Field Measurements for Passive Environmental Remote Sensing

Field Measurements for Passive Environmental Remote Sensing
Instrumentation, Intensive Campaigns, and Satellite Applications

Edited by

Nicholas R. Nalli
I.M. Systems Group, Inc. at NOAA/NESDIS Center for Satellite Applications and Research (STAR), College Park, MD, United States

ELSEVIER

Elsevier
Radarweg 29, PO Box 211, 1000 AE Amsterdam, Netherlands
The Boulevard, Langford Lane, Kidlington, Oxford OX5 1GB, United Kingdom
50 Hampshire Street, 5th Floor, Cambridge, MA 02139, United States

Copyright © 2023 Elsevier Inc. All rights reserved.

No part of this publication may be reproduced or transmitted in any form or by any means, electronic or mechanical, including photocopying, recording, or any information storage and retrieval system, without permission in writing from the publisher. Details on how to seek permission, further information about the Publisher's permissions policies and our arrangements with organizations such as the Copyright Clearance Center and the Copyright Licensing Agency, can be found at our website: www.elsevier.com/permissions.

This book and the individual contributions contained in it are protected under copyright by the Publisher (other than as may be noted herein).

Notices

Knowledge and best practice in this field are constantly changing. As new research and experience broaden our understanding, changes in research methods, professional practices, or medical treatment may become necessary.

Practitioners and researchers must always rely on their own experience and knowledge in evaluating and using any information, methods, compounds, or experiments described herein. In using such information or methods they should be mindful of their own safety and the safety of others, including parties for whom they have a professional responsibility.

To the fullest extent of the law, neither the Publisher nor the authors, contributors, or editors, assume any liability for any injury and/or damage to persons or property as a matter of products liability, negligence or otherwise, or from any use or operation of any methods, products, instructions, or ideas contained in the material herein.

ISBN: 978-0-12-823953-7

For information on all Elsevier publications
visit our website at https://www.elsevier.com/books-and-journals

Cover Credits

1. *Top Left:* Scientific instrumentation on the NOAA Ship *Ronald H. Brown* during the 2015 CalWater/ACAPEX field campaign, with an overflight of the DOE G-1 Research Aircraft in the background (photo credit: Jonathan Gero).

2. *Top Center:* Ozonesonde launch [exact credit and description should be available from Shutterstock].

3. *Top Right:* Icebreaker RV *Polarstern* embedded in Arctic sea ice during the 2020 MOSAiC campaign (photo credit Matthew Shupe).

4. *Middle Left:* NASA high-altitude ER-2 Research Aircraft descending for landing over California's Mojave Desert (photo credit: Tony Landis, NASA).

5. *Middle Center:* PNLL lidar research buoy [exact credit and description should be available from Shutterstock].

6. *Middle Right:* Deployment of a Prediction and Research Moored Array in the Tropical Atlantic (PIRATA) buoy from the NOAA Ship *Ronald H. Brown* during the January-February 2021 PIRATA Northeast Extension cruise (photo credit: Greg Foltz).

7. *Bottom Left:* Bulgaria 2200 m Arctic radar station [credit should be available from Shutterstock].

8. *Bottom Center:* Low Earth orbit satellite [exact credit and description should be available from Shutterstock].

9. *Bottom Right:* Instrument installation on the 32-meter meteorological tower at the Howard University GRUAN site in Beltsville Maryland (photo credit: Jose Fuentes).

Publisher: Candice Janco
Acquisitions Editor: Jennette McClain
Editorial Project Manager: Maria Elaine Desamero
Production Project Manager: Bharatwaj Varatharajan
Cover Designer: Christian J. Bilbow

Typeset by STRAIVE, India

Contents

Contributors   xi
Preface   xv

1. Introduction: Field measurements and remote sensing

Nicholas R. Nalli and Satya Kalluri

1. Background   2
2. Measurement principles   3
      2.1 In situ versus remote sensing   4
3. Remote sensing primer   4
      3.1 Terrestrial radiation   7
      3.2 Radiative transfer equations   7
4. Satellite remote sensing   9
      4.1 Environmental satellites   10
      4.2 Satellite applications   11
5. Calibration/validation (cal/val)   16
      5.1 Measurement correlation   16
      5.2 Error analysis   16
      5.3 Intensive cal/val and long-term monitoring   17
6. Outline and structure of the book   17
      6.1 Part I: Instrumentation   17
      6.2 Part II: Intensive campaigns   18
      6.3 Part III: Satellite applications   18
Acknowledgments   18
References   18

Part I
Instrumentation

2. Balloon-borne radiosondes

Holger Vömel and Bruce Ingleby

1. Introduction   23
2. Measurement technology   23
      2.1 Radiosondes   23
      2.2 Receiving system and data formats   27
      2.3 Operations   27
3. Cal/val considerations   28
      3.1 Temperature and humidity profiling   29
      3.2 Winds   31
      3.3 Clouds   31
      3.4 Using NWP data assimilation fields   32
4. Summary   32
References   33
Further reading   35

3. Frost point hygrometers

Dale F. Hurst, Masatomo Fujiwara, and Samuel Oltmans

1. Historical development of frost point hygrometers   37
      1.1 Early work   37
      1.2 1970s and 1980s development of FPHs for balloons   39
      1.3 Improvements in FPHs since the 1980s   41
      1.4 Modern day FPH measurement uncertainties   44
2. FPH measurement records   46
      2.1 FPH sounding sites   46
      2.2 What FPH records show   47
3. Examples of FPH comparisons with satellite-based remote sensors   48
      3.1 General approaches and methods for the comparisons   48
      3.2 Examples of comparisons between satellite sensors and FPHs   49
4. Summary   53
Acknowledgments   53
References   53

4. Ozonesondes: Instrumentation and Data Applications

Anne M. Thompson, Herman G.J. Smit, Debra E. Kollonige, and Ryan M. Stauffer

1. The role of ozonesondes in the global ozone measurement framework   59
      1.1 Sondes in the context of a global ozone measurement strategy   59
      1.2 Chapter overview   60

2. The ozonesonde instrument, operation, and data quality control   60
   2.1 Electrochemical ozonesondes   60
   2.2 The ECC ozonesonde: Principles of operation and sources of uncertainty   61
   2.3 Quality assurance of ozonesondes: Approach and current status   63
3. Ozonesonde networks   64
   3.1 The global network: Long-term sites   64
   3.2 Strategic networks: Global and campaign operations   68
4. Applications of ozonesonde data with remote sensing observations   69
   4.1 Satellite ozone product evaluation using ozonesonde data   70
   4.2 Use of satellite ozone data to track the performance of the ozonesonde   72
5. Summary and conclusions   72
   5.1 Scientific perspective: On-going need for profiles from global ozonesondes   72
   5.2 Quality assurance: Need for sonde intercomparisons and a global ozone reference   74
   5.3 Conclusions   74
Acknowledgments   75
References   75

5. Oceanographic buoys: Providing ocean data to assess the accuracy of variables derived from satellite measurements

Renellys C. Perez, Gregory R. Foltz, Rick Lumpkin, Jianwei Wei, Kenneth J. Voss, Michael Ondrusek, Menghua Wang, and Mark A. Bourassa

1. Introduction   79
2. Oceanographic buoy platforms and requirements   80
   2.1 Surface drifting buoys   81
   2.2 Moored buoys   83
3. Sea surface temperature   85
4. Sea surface salinity   87
5. Wind speed and direction   89
6. Surface heat fluxes   90
7. Sea level pressure   91
8. Water-leaving radiance   92
   8.1 Selection of location   92
   8.2 Instruments and measurements   92
   8.3 Data processing   93

   8.4 Characterization, calibration, and maintenance   94
   8.5 Accomplishments and prospective   94
9. Parting thoughts   94
Acknowledgments   96
References   96

6. Surface-based thermal infrared spectrometers

Peter J. Minnett, Robert O. Knuteson, and Jonathan Gero

1. Introduction   101
2. FTIR principles   103
3. Calibration   103
4. M-AERI   104
   4.1 ARM M-AERI   105
5. Deployments   105
6. Geophysical variables   106
   6.1 Sea surface emissivity   106
   6.2 SST_{skin} 107
   6.3 Thermal skin layer profiles   109
   6.4 Near-surface temperatures and air-sea temperature differences   110
   6.5 Atmospheric profiles   110
7. Applications   112
   7.1 Validating satellite SST_{skin} retrievals   112
   7.2 Atmospheric profiles   113
   7.3 Greenhouse gas heating of the ocean   114
8. Summary and conclusions   115
Acknowledgments   116
References   116

7. Sun photometers

Carlos Toledano, Victoria E. Cachorro, David Mateos, Roberto Román, Ramiro González, Alexander Smirnov, Julian Gröbner, Stelios Kazadzis, and Natalia Kouremeti

1. Introduction   121
2. Basics of Sun photometry   122
   2.1 Theoretical background   122
   2.2 Instrument specifications   124
3. Measurement networks and data products   125
   3.1 Aerosol robotic network (AERONET)   125
   3.2 Maritime aerosol network (MAN) as a component of AERONET   126

3.3 Global atmosphere watch: Precision filter radiometer network (GAW-PFR) 127
　　3.4 Brewer networks 129
　　3.5 Other networks 130
4. Calibration and uncertainty 131
5. Aerosol data interpretation 132
　　5.1 Aerosol types 132
　　5.2 Quality assurance and quality control 132
6. Conclusions 135
Acknowledgments 135
References 135

8. The AirCore atmospheric sampling system

Bianca C. Baier, Colm Sweeney, and Huilin Chen

1. Background 139
　　1.1 Atmospheric trace gases 139
　　1.2 Trace gas vertical profiling 140
2. The AirCore atmospheric sampling system 141
　　2.1 Principles of operation 141
　　2.2 AirCore instrument package and accompanying flight string components 142
　　2.3 Profile retrieval and vertical resolution 144
　　2.4 AirCore sampler configurations and associated platforms 145
　　2.5 Retrieved profile uncertainty and reproducibility 146
3. AirCore balloon-borne measurement sites 149
4. AirCore evaluation of remote sensing retrievals 149
　　4.1 Calculation of AirCore total-column trace gas abundances 150
　　4.2 Case study: AirCore-calculated XCO_2 comparison to satellite XCO_2 150
　　4.3 Case study: Linking remote sensing total column trace gas abundances to WMO standard scales using AirCore 151
5. Conclusions and future outlook 152
　　5.1 Toward a global AirCore network 152
　　5.2 On the use of high-altitude uncrewed aircraft systems 152
Acknowledgments 153
References 153

Part II
Intensive campaigns

9. High-altitude aircraft radiometric calibration-validation campaigns

Joseph K. Taylor, David C. Tobin, Fred A. Best, Raymond K. Garcia, Michelle L. Loveless, Henry E. Revercomb, and William L. Smith, Sr.

1. Introduction 159
2. The Scanning High-resolution Interferometer Sounder 160
3. S-HIS radiometric calibration, verification, and traceability 163
4. S-HIS high-altitude airborne calibration validation results 166
　　4.1 Calibration validation of the Cross-track Infrared Sounder 166
　　4.2 Calibration validation of the Advanced Baseline Imager 175
5. Summary 180
Acknowledgments 180
References 180

10. Aircraft dropsonde campaigns

Holger Vömel and Jason Dunion

1. Introduction 185
2. Measurement technology 186
　　2.1 Dropsonde description 186
　　2.2 Launchers 187
　　2.3 Data processing and quality control 187
3. Cal val campaigns 189
　　3.1 Temperature and humidity 189
　　3.2 Surface winds 190
　　3.3 Wind profiles 192
　　3.4 Clouds and precipitation 192
4. Summary 193
Acknowledgments 193
References 193

11. Ship-based cal/val campaigns

Nicholas R. Nalli, Gregory R. Foltz, Jonathan Gero, Laura Gibson, Robert O. Knuteson, Rick Lumpkin, Peter J. Minnett, Vernon R. Morris, Michael Ondrusek, Renellys C. Perez, Menghua Wang, and Jianwei Wei

1. Introduction 195
2. Research vessel/ship platforms 196
3. At-sea instrument and sensor deployments 198

	3.1 Atmospheric deployments	198
	3.2 Oceanographic deployments	201
4.	Early infrared remote sensing campaigns	203
	4.1 Ocean temperature interferometric survey (OTIS)	204
	4.2 Combined sensor program	205
5.	Collaborative remote sensing campaigns	205
	5.1 PNE/AEROSE	205
	5.2 CalWater/ACAPEX	207
	5.3 SHEBA/MOSAiC	208
	5.4 EUREC^{4}A/ATOMIC	209
6.	Ocean color campaigns	210
7.	Summary	214
	Acknowledgments	215
	References	215

12. Land-based cal/val campaigns

Lori A. Borg, Ruud J. Dirksen, and Robert O. Knuteson

1.	Introduction	219
2.	DOE ARM campaigns	220
	2.1 ARM Water Vapor Intensive Operating Periods	220
	2.2 AIRS Cal/Val at ARM	221
	2.3 JPSS CrIS Cal/Val at ARM	222
3.	GRUAN	226
	3.1 GRUAN data products	227
	3.2 GRUAN sites	227
	3.3 RS92-RS41 radiosonde intercomparisons	228
	3.4 WMO radiosonde intercomparisons	228
4.	Links to other land-based campaigns	230
5.	Summary	231
	Acknowledgment	231
	References	231
	Further reading	233

13. Aircraft vertical profile measurements for evaluation of satellite retrievals of long-lived trace gases

Eric A. Kort and Kathryn McKain

1.	Introduction	235
2.	"Global" intensive profiling campaigns	236
	2.1 HIPPO: HIaper Pole-to-Pole Observations	236
	2.2 ATom: Atmospheric Tomography Mission	239
3.	Sustained observations	240
	3.1 Global Greenhouse Gas Reference Network's Aircraft Network	240
	3.2 Deeter et al. (2019): Example of Role 1	241
	3.3 TCCON: Total Carbon Column Observing Network—Example of Role 3	241
4.	Aggregating aircraft observations from regional intensives	243
	4.1 Chevallier et al. (2019): Example of (2)	243
5.	Summary	243
	Acknowledgments	243
	References	244

14. Campaign situational awareness from operational satellite sounding retrievals

Rebekah Esmaili and Christopher D. Barnet

1.	Introduction	245
2.	Data sets and methods	247
	2.1 Satellite soundings	247
	2.2 Radiosondes	248
	2.3 RAP model	249
	2.4 Aircraft	249
3.	Hazardous Weather Testbed	249
4.	Intensity Forecast EXperiment (IFEX)	251
5.	Fire Influence on Regional to Global Environments and Air Quality (FIREX-AQ)	254
6.	Summary	256
	Acknowledgments	257
	References	257
	Further reading	260

Part III
Satellite applications

15. On-orbit VIIRS sensor calibration and validation in reflective solar bands (RSB)

Taeyoung Choi, Changyong Cao, Slawomir Blonski, Xi Shao, Wenhui Wang, and Sirish Uprety

1.	Introduction	263
2.	The NOAA-20 VIIRS sensor	264
3.	Overview of the prelaunch calibration	266
4.	On-orbit radiometric calibration for reflective solar band	267

4.1 Primary VIIRS RSB calibration coefficient (SD H- and F-factors)	267
4.2 Scheduled lunar calibration collection	268
4.3 Lunar F-factor calculation algorithm	270
4.4 The DCC technique for VIIRS	271
4.5 Simultaneous nadir overpass (SNO) trend derivation algorithm	272
5. On-orbit calibration	272
5.1 On-orbit SD F-factors	272
5.2 On-orbit lunar F-factor trends	274
5.3 Daily DCC trending	274
5.4 Simultaneous nadir overpass (SNO) trending validation	276
5.5 Decisions on operational calibration coefficients for VIIRS SDR data production	277
6. Conclusions	277
Acknowledgments	278
Disclaimer	278
References	278

16. The NOAA sounding Products Validation System (NPROVS)

Bomin Sun, Anthony Reale, Michael Pettey, Ryan Smith, and Charlie Brown

1. Introduction	281
2. Data sources	282
2.1 Radiosonde data	282
3. Collocation compiling	287
4. Vertical statistics computation	288
5. Graphic applications	289
5.1 Orbital Display System	289
5.2 Profile Display	291
5.3 NPROVS Archive Summary (NARCS) Statistics	293
6. Summary	294
Acknowledgments	295
References	295

17. Satellite microwave sounder product retrieval validation

Christopher Grassotti, Quanhua Liu, Yong-Keun Lee, Shuyan Liu, and Yan Zhou

1. Overview	297
2. Microwave sounder data	297
3. Microwave integrated retrieval system (MiRS)	299
4. MiRS product validation	301
4.1 Validation background	301
4.2 Temperature profile validation	304
4.3 Water vapor profile validation	308
4.4 Total precipitable water validation	311
5. Summary	315
Acknowledgments	316
References	316
Further reading	317

18. Considerations for thermal satellite infrared sounder applications

Nicholas R. Nalli, Nadia Smith, Rebekah Esmaili, Bomin Sun, Anthony Reale, and Christopher D. Barnet

1. Satellite thermal infrared remote sounding	319
2. Algorithm development	322
3. Algorithm validation	323
3.1 Objectives	323
3.2 Strategies and methodologies	323
3.3 Validation highlights using field data	327
4. Applications	329
5. Summary and outlook	332
Acknowledgments	332
References	332

19. Sea surface temperature validation and blended analysis

Christopher J. Merchant, Owen Embury, Chelle Gentemann, John J. Kennedy, Elizabeth C. Kent, Peter J. Minnett, and James While

1. Sea surface temperature products: An overview	337
2. In situ measurements in SST validation	342
3. Blended analysis of SST	346
4. Outlook	348
Acknowledgments	349
References	349

20. Satellite ocean color validation

Jianwei Wei, Menghua Wang, Michael Ondrusek, Alexander Gilerson, Joaquim Goes, Chuanmin Hu, Zhongping Lee, Kenneth J. Voss, Sherwin Ladner, Veronica P. Lance, and Nicholas Tufillaro

1. Introduction	351
2. Satellite ocean color products	352
2.1 Remote sensing reflectance	352

- 2.2 Water bio-optical properties — 353
- 2.3 Quality measures — 355
3. Field measurements of the radiance distribution — 356
4. Field measurements of $R_{rs}(\lambda)$ — 357
   - 4.1 Above-water approach — 357
   - 4.2 In-water approach — 358
   - 4.3 On-water approach — 359
5. Field measurements of water inherent optical properties — 360
   - 5.1 Backscattering coefficient — 360
   - 5.2 Non-water absorption coefficient — 361
   - 5.3 Phytoplankton and CDOM absorption coefficients — 362
6. Field measurements of phytoplankton pigments and functional types — 363
   - 6.1 Phytoplankton pigments — 363
   - 6.2 Phytoplankton cell counts and cell sizes (>10 μm) — 363
   - 6.3 Phytoplankton groups of small sizes (<10 μm) — 365
   - 6.4 Phytoplankton quantum efficiency of carbon fixation — 365
7. Matchup analysis — 366
8. Concluding remarks — 368
Acknowledgments — 369
References — 369

21. Land surface temperature validation

Yuling Liu, Yunyue Yu, Heshun Wang, and Peng Yu

1. Land surface temperature product introduction — 375
2. Methodology of LST retrieval — 376
   - 2.1 LST algorithm derivation — 376
   - 2.2 Simulation study and theoretical analysis — 376
3. LST product validation — 377
   - 3.1 VIIRS LST product — 377
   - 3.2 Strategies and methodologies — 377
   - 3.3 Ground LST estimates — 378
   - 3.4 Validation results and highlights — 378
4. Summary — 387
Acknowledgments — 387
References — 388

22. Heterogeneity of smoke from fires: Evaluation of VIIRS smoke detection using FIREX-AQ field campaign data

Pubu Ciren and Shobha Kondragunta

1. Introduction — 391
2. NOAA JPSS enterprise processing system aerosol detection product — 393
   - 2.1 Deep-blue aerosol detection algorithm — 394
   - 2.2 IR-visible aerosol detection algorithm — 396
   - 2.3 eMAS measurements during FIREX-AQ 2019 — 398
   - 2.4 Application of EPS ADP algorithm to eMAS measurements — 401
3. Comparison of ADP_e with ADP_v — 404
4. Conclusions — 407
Acknowledgment — 408
Disclaimer — 408
References — 409

23. Downburst monitoring and prediction studies

Kenneth L. Pryor

1. Introduction — 411
2. Theoretical background of severe convective windstorms: Genesis and evolution — 412
3. Field measurement application methodology — 414
4. Case studies of field measurement applications — 416
   - 4.1 27 April 2020 South Texas severe thunderstorm downbursts — 416
   - 4.2 29 June 2012 North American derecho — 421
5. Summary — 427
Acknowledgments — 428
References — 428

Index — 431

Contributors

Numbers in paraentheses indicate the pages on which the authors' contributions begin.

Bianca C. Baier (139), Cooperative Institute for Research in Environmental Sciences (CIRES), University of Colorado-Boulder; NOAA Global Monitoring Laboratory (GML), Boulder, CO, United States

Christopher D. Barnet (245, 319), Science and Technology Corp. (STC), Columbia, MD, United States

Fred A. Best (159), Space Science and Engineering Center (SSEC), University of Wisconsin-Madison, Madison, WI, United States

Slawomir Blonski (263), Global Science & Technology, Inc, Greenbelt, MD, United States

Lori A. Borg (219), Space Science and Engineering Center (SSEC); CIMSS, University of Wisconsin-Madison, Madison, WI, United States

Mark A. Bourassa (79), Florida State University, COAPS, Tallahassee, FL, United States

Charlie Brown (281), I.M. Systems Group, Inc.; NOAA/NESDIS Center for Satellite Applications and Research (STAR), College Park, MD, United States

Victoria E. Cachorro (121), Group of Atmospheric Optics, University of Valladolid, Valladolid, Spain

Changyong Cao (263), NOAA/NESDIS Center for Satellite Applications and Research (STAR), College Park, MD, United States

Huilin Chen (139), Centre for Isotope Research, University of Groningen, Groningen, The Netherlands; Joint International Research Laboratory of Atmospheric and Earth System Sciences, School of Atmospheric Sciences, Nanjing University, Nanjing, China

Taeyoung Choi (263), Global Science & Technology, Inc, Greenbelt, MD, United States

Pubu Ciren (391), I.M. Systems Group, Inc.; NOAA/NESDIS Center for Satellite Applications and Research (STAR), College Park, MD, United States

Ruud J. Dirksen (219), GRUAN Lead Centre, Deutscher Wetterdienst, Meteorologisches Observatorium Lindenberg, Lindenberg, Germany

Jason Dunion (185), NOAA/AOML Hurricane Research Division (HRD), Miami, FL, United States

Owen Embury (337), Department of Meteorology and National Centre for Earth Observation, University of Reading, Reading, United Kingdom

Rebekah Esmaili (245, 319), Science and Technology Corp. (STC), Columbia, MD, United States

Gregory R. Foltz (79, 195), NOAA/Atlantic Oceanographic and Meteorological Laboratory, Miami, FL, United States

Masatomo Fujiwara (37), Faculty of Environmental Earth Science, Hokkaido University, Sapporo, Japan

Raymond K. Garcia (159), Space Science and Engineering Center (SSEC), University of Wisconsin-Madison, Madison, WI, United States

Chelle Gentemann (337), Farallon Institute, Petaluma, CA, United States

Jonathan Gero (101, 195), Space Science and Engineering Center (SSEC); CIMSS, University of Wisconsin-Madison, Madison, WI, United States

Laura Gibson (195), NOAA/NESDIS Center for Satellite Applications and Research (STAR), College Park; NOAA/OMAO, Silver Spring, MD, United States

Alexander Gilerson (351), The City College of New York, New York, NY, United States

Joaquim Goes (351), Lamont-Doherty Earth Observatory, Palisades, NY, United States

Ramiro González (121), Group of Atmospheric Optics, University of Valladolid, Valladolid, Spain

Christopher Grassotti (297), Cooperative Institute for Satellite and Earth System Studies (CISESS), Earth System Science Interdisciplinary Center, University of Maryland, College Park, MD, United States

Julian Gröbner (121), Physikalisch-Meteorologisches Observatorium Davos, World Radiation Center (PMOD/WRC), Davos, Switzerland

Chuanmin Hu (351), University of South Florida, St. Petersburg, FL, United States

Dale F. Hurst (37), Cooperative Institute for Research in Environmental Sciences (CIRES), University of Colorado-Boulder; NOAA Global Monitoring Laboratory (GML), Boulder, CO, United States

Bruce Ingleby (23), European Centre for Medium Range Weather Forecasts (ECMWF), Reading, United Kingdom

Satya Kalluri (1), Joint Polar Satellite System (JPSS) Program Office, Greenbelt, MD, United States

Stelios Kazadzis (121), Physikalisch-Meteorologisches Observatorium Davos, World Radiation Center (PMOD/WRC), Davos, Switzerland

John J. Kennedy (337), Met Office, Exeter, United Kingdom

Elizabeth C. Kent (337), National Oceanography Centre, Southampton, United Kingdom

Robert O. Knuteson (101, 195, 219), Space Science and Engineering Center (SSEC); CIMSS, University of Wisconsin-Madison, Madison, WI, United States

Debra E. Kollonige (57), Earth Sciences Division, NASA-Goddard Space Flight Center, Greenbelt; SSAI, Lanham, MD, United States

Shobha Kondragunta (391), NOAA/NESDIS Center for Satellite Applications and Research (STAR), College Park, MD, United States

Eric A. Kort (235), University of Michigan, Ann Arbor, MI, United States

Natalia Kouremeti (121), Physikalisch-Meteorologisches Observatorium Davos, World Radiation Center (PMOD/WRC), Davos, Switzerland

Sherwin Ladner (351), Naval Research Laboratory, Stennis Space Center, MS, United States

Veronica P. Lance (351), NOAA/NESDIS Center for Satellite Applications and Research (STAR), College Park, MD, United States

Yong-Keun Lee (297), Cooperative Institute for Satellite and Earth System Studies (CISESS), Earth System Science Interdisciplinary Center, University of Maryland, College Park, MD, United States

Zhongping Lee (351), University of Massachusetts Boston, Boston, MA, United States

Quanhua Liu (297), NOAA/NESDIS Center for Satellite Applications and Research (STAR), College Park, MD, United States

Shuyan Liu (297), Cooperative Institute for Research in the Atmosphere (CIRA), Colorado State University, Fort Collins, CO, United States

Yuling Liu (375), Earth System Science Interdisciplinary Center of University of Maryland, College Park, MD, United States

Michelle L. Loveless (159), Space Science and Engineering Center (SSEC), University of Wisconsin-Madison, Madison, WI, United States

Rick Lumpkin (79, 195), NOAA/Atlantic Oceanographic and Meteorological Laboratory, Miami, FL, United States

David Mateos (121), Group of Atmospheric Optics, University of Valladolid, Valladolid, Spain

Kathryn McKain (235), Cooperative Institute for Research in Environmental Sciences, University of Colorado; NOAA Global Monitoring Laboratory (GML), Boulder, CO, United States

Christopher J. Merchant (337), Department of Meteorology and National Centre for Earth Observation, University of Reading, Reading, United Kingdom

Peter J. Minnett (101, 195, 337), Department of Ocean Sciences, Rosenstiel School of Marine and Atmospheric Science, University of Miami, Miami, FL, United States

Vernon R. Morris (195), School of Mathematical and Natural Sciences, Arizona State University, Phoenix, AZ, United States

Nicholas R. Nalli (1, 195, 319), I.M. Systems Group, Inc.; NOAA/NESDIS Center for Satellite Applications and Research (STAR), College Park, MD, United States

Samuel Oltmans (37), NOAA Global Monitoring Laboratory (GML), Boulder, CO, United States

Michael Ondrusek (79, 195, 351), NOAA/NESDIS Center for Satellite Applications and Research (STAR), College Park, MD, United States

Renellys C. Perez (79, 195), NOAA/Atlantic Oceanographic and Meteorological Laboratory, Miami, FL, United States

Michael Pettey (281), I.M. Systems Group, Inc.; NOAA/NESDIS Center for Satellite Applications and Research (STAR), College Park, MD, United States

Kenneth L. Pryor (411), NOAA/NESDIS Center for Satellite Applications and Research (STAR), College Park, MD, United States

Anthony Reale (281, 319), NOAA/NESDIS Center for Satellite Applications and Research (STAR), College Park, MD, United States

Henry E. Revercomb (159), Space Science and Engineering Center (SSEC), University of Wisconsin-Madison, Madison, WI, United States

Roberto Román (121), Group of Atmospheric Optics, University of Valladolid, Valladolid, Spain

Xi Shao (263), Cooperative Institute for Satellite and Earth System Studies (CISESS), Earth System Science

Interdisciplinary Center, University of Maryland, College Park, MD, United States

Alexander Smirnov (121), NASA Goddard Space Flight Center, Greenbelt; Science Systems and Applications, Inc., Lanham, MD, United States

Herman G.J. Smit (57), Forschungszentrum-Jülich, Jülich, Germany

Nadia Smith (319), Science and Technology Corp. (STC), Columbia, MD, United States

Ryan Smith (281), I.M. Systems Group, Inc.; NOAA/NESDIS Center for Satellite Applications and Research (STAR), College Park, MD, United States

William L. Smith, Sr. (159), Space Science and Engineering Center (SSEC), University of Wisconsin-Madison, Madison, WI, United States

Ryan M. Stauffer (57), Earth Sciences Division, NASA-Goddard Space Flight Center, Greenbelt, MD, United States

Bomin Sun (281, 319), I.M. Systems Group, Inc.; NOAA/NESDIS Center for Satellite Applications and Research (STAR), College Park, MD, United States

Colm Sweeney (139), NOAA Global Monitoring Laboratory (GML), Boulder, CO, United States

Joseph K. Taylor (159), Space Science and Engineering Center (SSEC), University of Wisconsin-Madison, Madison, WI, United States

Anne M. Thompson (57), Earth Sciences Division, NASA-Goddard Space Flight Center, Greenbelt; Joint Center for Environmental Technology, University of Maryland, Baltimore County, MD, United States

David C. Tobin (159), Space Science and Engineering Center (SSEC), University of Wisconsin-Madison, Madison, WI, United States

Carlos Toledano (121), Group of Atmospheric Optics, University of Valladolid, Valladolid, Spain

Nicholas Tufillaro (351), Oregon State University, Corvallis, OR, United States

Sirish Uprety (263), Cooperative Institute for Satellite and Earth System Studies (CISESS), Earth System Science Interdisciplinary Center, University of Maryland, College Park, MD, United States

Holger Vömel (23, 185), National Center for Atmospheric Research, Boulder, CO, United States

Kenneth J. Voss (79, 351), University of Miami, Coral Gables, FL, United States

Heshun Wang (375), Earth System Science Interdisciplinary Center of University of Maryland, College Park, MD, United States

Menghua Wang (79, 195, 351), NOAA/NESDIS Center for Satellite Applications and Research (STAR), College Park, MD, United States

Wenhui Wang (263), Cooperative Institute for Satellite and Earth System Studies (CISESS), Earth System Science Interdisciplinary Center, University of Maryland, College Park, MD, United States

Jianwei Wei (79, 195, 351), NOAA/NESDIS Center for Satellite Applications and Research (STAR), College Park; Global Science and Technology, Inc., Greenbelt, MD, United States

James While (337), Met Office, Exeter, United Kingdom

Peng Yu (375), Earth System Science Interdisciplinary Center of University of Maryland, College Park, MD, United States

Yunyue Yu (375), NOAA/NESDIS Center for Satellite Applications and Research (STAR), Silver Spring, MD, United States

Yan Zhou (297), Cooperative Institute for Satellite and Earth System Studies (CISESS), Earth System Science Interdisciplinary Center, University of Maryland, College Park, MD, United States

Preface

Our interpretation and understanding of remote sensing measurements from satellites is firmly rooted in the knowledge and experience we gained studying in a laboratory or field setting how Earth and its atmosphere interact with electromagnetic radiation. There are plenty of excellent books on remote sensing of land, oceans, and the atmosphere that provide a wealth of knowledge on the techniques and algorithms for processing the sensor measurements to geophysical variables that are used in various applications such as meteorology. This book does not intend to be a treatise on remote sensing techniques, but it attempts to bridge the knowledge gap between what we observe from satellites with what we measure on Earth.

Comparing satellite remote sensing measurements and derived geophysical parameters such as temperature, water vapor, and albedo, with in situ observations remains the gold standard of verification and validation. A vast majority of remote sensing scientists and algorithm developers spend most, if not all, of their time in front of a computer terminal processing the data and comparing the outputs with in situ measurements gathered in the field by someone else. Many, if not most, remote sensing scientists have limited experience with data collection in the field using sensors and data loggers. Thus, the intent here is to provide the reader with a background of Earth-based sensors that are independent of the satellite observing systems, and remote sensing products, highlighted herein. The contents of the book have been organized as ground, air, and ocean observing sensors such as radiosondes, hygrometers, and sunphotometers, along with the approaches to validate associated satellite products using these in situ observations.

Almost all Earth remote sensing missions have a calibration and validation plan that uses some type of in situ data for assuring the accuracy and integrity of their products. Having dedicated field campaigns to collect in situ measurements for satellite missions is an expensive endeavor that is typically done one time at the beginning of mission science operations. It is more common practice to routinely compare and verify satellite products with those measurements that are routinely collected by various organizations that are not primarily space agencies. Remote sensing measurements constitute more than 90% in data volume of measurements that are ingested into operational numerical weather prediction models (NWP). Nevertheless, routine observations from radiosondes, aircraft, and surface observations still play a critical role in NWP and forecasting. According to the World Meteorological Organization, there are more than 10,000 manned and automatic weather stations as well as several thousand sensors that are deployed on buoys, aircraft, and ships that operationally provide ocean, land, and atmosphere observations to NWP centers every day. Several of the sensors discussed in this book are part of the vital network. Exhibition floors at popular Earth science conferences such as the annual meeting of the American Meteorological Society and the American Geophysical Union have manufacturers of meteorological measurement equipment as well as manufacturers of remote sensing instruments and satellites on the same floor because these measurements complement each other.

The technology of field sensors has become very sophisticated due to miniaturization, advanced electronics, improved communication, automation, and new materials. This book does not attempt to cover every aspect of in situ observations, but rather to provide the reader with a brief history and status of some of the operational observations that are also used for validation of remote sensing data. The authors who contributed to the book have vast experience in sensors, field measurements, and remote sensing. We hope that the book will serve an important resource on remote sensing due to the unique aspects covered here.

Satya Kalluri
Greenbelt, MD

Chapter 1

Introduction: Field measurements and remote sensing

Nicholas R. Nalli[a,b] and Satya Kalluri[c]

[a]*I.M. Systems Group, Inc., College Park, MD, United States,* [b]*NOAA/NESDIS Center for Satellite Applications and Research (STAR), College Park, MD, United States,* [c]*Joint Polar Satellite System (JPSS) Program Office, Greenbelt, MD, United States*

Chapter outline

1. **Background**	2	5.1 Measurement correlation	16
2. **Measurement principles**	3	5.2 Error analysis	16
2.1 In situ versus remote sensing	4	5.3 Intensive cal/val and long-term monitoring	17
3. **Remote sensing primer**	4	6. **Outline and structure of the book**	17
3.1 Terrestrial radiation	7	6.1 Part I: Instrumentation	17
3.2 Radiative transfer equations	7	6.2 Part II: Intensive campaigns	18
4. **Satellite remote sensing**	9	6.3 Part III: Satellite applications	18
4.1 Environmental satellites	10	**Acknowledgments**	18
4.2 Satellite applications	11	**References**	18
5. **Calibration/validation (cal/val)**	16		

My Design in this Book is not to explain the Properties of Light by Hypotheses, but to propose and prove them by Reason and Experiments…

Sir Isaac Newton

Abbreviations

ABI	Advanced Baseline Imager (GOES-R series)
AERONET	Aerosol Robotic Network
AIRS	Atmospheric Infrared Sounder (NASA)
AMF	ARM Mobile Facility
AMSU	Advanced Microwave Sounding Unit
AOD	aerosol optical depth
ARM	Atmospheric Radiation Measurement program (US Department of Energy)
ATMS	Advanced Technology Microwave Sounder (JPSS series)
AVHRR	Advanced Very High-Resolution Radiometer (POES series)
cal/val	calibration/validation
CrIS	Cross-track Infrared Sounder (JPSS series)
CRTM	Community Radiative Transfer Model
DMW(s)	derived motion wind(s)
ECMWF	European Center for Medium Range Weather Forecasts
EDR(s)	environmental data record(s)
EM	electromagnetic
EUMETSAT	European Organization for the Exploitation of Meteorological Satellites
EXT	equator crossing time
FOV	field-of-view
FTS	Fourier transform spectrometer
GEO	geostationary (or geosynchronous) orbit

GOES	Geostationary Operational Environmental Satellite (NOAA)
IASI	Infrared Atmospheric Sounding Interferometer
IR	infrared spectrum
JPSS	Joint Polar Satellite System
LEO	low-earth orbit
(M)AERI	(Marine) Atmospheric Emitted Radiance Interferometer
MHS	Microwave Humidity Sounder
MODIS	Moderate resolution Imaging Spectroradiometer (NASA)
MW	microwave spectrum
NIR	near-IR spectrum
NOAA	National Oceanic and Atmospheric Administration
NUCAPS	NOAA-Unique Combined Atmospheric Processing System
NWP	numerical weather prediction
OE	optimal estimation
OISST	optimum interpolation SST
OMPS	Ozone Mapping and Profiler Suite
PDE(s)	partial differential equation(s)
POES	Polar Operational Environmental Satellite (NOAA)
RDR(s)	raw data record(s)
RMSE	root mean square error
RTA	radiative transfer algorithm (alternatively, rapid transmittance algorithm)
RTE	radiative transfer equation
RTM	radiative transfer model
SDR(s)	sensor data record(s)
SST(s)	sea surface temperature(s)
TOA	top of atmosphere
UV	ultraviolet spectrum
VIIRS	Visible Infrared Imaging Radiometer Suite (JPSS series)
VIS	visible (solar) spectrum

1. Background

Although environmental *remote sensing* is a relatively mature science, its importance has grown exponentially since the advent and proliferation of operational *environmental satellites*. As opposed to more traditional forms of Earth observation, these platforms flown in low-Earth and geostationary orbit have the distinct advantage of observing the Earth over its entire surface area. Surface-based remote sensing has also complemented satellites given their closer proximity to the boundary layer. Since their inception, these measurements have demonstrably improved short- and mid-term weather forecasts (Le Marshall et al., 2006, 2008; Noh et al., 2020), and they have been used for a host of other applications critical to the advancement of the atmospheric and oceanic sciences.

However, while environmental satellites are now indispensable, their success remains largely dependent on what this book refers to as *field measurements*, namely the well-designed acquisition of independent data with known error characteristics in a manner capable of specifying the geophysical state over the multidimensional space observed by the satellites. A heretofore understated common denominator in environmental remote sensing is the reliance on these independent, well-characterized field measurements, often managed and deployed by science teams with expertise and objectives separate from those in satellite remote sensing.

In spite of this, documentation on satellite applications of such data is typically found at either a somewhat vague and nonspecific overview level, or at a highly specialized and specific technical-report level, with little to no systematic treatment as an emergent subdiscipline of remote sensing. With the proliferation of numerous satellite remote sensing instruments on different platforms covering the electromagnetic spectrum, well-established methodologies have emerged that can be summarized within their own chapters, yet fall under the same general paradigm.

Of particular importance for environmental remote sensing are what we shall refer to as *intensive field campaigns*, which may be considered as a form of *closure experiment* (e.g., Quinn and Coffman, 1998). Closure experiments, as the name suggests, are an attempt to contend with the *closure problem*, namely the situation where we have more variables or unknown parameters than we do governing equations, for example, as in the case of turbulence (e.g., Stull, 1988, p. 197).

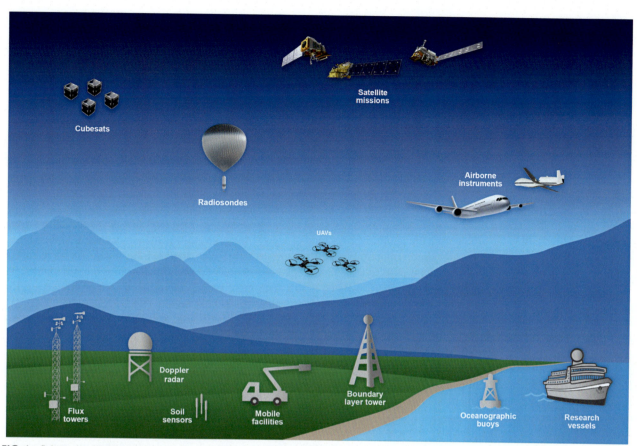

FIG. 1 Schematic of field measurements typically used in support of satellite passive remote sensing. Depicted are environmental satellites (including SmallSats/CubeSats), aircraft, balloon-borne sensors (e.g., radiosondes), uncrewed aerial vehicles (UAVs or "drones"), meteorological towers (flux and boundary layer), surface radar, mobile facilities (e.g., US Department of Energy [DOE] Atmospheric Radiation Measurement [ARM] Mobile Facilities [AMF]), soil sensors, oceanographic buoys, and research vessels or ships.

This is also the case for remote sensing problems, where we often find ourselves with far more unknowns than equations (discussed more below). Intensive campaigns are thus designed to attempt acquisition of multiple independent state parameter data, including coincidental remotely sensed measurements, in an effort to attain "closure" through a complete description of the environmental state under consideration. Intensive campaigns typically employ ground-based observing networks, as well as coordinated aircraft, balloon-borne, and ship-based instrumentation coordinated with satellite observations (see Fig. 1). Some intensive field campaigns of note that have supported the advancement of passive remote sensing are highlighted more in Part II (Chapters 9–14) of the book.

This book thus seeks to bring together a wide range of well-established instrument measurement techniques and observing strategies (ranging from intensive campaigns to observations of opportunity) widely utilized in remotely sensed environmental satellite data research and applications, compiling within a single volume an academic synthesis of the wealth of invaluable field measurements and techniques leveraged by the science of environmental remote sensing. As such, the book is envisioned to serve both as a practical textbook for students (upper-level undergraduate to graduate level) as well as a reference book for practitioners and researchers (in both remote sensing and conventional measurements/instrumentation). To this end, the remainder of this chapter will highlight the basic principles common to the topics covered within the book, including in situ and remote sensing measurements (both defined in Section 2), radiative transfer, environmental satellites, error analysis and calibration/validation (cal/val), and other miscellaneous applications of field measurements for passive remote sensing. For convenience, the definitions for abbreviations may be found at the beginning of the chapter.

2. Measurement principles

Measurement is the process of quantifying a physical reality, usually an object or phenomenon of interest. The quantification is obtained via a comparison against a known standard using an instrument or device that has been calibrated to that standard. Measurements may be obtained under controlled conditions in the laboratory or under less-controlled conditions in the field. An elementary example of a classical measurement in the laboratory would be taking the mass of an object using a triple-beam balance instrument. The balance is calibrated using known masses and levers, which then allow for a precise, unbiased quantification of the mass. In environmental sciences, laboratory measurements typically provide the basis for foundational relationships (e.g., fundamental constants, ideal gas laws, etc.), but field measurements provide the quantitative characterization of the geophysical state of interest.

2.1 In situ versus remote sensing

A measurement is considered in situ (Latin for "in place") when the sensor is in direct material contact with the medium being measured. The measurement is usually considered "direct" in the sense that a mathematical inversion scheme is not required for extracting the parameter of interest from the measurement. Examples of in situ instruments in environmental sciences include thermometers, thermistors, barometers, anemometers, hygrometers, salinometers, cascade impactors, among others. In the context of *satellite* remote sensing, an in situ measurement may also refer to a remotely sensed measurement acquired below the orbit within the satellite sensor field of view (FOV) (i.e., within the atmosphere). In this sense, ground-based, ship-based, or aircraft-based remotely sensed measurements are also acquired "in place" (relative to the satellite) and serve the same function or purpose to the satellite measurement as do traditional in situ measurements.

Generally speaking, measurements from a single, isolated in situ instrument such as thermometer are of limited use for macroscale environmental observation. First, routine acquisition of such data usually requires a *platform* (which includes the actual instrumentation, protective housing, power, data storage, and/or transmission, etc.). But perhaps more importantly, such measurements must also usually be coordinated in a manner that facilities large-scale state characterization; such approaches may be referred to as *observing systems*. Example in situ observing systems include radiosonde and surface station networks, oceanographic buoys, Aerosol Robotic Network (AERONET), sunphotometer networks (Holben et al., 1998), among others, which will be covered in Part I.

A *remotely sensed* measurement, on the other hand, is one that is obtained through electromagnetic (EM) radiative transfer. Remote sensing is something that we take for granted in our everyday life. Our eyes, for example, are sophisticated natural remote sensors that are sensitive to the solar spectrum, which is also referred to as the "visible spectrum" for this very reason. But how is it that we are able to measure something from a distance? To understand this, we must briefly review the nature of EM radiation; because this book is focused on macroscale measurements and applications, we may for the most part limit ourselves to the classical domain without having to delve into quantum effects.

3. Remote sensing primer

As mentioned earlier, remote sensing may to a large extent be understood from classical electrodynamics based on Maxwell's equations (see Box 1). Oscillating charges (e.g., electrons) found in everyday matter produce time-varying electric fields that give rise to self-sustaining EM waves called radiation (e.g., light). Both energy and momentum are transferred within EM waves (defined by the *Poynting vector*) as illustrated in Fig. 2 and discussed more in Box 2, where it is seen that the energy is proportional to the square of the electric field amplitude. Because EM waves do not depend upon any physical medium for propagation (as established by Albert Einstein in the Special Theory of Relativity; e.g., McMahon, 2006), EM radiation is thus capable of moving indefinitely through empty space, and thus the measurement of radiant energy (or "radiances") can be obtained "remotely," that is, not in direct contact.

BOX 1 Electromagnetic waves.

The *Maxwell equations* (named after 19th century physicist James Clerk Maxwell) consist of a system of four coupled, linear partial differential equations (PDEs), given by (e.g., Griffiths, 1989)

Characteristics of **E** and **B**:
$$\nabla \cdot \mathbf{E} = \frac{1}{\epsilon_0} \rho,$$
$$\nabla \cdot \mathbf{B} = 0,$$

Relationships between **E** and **B**:
$$\nabla \times \mathbf{B} = \mu_0 \epsilon_0 \frac{\partial \mathbf{E}}{\partial t} + \mu_0 \mathbf{J},$$
$$\nabla \times \mathbf{E} = -\frac{\partial \mathbf{B}}{\partial t},$$

where ρ is the electric charge, **E** and **B** are the electric and magnetic fields, respectively, $\mathbf{J} \equiv \rho \mathbf{v}$ is the electric current density (i.e., the flow of charge per unit area), and μ_0 and ϵ_0 are fundamental constants (specifically the permeability and permittivity of free space, respectively).

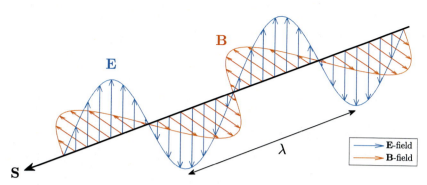

FIG. 2 Electromagnetic (EM) radiation schematic. The instantaneous space-time-varying electric and magnetic fields, **E** and **B**, are illustrated with *oscillating blue* and *red field lines*, respectively. The EM waves depicted have wavelength λ and are propagating from upper-right to lower-left (obliquely out of the page), with energy transfer indicated by the Poynting vector, $\mathbf{S} \equiv (1/\mu_0)\,\mathbf{E}\times\mathbf{B}$ (see Box 2).

BOX 1 Electromagnetic waves—cont'd

One may readily see that there is certain aesthetic symmetry to these equations. The two-left two equations describe the nature of the electric and magnetic fields. The top left equation is Gauss's law, which expresses that an electric field **E** originates from the presence of electric point charges ρ (i.e., it diverges from them), whereas the bottom-left equation states that a magnetic field **B** neither diverges nor converges, which is basically the same as saying that they do not result from point charges (unlike the electric fields).

The right two equations, on the other hand, describe the relationships between the **E**- and **B**-fields. The top-right equation (Ampère-Maxwell law) states that an electric current (moving charges), **J**, or a time-varying electric field, $\partial \mathbf{E}/\partial t$, create the curl (orthogonal curvature) of a magnetic field, $\nabla \times \mathbf{B}$; similarly, the bottom-right equation (Faraday's law) expresses that a time-varying magnetic field, $\partial \mathbf{B}/\partial t$, gives rise to electric field curl, $\nabla \times \mathbf{E}$.

Now, let us consider Maxwell's equations in empty space devoid of matter, and thus devoid of electric charge ρ or current **J**:

$$\nabla \cdot \mathbf{E} = 0,\quad \nabla \cdot \mathbf{B} = 0,\quad \nabla \times \mathbf{B} = \mu_0 \epsilon_0 \frac{\partial \mathbf{E}}{\partial t},\quad \nabla \times \mathbf{E} = -\frac{\partial \mathbf{B}}{\partial t}.$$

These can be reduced from four first-order PDEs to two decoupled, second-order linear PDEs by taking the second vector products (the curl of the curl) of the latter two equations, $\nabla \times (\nabla \times \mathbf{B})$ and $\nabla \times (\nabla \times \mathbf{E})$, then making the substitutions $\nabla \cdot \mathbf{E} = 0$ and $\nabla \cdot \mathbf{B} = 0$, which results in

$$\nabla^2 \mathbf{E} = \mu_0 \epsilon_0 \frac{\partial^2 \mathbf{E}}{\partial t^2} \quad \text{and} \quad \nabla^2 \mathbf{B} = \mu_0 \epsilon_0 \frac{\partial^2 \mathbf{B}}{\partial t^2}.$$

We now see that Maxwell's equations for **E** and **B** in free space are of the same form as the *classical wave equation*, which in 1D is given by (e.g., French, 1971; McMahon, 2006)

$$\frac{\partial^2 y(x,t)}{\partial x^2} = \frac{1}{v^2}\frac{\partial^2 y(x,t)}{\partial t^2},$$

where v is the wave speed with magnitude defined by $v \equiv \sqrt{1/\mu_0 \epsilon_0}$. Given the known values of μ_0 and ϵ_0, one finds (amazingly) that $v = c \approx 3.00 \times 10^8$ m s^{-1} (i.e., the observed speed of light in a vacuum).

BOX 2 Radiative energy transfer.

As already mentioned, EM waves carry energy with them away from their source at light speed, c (cf. Fig. 2). In empty space, these waves can propagate indefinitely, which is the basis behind remote sensing (cf. Box 1).

The radiative transfer of energy is derived from *Poynting's theorem*, which is the work-energy theorem of electrodynamics (e.g., Griffiths, 1989). In classical mechanics, we recall that work is defined as a force applied over distance, $dW = \mathbf{F}\cdot d\mathbf{l} = \mathbf{F}\cdot \mathbf{v}\, dt$, where **l** and **v** are the distance and velocity vectors, respectively. Integrated over a finite distance, the total work performed equals the negative of the potential energy (e.g., Barger and Olsson, 1995).

Applying the Lorentz force law, which specifies the force imposed on electric charge q as $\mathbf{F} = q\,[\mathbf{E} + (\mathbf{v}\times\mathbf{B})]$, and noting that $(\mathbf{v}\times\mathbf{B})\cdot\mathbf{v} = 0$, it can be shown that the work performed per unit time (i.e., the *power*) by the electric and magnetic fields on charges within a volume V is given by

Continued

BOX 2   Radiative energy transfer—cont'd

$$\frac{dW}{dt} = \iiint_V (\mathbf{E} \cdot \mathbf{J})\, dx\, dy\, dz.$$

This can be expressed solely in terms of the **E** and **B** fields (eliminating **J**) by applying the Ampère-Maxwell law (top-right equation in Box 1). We end up with a two-term equation, with one term being the energy stored in the fields, and the second term (the *Poynting vector* term) being the rate at which energy is carried out of the volume, which is given by

$$\mathbf{S} = \frac{1}{\mu_0} \mathbf{E} \times \mathbf{B},$$

where **S** quantifies the instantaneous energy flux density (power), which is the energy per unit area, per unit time.

If we consider the idealized case of monochromatic wavelength, $d\lambda$, and confine ourselves to a directional solid angle, $d\Omega$, the average power transferred over multiple wave cycles is referred to as the *intensity* (or *radiance*), denoted $I_\lambda(\Omega)$, which has units of power per area, per wavelength, per steradian (steradians, abbreviated "sr," are the SI unit of solid angle, basically the equivalent of radians for planar angles). For a plane wave originating from a distant source, I_λ is proportional to the square of the source **E**-field amplitude

$$I_\lambda \equiv \bar{S} = \frac{1}{2}\epsilon_0 c\, E_0^2.$$

Although this description of EM radiation is from the classical point of view, in reality the energy transferred within the waves are contained within discrete light quanta called *photons*, and is proportional to the frequency of the radiation, that is, $E = h\nu$, where h is Planck's constant.

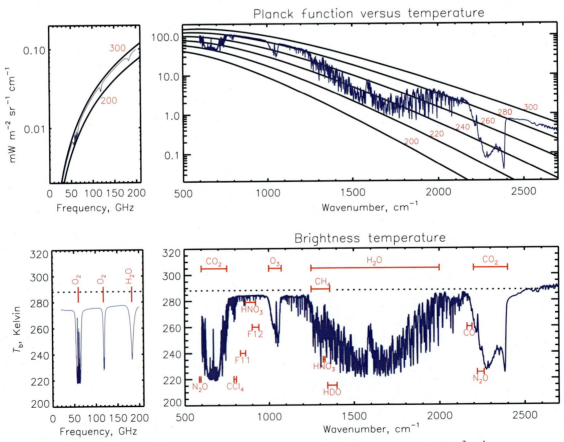

FIG. 3 Earth-emission spectra (*left*) MW and (*right*) thermal IR. The *top plots* show the radiance spectra (in mWm^{-2} sr^{-1} cm) along with the Planck blackbody curves, $B_\nu(T)$, terrestrial temperatures (200–300 K, *black lines*). The *bottom plots* show the equivalent brightness temperatures, T_B, obtained by taking the inverse Planck function defined by Eq. (2). Regions of optically active gases are denoted in *red fonts*. (*Courtesy of Chris Barnet, STC.*)

3.1 Terrestrial radiation

The sun is the primary source of incident energy on the Earth, with incoming solar radiation (insolation) either being reflected or absorbed and eventually reemitted back to space. In the microwave (MW) and thermal infrared (IR) regions, radiances emitted directly by Earth dominate over reflected solar contributions in these spectral regions. Fig. 3 shows Earth-emission spectra for both the IR and MW spectral regions in preferred spectral units of frequency (s^{-1}) and wavenumber (cm^{-1}), respectively. These spectral units are commonly denoted with Greek letter v and are inversely proportional to wavelength; for frequency, c/λ, and for wavenumber, $1/\lambda$ (hereafter in this chapter, v will implicitly refer to wavenumber, unless explicitly stated otherwise). It can be seen that different spectral regions are sensitive to different geophysical parameters. However, in this case, although surface variations (e.g., land vs. ocean) play a role, much of the so-called "action" occurs within the intervening atmosphere. The top plots show *blackbody* curves (denoted with black lines) defined by the *Planck function*, which specifies the thermally emitted radiance for an idealized "perfect emitter" (i.e., a "blackbody") given by

$$B_v(T) = \frac{2hc^2v^3}{\exp(hcv/kT) - 1}, \tag{1}$$

where B_v is the blackbody radiance typically expressed in $mWm^{-2}\, sr^{-1}\, cm$, T is the blackbody temperature in Kelvins (K), and h and k are the Planck and Boltzmann constants, respectively. Note that the Planck function can be inverted to obtain the effective emission temperature, or equivalent *brightness temperature*, T_B (also in K)

$$T_B(v) \equiv B_v^{-1}[B_v(T)] = \frac{hcv/k}{\ln\left(\frac{2hc^2v^3}{B_v(T)} + 1\right)}. \tag{2}$$

The Planck curves are shown alongside observed Earth emission for a standard atmosphere with a surface temperature of 288 K (blue lines), where it appears that large depressions (or "bites") have eaten into the curve relative to the idealized Planck curve at 288 K. These are the result of atmospheric absorption bands, whereby upwelling energy from the surface is absorbed, and as a result of the standard lapse rate (temperature decreasing with height in the troposphere) are reemitted at a cooler temperature. This is especially apparent in the thermal IR spectra (right-hand plots), and this phenomenon of net IR energy absorption by the atmosphere is referred to as the *greenhouse effect*.

3.2 Radiative transfer equations

Now that we understand that EM radiation carries with it energy (and momentum) from its source, the question is, given a passive remote measurement (cf. Section 4 for definitions of "passive" versus "active" remote sensing) of radiant intensity (or radiance), how do we extract from it the desired information, namely a measurement of the geophysical state parameter(s) of interest? To understand this, it is necessary to consider the relevant radiative transfer equation (RTE) for the given remotely sensed measurement. To illustrate this, we will consider the cases of passive solar spectrum ("visible," or VIS for short), thermal IR, and passive microwave (MW), spectral radiance measurements.

3.2.1 Solar spectrum (VIS)

In the VIS region of the electromagnetic spectrum, the observed Earth scene is illuminated by the sun and the incident solar radiation is *reflected* or *scattered* back to space. Unlike thermal IR and passive MW, where the radiance consists primarily of thermal absorption/reemission (discussed more in Sections 3.2.2 and 3.2.3), solar spectrum radiative transfer is primarily dominated by *scattering*, namely the process of radiation (whether conceived as waves or photons) is deflected from its initial path into or out of the sensor FOV. For environmental satellite applications that allow the assumption of a *single-scattering approximation* and an *isotropic* atmosphere (i.e., uniform in all directions), the electromagnetic radiation measured by a spaceborne VIS sensor located above the top of atmosphere (TOA) can be represented by the following equation (Gordon, 2003; Gordon and Castaño, 1987; Gordon and Clark, 1981; Gordon and Wang, 1994):

$$\underbrace{I_v(\theta)}_{\text{Radiance at TOA}} = \underbrace{I_{vs}(\theta)\, \mathcal{T}_v(\theta)}_{\text{Surface term}} + \underbrace{\left[\overbrace{I_{vr}(\theta)}^{\text{Rayleigh scattering}} + \overbrace{I_{va}(\theta)}^{\text{Aerosol scattering}}\right]}_{\text{Atmosphere term}}, \tag{3}$$

where $\mathcal{T}_v$ is the diffuse path transmittance, and I_{vr}, I_{va}, and I_{vs} are the radiances scattered from Rayleigh (molecular), aerosol, and surface scattering (reflectance), respectively, into the sensor FOV observing at zenith angle, θ. Note that the single-scattering approximation implies no interactions between the Rayleigh and aerosol contribution terms (Gordon and Wang, 1994).

The RTE may be simplified depending on the specific application or parameter of interest. For example, for ocean color, one may note that the term I_{vs} includes the contributions from both within water and from the interface reflectance; more details on the application of Eq. (3) toward ocean color can be found in Chapter 20.

In the case of aerosol remote sensing, the aerosol scattering term in Eq. (3), again assuming a single-scattering approximation, may be approximated as (Gordon and Castaño, 1987)

$$\underbrace{I_{va}(\theta)}_{\text{Aerosol radiance at TOA}} = \underbrace{\frac{1}{4\pi} F_{v\odot}}_{\text{Solar source term}} \cdot \underbrace{P_a(\Delta\theta)\,\omega_{va}}_{\text{Scattering properties}} \cdot \underbrace{\tau_{va} \sec(\theta)}_{\text{Optical thickness}}, \qquad (4)$$

where ω_{va} is the aerosol single-scattering albedo (i.e., the ratio of aerosol scattering to extinction), P_a is the aerosol scattering phase function (a measure of scattering of radiation from the solar zenith angle $\theta_\odot$ to θ, denoted by the relative angle $\Delta\theta$), τ_{va} is the vertical aerosol optical depth (AOD), and $F_{v\odot}$ is the attenuated solar flux after passage through the ozone layer. In a nutshell, Eq. (4) describes the radiation scattered by aerosols into a sensor FOV (observing at angle θ) as the product of the solar source radiance, the aerosol scattering properties, and the aerosol layer optical thickness (which is proportional to the aerosol concentration). Given assumptions about the optical properties of the aerosols, along with estimates of the magnitudes of the surface reflectance and Rayleigh scattering terms, Eqs. (3), (4) may be used to retrieve the AOD (discussed more in Chapter 22).

3.2.2 Thermal IR

In the thermal IR, terrestrial surfaces (e.g., land, water, snow, ice) are usually opaque (i.e., nontransparent) and absorbing (i.e., nonreflective, especially at small observing angles, $\theta < 30°$). Conservation of energy and thermodynamic equilibrium at the air-surface interface then imply that the surface emissivity is relatively high (i.e., somewhat close to unity). Thus, to simplify things here, one may consider the idealized case of a blackbody surface, where the surface emissivity is unity and the reflectance is zero. It can be shown in this case that the upwelling monochromatic IR radiance (or intensity) measured by a downlooking satellite sensor (located above the TOA, where pressure $p = 0$ hPa), viewing within a solid angle Ω, may be expressed as (e.g., Kidder and Vonder Haar, 1995, p. 63)

$$\underbrace{I_v(\Omega, p = 0)}_{\text{Radiance at TOA}} = \underbrace{B_v(T_s)\,\mathcal{T}_v(p_s, \Omega, X)}_{\text{Surface source term}} + \underbrace{\int_{p_s}^{0} B_v[T(p)] \frac{\partial \mathcal{T}_v(p, \Omega, X)}{\partial p} dp}_{\text{Atmosphere source term}}, \qquad (5)$$

where p_s is the surface pressure (hPa), B_v is the Planck function, $\mathcal{T}_v(p) \equiv \exp[-\tau_{va}\sec(\theta)]$, is the atmospheric path transmittance (a number between 0 and 1) from pressure p to the TOA, $T(p)$ is the atmospheric temperature profile, and X is the atmospheric absorbing gas concentration.

The first term on the right-hand side is the surface emission term (again, assuming a blackbody surface), which is attenuated by $\mathcal{T}_v$; in the case of thermal IR, the attenuation in question is primarily in the form of *absorption*. The second term on the right-hand side is the atmospheric contribution term, which accumulates the upwelling atmospheric radiance through the atmosphere; the vertical gradient in transmittance, $\partial \mathcal{T}_v/\partial p$, is referred to as the *weighting function*, which is a measure of how much a given infinitesimal layer dp contributes to upwelling radiance.

From Eq. (5), one may gain physical insight into the Earth-emission spectra shown in Fig. 3 (blue lines). Within atmospheric *window regions* (e.g., the region between 800 and 1000 cm^{-1} wavenumbers), the intervening atmosphere interacts minimally with the surface-leaving radiance. This implies that the total transmittance at the surface approaches unity, $\mathcal{T}_v \to 1$, and the gradient in transmittance from space goes to zero, $\partial \mathcal{T}_v/\partial p \to 0$, thus leaving only the surface term, $B_v(T_s)$.

However, for $0 < \mathcal{T}_v < 1$, the atmosphere term contributes in greater proportion, with the peak contributions coming from the layers where the weighting function peaks (i.e., where the gradient of transmittance peaks), while the surface term is attenuated. Because the temperature lapse rate through the troposphere (where the bulk of the optically active gases, X, are located) is a negative gradient in $T(p)$, the atmosphere term will generally be less than surface term, thereby resulting in a reduced radiance measured by the sensor at the TOA.

Note again that the assumption of a blackbody surface is a first-order approximation for small zenith-observing angles, $\theta < 30$ degrees; in general the surface has a spectral emissivity less than 1, and in practice it is not valid to assume a blackbody surface (given today's high-accuracy sensors). We consider the case of a nonblack surface with MW radiative transfer in Section 3.2.3.

3.2.3 Passive MW

The energy associated with Earth-emitted passive MW radiation is very weak, one million times weaker than solar visible and thermal IR radiation (Liu and Boukabara, 2014). MW wavelengths are much longer than that of VIS and IR radiation. Because of its long wavelength, MW radiation can penetrate most clouds, which make acquisition of atmospheric temperature and moisture under all-sky (i.e., clear and cloudy sky) conditions possible. Without significant scattering, the RTE of outgoing MW radiation at the TOA is very similar to Eq. (5), but MW variability of surface emissivity (ϵ_v) is quite large and must be explicitly included. Assuming an isotropic atmosphere with azimuthal symmetry, and *specular* (directional) reflectance, the passive MW RTE may be written as

$$\underbrace{I_v(\theta, p=0)}_{\text{Radiance at TOA}} = \underbrace{\epsilon_v(\theta) B_v(T_s) \mathcal{T}_v(p_s, \theta, X)}_{\text{Surface source term}} + \underbrace{\int_{p_s}^{0} B_v[T(p)] \frac{\partial \mathcal{T}_v(p, \theta, X)}{\partial p} dp}_{\text{Atmosphere source term}}$$
$$+ \underbrace{[1 - \epsilon_v(\theta)] \mathcal{T}_v(p_s, \theta, X) \int_{0}^{p_s} B_v[T(p)] \frac{\partial}{\partial p} \left[\frac{\mathcal{T}_v(p_s, \theta, X)}{\mathcal{T}_v(p, \theta, X)} \right] dp}_{\text{Surface reflectance term}} ,$$

(6)

where $\frac{\mathcal{T}_v(p_s, \theta, X)}{\mathcal{T}_v(p, \theta, X)}$ is the atmospheric path transmittance from pressure p to the surface.

On the right side of Eq. (6), the first two terms are the contributions from the surface and upward atmospheric radiation, respectively, similar to Eq. (5). The third term, however, is the surface-reflected downward atmospheric radiation, which is now a contributor given the nonunity surface emissivity. The MW radiative transfer becomes more complicated when scattering and polarization effects are significant and taken into account.

Both RTEs (5) and (6) can be *inverted* to obtain atmospheric profiles of temperature and water vapor (one of the radiatively active gases implicit in X), as will be discussed more in Section 4.2.

4. Satellite remote sensing

Environmental satellite remote sensing measurement systems are designed to be sensitive to specific portions of the EM spectrum, including ultraviolet (UV), VIS, thermal IR, and MW (see Table 1), and can be classified as being either *active* or *passive* systems. Active systems provide their own source energy (or "illumination") and then measure the backscattered energy; examples of active systems include radar and lidar. Passive systems, on the other hand, measure natural emitted or reflected energy from the scene of interest. Emission-based passive systems measure radiation emitted by the Earth at typical terrestrial temperatures (i.e., ≈200–320 K), whereas reflectance-based passive systems measure radiation emitted by the Sun at photosphere temperatures (i.e., ≈5500 K) that are then reflected back into the observer. Examples of passive sensors include radiometers and spectrometers. Observing systems may also be satellite based (downlooking) or surface

TABLE 1 Electromagnetic spectral regions for environmental passive remote sensing.

Region	λ range (μm)	Example applications	Example sensors
MUV (UV-B)	0.280–0.315	O_3 vertical profiles	OMPS, OMI, GOME-2, TOMS
NUV (UV-A)	0.315–0.400	Total column O_3, SO_2, aerosols	Same as UV-B
VIS	0.400–0.700	Clouds, aerosols, ocean color, vegetation	VIIRS, MODIS, GOES-ABI, AVHRR, MERSI/MERSI-II
NIR	0.700–3.50	Vegetation, soils, soil moisture, geology, burnt area	Same as VIS
IR	3.5–20	$T(p)$, $Q(p)$, $X(p)$, T_s, clouds, fires	CrIS, IASI, AIRS, VIIRS, GOES-ABI, AVHRR, MODIS, AERI, MAERI
MW	$0.16–1.5 \times 10^4$	$T(p)$, $Q(p)$, T_s, U_s	ATMS, AMSU-A, MHS

Spectral regions are listed in order from shorter to longer wavelengths for mid- and near-ultraviolet (MUV and NUV, or UV-B and UV-A, respectively), visible or solar spectrum (VIS), near-infrared (NIR), thermal infrared (IR), and finally, microwave (MW). Parameters $T(p)$, $Q(p)$, and $X(p)$ denote temperature, moisture, and gas concentration profiles as a function of atmospheric pressure, p, respectively; and T_s and U_s denote surface temperature and mean windspeeds, respectively.

based (uplooking). For the remainder of this book, our focus will be confined to passive remote sensing, given its prominence within satellite applications and its reliance on in situ field measurements.

4.1 Environmental satellites

Most English-speaking people have probably heard of the common phrase, "what goes up must come down," which is basically a statement about the everyday reality of living under the Earth's gravitational field. Throw a ball into the air, and it returns to the ground. Throw the ball harder, or better yet, hit it with a baseball bat or two-wood driver, and it takes longer to return, but return it still does. But in the case of using artificial satellites for observing the Earth from space, what goes up must decidedly *not* come down.

Remote sensing of the Earth from space requires satellites flying within stable, periodic, geocentric orbits, these providing platforms of practical use for Earth observing systems. Generally speaking, such orbits are determined based upon fundamental orbital mechanics, roughly following near-circular elliptical paths and behaving according to Kepler's laws. Kepler's third law for the satellite orbital period is given by (e.g., Barger and Olsson, 1995, pp. 151–152)

$$\tau = 2\pi \sqrt{\frac{(R_e + z_s)^3}{G M_e}}, \tag{7}$$

where z_s is the satellite altitude, G is the Newtonian gravitational constant ($G \approx 6.67430 \times 10^{-11}$ m^3 kg^{-1} s^{-2}), and R_e and M_e are the mean Earth radius and mass, respectively.

From Eq. (7), we may readily see that the orbital period is independent of the satellite mass and is proportional to $(R_e+z_s)^{3/2}$; thus the lower the satellite altitude, the shorter the orbital period and faster it must move to stay in orbit. There are a number of different orbital formations, but there are two basic orbits of particular interest for environmental applications (e.g., Kidder and Vonder Haar, 1995; Martin, 2004), namely the low-earth orbit (LEO) and geostationary or geosynchronous orbit (GEO), as illustrated in Fig. 4.

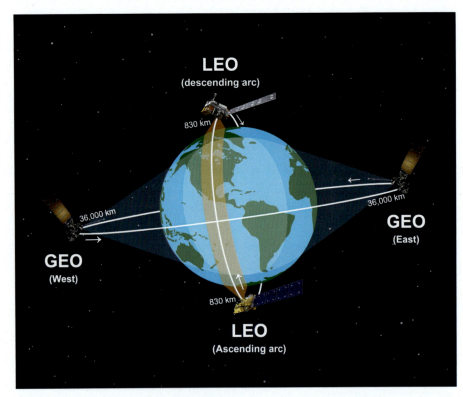

FIG. 4 Environmental satellite orbit schematic (not drawn to scale): low-earth orbit (LEO) and geostationary orbit (GEO). The GEO satellites are positioned depicting the NOAA GOES-East and GOES-West satellites, and revolve from west to east along the equator with a 24-h period. The LEO satellites roughly depict the JPSS series satellites, and are shown beginning their ascending (daytime) and descending (nighttime) orbital nodes, with the sensor scanning-swath coverage shown at the surface.

4.1.1 LEO satellites

As their name suggests, LEO satellites are flown at a very low altitude ($\approx$830 km), which from Eq. (7), allows for a complete orbital pass every $\approx$101.4 min. Because of this, one of the prime benefits of LEO satellites is a relatively high spatial resolution. The legacy Advanced Very High-Resolution Radiometer (AVHRR) instrument, for example, has an approximate 1.3 mrad FOV, which, assuming a locally flat surface, yields approximately $2 \times 830 \times \tan(0.0013/2) \approx 1.1$ km diameter nadir "footprint" at the surface.

For environmental remote sensing purposes, LEO satellites are typically flown in a polar sun-synchronous orbit that precesses throughout the year to maintain the same local equator crossing times (EXT) twice per day. These orbits pass near both poles several times throughout the day, and thus LEOs are often referred to as "polar orbits." Northbound portions of these orbits are typically during the daylight hours and referred to as "ascending node," whereas the southbound portions are during nighttime and called "descending node" orbits. Because the Earth rotates from west to east on its polar axis, the ascending passes proceed from south to north with a slight westward inclination and vice versa for the descending orbits, which facilitate a twice daily (1 day, 1 night) global precession per satellite (Fig. 4). To obtain near-global coverage of the earth during these orbits, satellite sensors typically employ a cross-track "whisk broom" scanning pattern, whereby a rotating mirror is used to obtain measurements from off-nadir FOV from the subsatellite point (e.g., Martin, 2004) (also, see Chapter 18, Fig. 2). However, these larger zenith angles can present challenges in the radiative transfer equation arising from more complicated geometry, notable examples including surface emissivity and reflectance (e.g., Dozier and Warren, 1982; Kornfield and Susskind, 1977; Nalli et al., 2001, 2008; Saunders, 1968; Smith et al., 1996; Wu and Smith, 1996), as well as apparent cloud cover and increased atmospheric path length (Naber and Weinman, 1984; Nalli et al., 2012, 2016).

Due to the limited mission lifetimes of these satellites, weather and climate agencies such as the US National Oceanic and Atmospheric Administration (NOAA) operate 2+ LEO satellites at any given time, with one serving as the primary operational satellite and older ones serving for overlapping continuity and redundant backup. But to increase the observation frequency, additional LEO satellites may be flown in different equator crossing time orbits to obtain >2 observations a day. NOAA has typically flown operational LEO satellites in the so-called "PM" or "afternoon" orbit, meaning that the daytime EXT occurs after 12:00. On the other hand, the European Organization for the Exploitation of Meteorological Satellites (EUMETSAT) flies operational LEO satellites in the AM orbit (daytime EXT before 12:00), which when combined with NOAA's Joint Polar Satellite System (JPSS) satellites (e.g., Goldberg et al., 2013), thus provides four EXT per day.

4.1.2 GEO satellites

GEO satellites, on the other hand, are flown over the equator with a 24-h orbital period that allows the satellite to remain "stationary" relative to the earth's rotating reference frame, hence the name "geostationary." However, this requires the orbit to be at a much higher altitude to support a much longer orbital period of 24 h (as opposed to $\approx$100 min). Solving Eq. (7) for z_s and setting $\tau = 24$ h, we find that the altitude needed to support such an orbit is $\approx$35,800 km.

GEO orbits have the advantage of allowing sensors to view the earth in a near-continuous fashion, thereby enabling sampling with high temporal resolution ($\leq$1 hourly) such that "movies" (on relevant meteorological timescales) can even be assembled. Most of us today are familiar with seeing satellite movie loops of weather systems such as hurricanes and synoptic-scale mid-latitude cyclones (often broadcast on television news programs), but such imagery only became possible with the introduction of GEO satellites.

The primary drawback of GEO satellites is that they are limited to quasihemispherical coverage (per satellite), along with greater difficulty in maintaining high spatial resolution (the FOV must be much smaller to retain the same footprint diameter at the surface) and the necessity of having to observe nontropical zones with increasingly oblique zenith-observing angles. These drawbacks are distinct from those posed from the LEO satellites; thus, GEO and satellites are generally complementary for environmental remote sensing. The quasihemispherical coverage can be compensated by positioning multiple GEO satellites around the globe. NOAA currently operates two R-Series Geostationary Operational Environmental Satellites (GOES-R) over each coastline (e.g., Goodman et al., 2020; Schmit et al., 2005), and EUMETSAT operates Meteosat Second Generation (MSG) satellites over Europe, Africa, and the Indian Ocean (with Meteosat Third Generation [MTG] in the works).

4.2 Satellite applications

A remote sensing radiometer measures the reflected and emitted energy in discrete wavelengths for which it is designed to measure. The measured radiation (i.e., radiance) is converted to a physical variable such as temperature, gas concentration, surface property (e.g., spectral reflectance), etc. that provides information about the state of the target at a given time and

12 Introduction: Field measurements and remote sensing

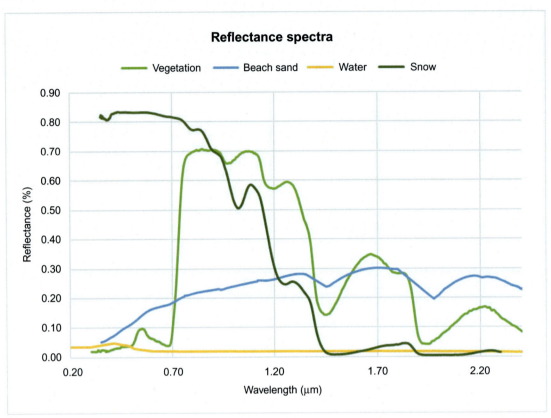

FIG. 5 Field measurements of the spectral surface reflectances (VIS to NIR) for four different surface types: green vegetation (*light green line*), beach sand (*blue line*), water (*yellow line*), and snow (*dark green line*). The measurements were obtained from the ground using a handheld spectroradiometer. *(From Kokaly, R., et al., 2017. USGS Spectral Library Version 7. Data Series 1035. US Geological Survey, Reston, VA, 61 pp. https://doi.org/10.3133/ds1035.)*

location. In order to interpret these measurements, it is important to understand the spectral properties of the target, which is typically measured in the field or in the laboratory.

For example, Fig. 5 shows field measurements (taken from a handheld spectroradiometer on the ground) of the spectral reflectances of green vegetation, sand, water, and snow surfaces. The spectral region encompasses the VIS to near-IR (NIR), which is highly transparent (i.e., $\mathcal{T}_v \to 1$) under noncloudy atmospheric conditions, thus allowing a relatively clear view of the surface. It can be seen that the spectral signatures of different surfaces (as shown in Fig. 5) vary a great deal, which means that in principle they can be distinguished from one another given such measurements from satellite. Green vegetation has a low reflectance in the red region of the spectrum due to absorption of chlorophyll during photosynthesis, with higher reflectance in the green, and very high reflectance in the NIR region due to the internal structure of the leaf. Clear, nonturbid water has a relatively low reflectance throughout the spectrum, and it is quasispecular (i.e., somewhat mirror-like), which means that it also depends on observing geometry. Slightly higher water reflectance can be seen in the more transparent blue region ($\lambda \leq 0.5\mu m$) compared to the rest of the spectrum, where the incident light is mostly absorbed. Ice and snow have a high reflectance throughout the VIS similar to clouds and therefore appear white. The reflectance of soil (e.g., sand) depends on a variety of factors such as the mineral composition, which gives different soils unique colors, moisture content, organic matter, and roughness. The spectral properties of various targets observed on the ground, therefore, provide vital information on the choice of wavelengths in which observations need to be made for studying the properties of ocean, land, and atmosphere features.

Sensors designed for land and ocean color remote sensing therefore are designed to collect reflectances in the visible (VIS) and NIR region of the spectrum, and the number of bands along with the band width defines the spectral resolution of the sensor. The AVHRR, which is an imaging radiometer that has been flown on NOAA Polar Operational Environmental Satellites (POES) since 1970s, only has five relatively narrow spectral channels covering the VIS to thermal-IR window regions. Its successor, the JPSS Visible Infrared Imaging Radiometer Suite (VIIRS), has 20 narrower bands making discrete spectral measurements in the 0.41–12.5 μm region. The better spectral resolution of VIIRS, resulting from a greater number of narrower bands, allows improved discrimination and more detailed analysis of the Earth's features compared to AVHRR.

Introduction: Field measurements and remote sensing **Chapter | 1** 13

Both AVHRR and VIIRS (as well as their counterpart on the GOES-R series, namely the Advanced Baseline Imager [ABI]) are examples of satellite passive VIS and IR *imagers*, which are primarily designed for high-resolution horizontal spatial mapping of a geophysical parameter. Passive IR and MW *sounders*, on the other hand, are instruments designed oriented toward three-dimensional (3D) sampling (vertical and horizontal), with channel selections sensitive to different pressure altitudes based upon the varying degrees of interaction of the radiation to the parameter of interest (i.e., the *sensitivity*). Today's IR sounders are *hyperspectral*, having thousands of such channels obtained from dispersive grating spectrometer or Fourier transform spectrometer (FTS) technologies (e.g., Chahine et al., 2006; Smith et al., 2009). Satellite sounder applications are discussed in more detail in Chapter 18. Figs. 6 and 7 highlight examples of operational satellite-derived product imagery from LEO and GEO platforms, respectively.

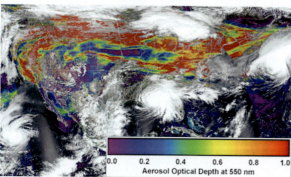

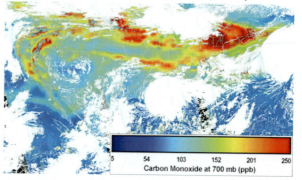

FIG. 6 NOAA-20 composite images of CONUS on September 15, 2020: (*top*) VIIRS true-color composite image, (*lower-left*) VIIRS retrieved solar spectrum (5.5 μm) AOD and (*lower-right*) 700 hPa carbon monoxide (CO) retrieved from the NUCAPS algorithm. Notable in these images are three different named tropical storms, as well as the large, transcontinental smoke plumes from fires over the western United States, especially California. The aerosol plumes are clearly evident in the true-color image (*top*) and retrieved AOD (*bottom left*), as well as the CO emissions from the fires (*lower right*).

14 Introduction: Field measurements and remote sensing

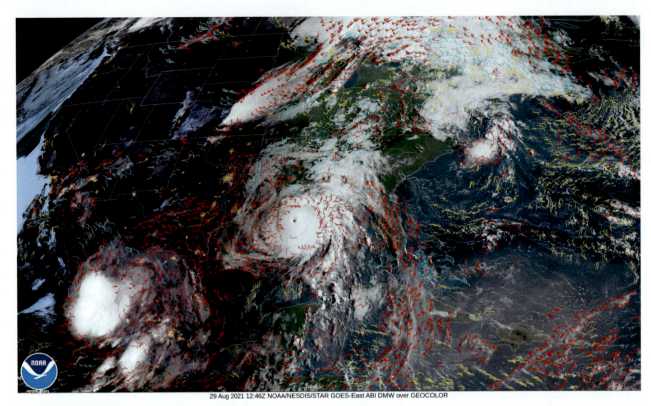

FIG. 7 GOES-16 (East) true-color image overlaid with derived motion winds (horizontal-component wind vectors) taken at 12:46 UTC, August 29, 2021. The day-night terminator is visible near the center of the image, and Hurricane Ida is about to make morning landfall with the Louisiana coast. Wind vectors are expressed using standard meteorological wind barbs, with speeds expressed in SI units using half, full, and pennant feathers for increments of 2.5, 5.0, and 25 ms^{-1}, respectively.

Fig. 6 shows an example of composite LEO images and derived products obtained from the NOAA-20 VIIRS (top and bottom left), and combined Cross-track Infrared Sounder (CrIS) and Advanced Technology Microwave Sounder (ATMS) sensors, during extensive fires (under drought conditions) over the western United States on September 15, 2020. The top plot shows a true-color (multispectral solar-reflected radiance) image over the continental United States (CONUS), with derived fire locations indicated by yellow to red dots. The smoke plumes from these fires have advected across the entire United States, extending into the northeast. Three named tropical storms are also visible in the image, as are different cloud formations and surface types. The lower-left plot is for the same data but with the retrieved AOD derived from solar backscattering in the 5.5 μm band, quantifying the optical thickness of smoke aerosol plumes from the fires. The lower right plot shows the retrieved carbon monoxide (CO) concentration at 700 hPa pressure altitude derived from the CrIS/ATMS NOAA-Unique Combined Atmospheric Processing System (NUCAPS) (e.g., Nalli et al., 2020; Smith and Barnet, 2019; Susskind et al., 2003, 2011), showing large-scale plumes of CO emissions from the fires.

Imagery from GEO platforms is highlighted in Fig. 7, which shows true-color images from the GOES-East (GOES-16) ABI instrument on August 29, 2021, along with derived motion winds (DMWs, also variously called in the past "atmospheric motion winds" and "cloud drift winds") overlaid using standard meteorological wind-barb symbols. The DMWs are inferred from cloud and moisture-gradient feature tracking between successive GEO images (Velden et al., 2005). This image captures both day and nighttime portions of the earth, with the day-night terminator running north to south roughly down the center. City lights visible over the western United States, and *sunglint*, basically the quasispecular reflection of the solar disk, is visible in the lower right over the tropical Atlantic Ocean. Hurricane Ida, a Category 4 Tropical Storm, is visible on the eastern edge of the terminator and about to make landfall with the Louisiana coast. Because of the oblique solar geometry, shadows are cast by vertically developed clouds. These shadows can provide a unique 3D perspective that includes the cloud vertical extents, which here clearly reveal the eye of the hurricane to be well developed. With a GEO satellite, multiple such images can be obtained and assembled to create "movie loops," much like a conventional video or movie camera, for monitoring the movement and progress of such storms; an animated movie loop of Hurricane Ida making landfall can be found in the online supplement at https://doi.org/10.1016/B978-0-12-823953-7.00002-2. (The MP4 animation of Hurricane Ida is courtesy of Matthew Jochum and the NESDIS/STAR GOES Imagery Team.)

4.2.1 Satellite remote sensing methods

The process of transforming satellite measurements from the sensor raw data to geophysical state measurements (e.g., as highlighted in Figs. 6 and 7) is comprised of three primary steps (cf. Chapter 18, Fig. 2), namely

1. *Calibration* of the sensor raw data (e.g., the digital counts registered by the detector) in the conversion to physical units (i.e., spectral radiance observations). The raw data records (RDRs) are also referred to as Level 1A (L1A for short).
2. *Geolocation* and mapping of the calibrated radiance observations to a fixed Earth coordinate grid (e.g., a geographic map projection); these are referred to as sensor data records (SDRs) or Level 1B (L1B for short).
3. *Retrieval* of the target geophysical state parameter from the L1B radiances such as sea surface temperature (SST) or atmospheric vertical temperature profile; these parameter *retrievals* have also been called environmental data records (EDRs) or Level 2 (L2 for short).

Each of these steps involves multiple tasks with uncertainties, assumptions, and/or approximations in the transformation from RDRs to geophysical parameters (EDRs). In particular, the last step described earlier is typically achieved using a *retrieval algorithm*. Retrieval algorithms may be based upon statistical relationships between the radiances and the parameter of interest, or inversion of a system of RTEs such as Eq. (5) or (6) given multiple spectral channels (Kidder and Vonder Haar, 1995). The former approaches are called *statistical algorithms* and they typically involve using traditional regression schemes (linear or nonlinear), or newer nonlinear machine learning methods such as neural networks, based upon empirical, model, and/or simulated training data. We may here note that empirically based statistical retrieval algorithms, by their very nature, require collocated independent field measurements.

The latter approaches, on the other hand, are referred to as *physical retrieval methods* (e.g., Chahine, 1972; Nalli and Smith, 1998; Smith, 1967; Susskind et al., 2003, 2011) or *inversion methods* (e.g., Chahine, 1970; Rodgers, 2000; Twomey, 1977). Eqs. (5), (6) are usually nonlinear and under-determined, the latter meaning that there are more unknowns than equations, so the problem is inherently *ill-posed*. The problem may be simplified via linearization of the RTE, along with subsequent regularization, typically in the form of an optimal estimation (OE) algorithm (e.g., Maahn et al., 2021; Rodgers, 1976; Smith and Barnet, 2019), which seeks to minimize a *cost function* between calculated or *calc* (via the RTE) and observed or *obs* (via the measurement) radiances. This process may be thought of as optimally minimizing "*obs minus calc*" (or *obs − calc*), and it is employed both by direct retrieval systems as well as within radiance assimilation systems for numerical weather prediction (NWP) models.

But note here that the OE approach requires the introduction of a priori information (i.e., a "virtual measurement") that is beforehand considered representative of the state (e.g., the mean or "best estimate") (Rodgers, 1976, 2000). It thus follows that a convenient and common a priori constraint would be a climatological value, which typically relies either directly or indirectly upon the acquisition and application of high-quality field data (e.g., Maddy et al., 2009). Because the *calc* is forward calculated from numerical integration of an RTE within a radiative transfer model or algorithm (RTM or RTA), it is referred to the *forward problem*, whereas the retrieval of the geophysical state parameters is referred to as the *inverse problem*. An example operational RTM designed for both direct NWP radiance assimilation and the retrieval of EDRs is the US Joint Center for Satellite Radiance Assimilation (JCSDA) Community Radiative Transfer Model (CRTM) (Han et al., 2006); the CRTM package is publicly available from GitHUB at https://github.com/JCSDA/crtm.

Other uses of field data within satellite remote sensing include radiance bias corrections and blended analyses. *Bias corrections* (also sometimes called "tuning") involve deriving an empirical adjustment to remove systematic differences between spectral calculations (*calc*) inherent in a given RTM versus concurrent measurements (*obs*) from a given sensor. As such, bias corrections address residual issues in the forward problem not accounted for by the RTM, and are thus important for inversion systems (assimilation or retrieval) given the dependence on the minimization of *obs − calc*. But note that to achieve this, bias correction algorithms rely on accurate state parameter specification to perform the necessary calculations for the atmospheric/oceanic state being observed. *Blended analyses*, on the other hand, involve bias corrections of a retrieved parameter in a postprocessing (i.e., postinversion) step. A prime example of blended analysis in passive remote sensing is the optimal interpolation SST (OISST) produced operationally by NOAA going back to 1981 (Reynolds, 1988; Reynolds and Smith, 1994; Reynolds et al., 2007). In a nutshell, the OISST is derived from thermal IR imagers onboard LEO satellites, but it uses concurrent in situ buoy data (cf. Chapters 5 and 19) to remove known systematic errors (biases) in the data caused by clouds, aerosols, and other atmospheric contaminants. This allows for the advantage of having globally analyzed SST fields, but without the biases that are sometimes present in the satellite data.

Because of the inherent variability of the Earth system in space and time, as well as the differences in the engineering design, specifications, and capabilities of remote sensing instruments, it is vital to compare remote sensing measurements with those that are collected in situ to verify and validate the observations (discussed in more detail in Section 5). Field measurements are therefore a critical component of routine verification and validation of remote sensing measurements.

5. Calibration/validation (cal/val)

Perhaps two of the most critical and widespread applications of field data in remote sensing are *calibration* and *validation* (or "cal/val" for short). These are two closely related endeavors involving "the process of ascribing uncertainties...through comparison with correlative observations" (Fetzer et al., 2003). Typically, "calibration" refers to the specification of uncertainties in SDRs (Level 1 radiances), with attention given to "zeroing" the measurement (eliminating bias) relative to a traceable standard, whereas "validation" refers to error specification of the ensuing geophysical EDRs (Level 2 retrievals). As with any assessment exercise, the validation task is typically performed with respect to some sort of established criteria that serves as the metric for mission success. Perhaps more fundamentally, rigorous validation comprises a key component of the algorithm development process (e.g., see Chapter 18).

5.1 Measurement correlation

As mentioned earlier, in situ field measurements are fundamentally different in nature from remotely sensed measurements. Because in situ measurements are by definition in direct physical contact with the medium, they represent what may be characterized as *point measurements*. Remotely sensed measurements, on the other hand, measure the integral, volumetric radiative contributions from within the finite FOV of a given sensor.

From an LEO or GEO satellite, the nadir FOV at the ground can be on the order of several kilometers or more in diameter horizontally. For multispectral sensors sensitive to different atmospheric layers (e.g., hyperspectral IR sounders), the sensor vertical resolution (which is also on the order of kilometers) must also be taken into account (see Chapter 18). Thus, in addition to space-time collocation of the two measurements, any variability within the FOV (horizontal and vertical) contributes to what we call *mismatch error*, discussed more below.

5.2 Error analysis

Validation is fundamentally an exercise in error analysis (cf. Taylor, 1997). For a remotely sensed (or retrieved) environmental parameter, $\hat{x}$, the *systematic error* (or *bias*) relative to the correlative truth measurement x is given by the mean difference for the sample

$$\overline{\Delta x} \equiv \frac{1}{n}\sum_{j=1}^{n}\Delta x_j, \tag{8}$$

where $\Delta x_j = \hat{x}_j - x_j$ and n is the sample size. The *random error* (or *variability*) is given by the standard deviation of the sample, denoted σ

$$\sigma(\Delta x) \equiv \sqrt{\frac{1}{n-1}\sum_{j=1}^{n}\left(\Delta x_j - \overline{\Delta x}\right)^2}. \tag{9}$$

Given a well-calibrated measurement traceable to a reference standard, the bias or systematic error may be assumed to be negligible, in which case the uncertainty of the measurement, denoted u, is simply given by the standard deviation (i.e., $u \equiv \sigma$). The root mean square error (RMSE) provides a single metric that includes both systematic and random errors in the collocation samples

$$\text{RMSE}(\Delta x) \equiv \sqrt{\overline{\Delta x}^2 + \sigma^2(\Delta x)}. \tag{10}$$

The earlier equations assume a perfect correlative truth measurement x, whereas in reality we only have an accepted reference measurement (e.g., a temperature profile from a radiosonde) that has its own uncertainty u along with the uncertainties arising from the different measurement types; the latter type of error is called *mismatch error*. In the case of passive sensor validation, mismatch error can arise from the difference between the times of measurement (the satellite overpass vs. the in situ measurement time) as well as the spatial sampling (a point measurement vs. a volume average over a sensor FOV). The *combined standard uncertainty*, u_c (Taylor and Kuyatt, 1994), is the expected uncertainty of the measurement comparison that takes into account the uncertainties in the two measurement systems, remotely sensed and in situ ($\hat{u}$ and u, respectively), along with the measurement mismatch error (u_m), arrived at by taking the quadratic sum

$$u_c = \sqrt{\hat{u}^2 + u^2 + u_m^2}. \tag{11}$$

The *expanded uncertainty* for a desired confidence level is obtained by multiplying by a *coverage factor*, k

$$U = k\, u_c, \quad (12)$$

where $k = \{1,2,3,...\}$ are for 68.2%, 95.4%, 99.6%, ...confidence levels, respectively.

Given a representative sample of $\hat{x}$ and x, we can empirically determine the standard deviation, σ, from Eq. (9), which could then be checked for consistency against the theoretical expanded uncertainty, that is, $\sigma \leq U$. Alternatively, given estimates of u and u_m (e.g., u from instrument manufacturer specifications and u_m determined experimentally), one may theoretically estimate the "true" uncertainty (or error) of the remotely sensed measurement, $\hat{u}$ from the empirically determined $\sigma \equiv u_c$ as

$$\hat{u} = \sqrt{\sigma^2 - u^2 - u_m^2}. \quad (13)$$

Because of the difficulty of quantifying the mismatch error term, u_m, application of Eq. (12) to passive remote sensing validation, for example, using atmospheric state best estimates from field campaigns (Tobin et al., 2006) remains an area of current active research (cf. Chapter 12).

5.3 Intensive cal/val and long-term monitoring

Toward these ends (environmental satellite cal/val), collaborative agencies have developed and sustained observing networks and cal/val strategies. For example, the JPSS Program has planned and supported continued validation over the various phases of satellite mission lifetimes, including prelaunch, early orbit checkout, intensive cal/val, and long-term monitoring phases (Zhou et al., 2016), and in response NOAA/NESDIS has established integrated cal/val and long-term monitoring (ICVS-LTM). (For more details on the JPSS ICVS-LTM, see https://www.star.nesdis.noaa.gov/icvs/.) The GOES-R Algorithm Working Group (AWG) (Goodman et al., 2020) has likewise supported cal/val of the GOES series. (For more details on GOES-R AWG cal/val, see https://www.star.nesdis.noaa.gov/goesr/docs_reports_ValWksp2.php.) The Committee on Earth Observation Satellites (CEOS) (Belward, 1999) has coordinated multiagency efforts, which include over 50 agencies operating over 100 satellites (e.g., https://ceos.org), and global community cal/val efforts within their online cal/val portal (http://calvalportal.ceos.org/home). Some of these missions and observing networks will be discussed or alluded to in chapters that follow.

6. Outline and structure of the book

This book is arranged in three parts, roughly proceeding from lower- to higher-level topics pertaining to the book's subject. The book content is not meant to be fully inclusive of all aspects of these topics (indeed, it would not be possible to attain this within a single volume), but we have tried to feature some of the more relevant and current subject areas.

6.1 Part I: Instrumentation

Part I features seven chapters on fundamental instrumentation and Earth observing systems as follows.

- Chapter 2 discusses balloon-borne radiosondes, in situ instrument packages that measure atmospheric profiles of state parameters up through the tropopause.
- Chapter 3 discusses frost-point hygrometers, which are instruments also typically deployed on balloons, but these providing high-accuracy water vapor reference measurements into the stratosphere.
- Chapter 4 similar to the previous two chapters, this chapter covers ozonesondes, in situ instruments that measure ozone profiles up into the stratosphere.
- Chapter 5 is on oceanographic buoys, floating platforms for obtaining routine in situ marine surface measurements.
- Chapter 6 covers surface-based thermal IR spectrometers, which are similar to comparable satellite-based instruments, but located at the surface.
- Chapter 7 provides a technical overview of ground-based sunphotometers (handheld and otherwise) for measuring solar spectrum AOD.
- Chapter 8 overviews the AirCore balloon-borne in situ sampling system, a relatively new technology for collecting ambient air samples to perform ground-based analysis allowing a derivation of atmospheric trace gas profiles.

6.2 Part II: Intensive campaigns

Part II follows up with six chapters discussing how these basic measurements and observations are utilized within macroscale observing strategies (e.g., closure experiments) in the support of passive remote sensing applications.

- Chapter 9 provides an overview of intensive campaigns involving high-altitude aircraft (e.g., ER-2) with high-accuracy radiometric sensors (e.g., S-HIS) for satellite sensor cal/val.
- Chapter 10 discusses aircraft-based campaigns focused on acquiring in situ profiles using dropsondes (radiosondes "dropped" from aircraft) strategically deployed in the vicinity of inclement weather events (namely hurricanes).
- Chapter 11 covers ship-based campaigns designed to establish, acquire, and maintain critical in situ data over remote ocean locations.
- Chapter 12 describes coordinated, land-based intensive observing periods (IOPs) designed to acquire special datasets over a study region.
- Chapter 13 covers aircraft-based campaigns designed to acquire profiles of atmospheric trace gases, especially including carbon greenhouse gases.
- Chapter 14 presents a reverse perspective on the use of satellite data for the benefit of providing near real-time guidance for intensive campaigns.

6.3 Part III: Satellite applications

Finally, Part III consists of nine chapters covering some of the more prominent applications of these measurements and observing strategies for environmental passive remote sensing.

- Chapter 15 provides a technical overview example on the use of in situ datasets for the purpose of satellite sensor calibration/validation.
- Chapter 16 discusses the NOAA Products Validation System (NPROVS) for routine collocation of global radiosonde data with multiple satellite sounder products for convenient, user-friendly statistical analyses, and intercomparisons.
- Chapter 17 overviews the validation of satellite passive microwave sounder profile retrievals.
- Chapter 18 provides a technical overview of the methodologies used for validating thermal infrared satellite sounder profile retrievals along with example applications.
- Chapter 19 discusses the use of in situ data (e.g., oceanographic buoys) for bias-correcting satellite-derived SST and generating climate-quality blended analyses.
- Chapter 20 overviews the vicarious calibration and validation of satellite ocean color retrievals.
- Chapter 21 similar to Chapter 19, but instead focuses on land surface temperature validation.
- Chapter 22 similar to Chapter 21, but focused on atmospheric aerosol product validation.
- Chapter 23 closes the book with an example of the application of satellite data and surface-based field data for studying mesoscale severe weather phenomena and improving forecasts (namely downbursts).

Acknowledgments

The authors acknowledge the support of the JPSS Program Office and Proving Ground and Risk Reduction (PGRR) initiatives, the NOAA/NESDIS/STAR Satellite Meteorology and Climatology Division (SMCD), and the STAR IR Soundings Team (K. Pryor et al.). We also express our appreciation to Alexander Smirnov (NASA/GSFC), and Quanhua (Mark) Liu, Pubu Ciren, Jianwei Wei, and Chris Grassotti (NOAA/NESDIS/STAR), for their contributions to Section 3.2; Matt Jochum and Lori Brown (STAR) for assistance with Fig. 7, and Chris Barnet (STC) for kindly providing Fig. 3; and finally to Renellys Perez (NOAA/AOML) for carefully reviewing the manuscript and providing constructive feedback that has improved the quality of the chapter. The scientific results and conclusions, as well as any views or opinions expressed herein, are those of the author(s) and do not necessarily reflect those of NOAA or the Department of Commerce.

References

Barger, V.D., Olsson, M.G., 1995. Classical Mechanics: A Modern Perspective, second ed. McGraw-Hill, Inc., New York, NY.
Belward, A.S., 1999. International co-operation in satellite sensor calibration; the role of the CEOS working group on calibration and validation. Adv. Space Res. 23 (8), 1443–1448. https://doi.org/10.1016/S0273-1177(99)00296-3.
Chahine, M.T., 1970. Inverse problems in radiative transfer: determination of atmospheric parameters. J. Atmos. Sci. 27 (6), 960–967.
Chahine, M.T., 1972. A general relaxation method for inverse solution of the full radiative transfer equation. J. Atmos. Sci. 29 (4), 741–747.
Chahine, M.T., et al., 2006. AIRS: improving weather forecasting and providing new data on greenhouse gases. Bull. Am. Meteorol. Soc. 87 (7), 911–926.

Dozier, J., Warren, S.G., 1982. Effect of viewing angle on the infrared brightness temperature of snow. Water Resour. Res. 18 (5), 1424–1434.
Fetzer, E., et al., 2003. AIRS/AMSU/HSB validation. IEEE Trans. Geosci. Remote Sens. 41 (2), 418–431.
French, A.P., 1971. Vibrations and waves. In: MIT Introductory Physics, W.W. Norton & Company, New York.
Goldberg, M.D., Kilcoyne, H., Cikanek, H., Mehta, A., 2013. Joint polar satellite system: the United States next generation civilian polar-orbiting environmental satellite system. J. Geophys. Res. Atmos. 118, 13463–13475. https://doi.org/10.1002/2013JD020389.
Goodman, S.J., Schmit, T.J., Daniels, J., Redmon, R.J., 2020. The GOES-R Series: A New Generation of Geostationary Environmental Satellites. Elsevier, Cambridge, MA, https://doi.org/10.1016/B978-0-12-814327-8.09988-1. 283 pp.
Gordon, H.R., 2003. Pitfalls in atmospheric correction of ocean color imagery: how should aerosol optical properties be computed?: Comment. Appl. Opt. 42 (3), 542–544. https://doi.org/10.1364/AO.42.000542.
Gordon, H.R., Castaño, D.J., 1987. Coastal Zone Color Scanner atmospheric correction algorithm: multiple scattering effects. Appl. Opt. 26 (11), 2111–2122. https://doi.org/10.1364/AO.26.002111.
Gordon, H.R., Clark, D.K., 1981. Clear water radiances for atmospheric correction of coastal zone color scanner imagery. Appl. Opt. 20 (24), 4175–4180.
Gordon, H.R., Wang, M., 1994. Retrieval of water-leaving radiance and aerosol optical thickness over the oceans with SeaWiFS: a preliminary algorithm. Appl. Opt. 33 (3), 443–452. https://doi.org/10.1364/AO.33.000443.
Griffiths, D.J., 1989. Introduction to Electrodynamics, second ed. Prentice-Hall, Inc., Englewood Cliffs, NJ. 532 pp.
Han, Y., van Delst, P., Liu, Q., Weng, F., Yan, B., Treadon, R., Derber, J., 2006. JCSDA Community Radiative Transfer Model (CRTM)—Version 1. NOAA Technical Report NESDIS 122. National Oceanic and Atmospheric Administration (NOAA), Washington, DC.
Holben, B.N., et al., 1998. AERONET—a federated instrument network and data archive for aerosol characterization. Remote Sens. Environ. 66, 1–16.
Kidder, S.Q., Vonder Haar, T.H., 1995. Satellite Meteorology: An Introduction. Academic Press, San Diego. 466 pp.
Kornfield, J., Susskind, J., 1977. On the effect of surface emissivity on temperature retrievals. Mon. Weather Rev. 105, 1605–1608.
Le Marshall, J., et al., 2006. Improving global analysis and forecasting with AIRS. Bull. Am. Meteorol. Soc. 87 (7), 891–894.
Le Marshall, J., Jung, J., Goldberg, M., Barnet, C., Wolf, W., Derber, J., Treadon, R., Lord, S., 2008. Using cloudy AIRS fields of view in numerical weather prediction. Aust. Meteorol. Mag. 57, 249–254.
Liu, Q., Boukabara, S., 2014. Community radiative transfer model (CRTM) applications in supporting the Suomi national polar-orbiting partnership (SNPP) mission validation and verification. Remote Sens. Environ. 140, 744–754. https://doi.org/10.1016/j.rse.2013.10.011.
Maahn, M., Turner, D.D., Löhnert, U., Posselt, D.J., Ebell, K., Mace, G.G., Comstock, J.M., 2021. Optimal estimation retrievals and their uncertainties: what every atmospheric scientist should know. Bull. Am. Meteorol. Soc. 101 (9), E1512–E1523. https://doi.org/10.1175/BAMS-D-19-0027.1.
Maddy, E.S., Barnet, C.D., Gambacorta, A., 2009. A computationally efficient retrieval algorithm for hyperspectral sounders incorporating a priori information. IEEE Geosci. Remote Sens. Lett. 6, 802–806. https://doi.org/10.1109/LGRS.2009.2025780.
Martin, S., 2004. An Introduction to Ocean Remote Sensing, first ed. Cambridge University Press, New York. 416 pp.
McMahon, D., 2006. Relativity Demystified. "Demystified" series, McGraw-Hill, New York. 345 pp.
Naber, P.S., Weinman, J.A., 1984. The angular distribution of infrared radiances emerging from broken fields of cumulus clouds. J. Geophys. Res. 89 (D1), 1249–1257.
Nalli, N.R., Smith, W.L., 1998. Improved remote sensing of sea surface skin temperature using a physical retrieval method. J. Geophys. Res. 103 (C5), 10527–10542.
Nalli, N.R., Smith, W.L., Huang, B., 2001. Quasi-specular model for calculating the reflection of atmospheric emitted infrared radiation from a rough water surface. Appl. Opt. 40 (9), 1343–1353.
Nalli, N.R., Minnett, P.J., van Delst, P., 2008. Emissivity and reflection model for calculating unpolarized isotropic water surface leaving radiance in the infrared. 1: Theoretical development and calculations. Appl. Opt. 47 (21), 3701–3721.
Nalli, N.R., Barnet, C.D., Maddy, E.S., Gambacorta, A., 2012. On the angular effect of residual clouds and aerosols in clear-sky infrared window radiance observations: sensitivity analyses. J. Geophys. Res. 117, D12208. https://doi.org/10.1029/2012JD017667.
Nalli, N.R., Smith, W.L., Liu, Q., 2016. Angular effect of undetected clouds in infrared window radiance observations: aircraft experimental analyses. J. Atmos. Sci. 73 (5), 1987–2010. https://doi.org/10.1175/JAS-D-15-0262.1.
Nalli, N.R., et al., 2020. Validation of carbon trace gas profile retrievals from the NOAA-unique combined atmospheric processing system for the cross-track infrared sounder. Remote Sens. 12 (19), 3245. https://doi.org/10.3390/rs12193245.
Noh, Y.-C., Lim, A.H.N., Huang, H.-L., Goldberg, M.D., 2020. Global forecast impact of low data latency infrared and microwave sounders observations from polar orbiting satellites. Remote Sens. 12, 14. https://doi.org/10.3390/rs12142193.
Quinn, P.K., Coffman, D., 1998. Local closure during the first aerosol characterization experiment (ACE 1): aerosol mass concentration and scattering and backscattering coefficients. J. Geophys. Res. 103 (D13), 16575–16596.
Reynolds, R.W., 1988. A real-time global sea surface temperature analysis. J. Clim. 1, 75–76.
Reynolds, R.W., Smith, T.M., 1994. Improved global sea surface temperature analyses using optimum interpolation. J. Clim. 7, 929–948.
Reynolds, R.W., Smith, T.M., Liu, C., Chelton, D.B., Casey, K.S., Schlax, M.G., 2007. Daily high-resolution-blended analyses for sea surface temperature. J. Clim. 20, 5473–5496.
Rodgers, C.D., 1976. Retrieval of atmospheric temperature and composition from remote measurements of thermal radiation. Rev. Geophys. Space Phys. 14 (4), 609.
Rodgers, C.D., 2000. Inverse Methods for Atmospheric Sounding: Theory and Practice, Atmospheric, Oceanic and Planetary Physics, vol. 2 World Scientific, Singapore.
Saunders, P.M., 1968. Radiance of sea and sky in the infrared window 800–1200 cm^{-1}. J. Opt. Soc. Am. 58 (5), 645–652.

Schmit, T.J., Gunshor, M.M., Menzel, W.P., Gurka, J.J., Li, J., Bachmeier, A.S., 2005. Introducing the next-generation advanced baseline imager on GOES-R. Bull. Am. Meteorol. Soc. 86 (8), 1079–1096.

Smith, W.L., 1967. An iterative method for deducing tropospheric temperature and moisture profiles from satellite radiation measurements. Mon. Weather Rev. 95 (6), 363–369.

Smith, N., Barnet, C.D., 2019. Uncertainty characterization and propagation in the community long-term infrared microwave combined atmospheric product system (CLIMCAPS). Remote Sens. 11 (10), 1227. https://doi.org/10.3390/rs11101227.

Smith, W.L., et al., 1996. Observations of the infrared properties of the ocean: implications for the measurement of sea surface temperature via satellite remote sensing. Bull. Am. Meteorol. Soc. 77, 41–51.

Smith, W.L., et al., 2009. Technical note: Evolution, current capabilities, and future advance in satellite nadir viewing ultra-spectral IR sounding of the lower atmosphere. Atmos. Chem. Phys. 9, 5563–5574.

Stull, R.B., 1988. An Introduction to Boundary Layer Meteorology. Kluwer Academic Publishers, Dordrecht, The Netherlands.

Susskind, J., Barnet, C.D., Blaisdell, J.M., 2003. Retrieval of atmospheric and surface parameters from AIRS/AMSU/HSB data in the presence of clouds. IEEE Trans. Geosci. Remote Sens. 41 (2), 390–409.

Susskind, J., Blaisdell, J., Iredell, L., Keita, F., 2011. Improved temperature sounding and quality control methodology using AIRS/AMSU data: the AIRS science team version 5 retrieval algorithm. IEEE Trans. Geosci. Remote Sens. 49 (3), 883–907. https://doi.org/10.1109/TGRS.2010.2070508.

Taylor, J.R., 1997. An Introduction to Error Analysis: The Study of Uncertainties in Physical Measurements, second ed. University Science Books, Sausalito, CA.

Taylor, B.N., Kuyatt, C.E., 1994. Guidelines for Evaluating and Expressing the Uncertainty of NIST Measurement Results. NIST Technical Note 1297. US National Institute of Standards and Technology, Gaithersburg, MD.

Tobin, D.C., et al., 2006. Atmospheric radiation measurement site atmospheric state best estimates for atmospheric infrared sounder temperature and water vapor retrieval validation. J. Geophys. Res. 111, D09S14. https://doi.org/10.1029/2005JD006103.

Twomey, S., 1977. Introduction to the Mathematics of Inversion in Remote Sensing and Indirect Measurements. Developments in Geomathematics, vol. 3 Elsevier Scientific, Mineola, NY. Dover Phoenix 2002 ed.

Velden, C., et al., 2005. Recent innovations in deriving tropospheric winds from meteorological satellites. Bull. Am. Meteorol. Soc. 86 (2), 205–224.

Wu, X., Smith, W.L., 1996. Sensitivity of sea surface temperature retrieval to sea surface emissivity. ACTA Meteorol. Sin. 10, 376–384 (Engl. Transl.).

Zhou, L., Divakarla, M., Liu, X., 2016. An overview of the joint polar satellite system (JPSS) science data product calibration and validation. Remote Sens. 8, 139. https://doi.org/10.3390/rs8020139.

Part I

Instrumentation

Chapter 2

Balloon-borne radiosondes

Holger Vömel[a] and Bruce Ingleby[b]

[a]*National Center for Atmospheric Research, Boulder, CO, United States,* [b]*European Centre for Medium Range Weather Forecasts (ECMWF), Reading, United Kingdom*

Chapter outline

1. Introduction	23	3.2 Winds	31
2. Measurement technology	23	3.3 Clouds	31
2.1 Radiosondes	23	3.4 Using NWP data assimilation fields	32
2.2 Receiving system and data formats	27	4. Summary	32
2.3 Operations	27	References	33
3. Cal/val considerations	28	Further reading	35
3.1 Temperature and humidity profiling	29		

The icy layers of the upper atmosphere contain conundrums enough to be worthy of humanity's greatest efforts.

Hugo Hergesell

1. Introduction

In situ profiling of our atmosphere has been the mainstay of meteorology since the late 1890s. Early observations used kites as platform and mechanical chart recorders, which were replaced by radiosondes around the middle of the 20th century. These instruments have been a leading technique for vertically resolved observations of the atmospheric state for over 80 years.

Radiosondes are launched globally, and their data are used in numerical weather prediction models, in the prediction of severe storms, for aviation meteorology and marine forecasts, for climate change research, and as ground truth for satellite observations. For example, Fig. 1 shows an operator launching a radiosonde at the meteorological station at San Cristobal, Galapagos, Ecuador.

Most radiosonde observations are taken at about 800 meteorological observatories (Ingleby et al., 2016a). In addition, science projects conducted by research institutions often supplement operational observations with higher frequency launches and launches in regions away from operational soundings.

Before the advent of satellite meteorology in the late 1970s, radiosonde observations were the most important technique to measure the state of the atmospheric column. Meanwhile, satellite observations provide near-global observations of the entire atmosphere, but radiosondes are still of great importance. While aircraft observations cover the lowest 10 km or so of the atmosphere, radiosondes are still the only technique that provides direct and simultaneous observations of pressure, temperature, relative humidity, and horizontal winds reaching into the middle stratosphere. Furthermore, satellites rely on these observations for calibration and validation of retrieved parameters. Therefore, it is essential to understand the capabilities and limitations of radiosondes to interpret observations by satellites correctly.

2. Measurement technology

2.1 Radiosondes

Modern radiosondes, such as the sonde shown in Fig. 2, are small, lightweight devices, which contain sensors to measure pressure, temperature, relative humidity, and winds with remarkable accuracy. Various balloon sizes are used to reach different target altitudes. Routine climatological observations cover the altitude region between the surface and the middle

FIG. 1 Meteorologist launching a radiosonde at San Cristobal, Galapagos, Ecuador.

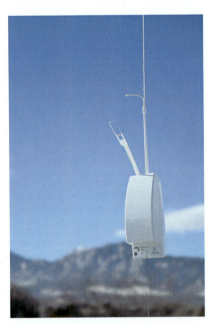

FIG. 2 Radiosonde with temperature and humidity sensor boom extending up and out.

stratosphere up to around 35 km, and routine meteorological observations reach 20–25 km. Operators typically inflate balloons to achieve an average rise rate of about 5 m/s, which then cover a complete tropospheric profile in roughly 45 min.

A typical radiosonde profile measured at San Jose, Costa Rica, is shown in Fig. 3. The skew-T plot used in this figure is a widespread tool in the meteorological community to represent the atmospheric state. It allows immediate identification of atmospheric parameters used in analysis and forecasting.

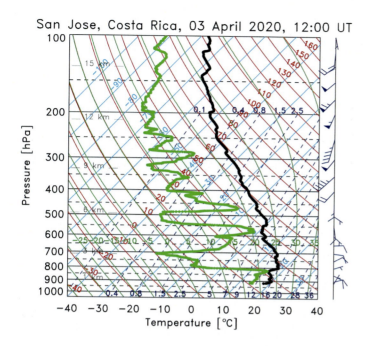

FIG. 3 Skew-T plot for a radiosonde ascent at San Jose, Costa Rica. Pressure is used as a vertical coordinate; air *(black)* and dew point *(green)* temperature are shown as *solid lines*, wind profile is displayed as wind barbs.

Although radiosondes are small, low-cost instruments, their technology and level of sophistication are remarkable. Here we describe the measurement of temperature, humidity, winds, and pressure separately.

2.1.1 Temperature

Atmospheric temperature is measured using a small sensor mounted on an arm extending out of the radiosonde package to reduce contamination by the package itself. Most manufacturers use either a small bead thermistor or a small wire resistance thermometer. These sensors must be calibrated over the entire range of temperatures for which radiosondes are designed, typically between −90°C and +30°C. Thermistors are highly non-linear elements that require a slightly more complex calibration than wire resistance sensors, which benefit from the higher linearity of these sensing elements. For atmospheric measurements, the calibration accuracy of both techniques is generally sufficient.

The most challenging environmental influence for radiosonde temperature measurements is solar heating of the sensing element and mount. To minimize this effect, manufacturers use sensors as small as possible and apply coatings that are highly reflective over most of the solar and infrared spectrum. In addition, the rising balloon provides forced ventilation, which supports a fast equilibration of the sensor with the ambient temperature and further reduces the effect of solar heating. However, at lower pressures in the upper troposphere and particularly in the stratosphere, solar radiative heating must be corrected for in data processing. This solar radiation correction is typically based on empirical corrections, laboratory measurements of the coatings, and comparisons of day and night time soundings. In the stratosphere, the solar radiation correction is the largest source of measurement uncertainty (e.g., Dirksen et al., 2014). Furthermore, the solar radiation correction applied by all manufacturers is a climatological correction without knowledge of the instantaneous cloud field. Thus, individual temperature profiles may have some level of bias due to the variable cloud cover and resulting radiative flux onto the radiosonde temperature sensor.

Other, generally transient environmental effects exist that may degrade radiosonde temperature measurements. For example, sensor icing may happen while a radiosonde passes through a supercooled liquid cloud. An ice layer covering the sensor decouples it from the atmosphere, and its evaporation in the drier layer above a cloud may cool the sensor well below ambient temperature. While this condition does generally not last long, it may lead to artificially low temperature readings at the cloud tops.

The temperature of the fill gas inside the balloon is significantly warmer than the ambient temperature during the day and colder during the night. Air passing over the balloon as it rises creates a wake of warmer air during the day and colder air during the night. Especially in the stratosphere, the radiosonde passing through these wakes may report temperature spikes that are not representative of ambient conditions. Instead of smoothing over these spikes, the processing software should filter them to report unbiased profiles.

Temperature sensors from different manufacturers vary considerably in size and geometry. In addition, the mass of the sensor requires a minor lag correction, which strengthens vertical gradients.

2.1.2 Humidity

Most modern radiosondes measure humidity using thin-film polymer sensors. These sensors are small capacitive elements, in which a thin film of select polymer materials forms the humidity-sensitive dielectric, which is sandwiched between permeable electrodes. As atmospheric humidity changes, voids in the thin polymer film take up or release water molecules, thereby changing its dielectric constant. As a result, the capacitance of these elements is a nearly linear function of the relative humidity and only a weak function of atmospheric temperature. Nevertheless, for accurate measurements, these elements must be calibrated over the entire humidity and temperature range encountered in the atmosphere.

Solar radiative heating also changes the temperature of the humidity sensor. Although it still measures relative humidity, the sensor now does so at an elevated temperature, leading to a strong dry bias at ambient temperature. Manufacturers have historically put a cap over the humidity sensor to avoid this artifact and in addition to shield it from precipitation. While a cap reduces the solar heating effect, it cannot eliminate it. As a result, manufacturers have started adding a dedicated temperature sensor into the humidity-sensing element, to measure its temperature directly and to properly compensate for the solar heating effect.

Icing in clouds can also degrade humidity measurements by decoupling the sensor from the atmosphere and by adding a source of water vapor near the sensing element. Caps and hydrophobic coatings may reduce this problem, and some manufacturers have started to include heating elements into the humidity-sensing element. A heater allows the humidity sensor to operate well above ambient temperature, eliminating the risk of icing. Heated sensors require a sensor temperature measurement, which allows calculating the correct ambient vapor pressure and subsequently the ambient relative humidity.

The response time of polymer sensors is highly temperature dependent. Fast sensors react to step changes in humidity at room temperature in well under 1 s. At −40°C, the response time increases to many seconds, and, at −60°C, it exceeds 1 min. Manufacturers have implemented time lag corrections to compensate for the slow response at cold temperatures; however, the uncertainty of the correction dominates the uncertainty budget at cold temperatures. Humidity sensors used by some manufacturers respond to humidity down to temperatures of −80°C, while those used by other manufacturers stop responding to humidity at around −40°C (Nash et al., 2011).

Some trace gases, especially outgassing of plastic materials, may contaminate the polymer film and take up space in the voids that would normally be taken up by water molecules. This contamination of the material may lead to a dry bias. Some manufacturers have started defining dedicated reconditioning procedures prior to the launch of a radiosonde, in which the sensor is heated to remove these contaminants and to restore the factory calibration.

Many different humidity sensors exist on the market, and only a detailed knowledge of the sensor allows a proper interpretation of the measurement quality. Careful treatment and preparation of the humidity sensor, its placement on the balloon payload, and the proper processing are additional factors controlling the quality of humidity measurements (Ingleby, 2017).

2.1.3 Winds

In most modern radiosondes, radio signals from Global Navigation Satellite Systems (GNSS), such as the American Global Positioning Satellite (GPS), the European Galileo, the Russian Global Navigation Satellite System (GLONASS), or the Chinese BeiDou Navigation Satellite System, provide the necessary information to calculate position and speed of the balloon. Parts of the world still use radio theodolite and tracking radar to measure the winds; however, with the low cost and ubiquity of GNSS receivers, this technology dominates in most parts of the world.

The balloon's motion is a good approximation of the horizontal winds; however, the radiosonde motion underneath the balloon includes a pendulum motion on a long string. The radiosonde processing software must filter this pendulum motion, which degrades the vertical resolution of horizontal wind measurements. Especially in the lower boundary layer, the vertical resolution of radiosonde wind measurements may often be too limited for detailed process studies.

2.1.4 Pressure

Modern radiosondes either measure the atmospheric pressure using small solid-state pressure sensors or derive it from the geometric height measured by GNSS. Russian radiosondes also derive pressure from geometric height measured by tracking radar but with much larger uncertainty.

The geometric height measured by the GNSS receiver onboard the radiosonde for wind measurements can be inverted using the hypsometric equation to calculate a vertical profile of pressure. This calculation requires an independent and accurate measurement of the surface pressure and accurate measurements of the temperature and humidity profiles.

Upper tropospheric and stratospheric pressures derived from GNSS have significantly lower uncertainty and lower systematic error than those measured by a pressure sensor. On the other hand, the uncertainty budget of the pressure sensor in the lower to middle troposphere is superior to that of the GNSS-derived values (Dirksen et al., 2014). Programs such as the Global Climate Observing System (GCOS) Reference Upper Air Network (GRUAN) make use of both techniques to provide optimal pressure observations over the entire tropospheric and stratospheric profile.

2.2 Receiving system and data formats

Processing of the radiosonde data typically happens within the sounding system. Larger meteorological services may use additional quality control procedures based on additional information not available to the sounding system. Quality control procedures filter out bad data due to telemetry errors, sensor malfunction, or other causes. Sounding systems then apply all necessary corrections and adjustments defined by the manufacturer before generating properly formatted output files.

Historically, radiosonde profiles were stored in highly compressed formats defined by the World Meteorological Organization (WMO) to minimize the transmission bandwidth required to send the data from remote stations to the Global Telecommunication System (GTS). Data formats such as FM-35 TEMP and similar formats (WMO, 2019) contain only a limited number of vertical levels and only the location of the launch site. To satisfy the need for higher resolution in current Numerical Weather Prediction (NWP) models, WMO decided to migrate data reporting to the Binary Universal Form for the Representation of meteorological data (BUFR; Ingleby et al., 2016a). This format allows storing observations with up to 1 s resolution and includes the location and time for each data point in addition to significantly more metadata.

For research purposes, NetCDF formatted data files for radiosondes are commonly used, which implement the climate and forecasting (CF) metadata conventions to improve interoperability of these data. Most recently, researchers are working on using radiosonde data on descent after the balloon has burst. These data provide a second profile, which may benefit a forecast model, especially if the balloon has covered a significant distance. However, measurement artifacts due to the different descent rates and ambiguous sensor orientation reduce the quality of these observations (Ingleby et al., 2021).

2.3 Operations

Meteorological observatories launch radiosondes typically once or twice daily. Globally, about 800 stations report regular radiosonde observations, which are shown in Fig. 4.

Reporting times for balloon observations at most stations are 0:00 UTC and 12:00 UTC, regardless of the geographic location. In Europe, balloons are typically launched 40–90 min before local noon and before local midnight to collect a complete profile by the nominal reporting time. In contrast, in the Americas, balloons are more likely to be released around local sunrise and sunset. Therefore, validation of sun-synchronous satellite observations using radiosondes may have a geographic bias, where time of day is important. Stations launching only once per day select the synoptic time to maximize the number of successful observations and may choose to avoid the time with a higher incidence of strong convection.

Most balloons are still launched manually; however, automated systems, which can launch balloons at pre-programmed times without human intervention, are becoming more common (e.g., Madonna et al., 2020). These systems operate autonomously and require station visits only to restock supplies and for maintenance. Radiosonde balloons are launched in all weather conditions as long as the safety of the operator and the safety of the launch system can be guaranteed. Climate records based on radiosonde launches will have little bias based on weather conditions but may show a slight difference between day and night observations. Balloons launched in stormy conditions are susceptible to early balloon bursts and loss of parts of the profile.

Balloon observations in the polar winter are particularly challenging, as the extreme stratospheric cold and darkness frequently lead to early balloon bursts. Consequently, stations may employ special treatments of the sounding balloon to maximize the ceiling altitude attainable under these conditions.

The number of global radiosonde providers varies between 10 and 15. Manufacturers implement technological advances into their radiosondes and introduce new models into the market as they become available. Therefore, the global network of radiosonde stations is heterogeneous. Knowledge about the sondes launched at different stations and their performance is needed to evaluate the quality of reference observations that a station can provide.

28 PART | I Instrumentation

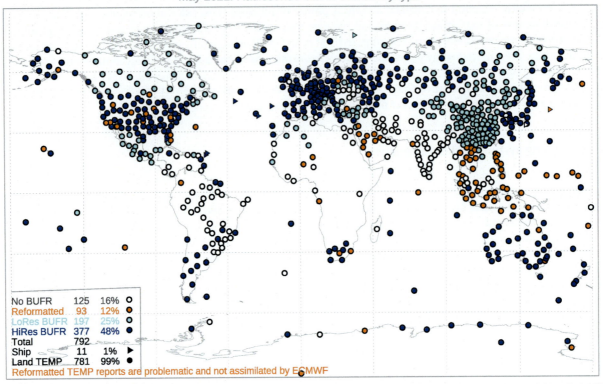

FIG. 4 Radiosonde stations reporting in May 2021. Different colors indicate the different vertical resolution of the reports shared through the GTS.

Some of today's best available radiosonde technology shows only minimal bias against satellite observations under cloud-free conditions. These observations may be taken as reference (Sun et al., 2021). Although the radiosonde network is under continuous financial pressure and with it the ability to validate satellite observations (Ingleby et al., 2016a), radiosonde observations, in particular those of high quality, such as those launched within GRUAN, remain an essential tool for the ongoing evaluation of satellite radiance observations in numerical weather prediction models (Eyre, 2016).

3. Cal/val considerations

High-quality radiosondes provide reference profiles with high vertical resolution in the troposphere and lower to middle stratosphere; however, their spatial coverage is poor compared to satellite observations, and only about 50% of all stations report their data at high resolution. Radiosonde observation data can be compared to satellite observations in two distinct approaches. First, the atmospheric state observed by radiosondes can be converted to radiances observed by the satellite instrument. This approach is less dependent on retrieval algorithms and a priori assumptions. It may also provide more insight on the technical performance of the radiometers and that of the radiosondes (Soden and Lanzante, 1996; John and Buehler, 2005; Moradi et al., 2013). Alternatively, atmospheric profiles can be retrieved from satellite observations and directly compared to radiosonde observations. This approach allows a more straightforward interpretation of the performance of the satellite observations in physical quantities measured by the radiosondes. Known biases in radiosonde data must be removed and collocation errors must be minimized as much as possible (Calbet et al., 2011) to provide the best reference for satellite observations. The vertical resolution of the in situ observations needs to be reduced to that of the satellite retrieval ideally by using the appropriate averaging kernels. However, the retrieval requires a set of assumptions and prior information, which can significantly affect the comparison.

Errors in the a priori information and retrieval algorithms are more difficult to characterize than errors in radiosonde observations. As a result, there is a trend toward comparing satellite and radiosonde observations in the quantities measured by the satellite, i.e., radiances or brightness temperatures.

Collocation and spatial variability are significant issues between radiosondes and satellite observations (e.g., Bruce et al., 1977; Sun et al., 2010). In particular, nadir-looking satellite instruments with a small footprint need to consider the horizontal drift of radiosondes, which may vary dramatically with the season. Geostationary satellites provide higher temporally resolved observations but poorer horizontal resolution than polar-orbiting satellites. They may need to consider the relatively slow balloon ascent to create a complete radiosonde profile.

Satellite validation efforts have made use of data from dedicated radiosonde campaigns as well as from operational networks. Dedicated campaigns can provide a more homogeneous data set, which is well-timed with satellite overpasses, but provide fewer profiles. Validation efforts using operational networks need to consider the performance differences of radiosondes from different manufacturers. The timing of operational soundings may also lead to biased comparisons due to the systematically different overpass times, particularly for polar orbiting satellites.

Historically, operational radiosonde profiles have been treated as a vertical snapshot of the atmospheric state profile, ignoring the ascent and horizontal drift. The use of operational data will further improve as complete time and position data are reported and assimilation systems make full use of the vertical resolution (Andersson and Sato, 2012).

Using global analysis systems such as that at the European Center for Medium-Range Weather Forecasting (ECMWF; Hersbach et al., 2020) can reduce some collocation errors by explicitly considering the dynamics of the atmosphere. This approach significantly expands the available data against which to compare satellite observations. High-quality operational networks, particularly GRUAN, can be used to evaluate the biases in analysis systems and thereby allow for a more extensive set of reference observations for satellite validation (Carminati et al., 2019). Nalli et al. (2013) describe a validation hierarchy for satellite sounders, using NWP model global comparisons, global observations by other satellite instruments, operational radiosonde matchups, dedicated reference sounding matchups, and intensive field campaigns. Each of these comparisons has its strengths and justification at different stages of the lifetime of a satellite instrument.

Lastly, observations in the infrared spectrum are very sensitive to clouds in the satellite field of view. Validation observations under clear sky provide the best reference; however, a lot of effort has been made to develop cloud-clearing algorithms from remote sensing observation (Rodgers, 1970; Susskind et al., 2003). Observations with all but the most extreme clouds in the field of view have only slightly degraded errors.

3.1 Temperature and humidity profiling

Meteorological observations from space started in the 1960s. Their expansion was driven by newly developed spaceborne radiometers and retrieval algorithms for atmospheric state variables from spectral observations. Rodgers (1976) gives a review of the theory of remote sensing developed in this period, Menzel et al. (2018) give a review of the developments of spaceborne infrared sounders, and Eyre et al. (2020) provide a review of the assimilation of early satellite observations in NWP models. Measurements of temperature and trace gases (water vapor, ozone, and others) from space required dedicated retrieval algorithms and, most importantly, validation. The series of Nimbus experimental satellites launched between 1964 and 1978 carried different infrared and later microwave spectrometers, which provided a test platform for a line of infrared and microwave sounders. For example, Smith (1968) and Smith et al. (1970) used observations from the Medium Resolution Infrared Radiometer (MRIR) and the Satellite Infrared Spectrometer (SIRS) to retrieve profiles of temperature and humidity. Comparing these profiles with those by measured radiosondes and rocketsondes clearly showed the validity of these early efforts and highlighted some challenges. Methods to account for the influence of clouds, mountains, and hot terrain had to be developed and validated. Nevertheless, these retrievals showed that temperature profiles with accuracies reaching that of radiosondes could be measured globally. Barnett et al. (1975) compared radiance measurements by the selective chopper radiometers (SCR) with those derived from radiosondes and rocketsondes to demonstrate that atmospheric temperatures from satellites could be extended into the upper stratosphere and mesosphere.

These satellite instruments used very broad spectral regions to derive temperature and water vapor. The spectral regions include absorption features of other trace gases such as CO_2, ozone, etc., which must be considered in the retrieval. Validation is essential to confirm that these effects have been adequately taken into account and to evaluate the errors in the presence of other absorption features.

Clouds represent a significant limitation for the direct retrieval of atmospheric profiles from spectra in the infrared. Two different approaches were developed to overcome this challenge. First, infrared signals from several contiguous fields of view can define a single cloud-free profile. The influence from clouds can be detected in the variability of the signals, and a cloud-free measurement can be extracted. Alternatively, measurements from spectrometers in the microwave bands, which are not strongly affected by cloud droplets, can be used to derive profiles in most but the cloudiest scenes (Smith et al., 1979).

The Television InfraRed Observation Satellite (TIROS) Operational Vertical Sounder (TOVS) suite of instruments combined the High-Resolution Infrared Radiation Sounder (HIRS), the Microwave Sounding Unit (MSU), and the Stratospheric Sounding Unit (SSU), which forms the basis for all operational satellites until the present. The first suite of TOVS instruments was launched on the TIROS-N satellite in 1978, and the first Advanced TOVS (ATOVS) suite, including the

Advanced MSU (AMSU), was launched in 1998 on the NOAA-15 satellite. Phillips et al. (1979) compared the operational satellite retrievals for temperature with operational radiosondes. Unlike radiosondes, satellite observations have global coverage. However, their vertical resolution is much poorer, and the consistency of errors in the satellite observations over larger geographic regions can cause problems in numerical weather forecasting. Gruber and Watkins (1982) highlighted that satellite soundings made under cloudy conditions, which depend heavily on radiances in the microwave channels, have larger temperature biases relative to radiosondes than soundings made in clear air.

Temperature retrievals in the stratosphere using limb scanning allow a much higher vertical resolution and avoid cloud effects near the tropopause, where cloud contamination becomes noticeable. Gille et al. (1984) used radiosonde data from the WMO global observing system to validate the limb retrievals of temperature from the Limb Infrared Monitor of the Stratosphere (LIMS). They showed that limb retrievals can reach similar levels of precision as that of radiosondes and rocketsondes. Seasonal and regional differences may possibly be attributed to the different radiosonde systems used in this study and their different methods of processing raw measurements, particularly radiation corrections. Near the tropopause, particularly in the tropics, satellite limb observations may be affected by clouds and are flagged in the processing. Above that, clouds do not affect retrievals and allow a true global determination of stratospheric temperature with much higher vertical resolution than nadir-looking sounders. On the other hand, the horizontal footprint of limb retrievals is much larger, and some of the variability in comparison with radiosondes may be explained by spatial inhomogeneities, which are resolved differently between the two observing systems.

The differences in radiosonde systems and processing algorithms have not played a significant role in the early periods of validating satellite observations. However, in building long-term climate records, changes in radiosonde systems and processing algorithms are as crucial as drifts of single satellite instruments and changes in satellite sensors. Christy et al. (2007) studied two homogenized data sets of channel 2 of the MSU and channel 5 of the AMSU, which represent a mean temperature in a layer from roughly the surface to 350 hPa. At 58 tropical stations, radiosonde data contain numerous biases and require adjustments relative to these satellite data to derive a temperature trend from these radiosondes. Therefore, while operational uncertainties of radiosondes are sufficiently small for the validation of satellite instruments, sporadic changes in these systems may be larger than drifts in satellite instruments, making radiosonde data challenging to use for climate trend estimates. High-quality networks such as GRUAN attempt to minimize the impact of system change on long-term records through rigorous uncertainty characterization, validation, and metadata tracking.

Hyperspectral infrared (IR) sounders such as the Atmospheric Infrared Sounder (AIRS) onboard Aqua, the Infrared Atmospheric Sounding Interferometer (IASI) onboard the Metop satellites, and the Cross-track Infrared Sounder (CrIS) onboard the Suomi-NPP (SNPP) and Joint Polar Satellite System (JPSS) series provide retrievals with a much higher vertical resolution than the infrared sounders as part of the TOVS and ATOVS suite of instruments. These instruments use many more spectral features in the IR band and provide data for the retrieval of additional trace species, particularly ozone and CO_2. As all previous sounders, these instruments require special retrieval algorithms to handle observations over land and in the presence of clouds (Susskind et al., 2003). In addition, these instruments can resolve the upper tropospheric water vapor with higher fidelity, which places additional requirements on the validation observations.

Fetzer et al. (2003) described the basic approach for validating the infrared and microwave remote sensing spectrometer AIRS, AMSU, and HSB onboard the Aqua satellite. Divakarla et al. (2006) compared operational radiosondes with AIRS and confirmed that AIRS can retrieve the tropospheric temperature profile over oceans in cloud free scenes to within 1 K and with a 1 km vertical resolution. AIRS can retrieve relative humidity to within 15% with a 2 km vertical resolution. In cloudy observations and over land, the instrument performance almost reaches this level of accuracy, possibly with a slight degradation near the surface. Gettelman et al. (2006) compared dedicated radiosonde launches from the Dome-C Antarctic research station with overpass observations by AIRS and climate models. This validation campaign again showed good agreement in temperature but some larger bias in humidity. The data from the three different radiosonde models used in this campaign had to be homogenized before comparison with the satellite observations, which may have contributed to some of the bias. The radiosonde profiles were also combined with spectral radiance measurements from the surface to calculate the top of the atmosphere radiances in a direct validation of the AIRS spectral measurements. Due to the high latitude, overpasses close to the station happened more frequently than near the equator, and finding suitable co-located profiles was less of a challenge. Some of the variability found in the comparison is attributed to the spatial variability of water vapor.

To avoid some of the limitations of radiosondes, mainly the limited sampling frequency and balloon drift, (Tobin, 2006) constructed site atmospheric state best estimates for temperature and humidity. High-quality radiosonde data are adjusted using remote sensing instrumentation at highly instrumented research sites. These adjustments can compensate for calibration and contamination biases and reduce the co-location errors due to the slow drift of the radiosonde balloons while at the same time maintaining the high resolution of the radiosonde profile.

In the transition from the troposphere to the stratosphere, most operational radiosondes humidity sensors lose sensitivity to water vapor. Their reported values should no longer be used for the validation of satellite measurements of stratospheric water vapor. Detailed knowledge about the capabilities of the respective relative humidity is needed for use in the validation of relative humidity in the tropopause region. For the validation of water vapor measurements in the stratosphere, specialized sensors (see Chapter 3: Frost Point Hygrometers) are required.

3.2 Winds

Measurements of wind from satellites are very challenging, since there are only few spectral signatures that can be detected by radiometers. The earliest method for the detection of atmospheric winds from geostationary satellites is through tracking of cloud features (Menzel et al., 1983), and soon atmospheric motion vectors (AMV) have become an indispensable tool in NWP (Schmetz et al., 1993). This method requires a reasonably high temporal resolution of observations to be able to track the movement of atmospheric features, which can only be done from geostationary satellites. Simultaneously, algorithms assign altitudes to the cloud features and distinguish a high, middle, or low altitude based on radiometric measurements. These algorithms are implemented in numerical data analysis models and use forecast fields in the processing, which most likely contribute to highly correlated errors (Bormann et al., 2003). The characterization of errors of AMVs and their correlation using radiosondes uses a dense network of radiosondes such as over Europe, the Eastern US, or Eastern Asia. Errors in wind measurements from radiosondes between different stations are assumed uncorrelated, i.e., error correlations over these horizontal distances are due to the satellite wind derivations. Over oceans, AMV-derived winds have a stronger influence on NWP since radiosonde winds dominate over land.

In addition to cloud features, AMVs can be derived from spatial features of trace gases. Velden et al. (1997) developed an algorithm to determine wind vectors from water vapor fields measured by geostationary satellites. In contrast to clouds, water vapor fields are continuous fields and can provide denser wind fields.

Although in earlier studies where the errors of radiosonde winds contributed significantly to the overall comparison, errors are minor in modern radiosondes using GNSS-derived winds. Velden and Bedka (2009) compared GPS radiosonde winds with AMVs at research sites of the Department of Energy Atmospheric Radiation Measurements (DOE/ARM) program. Here, the height assignment of the AMV was the most significant error source compared to radiosondes. Furthermore, radiosonde winds need to be averaged over specific layers to minimize the errors in the comparisons with AMVs.

Surface winds over the ocean can be measured from satellites measuring spectral signatures in the microwave emissions from the sea surface. These scatterometers detect the wind information directly at the surface. However, instead of radiosondes, which require a land-based launch site, these instruments are, among others, validated using dropsondes (see Chapter 10: Airborne Dropsonde Campaigns).

The first direct measurements of atmospheric wind components from satellites started with the deployment of the Atmospheric Laser Doppler Instrument (ALADIN) onboard the Atmospheric Dynamics Mission Aeolus satellite. This instrument measures the line of sight wind component, which can be assimilated and used in NWP. Detailed analysis and validation of these measurements require radiosondes to measure the complete wind profile over the entire altitude range of ALADIN. The co-location and time mismatch between radiosondes and satellites is corrected in NWP models. Baars et al. (2020) used dedicated radiosonde launches over the Atlantic from the R/V Polarstern, and Martin et al. (2021) used collocated radiosonde profiles from the operational network and NWP models to evaluate the error characteristics of ALADIN. The horizontal drift of some radiosonde launches makes these unsuitable for the comparison; however, the minimal errors of GNSS winds make them nearly ideal reference observations.

3.3 Clouds

Although radiosondes do not detect clouds directly, the structure of the relative humidity profile can provide some information about the likelihood of clouds. In particular, at the cloud top, relative humidity typically shows a distinct and sudden drying. This feature can be used to detect cloud tops and cloud top temperatures and is often used in conjunction with other remote sensing profiling instruments (Smith and Platt, 1978; Pavolonis et al., 2005; Liu et al., 2020). Wylie and Menzel (1989) used this signal to validate the cloud statistics of the Visible Infrared Spin-Scan Radiometer (VISSR) Atmospheric Sounder (VAS). However, they noted that the humidity sensor of these earlier radiosondes sometimes stopped responding somewhere inside the cloud. Wang and Rossow (1995) derived a more detailed cloud vertical structure from the structure of the radiosonde humidity profile, generally agreeing well with the International Climate Cloud Project (ISCCP) data. High clouds are not well detected by radiosondes, in parts due to the difficulty of some sensors to accurately measure relative humidity in the upper troposphere and in parts due to the weaker correlation of high relative humidity and the presence of clouds. Furthermore, upper tropospheric cirrus clouds may be optically thin and may not be detected by satellite instruments, leading to large differences in cloud top temperatures relative to those derived from radiosondes (Hanna et al., 2008). Garay et al. (2008) investigated biases in satellite cloud top heights in marine cumulus layers, using cloud radars and radiosondes. In radiosondes using unheated humidity sensors, cloud top temperatures may be impacted by icing and evaporative cooling, depending on radiosonde type (Nash et al., 2011). Heated humidity sensors, which are used on some radiosondes, alleviate this measurement artifact.

3.4 Using NWP data assimilation fields

The sparsity of reference data is a significant problem in the validation of satellite data, and only a tiny fraction of the available satellite data can be used in direct comparisons. Similarly, high-quality reference data are unlikely to sample the full range of environmental conditions from the tropics to the poles and from the oceans to the deserts. Furthermore, operational radiosonde measurements require additional quality screening since some radiosonde types have larger uncertainty than others (Ingleby, 2017).

One approach to address these challenges is to use data assimilation fields from NWP models (e.g., Loew et al., 2017, Section 4.6). This way, all satellite data can be compared and some patterns such as scan-angle dependent errors or geolocation errors may become apparent with observations on a single day, whereas they may be very difficult to detect using only the sparse reference data. More subtle effects such as shifts, drifts, and uncertainties in pass-band center frequencies can also be detected (Lu and Bell, 2014).

NWP fields used in data assimilation have their own errors and need to be understood as well as possible. The GAIA-CLIM project (2015-18, www.gaia-clim.eu) aimed to establish sound methods for the characterization of satellite-based Earth Observation (EO) data by surface-based measurements. For example, radiosonde data from GRUAN can be used to characterize errors in NWP fields, even though this is somewhat limited by their spatial sampling. Carminati et al. (2019) expanded this approach and projected resulting uncertainty estimates into radiance space and Newman et al. (2020) used these tools to assess new satellite missions. Nevertheless, even though NWP biases may be long-term, they are not fixed and there are ongoing efforts to reduce them (e.g., Laloyaux et al., 2020).

4. Summary

Satellite observations of the atmosphere require validation and calibration observations from in situ profiling observations. Modern high-quality radiosondes are a suitable technology for this purpose and provide high-resolution vertical profiles of the atmospheric state parameter pressure, temperature, relative humidity, and winds. The satellite and radiosonde measurements can be compared in terms of the variables measured by the radiosondes (temperature, humidity, and wind) or measured by the satellite instrument (usually radiance or brightness temperature). Comparing these systems in variables measured by the radiosonde introduces ambiguity due to the a priori assumptions and retrieval algorithms but allows a more straightforward interpretation. Using a radiative transfer forward model translates the radiosonde data into the variables observed by the satellites. This method has fewer assumptions and is often the preferred method of comparison.

Validation observations are essential as soon as a satellite instrument is placed into orbit to characterize its performance and limitations. Ongoing validation is vital to characterize drifts and other problems in satellite instruments. Typically, the performance of satellite instruments is assumed to drift slowly, if at all. However, long-term climate records from satellites are affected by the limited lifetime of any instrument and the necessary management of continuity between different satellites. Radiosonde systems, on the other hand, may change more frequently across the network, but largely uncorrelated between different stations.

Radiosondes have seen large improvements in their performance and reduced systematic and random uncertainties since their introduction over 80 years ago. The WMO coordinates intercomparisons between the most common radiosonde systems at regular intervals, the last of which took place in 2010 at Yangjiang, China (Nash et al., 2011) and the next scheduled to take place at Lindenberg, Germany in 2022. These intercomparisons evaluate the capabilities of the operational radiosondes in use at that time. This information is essential for the interpretation of radiosonde data in satellite validation and calibration efforts.

The dominant source of uncertainty in comparing radiosonde and satellite observations directly is due to co-location and overpass time mismatch. Other systematic and random sources of uncertainty in modern, high-quality radiosondes have been reduced such that they are secondary to co-location and overpass timing.

Some NWP centers now perform satellite calibration and validation against NWP models, which are "kept on the right track" by anchor observations: principally radio occultation and radiosonde observations. At these institutes, most validation is indirect via the NWP fields rather than via direct collocations. One significant advantage is that all satellite observations can be compared with the NWP fields giving samples many times larger than from direct collocations. Furthermore, radiative transfer calculations need a full state vector, which includes the skin temperature of Earth that is not available from radiosondes.

Operational radiosondes are an essential source of observational data for NWP model analyses, which may be used to evaluate the performance of satellite observations. Like satellites, these analyses provide global coverage and parameters not measured by radiosondes to evaluate satellite observations. However, bias adjustments, model errors, and the errors of poorer performing operational radiosondes may introduce additional uncertainty into the evaluation. Dedicated high-quality radiosonde launches, timed with overpasses at select locations, provide higher fidelity data to evaluate the performance of new spaceborne instruments. These in situ observations are usually part of larger projects, which include additional ground-based remote sensing observations. The cross-validation of all platforms provides the most consistent set of data to evaluate the performance of satellite observations.

References

Andersson, E., Sato, Y. (Eds.), 2012. Final report of the Fifth WMO Workshop on the Impact of Various Observing Systems on Numerical Weather Prediction (Sedona, Arizona, USA, 22–25 May 2012), WMO Integrated Global Observing System. Technical Report No. 2012-1. library.wmo.int/index.php?lvl=notice_display&id=13475 (last Accessed 12 June 2021).

Baars, H., Herzog, A., Heese, B., Ohneiser, K., Hanbuch, K., Hofer, J., Yin, Z., Engelmann, R., Wandinger, U., 2020. Validation of Aeolus wind products above the Atlantic Ocean. Atmos. Meas. Tech. 13, 6007–6024. https://doi.org/10.5194/amt-13-6007-2020.

Barnett, J.J., Harwood, R.S., Houghton, J.T., Morgan, C.G., Rodgers, C.D., Williamson, E.J., 1975. Comparison between radiosonde, rocketsonde, and satellite observations of atmospheric temperatures. Q. J. Roy. Meteorol. Soc. 101, 423–436. https://doi.org/10.1002/qj.49710142903.

Bormann, N., Saarinen, S., Kelly, G., Thépaut, J., 2003. The spatial structure of observation errors in atmospheric motion vectors from geostationary satellite data. Mon. Weather Rev. 131 (4), 706–718. https://doi.org/10.1175/1520-0493(2003)131<0706:TSSOOE>2.0.CO;2.

Bruce, R.E., Duncan, L.D., Pierluissi, J.H., 1977. Experimental study of the relationship between radiosonde temperatures and satellite-derived temperatures. Mon. Weather Rev. 105 (4), 493–496. https://doi.org/10.1175/1520-0493(1977)105<0493:ESOTRB>2.0.CO;2.

Calbet, X., Kivi, R., Tjemkes, S., Montagner, F., Stuhlmann, R., 2011. Matching radiative transfer models and radiosonde data from the EPS/Metop Sodankylä campaign to IASI measurements. Atmos. Meas. Tech. 4, 1177–1189. https://doi.org/10.5194/amt-4-1177-2011.

Carminati, F., Migliorini, S., Ingleby, B., Bell, W., Lawrence, H., Newman, S., Hocking, J., Smith, A., 2019. Using reference radiosondes to characterise NWP model uncertainty for improved satellite calibration and validation. Atmos. Meas. Tech. https://doi.org/10.5194/amt-2018-219.

Christy, J.R., Norris, W.B., Spencer, R.W., Hnilo, J.J., 2007. Tropospheric temperature change since 1979 from tropical radiosonde and satellite measurements. J. Geophys. Res. 112, D06102. https://doi.org/10.1029/2005JD006881.

Dirksen, R.J., Sommer, M., Immler, F.J., Hurst, D.F., Kivi, R., Vömel, H., 2014. Reference quality upper-air measurements: GRUAN data processing for the Vaisala RS92 radiosonde. Atmos. Meas. Tech. 7, 4463–4490. https://doi.org/10.5194/amt-7-4463-2014.

Divakarla, M.G., Barnet, C.D., Goldberg, M.D., McMillin, L.M., Maddy, E., Wolf, W., Zhou, L., Liu, X., 2006. Validation of atmospheric infrared sounder temperature and water vapor retrievals with matched radiosonde measurements and forecasts. J. Geophys. Res. 111, D09S15. https://doi.org/10.1029/2005JD006116.

Eyre, J.R., 2016. Observation bias correction schemes in data assimilation systems: a theoretical study of some of their properties. Q. J. Roy. Meteorol. Soc. 142, 2284–2291. https://doi.org/10.1002/qj.2819.

Eyre, J.R., English, S.J., Forsythe, M., 2020. Assimilation of satellite data in numerical weather prediction. Part I: the early years. Q. J. Roy. Meteorol. Soc. 146, 49–68. https://doi.org/10.1002/qj.3654.

Fetzer, E., et al., 2003. AIRS/AMSU/HSB validation. IEEE Trans. Geosci. Remote Sens. 41 (2), 418–431. https://doi.org/10.1109/TGRS.2002.808293.

Garay, M.J., de Szoeke, S.P., Moroney, C.M., 2008. Comparison of marine stratocumulus cloud top heights in the southeastern Pacific retrieved from satellites with coincident ship-based observations. J. Geophys. Res. 113, D18204. https://doi.org/10.1029/2008JD009975.

Gettelman, A., Walden, V.P., Miloshevich, L.M., Roth, W.L., Halter, B., 2006. Relative humidity over Antarctica from radiosondes, satellites, and a general circulation model. J. Geophys. Res. 111, D09S13. https://doi.org/10.1029/2005JD006636.

Gille, J.C., Russell, J.M., Bailey, P.L., Gordley, L.L., Remsberg, E.E., Lienesch, J.H., Planet, W.G., House, F.B., Lyjak, L.V., Beck, S.A., 1984. Validation of temperature retrievals obtained by the Limb Infrared Monitor of the Stratosphere (LIMS) Experiment on NIMBUS 7. J. Geophys. Res. 89 (D4), 5147–5160. https://doi.org/10.1029/JD089iD04p05147.

Gruber, A., Watkins, C.D., 1982. Statistical assessment of the quality of TIROS-N and NOAA-6 satellite soundings. Mon. Weather Rev. 110 (7), 867–876. https://doi.org/10.1175/1520-0493(1982)110<0867:SAOTQO>2.0.CO;2.

Hanna, J.W., Schultz, D.M., Irving, A.R., 2008. Cloud-top temperatures for precipitating winter clouds. J. Clim. Appl. Meteorol. 47 (1), 351–359. https://doi.org/10.1175/2007JAMC1549.1.

Hersbach, H., Bell, B., Berrisford, P., et al., 2020. The ERA5 global reanalysis. Q. J. Roy. Meteorol. Soc. 146, 1999–2049. https://doi.org/10.1002/qj.3803.

Ingleby, B., 2017. An assessment of different radiosonde types 2015/2016. ECMWF Technical Memorandum 807, August 2017. https://www.ecmwf.int/sites/default/files/elibrary/2017/17551-assessment-different-radiosonde-types-20152016.pdf (Accessed 6 July 2021).

Ingleby, B., Pauley, P., Kats, A., Ator, J., Keyser, D., Doerenbecher, A., Fucile, E., Hasegawa, J., Toyoda, E., Kleinert, T., Qu, W., St. James, J., Tennant, W., Weedon, R., 2016a. Progress toward high-resolution, real-time radiosonde reports. Bull. Am. Meteorol. Soc. 97 (11), 2149–2161. https://doi.org/10.1175/BAMS-D-15-00169.1.

Ingleby, B., Motl, M., Marlton, G., Edwards, D., Sommer, M., von Rohden, C., Vömel, H., Jauhiainen, H., 2021. On the quality of RS41 radiosonde descent data. Atmos. Meas. Tech. Discuss. https://doi.org/10.5194/amt-2021-183. [preprint]. accepted.

John, V.O., Buehler, S.A., 2005. Comparison of microwave satellite humidity data and radiosonde profiles: a survey of European stations. Atmos. Chem. Phys. 5, 1843–1853. https://doi.org/10.5194/acp-5-1843-2005.

Laloyaux, P., Bonavita, M., Dahoui, M., Farnan, J., Healy, S.B., Hólm, E.V., Lang, S.T.K., 2020. Towards an unbiased stratospheric analysis. Q. J. Roy. Meteorol. Soc. 146, 2392–2409. https://doi.org/10.1002/qj.3798.

Liu, C.-.Y., Chiu, C.-.H., Lin, P.-.H., Min, M., 2020. Comparison of cloud-top property retrievals from Advanced Himawari Imager, MODIS, CloudSat/CPR, CALIPSO/CALIOP, and radiosonde. J. Geophys. Res. 125, e2020JD032683. https://doi.org/10.1029/2020JD032683.

Loew, A., Bell, W., Brocca, L., Bulgin, C.E., Burdanowitz, J., Calbet, X., et al., 2017. Validation practices for satellite-based Earth observation data across communities. Rev. Geophys. 55, 779–817. https://doi.org/10.1002/2017RG000562.

Lu, Q., Bell, W., 2014. Characterizing channel center frequencies in AMSU-A and MSU microwave sounding instruments. J. Atmos. Oceanic Tech. 31, 1713–1732. https://doi.org/10.1175/JTECH-D-13-00136.1.

Madonna, F., Kivi, R., Dupont, J.-C., Ingleby, B., Fujiwara, M., Romanens, G., Hernandez, M., Calbet, X., Rosoldi, M., Giunta, A., Karppinen, T., Iwabuchi, M., Hoshino, S., von Rohden, C., Thorne, P.W., 2020. Use of automatic radiosonde launchers to measure temperature and humidity profiles from the GRUAN perspective. Atmos. Meas. Tech. 13, 3621–3649. https://doi.org/10.5194/amt-13-3621-2020.

Martin, A., Weissmann, M., Reitebuch, O., Rennie, M., Geiß, A., Cress, A., 2021. Validation of Aeolus winds using radiosonde observations and numerical weather prediction model equivalents. Atmos. Meas. Tech. 14, 2167–2183. https://doi.org/10.5194/amt-14-2167-2021.

Menzel, W.P., Smith, W.L., Stewart, T.R., 1983. Improved cloud motion wind vector and altitude assignment using VAS. J. Appl. Meteorol. Climatol. 22 (3), 377–384. https://doi.org/10.1175/1520-0450(1983)022<0377:ICMWVA>2.0.CO;2.

Menzel, W.P., Schmit, T.J., Zhang, P., Li, J., 2018. Satellite-based atmospheric infrared sounder development and applications. Bull. Am. Meteorol. Soc. 99 (3), 583–603. https://doi.org/10.1175/BAMS-D-16-0293.1.

Moradi, I., Soden, B., Ferraro, R., Arkin, P., Vömel, H., 2013. Assessing the quality of humidity measurements from global operational radiosonde sensors. J. Geophys. Res. 118, 8040–8053. https://doi.org/10.1002/jgrd.50589.

Nalli, N.R., Barnet, C.D., Reale, A., Tobin, D., Gambacorta, A., Maddy, E.S., Joseph, E., Sun, B., Borg, L., Mollner, A.K., Morris, V.R., Liu, X., Divakarla, M., Minnett, P.J., Knuteson, R.O., King, T.S., Wolf, W.W., 2013. Validation of satellite sounder environmental data records: application to the Cross-track Infrared Microwave Sounder Suite. J. Geophys. Res. 118, 13628–13643. https://doi.org/10.1002/2013JD020436.

Nash, J., Oakley, T., Vömel, H., Li, W., 2011. WMO Intercomparison of High Quality Radiosonde Systems, Yangjiang, China, 12 July–3 August 2010. Instruments and Observing Methods Report No. 107, World Meteorological Organization. 248 p.

Newman, S., Carminati, F., Lawrence, H., Bormann, N., Salonen, K., Bell, W., 2020. Assessment of new satellite missions within the framework of numerical weather prediction. Remote Sens. (Basel) 12, 1580. https://doi.org/10.3390/rs12101580.

Pavolonis, M.J., Heidinger, A.K., Uttal, T., 2005. Daytime global cloud typing from AVHRR and VIIRS: algorithm description, validation, and comparisons. J. Appl. Meteorol. 44 (6), 804–826. https://doi.org/10.1175/JAM2236.1.

Phillips, N., McMillin, L., Gruber, A., Wark, D., 1979. An evaluation of early operational temperature soundings from TIROS–N. Bull. Am. Meteorol. Soc. 60 (10), 1188–1197.

Rodgers, C.D., 1970. Remote sounding of the atmospheric temperature profile in the presence of cloud. Q. J. Roy. Meteorol. Soc. 96, 654–666. https://doi.org/10.1002/qj.49709641008.

Rodgers, C.D., 1976. Retrieval of atmospheric temperature and composition from remote measurements of thermal radiation. Rev. Geophys. 14 (4), 609–624. https://doi.org/10.1029/RG014i004p00609.

Schmetz, J., Holmlund, K., Hoffman, J., Strauss, B., Mason, B., Gaertner, V., Koch, A., Van De Berg, L., 1993. Operational cloud-motion winds from meteosat infrared images. J. Clim. Appl. Meteorol. 32 (7), 1206–1225. https://doi.org/10.1175/1520-0450(1993)032<1206:OCMWFM>2.0.CO;2.

Smith, W.L., 1968. An improved method for calculating tropospheric temperature and moisture from satellite radiometer measurements. Mon. Weather Rev. 96 (6), 387–396. https://doi.org/10.1175/1520-0493(1968)096<0387:AIMFCT>2.0.CO;2.

Smith, W.L., Platt, C.M.R., 1978. Comparison of satellite-deduced cloud heights with indications from radiosonde and ground-based laser measurements. J. Clim. Appl. Meteorol. 17 (12), 1796–1802. https://doi.org/10.1175/1520-0450(1978)017<1796:COSDCH>2.0.CO;2.

Smith, W.L., Woolf, H.M., Jacob, W.J., 1970. A regression method for obtaining real-time temperature and geopotential height profiles from satellite spectrometer measurements and its application to Nimbus 3 "SIRS" observations. Mon. Weather Rev. 98 (8), 582–603. https://doi.org/10.1175/1520-0493(1970)098<0582:ARMFOR>2.3.CO;2.

Smith, W., Woolf, H., Hayden, C., Wark, D., McMillin, L., 1979. The TIROS–N operational vertical sounder. Bull. Am. Meteorol. Soc. 60 (10), 1177–1187.

Soden, B.J., Lanzante, J.R., 1996. An assessment of satellite and radiosonde climatologies of upper-tropospheric water vapor. J. Climate 9 (6), 1235–1250. https://doi.org/10.1175/1520-0442(1996)009<1235:AAOSAR>2.0.CO;2.

Sun, B., Reale, A., Seidel, D.J., Hunt, D.C., 2010. Comparing radiosonde and COSMIC atmospheric profile data to quantify differences among radiosonde types and the effects of imperfect collocation on comparison statistics. J. Geophys. Res. 115, D23104. https://doi.org/10.1029/2010JD014457.

Sun, B., Calbet, X., Reale, A., Schroeder, S., Bali, M., Smith, R., Pettey, M., 2021. Accuracy of Vaisala RS41 and RS92 upper tropospheric humidity compared to satellite hyperspectral infrared measurements. Remote Sens. (Basel) 13, 173. https://doi.org/10.3390/rs13020173.

Susskind, J., Barnet, C.D., Blaisdell, J.M., 2003. Retrieval of atmospheric and surface parameters from AIRS/AMSU/HSB data in the presence of clouds. IEEE Trans. Geosci. Remote Sens. 41 (2), 390–409. https://doi.org/10.1109/TGRS.2002.808236.

Tobin, D.C., et al., 2006. Atmospheric Radiation Measurement site atmospheric state best estimates for Atmospheric Infrared Sounder temperature and water vapor retrieval validation. J. Geophys. Res. 111, D09S14. https://doi.org/10.1029/2005JD006103.

Velden, C.S., Bedka, K.M., 2009. Identifying the uncertainty in determining satellite-derived atmospheric motion vector height attribution. J. Clim. Appl. Meteorol. 48 (3), 450–463. https://doi.org/10.1175/2008JAMC1957.1.

Velden, C.S., Hayden, C.M., Nieman, S.J.W., Paul Menzel, W., Wanzong, S., Goerss, J.S., 1997. Upper-tropospheric winds derived from geostationary satellite water vapor observations. Bull. Am. Meteorol. Soc. 78 (2), 173–196. https://doi.org/10.1175/1520-0477(1997)078<0173:UTWDFG>2.0.CO;2.

Wang, J., Rossow, W.B., 1995. Determination of cloud vertical structure from upper-air observations. J. Clim. Appl. Meteorol. 34 (10), 2243–2258. https://doi.org/10.1175/1520-0450(1995)034<2243:DOCVSF>2.0.CO;2.

WMO, 2019. Manual on Codes—International Codes, Volume I.1, Annex II to the WMO Technical Regulations: Part A-Alphanumeric Codes, WMO-No. 306, 2011 Edition, Updated in 2019. World Meteorological Organization, Geneva. library.wmo.int/doc_num.php?explnum_id=4157.

Wylie, D.P., Menzel, W.P., 1989. Two years of cloud cover statistics using VAS. J. Climate 2 (4), 380–392. https://doi.org/10.1175/1520-0442(1989)002<0380:TYOCCS>2.0.CO;2.

Further reading

Hanna, J.W., Schultz, D.M., Irving, A.R., 2008. Cloud-top temperatures for precipitating winter clouds. J. Clim. Appl. Meteorol. 47 (1), 351–359. https://doi.org/10.1175/2007JAMC1549.1.

Ingleby, B., Rodwell, M., Isaksen, L., 2016b. Global radiosonde network under pressure. ECMWF Newsletter 149. www.ecmwf.int/en/newsletter/149/meteorology/global-radiosonde-network-under-pressure (Accessed 25 February 2021).

Kitchen, M., 1989. Representativeness errors for radiosonde observations. Q. J. Roy. Meteorol. Soc. 115, 673–700. https://doi.org/10.1002/qj.49711548713.

Oort, A.H., 1978. Adequacy of the rawinsonde network for global circulation studies tested through numerical model output. Mon. Weather Rev. 106 (2), 174–195. https://doi.org/10.1175/1520-0493(1978)106<0174:AOTRNF>2.0.CO;2.

Pougatchev, N., August, T., Calbet, X., Hultberg, T., Oduleye, O., Schlüssel, P., Stiller, B., Germain, K.S., Bingham, G., 2009. IASI temperature and water vapor retrievals—error assessment and validation. Atmos. Chem. Phys. 9, 6453–6458. https://doi.org/10.5194/acp-9-6453-2009.

Schlatter, T.W., 1981. An assessment of operational TIROS–N temperature retrievals over the United States. Mon. Weather Rev. 109 (1), 110–119. https://doi.org/10.1175/1520-0493(1981)109<0110:AAOOTT>2.0.CO;2.

Seidel, D.J., Angell, J.K., Christy, J., Free, M., Klein, S.A., Lanzante, J.R., Mears, C., Parker, D., Schabel, M., Spencer, R., Sterin, A., Thorne, P., Wentz, F., 2004. Uncertainty in signals of large-scale climate variations in radiosonde and satellite upper-air temperature datasets. J. Climate 17 (11), 2225–2240. https://doi.org/10.1175/1520-0442(2004)017<2225:UISOLC>2.0.CO;2.

Smith, W.L., 1970. Iterative solution of the radiative transfer equation for the temperature and absorbing gas profile of an atmosphere. Appl. Optics 9, 1993–1999. https://doi.org/10.1364/AO.9.001993.

Spencer, R.W., Christy, J.R., 1993. Precision lower stratospheric temperature monitoring with the MSU: technique, validation, and results 1979–1991. J. Climate 6 (6), 1194–1204. https://doi.org/10.1175/1520-0442(1993)006<1194:PLSTMW>2.0.CO;2.

Wark, D.Q., Lienesch, J.H., Weinreb, M.P., 1974. Satellite observations of atmospheric water vapor. Appl. Optics 13, 507–511. https://doi.org/10.1364/AO.13.000507.

Waugh, S., Schuur, T.J., 2018. On the use of radiosondes in freezing precipitation. J. Atmos. Oceanic Tech. 35 (3), 459–472. https://doi.org/10.1175/JTECH-D-17-0074.1.

Zuidema, P., Painemal, D., De Szoeke, S., Fairall, C., 2009. Stratocumulus cloud-top height estimates and their climatic implications. J. Climate 22 (17), 4652–4666. https://doi.org/10.1175/2009JCLI2708.1.

Chapter 3

Frost point hygrometers

Dale F. Hurst[a,b], Masatomo Fujiwara[c], and Samuel Oltmans[b]
[a]Cooperative Institute for Research in Environmental Sciences (CIRES), University of Colorado-Boulder, Boulder, CO, United States, [b]NOAA Global Monitoring Laboratory (GML), Boulder, CO, United States, [c]Faculty of Environmental Earth Science, Hokkaido University, Sapporo, Japan

Chapter outline

1. Historical development of frost point hygrometers 37
   1.1 Early work 37
   1.2 1970s and 1980s development of FPHs for balloons 39
   1.3 Improvements in FPHs since the 1980s 41
   1.4 Modern day FPH measurement uncertainties 44
2. FPH measurement records 46
   2.1 FPH sounding sites 46
   2.2 What FPH records show 47
3. Examples of FPH comparisons with satellite-based remote sensors 48
   3.1 General approaches and methods for the comparisons 48
   3.2 Examples of comparisons between satellite sensors and FPHs 49
4. Summary 53
Acknowledgments 53
References 53

We forget that the water cycle and the life cycle are one.

Jacques Cousteau

1. Historical development of frost point hygrometers

1.1 Early work

In the 1810s, British chemist John Daniell developed a dew point hygrometer that was widely adopted by meteorologists to measure atmospheric humidity in the moist lower atmosphere (Daniell, 1820). The high-quality dew point measurements helped with weather forecasting, especially when estimating rainfall probabilities. Two centuries later, dew and frost point hygrometers (FPHs) are technologically advanced instruments that utilize the same traditional measurement principle. Today's hygrometers provide environmental humidity information not only to weather forecasting and other atmospheric science but also to strict moisture-controlled processes in industries like microelectronic manufacturing, food processing and packaging, and pharmaceutical production.

One may question the need for FPHs when there are thousands of radiosondes launched every day around the world that make direct measurements of relative humidity (RH) profiles in the atmosphere. Radiosonde RH sensors perform well from the surface to the middle troposphere, but above ~8 km their performance begins to degrade due to some well-documented problems (see Chapter 1 for more details). In general, radiosonde RH sensors lack the sensitivity of FPHs and have slow response times in the cold and dry upper troposphere and stratosphere, requiring corrections to reduce measurement lags and biases, including one caused by solar radiation heating the sensor (e.g., Dirksen et al., 2014). Frost point hygrometer measurements require no such corrections, hence their ability to measure water vapor profiles with high accuracy from the surface to the middle stratosphere (~30 km).

Dew point and FPHs function similarly, cooling a surface just enough to condense airborne water vapor into a small deposit of liquid water (dew) or ice (frost) on the surface. The temperature at which this occurs is known as the dew point (frost point) temperature. Well-documented equations relating the saturation vapor pressure over liquid water (e.g., Hyland and Wexler, 1983) and ice (e.g., Goff, 1957) to the temperature of the condensed phase facilitate accurate conversions of dew and frost point temperatures to water vapor partial pressures. Relative humidity is determined by dividing water vapor partial pressures by the saturation vapor pressure at the ambient air temperature. A volume mixing ratio (i.e., mole fraction in dry air) is calculated by dividing the water vapor partial pressure by the pressure of dry air (i.e., ambient air pressure minus the water vapor partial pressure).

There are two primary challenges in dew and frost point hygrometry: maintaining a stable deposit of condensed water on the chilled surface, and accurately measuring the temperature of the surface when and where there is a stable condensate layer. Dew point measurements are achievable using a variety of cold sources because they require only moderate cooling power since dew point temperatures never dip below the freezing point of super-cooled liquid water, typically −30°C. On the other hand, frost point measurements demand much greater cooling power, as frost point temperatures can range as low as −95°C in very dry regions of the atmosphere (tropical upper troposphere and stratosphere). With this requirement of stronger cooling comes the need for a robust and efficient method to rapidly heat the chilled surface so that a stable layer of frost can be maintained against the powerful cooling.

The earliest use of FPHs to measure atmospheric water vapor from aircraft was over southern England during World War II. These early measurements (Dobson et al., 1946; Brewer et al., 1948), conducted by the Meteorological Research Flight (MRF) project, typically reached altitudes of ~12 km and revealed the extreme dryness of the lowermost stratosphere over the Northern Hemisphere mid-latitudes (Fig. 1). Two versions of FPH instruments were used: one in which an observer viewed and adjusted the frost deposit on a chilled polished surface (thimble) then recorded the temperature at which frost appeared, and another where the frost deposited on the mirror was detected photoelectrically and the temperature of the mirror face was recorded. Both versions are described in detail, including schematics and photographs, in a handbook of meteorological instruments for upper air observations (United Kingdom Meteorological Office, 1961).

In 1954 and 1955, a Canberra jet aircraft carried out 339 more MRF flights, with 60 reaching the ~18 km flight ceiling (Tucker, 1957). The ubiquitous dryness of the lower stratosphere was confirmed through measurements of near-constant water vapor mixing ratios of ~3 parts per million by volume (ppmv), <0.03% of typical mixing ratios at the surface. It was shown that the transport of air masses through the very cold equatorial tropopause, with subsequent poleward advection, was the source of the low water vapor mixing ratios observed in the lower stratosphere over temperate latitudes (Brewer, 1949). Data obtained from MRF flights during 1972–76 prompted a re-calculation of data from the 1954 and 1955 flights using newer calibration procedures, and this adjusted the earlier mixing ratios upward by ~0.5 ppmv (Cluley and Oliver, 1978). Despite these adjustments, it was evident that stratospheric water vapor had increased over the 17-year period between the two sets of measurements (Cluley and Oliver, 1978).

Early attempts to measure water vapor above the flight ceilings of ordinary research aircraft using balloon-borne instruments, including profile measurements by FPHs of varying designs (Barrett et al., 1950; Kobayashi, 1960; Hayashi, 1961; Mastenbrook and Dinger, 1961), showed markedly increasing mixing ratios with altitude in the stratosphere. Unfortunately, many of these profiles obtained from balloons were biased wet (high) above 18 km and therefore inconsistent with the highest altitude data from the MRF flights. Consequently, it was determined that the balloon-borne measurements were contaminated at high altitudes by moisture that outgassed from the balloon skin, parachute, tether and harness strings, and possibly the insulation surrounding the instrument.

Mastenbrook and Dinger (1961) developed an FPH capable of being deployed on meteorological balloons (Mastenbrook, 1966, 1968) while working at the U.S. Naval Research Laboratory (NRL) in Washington, D.C. The basic configuration of the NRL instrument employed a polished mirror that was interposed between a coil heater and a heat sink

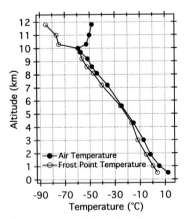

FIG. 1 Profiles of frost-point temperature and air temperature measured from aircraft over southern England on May 30, 1945, as part of the Meteorological Research Flights (MRF) project.

of chlorofluorocarbon-13 (CFC-13, CClF$_3$). A light source was pointed at the chilled mirror, and the measured intensity of light reflected by the dew- or frost-covered mirror was used to adjust the mirror heating so that a constant layer of dew/frost was maintained. This early photodetection system used a visible light source, a photomultiplier, and analog electronics. A calibrated thermistor mounted in the mirror face measured the mirror temperature at the edge of the dew or frost deposit.

An important source of error in detecting the true frost point temperature is how proximate the mirror thermistor is to the deposit of condensate. For the NRL hygrometer, this distance was minimized before flight by visual inspection and adjustment, but after launch there was no means for further adjustment. The mirror temperature, along with standard meteorological radiosonde measurements of air temperature and pressure, was telemetered by the radiosonde to a ground-based receiving station and recorded.

Also developed at NRL during these early years was a novel valve system that transitions an ascending balloon to a balloon descending at a controlled rate of $\sim 5\,\mathrm{m\,s^{-1}}$. The concept relied on helium being vented from the balloon before it bursts and rapidly plummets through the stratosphere, where air density and pressure are too low to sufficiently slow the scientific payload with a parachute (Mastenbrook, 1966). This was achieved by inserting a simple valve in the neck of the balloon that opens and releases a portion of the helium from the balloon when a pre-set atmospheric pressure is reached. This innovative, autonomous balloon valve greatly reduced the impact of water contamination on flight profiles because it enabled high-quality measurements during controlled balloon descent, when the FPH leads instead of follows the moisture-shedding balloon, parachute and string harness. These flight train components collect moisture in the lower atmosphere and eventually release it at higher altitudes, especially in the dry and sunbaked stratosphere. These moisture sources always have the potential to contaminate measurements during ascent when the FPH samples air in their wake. In terms of other measurement contaminants, the instrument's mirror and inlet tubes were always carefully cleaned with high-purity ethanol before launch to minimize any undesirable instrumental artifacts by materials that had adsorbed on these critical surfaces.

Results of the soundings by the NRL balloon hygrometer in Washington, D.C. (Fig. 2) and at several other sites (Mastenbrook, 1968) firmly established that water vapor mixing ratios in the 20–28 km altitude range were $\sim$3.2–4.5 ppmv and increased only gradually with height due to the strengthening in situ production of water vapor by the oxidation of methane (CH$_4$) and hydrogen (H$_2$). In the mid-latitudes, the seasonal variations in water vapor, driven by the annual cycle of the coldest temperatures within the tropical tropopause layer (TTL), generally produced a positive gradient with altitude in the lowermost stratosphere during late winter and spring when TTL temperatures are coldest. This gradient becomes negative during the summer and fall when TTL temperatures are warmest, allowing more water vapor to enter the lower stratosphere (Mastenbrook, 1971). The measurements above Washington, D.C., confirmed these basic characteristics of water vapor vertical profiles and their seasonal variations in the lower stratosphere over the Northern Hemisphere mid-latitudes (Mastenbrook, 1968, 1971, 1974). Stratospheric water vapor profiles obtained in the tropics at Trinidad, West Indies (10.7° N), featured extremely dry air with <1 ppmv water vapor just above the tropopause and mixing ratios that gradually increased with altitude (Mastenbrook, 1968).

1.2 1970s and 1980s development of FPHs for balloons

1.2.1 Progress and results of Mastenbrook and Oltmans

With the goal of developing an instrument capable of deployment for routine monitoring of water vapor in the upper troposphere and stratosphere, a program of upgrades to the NRL frost point hygrometer was undertaken in 1977 (Mastenbrook and Oltmans, 1983; Oltmans, 1985). Features of the redesigned instrument included the sampling of ram air flow instead of pumped flow, replacement of the temperature gradient mirror described above with a design having a uniform temperature across the mirror face, and the implementation of solid-state electronics throughout the instrument. As with all new FPHs, the historic measurement principle was retained, but the instrument's ability to maintain a stable frost layer was improved. The solid-state electronics enabled better control of the mirror temperature by providing well-regulated, intermittent pulses of electrical current to a resistive heating coil wrapped around the mirror stem just below the rhodium- (or later gold-) plated copper mirror. High precision heating pulses applied against the constant cooling by the cold CFC-13 reservoir better maintained a stable layer of condensate on the mirror.

During these early years, the amount of mirror heating needed to maintain a stable condensate layer was determined by two pairs of temperature-matched light emitting diodes (LEDs) and photodetectors. The "reference" pair LED and detector were pointed directly at each other while the "condensate detection" pair were pointed at the mirror to measure the LED light it reflected. The signal difference between the condensate detection pair and the reference pair determined the amount of heat that needed to be applied to the mirror. Both photodetectors were mounted closely together in an aluminum block to help compensate for temperature drifts in their responses. This design limited differences between the

40 PART | I Instrumentation

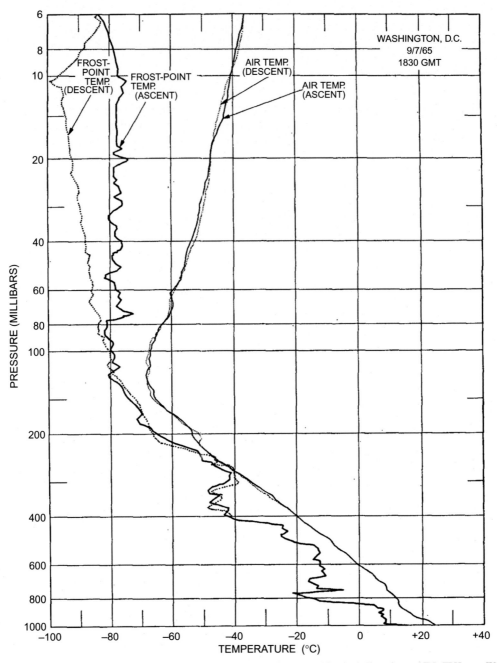

FIG. 2 Ascent and descent profiles of air temperature and frost point temperature measured by the balloon-borne NRL FPH over Washington, D.C., United States (38.9° N) on September 7, 1965. The ascent and descent air temperature profiles are generally in good agreement. Significant differences between the two frost point temperature profiles exist above 80 hPa and are likely due to the ascent measurements being contaminated by moisture (see text) and an initial delay in the recovery of frost control logic during balloon descent above ~12 hPa. *(Reproduced with permission from Mastenbrook, H.J., 1968. Water vapor distribution in the stratosphere and high troposphere. J. Atmos. Sci. 25, 299–311.)*

temperature-dependent response functions of the two LED/detector pairs to <3% during a sounding. Solid-state light sources and detectors were chosen for their reliability, while the servo system control circuitry also utilized solid-state electronic components. To ensure that the condensate layer was steadily controlled by the servo system, and as a check of the instrument response time during a flight, a heat pulse was applied to the mirror once per flight to remove a small amount of the condensate, then the return to equilibrium control was observed.

In the redesigned NRL instrument, temperature variations across the mirror face were <0.1°C, while in the earlier version the gradient was approximately 3°C. The change to a uniform temperature mirror revealed that the largest source of error in determining the frost point temperature during flights using the gradient mirror was any mis-positioning of the condensate layer's edge relative to the mirror thermistor. Assuming that the redesigned hygrometer was controlling the temperature of the mirror at or near the frost point temperature, the maximum combined measurement uncertainty from all sources, including thermistor calibration errors, non-uniform mirror temperatures, and errors associated with radiosonde measurements, data transmission and data reception on the ground, was estimated to be ±0.55°C, or ±10% of stratospheric mixing ratios.

While the soundings performed by the redesigned NRL instrument at Washington, D.C., were considered part of the balloon-borne FPH's development phase, these profiles met the criteria for reliable measurements that were adopted when the National Oceanic and Atmospheric Administration (NOAA) program of routine water vapor soundings began at Boulder, Colorado (40° N). The first Boulder FPH sounding was in April 1980, and a quasi-monthly launch schedule was initiated in January 1981. By that time, the earliest soundings at Washington, D.C. (1964–76), along with the 1977–80 profiles over D.C. and the first few flights at Boulder with the NOAA FPH, comprised a 17-year measurement record. The continuity of measurement accuracy from the earliest soundings to those performed today is assured by the consistent procedure used to calibrate mirror thermistors over the last 57 years (see Section 1.2.2).

From the initial time series of water vapor profiles at Washington, D.C., to the 1980 start-up of the NOAA program of routine FPH soundings at Boulder, the observed secular increase in stratospheric water vapor at both sites was of particular interest (Mastenbrook, 1971, 1974; Mastenbrook and Oltmans, 1983; Oltmans and Hofmann, 1995; Oltmans et al., 2000). Based on the first 6 years of measurements at Washington, D.C. (Mastenbrook, 1971), a strong positive trend in stratospheric water vapor mixing ratios was noted. However, as the record lengthened, the growth rate was found to have slowed (Mastenbrook, 1974; Mastenbrook and Oltmans, 1983). A later analysis of the stratospheric record determined an overall net increase of 0.05–0.08 ppmv year^{-1} (1%–2% year^{-1}) during 1964–76 at Washington, D.C. (Oltmans et al., 2000). By 1994, the nearly 15-year time series at Boulder also showed a significant net increase of 0.04–0.05 ppmv year^{-1} (~1% year^{-1}) in the 16–26 km altitude interval (Oltmans and Hofmann, 1995). Though this first report of a trend over Boulder was based on soundings performed 11 years prior to the 1991 introduction of digital radiosondes and the first 3 years after that change, the increase was confirmed by a subsequent trend analysis that included another 6 years of Boulder FPH data (Oltmans et al., 2000). More details about the long-term trends in stratospheric water vapor observed over Boulder are presented below in Section 2.2. Improvements in the NOAA FPH and its data quality since the 1980s are described in Section 1.3, while greater details can be found in Hall et al. (2016).

1.2.2 Accuracy and stability of accuracy—1970s and 1980s

Accurate and repeatable calibrations of mirror thermistors were, and will always be, of utmost importance to FPH measurement records. Until 2014, each thermistor was individually calibrated at three temperatures (0°C, −45°C, and −79°C) in a well-mixed bath of ethanol and dry ice. A National Institute of Standards and Technology (NIST)-traceable temperature probe was used to measure the bath temperature with high accuracy. For colder temperatures (<−79°C), an empirical correction was applied to the extrapolated polynomial curves fit to the measured thermistor resistances at the three temperatures. The correction was determined from a set of thermistor resistance measurements down to −93°C (Scherer et al., 2008). Each batch of thermistor calibrations included at least one archived thermistor used as a control to ensure that the calibration results duplicated those of previous calibrations. Details of this procedure, including its evolution as technology advanced after the 1980s, are described below (Section 1.3.3) and discussed in greater detail by Hall et al. (2016).

Four profile measurements made over 2 days in February 1991 at Edwards Air Force Base, CA (34.9° N), near the end of the analog radiosonde era, established the measurement precision for NOAA FPH soundings during the 1980s as ±8% of stratospheric mixing ratios (Fig. 3). This estimate was based on the variability of frost point temperatures measured during a 30-h period in specific stratospheric layers. Errors that were more difficult to quantify, including potential contamination and detector drift, did not exceed ±0.5°C in frost point temperature, or ±9% of stratospheric mixing ratios. The overall FPH measurement uncertainties from the 1970s to the early 1980s were estimated to be ±1°C in frost point temperature and ±0.9 ppmv (±17%) in stratospheric mixing ratios.

1.3 Improvements in FPHs since the 1980s

Both the reliability and quality of balloon-borne FPH measurements of atmospheric water vapor have improved significantly since the 1980s, largely in conjunction with the development of radiosondes (see Chapter 1) that enabled data from peripheral instruments like ozonesondes (see Chapter 2) and FPHs, as well as their own sensor outputs, to be telemetered to a ground-based receiving and recording system. As the bandwidth of radiosonde telemetry increased, permitting larger

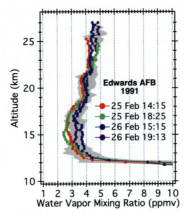

FIG. 3 Four water vapor profiles obtained by the balloon-borne NOAA FPH on February 25 and 26, 1991, at Edwards Air Force Base (AFB), California (34.9° N). Launch times are local (PST). Colored markers depict mixing ratio averages in 250-m altitude bins. *Gray shading* shows the 95% confidence intervals of mean mixing ratios for the final flight. Despite differences in the profiles below 19 km, they demonstrated the repeatability of stratospheric water vapor measurements by the NOAA FPH.

quantities of data from instruments to be received and recorded, so did the quality of radiosonde sensors. Today's radiosondes have adequate bandwidth to telemeter not only measurement data, but also instrumental engineering data that are highly useful in performing quality assurance of the FPH profiles.

Since the 1980 deployment of the NOAA FPH in Boulder, several other balloon-borne FPHs have been developed and employed for water vapor soundings. The French Laboratoire de Météorologie Dynamique (LMD) developed a balloon-borne FPH in the early 1990s (Ovarlez, 1991) and its water vapor measurements in the upper atmosphere were compared to profiles from various satellite-based instruments (Pruvost et al., 1993; Berthet et al., 2013). Starting in 1996, the Swiss company Meteolabor AG offered a thermoelectric- (Peltier-) cooled "Snow White" FPH for balloon-borne measurements (Fujiwara et al., 2003; Vömel et al., 2003). The cryogenic FPH (CFH), a modern derivative of the NOAA FPH, became a commercial off-the-shelf instrument in 2007 (Vömel et al., 2007a). All of these instruments rely on the same measurement technique but their designs differ in subtle but important ways.

In the next sections, the most significant advances made over the last four decades in balloon-borne FPH measurements are described. For the most part, the improvements described are generally applicable to all balloon-borne FPHs, but the specific details provided pertain almost exclusively to the NOAA instrument.

1.3.1 More stable frost control

Technical advances since the 1980s improved the ability of FPHs to maintain stable frost layers on their temperature-controlled mirrors. Older analog frost control systems were slow to respond to changes in frost thickness and often caused oscillations ("ringing") in the measured frost point temperatures. Digital systems allow the use of proportional-integral-derivative (P-I-D) logic to control frost on the mirror through a finely-tuned feedback loop with fast response times to frost layer changes. Modern components allow the resistive heating of the mirror to be applied in highly controlled electrical pulses, minimizing both the amplitude and frequency of oscillations in measured frost point temperatures. The use of smaller mirrors and low mass bead thermistors embedded in them reduces the latency of the resistive mirror heating by providing rapid temperature responses.

1.3.2 Update of the photodetection system and the exclusion of false signals

The use of two pairs of LED light sources and photodetectors ("reference" and "condensate detection") to compensate for thermal drift in their response functions was discussed earlier. With the older instruments, differences between the temperature coefficients of the source/detector pairs were a potential source of measurement uncertainty. An improved design of the photodetection system, implemented in the 1990s, removed the reference pair and added heated temperature stabilization of the condensate detection pair, as well as warming the beam-focusing lens to prevent fogging (Fig. 4). This temperature-controlled "optics block" in the modern NOAA FPH is resistively heated to 32°C and temperature controlled to ±0.1°C during flights to minimize thermal changes in both the LED output intensity and the photodiode sensitivity (Hall et al., 2016).

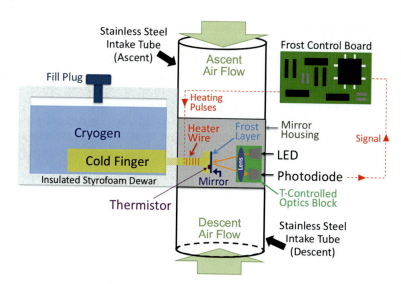

FIG. 4 Schematic of the modern version of the NOAA FPH. The direction of air flow through the instrument is dictated by the ascending or descending motion of the balloon. The mirror is continually cooled by a cold finger immersed in liquid cryogen and accumulates condensate (dew or frost) from moisture in the air flowing past it. The intensity of LED light reflected by the mirror and frost layer is monitored by a photodiode to control the frost layer. The frost control board adjusts mirror heating by sending electrical pulses to the resistive heater wire (coiled around the mirror stem) in response to deviations in the LED light signal at the photodiode to maintain a stable frost layer.

The paired LED and photodiode used for frost detection operate at wavelengths emitted by the sun, so any solar radiation entering the instrument is cause for concern. Sunlight entering the upward-pointing ascent intake tube and reaching the photodiode can be falsely interpreted as inadequate frost on the mirror, temporarily disabling mirror heating. Over the years, various attempts were made to physically block sunlight from entering the instrument through the intake tubes using physical "sun shields." Several materials and designs were tried, but all substantially increased the potential for measurement contamination through their initial collection and subsequent shedding of moisture into the sample flow. The "Snow White" FPH (Fujiwara et al., 2003; Vömel et al., 2003) utilized a non-linear flow path into the instrument to block sunlight, but the resulting turbulent air flow enhanced its contact with the instrument's internal surfaces, sometimes adversely affecting the measurements.

One solution that significantly minimized false sunlight signals while retaining laminar air flow through the instrument was to rapidly and repeatedly cycle the LED power. In this way, any "dark signals" measured at the photodiode while the LED power is off are those caused by stray sunlight in the hygrometer. Dark signals are simply subtracted from an interpolation of the previous and subsequent frost detection signals at the photodiode when the LED is powered on. With modern technology, the LED blinks frequently enough (24 Hz) to produce a very short time lag between consecutive light and dark signals. Using this method, false signals from sunlight are digitally removed from the light signals measured at the photodiode to provide a true mirror reflectance signal for the frost control logic.

1.3.3 Advanced thermistor calibration technique and curve fitting

Calibration standards used for in situ measurements of atmospheric gases are often prepared in a laboratory setting and stored under pressure in some type of container, such as a gas cylinder. For water vapor, these types of calibration standards are notoriously difficult to prepare with accuracy because water vapor adsorbs onto most surfaces it contacts. Hydrophilic surfaces can degrade the long-term stability of low concentration water vapor standards stored in gas cylinders, even at high pressures. Frost point hygrometry eliminates the need for gas standards of water vapor because the technique requires only highly accurate temperature measurements. Temperature is not only easy to measure with high accuracy, it is straightforward to measure with sustained accuracy if calibrations are performed in a very repeatable manner.

Before each thermistor is embedded in an FPH mirror, it is independently calibrated in a cold, well-stirred ethanol bath along with 35–40 other thermistors. As mentioned previously, every calibration of new thermistors includes the re-calibration of a small archive of previously calibrated thermistors plus a reference thermometer that is routinely checked against NIST temperature standards. During 1980–98, a quartz thermometer was employed as this NIST-traceable reference thermometer, and since 1998 a platinum resistance thermometer (PRT) is used. Re-calibrations of the PRT every 6 years indicate that the reference thermometer's calibration has not changed significantly over time. The inclusion of a reference thermometer and a set of archived thermistors in each and every calibration of new thermistors provides a traceable, quantitative link between all new and past calibrations.

Until 2014, the ethanol bath used for thermistor calibrations was cooled to three temperatures (−79°C, −45°C, and 0°C) by manual additions of finely ground dry ice powder. The target temperatures were stably maintained (±0.005°C) for several minutes, while the resistance of each thermistor was recorded. The 3-point calibration curve derived for each thermistor required extrapolation to temperatures above 0°C and below −79°C.

In 2008, a set of measurements of thermistor resistances down to −93°C provided corrections to the low temperature extrapolations, significantly reducing small systematic errors in the historical frost point temperature data through reprocessing (Scherer et al., 2008). The well-stirred ethanol calibration bath was initially cooled to near −93°C using liquid nitrogen and then left for ~2 days to warm naturally to +19°C. The resistance of each thermistor was autonomously measured and recorded throughout the process, and a unique calibration curve was determined for each thermistor using its resistance values at six temperatures (−91°C, −80°C, −59°C, −39°C, −15°C, and +19°C). The wider range of temperatures eliminated the need for calibration curve extrapolations and corrections, and further reduced systematic errors when converting thermistor resistances into temperatures.

1.3.4 Radiosonde improvements

FPHs need simultaneous measurements of atmospheric temperature (T) and pressure (P) to convert water vapor partial pressures to relative humidity values and volume mixing ratios, respectively. These T and P measurements are almost always made by a radiosonde launched on the same balloon as the hygrometer. Radiosonde temperature and pressure sensors have improved over the years to better measure the wide dynamic ranges of T and P they encounter during a typical sounding. The calibration of T and P sensors, usually performed in bulk by manufacturers within a laboratory setting, has benefitted from automation and advanced environmental control measures (see Chapter 1). Nevertheless, sensors are still prone to systematic biases driven by the wide ranges of environmental conditions in which they must operate. One example of this is the radiative heating of T sensors by the strong and ubiquitous solar flux in the stratosphere, despite the sensors being coated with reflective material. Ample testing has led to the development and application of corrections to the T and P sensors that minimize systematic errors in their profile data. However, even with these corrections, the measurement errors created by calibration uncertainties and imperfect correction algorithms for the T and P sensors add non-negligible biases to the mixing ratios and relative humidity values calculated from FPH measurements.

Radiosondes also perform the critical function of telemetering data from FPHs and other instruments to a ground-based receiving unit. This telemetry allows the flight data to be recorded in real time and eliminates the requirement that the payload be recovered after the flight, which in some cases is impossible. The amount of data that can be telemetered varies with the radiosonde make and model, but has increased over time. The increasing bandwidth has allowed for more sophisticated instruments to be carried by balloons because they generally output larger amounts of data. Most modern radiosondes are also equipped with a GPS (GNSS) receiver that reports the three-dimensional position of the payload, making it easier to find and recover after the flight if desired. Care must be taken to ensure the telemetered radio frequency (RF) waves do not interfere with the FPH electronics and produce noise in the frost point measurements. This is typically achieved by shielding the FPH electronics from RF interference and/or physically separating the radiosonde and FPH by ≥1 m using a Styrofoam or other lightweight boom.

1.3.5 New cryogen implementation

One significant improvement in the NOAA FPH occurred when the ozone-depleting cryogen (CFC-13) used in early instruments was replaced by HFC-23 (CHF_3). This change did not decrease measurement uncertainties or improve instrument reliability, but it immediately reduced the instrument's impact on Earth's stratospheric ozone layer. Given the high GWP (~12,000) of HFC-23 and the recent Kigali Amendment to the Montreal Protocol that phases down HFC production and consumption, there is now a pressing need to replace HFC-23 with a more environmentally friendly cooling mechanism. Work is currently ongoing to develop new FPH instruments with thermoelectric cooling as well as to modify existing FPHs to use a cold dry ice and ethanol mixture instead of HFC-23.

1.4 Modern day FPH measurement uncertainties

Uncertainties in frost point temperature measurements fall into three distinct categories: those associated with maintaining a stable layer of frost on the chilled mirror, those that propagate from the calibration of mirror thermistors, and those that derive from manufacturing differences between instruments.

1.4.1 Frost control uncertainties

Uncertainties associated with frost control dominate the uncertainty budget for frost point temperature measurements, especially in the troposphere where they account for >95% of the total measurement uncertainties (Vömel et al., 2016;

Hall et al., 2016). The assumption that the mirror temperature is equal to the frost point temperature is based on the frost layer being stable. When that condition is met, the frost layer is in equilibrium with the moisture in the air stream passing over it, and the frost layer (mirror) temperature is the frost point temperature. The frost point temperature is then used in the Goff-Gratch equation (Goff, 1957) that relates water vapor saturation pressures over ice to the ice surface temperature. Without stability of the frost layer, there can be no argument of equality between the mirror temperature and the frost point temperature. Small instabilities lead to random errors in frost point temperatures that can be averaged out over time, but large instabilities can drive substantial oscillatory behavior of the frost layer and a significant degradation of the quantitative connection between mirror temperatures and frost point temperatures.

Extensive work has gone into determining the best P-I-D parameters to achieve stable frost control over a wide range of moisture conditions, from the wet lower troposphere to the very dry stratosphere. Most modern hygrometers use a progressive set of parameters based on the measured frost point temperature, with unique parameter sets for instruments cooled by different methods, e.g., liquid cryogen versus thermoelectric cooling. Hall et al. (2016) estimate that typical frost control instabilities lead to frost point temperature uncertainties of ±1.5°C in the lower troposphere, decreasing to ±0.5°C in the upper troposphere, then further reducing to ±0.15°C in the lower stratosphere (Fig. 5). These translate to mixing ratio uncertainties of ±10%, ±6%, and ±3% for the three altitude regions, respectively. In the dry stratosphere, frost control uncertainties account for ~50% of the total measurement uncertainties.

1.4.2 Calibration uncertainties

Calibration uncertainties for the FPH are small systematic biases that reflect how well a parametric equation fits the thermistor calibration data. Residuals from parametric fits generally decrease as the order of the fitting equation increases. The change in 2014 from 3-point to 6-point fits to produce the calibration curve for each thermistor significantly reduced calibration biases. This change was most significant for frost point temperatures below −79°C since the 3-point calibration curves were historically extrapolated to these cold temperatures and then corrected. The 6-point curves require no extrapolation. Hall et al. (2016) estimate calibration uncertainties for the NOAA FPH to be ±0.1°C before 1990, ±0.07°C between 1990 and 2014, and ±0.03°C since 2014. Calibration uncertainties are smaller than frost control uncertainties in both the troposphere and stratosphere, contributing 0.5%–1.2% to the uncertainties of FPH water vapor mixing ratios (Fig. 5).

1.4.3 Uncertainties due to radiosonde measurement errors

Most radiosonde manufacturers provide technical data sheets reporting measurement uncertainties, mainly accuracy and repeatability limits for the sensors deployed on their specific models. The estimated accuracy of T measurements is typically in the range of 0.3–1.0°C, with a step-change in accuracy at a specified altitude or atmospheric pressure level. Systematic errors in radiosonde T measurements can therefore induce biases as large as 1%RH in RH values calculated from FPH measurements. Uncertainty estimates for radiosonde P measurements are also commonly provided with a step change above a specified altitude or atmospheric pressure level. Some manufacturers claim smaller absolute uncertainties at

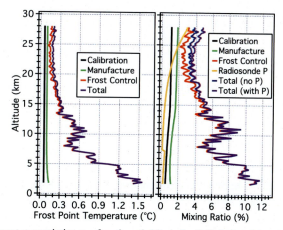

FIG. 5 Frost point hygrometer measurement uncertainties as a function of altitude for (left) frost point temperatures in absolute units (°C) and (right) water vapor mixing ratios in relative units (%). Frost control errors contribute the largest uncertainties for both frost point temperatures and mixing ratios, although above 25 km radiosonde pressure measurement uncertainties become nearly as important for mixing ratio uncertainties.

$P < 100$ hPa, while others report absolute uncertainties that increase with altitude. Measurement uncertainty estimates for $P > 100$ hPa are typically ≤ 1 hPa, but at $P < 100$ hPa range from 0.6 to 2 hPa. Hall et al. (2016) demonstrate how P measurement uncertainties for International Met Systems model iMet-1-RSB radiosondes contribute uncertainties of $<1\%$ to tropospheric water vapor mixing ratios, while stratospheric uncertainties increase from $<1\%$ near the tropopause to as large as 3% at 28 km due to decreasing atmospheric pressure (Fig. 5).

The use of GPS (GNSS) altitude as a replacement for radiosonde P sensors has become increasingly common since the early 2000s. Atmospheric P is calculated from GPS-based geometric altitude data using a well-calibrated P sensor at the surface and standard atmospheric P versus altitude equations. Hence, the method is free from uncertainties normally associated with radiosonde P measurements. Some long-term FPH data records pre-date the widespread availability of GPS technology in radiosondes. These records have traditionally reported geopotential height as their vertical coordinate, calculated using the hypsometric equation and measured incremental changes in P, T and relative humidity as the balloon ascends. Since GPS receivers provide values of the geometric altitude instead of geopotential altitude, a simple latitude-dependent mathematical conversion between the two is required. A comparison between GPS altitudes and geopotential heights for each flight can be used to calculate and apply an optimal constant correction to the measured atmospheric pressures that best aligns GPS-based geopotential altitudes with those calculated from radiosonde sensor data. In some cases, a constant P correction based on a pre-flight comparison with an accurate P sensor at the surface is applied to the radiosonde's entire profile of atmospheric P.

1.4.4 Environmental uncertainties

Other sources of uncertainties in FPH measurements exist, but these are quite difficult to estimate because they are episodic in nature. For example, and as discussed previously, all flight train surfaces, including the balloon skin, parachute, payload tethers and hygrometer intake tubes, can collect and subsequently shed moisture, potentially moistening the air stream measured by the instrument during ascent. This "contamination" is likely episodic because the pendulum or rotational motion of the instrument payload moves it in and out of the wake of the ascending balloon. Moisture shedding by the payload tethers or intake tube can provide a more proximate and therefore a more persistent source of contamination. The most fatal location for moisture shedding is within an intake tube, where even the tiniest spot of condensed water will continuously contaminate the air stream flowing through the FPH. Jorge et al. (2021) demonstrated how this type of contamination occurred during several soundings where the balloon-borne FPH ascended through mixed-phase clouds. A liquid water droplet collided with and froze on the interior wall of the intake tube to become a source of long-lasting, uncorrectable contamination. Intake tube contamination is fairly easy to detect in an FPH profile because it is substantial and persistent, unlike the more intermittent and weaker contamination from the balloon skin. To minimize the potential for moisture contamination of the measurements, FPHs are seldom launched when there is precipitation or a thick cloud layer overhead.

2. FPH measurement records

FPHs have been routinely launched from Boulder, Colorado, since April 1980 by NOAA's Global Monitoring Laboratory. The program was initiated to monitor water vapor abundance in the upper troposphere and stratosphere over decadal time scales. As mentioned earlier, hygrometers of various designs from a number of different manufacturers have been launched over the last three decades at many locations around the globe—some for short-duration intensive "campaigns" and others for longer-term studies of the behavior of upper atmospheric water vapor. Though also developed in Boulder, the CFH (Vömel et al., 2007a) is manufactured and calibrated independently from the NOAA FPH. Together, the CFH and NOAA FPH are the instruments behind most of the long-term FPH records in existence. Here we describe several of those records that have been used or are available for use to evaluate atmospheric water vapor retrievals from satellite-based instruments.

2.1 FPH sounding sites

Balloon-borne FPHs have been launched for at least 5 years at only 10 sites around the globe (Table 1) during the last ~20 years. The 5-year minimum length is relevant here because shorter records have far less statistical value when used to evaluate biases and drifts in the remote-sensing records of satellite-based instruments. The historical timelines of FPH launches at the 10 sites are shown in Fig. 6. The sites at San Cristóbal, Ecuador, and Biak, Indonesia performed their soundings with CFHs or a blend of NOAA FPHs and CFHs during annual campaigns of the Soundings of Ozone and Water in the Equatorial Region (SOWER) project (Fujiwara et al., 2010; Hasebe et al., 2013). The other records listed in Table 1 are generally more temporally uniform and have been compiled primarily for longer term climate studies. Most sites are

TABLE 1 Modern frost point hygrometer sounding sites with 5+ year records.

Site location	Site code	Affiliated networks[a]	Latitude	Longitude	Start year	Number of soundings[b]
Ny-Ålesund, Norway	NYA	N, G	78.9° N	11.9° E	2013	51
Sodankylä, Finland	SOD	N, G	67.4° N	26.6° E	2004	134
Lindenberg, Germany	LIN	N, G	52.2° N	14.1° E	2006	277
Boulder, Colorado, United States	BLD	N, G	40.0° N	105.3° W	1980	536
Beltsville, Maryland, United States	BEL	N, G	39.1° N	76.9° W	2006	69
Hilo, Hawaii, United States	HIL	N	19.7° N	155.1° W	2002	116
San José, Costa Rica	SJC	N	9.9° N	84.0° W	2005	252
San Cristóbal, Ecuador	SAC		0.9° S	89.4° W	2000	13
Biak, Indonesia	BIK		1.2° S	136.0° E	2005	54
Lauder, New Zealand	LDR	N, G	45.0° S	169.7° E	2003	173

[a]Affiliated networks are (N) NDACC and (G) GRUAN.
[b]Through 2020 or early 2021.

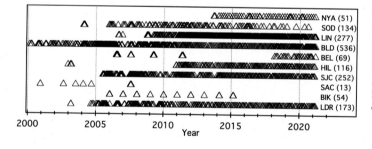

FIG. 6 Records of balloon-borne frost point hygrometer soundings at 10 sites with >5 years of measurements. Each marker represents an individual sounding, and the total number of soundings in the records appears in parentheses next to the site codes (see Table 1). The sites are arranged from most northerly (Ny-Ålesund) to most southerly (Lauder). The start date of January 2000 excludes the first 20 years of the Boulder record so that the other records can be viewed with greater temporal resolution.

affiliated with at least one of two climate observing networks that include FPHs; the Network for the Detection of Atmospheric Composition Change (NDACC) and the Global Climate Observing System (GCOS) Reference Upper Air Network (GRUAN).

2.2 What FPH records show

FPHs provide water vapor profile measurements from the surface to the middle stratosphere at a vertical resolution of tens of meters. They are capable of measuring over the entire range of atmospheric water vapor mixing ratios, from as much as 40,000 ppmv at the warm surface to as little as 1 ppmv near the cold tropical tropopause. These values depict a dynamic range of $>10^6$ in water vapor number density that is too wide for the measurement capabilities of most high-quality instruments. Water vapor abundance in the troposphere is highly variable, such that profiles below the tropopause can change significantly in less than an hour. As a result, meaningful comparisons of tropospheric water vapor measurements by FPHs and satellite instruments require that they are highly coincident in both time and space (e.g., Read et al., 2022).

Water vapor mixing ratios above the tropopause are much less variable, such that profiles measured a few days or 500 km apart can be considered "coincident." However, lower stratospheric mixing ratios vary seasonally due to the annual cycle of temperatures in the TTL, and vary inter-annually because of dynamical phenomenon like the quasi-biennial

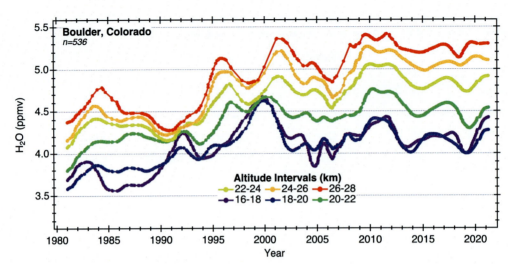

FIG. 7 The 41-year record of stratospheric water vapor over Boulder measured by balloon-borne NOAA frost point hygrometers. *Filled circles* depict the average mixing ratios in 2-km altitude intervals for the 536 soundings. The record shown is smoothed using a ±1 year boxcar filter. The average net increase in stratospheric water vapor (16–26 km) between 1980 and 2010 was ~1 ppmv or ~25%. For most 2-km altitude intervals, mixing ratio averages in 2021 are not significantly different from those in 2010. *(Modified and updated from Hurst, D.F., Oltmans, S.J., Vömel, H., Rosenlof, K.H., Davis, S.M., Ray, E.A., et al., 2011. Stratospheric water vapor trends over Boulder, Colorado: analysis of the 30 year Boulder record. J. Geophys. Res. 116, D02306. https://doi.org/10.1029/2010JD015065.)*

oscillation (QBO; e.g., Fujiwara et al., 2010; Konopka et al., 2010) and the El Niño Southern Oscillation (ENSO; e.g., Liang et al., 2011; Garfinkel et al., 2018). The timescales of the QBO and ENSO imply that several years of overlapping FPH and satellite measurements of stratospheric water vapor are needed to conduct a comparison that is statistically meaningful.

Vertical profiles from FPHs show a distinct seasonality in the lowermost stratosphere, with the driest air present during the NH winter months when the Brewer-Dobson circulation is strongest and temperatures in the TTL are coldest. In the tropics, water vapor mixing ratios are largely conserved as stratospheric air masses ascend, and the seasonal cycle signal in water vapor can be followed upward from the tropical tropopause, an observation known as the "tape recorder" (Mote et al., 1996; Fujiwara et al., 2010). Quantification of the "transit times" of air masses from the tropical tropopause to higher altitudes based on this tape recorder provides information about ascent rates in the tropical stratosphere. The seasonal signatures in water vapor diminish with altitude as air masses mix and in situ water vapor sources (i.e., the oxidation of CH_4 and H_2) become stronger.

The 41-year FPH record over Boulder (Fig. 7) shows a statistically significant net water vapor increase of 0.6–1.2 ppmv (15%–30%) in the stratosphere (16–26 km) from 1980 to the year 2000 or 2010 (Oltmans et al., 2000; Rosenlof et al., 2001; Hurst et al., 2011). Some of this increase occurred during 1980–92, before any continuous, long-term satellite-based records of stratospheric water vapor. Some studies have merged together the records of different satellite sensors to create multi-decadal timeseries of stratospheric water vapor for trend analysis (Hegglin et al., 2014; Froidevaux et al., 2015; Davis et al., 2016). Each employed a different method to reduce biases between the different satellite records before combining them. Using such a "homogenized," merged satellite record, Hegglin et al. (2014) reported finding no significant trend in stratospheric water vapor over Boulder between 1989 and 2010. The lack of a continuous, >5-year satellite-based measurement record of stratospheric water vapor before 1993 hinders direct comparisons of satellite trends with the Boulder FPH record that started in 1980. In general, stratospheric mixing ratios over Boulder in 2021 are not significantly different from those measured back in 2010.

3. Examples of FPH comparisons with satellite-based remote sensors

3.1 General approaches and methods for the comparisons

After the launch of a new satellite instrument for stratospheric and upper tropospheric water vapor, special short-term campaigns and longer-term programs of FPH soundings have been established to evaluate its measurements. Spatiotemporal coincidence criteria used to match satellite and FPH profile data include a time window for temporal matching (e.g., satellite measurements taken within ±X hours of a balloon launch) and latitude/longitude or distance limits for spatial matching (e.g., within ±Y degrees of latitude and ±Z degrees of longitude, or within N km from the balloon launch site).

In some cases, balloon launches have been scheduled such that FPH measurements in the lower-to-middle stratosphere are made as close in time as possible to the satellite measurements over that location. Depending on the frequency of satellite overpasses of the FPH launch site, the criteria applied to match the two profiles can have a wide range of tolerances. For satellites with 1–3 day return intervals, temporal tolerances for coincidence are often within ±24 h and spatial criteria are within a few degrees of latitude and 8–10 degrees of longitude. Coincidences for satellites with less frequent overpasses require more relaxed criteria, up to ±7 days and within 10–15 degrees of latitude and several tens of degrees of longitude. For instruments with weekly or longer return intervals, the only way to apply more stringent coincidence criteria without excluding too much data is to temporally coordinate FPH launches with satellite overpasses of the site.

Most FPH launch sites where measurement programs have been running for at least 5 years (Table 1) have adequate numbers of soundings to evaluate satellite retrievals if the coincidence criteria are judiciously chosen. In practice, when using FPH data to evaluate satellite retrievals, all available sounding data are compiled and coincidence criteria are defined to identify an adequate number of FPH profiles to produce reliable statistics. More stringent criteria (smaller spatiotemporal windows) may provide more robust comparisons by ensuring the FPH and satellite are measuring the same airmass. Coincidence criteria must also depend on the variability of the measured species in the region of interest. More stringent coincidence criteria reduce the number of comparison cases, often producing results with lower statistical significance. Note that the coincidence criteria for water vapor comparisons must be far more stringent below the tropopause than above it because of much greater spatiotemporal variability in the troposphere. The best criteria for such a study are determined through consideration of these and other factors to produce meaningful comparison statistics.

The vertical resolution of water vapor measurements in the UT and stratosphere by satellite instruments ranges from 1 to 5 km and is typically 2–3 km. This is relatively coarse compared to the vertical resolution of FPH measurements, even if FPH response times approach 30 s (Vömel et al., 2016) and balloon ascent/descent rates of $\sim$5 m s^{-1} are considered. It is therefore necessary to convolve each FPH profile with a vertical weighting function or averaging kernel specific to the satellite instrument being evaluated before making a comparison. In this way, satellite measurements are evaluated at their theoretical upper limit of vertical resolution, and FPH data are convolved at that resolution and placed on the satellite's grid of retrieval pressures or altitudes.

Large sets of differences between satellite retrievals and vertically convolved FPH profiles are typically analyzed to evaluate biases using standard statistical measures such as means, medians, standard deviations, probability distribution functions, and others. Normally, biases are attributed to the satellite retrievals under the general assumption that FPH measurements are bias-free or have negligible biases. The biases determined in this way may depend on altitude, latitude, and time (e.g., season, number of years from the satellite launch, etc.), and investigations of biases are normally made as functions of these variables.

3.2 Examples of comparisons between satellite sensors and FPHs

In this section, we discuss the comparison methods and results of stratospheric water vapor measurements by the NOAA FPH and two satellite-based water vapor instruments: The Microwave Limb Sounder (MLS) on NASA's *Aura* satellite and the 3rd generation Stratospheric Aerosol and Gas Experiment (SAGE III) spectrometer aboard the International Space Station (ISS). Normally, satellite data products improve with progressive updates after the initial data release, as enhancements are made to data processing software, radiative transfer models, and other procedures. In some cases, more than one data processing center is given the opportunity to produce their own satellite data products from the same raw radiance data, with unique ownership and version numbers assigned to their products. This is not the case for either MLS or SAGE III/ISS since both instruments have their own dedicated teams who have continued to develop and refine their retrieval algorithms. For the case of MLS water vapor data, versions 1.5, 2.2, 3.3, 4.2, and 5 have been released since the instrument was deployed in 2004. Data versions 5.1 and 5.2 for SAGE III/ISS retrievals have been made available since the instrument became operational on the ISS in 2017.

Re-evaluations of satellite retrievals through these types of comparisons are essential when new versions of satellite data products are released, especially when the retrieved values are expected to change significantly. For MLS water vapor, Vömel et al. (2007b) evaluated primarily version 1.5 (along with limited cases for version 2.2) using NOAA FPH profile data between February 2005 and January 2007 at 10 stations worldwide. Read et al. (2007) evaluated MLS version 2.2, Hurst et al. (2014) evaluated version 3.3 at three NOAA FPH sites between August 2004 and December 2012, and Hurst et al. (2016) evaluated version 3.3 and version 4.2 between 2004 and mid-2015 at five sites launching CFHs and NOAA FPHs. Most recently, Livesey et al. (2021) evaluated MLS versions 4.2 and 5 for 2004–20 at three NOAA FPH sites. To evaluate the much shorter record of SAGE III/ISS water vapor retrievals, Davis et al. (2021) utilized several different data sets, including FPH profiles from five different sites.

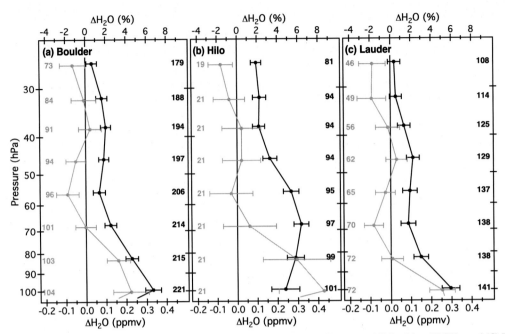

FIG. 8 Mean MLS(v4.2)-FPH differences at eight pressure levels from 100 to 26 hPa (16–26 km) over (A) Boulder, (B) Hilo, and (C) Lauder. Top and bottom axes provide separate scales for water vapor differences in % and ppmv, respectively. *Gray markers* depict mean differences for the period 2004–12 for Boulder and Lauder, and the period 2010–12 for Hilo, as reported in Hurst et al. (2014). *Black markers* show updated mean differences for the period 2004–21 (2010–21 for Hilo). Error bars represent the 95% confidence intervals of mean differences. The number of MLS-FPH differences at each pressure level are listed for both date ranges.

Figs. 8–10 present the results of comparison studies between MLS version 4.2 water vapor retrievals and the FPH profile data routinely obtained at three to five sounding sites. MLS makes ~3500 near-global (82° S–82° N) measurements each day, therefore even randomly-timed balloon launches from any site are likely spatiotemporally coincident with at least one MLS profile. The MLS-FPH comparison results shown here are updated and slightly modified from those described in Hurst et al. (2014, 2016), with temporal and spatial coincidence criteria of ±18 h, ±2° latitude and ±8° longitude. These tolerances typically identified four to six MLS profiles that were coincident with each FPH profile. When multiple coincident MLS profiles were identified for a single FPH profile, the median MLS mixing ratio at each of eight MLS retrieval pressures (100–26 hPa) was determined from each coincidence "cluster." This method allowed for a single MLS mixing ratio (at each pressure level) per FPH sounding to be compared to the convolved FPH profile. Median values were used instead of averages because the former are far less sensitive to anomalous retrievals than the latter.

As explained in the previous section, the vertical resolution of FPH measurements, natively recorded at 5–10 m and typically reported as averages in 250 m altitude bins, was degraded to that of the MLS retrievals (~3 km in the lower stratosphere) to ensure fair and robust comparisons. This convolution was done using a water vapor a priori profile, a forward model smoothing function, and averaging kernels (for non-polar latitudes here) for the MLS measurements. For more details see Equation (1) of Read et al. (2007). For simplicity, the median MLS profiles (described earlier) were used as the a priori profiles instead of those available from the MLS database because they have been shown to produce nearly identical convolved FPH profiles (Hurst et al., 2014).

Fig. 8 shows the profiles of mean MLS-FPH differences at the Boulder, Hilo and Lauder sounding sites for two different periods, 2004–12 (2010–12 for Hilo) and 2004–21 (2010–21 for Hilo). For the shorter periods through 2012, which is the same date range used in the Hurst et al. (2014) analysis of MLS version 3.3, differences in the newer version 4.2 MLS retrievals were calculated. Despite the use of these more contemporary MLS retrievals and some minor differences in analysis methods used for the comparisons, the results presented here are basically the same as those for MLS version 3.3 in Hurst et al. (2014). Over the pressure range 68–26 hPa (~19–25 km), the mean differences are either not statistically different from zero or within ±0.1 ppmv (±2%). At 83 hPa over Boulder and Hilo and at 100 hPa over all three sites, MLS v4.2 retrieved values are ~0.15 to 0.45 ppmv (~3%–8%) greater than FPH mixing ratios. The differences at 83 and 100 hPa are likely due to difficulties in MLS trying to resolve the steep vertical gradient of water vapor across the tropopause given the instrument's vertical resolution of ~3 km (Hurst et al., 2014). However, for the longer periods extending into 2021, the mean MLS-FPH differences are almost all positive (0.1–0.3 ppmv) and statistically significant at the eight stratospheric

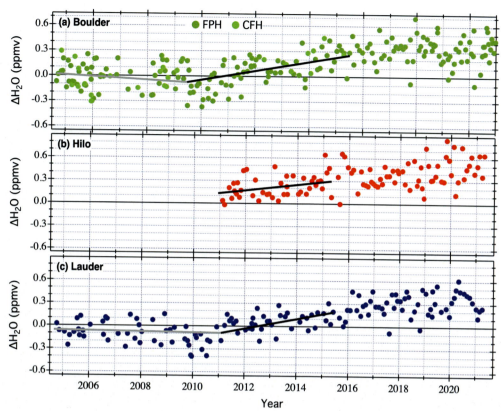

FIG. 9 Timeseries of MLS(v4.2)-FPH differences at 68 hPa (~19 km) over (A) Boulder, (B) Hilo, and (C) Lauder. All FPH data are from NOAA FPH soundings except at Boulder, where 24 flights were performed with CFHs *(light green)*. For Boulder and Lauder, *gray lines* represent the linear trends in differences before their statistically significant changepoints, while *black lines* depict the linear trends from the changepoint at each site to mid-2015, as described in Hurst et al. (2016). The record of MLS-FPH differences at Hilo has no statistically significant changepoint so only the linear trend from record start to mid-2015 *(black line)* is shown.

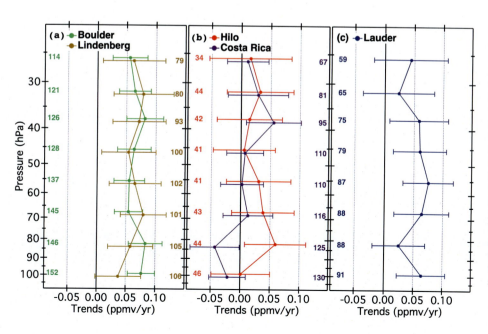

FIG. 10 Trends in MLS(v4.2)-FPH differences for five different launch sites for the time period starting at the statistical changepoint for each record to mid-2015, as described in Hurst et al. (2016). Error bars depict the 95% confidence interval of the trends. The numbers of FPH launches coincident with MLS profiles at each site are listed for every pressure level. NOAA FPHs were launched at Boulder, Hilo, and Lauder, while CFHs were launched at Lindenberg, Germany and San José, Costa Rica. Trends in MLS-FPH differences at nearly every pressure level above the extra-tropical sites are statistically significant, but very few trends at tropical sites Hilo and San José are statistically different from zero.

pressure levels above each of the three sites. This result confirms the previous finding that MLS v4.2 mixing ratios were consistently greater than FPH mixing ratios after ~2012, at all three sites, possibly because of drifts or step changes in the MLS v4.2 water vapor measurements.

Fig. 9 shows time series of the MLS-FPH differences at 68 hPa over these three NOAA FPH sites. The differences are not statistically significant from record start to ~2009, but after that, MLS values diverged from FPH values, as indicated by the black linear trend line in each panel. During 2016–21, MLS mixing ratios continued to be ~0.2–0.6 ppmv greater than NOAA FPH mixing ratios at all three sites. Fig. 10 summarizes the linear trends in MLS-FPH differences for the time period starting on the date of the detected statistical changepoint in each time series (calculated independently for each site/pressure level combination) and ending in mid-2015, as was done in Hurst et al. (2016).

Statistical changepoints in the timeseries of MLS-FPH differences at eight pressure levels above the three NOAA FPH sites, plus two CFH sounding sites at Lindenberg, Germany (52.2° N) and San José, Costa Rica (10.0° N), all occurred between 2008 and 2011. Hurst et al. (2016) provide greater details about the changepoint determinations and piecewise continuous linear regression fitting routines that were used in the analysis. The common features for all retrieval pressure levels, at least at the extratropical sites, is an increase in MLS-FPH differences after their changepoints followed by persistent positive differences for at least the last 5 years. Hurst et al. (2016) first reported this potential problem with MLS v4.2 retrievals, and in response, the MLS team recently produced a new version 5 of water vapor retrievals (Livesey et al., 2021). The new version corrects for a drift in the relative sensitivity of the MLS 190-GHz receiver to the two different parts of the microwave spectrum it observes. These corrections reduce the rates of drift in MLS v5 retrievals relative to FPH measurements at Boulder, Hilo, and Lauder by an average of 46 ± 28% (Livesey et al., 2021).

Another satellite-based water vapor instrument that was recently compared to FPH profiles is the SAGE III spectrometer aboard the ISS. Operational since June 2017, the instrument makes vertical profile measurements of stratospheric water vapor (and ozone) with ~0.5 km vertical resolution using the technique of solar (and lunar) occultation. With the requisite need to observe the rising or setting sun (moon) at Earth's limb to make a measurement, SAGE III/ISS provides only ~30 stratospheric water vapor profiles per day. The initial release of SAGE III/ISS water vapor data (version 5.1) was compared to FPH profiles in late 2020 by Davis et al. (2021), and since that work was published a newer version 5.2 that reduces measurement biases has been released.

The FPH sites at Boulder and Lauder have been coordinating FPH soundings with ISS overpasses since August 2017 for the purpose of validating the SAGE III stratospheric water vapor (and ozone) profiles. To date, there have been 40 (20) FPH soundings at Boulder (Lauder) that were spatiotemporally coincident with SAGE III/ISS measurements. Davis et al. (2021) applied coincidence criteria of ±1 day, ±2° latitude and ±1113 km in east-west distance to match SAGE III/ISS overpass data (v5.1) with FPH soundings from five sites, including Boulder and Lauder. The study concluded that SAGE III v5.1 stratospheric retrievals are low (dry) biased by 0.2–1.2 ppmv (5%–18%) relative to the FPH profiles. Fig. 11, produced recently to include several more coincident profiles that were not available for the earlier comparison (Davis et al., 2021),

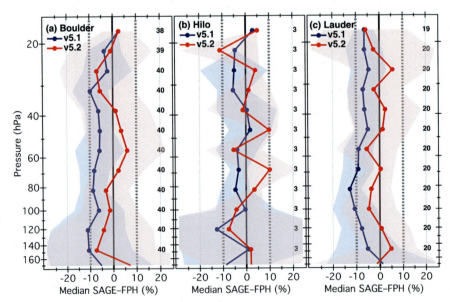

FIG. 11 Median relative differences (%) between water vapor profiles from the SAGE III spectrometer on the International Space Station (ISS) and spatiotemporally coincident NOAA FPH profiles at (A) Boulder, (B) Hilo and (C) Lauder, along with their 95% confidence intervals (shading). For each site, the numbers of NOAA FPH soundings coincident with SAGE III/ISS profiles at each retrieval level are given in *black text*. The *dotted vertical lines* at ±10% differences are provided as visual guides. Current version 5.2 SAGE III/ISS retrievals (*red*) show no statistically significant biases with the FPH, while the previous version 5.1 retrievals (*blue*) were typically low (dry) biased by 5%–10% at the two mid-latitude sites.

clearly shows these dry biases in SAGE III v5.1 data between 150 and 18 hPa over Boulder and Lauder. The release of version 5.2 retrievals significantly reduced these differences to $<\pm 5\%$ at Boulder and Lauder, with median differences at the various pressure levels randomly distributed around zero (Fig. 11). For both data versions, the results at Hilo are noisier and less suggestive of biases because far fewer coincident FPH profiles are available for the site where balloon launches with SAGE III overpasses are not temporally coordinated.

4. Summary

Pioneering efforts began nearly 80 years ago to measure water vapor in the very dry stratosphere, first by aircraft-based FPHs and then by balloon-borne FPHs. Though the reliability and quality of instrument data improved as technology advanced, the measurement principle and calibration procedure for FPHs has always remained the same. Evolving from analog to digital components enabled better control of the frost layer deposited on the chilled mirror, significantly improving measurement precision. Modern electronics provided the FPH with an enhanced ability to differentiate between real measurement signals and stray sunlight within the instrument, preventing errant frost control due to false signals. The quality of radiosonde measurements of pressure and temperature also greatly improved, reducing the uncertainties in the water vapor mixing ratios and relative humidity values derived from the FPH measurements of frost point temperatures. Two examples of comparisons between remotely-sensed satellite measurements and in situ FPH measurements of water vapor in the stratosphere demonstrate the critical need for reference-quality FPH measurements in the ongoing efforts to improve satellite-based water vapor data products in the upper atmosphere. Not only do FPH soundings provide key validation information for newly launched satellite instruments, they also help satellite instrument teams continually monitor retrievals for potential response drifts over the complete lifetimes of their instruments.

Acknowledgments

The authors would like to thank Bianca Baier and Ryan Stauffer for their thorough reviews of this chapter and their helpful comments that improved it.

References

Barrett, E.W., Herndon, T.R., Carter, H.J., 1950. Some measurements of the distribution of water vapor in the stratosphere. Tellus 2, 302–311.

Berthet, G., Renard, J.-B., Ghysels, M., Durry, G., Gaubicher, B., Amarouche, N., 2013. Balloon-borne observations of mid-latitude stratospheric water vapour: comparisons with HALOE and MLS satellite data. J. Atmos. Chem. 70, 197–219. https://doi.org/10.1007/s10874-013-9264-7.

Brewer, A.W., 1949. Evidence for a world circulation provided by the measurements of helium and water vapor distribution in the stratosphere. Q. J. Roy. Meteorol. Soc. 75, 351–363.

Brewer, A.W., Cwilong, B.M., Dobson, G.M.B., 1948. Measurements of absolute humidity in extremely dry air. Proc. Phys. Soc. A60, 52–70.

Cluley, A.P., Oliver, M.J., 1978. Aircraft measurements of humidity in low stratosphere over southern England 1972–1976. Q. J. Roy. Meteorol. Soc. 104, 511–526.

Daniell, J.F., 1820. On the new hygrometer. Q. J. Sci. Lit. Arts 9, 128–134.

Davis, S.M., Rosenlof, K.H., Hassler, B., Hurst, D.F., Read, W.G., Vömel, H., et al., 2016. The stratospheric water and ozone satellite homogenized (SWOOSH) database: a long-term database for climate studies. Earth Syst. Sci. Data 8, 461–490. https://doi.org/10.5194/essd-8-461-2016.

Davis, S.M., Damadeo, R., Flittner, D., Rosenlof, K.H., Park, M., Randel, W.J., et al., 2021. Validation of SAGE III/ISS solar water vapor data with correlative satellite and balloon-borne measurements. J. Geophys. Res. Atmos. 126. https://doi.org/10.1029/2020JD033803.

Dirksen, R.J., Sommer, M., Immler, F.J., Hurst, D.F., Kivi, R., Vömel, H., 2014. Reference quality upper-air measurements: GRUAN data processing for the Vaisala RS92 radiosonde. Atmos. Meas. Tech. 7, 4463–4490. https://doi.org/10.5194/amt-7-4463-2014.

Dobson, G.M.B., Brewer, A.W., Cwilong, B.M., 1946. Meteorology of the lower stratosphere. Proc. R. Soc. Med. A185, 144–175.

Froidevaux, L., Anderson, J., Wang, H.-J., Fuller, R.A., Schwartz, M.J., Santee, M.L., et al., 2015. Global ozone chemistry and related trace gas data records for the stratosphere (GOZCARDS): methodology and sample results with a focus on HCl, H_2O, and O_3. Atmos. Chem. Phys. 15, 10471–10507. https://doi.org/10.5194/acp-15-10471-2015.

Fujiwara, M., Shiotani, M., Hasebe, F., Vömel, H., Oltmans, S.J., Ruppert, P.W., et al., 2003. Performance of the Meteolabor "Snow White" chilled-mirror hygrometer in the tropical troposphere: comparisons with the Vaisala RS80 A/H-Humicap sensors. J. Atmos. Oceanic Tech. 20 (11), 1534–1542. https://doi.org/10.1175/1520-0426(2003)020<1534:POTMSW>2.0.CO;2.

Fujiwara, M., Vömel, H., Hasebe, F., Shiotani, M., Ogino, S.-Y., Iwasaki, S., et al., 2010. Seasonal to decadal variations of water vapor in the tropical lower stratosphere observed with balloon-borne cryogenic frostpoint hygrometers. J. Geophys. Res. 115, D18304. https://doi.org/10.1029/2010JD014179.

Garfinkel, C.I., Gordon, A., Oman, L.D., Li, F., Davis, S., Pawson, S., 2018. Nonlinear response of tropical lower-stratospheric temperature and water vapor to ENSO. Atmos. Chem. Phys. 18, 4597–4615. https://doi.org/10.5194/acp-18-4597-2018.

Goff, J.A., 1957. Saturation pressure of water on the new Kelvin temperature scale. In: Transactions of the American Society of Heating and Ventilating Engineers, presented at the Semi-Annual Meeting of the American Society of Heating and Ventilating Engineers, Murray Bay, Quebec, Canada, pp. 347–354.

Hall, E.G., Jordan, A.F., Hurst, D.F., Oltmans, S.J., Vömel, H., Kühnreich, B., et al., 2016. Advancements, measurement uncertainties and recent comparisons of the NOAA frost point hygrometer. Atmos. Meas. Tech. 9, 4295–4310. https://doi.org/10.5194/amt-9-4295-2016.

Hasebe, F., Inai, Y., Shiotani, M., Fujiwara, M., Vömel, H., Nishi, N., et al., 2013. Cold trap dehydration in the tropical tropopause layer characterised by SOWER chilled-mirror hygrometer network data in the tropical Pacific. Atmos. Chem. Phys. 13, 4393–4411. https://doi.org/10.5194/acp-13-4393-2013.

Hayashi, E., 1961. Water vapor distribution in the stratosphere by the dew-point radiosonde observation. J. Meteor. Res. 13, 905–916.

Hegglin, M.I., Plummer, D.A., Shepherd, T.G., Scinocca, J.F., Anderson, J., Froidevaux, L., et al., 2014. Vertical structure of stratospheric water vapor trends derived from merged satellite data. Nat. Geosci. 7, 1–9. https://doi.org/10.1038/NGEO2236.

Hurst, D.F., Oltmans, S.J., Vömel, H., Rosenlof, K.H., Davis, S.M., Ray, E.A., et al., 2011. Stratospheric water vapor trends over Boulder, Colorado: analysis of the 30 year Boulder record. J. Geophys. Res. 116, D02306. https://doi.org/10.1029/2010JD015065.

Hurst, D.F., Lambert, A., Read, W.G., Davis, S.M., Rosenlof, K.H., Hall, E.G., et al., 2014. Validation of Aura Microwave Limb Sounder stratospheric water vapor measurements by the NOAA frost point hygrometer. J. Geophys. Res. Atmos. 119. https://doi.org/10.1002/2013JD020757.

Hurst, D.F., Read, W.G., Vömel, H., Selkirk, H.B., Rosenlof, K.H., Davis, S.M., et al., 2016. Recent divergences in stratospheric water vapor measurements by frost point hygrometers and the Aura Microwave Limb Sounder. Atmos. Meas. Tech. 9, 4447–4457. https://doi.org/10.5194/amt-9-4447-2016.

Hyland, R.W., Wexler, A., 1983. Formulations for the thermodynamic properties of the saturated phases of H_2O from 173.15 to 473.15K. ASHRAE Trans. 89, 500–519.

Jorge, T., Brunamonti, S., Poltera, Y., Wienhold, F.G., Luo, B.P., Oelsner, P., et al., 2021. Understanding balloon-borne frost point hygrometer measurements after contamination by mixed-phase clouds. Atmos. Meas. Tech. 14, 239–268. https://doi.org/10.5194/amt-14-239-2021.

Kobayashi, J., 1960. Investigations on hygrometry. Pap. Meteorol. Geophys. 11 (2–4), 213–338.

Konopka, P., Grooß, J.U., Günther, G., Plöger, F., Pommrich, R., Müller, R., et al., 2010. Annual cycle of ozone at and above the tropical tropopause: observations versus simulations with the Chemical Lagrangian Model of the Stratosphere (CLaMS). Atmos. Chem. Phys. 10 (1), 121–132. https://doi.org/10.5194/acp-10-121-2010.

Liang, C.K., Eldering, A., Gettelman, A., Tian, B., Wong, S., Fetzer, E.J., et al., 2011. Record of tropical interannual variability of temperature and water vapor from a combined AIRS-MLS data set. J. Geophys. Res. 116, D06103. https://doi.org/10.1029/2010JD014841.

Livesey, N.J., Read, W.G., Froidevaux, L., Lambert, A., Santee, M.L., Schwartz, M.J., et al., 2021. Investigation and amelioration of long-term instrumental drifts in water vapor and nitrous oxide measurements from the Aura Microwave Limb Sounder (MLS) and their implications for studies of variability and trends. Atmos. Chem. Phys. 21, 15409–15430. https://doi.org/10.5194/acp-21-15409-2021.

Mastenbrook, H.J., 1966. A control system for ascent-descent soundings of the atmosphere. J. Appl. Meteorol. 5, 737–740.

Mastenbrook, H.J., 1968. Water vapor distribution in the stratosphere and high troposphere. J. Atmos. Sci. 25, 299–311.

Mastenbrook, H.J., 1971. The variability of water vapor in the stratosphere. J. Atmos. Sci. 28, 1495–1501.

Mastenbrook, H.J., 1974. Water-vapor measurements in the lower stratosphere. Can. J. Chem. 52, 1527–1531.

Mastenbrook, H.J., Dinger, J.E., 1961. Distribution of water vapor in the stratosphere. J. Geophys. Res. 66, 1437–1444.

Mastenbrook, H.J., Oltmans, S.J., 1983. Stratospheric water vapor variability for Washington DC/Boulder, CO: 1964–82. J. Atmos. Sci. 40, 2157–2165.

Mote, P.W., Rosenlof, K.H., McIntyre, M.E., Carr, E.S., Gille, J.C., Holton, J.R., et al., 1996. An atmospheric tape recorder: the imprint of tropical tropopause temperatures on stratospheric water vapor. J. Geophys. Res. 101, 3989–4006.

Oltmans, S.J., 1985. Measurements of water vapor in the stratosphere with a frost-point hygrometer. In: Proceedings of the 1985 International Symposium on Moisture and humidity. Instrument Society of America, Research Triangle Park, North Carolina, pp. 251–258.

Oltmans, S.J., Hofmann, D.J., 1995. Increase in lower-stratospheric water vapour at a mid-latitude Northern Hemisphere site from 1981 to 1994. Nature 374 (6518), 146–149. https://doi.org/10.1038/374146a0.

Oltmans, S.J., Vömel, H., Hofmann, D.J., Rosenlof, K.H., Kley, D., 2000. The increase in stratospheric water vapor from balloon-borne, frostpoint hygrometer measurements at Washington, D.C., and Boulder, Colorado. Geophys. Res. Lett. 27 (21), 3453–3456. https://doi.org/10.1029/2000GL012133.

Ovarlez, J., 1991. Stratospheric water vapor measurement in the tropical zone by means of a frost point hygrometer on board long-duration balloons. J. Geophys. Res. 96 (D8), 15541–15545.

Pruvost, P., Ovarlez, J., Lenoble, J., Chu, W.P., 1993. Comparison of stratospheric aerosol and gas experiment and balloon-borne stratospheric water vapor measurement. J. Geophys. Res. 98 (D3), 4889–4896.

Read, W.G., Lambert, A., Bacmeister, J., Cofield, R.E., Christensen, L.E., Cuddy, D.T., et al., 2007. Aura Microwave Limb Sounder upper tropospheric and lower stratospheric H_2O and relative humidity with respect to ice validation. J. Geophys. Res. 112 (D24), D24S35. https://doi.org/10.1029/2007JD008752.

Read, W.J., Stiller, G., Lossow, S., Kiefer, M., Khosrawi, F., Hurst, D., et al., 2022. The SPARC Water Vapor Assessment II: assessment of satellite measurements of upper tropospheric humidity. Atmos. Meas. Tech. 15, 3377–3400. https://doi.org/10.5194/amt-15-3377-2022.

Rosenlof, K.H., Oltmans, S.J., Kley, D., Russel III, J.M., Chiou, E.-W., Chu, W.P., et al., 2001. Stratospheric water vapor increases over the past half-century. Geophys. Res. Lett. 28 (7), 1195–1198. https://doi.org/10.1029/2000GL012502.

Scherer, M., Vömel, H., Fueglistaler, S., Oltmans, S.J., Staehelin, J., 2008. Trends and variability of midlatitude stratospheric water vapour deduced from the re-evaluated Boulder balloon series and HALOE. Atmos. Chem. Phys. 8, 1391–1402.

Tucker, G.B., 1957. An Analysis of Humidity Measurements in the Upper Troposphere and Lower Stratosphere Over Southern England. MRP 1052, Meteorological Research Committee, London. 35 pp.

United Kingdom Meteorological Office, 1961. Handbook of Meteorological Instruments, Part II: Instruments for Upper Air Observations. Her Majesty's Stationery Office, London, England. https://digital.nmla.metoffice.gov.uk/IO_a72e5bab-d45e-4efd-9fa5-9cbd18d43c16.

Vömel, H., Fujiwara, M., Shiotani, M., Hasebe, F., Oltmans, S.J., Barnes, J.E., 2003. The behavior of the Snow White chilled-mirror hygrometer in extremely dry conditions. J. Atmos. Oceanic Tech. 20 (11), 1560–1567. https://doi.org/10.1175/1520-0426(2003)020<1560:TBOTSW>2.0.CO;2.

Vömel, H., David, D.E., Smith, K., 2007a. Accuracy of tropospheric and stratospheric water vapor measurements by the cryogenic frost point hygrometer: instrumental details and observations. J. Geophys. Res. 112, D08305. https://doi.org/10.1029/2006JD007224.

Vömel, H., Barnes, J.E., Forno, R.N., Fujiwara, M., Hasebe, F., Iwasaki, S., et al., 2007b. Validation of Aura MLS water vapor by balloon-borne cryogenic frostpoint hygrometer measurements. J. Geophys. Res. 112 (D24), D24S37. https://doi.org/10.1029/2007JD008698.

Vömel, H., Naebert, T., Dirksen, R., Sommer, M., 2016. An update on the uncertainties of water vapor measurements using cryogenic frostpoint hygrometers. Atmos. Meas. Tech. 9, 3755–3768. https://doi.org/10.5194/amt-9-3755-2016.

Chapter 4

Ozonesondes: Instrumentation and Data Applications

Anne M. Thompson[a,b], Herman G.J. Smit[c], Debra E. Kollonige[a,d], and Ryan M. Stauffer[a]

[a]*Earth Sciences Division, NASA-Goddard Space Flight Center, Greenbelt, MD, United States,* [b]*Joint Center for Environmental Technology, University of Maryland, Baltimore County, MD, United States,* [c]*Forschungszentrum-Jülich, Jülich, Germany,* [d]*SSAI, Lanham, MD, United States*

Chapter outline

1. The role of ozonesondes in the global ozone measurement framework — 59
   - 1.1 Sondes in the context of a global ozone measurement strategy — 59
   - 1.2 Chapter overview — 60
2. The ozonesonde instrument, operation, and data quality control — 60
   - 2.1 Electrochemical ozonesondes — 60
   - 2.2 The ECC ozonesonde: Principles of operation and sources of uncertainty — 61
   - 2.3 Quality assurance of ozonesondes: Approach and current status — 63
3. Ozonesonde networks — 64
   - 3.1 The global network: Long-term sites — 64
   - 3.2 Strategic networks: Global and campaign operations — 68
4. Applications of ozonesonde data with remote sensing observations — 69
   - 4.1 Satellite ozone product evaluation using ozonesonde data — 70
   - 4.2 Use of satellite ozone data to track the performance of the ozonesonde — 72
5. Summary and conclusions — 72
   - 5.1 Scientific perspective: On-going need for profiles from global ozonesondes — 72
   - 5.2 Quality assurance: Need for sonde intercomparisons and a global ozone reference — 74
   - 5.3 Conclusions — 74

Acknowledgments — 75
References — 75

Acronyms

AASE II	Airborne Arctic Stratospheric Expedition II
ACE-FTS	Atmospheric Chemistry Experiment—Fourier Transform Spectrometer on Canadian SCISAT satellite
ASOPOS	Assessment of Standard Operating Procedures for OzoneSondes
BESOS	Balloon Experiment on Standards for OzoneSondes
BORTAS	Quantifying the impact of BOReal forest fires on Tropospheric oxidants over the Atlantic using Aircraft and Satellites
DU	Dobson Unit, the unit to express vertical ozone column abundances, 1 DU $= 2.69 \times 10^{16}$ molecules per cm^2 at STP 1×10^{-3} atm cm at STP
EASOE	European Arctic Stratospheric Ozone Experiment
ECC	Electrochemical Concentration Cell
EN-SCI	Environmental Science Corporation; ECC ozonesonde manufacturer
ESRL	Earth System Research Laboratories
GAW	Global Atmospheric Watch
GCOS	Global Climate Observing System
GEMS	Geostationary Environment Monitoring Spectrometer
GML	Global Monitoring Laboratory (division of NOAA's ESRL; formerly GMD)
GOES	Geostationary Operational Environmental Satellites
GOME	Global Ozone Monitoring Experiment (onboard MetOp satellites)
GNSS	Global Navigational Satellite System
GRUAN	GCOS Reference Upper Air Network
IAP	Institute of Atmospheric Physics, Beijing, China

Field Measurements for Passive Environmental Remote Sensing. https://doi.org/10.1016/B978-0-12-823953-7.00011-3
Copyright © 2023 Elsevier Inc. All rights reserved.

IGACO	Integrated Global Atmospheric Chemistry Observations
IOC	International Ozone Commission
IONS	Intensive Ozonesonde Network Study
IPCC	Intergovernmental Panel on Climate Change
ISS	International Space Station
JOSIE	Jülich OzoneSonde Intercomparison Experiment
KI	Potassium Iodide
LEO	Low Earth orbit
MLS	Microwave Limb Sounder (on Aura satellite)
NASA	National Aeronautics and Space Administration
NDACC	Network for the Detection of Atmospheric Composition Change
NOAA	National Oceanic and Atmospheric Administration
OMI	Ozone Monitoring Instrument (on Aura satellite)
OMPS-LP	Ozone Mapping and Profiler Suite—Limb Profiler (onboard Suomi-NPP and JPSS satellites)
OPM	Ozone PhotoMeter Instrument (used as UV-reference)
OSIRIS	Optical Spectrograph and InfraRed Imaging System, on Odin satellite
O3S-DQA	Ozone Sonde Data Quality Assessment
QA	Quality Assurance
RECONCILE	Reconciliation of essential process parameters for an enhanced predictability of Arctic stratospheric ozone loss and its climate interactions
SAGE III	Stratospheric Aerosol and Gas Experiment (fourth generation on ISS)
SBUV	Solar Backscatter UltraViolet (referring to instrument type on satellites measuring ozone)
SCIAMACHY	SCanning IMaging Absorption SpectroMeter for Atmospheric CHartographY
SCIATRAN	radiative transfer and retrieval code used by Univ. Bremen SCIAMACHY and TROPOMI algorithm group
SCOUT-O3	Stratospheric-Climate links with emphasis On the Upper Troposphere and lower stratosphere
SEACIONS	Southeast America Consortium for Intensive Ozonesonde Network Study
SESAME	Second European Stratospheric Arctic and Mid-latitude Experiment
SHADOZ	Southern Hemisphere ADditional OZonesondes
SI^{2}N	Ozone trend assessment study supported by SPARC, IOC, IGACO, and NDACC
SMILES	Submillimeter-Wave Limb Emission Sounder onboard ISS
SOLVE	SAGE III Ozone Loss and Validation Experiment
SOP	Standard Operating Procedure
SPARC	Stratosphere-troposphere Processes And their Role in Climate
SPC	Science Pump Corporation; ECC ozonesonde manufacturer
SST	Sensing Solution Type
STP	Standard Temperature ($= 273.15\,K$) and Pressure ($= 1013.25\,hPa$) conditions
StratoClim	Stratospheric and upper tropospheric processes for better climate predictions
TCO	Total Column Ozone
TEMPO	Tropospheric Emissions: Monitoring of Pollution
THESEO	Third European Stratospheric Experiment on Ozone
TOMS	Total Ozone Mapping Spectrometer
TOPAS	Tikhonov regularized Ozone Profile retrievAl with SCIATRAN
TROPOMI	TROPOspheric Monitoring Instrument
TrCO	Tropospheric Column Ozone
UNEP	United Nations Environment Programme
UV	UltraViolet
VINTERSOL	Validation of INTERnational satellites and Study of Ozone Loss
WCCOS	World Calibration Center for OzoneSonde
WDCRG	World Data Centre for Reactive Gases
WMO	World Meteorological Organization
WOUDC	World Ozonesonde and Ultraviolet Data Centre

Nobody can be uncheered with a balloon.

A. A. Milne, *Winnie the Pooh*

1. The role of ozonesondes in the global ozone measurement framework

1.1 Sondes in the context of a global ozone measurement strategy

The ozonesonde instrument, although more than 50 years old in design, and simple to operate, remains an essential component of the global observing strategy for stratospheric and tropospheric ozone. The profiles from ozonesondes are foundational in the development of satellite ozone retrievals and are used for validating satellite products from a growing constellation of ozone-measuring sensors. The ozonesonde instrument is unique in providing readings at (5–10)% uncertainty or better throughout the troposphere to the mid-stratosphere at 100–150 m resolution independent of conditions of cloudiness or precipitation (Fig. 1). Because it is relatively inexpensive and easy to operate—launching with a standard radiosonde instrument—the ozonesonde can be used virtually anywhere. Ozone sounding records provide the longest record of the vertical distribution of ozone and thus play a key role in monitoring changes in stratospheric ozone in accordance with the Montreal Protocol (WMO/UNEP, 2019).

Fig. 2 illustrates how ozonesondes fit into the global ozone observing strategy that employs various ground-based spectroscopic and lidar techniques, ozone instruments on aircraft and balloons as well as from space-borne platforms. The altitude ranges of sonde operation, aircraft, and Low Earth Orbit (LEO) satellites are illustrated. Note that ozone-measuring instruments have been hosted on the International Space Station (SAGE III is currently operational). Geostationary satellites (e.g., the Korean GEMS, NOAA's GOES series) also carry ozone measuring instruments; these are typically 36,000 km above earth. The tropospheric and stratospheric segments of the atmosphere are usually measured by two separate lidar instruments (McDermid et al., 1990; McGee et al., 1990) An advantage of ozonesondes is that a single sounding encompasses the troposphere and lower and middle stratosphere.

In addition to monitoring and validation of other sensors, ozonesonde data are important in understanding atmospheric dynamics, lifetimes, and sources and sinks of ozone. Above the atmospheric boundary layer, the ozone lifetime is weeks to months. Thus, in the troposphere, sonde data are used to study the transport of pollution throughout the troposphere and lowermost stratosphere. Pollution from biomass fires in the tropics (Thompson et al., 1996, 2001, 2003a,b), throughout mid-latitudes by intercontinental transport (Stauffer et al., 2017) and from boreal fires (Moeini et al., 2020), has been

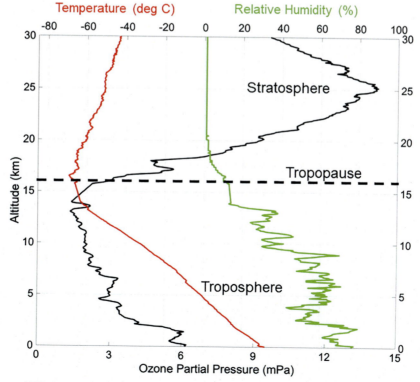

FIG. 1 Ozone profile from an ECC ozonesonde with the temperature and humidity recorded by the accompanying radiosonde. The radiosonde also measures wind speed and direction. Data from a launch at Wallops Island, VA (37.9 N, 75.5 W) on July 17, 2019.

60 PART | I Instrumentation

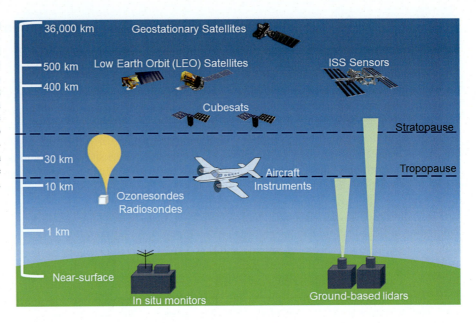

FIG. 2 Altitude ranges of techniques used to measure ozone, ground-based, airborne, and satellites. Other ground-based instrumentation (lidar, surface monitors) show context for the ozonesonde measurement. The schematic shows lidar that measure in the troposphere only (Sullivan et al., 2015) and that cover troposphere and stratosphere. In fact, only one or two of the most widely used ozone lidar instruments, e.g., within NDACC, detect both troposphere and stratosphere; most ozone lidars report data only in the stratosphere.

investigated. Recently, sonde data across the mid-latitude northern hemisphere quantified a significant drop in tropospheric ozone due to the global economic crisis instigated by the 2020 COVID-19 pandemic (Steinbrecht et al., 2021).

1.2 Chapter overview

The purpose of this chapter is to present the capabilities and applications of the ozonesonde measurement as they relate to remote sensing (Sections 3 and 4). We begin with a description of the ozonesonde instrument and ongoing research related to the quality assurance (QA) of the data (Section 2).

2. The ozonesonde instrument, operation, and data quality control

2.1 Electrochemical ozonesondes

Ozonesondes are small, light-weight instruments that are flown on weather balloons coupled via interfacing electronics to radiosondes for data transmission and measurements of meteorological parameters: pressure, temperature, humidity, wind, and position. The total weight of the ozonesonde-radiosonde flight package is ∼1 kg so the payload can be flown on relatively small balloons (typically 1200–1500 g). Using the telemetry of the radiosonde, the measured data are transmitted to the ground station for further processing. Normally, data are taken during ascent at a rise rate of about 5 m/s to a balloon burst altitude of 30–33 km altitude. The inherent response time of the chemical measurement of the ozonesonde is 20–30 s, which provides an effective height resolution in the ozone profile data of 100–150 m.

Since their first design in the 1960s, the most commonly used ozonesonde instruments are based on electrochemical detection methods that convert the sampled ozone into an electrical current. Smit (2014) describes the common ozonesonde types in use over the past 50 years. At the present time, the most widely used ozonesonde type is the electrochemical concentration cell (ECC). Although widely deployed in the past, the Brewer Mast sonde is presently only launched at the Meteorological Observatory Hohenpeissenberg in Germany in a time series that started in 1967. Two other major electrochemical sonde types, developed by the India Meteorological Department and the Japan Meteorological Agency, are no longer used.

Each ozonesonde instrument is unique and is prepared and provisionally calibrated prior to launch. It is important for remote sensing researchers to understand operational aspects of the ozonesonde and the procedures that sonde data providers take to minimize uncertainties within an individual profile and to ensure consistency of the global ozonesonde record over time. The instrument and data treatment are described in the following sections.

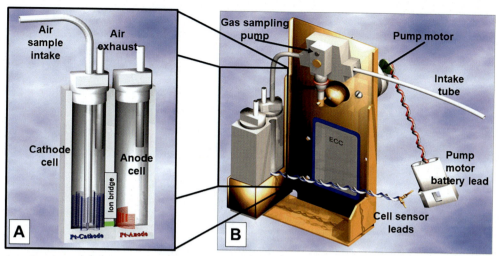

FIG. 3 (A) Cross-section of the electrochemical concentration cells (ECC) in (B) the ozonesonde sensor. There are two widely used ECC ozonesonde types, manufactured by Science Pump Corporation and the EN-SCI Corporation, producing the SPC-6A and EN-SCI instrument, respectively. The design of both ECCs is similar but there is a consistent 4%–5% difference in their performance (Fig. 4A and B) when launched under the same conditions (Smit et al., 2007; Thompson et al., 2007c; Smit, 2014). Since 2014, a third ECC-type instrument manufactured at the Institute of Atmospheric Physics (IAP), Beijing, China, has been flown at several East Asian stations; the new instrument has not been extensively intercompared with the SPC-6A or EN-SCI in laboratory or field tests.

2.2 The ECC ozonesonde: Principles of operation and sources of uncertainty

The ECC ozonesonde (Fig. 3) developed by Komhyr (1969) consists of two cells, made of Teflon or molded plastic, which serve as a cathode and anode chamber. There are two widely used ECC ozonesonde types, manufactured by Science Pump Corporation and the EN-SCI Corporation, producing the SPC-6A and EN-SCI instrument, respectively. The design of both ECCs resembles Fig. 3, but there is a consistent 4%–5% difference in their performance (Fig. 4A and B) when the different instrument types are operated under the same conditions (Smit et al., 2007; Thompson et al., 2007c; Smit, 2014). Both cells contain platinum mesh electrodes. They are immersed in aqueous potassium iodide (KI) solutions of different concentrations, whereby the cathode cell is charged with a solution of low KI concentration and the anode cell with a solution saturated with KI. The two chambers are linked together by an ion-bridge to provide an ion-pathway and to prevent mixing of the cathode and anode electrolytes. The detection is based on the titration of ozone in KI according to the redox reaction:

$$2KI + O_3 + H_2O \rightarrow I_2 + O_2 + 2KOH \tag{R-1}$$

In the cathode cell, the iodine (I_2) is converted back into two iodide ions (I^-) by the uptake of two electrons from the platinum electrode surface. Continuous sampling is achieved by a small battery-driven gas pump made of Teflon that bubbles ambient air through the sensing solution of the electrochemical cell. The iodine molecules that are produced by the reaction are transported toward the cathode electrode to be converted back to I^-; this process generates an electrical current in an external circuit that is proportional to the sampled ozone per unit time. Given the pump flow rate (Φ_P in $cm^3\,s^{-1}$), the pump temperature (T_P in K), the overall efficiency (η_T) of the sensor cell, the measured electrical current (I_M in μA), after a correction for a background current (I_B in μA), is converted to the ozone partial pressure (P_{O_3} in mPa):

$$P_{O_3} = 0.043085 \times \frac{T_P}{(\eta_T \times \Phi_P)} \times (I_M - I_B) \tag{E-1}$$

The constant 0.043085 is determined by the ratio of the gas constant (R) to two times the Faraday constant (for each O_3 molecule two electrons flow in the electrical circuit from reaction R-1). The overall efficiency, η_T, includes the absorption efficiency η_A of O_3 into the sensing solution (usually 1.00), the pressure dependent pump efficiency η_P, and the conversion efficiency η_C of the ECC sensor cell. The last efficiency is predominantly determined by the stoichiometry of redox reaction (R-1) followed by the conversion of the produced iodine into the measured electrical current I_M. In practice, most operators add a sodium-hydrogen phosphate buffer to the cathode KI-solution to maintain the pH at 7.0 to keep the stoichiometry of the redox reaction (R-1) close to one.

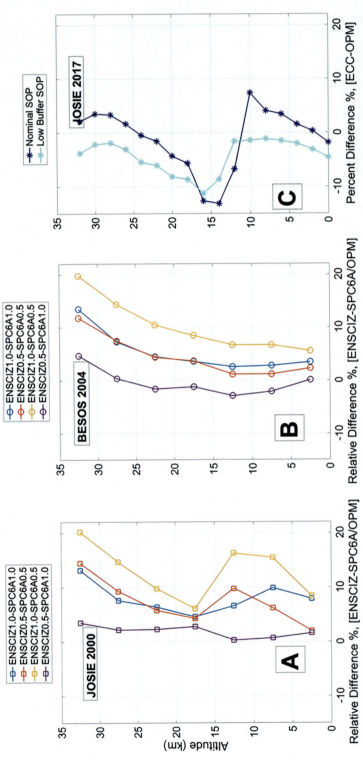

FIG. 4 (A) JOSIE 2000 & BESOS (B): Relative differences between measurements of ozone by EN-SCI and SPC-6A using different combinations of 1% KI & full buffer and 0.5%KI & half buffer sensing solution strength. Data are averaged over 5km altitude. All profiles were first referenced to the WMO/GAW standard ozone photometer (OPM). In JOSIE-2000, the OPM was in the Jülich (Germany) WCCOS facility; in BESOS, the OPM flew on a gondola with 18 ozonesonde instruments in Laramie, Wyoming (United States). (C) Mean percent differences between ozone measured by EN-SCI and SPC-6A sondes following (Smit and ASOPOS Panel, 2014) recommendations and sondes using 1% KI and 0.1 buffer, during JOSIE-2017. Both sets of measurements were referenced to the OPM.

The uncertainty of the ECC sonde measurements of the ozone partial pressure (P_{O_3}) is a composite of the contributions of the individual uncertainties of the instrumental parameters (I_M, I_B, T_P, F_P, $\eta_T = \eta_{A*}\eta_{P*}\eta_C$), as described in detail by Tarasick et al. (2021). Tarasick et al. (2021) assumed that all systematic uncertainty components are known and corrected for. All instrumental uncertainties are assumed to be random and uncorrelated such that they follow Gaussian statistics to determine the overall uncertainty of the measured P_{O_3}. In the troposphere, the background current I_B is the dominant uncertainty, particularly in the upper troposphere where the ozone concentration is generally low (mid-latitudes) to very low (near the tropical tropopause).

In the stratosphere, uncertainties of pump characteristics (Johnson et al., 2002) and conversion efficiencies are the major contributors to the overall uncertainty (Smit et al., 2021). Since 2000–10, the radiosondes flown with the ozonesondes are equipped to measure GNSS altitude. This means that the ambient air pressure is determined from the altitude measurement (e.g., Stauffer et al., 2014) in which case the pressure uncertainty is better than 0.05–0.10 hPa above 50 hPa, making only a minor contribution to the overall uncertainty. However, in case of ozonesondes flown with non-GNSS radiosondes, generally those prior to ~2000, the uncertainty of the radiosonde pressure sensor measurement above 50 hPa could be the dominant source of error.

2.3 Quality assurance of ozonesondes: Approach and current status

There has been considerable research activity to understand the performance of the ozonesonde instrument and to establish standard operating procedures (SOP). Twenty-five years ago, the ozonesonde measurement was assigned a 15%–20% accuracy (SPARC/IOC/GAW, 1998). The total column ozone (TCO) amount is now typically accurate to within 2%–3% when evaluated against co-located ground-based instruments. Accuracy throughout the column, when best practices are followed, is ~5%–10%, with the potential to improve to 3%–5%.

2.3.1 Overview of ozonesonde community quality assurance activities

The ozonesonde community, working together under the auspices of World Meteorological Organization/Global Atmospheric Watch (WMO/GAW) and groups like NDACC, the International Ozone Commission (IO3C), and, in the past decade, the GCOS Reference Upper Air Network (GRUAN), has organized quality assurance (QA) research around three important activities. The first of these was the creation of a testing facility for ozonesondes. In the mid-1990s, as part of the WMO/GAW Quality Assurance plan (World Meteorological Organization WMO, 1996), a World Calibration Centre for OzoneSondes (WCCOS) was established at Germany's Forschungszentrum-Jülich (Smit et al., 2000). The heart of the WCCOS is an environmental simulation chamber in which up to four ozonesondes can be intercompared and calibrated against a dual beam UV-photometer (OPM; Proffitt and McLaughlin, 1983) that is traceable to the NIST standard for ozone. During testing, pressure, temperature, and ozone concentration are varied at the rate of an actual ascent from the surface until burst altitude at 33–35 km altitude. In its first 5 years of operation, a set of campaigns, each referred to as a Jülich Ozone Sonde Intercomparison Experiment (JOSIE; Smit and Kley, 1998; Smit and Straeter, 2004a,b), quantified biases among ozonesonde types, ECC or otherwise, between the two major ECC types of instruments, among different sensing solution types (SST). Smit et al. (2007) summarized a JOSIE-2000 in which eight groups compared instruments and preparation methods over 10 simulations of various environments: polar, tropical, mid-latitude.

The second ozonesonde QA activity has been intercomparisons of ECC ozonesondes in the field. For example, JOSIE-2000 results on biases were confirmed in the field during the Balloon Experiment on Standards for Ozone (BESOS) campaign in 2004 (Deshler et al., 2008), with 18 sondes flown on a single gondola along with the WCCOS standard OPM.

Examples from laboratory and field comparisons appear in Fig. 4. In Fig. 4A and B, offsets in the measurement of ozone between the two instruments from JOSIE-2000 and BESOS, respectively, are shown. The OPM was the absolute reference in both experiments.

2.3.2 Development of consensus-based standard operating procedures (ASOPOS)

The third component of enhancing QA was the establishment in 2004 of an international team of 15–20 sonde experts to review laboratory and field tests in an Assessment of Standard Operating Procedures (SOPs) for OzoneSondes (ASOPOS). The first ASOPOS led to a community consensus for SOPs. Largely based on the 1996–2000 JOSIE campaigns and BESOS, the recommended SOPs were published as Smit and ASOPOS Panel (2014).

The 2017 JOSIE campaign, with simulations of only tropical conditions (Thompson et al., 2019), was the basis for an ASOPOS 2.0 evaluation (Smit et al., 2021). The ASOPOS 2.0 report outlines (1) an improved treatment to correct the pump flow rate that falls off at low pressures; (2) a correction of the ozone exposure-dependent stoichiometry of the $O_3 + KI$ redox reaction (R-1) to account for both slow (≅20–25 min) and fast (≅20–25 s) reactions that take place in the ECC during an ascent (Vömel et al., 2020); (3) a new conversion efficiency in Eq. (E-1) that relates the final calculation of ozone amount to the OPM used at the WCCOS, making every reported sounding traceable to a common standard; (4) an extended list of metadata to be collected at launch time so data can be reprocessed; (5) continuous monitoring of station QA by comparing sonde ozone amounts to ground-based and satellite overpass measurements for detecting problems like the post-2013 total ozone "dropoff" observed at a number of stations (Stauffer et al., 2020; see Section 4.2). Fig. 4C displays some JOSIE-2017 results. Operators prepared their sondes used for determining the average labeled "nominal SOP" according to their home station practices; for seven of eight stations tested, the preparation followed the first ASOPOS Report (Smit and ASOPOS Panel, 2014). For the "Low Buffer" tests all operators used a sensing solution with 1% KI and 10% of the standard buffer solution. Ozone measured with the low-buffer solution, irrespective of instrument type, measured closer to the OPM near the simulated tropopause altitude (~15 km) but always lower than the OPM elsewhere in the profile.

2.3.3 Homogenization of long ozonesonde time-series

The bias effects, i.e., discontinuities and trends introduced by instrumental artifacts, as described in the first ASOPOS Report (Smit and ASOPOS Panel, 2014), need to be accounted for in calculating reliable ozone profile trends. ECC ozonesondes were first manufactured 50 years ago and have undergone modifications of the instrument and in some cases, operational procedures, resulting in inhomogeneities in some station records and biases among stations. Discontinuities in total ozone or profile segments have appeared in the time series at various stations. This phenomenon was recognized in a 2011/2012 Ozone Sonde Data Quality Assessment (O3S-DQA) that reviewed 40 years of ozonesonde records from a number of stations. The O3S-DQA activity led to guidelines for data providers to resolve inhomogeneities in long-term sonde records (Smit and O3S-DQA, 2012; https://www.wccos-josie.org/o3s-dqa). Generic transfer functions were developed (Deshler et al., 2017) to aid the process of harmonizing sonde records to the common standard of the combinations recommended in the Smit and ASOPOS Panel (2014).

Since 2015, ~40 of the long-term ozonesonde records within the global network have been re-processed following the O3S-DQA guidelines, removing known inhomogeneities to achieve overall uncertainties of 5%–10%. These include the Canadian stations (Tarasick et al., 2016), several European stations (Van Malderen et al., 2016), those of the SHADOZ network (Witte et al., 2017, 2018; Thompson et al., 2017), Wallops Island, VA (Witte et al., 2019), and eight stations in the NOAA network (Sterling et al., 2018). Fig. 5 shows the result of the homogenization effort of the ozonesonde time series at Boulder, CO (cyan triangle on the Fig. 6 map), by comparing the total ozone column (TCO) derived from the sondes with TCO measured by the Dobson spectrophotometer before (Fig. 5A) and after the reprocessing (Fig. 5B).

3. Ozonesonde networks

3.1 The global network: Long-term sites

Stations launching ozonesondes on a regular basis are displayed in Fig. 6. All except one launch ECC-type ozonesonde instruments. WOUDC archives the sonde profiles along with co-located total column ozone amounts from Dobson, Brewer, and SAOZ spectrometers where these are available. NDACC is another repository for ozonesonde data. Other oft-used archives are NOAA/GML (https://gml.noaa.gov/aftp/data/ozwv/Ozonesonde/) and NASA's SHADOZ (https://tropo.gsfc.nasa.gov/shadoz). Surface ozone concentrations are archived with other reactive gases at the WDCRG.

The global ozonesonde network, consisting of stations operated by meteorological services, space agencies, and several universities, has evolved over more than 80 years. A number of stations originated in the 1950s during the International Geophysical Year. Other sounding stations became operational as the number of ozone-measuring satellites increased after 1990 (Fig. 7). Because most Antarctic ozonesonde stations began operating before the 1980s, a robust record exists of the lower stratospheric ozone depletion associated with the Antarctic "ozone hole" in the Austral winter to early spring when UV-based satellites have limited views. The discovery of extreme Antarctic ozone loss was first reported at the 1984 Quadrennial Ozone Symposium (Chubashi, 1985) based on soundings from the Japanese Syowa station (*black triangle* on Fig. 6) and on column ozone losses at the British Halley Bay station in 1985 (Farman et al., 1985). Fig. 8A displays an example from South Pole station (*magenta triangle* on the Fig. 6 map) in 2018 of the morphology of low-ozone profiles that occur during September and October when there is a sustained Antarctic polar vortex. The contrasting profiles are from July 2018 at South Pole.

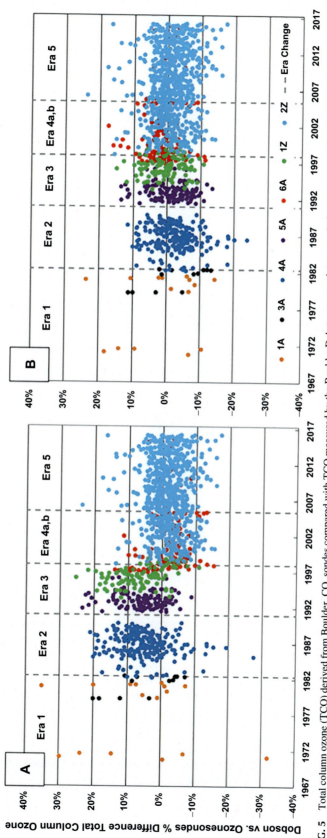

FIG. 5 Total column ozone (TCO) derived from Boulder, CO, sondes compared with TCO measured by the Boulder Dobson spectrophotometer before (A) and after (B) reprocessing of sonde data. An artifact step-function drop has been eliminated with the reprocessing. (From Sterling, C.W., Johnson, B.J., Oltmans, S.J., Smit, H.G.J., Jordan, A.F., Cullis, P.D., Hall, E.G., Thompson, A.M., Witte, J.C., 2018. Homogenizing and estimating the uncertainty in NOAA's long-term vertical ozone profile records measured with the electrochemical concentration cell ozonesonde. Atmos. Meas. Tech. 11, 3661–3687. https://doi.org/10.5194/amt-11-3661-2018.)

66 PART | I Instrumentation

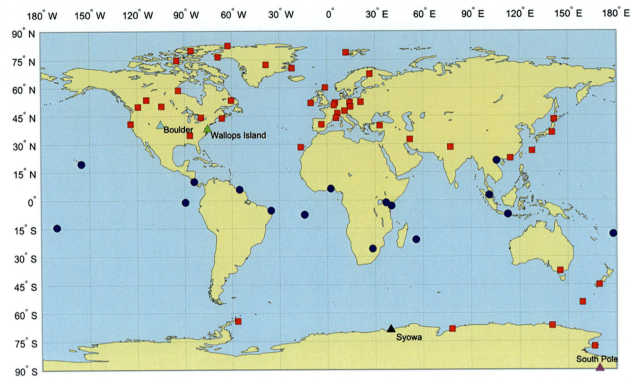

FIG. 6 Distribution of 64 most active ozone sounding stations in the global network (after Smit et al., 2021). These stations deposit data in major public archives. The latter include the archive WOUDC (World Ozone and Ultraviolet Data Center) sponsored by the World Meteorological Organization Global Atmospheric Watch (WMO/GAW; see Acronyms section). Other commonly used archives are those of the Network for Detection of Atmospheric Composition Change (NDACC; DeMazière et al., 2018), at the websites of NASA for the Southern Hemisphere ADditional OZonesonde Network (SHADOZ; Thompson et al., 2012, 2017) or at the NOAA/Global Monitoring Laboratory (GML).

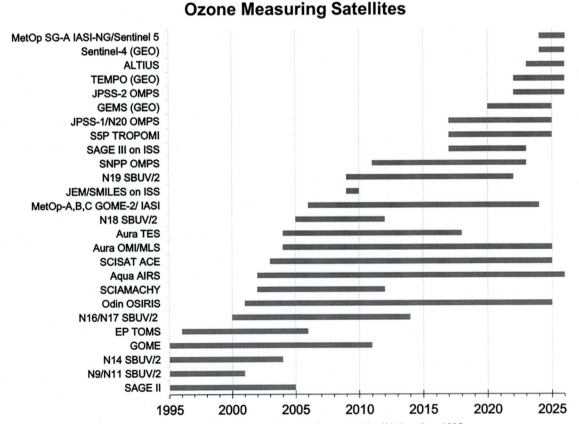

FIG. 7 Ozone-measuring satellites that have used sonde data for algorithm development and validation since 1995.

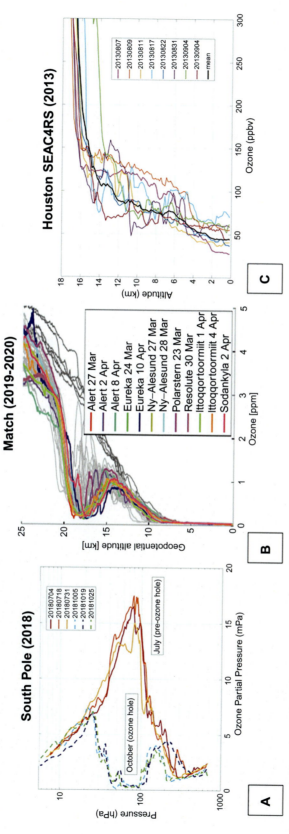

FIG. 8 Examples of dynamic and/or chemical processes affecting the ozone profile, as captured by soundings. (A) Ozonesonde profiles over NOAA's South Pole station that illustrate extreme ozone loss due to catalytic chemical destruction in the region ~15–20 km (above 100 hPa) in October of 2018, compared to July 2018 (pre-ozone hole); (B) 2019–20 winter-spring season Match ozone soundings over Greenland, Ny-Ålesund (Svalbard, Norway), Canada, and Finland (from Wohltmann, I., von der Gathen, P., Lehmann, R., Maturilli, M., Deckelmann, H., Manney, G. L., Davies, J., Tarasick, D., Jepsen, N., Kivi, R., Lyall, N., Rex, M., 2020. Near complete local reduction of Arctic stratospheric ozone by record chemical loss in spring 2020. Geophys. Res. Lett. 47(20), doi:https://doi.org/10.1029/2020GL089547); Used by permission from AGU. (C) A series of ozone profiles during the 2013 SEACIONS campaign (https://tropo.gsfc.nasa.gov/seacions/) at Ellington Field, Texas (29.6 N, 95.2 W). STE influences appear in profiles of 7, 9 August and 4 September (green line) 2013. An example of low-ozone air lofted in convection appears in the profile of 4 September (maroon).

3.2 Strategic networks: Global and campaign operations

Ozonesondes have been organized for targeted purposes in what are referred to as strategic ozonesonde networks (Thompson et al., 2011). The global SHADOZ network (*blue circles* in Fig. 6), organized in 1998 (Thompson et al., 2003a), consists of tropical and subtropical stations that launch 2–5 sondes monthly, generally coordinated with a midday overpass of one or more instruments on a polar-orbiting satellite. The zonal distribution of SHADOZ stations (Thompson et al., 2003b) was chosen to investigate the wave-one pattern in tropical total column ozone (Fig. 9) first reported in the 1980s by (Fishman and Larsen, 1987) An important contribution of SHADOZ has been the characterization of a distinct tropical tropopause layer (TTL, sometimes referred to as a tropopause transition layer (Gettelman and Forster, 2002; Fuglistaler et al., 2009; Thompson et al., 2012). This region is typically given as between 13 and 18 km; note steep ozone gradients at ~13 km in Fig. 9.

Other strategic ozonesonde networks operate on a campaign basis (Thompson et al., 2011); a list of major campaigns is given in Table 1. These soundings provide fixed-site ozone profiles to complement the multispecies payloads that aircraft deploy to study chemical and meteorological processes influencing ozone in the stratosphere and/or troposphere. The Match campaigns (von der Gathen et al., 1995; Rex et al., 1999) have coordinated polar and mid-latitude soundings to study in situ ozone losses during two Antarctic and 19 Arctic springs since the 1991–92 Arctic winter (Table 1). Using forecast trajectories to predict where layers of depleted ozone observed in one sounding will travel, the projected arrival of such a parcel over another station triggers a timed launch. Match has also supported a number of international aircraft experiments (Table 1). For the first time, in the 2019–20 winter-spring season, Match showed that the magnitude of Arctic ozone profile loss, recorded by soundings over Greenland, Ny-Ålesund (Svalbard, Norway), Canada, and Finland, could approach the magnitude of Antarctic "ozone hole" loss, with ozone mixing ratio values at ≤0.2 ppmv at 18 km (Fig. 8B; Wohltmann et al., 2020).

Over North America, a series of Intensive Ozonesonde Network Studies (IONS), supported multiaircraft and satellite validation studies from 2004 through 2013. For four IONS campaigns, sondes were coordinated at 6 to as many as 23 sites (August 2006) for midday satellite overpasses from 3 to 7 times/week. The IONS experiments led to a deeper understanding of tropospheric ozone during North American summers and have been especially useful in identifying stratosphere-troposphere exchange (STE) episodes. STE turns out to be more prevalent than previously thought, with significant intrusions of stratospheric air taking place after April–May, the typical "springtime" maximum in STE activity (Ott et al., 2016; Kuang et al., 2017; Tarasick et al., 2019). During the July–August 2004 IONS, ozonesonde observations along with satellite data showed that ~1/4 of the free tropospheric ozone budget from mid-Atlantic states to southeastern Canada originated

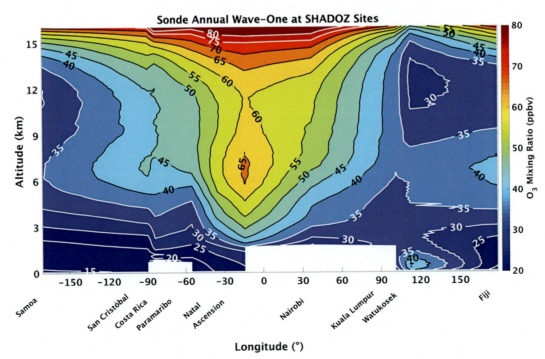

FIG. 9 Composite data from a strategic global network, SHADOZ, displaying the zonal ozone structure (mixing ratios) that gives rise to the wave-one pattern in satellite TCO. The contours are based on annually averaged profile data over 1998–2020.

TABLE 1 Strategic ozonesonde networks and related campaigns.

Observation Years	Campaign	Observation years	Campaign
1991–1992	Match (Arctic Spring) with EASOE & *AASE II*	July–Aug 2004	IONS-04 (Intensive Ozonesonde Network Study, INTEX-A, ICARTT)
1992–1993	Match (Arctic Spring)		
1993–1994	Match (Arctic Spring) with SESAME		
1994–1995	Match (Arctic Spring) with SESAME	March, May, Aug–Sept 2006	IONS-06 (Intensive Ozonesonde Network Study, INTEX-B, MILAGRO)
1995–1996	Match (Arctic Spring)		
1996–1997	Match (Arctic Spring)		
1997–1998	Match (Arctic Spring)		
1998–1999	Match (Arctic Spring) with THESEO	April 2008, June–July 2008	ARCIONS (ARCTAS IONS)
1999–2000	Match (Arctic Spring) with THESEO 2000 & *SOLVE*		
2002–2003	Match (Arctic Spring) with VINTERSOL & *SOLVE II*		
2004–2005	Match (Arctic Spring) with SCOUT-O3		
2006–2007	Match (Arctic Spring) with SCOUT-O3	July–Aug 2010, 2011	BORTAS
2007–2008	Match (Arctic Spring) with SCOUT-O3		
2009–2010	Match (Arctic Spring) with RECONCILE		
2010–2011	Match (Arctic Spring) with RECONCILE		
2013–2014	Match (Arctic Spring) with StratoClim		
2015–2016	Match (Arctic Spring) with StratoClim		
2017–2018	Match (Arctic Spring) with StratoClim		
2019–2020	Match (Arctic Spring)		
2003	Match (Antarctic Spring)	Aug–Sept 2013	SEACIONS (SEAC4RS IONS)
2007			

Campaigns aligned with Match are in *black* (European-sponsored) and *red* (NASA-sponsored).

from the stratosphere (Thompson et al., 2007a,b). Fig. 8C illustrates ozone profiles below 18 km at a Houston site during SEACIONS (2013). Varying ozone concentrations in the upper troposphere reflect stratospheric influences as well as lightning, as Thompson et al. (2008) showed with the identification of ozone laminae and satellite data analysis with IONS-06 summertime soundings over Houston. These same influences are reflected in the 2013 SEACIONS profiles (Fig. 8C).

4. Applications of ozonesonde data with remote sensing observations

Ozonesonde observations and remote sensing observations have a symbiotic relationship in that they are both useful to each other for producing high-quality data sets. The simple satellite retrieval flowchart of Fig. 10 demonstrates that climatologies based on ozonesonde profiles (e.g., McPeters and Labow, 2012) are used in satellite algorithms as a priori or first guess information. Limb-measuring satellites rely on comparisons with sonde ozone profiles for validation of their products. With a number of ozone-measuring satellites lasting a decade or more (Fig. 7), ozonesonde data are being used to evaluate drift in the satellite instruments (Hubert et al., 2016). The latter application has been an important factor in increasing demand for sonde data with reduced uncertainty and more rapid data delivery. Total column ozone (TCO) or tropospheric column ozone (TrCO) from sondes, as well as ground-based spectrometers, are routinely compared with the satellite TCO or TrCO. Examples are given in the next section.

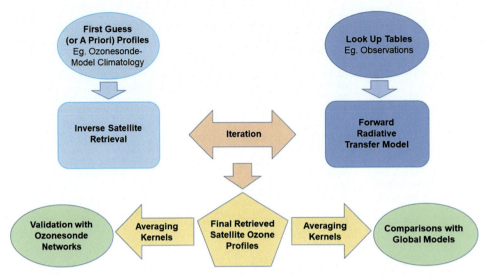

FIG. 10 Generalized flowchart indicating how ozonesonde data are used for a first guess or a priori profile in the retrieval process and for validation of the final satellite product.

4.1 Satellite ozone product evaluation using ozonesonde data

Ozonesonde data are typically used to evaluate two types of satellite products: profiles and column amounts. For example, stratospheric ozone profiles from the SAGE III instrument on the International Space Station (ISS/SAGE III) were recently examined by Wang et al. (2020). The satellite profiles are based on limb-viewing observations at sunrise and sunset. Twenty ozonesonde stations (between ±55 degrees latitude) provided the statistics, using a total of 273 profiles. Wang et al. (2020) also compared the SAGE III data to ozone from four other limb-measuring satellites, OSIRIS, Aura/MLS, ACE-FTS, and OMPS-LP. Agreement of the satellites as a whole was somewhat better at mid-latitudes than in the tropics.

Extracting profiles from nadir-viewing UV-measuring satellites is challenging. Huang et al. (2017) present a 10-year record of tropospheric profiles derived from OMI. The record is somewhat compromised due to a partial detector failure in 2009, which introduced a sampling bias into the ozone readings. For the newer TROPOMI (2017-), Mettig et al. (2021) employed a novel technique (TOPAS, Tikhonov regularized Ozone Profile retrievAl with SCIATRAN) to nadir retrievals in tropical and mid-latitudes to estimate ozone throughout the troposphere and lower-mid stratosphere; the method follows the simple flowchart in Fig. 10. The vertical resolution of the TOPAS method is fairly coarse (~9km on average) based on the averaging kernels reported with only 1–2 degrees of freedom (DOFs) in the troposphere, which is not unlike other UV-only satellite instruments. This indicates that similar instruments are highly dependent on the a priori profile (e.g., an ozonesonde climatology) in the troposphere. However, agreement between the TROPOMI-retrieved ozone profiles and ozonesonde measurements is generally within 20% (Fig. 11). New retrievals that combine observations from UV-satellite instruments and IR instruments (e.g., NOAA's CrIS) can improve both tropospheric and stratospheric comparisons with ozonesondes due to increased sensitivity throughout the ozone profile (Mettig et al., 2022).

Other techniques for estimating tropospheric ozone are based on column amounts, following the heritage of Fishman et al. (1991, 1996). Their "residual" approach to tropospheric ozone consists of subtracting the stratospheric column extracted from one satellite sensor from a highly accurate TCO from a backscattered UV instrument, initially from TOMS (several instruments from 1978 to 2005). The OMI/MLS series (Ziemke et al., 2006, 2019) is one of the most-used tropospheric column ozone (TrCO) data sets based on a residual technique. Fig. 12 shows the monthly mean TrCO from SHADOZ sondes from 10 tropical sites (latitude within ±20 degrees) (Thompson et al., 2021) compared to the corresponding monthly average OMI/MLS estimated tropospheric column. The offset is ~25% where the sonde TrCO is 40 DU although the correlation ($r^2 = 0.66$) is reasonably good. Part of the offset may be sampling differences (daily satellite data, with averaging over several pixels, vs 2–4 sondes/month). The satellite measurements do not typically capture the full range of ozone extremes measured by the sondes.

Cloud-slicing techniques (Ziemke et al., 2001; Heue et al., 2016) constitute an alternative approach to estimating upper and lower tropospheric column amounts; this has been applied to TROPOMI (Hubert et al., 2021). Agreement with ozonesonde-based totals is ~15%. A shortcoming of both cloud slicing and residual methods is incomplete knowledge

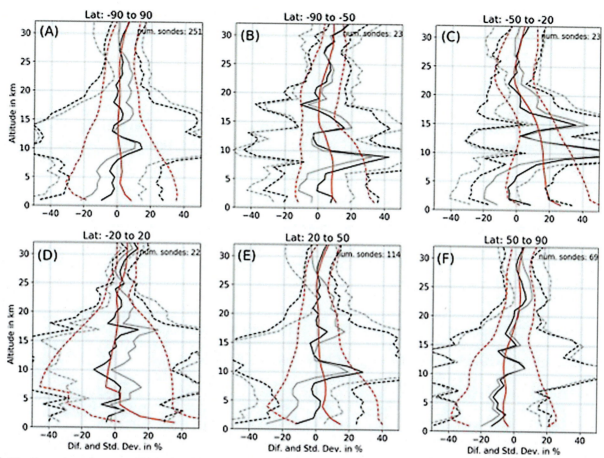

FIG. 11 Comparison of ozone profiles retrieved from TROPOMI and those from ozonesondes for different zonal bands. The relative mean difference between the retrieval results and the high-resolution sonde data *(solid line)*, as well as the standard deviation of the differences *(dashed line)*, is shown in *black*. The comparison with the sonde profiles convolved with the averaging kernels is shown in *red*. In *gray*, the relative difference between the a priori ozone profiles and high-resolution ozonesonde profiles is displayed, along with the corresponding standard deviations. *(From Mettig, N., Weber, M., Rozanov, A., Arosio, C., Burrows, J.P., Veefkind, P., Thompson, A.M., Querel, R., Leblanc, T., Godin-Beekman, S., Kivi, R., Tully, M.B., 2021. Ozone profile retrieval from nadir TROPOMI measurements in the UV range. Atmos. Meas. Tech. 14, 6057–6082. https://doi.org/10.5194/amt-14-6057-2021.)*

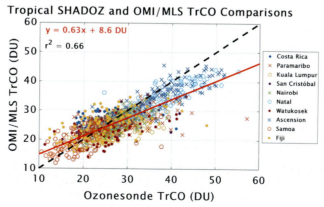

FIG. 12 Scatterplot of monthly mean TrCO estimated by the tropospheric residual OMI/MLS product (Ziemke et al., 2019) vs the corresponding TrCO from 10 SHADOZ sites (Thompson et al., 2021), the latter computed by integrating ozone from surface to tropopause determined from the coupled radiosonde. Comparisons are for SHADOZ stations with latitude within ±20 degrees.

of the tropopause height, i.e., what the column actually represents. This limitation is particularly relevant in the extra tropics where the tropopause height can vary greatly and change from <10 km to more than 15 km within hours. Time series with residual products (Ziemke et al., 2019) capture seasonal variability and oscillations like the ENSO, but caution is warranted for trends.

Fig. 13 shows examples of ozonesonde comparisons from two instruments on the Aura satellite (OMI and MLS) that has operated for 17 years. The comparisons are for soundings taken at the Wallops Island, VA (*green triangle marks* location in Fig. 6). Good agreement between the ozonesondes and MLS (Fig. 13A) is observed throughout the stratosphere (Witte et al., 2019). Dobson spectrophotometer measurements at Wallops Island are within ±5% of the ozonesonde TCO over the 25-year record illustrated (1995–2020), demonstrating the stability and high-quality of the sounding record); the Dobson is calibrated regularly against the world reference instrument at Boulder, CO. Fig. 13B shows that agreement between OMI (October 2004-) and ozonesonde TCO also averages 5% or better to 2020.

4.2 Use of satellite ozone data to track the performance of the ozonesonde

The examples above illustrate how ozonesonde data are used for evaluation of satellite products. Conversely, because several satellite records have been processed and improved multiple times, high-accuracy satellite data can be useful in monitoring the quality of sonde data. The ozonesonde community has been systematically reprocessing long-term sonde records over the past decade. Comparisons in total column ozone between integrated total ozone from soundings and coincident satellite overpasses may show a discontinuity that signifies a problem in the sonde measurements. For example, Witte et al. (2017, 2018) showed that an inadvertent change in the sensing solution in soundings at La Réunion led to an artificial 18 DU increase in the mean TCO from 2007 to 2016 compared to the average TCO from 1998 to 2006. Witte et al. (2017, 2018) corrected the affected ozone profiles to remove the discontinuities, using the homogenization procedures recommended by ASOPOS in Deshler et al. (2017).

In the past 5 years, there have been concerns about drifts or discontinuities in the ozonesonde TCO at ~20% of the global ozonesonde record since 2005. The direction of change is a loss of 3% or more in TCO since 2013. Fig. 14 illustrates how data from five operational satellite instruments, MLS (stratosphere), OMI, OMPS, and two GOME-2 instruments (TCO) are used to evaluate the ozonesonde data quality in the Aura era. In the upper panels of Fig. 14A and B, comparisons of sonde stratospheric ozone are made with ozone at standard MLS pressure levels. The lower panels show TCO comparisons with the 4 UV-based satellite instruments. The Wallops Island record (Figs. 13 and 14A) is stable in both TCO and stratospheric ozone above 50 hPa, whereas after 2013, the Samoa data (Fig. 14B) display more variability and an overall TCO decline (lower panel in Fig. 14B) that averages 3%–4% (Stauffer et al., 2020); the cause is partially due to changes in one sonde instrument type. The ASOPOS 2.0 Report (Smit et al., 2021), in which procedures are detailed to maximize quality in ozonesonde measurements, recommends ongoing comparisons of both the TCO and the stratospheric profile. The goal is to detect any change in procedure or instrument performance as quickly as possible.

5. Summary and conclusions

5.1 Scientific perspective: On-going need for profiles from global ozonesondes

The vertical profiles of the ozonesonde instrument provide unique information in the global ozone observing system for several reasons. First, no other widely used method is as free of weather effects. Second, although lidar has high vertical resolution, there are many fewer lidar stations compared to ozonesonde monitoring sites.

The near-real time measurement of the ozonesonde is ideal for tracking layers of stratospheric ozone (Match campaigns) and ozone pollution in the troposphere (IONS campaigns). Interest in ingesting sonde profiles into regional air-quality forecasts in near-real time and global chemistry-climate models is another motivator for adding to the number of ozonesonde stations. Unfortunately, numbers of sonde records have been declining in the past years. The combined WOUDC, NDACC, SHADOZ, and NOAA/GML archives include >2800 soundings for 2017 but fewer than 2400 records in 2019. Key Arctic and mid-latitude stations have reduced or eliminated soundings.

The satellite community continues to be an important user of ozonesonde data as well as a driver for faster data delivery and more stringent QA (SPARC/IO3C/GAW, 2019). With 5% uncertainty in TCO now achievable, ozonesonde data can be used to detect drifts of profiling ozone monitoring satellites and to evaluate new algorithms and satellite ozone products in a timely manner. Conversely, satellite data have been shown to be an important component in ensuring continuous evaluation of ozonesonde instrument and operational QA.

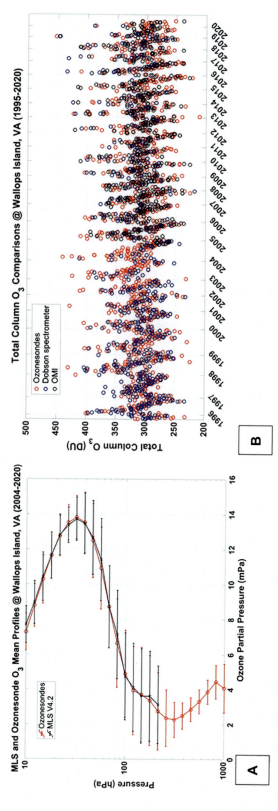

FIG. 13 (A) Comparison of ozone from Wallops Island, VA, United States, ozonesondes (*red*) and Aura/MLS data (*black*) at the standard levels of the MLS measurement (mean over 2004–20) with standard deviations indicated by horizontal bars; (B) TCO from Wallops sondes (*red*) compared to TCO from the Aura/OMI (*black*), 2004–20, and Dobson spectrophotometer (*blue*), 1995–2020.

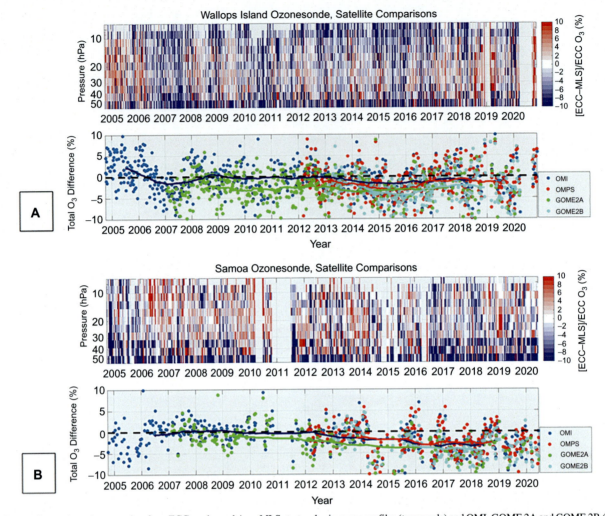

FIG. 14 Comparisons between data from ECC sondes and Aura MLS stratospheric ozone profiles (top panels) and OMI, GOME 2A and GOME 2B *(blue dots)*, and OMPS *(red dots)* TCO (bottom panels). (A) Wallops Island, VA, record; (B) Samoa SHADOZ record. *Red (blue)* colors in the top panels indicate where the ECC ozone is greater (less) than MLS. *Horizontal dashed lines* in the lower panels indicate the 0% line for TCO differences. Note a post-2014 drop in Samoa TCO relative to satellite measurements.

5.2 Quality assurance: Need for sonde intercomparisons and a global ozone reference

Changes in ozonesonde instrumentation are unavoidable as individual components may be modified by manufacturers. Operational and data processing practices may also change at individual stations. Accordingly, there is an ongoing need for periodic evaluation of ozonesonde performance and intercomparisons with a global ozone reference as the ASOPOS process has demonstrated. Essential elements of QA assessments are (1) regular laboratory evaluation of instruments and operational practices, such as the JOSIE experiments; (2) field tests; and (3) a process whereby global data and SOPs are continuously evaluated by a broad team of ozonesonde experts. These assessments must be supported by maintaining a world ozone standard photometer and one or more environmental test centers, e.g., the WCCOS. A strength of the ASOPOS process has been the inclusion of dedicated researchers who provide and archive ozone profiles, data users and instrument manufacturers. The recommendations, supported by analyses in the peer-reviewed literature, are consensus based. The ASOPOS Reports are themselves peer reviewed and are publicly available through the WMO/GAW website.

5.3 Conclusions

The ozonesonde instrument is unmatched in producing profiles of ozone with high vertical resolution throughout the troposphere and lower-mid stratosphere. Over the past 25 years, dedicated attention to ozonesonde QA has led to significant advances. This in turn led to new laboratory and field experiments to further refine SOP and guidelines for traceable ozonesonde records, bringing the target of 5% uncertainty throughout the ozone profile within reach. With reprocessed data, it has been possible to reduce residual uncertainties, biases, and discontinuities in ozonesonde time series. We can expect that there will be further homogenization efforts of ozonesonde data and evaluation of the new data within the global network in the coming years.

Acknowledgments

Valuable comments were received from reviewer Holger Vömel (NCAR). Thanks to Peter von der Gathen (Alfred Wegener Institute, Potsdam) for information on the Match and related aircraft and ground campaigns.

References

Chubashi, S., 1985. A special ozone observation a Syowa station, Antarctica, from February 1982 to January, 1983. In: Zerefos, C.S., Ghazi, A. (Eds.), Atmospheric Ozone. D. Reidel Publishers, Dordrecht, The Netherlands, pp. 285–289.

DeMazière, M., Thompson, A.M., Kurylo, M.J., Wild, J., Bernhard, G., Blumenstock, T., Hannigan, J., Lambert, J.-C., Leblanc, T., McGee, T.J., Nedoluha, G., Petropavlovskikh, I., Seckmeyer, G., Simon, P.C., Steinbrecht, W., Strahan, S., Sullivan, J.T., 2018. The Network for the Detection of Atmospheric Composition Change (NDACC): history, status and perspectives. Atmos. Chem. Phys. 18, 4935–4964. https://doi.org/10.5194/acp-18-4935-2018.

Deshler, T., Mercer, J., Smit, H.G.J., Stubi, R., Levrat, G., Johnson, B.J., Oltmans, S.J., Kivi, R., Thompson, A.M., Witte, J., Davies, J., Schmidlin, F.J., Brothers, G., Sasaki, T., 2008. Atmospheric comparison of electrochemical cell ozonesondes from different manufacturers, and with different cathode solution strengths: the Balloon Experiment on Standards for Ozonesondes. J. Geophys. Res. 113, D04307. https://doi.org/10.1029/2007JD008975.

Deshler, T., Stübi, R., Schmidlin, F.J., Mercer, J.L., Smit, H.G.J., Johnson, B.J., Kivi, R., Nardi, B., 2017. Methods to homogenize ECC ozonesonde measurements across changes in sensing solution concentration or ozonesonde manufacturer. Atmos. Meas. Tech. 10, 2012–2043. https://doi.org/10.5194/amt-10-2021-2017.

Farman, J., Gardiner, B., Shanklin, J., 1985. Large losses of total ozone in Antarctica reveal seasonal ClO_x/NO_x interaction. Nature 315, 207–210. https://doi.org/10.1038/315207a0.

Fishman, J., Larsen, J.C., 1987. Distribution of total ozone and stratospheric ozone in the tropics: implications for the distribution of tropospheric ozone. J. Geophys. Res. 92, D6. https://doi.org/10.1029/JD092iD06p06627.

Fishman, J., Fakhruzzaman, K., Cros, B., Nganga, D., 1991. Identification of widespread pollution in the Southern Hemisphere deduced from satellite analyses. Science 252 (5013), 1693–1696. https://doi.org/10.1126/science.252.5013.1693.

Fishman, J., Brackett, V.G., Browell, E.V., Grant, W.B., 1996. Tropospheric ozone derived from TOMS/SBUV measurements during TRACE-A. J. Geophys. Res. 101, D19. https://doi.org/10.1029/95JD03576.

Fuglistaler, S., Dessler, A.E., Dunkerton, T.J., Folkins, I., Fu, Q., Mote, P.W., 2009. Tropical tropopause layer. Rev. Geophys. https://doi.org/10.1029/2008RG000267.

Gettelman, A., Forster, P.M.F., 2002. A climatology of the tropical tropopause layer. J. Meterol. Soc. Jpn. 80 (4B), 911–924. https://doi.org/10.2151/jmsj.80.911.

Huang, G., Liu, X., Chance, K., Yang, K., Bhartia, P.K., Cai, Z., Allaart, M., Acellet, G., Calpini, B., Coetzee, G.J.R., Cuevas-Agulló, E., Cupeiro, M., De Backer, H., Dubey, M.K., Fuelberg, H.E., Fujiwara, M., Godin-Beekmann, S., Hall, T.J., Johnson, B., Joseph, E., Kivi, R., Kois, B., Komala, N., König-Langlo, G., Laneve, G., Leblanc, T., Marchand, M., Minschwaner, K.R., Morris, G., Newchurch, M.J., Ogino, S.-Y., Ohkawara, N., Piters, A.J.M., Posny, F., Querel, R., Scheele, R., Schmidlin, F.J., Schnell, R.C., Schrems, O., Selkirk, H., Shiotani, M., Skrivánková, P., Stübi, R., Taha, G., Tarasick, D.W., Thompson, A.M., Thouret, V., Tully, M.B., van Malderen, R., Vömel, H., von der Gathen, P., Witte, J.C., Yela, M., 2017. Validation of 10-year SAO OMI Ozone Profile (PROFOZ) product using ozonesonde observations. Atmos. Meas. Tech. 10, 2455–2475. https://doi.org/10.5194/amt-10-2455-2017.

Hubert, D., Lambert, J.-C., Verhoelst, T., Granville, J., Keppens, A., Baray, J.-L., Bourassa, A.E., Cortesi, U., Degenstein, D.A., Froidevaux, L., Godin-Beekmann, S., Hoppel, K.W., Johnson, B.J., Kyrölä, E., Leblanc, T., Lichtenberg, G., Marchand, M., McElroy, C.T., Murtagh, D., Nakane, H., Portafaix, T., Querel, R., Russell III, J.M., Salvador, J., Smit, H.G.J., Stebel, K., Steinbrecht, W., Strawbridge, K.B., Stübi, R., Swart, D.P.J., Taha, G., Tarasick, D.W., Thompson, A.M., Urban, J., van Gijsel, J.A.E., Van Malderen, R., von der Gathen, P., Walker, K.A., Wolfram, E., Zawodny, J.M., 2016. Ground-based assessment of the bias and long-term stability of 14 limb and occultation ozone profile data records. Atmos. Meas. Tech. 9, 2497–2534. https://doi.org/10.5194/amt-9-2497-2016.

Hubert, D., Heue, K.-P., Lambert, J.-C., Verhoelst, T., Allaart, M., Compernolle, S., Cullis, P.D., Dehn, A., Félix, C., Johnson, B.J., Keppens, A., Kollonige, D.E., Lerot, C., Loyola, D., Maata, M., Mitro, S., Mohamad, M., Piters, A., Romahn, F., Selkirk, H.B., da Silva, F.R., Stauffer, R.M., Thompson, A.M., Veefkind, J.P., Vömel, H., Witte, J.C., Zehner, C., 2021. TROPOMI tropospheric ozone column data: geophysical assessment and comparison to ozonesondes, GOME-2B and OMI. Atmos. Meas. Tech 14, 7405–7433. https://doi.org/10.5194/amt-14-7405-2021.

Heue, K.-P., Coldewey-Egbers, M., Delcloo, A., Lerot, C., Loyola, D., Valks, P., van Roozendael, M., 2016. Trends of tropical tropospheric ozone from 20 years of European satellite measurements and perspectives for the Sentinel-5 Precursor. Atmos. Meas. Tech. 9, 5037–5051. https://doi.org/10.5194/amt-9-5037-2016.

Johnson, B.J., Oltmans, S.J., Vömel, H., Smit, H.G.J., Deshler, T., Kroeger, C., 2002. ECC ozonesonde pump efficiency measurements and tests on the sensitivity to ozone of buffered and unbuffered ECC sensor cathode solutions. J. Geophys. Res. 107, D19. https://doi.org/10.1029/2001JD000557.

Komhyr, W.D., 1969. Electrochemical concentration cells for gas analysis. Ann. Geophys. 25, 203–210.

Kuang, S., Newchurch, M.J., Thompson, A.M., Stauffer, R.M., Johnson, B.J., Wang, L., 2017. Ozone variability and anomalies observed during SENEX and SEAC4RS campaigns in 2013. J. Geophys. Res. 122 (20), 11227–11241. https://doi.org/10.1002/2017JD027139.

McDermid, I.S., Godin, S.M., Lindqvist, L.O., 1990. Ground-based laser DIAL system for long-term measurements of stratospheric ozone. Appl. Opt. 29, 3603–3612. https://doi.org/10.1364/AO.29.003603.

McGee, T.J., Newman, P., Ferrare, R., Whiteman, D., Butler, J.J., Burris, J., Godin, S.M., McDermid, I.S., 1990. Lidar observations of ozone changes induced by sub-polar airmass motion over Table Mountain (34.4N). J. Geophys. Res. 95, 20527–20530.

McPeters, R.D., Labow, G.J., 2012. Climatology 2011: an MLS and sonde derived ozone climatology for satellite retrieval algorithms. J. Geophys. Res. 117, D10303. https://doi.org/10.1029/2011JD017006.

Mettig, N., Weber, M., Rozanov, A., Arosio, C., Burrows, J.P., Veefkind, P., Thompson, A.M., Querel, R., Leblanc, T., Godin-Beekman, S., Kivi, R., Tully, M.B., 2021. Ozone profile retrieval from nadir TROPOMI measurements in the UV range. Atmos. Meas. Tech. 14, 6057–6082. https://doi.org/10.5194/amt-14-6057-2021.

Mettig, N., Weber, M., Rozanov, A., Burrows, J.P., Veefkind, P., Barnet, C., Thompson, A.M., Stauffer, R.M., Leblanc, T., Ancellet, G., Newchurch, M., Kivi, R., Tully, M.B., Van Malderen, R., Steinbrecht, W., Piters, A., Allaart, M., Kois, B., Stübi, R., Davies, J., Skrivankova, P., 2022. Combined UV and IR ozone profile retrieval from TROPOMI and CrIS measurements. Atmos. Meas. Tech. Discuss. Atmos. Meas. Tech. 15, 2955–2978. https://doi.org/10.5194/amt-15-2955-2022.

Moeini, O., Tarasick, D.W., McElroy, C.T., Liu, J., Osman, M.K., Thompson, A.M., Parrington, M., Palmer, P.I., Johnson, B., Oltmans, S.J., Merrill, J., 2020. Estimating boreal fire-generated ozone over North American using ozonesonde profiles and a differential back trajectory technique. Atmos. Environ. https://doi.org/10.1016/j.aeaoa.2020.100078.

Ott, L.E., Duncan, B.N., Thompson, A.M., Diskin, G., Fasnacht, Z., Langford, A.O., Lin, M., Molod, A.M., Nielsen, J.E., Pusede, S.E., Wargan, K., Weinheimer, A.J., Yoshida, Y., 2016. Frequency and impact of summertime stratospheric intrusions over Maryland during DISCOVER-AQ (2011): new evidence from NASA's GEOS-5 simulations. J. Geophys. Res. 121 (7), 3687–3706. https://doi.org/10.1002/2015JD024052.

Proffitt, M.H., McLaughlin, R.J., 1983. Fast response dual-beam UV-absorption photometer suitable for use on stratospheric balloons. Rev. Sci. Instrum. 54, 1719–1728.

Rex, M., von der Gathen, P., Braathen, G., Harris, N.R.P., Reimer, E., Beck, A., Alfier, R., Krüger-carstensen, R., Chipperfield, M., De Backer, H., Balis, D., O'Connor, F., Dier, H., Dorokhov, V., Fast, H., Gamma, A., Gil, M., Kyrö, E., Litynska, Z., Mikkelsen, I.S., Molyneux, M., Murphy, G., Reid, S.J., Rummukainen, M., Zerefos, C., 1999. Chemical ozone loss in the Arctic winter 1994/95 as determined by the match technique. J. Atmos. Chem. 32, 35–59. https://doi.org/10.1023/A:1006093826861.

Smit, H.G.J., 2014. Ozone sondes. In: North, G.R., Pyle, J.A., Zhang, F. (Eds.), Encyclopedia of Atmospheric Sciences, second ed. vol. 1. Academic Press, London, pp. 372–378.

Smit, H.G.J., ASOPOS Panel, 2014. Quality assurance and quality control for ozonesonde measurements in GAW. WMO Global Atmosphere Watch report series, No. 201, World Meteorological Organization, Geneva. Available from: https://library.wmo.int/doc_num.php?explnum_id=7167.

Smit, H.G.J., Kley, D., 1998. JOSIE: the 1996 WMO international intercomparison of ozonesondes under quasi flight conditions in the environmental simulation chamber at Jülich. WMO Global Atmosphere Watch Report Series, No. 130, WMO/TD No. 926, World Meteorological Organization, Geneva.

Smit, H.G.J., Straeter, W., 2004a. JOSIE-1998, performance of ECC ozone sondes of SPC-6A and ENSCI-Z type. WMO Global Atmosphere Watch Report Series, No. 157, WMO/TD No. 1218, World Meteorological Organization, Geneva. Available from: https://library.wmo.int/index.php?lvl=notice_display&id=11089#.Ya-MV1MxlUM.

Smit, H.G.J., Straeter, W., 2004b. JOSIE-2000, Jülich Ozone Sonde Intercomparison Experiment 2000, the 2000 WMO international intercomparison of operating procedures for ECC ozonesondes at the environmental simulation facility at Jülich. WMO Global Atmosphere Watch Report Series, No. 158, WMO TD No. 1225, World Meteorological Organization, Geneva. Available from: https://library.wmo.int/index.php?lvl=notice_display&id=11090#.Ya-M6FMxlUM.

Smit, H.G.J., Sträter, W., Helten, M., Kley, D., 2000. Environmental simulation facility to calibrate airborne ozone and humidity sensors. Jül Berichte Nr 3796, Forschungszentrum Jülich.

Smit, H.G.J., Straeter, W., Johnson, B.J., Oltmans, S.J., Davies, J., Tarasick, D.W., Hoegger, B., Stubi, R., Schmidlin, F.J., Northam, T., Thompson, A.M., Witte, J.C., Boyd, I., Posny, F., 2007. Assessment of the performance of ECC-ozonesondes under quasi-flight conditions in the environmental simulation chamber: insights from the Jülich Ozone Sonde Intercomparison Experiment (JOSIE). J. Geophys. Res. 112, D19306. https://doi.org/10.1029/2006JD007308.

Smit, H.G.J., O3S-DQA, 2012. Guidelines for Homogenization of Ozonesonde Data, SI2N/O3S-DQA Activity as Part of "Past Changes in the Vertical Distribution of Ozone Assessment". Available from: https://www.wccos-josie.org/o3s-dqa/.

Smit, H.G.J., Thompson, A.M., ASOPOS Panel, 2021. Ozonesonde measurement principles and best operational practices, ASOPOS (Assessment of standard operating procedures for ozonesondes) 2.0. WMO Global Atmosphere Watch report series, No. 268, World Meteorological Organization, Geneva. Available from: https://library.wmo.int/index.php?lvl=notice_display&id=21986#.YaFNSbpOlc8.

SPARC/IOC/GAW, 1998. Assessment of trends in the vertical distribution of ozone. SPARC Report No.1, WMO Global Ozone Research and Monitoring Project Report No. 43, World Meteorological Organization, Geneva.

SPARC/IO3C/GAW, 2019. In: Petropavlovskikh, I., Godin-Beekmann, S., Hubert, D., Damadeo, R., Hassler, B., Sofieva, V. (Eds.), Report on Long-Term Ozone Trends and Uncertainties in the Stratosphere., https://doi.org/10.17874/f899e57a20b. SPARC Report No. 9, GAW Report No. 241, WCRP-17/2018.

Stauffer, R.M., Morris, G.A., Thompson, A.M., Joseph, E., Coetzee, G.J.R., Nalli, N.R., 2014. Propagation of radiosonde pressure sensor errors to ozonesonde measurements. Atmos. Meas. Tech. 7, 65–79. https://doi.org/10.5194/amt-7-65-2014.

Stauffer, R.M., Thompson, A.M., Oltmans, S.J., Johnson, B.J., 2017. Tropospheric ozonesonde profiles at long-term US monitoring sites: 2. Links between Trinidad, CA, profile clusters and inland surface ozone measurements. J. Geophys. Res. 122. https://doi.org/10.1002/2016JD025254.

Stauffer, R.M., Thompson, A.M., Kollonige, D.E., Witte, J.C., Tarasick, D.W., Davies, J.M., Vömel, H., Morris, G.A., Van Malderen, R., Johnson, B.J., Querel, R.R., Selkirk, H.B., Stübi, R., Smit, H.G.J., 2020. A post-2013 drop-off in total ozone at a third of global ozonesonde stations: electrochemical concentration cell instrument artifacts? Geophys. Res. Lett. 47 (11). https://doi.org/10.1029/2019/GL086791.

Steinbrecht, W., Kubistin, D., Plass-Dülmer, C., Davies, J., Tarasick, D.W., von der Gathen, P., Deckelmann, H., Jepsen, N., Kivi, R., Lyall, N., Palm, M., Notholt, J., Kois, B., Oelsner, P., Allaart, M., Piters, A., Gill, M., Van Malderen, R., Delcloo, A.W., Sussmann, R., Mahieu, E., Servais, C., Romanens, G., Stübi, R., Ancellet, G., Godin-Beekmann, S., Yamanouchi, S., Strong, K., Johnson, B., Cullis, P., Petropavlovskikh, I., Hannigan, J.W., Hernandez, J.-L., Rodriguez, A.D., Nakano, T., Chouza, F., Leblanc, T., Torres, C., Garcia, O., Röhling, A.N., Schneider, M., Blumenstock, T., Tully, M., Paton-Walsh, C., Jones, N., Querel, R., Strahan, S., Stauffer, R.M., Thompson, A.M., Inness, A., Engelen, R., Chang, K.-L., Cooper, O.R., 2021. Did the COVID-19 crisis reduce free tropospheric ozone across the Northern Hemisphere? Geophys. Res. Lett. 48, e2020GL091987. https://doi.org/10.1029/2020GL091987.

Sterling, C.W., Johnson, B.J., Oltmans, S.J., Smit, H.G.J., Jordan, A.F., Cullis, P.D., Hall, E.G., Thompson, A.M., Witte, J.C., 2018. Homogenizing and estimating the uncertainty in NOAA's long-term vertical ozone profile records measured with the electrochemical concentration cell ozonesonde. Atmos. Meas. Tech. 11, 3661–3687. https://doi.org/10.5194/amt-11-3661-2018.

Sullivan, J.T., McGee, T.J., Leblanc, T., Sumnicht, G.K., Twigg, L.W., 2015. Optimization of the GSFC TROPOZ DIAL retrieval using synthetic lidar returns and ozonesondes—part 1: algorithm validation. Atmos. Meas. Tech. 8, 4133–4143. https://doi.org/10.5194/amt-8-4133-2015.

Tarasick, D.W., Davies, J., Smit, H.G.J., Oltmans, S.J., 2016. A re-evaluated Canadian ozonesonde record: measurements of the vertical distribution of ozone over Canada from 1966 to 2013. Atmos. Meas. Tech. 9, 195–214. https://doi.org/10.5194/amt-9-195-2016.

Tarasick, D.W., Carey-Smith, T.K., Hocking, W.K., Moeini, O., He, H., Liu, J., Osman, M.K., Thompson, A.M., Johnson, B.J., Oltmans, S.J., Merrill, T.J., 2019. Quantifying stratosphere-troposphere transport of ozone using balloon-borne ozonesondes, radar windprofilers and trajectory models. Atmos. Environ. 198 (2019), 496–509. https://doi.org/10.1016/j.atmosenv.2018.10.040.

Tarasick, D.W., Smit, H.G.J., Thompson, A.M., Morris, G.A., Witte, J.C., Davies, J., Nakano, T., Van Malderen, R., Stauffer, R.M., Johnson, B.J., Stübi, R., Oltmans, S.J., Vömel, H., 2021. Improving ECC ozonesonde data quality: assessment of current methods and outstanding issues. Earth Space Sci. 8, e2019EA000914. https://doi.org/10.1029/2019EA000914.

Thompson, A.M., Pickering, K.E., McNamara, D.P., Schoeberl, M.R., Hudson, R.D., Kim, J.H., Browell, E.V., Kirchhoff, V.W.J.H., Nganga, D., 1996. Where did tropospheric ozone over southern Africa and the tropical Atlantic come from in October 1992? Insights from TOMS, GTE TRACE A, and SAFARI 1992. J. Geophys. Res. 101 (D19), 24251–24278. https://doi.org/10.1029/96JD01463.

Thompson, A.M., Witte, J.C., Hudson, R.D., Guo, H., Herman, J.R., Fujiwara, M., 2001. Tropical tropospheric ozone and biomass burning. Science 291, 2128–2132. https://doi.org/10.1126/science.291.5511.2128.

Thompson, A.M., Witte, J.C., McPeters, R.D., Oltmans, S.J., Schmidlin, F.J., Logan, J.A., Fujiwara, M., Kirchhoff, V.W.J.H., Posny, F., Coetzee, G.J.R., Hoegger, B., Kawakami, S., Ogawa, T., Johnson, B.J., Vömel, H., Labow, G., 2003a. Southern hemisphere additional ozonesondes (SHADOZ) 1998–2000 tropical ozone climatology 1. Comparison with Total Ozone Mapping Spectrometer (TOMS) and ground-based measurements. J. Geophys. Res. 108, 8238. https://doi.org/10.1029/2001JD000967.

Thompson, A.M., Witte, J.C., Oltmans, S.J., Schmidlin, F.J., Logan, J.A., Fujiwara, M., Kirchhoff, V.W.J.H., Posny, F., Coetzee, G.J.R., Hoegger, B., Kawakami, S., Ogawa, T., Fortuin, J.P.F., Kelder, H.M., 2003b. Southern hemisphere additional ozonesondes (SHADOZ) 1998–2000 tropical ozone climatology. 2. Tropospheric variability and the zonal wave-one. J. Geophys. Res. 108, 8241. https://doi.org/10.1029/2002JD002241.

Thompson, A.M., Stone, J.B., Witte, J.C., Miller, S.K., Pierce, R.B., Chatfield, R.B., Oltmans, S.J., Cooper, O.R., Loucks, A.L., Taubman, B.F., Johnson, B.J., Joseph, E., Kucsera, T.L., Merrill, J.T., Morris, G.A., Hersey, S., Forbes, G., Newchurch, M.J., Schmidlin, F.J., Tarasick, D.W., Thouret, V., Cammas, J.-P., 2007a. Intercontinental chemical transport experiment ozonesonde network study (IONS) 2004: 1 Summertime upper troposphere/lower stratosphere ozone over northeastern North America. J. Geophys. Res. 112, D12S12. https://doi.org/10.1029/2006JD007441.

Thompson, A.M., Stauffer, R.M., Wargan, K., Witte, J.C., Kollonige, D.E., Ziemke, J.R., 2021. Regional and seasonal trends in tropical ozone from SHADOZ profiles: reference for models and satellite products. J. Geophys. Res. Atmos. 126. https://doi.org/10.1029/2021JD034691.

Thompson, A.M., Stone, J.B., Witte, J.C., Miller, S.K., Oltmans, S.J., Ross, K.L., Kucsera, T.L., Merrill, J.T., Forbes, G., Tarasick, D.W., Joseph, E., Schmidlin, F.J., McMillan, W.W., Warner, J., Hintsa, E.J., Johnson, J.E., 2007b. Intercontinental transport experiment ozonesonde network study (IONS, 2004): 2. Tropospheric ozone budgets and variability over Northeastern North America. J. Geophys. Res. 112, D12S13. https://doi.org/10.1029/2006JD007670.

Thompson, A.M., Witte, J.C., Smit, H.G.J., Oltmans, S.J., Johnson, B.J., Kirchhoff, V.W.J.H., Schmidlin, F.J., 2007c. Southern hemisphere additional ozonesondes (SHADOZ) 1998–2004 tropical ozone climatology. 3. Instrumentation, station variability, evaluation with simulated flight profiles. J. Geophys. Res. 112, D03304. https://doi.org/10.1029/2005JD007042.

Thompson, A.M., Yorks, J.E., Miller, S.K., Witte, J.C., Dougherty, K.M., Morris, G.A., Baumgardner, D., Ladino, L., Rappenglueck, B., 2008. Tropospheric ozone sources and wave activity over Mexico City and Houston during Milagro/Intercontinental Transport Experiment (INTEX-B) Ozonesonde Network Study, 2006 (IONS-06). Atmos. Chem. Phys. 8, 5113–5126.

Thompson, A.M., Oltmans, S.J., Tarasick, D.W., von der Gathen, P., Smit, H.G.J., Witte, J.C., 2011. Strategic ozone sounding networks: review of design and accomplishments. Atmos. Environ. 45, 2145–2163. https://doi.org/10.1016/j.atmosenv.2010.05.002.

Thompson, A.M., Miller, S.K., Tilmes, S., Kollonige, D.W., Witte, J.C., Oltmans, S.J., Johnson, B.J., Fujiwara, M., Schmidlin, F.J., Coetzee, G.J.R., Komala, N., Maata, M., bt Mohamad, M., Nguyo, J., Mutai, C., Ogino, S.-Y., Raimundo Da Silva, F., Paes Leme, N.M., Posny, F., Scheele, R., Selkirk, H.B., Shiotani, M., Stübi, R., Levrat, G., Calpini, B., Thouret, V., Tsuruta, H., Valverde Canossa, J., Vömel, H., Yonemura, S., Andrés Diaz, J., Tan

Thanh, H.T., Thuy Ha, H.T., 2012. Southern Hemisphere Additional Ozonesondes (SHADOZ) ozone climatology (2005–2009): tropospheric and tropical tropopause layer (TTL) profiles with comparisons to OMI-based ozone products. J. Geophys. Res. 117, D23301. https://doi.org/10.1029/2011JD016911.

Thompson, A.M., Witte, J.C., Sterling, C., Jordan, A., Johnson, B.J., Oltmans, S.J., Fujiwara, M., Vömel, H., Allaart, M., Piters, A., Coetzee, G.J.R., Posny, F., Corrales, E., Diaz, J.A., Félix, C., Komala, N., Lai, N., Ahn Nguyen, H.T., Maata, M., Mani, F., Zainal, Z., Ogino, S.-Y., Paredes, F., Penha, T.L.B., da Silva, F.R., Sallons-Mitro, S., Selkirk, H.B., Schmidlin, F.J., Stübi, R., Thiongo, K., 2017. First reprocessing of Southern Hemisphere ADditional OZonesondes (SHADOZ) Ozone Profiles (1998–2016). 2. Comparisons with satellites and ground-based instruments. J. Geophys. Res. 122. https://doi.org/10.1002/2017JD027406.

Thompson, A.M., Smit, H.G.J., Witte, J.C., Stauffer, R.M., Johnson, B.J., Morris, G., von der Gathen, P., Van Malderen, R., Davies, J., Piters, A., Allaart, M., Posny, F., Kivi, R., Cullis, P., Anh, N.T.H., Corrales, E., Machinini, T., da Silva, F.R., Paiman, G., Thiong'o, K., Zainal, Z., Brothers, G.B., Wolff, K.R., Nakano, T., Stübi, R., Romanens, G., Coetzee, G.J.R., Diaz, J.A., Mitro, S., Mohamad, M., Ogino, S.-Y., 2019. Ozonesonde quality assurance: the JOSIE-SHADOZ (2017) experience. Bull. Am. Meteor. Soc. 100 (1). https://doi.org/10.1175/BAMS-D-17-0311.1.

Van Malderen, R., Allaart, M.A.F., De Backer, H., Smit, H.G.J., De Muer, D., 2016. On instrumental errors and related correction strategies of ozonesondes: possible effect on calculated ozone trends for the nearby sites Uccle and De Bilt. Atmos. Meas. Tech. 9, 3793–3816. https://doi.org/10.5194/amt-9-3793-2016.

Vömel, H., Smit, H.G.J., Tarasick, D., Johnson, B., Oltmans, S.J., Selkirk, H., Thompson, A.M., Stauffer, R.M., Witte, J.C., Davies, J., van Malderen, R., Morris, G.A., Nakano, T., Stübi, R., 2020. A new method to correct the ECC ozone sonde time response and its implications for "background current" and pump efficiency. Atmos. Meas. Tech. 13 (10), 5667–5680. https://doi.org/10.5194/amt-13-5667-2020.

von der Gathen, P., Rex, M., Harris, N.R.P., Lucic, D., Knudsen, B.M., Braathen, G.O., De Backer, H., Fabian, R., Fast, H., Gil, M., Kyrö, E., Mikkelsen, I.S., Rummukainen, M., Stähelin, J., Varotsos, C., 1995. Observational evidence for chemical ozone depletion over the Arctic in winter 1991–92. Nature 375, 131–134. https://doi.org/10.1038/375131a0.

Wang, H.J.R., Damadeo, R., Flittner, D., Kramarova, N., Taha, G., Davis, S., Thompson, A.M., Strahan, S., Wang, Y., Froidevaux, L., Degenstein, D., Bourassa, A., Steinbrecht, W., Walker, K., Querel, R., Leblanc, T., Godin-Beekman, S., Hurst, D., Hall, E., 2020. Validation of SAGE III/ISS solar occultation ozone products with correlative satellite and ground based measurements. J. Geophys. Res. 125. https://doi.org/10.1029/2020JD032430.

Witte, J.C., Thompson, A.M., Smit, H.G.J., Fujiwara, M., Posny, F., Coetzee, G.J.R., Northam, E.T., Johnson, B.J., Sterling, C.W., Mohamad, M., Ogino, S.-Y., Jordan, A., da Silva, F.R., 2017. First reprocessing of Southern Hemisphere ADditional OZonesondes (SHADOZ) profile records (1998–2015) 1: methodology and evaluation. J. Geophys. Res. 122. https://doi.org/10.1002/2016JD026403.

Witte, J.C., Thompson, A.M., Smit, H.G.J., Vömel, H., Posny, F., Stübi, R., 2018. First reprocessing of Southern hemisphere additional ozonesondes (SHADOZ) profile records. 3. Uncertainty in ozone profile and total column. J. Geophys. Res. 123 (6), 3243–3268. https://doi.org/10.1002/2017JD027791.

Witte, J.C., Thompson, A.M., Schmidlin, F.J., Northam, E.T., Wolff, K.R., Brothers, G.B., 2019. The NASA Wallops Flight Facility digital ozonesonde record: reprocessing, uncertainties, and dual launches. J. Geophys. Res. 124, 3565–3582. https://doi.org/10.1029/2018JD030098.

World Meteorological Organization (WMO), 1996. Report of the fourth WMO meeting of experts on the quality assurance/science activity centers (QA/SACs) of the global atmosphere watch. WMO Global Atmosphere Watch Report Series, No. 104, World Meteorological Organization, Geneva.

WMO/UNEP, 2019. Scientific assessment of ozone depletion: 2018, global ozone research and monitoring project. Report No. 58, World Meteorological Organization, Geneva. Available from: https://library.wmo.int/index.php?lvl=notice_display&id=20763#.Ya-O8lMxlUM.

Wohltmann, I., von der Gathen, P., Lehmann, R., Maturilli, M., Deckelmann, H., Manney, G.L., Davies, J., Tarasick, D., Jepsen, N., Kivi, R., Lyall, N., Rex, M., 2020. Near complete local reduction of Arctic stratospheric ozone by record chemical loss in spring 2020. Geophys. Res. Lett. 47 (20). https://doi.org/10.1029/2020GL089547.

Ziemke, J.R., Chandra, S., Bhartia, P.K., 2001. "Cloud slicing": a new technique to derive upper tropospheric ozone from satellite measurements. J. Geophys. Res. 106 (D9), 9853–9867. https://doi.org/10.1029/2000JD900768.

Ziemke, J.R., Chandra, S., Duncan, B.N., Froidevaux, L., Bhartia, P.K., Levelt, P.F., Waters, J.W., 2006. Tropospheric ozone determined from Aura OMI and MLS: evaluation of measurements and comparison with the Global Modeling Initiative's Chemical Transport Model. J. Geophys. Res. 111, D19303. https://doi.org/10.1029/2006JD007089.

Ziemke, J.R., Oman, L.D., Strode, S.A., Douglass, A.R., Olsen, M.A., McPeters, R.D., Bhartia, P.K., Froidevaux, L., Labow, G.J., Witte, J.C., Thompson, A.M., Haffner, D.P., Kramarova, N.A., Frith, S.M., Huang, L.K., Jaross, G.R., Seftor, C.J., Deland, M.T., Taylor, S.L., 2019. Trends in global tropospheric ozone inferred from a composite record of TOMS/OMI/MLS/OMPS satellite measurements and the MERRA-2 GMI simulation. Atmos. Chem. Phys. 19, 3257–3269. https://doi.org/10.5194/acp-19-3257-2019.

Chapter 5

Oceanographic buoys: Providing ocean data to assess the accuracy of variables derived from satellite measurements

Renellys C. Perez[a], Gregory R. Foltz[a], Rick Lumpkin[a], Jianwei Wei[b], Kenneth J. Voss[c], Michael Ondrusek[b], Menghua Wang[b], and Mark A. Bourassa[d]

[a]*NOAA/Atlantic Oceanographic and Meteorological Laboratory, Miami, FL, United States,* [b]*NOAA/NESDIS Center for Satellite Applications and Research (STAR), College Park, MD, United States,* [c]*University of Miami, Coral Gables, FL, United States,* [d]*Florida State University, COAPS, Tallahassee, FL, United States*

Chapter outline

1. Introduction	79	8. Water-leaving radiance	92
2. Oceanographic buoy platforms and requirements	80	8.1 Selection of location	92
2.1 Surface drifting buoys	81	8.2 Instruments and measurements	92
2.2 Moored buoys	83	8.3 Data processing	93
3. Sea surface temperature	85	8.4 Characterization, calibration, and maintenance	94
4. Sea surface salinity	87	8.5 Accomplishments and prospective	94
5. Wind speed and direction	89	9. Parting thoughts	94
6. Surface heat fluxes	90	Acknowledgments	96
7. Sea level pressure	91	References	96

Knowledge of the oceans is more than a matter of curiosity. Our very survival may hinge upon it.

President John F. Kennedy

1. Introduction

The collection of independent in situ field measurements of ocean and atmospheric parameters is a crucial element of passive environmental satellite calibration and validation (hereafter cal/val). Despite multigenerational efforts by scientists, engineers, and the crew aboard research and commercial vessels (see Chapter 11), it remains a daunting challenge to maintain an ocean observing system capable of assessing the accuracy of variables derived from satellite measurements on a global scale. Oceanographic buoys, defined here as floating, drifting, or anchored ocean observing platforms, have long been used to collect data in the ocean either during relatively short-term process studies that support science-based inquiries, or as part of sustained programs that contribute near-real-time data for operational and scientific use. The form and function of oceanographic buoys vary tremendously and includes anchored surface and subsurface moorings, and quasi-Lagrangian buoys that follow fluid motion in two or three-dimensions. Some of these buoys may spend their lifetime at the sea surface (e.g., surface drifting buoys); some transit the ocean following a fixed depth, pressure, or density surface (e.g., isobaric floats); some move three dimensionally through the water column collecting measurements as they ascend or descend (e.g., Argo profiling floats), some are affixed to the seafloor by an anchor, collecting measurements just above the seafloor, from the seafloor to an intermediate depth, or from the seafloor to the sea surface (e.g., moored buoys with subsurface or surface flotation).

Here we primarily focus on two specific types of oceanographic buoys that intersect with and sample near the air-sea interface with meteorological and/or oceanographic instrumentation in the marine boundary layer: surface drifting buoys and moored buoys with a surface float (hereafter drifters and moorings, respectively). Drifters and moorings provide data with only an hour or a day delay of the measurement time, and these "real-time" near-surface field measurements are invaluable for satellite cal/val. In Section 4, however, we will also discuss Argo profiling floats which routinely measure temperature and salinity from the surface down to 2000 m of the water column and provide key salinity measurements near the ocean surface. While there are many new types of autonomous or unpiloted platforms that can provide data at the air-sea interface (e.g., Saildrones and wavegliders), with an ever-expanding suite of physical and biogeochemical parameters, this technology is relatively nascent and presently lacks the requisite spatiotemporal coverage needed for routine satellite cal/val.

Here, we will discuss some of the current/best practices for deploying, servicing, maintaining, and calibrating different types of instrumentation on oceanographic buoys, and describe what spatial and temporal coverage these observing platforms provide. The oceanic or atmospheric variables that we focus on in this chapter are sea surface temperature and salinity, wind speed and direction, surface heat fluxes, sea-level pressure, and water-leaving radiance. While we do not provide an exhaustive list of all of the buoys used to calibrate and validate satellite measurements, nor the full scope of the effort required to deploy and maintain these buoys, our intent is to give a flavor of what is involved in such efforts.

2. Oceanographic buoy platforms and requirements

Most sustained global ocean observational networks are funded, managed, and maintained by partners from multiple national and/or international institutions. These networks are often coordinated by scientific steering groups and intergovernmental agencies such as the World Meteorological Organization (WMO), the Intergovernmental Oceanographic Commission (IOC) of the United Nations Educational, Scientific and Cultural Organization (UNESCO). They may also receive input from entities such as the Data Buoy Cooperation Panel (DBCP) of the WMO-IOC, or the Ocean Observations Panel for Climate (OOPC), which is sponsored by the WMO's Global Climate Observing System (GCOS), the Global Ocean Observing System (GOOS), and the World Climate Research Program (WCRP). These organizations set guidelines and best practices for making systematic field measurements of essential ocean and climate variables (parameters) through the GOOS and GCOS frameworks (e.g., Bojinski et al., 2014; Moltmann et al., 2019). Sustained observing systems with a more regional or coastal focus, as well as short-term field programs that are focused on a specific process study, may be managed by a single research group or a consortium of research groups within a specific country (e.g., Bailey et al., 2019). It is important to note that most marine observations are only useful for satellite calibration or validation when they are collected at least half of the satellite footprint away from land or ice. For microwave observations that typically have footprints of 25 km for wind and temperature and 40 km for salinity, many coastal buoys are unfortunately too close to the shore to be useful for satellite cal/val.

Observational networks require a significant amount of fieldwork in the form of regular and repeated buoy deployments. In the case of moorings, frequent recoveries and redeployments from research vessels and ships of opportunity are required to keep the instruments healthy and ensure a high-quality data stream (see Chapter 11). For observing systems that are basin-wide or global in scope, this can require international collaborations, memorandums of understanding (MOUs) which articulate the partner countries' roles and responsibilities, clearances to deploy instruments in a country's Exclusive Economic Zone (EEZ), and policies for what happens when floating oceanic buoys transit between EEZs and international waters (i.e., who takes custody of the equipment). EEZ boundaries extend up to 200 nautical miles from the territorial coast of a country.

The ocean is a harsh environment and after months to years deployed in the field, instruments can experience sensor drift, data loss, and/or damage due to battery failure; biofouling; deposition of dust and salt; water intrusion and corrosion; underwater implosion due to a collapse of the pressure casing; loss of connection to the drogue or the anchor (i.e., become untethered), wind and wave damage; vandalism; piracy; and, even accidental damages during the deployment and recovery process. Instrument loss can also occur when portions of the moorings fail to release from the seafloor, drifters and floats run aground, or the buoy experiences a communication or beacon failure which makes recovery in the middle of the ocean nearly impossible. While some of these occurrences are unusual and cannot be predicted or prevented, regular maintenance can minimize the impact of the more routine wear and tear that occurs in the field (e.g., battery failure, biofouling, dust deposition).

Sensors deployed on moorings can start to drift from their factory or pre-cruise calibrations, and sensors need to be routinely recovered and calibrated in the field or in the lab after the cruise, and are sometimes sent back to the manufacturer for calibration and refurbishment. Data analysts look for unexpected spikes and outliers in the time series data to ensure that the data are of the highest accuracy possible. For determining the accuracy of the mooring sensors, it is crucial to obtain concurrent and collocated in situ surface atmospheric and oceanic data, as well as vertical profile measurements (e.g., conductivity temperature depth "CTD" casts) from the ocean surface to a predetermined depth or to the sea floor. These measurements are obtained from the ship used to deploy and recover the instruments (see Chapter 11). Vertical CTD profiles obtained from the ship are often collected from dual sets of sensors (i.e., two of each type of sensor are used for cross-calibration of the CTD profile data). Additionally, sensor data is cross-calibrated by collecting water samples during each cast to verify that measurements conform to World Ocean Circulation Experiment (WOCE) accuracy standards (e.g., Hood et al., 2010). These samples are processed in research labs on the ship. When available, mooring data is also compared with data from prior mooring deployments at the same location to help ensure that there are no biases, spurious signals, or data spikes that deviate too far from climatological values. For drifters and other floating buoys, data analysts can verify that the measurements are compatible with nearby contemporaneous observations and that they are reasonable given historical measurements collected in the current vicinity of the buoy. Ocean buoys provide raw data at some fixed temporal increment (hourly, daily, etc.) in real-time. Data centers apply automated quality control procedures to the data and provide that data to the public with a short-time delay (near-real time). Higher frequency data (seconds, minutes, etc.) are routinely stored internally in the instruments and this crucial data are recovered during regular servicing cruises. Substantial work is then performed by data analysts to produce delayed-mode versions of the data sets, which are of higher quality and may be available with finer temporal resolution than is provided in real time.

2.1 Surface drifting buoys

Satellite-tracked surface drifters are buoys designed to move with ocean currents while they collect various upper ocean and air-sea boundary measurements. For a review of different drifter designs, see Lumpkin et al. (2016a). One of the most widely used drifter designs was standardized during the WOCE Surface Velocity Program (Lumpkin and Pazos, 2007), and are used in the global drifter array of the Global Drifter Program (GDP). This design includes a sea anchor, or drogue, centered at a depth of 15 m to follow mixed layer currents, attached by a tether to a surface float containing the electronics, sensors, batteries, and the transmitter to relay data to a data center via satellites (Fig. 1). Drifter measurements of velocity are derived by differentiation of their location information and indicate currents at a depth of 15 m while the drogue remains attached. Once the drogue is lost, the slip of the drifter with respect to currents at 15 m depth increases from ~0.1% of the wind speed to 1 to 2% of the wind speed (e.g., Laurindo et al., 2017).

Before 2017, the majority of GDP drifters transmitted data using the Argos satellite tracking system, which also provided location estimates with a mean accuracy of 400–450 m (Elipot et al., 2016). Early in the GDP's history, drifter locations were provided a few times per day, once every 3 days, a sampling scheme abandoned by the year 2000 in favor of continuous sampling (Elipot and Lumpkin, 2008). The number of location fixes per day then increased steadily as the size of the Argos constellation grew, with a significant increase in fixes starting in January 2005 when tracking switched from two satellites to a constellation of five or six satellites, yielding locations spaced 1 to 2 h apart (Elipot and Lumpkin, 2008). Since 2017, a majority of GDP drifters (>98% since mid-2020) acquire location via GPS with an accuracy of a few meters and transmit data using the Iridium constellation of satellites providing more frequent data transmissions. Modern iridium drifters transmit their GPS locations every hour, on the hour. As a consequence, errors in velocity estimates have generally reduced with time (for details, see Elipot et al., 2016); the median error in hourly velocity for a modern GPS drogued drifter is $1.4\,\mathrm{cm\,s^{-1}}$. For comparison, the GOOS spatial accuracy requirements for surface velocity observations were set in 1999 at $2\,\mathrm{cm\,s^{-1}}$ per 5° longitude × 5° latitude bin per month (Lumpkin et al., 2016b). Additional information about drifter data accuracy and access to real-time data can be obtained at https://www.aoml.noaa.gov/global-drifter-program/.

FIG. 1 Schematic of a global drifter program drifting buoy including drogue attached by a tether to a surface float. *(Created by Bertrand Dano (UM/CIMAS).)*

Maintaining the global array of drifters at approximately 5° longitude × 5° latitude resolution across the global ocean excluding the polar regions requires an array of ~1300 evenly distributed drifters (Fig. 2). Each drifter has a designed or optimal half-life of ~450 days (not counting externalities such as running aground or being picked up by boaters; Lumpkin et al., 2016b). The actual lifetime of a drifter may depend on manufacturer design, and the health and size of the drifter array is closely monitored by the GDP. Because drifters by definition will drift away from where they are deployed, regions need to be regularly reseeded in order to maintain the desired coverage. In practice, approximately 100 deployments per month

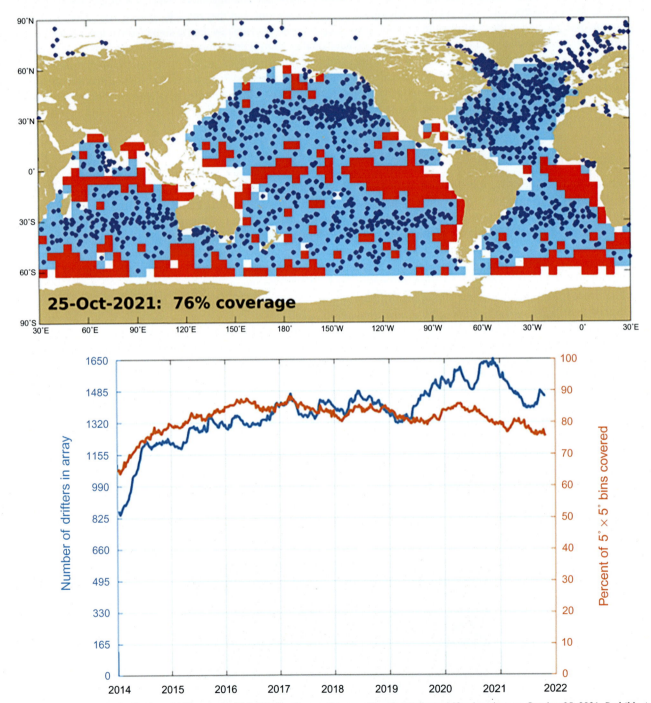

FIG. 2 *Upper panel*: Distribution of drifters in the Global Drifter Program's array. *Blue dots* indicate drifter locations on October 25, 2021. *Red (blue)* shading indicates open-ocean 5° longitude × 5° latitude bins without (with) a drifter. *Lower panel*: Number of drifters in the array *(blue)* and percent coverage of open-ocean 5° longitude × 5° latitude bins *(red)* as a function of time starting in the year 2014. *(Adapted from Lumpkin, R., Centurioni, L., Perez, R.C., 2016b. Fulfilling observing system implementation requirements with the global drifter array. J. Atmos. Ocean. Technol. 33, 685–695, https://doi.org/10.1175/JTECH-D-15-0255.1.)*

are needed worldwide, with deployments concentrated in regions such as the equator where it is hard to maintain a high density of drifters because of prevailing currents (Fig. 2; Lumpkin et al., 2016b).

All GDP drifters measure sea surface temperature (SST), as described in Section 3. Approximately half are standard barometric drifters that measure sea-level pressure (SLP; Section 7). As with any oceanographic research platform, efforts are always underway to innovate and maximize the benefit obtained from the GDP array through the addition of new sensors to expand the suite of observations collected. These added instruments can dramatically increase the expense of each individual buoy, and hence are not always possible to implement across the entire array. Here we briefly describe some of specialized drifters that have been developed for specific satellite cal/val missions. The Salinity Process in the Upper-ocean Regional Study (SPURS; Lindstrom et al., 2015; Lindstrom et al., 2019) and Elucidating the Role of Clouds-Circulation Coupling in Climate (EUREC^{4}A; Stevens et al., 2021) research campaigns included a considerable focus on salinity measurements in the tropical oceans (see Section 4). For these campaigns, a large number of drifters were deployed with additional sensors at one or more depth, including drifters measuring salinity and temperature at three upper-ocean depths during Atlantic Tradewind Ocean-Atmosphere Mesoscale Interaction Campaign (ATOMIC; Quinn et al., 2021), the US component of EUREC^{4}A, to measure barrier layer formation and evolution. In addition to the measurements of a standard barometric drifter mentioned above, hurricane drifters measure wind speed and direction using a sonic anemometer and can have the 15 m drogue replaced with a thermistor chain that extends to 15 m depth. Directional wave spectrum drifters designed at the Lagrangian Drifter Laboratory of the Scripps Institution of Oceanography (https://gdp.ucsd.edu/ldl/dwsd/) are a growing component of the GDP array, and measure the temporal evolution of the two-dimensional (2D) wave spectra following internationally-agreed upon standards. Hurricane and wave drifters can be air-deployed ahead of or within tropical storms and hurricanes from NOAA research and Air Force operational aircraft for improved hurricane and storm surge forecasts.

2.2 Moored buoys

In this section, we only consider measurements obtained from anchored buoys which have a surface float that measure variables at the ocean surface (as well as subsurface) and in the overlying atmosphere (Fig. 3). Sensors are typically attached at fixed depths to a cable or nylon rope connecting the surface float and the anchor. Extensive bottom surveys

FIG. 3 Schematic of a global tropical moored buoy including surface float with meteorological sensors. Oceanographic instruments are attached to the surface float and to a cable or nylon rope that descends from the surface float to the anchor. *(Schematic created by Bertrand Dano (UM/CIMAS).)*

using ship mounted echosounders are conducted prior to choosing a mooring deployment site to find flat terrain where the depth of the mooring can be well constrained (see Chapter 11 for mooring deployment details). Flat terrain is also important for ensuring that the depth of the mooring remains somewhat similar over successive deployments. Although the anchor is fixed and the buoy floats on the ocean surface, there is typically some slack or stretch in the nylon rope. As a result, the float and attached instruments can move in response to the currents and waves in "watch circle" (the diameter of the watch circle depends on the mooring design) which creates a temporary tilt in the nylon rope (i.e., it may not be perfectly vertical). Surface buoys typically collect and report latitude and longitude position data, so that their horizontal displacements can be tracked. Hydrostatic pressure is an oceanic parameter typically measured at a few subsurface levels along the mooring line to infer any vertical movement associated with the tilted nylon rope or wave/current motion. For some measurements, the time series of surface observations collected by moorings can be treated as fixed point measurements regardless of buoy motion. But, for other observations, the horizontal and vertical movement of the sensors is important metadata that must be recorded.

We will focus on a few different examples of moored buoy networks that provide observations that are useful for satellite cal/val or other satellite-related applications. These include OceanSITES flux reference buoys (Fig. 4; http://www.oceansites.org/), which make all the observations needed to calculate stress and radiative fluxes and turbulent heat fluxes via bulk formulas; buoys from regional/national arrays such as the National Data Buoy Center or NDBC buoys in the United States (https://www.ndbc.noaa.gov/), and the internationally maintained Global Tropical Moored Buoy Array (GTMBA; https://www.pmel.noaa.gov/gtmba/). The GTMBA moorings (Fig. 5) are deliberately located in regions in the Pacific (TAO; McPhaden et al., 1998), Indian (RAMA; McPhaden et al., 2009), and Atlantic (PIRATA; Bourlès et al., 2008, 2019) oceans that are otherwise sparsely sampled. GTMBA mooring observations can be used to categorize and understand coupled climate variations such as the well-known El Niño-Southern Oscillation. Many of the GTMBA moorings also serve as flux reference sites (Figs. 4 and 5). Although there are many key regions for which in situ data are unavailable for satellite measurement validation, moored networks such as the GTMBA serve admirably to address issues such as the impacts of rain contamination in the satellite observations in the Intertropical Convergence Zones (where data retrievals cannot be

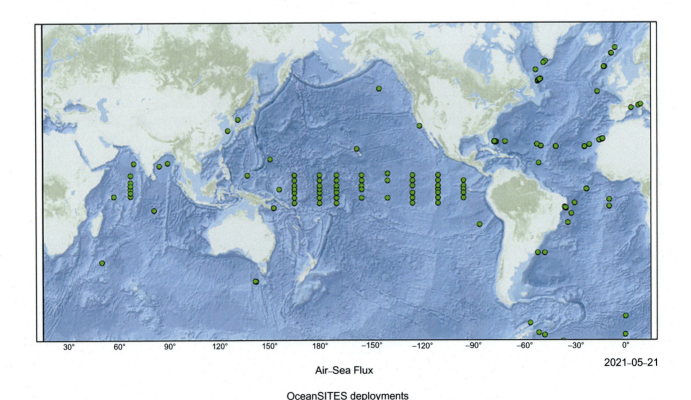

FIG. 4 Subset of OceanSITES flux reference moorings that collect the surface observations (selection criteria: a minimal sensor height of 0 m) needed to calculate stress and radiative fluxes and turbulent heat fluxes via bulk formulas and are active as of May 21, 2021. *(Image courtesy of Jiang Long and generated using the OceanOPS viewer (http://ocean-ops.org).)*

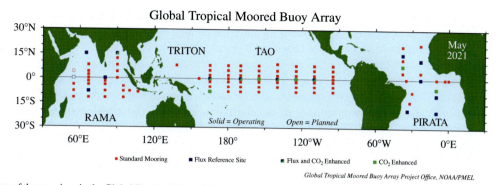

FIG. 5 Locations of the moorings in the Global Tropical Moored Buoy Array (GTMBA) as of May 2021. *Red squares* indicate standard moorings which sample basic meteorological and ocean variables, *blue squares* indicate flux reference sites, *green squares* indicate CO$_2$ enhanced sites, and a *blue/green square* indicates a combined flux reference and CO$_2$ enhanced site. *(Image obtained from https://www.pmel.noaa.gov/gtmba/.)*

made when it is raining or clouds are in the field of view of the satellite instrument) in all three ocean basins. Similarly, observations from arrays in extratropical latitudes such as the NDBC network are needed to cover the parameter space of marine observations (particularly temperature, humidity, wind speed, and wave state measurements) closer to the coastlines. A specialized type of mooring called the Marine Optical BuoY (MOBY) provides accurate surface determinations of water-leaving radiance that are used to calibrate ocean color satellite measurements in the clear oligotrophic waters near Hawaii (Wang and Franz, 2000; Clark et al., 2003). The unique features of this mooring will be discussed in Section 8.

It is important for researchers to keep track of how the sampling methodologies of long-term moored networks evolve so as to not introduce measurement biases. New instrumentation may become available that alter the mooring design for reasons of efficiency, accuracy, or cost. Sometimes a mooring location has to change for a variety of logistical reasons (i.e., accessibility, damage from fishing activities, piracy, etc.), and this can cause a change in the water depth, water mass properties, currents, and sea state observed by the mooring. These changes must be carefully introduced, and the details must be recorded as metadata and maintained or the quality of the dataset may diminish. For example, observations from the flux reference buoys are easier for the broader community to use if they are height adjusted to a standard height or converted to equivalent neutral winds (i.e., wind speed for which neutral atmospheric stratification would produce a stress consistent with the observations). The metadata must address these reference heights and any conversions. Although there is very little sensitivity to the choice of flux algorithm used to make these adjustments, so long as the adjustment considers atmospheric stability and the feedback of this stability on stress and surface roughness (Kara et al., 2008), complications occur when the engineering specifications of a buoy (i.e., changes in observation height) vary over the lifetime of the array. To verify that a spurious signal will not be introduced into the long-term time series, arrays often deploy next-generation instrumentation or moorings side-by-side with their older generation counterparts during a prototype testing stage.

3. Sea surface temperature

The temperature of the earth's surface controls how hospitable our planet is for life through its influence on our weather, climate, food abundance, and health. Since the oceans cover approximately 70% of earth's surface and approximately 90% of excess anthropogenic warming is being absorbed by the oceans (e.g., Rhein et al., 2014; Meyssignac et al., 2019), ocean temperature is arguably the most important parameter to observe regularly and accurately. Not surprisingly, SST is one of the longest observed parameters in the ocean both in situ and from space measurements (e.g., Kilpatrick et al., 2001; Reynolds et al., 2002; Zhang et al., 2009; Durack et al., 2018; Minnett et al., 2019; O'Carroll et al., 2019), with earliest satellite SST measurements being available in the 1970s and used to study large horizontal temperature gradients (i.e., fronts) associated with the Gulf Stream (e.g., Legeckis, 1975) and equatorial waves in the tropical Pacific (e.g., Legeckis, 1977). Satellites have provided some of the first images of the spatial complexity of surface temperature fronts that can be punctuated by eddies, rings, meanders, loops, or filaments. Alone, however, these beautiful remotely sensed images would not be useful for science without in situ observations to help characterize their accuracy. From satellite and in situ measurements, we know that seawater temperatures can range from −1.8 degrees Celsius (note: seawater freezes at −1.8°C) at high latitudes near the poles to a few degrees above 30°C in the tropical oceans and can exhibit a diurnal temperature range (SST day - SST night) of 3–4°C in some tropical (e.g., Fairall et al., 1996a; Soloviev and Lukas, 1997) and subtropical regions such as the Sargasso Sea (Stramma et al., 1986). SST diurnal variations can even exceed 6°C in some extreme instances (e.g., Flament et al., 1994; Gentemann et al., 2008). SSTs can also fluctuate on interannual, decadal, multidecadal, and longer timescales that could exceed the duration of any single multiyear satellite mission. Although, some science satellite missions have successfully lasted for a decade or even two decades.

The accuracy and spatial and temporal coverage of data obtained from satellites has changed over the years and decades, and as a result the error statistics of remotely sensed SST are non-stationary. The orbit of satellites can drift during their lifetime, which can cause the local overpass time to be later or earlier, and can lead to spurious trends in SST or a misinterpretation of long-term changes in the amplitude of the diurnal cycle of SST (Stuart-Menteth et al., 2003). The errors and uncertainties in the satellite-derived SSTs are also dependent on atmospheric conditions that can change rapidly and over short distances, which is why atmospheric correction algorithms are applied on a pixel-by-pixel basis. As technology advances and correction algorithms are refined, the quality and sampling characteristics of satellite measurements can improve from one satellite mission to the next (see review article by O'Carroll et al., 2019). Without timely follow-on missions, temporal gaps in the satellite record could occur making it difficult to cross-validate old and new satellite missions. Conversely, there may be several satellites orbiting the earth at the same time, each providing SST information with differing spatial coverage, resolution, and accuracy that could be combined together to create a multisatellite SST product. Thus, it is important to have well-calibrated and validated measurements from each individual satellite mission so that they can be merged to generate long-term, reliable, stable SST records that can be used for scientific research (e.g., Smith, 2000; Kilpatrick et al., 2001, 2015). This in turn requires accurate long-term in situ SST records (sometimes referred to as climate data records) from as many parts of the global ocean as possible (e.g., Bojinski et al., 2014; Centurioni et al., 2019; O'Carroll et al., 2019), as well as documentation in the form of metadata detailing any significant changes to the uncertainty associated with those measurements.

Depending on the type of radiometers on the satellite (i.e., infrared or microwave), some barriers to accurately observing SST from space include the contamination of measurements by the effects of aerosols (e.g., Saharan dust), clouds, rainfall, or land masses (e.g., Kilpatrick et al., 2001; Merchant et al., 2005; Zhang et al., 2009; Minnett et al., 2019; Luo et al., 2021). Sparse SST field measurements in regions of aerosol, cloud, rainfall, or land contamination complicate satellite cal/val efforts. Before the mid-2000s, when drifters began to routinely provide continuous high-frequency sampling with global data coverage, in situ SST measurements were primarily collected in the Northern hemisphere from ships and/or moorings (e.g., Zhang et al., 2009; O'Carroll et al., 2019; Wong et al., 2020). Moorings were strategically deployed to provide observations in otherwise undersampled regions such as the ITCZ where rain contamination was an issue for accurate retrieval of geophysical data from satellites. Observing system design studies have also been conducted to optimize the observing system coverage (e.g., number of buoys, spacing between buoys, etc.), minimize satellite bias errors, and generate accurate satellite-in situ SST-blended products (see Chapter 19; Zhang et al., 2009). Without making use of in situ observations, globally averaged satellite SST biases could be as large as 2°C, but with at least 2 buoys in a 10° longitude × 10° latitude area, average SST biases decrease to 0.5°C (Zhang et al., 2009). Operationally, drifter deployments have been optimized by routinely repeating the analyses of the SST in situ coverage for individual 10° longitude × 10° latitude grid boxes following the methods similar to Zhang et al. (2009). The performance of the SST in situ network can similarly be studied over time in the context of satellite cal/val. For example, as the number of drifters in the global ocean increases, satellite in situ bias errors have decreased over time (c.f. Fig. 6 in Zhang et al., 2009).

The disparate depths at which measurements of the "surface" temperature are taken from buoys have also complicated satellite in situ comparisons in the past (e.g., Donlon et al., 2002). Drifters measure bulk SST via a thermistor (i.e., a semiconductor that registers a resistance change in response to changing temperatures) on the bottom half of the surface float, located approximately 20 cm below the sea surface. Most moored oceanographic buoys measure a near-surface temperature in the upper 1 to 10 m of the water column rather than the actual temperature at the thin (molecular) layer between the air and sea. Commonly used definitions of surface or near-surface temperatures have been established, which include the air-sea interface SST, skin SST less than 100 μm below the air-sea interface and which is traditionally measured with an infrared radiometer (see Chapters 6 and 11), subskin SST about 1 mm below the air-sea interface, subsurface or "bulk" SST

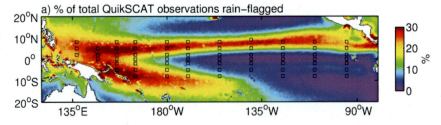

FIG. 6 Fraction of time that QuikSCAT overpasses were flagged as rain contaminated. This subset of observations is usually typically not used for satellite calibration, except for simultaneous retrievals of wind and rain. *(Graphic courtesy of Larry O'Neill (Oregon State University).)*

traditionally up to 5 m below the sea surface, and the foundation SST which is approximately 10 m below the sea surface and is considered to be free of diurnal SST variability (e.g., Donlon et al., 2002). Skin SSTs are almost always colder than the subskin temperatures by a fraction of a degree Celsius (on average by about −0.17°C), and the skin SSTs are only warmer than bulk and foundation SSTs if there is strong diurnal heating (e.g., Donlon et al., 2002; Minnett et al., 2019).

Pseudo-bulk SST measurements from satellites such as those obtained by the advanced very high resolution radiometer (AVHRR) are typically adjusted to be a representation of the temperature at a depth comparable to that of buoy measurements. As a result, these calibrated AVHRR pseudobulk SST measurements do indeed have low biases when compared to buoy temperatures. However, pseudobulk SSTs still exhibit temporal variability akin to skin SST variability as evidenced by small root mean square errors when they are compared against skin Marine-Atmospheric Emitted Radiance Radiometer (M-AERI) SST measurements (e.g., Kearns et al., 2000; Kilpatrick et al., 2001). Moreover, these pseudo-bulk SST corrections use algorithms derived by comparisons between the satellite brightness temperature measurements and collocated buoy data. The coefficients of the algorithm are estimated for a monthly time period, and the same coefficients are applied to both day and nighttime data, and hence the correction algorithms do not take into account potential diurnal variations in those coefficients (e.g., Stuart-Menteth et al., 2003).

Ocean buoy temperature measurements have themselves become more accurate over time, thanks to the efforts of groups of researchers, instrument manufacturers, and operational users working together to improve the accuracy of SST fields for a variety of applications. Examples of these collective efforts include the Group for High Resolution SST (GHRSST; Donlon et al., 2009; http://www.ghrsst.org) and the in situ SST Quality Monitor (iQuam; Xu and Ignatov, 2014; http://www.star.nesdis.noaa.gov/sod/sst/iquam/). GHRSST is trying to meet the SST requirements set forth by the Global Ocean Data Assimilation Experiment (GODAE) requiring global in situ SST coverage with horizontal resolution of 10 km or finer in 6 h increments with an accuracy of at least 0.2°C (e.g., Smith and Koblinsky, 2001; Bell et al., 2009; O'Carroll et al., 2019). In recent years, driven by input from GHRSST, drifter thermistors now typically provide SST at an accuracy of 0.05°C (previously they had an accuracy of 0.1°C), and their accuracy is being evaluated by comparison with high-resolution SST drifters (Poli et al., 2019; Le Menn et al., 2019). Note, because drifter measurements are hourly and follow near-surface currents, they can resolve SST diurnal and higher frequency fluctuations such as rapid frontal changes that satellites still cannot resolve.

NDBC maintains a network of over 100 coastal moored buoys which typically measure temperature with probes attached to the inside of the buoy hull (NDBC, 2009). As a result, the surface temperatures observed by NDBC moorings may more closely reflect the temperature of the water around the hull because of the high thermal conductivity of the hull. The NDBC Coastal Marine Automated Network (C-MAN) stations do make point measurements of water temperature, but the thermistors are at fixed distances relative to the bottom; therefore, the thermistor is at different depths throughout the day owing to changing water levels (NDBC, 2009). The reported water temperature accuracy of NDBC thermistors is 0.08°C which is below the WMO requirement of 0.1°C. GTMBA buoys typically measure temperature at 1 m (subsurface or bulk SST) and 10 m (foundation SST) depth, with hourly or daily averages available in real-time (and 10-min data stored internally) with an accuracy ranging from values on the order of ±0.002°C to ±0.02°C depending on which type/generation of GTMBA mooring and thermistors are being considered (e.g., A'Hearn et al., 2002; Freitag et al., 2005; https://www.pmel.noaa.gov/gtmba/sensor-specifications).

Each in situ observing network performs real-time and delayed-mode processing of the SST data. For example, for the GTMBA network, this process starts by detecting outliers (SSTs that are >33°C or <9°C) or spikes (excessively large hourly or daily SST changes). Visual inspection is also performed on 5-day running mean SST and wind vector time series to identify spurious signals (https://www.pmel.noaa.gov/gtmba/data-quality-control). On a weekly basis, SSTs are compared against typical values for that week of the year (i.e., climatology) detecting unusual weekly averages and fluctuations. Lastly, quality control can happen at the research stage by scientists and data analysts when data records are examined in the context of longer-term records and independent data sets (e.g., Foltz et al., 2018), often months to years after the data has been collected.

There are also larger community efforts, notably those led by iQuam that examine the quality of in situ SST collected by a variety of networks relative to gridded reference SST products. One example of a gridded reference SST product is the NOAA Optimum Interpolation Sea Surface Temperature (or daily OISST), which is available daily at 1/4° resolution (e.g., Reynolds and Smith, 1994; Reynolds et al., 2002, Reynolds et al., 2007; Banzon et al., 2020; https://www.ncdc.noaa.gov/oisst). OISST is produced by combining bias adjusted estimates of satellite, ship, buoy, and Argo float SST data and is used for a wide range of science and societal applications.

4. Sea surface salinity

Sea surface salinity (SSS) varies widely over the global ocean, from nearly zero close to river mouths to more than 38 parts per thousand in the saltiest area of the subtropical North Atlantic. Salinity, together with temperature and pressure, determines the density of seawater, and surface salinity is a good indicator of changes in Earth's hydrological cycle because it responds directly to variations in precipitation and evaporation. Satellites have been measuring SSS since the 2009 launch of the Soil Moisture and Ocean Salinity (SMOS) mission, which is still active. In addition, Aquarius/SAC-D (Satélite de Aplicaciones Científicas-D) measured global SSS between 2011 and 2015, and soil moisture active passive (SMAP) mission has been providing SSS data since 2015. The satellite sensors detect microwave radiation that is naturally emitted from the sea surface. There is a small but detectable dependence of the intensity of the radiation on SSS, and the dependence is stronger for higher SST, allowing more accurate SSS retrievals in the tropics than in high latitudes. Numerous studies have been performed to assess the accuracy of satellite SSS retrievals and modify retrieval algorithms, and many are underway (e.g., Vinogradova et al., 2019; Reul et al., 2020). In situ measurements of near-surface salinity are crucial for this, similar to the case for satellite SST discussed in the previous section.

Salinity is a measure of how much matter is dissolved in water and is defined as a mass ratio or fraction and has no units. It is typically determined in situ through measurement of seawater conductivity, temperature, and pressure. Temperature and pressure must be considered because for a given concentration of dissolved matter ("salt"), the water's conductivity depends on both. Conversion of temperature, conductivity, and pressure to salinity is complicated by the existence of many different dissolved elements in seawater, making it virtually impossible to determine the concentration of each. Each element makes a different contribution to the water's conductivity. Fortunately, the ratios of concentrations of different elements are approximately constant throughout the ocean. The most commonly applied method to estimate salinity takes advantage of this fact and makes use of the Practical Salinity Scale 1978 (PSS-78). Salinity in PSS-78 is defined based on the ratio of the conductivity of seawater at a temperature of 15°C and pressure of one standard atmosphere (760 mmHg) to the conductivity of a potassium chloride solution (KCl) at the same temperature and pressure and with a mass fraction of KCl of 32.4356×10^{-3}. The conversion of measured conductivity, temperature, and pressure to practical salinity (i.e., for values of temperature and pressure that are different than those of the reference sample) is complex and nonlinear. For a given value of conductivity, salinity is most strongly dependent on temperature. For example, for a typical value of conductivity at the surface of the ocean, increasing temperature from 20°C to 25°C will result in a decrease in measured salinity of about 4 PSS-78. However, for the same value of conductivity and a temperature of 22.5°C, increasing pressure from 1 to 100 dbar (corresponding to depths of approximately 1 m and 100 m), salinity decreases by only about 0.03 PSS-78. A newer and more comprehensive definition of salinity, termed Thermodynamic Equation of Seawater 2010 (TEOS-10; Pawlowicz et al., 2012), is starting to be used more widely in the oceanographic community. Among other improvements, it accounts for the different chemical compositions in seawater by adding a regionally dependent variable to the salinity equation.

Determination of seawater salinity is more difficult than temperature because of the added requirement of accurate conductivity measurements. Most oceanographic buoys in the open ocean are deployed for several months at a time. Biological growth on the conductivity cell eventually begins to decrease the measured conductivity, leading to a low bias in salinity that increases with time as growth accumulates. In addition, particles passing through the conductivity cell can generate "spikes" in conductivity and salinity that need to be filtered out during the quality-control process.

Salinity is not affected directly by changes in solar radiation, so diurnal variations of surface salinity are normally much smaller than those of temperature. Diurnal amplitudes of surface salinity (based on measurements at 0.4 m) are typically 0.01 PSS-78 (Anderson and Riser, 2014; Drushka et al., 2014; Dong et al., 2017; Volkov et al., 2019), compared to 0.2°C for temperature, sometimes reaching more than 2°C (Fairall et al., 1996a; Soloviev and Lukas, 1997; Flament et al., 1994; Gentemann et al., 2003, 2008). Therefore, in the absence of rain, bulk surface salinity measurements (in the upper 5–10 m) can be compared directly to satellite measurements of surface salinity, which typically represent the upper 1–2 cm. In the presence of rain, salinity in the upper meter can be as much as 2 PSS-78 fresher than salinity at a depth of 5 m (Reverdin et al., 2012; Volkov et al., 2019). In these cases, caution should be used when comparing in situ surface salinity to satellite-based measurements.

Sustained in situ near-surface salinity measurements are made primarily by Argo profiling floats (e.g., Wong et al., 2020) and moored buoys, most of which are part of the GTMBA mentioned in Section 2.2. After deployment, Argo floats move freely through the ocean. Their buoyancy is controllable, allowing them to remain in a certain depth range. Normally they are parked at a depth of about 1000 m, and every 10 days a float descends to 2000 m, and then rises to the surface while measuring temperature, conductivity, and pressure. Parking at 1000 m minimizes biological growth on the sensors so that the float's lifetime is dictated mainly by battery life, which is typically several years. The uppermost measurement is normally about 5 m below the surface, with approximately 5-m vertical resolution in the upper 100 m. At the surface, the floats transmit data to shore via satellite, and the real-time data are available for assimilation into numerical models and for scientific analysis. The accuracy of real-time Argo salinity is about 0.01 PSS-78 (Wong et al., 2020). Sometimes when a float rises through a region with a strong vertical temperature gradient, there can be a lag between the measurement of temperature and conductivity, resulting in a biases and spikes in derived salinity. These data are corrected before the delayed-mode data are released. As of April 2021, there were about 3800 Argo floats in the global ocean. The array is maintained through regular deployments from various research and commercial vessels (see Chapter 11).

The other main source of real time sustained near-surface salinity observations is the GTMBA (McPhaden et al., 1998, 2009; Bourlès et al., 2008, 2019), which extends globally in the tropics (Fig. 5). In April 2021, there were 67 moorings in the Pacific, 27 in the Indian Ocean, and 18 in the Atlantic. All moorings measure near-surface salinity at a depth of about 1 m and 10 m, and most have additional subsurface salinity measurements at other depths in the upper 100–200 m. Data are recorded internally in the buoys and are available as hourly averages after mooring recovery, typically about once per year. All data from the moorings are also transmitted in real time, either as hourly or daily averages, depending on the type of buoy. The accuracy of salinity from the moorings is about 0.02 PSS-78 (Freitag et al., 1999). The near-surface sensors are more prone to biological growth than the sensor on Argo floats because they are continuously in the high-light, high-nutrient mixed layer (Foltz et al., 2018). In addition to the GTMBA, there are several other OceanSITES moorings that provide real-time near-surface salinity data, such as the Kuroshio Extension Observatory (Cronin et al., 2005), Ocean Station Papa (Kamphaus et al., 2008), Northwest Tropical Atlantic Station for air-sea flux measurements at 15°N, 51°W (Bigorre and Plueddemann, 2021), and Stratus at 22°S, 85°W (Fig. 4).

Shorter-duration process-oriented studies are also a valuable source of upper-ocean in-situ salinity data. Process studies normally take place in one specific area for a limited duration, typically 1 year or less. Intensive measurements are acquired from numerous different platforms. Examples include the Salinity Process in the Upper-ocean Regional Study (SPURS; Lindstrom et al., 2015), SPURS-2 (Lindstrom et al., 2019), Elucidating the Role of Clouds-Circulation Coupling in Climate (EUREC^{4}A; Stevens et al., 2021), Atlantic Tradewind Ocean-Atmosphere Mesoscale Interaction Campaign (ATOMIC; Quinn et al., 2021), and the Air Sea Interaction Regional Initiative (ASIRI) in the Bay of Bengal (Mahadevan et al., 2016). These studies included near-surface salinity measurements from moored buoys, surface drifters, profiling floats, and/or other autonomous ocean systems. The near-surface salinity data have been useful for validating and calibrating satellite surface salinity retrievals (Bingham, 2019; Reverdin et al., 2021).

5. Wind speed and direction

Buoys typically measure wind speeds with anemometers (e.g., propeller anemometers for NDBC buoys and sonic anemometers on the GTMBA). Both approaches must account for the motion of the buoy and measurements must be averaged over sufficient time to obtain a stable mean. For very low wind speeds this time could be 20–60 min, which might exceed the sampling interval of the buoy observations. At high wind speeds, sampling faster than 10 min would be preferred but such high-frequency measurements are rarely available. Wind directions are typically determined from a wind vane and a magnetic compass. Anemometer heights on buoys have been trending downward with time. There are now very few buoys with anemometers at a height of 10 m, and many more buoys with anemometers at 3 or 4 m above sea level. Since wind speeds near the surface (often heights lower than 60 to 100 m above the surface) vary logarithmically with height, it is important to account for anemometer height and adjust to a common standard, which is typically 10 m (World Meteorological Organization, 1981). This is particularly true for extreme weather-related risk analysis, where not accounting for biases associated with this change combined with a downward trend in NDBC anemometer heights could cause a spurious downward trend in the return period (the average time between events) of high winds.

Winds derived from satellite measurements are calibrated to equivalent neutral winds (Ross et al., 1985; Liu and Tang, 1996) at a height of 10 m above the displacement height (the height at which the log-profile extrapolates to zero wind speed; Stull, 1988), where the displacement height is assumed to be negligibly different from the local mean sea surface height. This approach is applied because active satellite sensors (i.e., radars) respond to roughness on the scale of the wavelength of capillary waves or very short gravity waves (the specific wavelength depends on the radar wavelength, incidence angle of the radar, and arguably on the slope of swell) and passive satellite sensors respond to excess emissivity (i.e., excess brightness beyond what would be expected for a smooth water surface due to foam and whitecaps), both of which respond more directly to wind stress than to wind velocity. In concept, the roughness length and friction velocity are determined for non-neutral conditions, and then used to adjust the wind speeds to a height of 10 m where the stability term is set to zero (wind direction does not change with height in this log-layer). This approach has the advantage that the winds can be converted accurately to a stress using a neutral drag coefficient. A further consideration is that buoy winds are Earth-relative, whereas satellite winds are current relative (Cornillon and Park, 2001; Kelly et al., 2001; Plagge et al., 2012). Both types of satellite sensors are also sensitive to wind direction relative to the azimuthal and zenith look directions, thus winds can be retrieved more accurately when direction is solved simultaneously with speed. In other words, satellite measurements are simultaneously dependent on speed and direction, hence wind vectors are retrieved, in contrast to buoys which can measure speed independently from direction. NOAA and Air Force airborne observations bypass observational directional complications for hurricane applications by observing straight down. This makes retrievals much easier to calibrate, but provides only a single track of observations below the aircraft, whereas most satellites observe over a combination of angles to provide a wide swath of observations.

There are a few ways through which winds derived from satellite measurements are sensitive to air and sea surface temperatures, which is why collocated wind, SST, and air temperature buoy measurements are desired for more precise cal/val of satellite winds. Surface stress (Section 6) is proportional to air density, consequently the wind signal is roughly proportional to the square root of air density. For example, the globally averaged air density is roughly $1.2 \, kg \, m^{-3}$, whereas air density in the tropics is typically closer to $1.1 \, kg \, m^{-3}$, causing an underestimation of wind speed unless this is accounted for (Bourassa et al., 2010). Similarly, in very cold conditions, the air density can be as large as $1.4 \, kg \, m^{-3}$ resulting in a substantial over-estimate of satellite wind speeds. For active systems that respond to roughness related to short water waves, the impact of water temperature is greatest for very cold temperatures because it is hypothesized that cold temperatures inhibit the formation of ultra-gravity waves (Wang et al., 2017). Passive systems are also highly sensitive to SST because the emissivity of the water is a strong function of SST, and wind is retrieved based on a brightness relative to the brightness expected at the SST. Passive instruments have some dependency on air temperature because the emission and absorption of the atmosphere between the surface and the sensor must also be considered, as well as the reflection of downwelling radiation off the sea surface. Note, microwave radiometers take measurements at multiple frequencies with different sensitivities to SST and wind, so that SST and wind are derived simultaneously.

Wind speeds from buoys are considered relatively accurate, with uncertainties on the order of a few tenths of a $m \, s^{-1}$. There are suggestions of small systematic errors associated with flow distortion (Schlundt et al., 2020) and a dependency on sea state (Wright et al., 2021). However, these considerations are smaller than random errors due to differences in collocation of in situ and satellite measurements in space and time—i.e., when collocation differences exceed 5 min or 5 km based on wind variability observed from research vessels (May and Bourassa, 2011). Most calibration efforts avoid comparison data that are contaminated by rain because an insufficient fraction or quantity of uncontaminated data adversely impacts the retrieval. The threshold rain rate for serious impacts increases as the wind speed increases (Draper and Long, 2004), making observations from the GTMBA more challenging to use in calibration, but also providing a wealth of comparison data for calibrations that attempt to simultaneously retrieve wind and rain (Fig. 6).

6. Surface heat fluxes

The exchange of heat flux between the ocean surface and the overlying atmosphere is composed of a combination of radiative and turbulent heat fluxes. Radiative heat fluxes near the surface are typically split into four components: upwelling and downwelling electromagnetic observations and longwave (earth temperature) and shortwave (solar) observations. NDBC buoys measure accumulated downwelling solar radiation ($W \, m^{-2}$) with a resolution of $0.5 \, W \, m^{-2}$ (i.e., the mean global climate change signal) and an accuracy of 5%. Sampling is at 1 Hz, with an accumulation interval of 8 min. NDBC buoys do not measure longwave radiation. The GTMBA measures downwelling shortwave (resolution of $0.4 \, W \, m^{-2}$ and pre-deployment accuracy of 2%) and downwelling longwave ($0.1 \, W \, m^{-2}$ resolution and 1% accuracy). Accumulation of materials on the sensors inhibits transmission of the electromagnetic radiation (e.g., Foltz et al., 2018), artificially increasing the apparent optical thickness of the atmosphere.

Turbulent heat fluxes are not directly measured on operational buoys; however, they can be inferred through the use of bulk parameterizations (e.g., Liu et al., 1979; Fairall et al., 1996b; Bourassa et al., 1999; Edson et al., 2013). That is, they can be approximated from mean observations of wind speed (ideally relative to the surface current), SST, air temperature, humidity, and pressure. Some parameterizations (e.g., Hsu, 1974; Taylor and Yelland, 2001; Oost et al., 2002; Drennan et al., 2005; Bourassa, 2006) make use of sea state information; however, sea state is only observed on NDBC buoys and the coarse resolution of the observations (e.g., wave height is recorded to the nearest foot) makes the observations challenging to work with in this context. Consequently, sea state is not often used when estimating fluxes. Some parameterizations for stress have a large dependency on sea state (Hsu, 1974; Taylor and Yelland, 2001; Oost et al., 2002); whereas others (sometimes in comparison to the same data) find sea state to be a minor consideration (Bourassa, 2006; Edson et al., 2013). This remains a topic of considerable debate.

The three turbulent fluxes that are normally calculated are stress (τ, the flux density of horizontal momentum), sensible heat (H, the flux density of thermal energy that is transported without a phase change), and latent heat (L, the flux density of energy stored in a phase change due to transported water vapor). The moisture flux due to evaporation times the latent heat of vaporization is equal to the latent heat flux. There are two equivalent methods for using bulk formulas to estimate turbulent fluxes. One approach, the Coupled Ocean-Atmosphere Response Experiment (COARE) algorithm (e.g., Edson et al., 2013) estimates turbulent fluxes through measured air-sea differences and a drag coefficient.

$$\tau = \rho\, C_{D10} |U(z) - U_{sfc}| \left(U(z) - U_{sfc}\right) \quad (1)$$

$$H = \rho\, C_p\, C_{H10} |U(z) - U_{sfc}|(SST - \Theta(z)) \quad (2)$$

$$L = \rho\, L_v\, C_{L10} |U(z) - U_{sfc}| \left(q_{sfc} - q(z)\right) \quad (3)$$

where Θ is the time averaged potential temperature; q is the time averaged specific humidity; ρ is the density of air; C_p is the heat capacity of air at a constant pressure, and L_v is the latent heat of vaporization. However, the transfer coefficients (C_D, C_H, and C_L) are all calibrated to specific instrument heights and these heights are normally 10 m, which is well above the height of instruments on modern buoys. The second approach to estimate turbulent fluxes is to model the vertical profiles of wind velocity, potential temperature, and humidity and match the observations at the appropriate heights.

$$U(z) - U_{sfc} = \frac{u_*}{kv}\left[\ln(z/z_{oM}) - \psi_M(z,\, z_{oM},\, L)\right] \quad (4)$$

$$\Theta(z) - SST = \frac{\Theta_*}{kv}\left[\ln(z/z_{o\Theta}) - \psi_\Theta(z,\, z_{o\Theta},\, L)\right] \quad (5)$$

$$q(z) - q_{sfc}(z) = \frac{q_*}{kv}\left[\ln\left(z/z_{oq}\right) - \psi_q\left(z,\, z_{oq},\, L\right)\right] \quad (6)$$

And the fluxes are determined as

$$\tau = \rho |u_*| u_* \quad (7)$$

$$H = -\rho\, C_p |u_*| \Theta_* \quad (8)$$

$$L = -\rho\, L_v |u_*| q_* \quad (9)$$

where z is the observation height (which is usually the same for temperature and humidity, but is often different for wind speed); u_* is the friction velocity which is a scale factor for wind and is equal to the square root of the kinematic stress; q_* and Θ_* are scale factors for the moisture and potential temperature profiles, respectively; the z_o terms are the roughness lengths for momentum (wind), humidity, and potential temperature, respectively; and the ψ terms are the corresponding adjustments for boundary-layer stability. Roughness length is usually parameterized as a function of friction velocity (Liu et al., 1979; Clayson et al., 1996; Bourassa et al., 1999) and sometimes wave state although the functional dependence on wave state remains controversial. The course resolution of wave height makes it challenging to use these observations and evaluate flux parameterizations, particularly for low wind speeds where small wave heights might be rounded down to a height of zero.

Another concern is that the buoy values for the sea surface temperature are often measured 0.2 m up to a few meters below the surface by moorings (Section 3), and as with the atmosphere the behavior of temperature can change rapidly with distance from the ocean surface (Section 3). Early efforts to adjust observations to surface values demonstrated rather little skill, but recent efforts have proven much more effective (Castro et al., 2003; Gentemann et al., 2009).

The directional convention for stress is positive downward. Oceanographers treat latent and sensible heat fluxes as positive downward (appropriate for an ocean energy budget) and meteorologists treat sensible and latent heat fluxes as positive upward, which makes sense for an atmospheric energy budget. The equations above are in atmospheric convention.

Note that the same Eqs. (4)–(6) can be used to adjust the atmospheric measurements made by the buoys to a different observation height (Berry and Kent, 2011). The process of adjusting to a different height can also be used to solve for turbulent heat fluxes. The uncertainty in a height adjustment is very tiny compared to uncertainty in the observations and has very little dependence on the choice of flux algorithm provided that the roughness length adjusts to stability induced changes in friction velocity (Kara et al., 2008). However, the uncertainty in a flux can be quite large and non-Gaussian, particularly for large wind speeds and large air-sea temperature differences (Gulev et al., 2007a,b). Temporal averaging is usually needed to reduce this uncertainty.

7. Sea level pressure

The unit for SLP is a hectoPascal (hPa), which is equivalent to a millibar (mb), and the global average SLP is approximately 1013 hPa. While surface pressure is not currently measured by satellites, it is an essential climate variable and used for a variety of applications. SLP is an important parameter assimilated into numerical weather forecast models and modern (e.g., the Modern-Era Retrospective analysis for Research and Applications or the ECMWF Reanalysis 5th generation; Gelaro et al., 2017; Hersbach et al., 2020) and historical (e.g., the 20th Century Reanalysis; Slivinski et al., 2021) weather reconstructions. In the context of this chapter that focuses on in situ measurements for satellite cal/val, in situ measurements of pressure are often needed to provide atmospheric corrections to variables, or to help estimate quantities from in situ measurements that better correspond to a satellite measurement. For example, there is an inverse barometer effect (e.g., Wunsch and Stammer, 1997) whereby a long-term decrease of 100 hPa in SLP can lead to a 1 cm rise in sea level which could influence remotely sensed sea surface height.

SLP data are used to estimate other quantities for which there are not as many direct estimates in the open ocean. Scatterometer winds have a weak dependence on surface pressure (Bourassa et al., 2010). Horizontal SLP gradients can provide information about near-surface winds, and the surface turbulent fluxes described in the previous section are pressure dependent through the air density used in Eqs. (7)–(9), and calculations of near-surface moisture are dependent on pressure. Conversely, horizontal pressure gradients can be estimated from satellite observations of winds, and SLP measurements from buoys can be used to validate these pressure gradients (Patoux et al., 2003; Patoux and Foster, 2012).

Standard barometric drifters measure SLP at an accuracy of 1 hPa, with less than 1 hPa/year drift specified for the barometer (Sybrandy et al., 2009). The barometer port is located at the top of the surface float and is designed to reduce wind-induced Bernouilli effect pressure offsets and to prevent water from entering the internal pressure sensor, so that it can withstand frequent immersion with no loss of accuracy (Sybrandy et al., 2009). Currently, ~65% of the global drifter array is outfitted with barometers; deployments of these instruments are concentrated at high latitudes, where pressure signals are largest in magnitude. Many of the barometers are purchased by meteorological partners of the GDP via the DBCP's barometer upgrade program (see https://www.ocean-ops.org/dbcp/platforms/barometer.html). Of all components of the ocean observing system, drifter measurements of SLP are the most impactful on a per-observation basis at improving weather forecasts globally (Centurioni et al., 2017; Poli, 2018). For many years, the DBCP of the WMO-IOC has issued a standing recommendation that 100% of the global array of drifters be implemented with barometers.

About 40% of the GTMBA buoys measure SLP. The barometers are located at a height of 3 m on the buoys' towers, and the accuracy of the measurement is 0.01% (typically about 0.1 hPa). Of the measurements made by moored buoys, SLP and winds provide the largest benefit to global weather forecasts (Centurioni et al., 2017; Poli, 2018). Most of the NDBC moored buoys and C-MAN stations also measure surface pressure, typically at a height of about 2.4 m. These NDBC buoys are located in the western North Atlantic and central-eastern North Pacific, where marine weather systems (including tropical storms and hurricanes) can cause dramatic decreases in SLP.

8. Water-leaving radiance

Accurately measuring from a satellite the radiance coming out of the ocean is very difficult because approximately 90% of the top of the atmosphere (TOA) radiance originates from the atmosphere and the ocean surface. Thus, deriving the ocean signal to within 5% accuracy requires the TOA signal to be determined within ~0.5% of the total signal. To achieve this accuracy, all ocean color satellite instruments, to date, have found it necessary to use accurate surface measurements in a system vicarious calibration process. Most, if not all, of these satellite instruments, have used measurements from the Marine Optical BuoY (MOBY) (Wang and Franz, 2000; Clark et al., 2003).

8.1 Selection of location

MOBY is located about 14 km west of the island of Lanai, Hawaii, where the water depth is greater than 1000 m. Away from land, this site is free from contamination from the land adjacency effect and bottom reflectance. The clear oligotrophic waters have low chlorophyll-a concentration of ~0.07 mg m^{-3} and are relatively spatially homogeneous (Voss et al., 2010). As the mountains of the Hawaiian Islands provide a lee to this site, clear skies and low sea states are frequently encountered, which are ideal for ocean color observations. In addition, the maritime atmosphere at the MOBY site can be considered to be representative of the global open ocean.

8.2 Instruments and measurements

MOBY is tethered to a slack-line mooring buoy to prevent drifting (Fig. 7). The spar-like structure helps maintain MOBY in a vertical position to within a few degrees. MOBY collects the above-water downwelling irradiance ($E_s(\lambda)$), in-water downwelling irradiance ($E_d(\lambda)$), and upwelling radiance ($L_u(\lambda)$) from collectors mounted at the ends of the three standoff arms at nominal depths of 1 m, 5 m, and 9 m. These arms extend out from the main buoy frame to reduce the instrument's

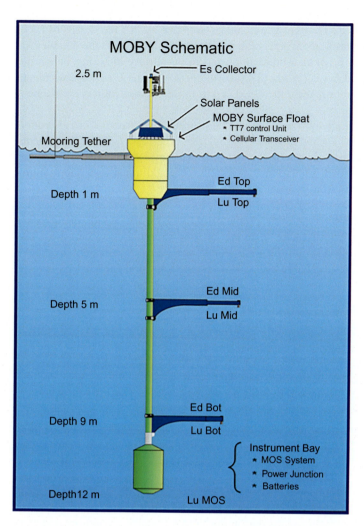

FIG. 7 Cartoon drawing of MOBY system. Three arms extend from the side to reduce instrument self-shadowing effects on the measurement of upwelling radiance (L_u) and downwelling irradiance (E_d). The downwelling surface irradiance (E_s) is measured at the top of the mast of MOBY. The system is self-powered (batteries and solar panels) and communicates to shore via a cellular data link. *(Credit: Drawing by MOBY Team.)*

self-shadowing effect. All the radiance and irradiance measurements are hyperspectral at <1 nm spectral intervals from 380 to 900 nm using two radiometers (one for the short-wavelength portion of the spectrum, one for the longer-wavelength portion of the spectrum, with a common cross-over region). Meteorological data, including wind speed and direction, air temperature, relative humidity, and barometric pressure, are recorded with sensors mounted on the MOBY surface float (Fig. 7).

8.3 Data processing

The radiometric measurements from MOBY allow deriving the water-leaving radiance ($L_w(\lambda)$) or normalized water-leaving radiance ($nL_w(\lambda)$), which are the desired quantity for satellite vicarious calibration. For instance, $L_w(\lambda)$ can be obtained from measurements at the first and second depths, z_1 and z_2:

$$L_w(\lambda) = L_u(\lambda z_1) \exp\left[K_{Lu}(\lambda, z_1, z_2) \times z_1\right] T(\lambda) / [n_w(\lambda)]^2, \tag{10}$$

where T is the transmission of radiance across the air-sea interface, n_w is the refractive index of water, and K_{Lu} is the water diffuse upwelling radiance attenuation coefficient, determined between the paired depth measurements:

$$K_{Lu}(\lambda, z_1, z_2) = \ln\left[L_u(\lambda z_1)/L_u(\lambda z_2)\right]/(z_2 - z_1). \tag{11}$$

The normalized water-leaving radiance, $nL_w(\lambda)$, can then be obtained by using the downwelling surface irradiance, $E_s(\lambda)$, as

$$nL_w(\lambda) = L_w(\lambda) F_0(\lambda) / E_s(\lambda), \tag{12}$$

where F_0 is a reference extra-terrestrial solar irradiance (e.g., Thuillier et al., 2003). Note that, with three depths of $L_u(\lambda)$ measurements, three different combinations of radiance measurement pairs of depths can be obtained, thus three values for $L_w(\lambda)$ and $nL_w(\lambda)$ can be estimated. $K_{Lu}(\lambda)$ propagates the $L_u(\lambda)$ measurement closest to the surface to be just below the surface. Since $K_{Lu}(\lambda)$ can change with depth, the preferred $L_w(\lambda)$ (or $nL_w(\lambda)$) measurement uses the $L_u(\lambda)$ measurements at 1 m and 5 m to determine $K_{Lu}(\lambda)$. $K_{Lu}(\lambda)$ is then used to propagate the $L_u(\lambda)$ at 1 m to the surface (this is the shortest propagation distance).

The $L_u(\lambda)$ measurements at one or more arms may not always be available due to instrument damage or calibration issues, hence having measurements at the three arms allows for redundancy. In rare cases when two $L_u(\lambda)$ measurements are not available, a historical, empirical monthly $K_{Lu}(\lambda)$ value is used to propagate the remaining $L_u(\lambda)$ measurement to the surface. The optical properties of the water at the measurement site are very stable, thus the additional uncertainty introduced by this approach is small (less than 1%), and the resulting product is still useful.

During quality control, each data set is inspected by an expert analyst during processing and categorized as either "good" (highest quality), "questionable" (something indicates a possible problem) or "bad." Only "good" quality data should be used for vicarious calibration, but "good" and "questionable" data can be used for monitoring purposes. The processed data are typically posted within 1–3 days of collection on the CoastWatch site (https://coastwatch.noaa.gov) and are available in both files of data weighted for specific satellite missions and complete spectral files.

8.4 Characterization, calibration, and maintenance

There are currently two MOBY systems that are alternatively deployed for approximately 4 months. The Marine Optical System (MOS) on MOBY has been extensively characterized and is calibrated pre- and post-deployment using National Institute of Standards and Technology (NIST) traceable standards (Clark et al., 2003). There are also calibration stability sources in the MOS as a system check during deployment. The characterizations include temperature effects, measurement bandwidth, cosine response of irradiance collectors, and straylight characterization. Routine pre- and post-deployment calibrations include absolute radiometric response and wavelength calibration. Other characterizations, such as checking the straylight, may take place if the optical system has been opened or otherwise been disturbed. The straylight characterization is a process to determine how light at a single wavelength affects light at other wavelengths in the optical system. The sources used for the radiometric calibration are maintained through regular recalibration at the NIST, in addition to custom stability radiometers used during each calibration. Calibrations are done on the complete optical buoy at the MOBY base site, located at the University of Hawaii Marine Center. During deployments, data are processed with the predeployment calibration and distributed through the NOAA CoastWatch site. After the post-deployment calibration is performed, this additional information is used in a reprocessing of the deployment and the data at the NOAA CoastWatch site are updated. During the deployments, monthly diver cleanings and calibrations are also performed. These diver calibrations are useful as gross system checks, but cannot be done with significant accuracy to replace the pre- and post-deployment calibrations.

8.5 Accomplishments and prospective

MOBY has continuously measured ocean radiance data since late 1996 (Clark et al., 2003). The quality-controlled and spectrally weighted time series data (Fig. 8) provide in situ observations to calibrate a fleet of ocean color satellite instruments, such as the Visible Infrared Imaging Radiometer Suite (VIIRS) onboard the Suomi National Polar-orbiting Partnership (SNPP) and NOAA-20, the Moderate Resolution Imaging Spectroradiometer (MODIS) onboard the Aqua and Terra, and the decommissioned Sea-viewing Wide Field-of-view Sensor (SeaWiFS).

MOBY is currently limited by the solar charging capacity to three measurement sets each day. The times of these measurements are determined to optimize alignment with the nominal satellite overpass times of the different satellite instruments using the data for vicarious calibration and satellite data monitoring. Raw data from each collection are transmitted via a cellular link, daily, to servers to be processed.

In the current instrument configuration, the different light measurements are done sequentially, which is not ideal for radiance measurements in dynamic environments. A new system that will provide simultaneous measurements of $E_s(\lambda)$ and $L_u(\lambda)$ is being tested and is expected to be operational in 2023.

9. Parting thoughts

Oceanographic buoys collecting measurements near the air–sea interface are useful for many different applications including weather, climate, and ecosystems research, numerical model validation, and operational weather forecasting.

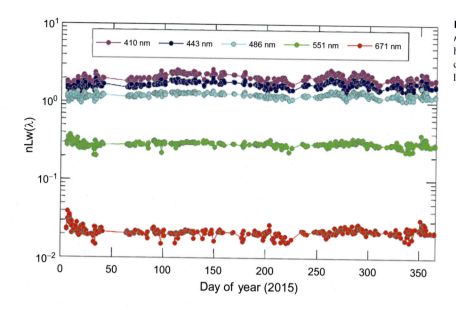

FIG. 8 MOBY-derived daily measurements of $nL_w(\lambda)$ in 2015. The original hyperspectral data have been weighted for the VIIRS/SNPP ocean color sensor. The individual wavelengths are labeled in the legend.

As described in this chapter, measurements of quantities such as SST, SSS, wind speed and direction, surface heat fluxes, SLP, and water leaving radiance are also invaluable for satellite calibration/validation. Field measurements can also be blended with satellite data to generate new products, and in situ data can also inform the design of new satellite missions, products, and applications. New observing platforms and instrumentation are allowing for the collection of new multidisciplinary (e.g., biogeochemistry, nutrients, pigments, and genomic information) and physical observations (e.g., wave spectrum), or improving our ability to observe the ocean with increased spatial or temporal coverage or enhanced resolution. Long-term observing networks strategically seek ways to optimize and innovate while not interrupting operational data streams.

When designing new in situ observing systems or adding new measurements to existing platforms, some consideration should be given to existing, planned, and future satellite missions that may require calibration and validation of derived variables. For example, a proposed satellite mission has the potential to monitor ocean currents in the near-future (the Sea surface KInematics Multiscale monitoring of ocean surface currents or SKIM mission, Ardhuin et al., 2019), and in situ velocity measurements will be needed to validate satellite-derived currents. Although all drifters are used to estimate near-surface velocity, the tropics are a region of surface divergence which means the ocean circulation causes drifters to quickly advect away (Fig. 2; e.g., Lumpkin et al., 2016b). While many moored buoys in the GTMBA are equipped with a point current meter in the upper 10–15 m of the water column, enhanced velocity measurements capabilities at all of the moorings could compensate for limited velocity information in this region (e.g., Foltz et al., 2019; Perez et al., 2019).

The oceanographic community comes together through meetings such as the OceanObs'19 workshop (e.g., Ardhuin et al., 2019; Bailey et al., 2019; Cronin et al., 2019; Centurioni et al., 2019; Foltz et al., 2019; Meinig et al., 2019; Meyssignac et al., 2019; Moltmann et al., 2019; O'Carroll et al., 2019; Vinogradova et al., 2019), and other working groups/activities coordinated through the National Academy of Sciences, national and international Climate and Ocean: Variability, Predictability and Change (CLIVAR) program offices, the Scientific Community on Oceanographic Research (SCOR), and the United Nations Decade of the Ocean to establish the aspirational long-term vision and strategic plans for ocean observing. One such new initiative that has emerged from these community consensus-building exercises is the Observing Air-Sea Interactions Strategy (OASIS, Cronin et al., 2019) program which is focused on observing air-sea exchanges of energy, moisture, and gases in the oceans. OASIS will hopefully lead to an expansion of ocean and atmospheric boundary layer measurements in the field that are suitable for satellite cal/val.

Buoy data are most useful for operational purposes if the data are of high quality and easily available with a short time delay. Ironically, this means using satellite transmission systems to bring the data "to shore" as quickly and cheaply as possible. Data from moored networks are typically served and made available to researchers and the public by the programs themselves. Organizations such as NDBC and OceanSITES have taken on the mission to provide additional infrastructure to collect, serve, and promote time series observations from various networks to support a variety of stakeholders and users. Data are also rapidly distributed by the Global Telecommunications System (GTS) and other data providers for immediate use by the operational forecast and satellite cal/val communities. As mentioned earlier, coordinated efforts to develop data quality standards and SI traceability for disparate data sets (such as for SST in Section 3) are a crucial part of this process. Similarly, increasing the findability, accessibility, interoperability, and reuse (FAIR; Wilkinson et al., 2016) of in situ data will continue to be important.

It can be difficult to obtain in situ measurements in remote or harsh ocean environments using traditional platforms such as buoys and ships. Although hurricane and wave drifters, Argo floats, and expendable probes can be deployed from airplanes to collect oceanic information to support hurricane track and intensity forecasts, it is not possible to adjust an autonomous drifters' sampling pattern in response to an unexpected, extreme weather event once they are deployed. In the future, moored arrays such as PIRATA in the tropical Atlantic may utilize adaptive temporal sampling to provide higher frequency (sub-hourly) data in real time during the passage of tropical storms and hurricanes. However, once deployed, moored buoys are fixed in space and can only provide limited data if a storm happens to pass by a mooring. New piloted platforms including Saildrones, wave gliders, and autonomous underwater vehicles are providing opportunities to collect a suite of multidisciplinary observations in extreme or remote environments with adaptive sampling strategies (Domingues et al., 2015; Meinig et al., 2019; Le Hénaff et al., 2021; Quinn et al., 2021). These platforms will play a key role in the future of ocean measurements and satellite cal/val as their use by the scientific research community expands. Until then, a combination of field measurements obtained from traditional oceanographic buoys, ships, floats, and piloted platforms will be required for satellite calibration and validation.

Acknowledgments

The authors also thank Nicholas Nalli for his guidance throughout the generation of the article and Marlos Goes and Peter Minnett for their insightful comments as reviewers. The authors also thank Bertrand Dano, Jiang Long, and Larry O'Neill for providing graphics.

References

A'Hearn, P.N., Freitag, H.P., McPhaden, M.J., 2002. ATLAS module temperature bias due to solar heating. In: NOAA Tech. Memo OAR PMEL-121. NOAA/Pacific Marine Environmental Laboratory, Seattle, WA. 24 pp.

Anderson, J.E., Riser, S.C., 2014. Near-surface variability of temperature and salinity in the near-tropical ocean: observations from profiling floats. J. Geophys. Res. Oceans 119, 7433–7448. https://doi.org/10.1002/2014JC010112.

Ardhuin, F., Brandt, P., Gaultier, L., Donlon, C., Battaglia, A., Boy, F., Casal, T., Chapron, B., Collard, F., Cravatte, S., Delouis, J.-M., De Witte, E., Dibarboure, G., Engen, G., Johnsen, H., Lique, C., Lopez-Dekker, P., Maes, C., Martin, A., Marié, L., Menemenlis, D., Nouguier, F., Peureux, C., Rampal, P., Ressler, G., Rio, M.-H., Rommen, B., Shutler, J.D., Suess, M., Tsamados, M., Ubelmann, C., van Sebille, E., van den Oever, M., Stammer, D., 2019. SKIM, a candidate satellite mission exploring global ocean currents and waves. Front. Mar. Sci. 6. https://doi.org/10.3389/fmars.2019.00209.

Bailey, K., Steinberg, C., Davies, C., Galibert, G., Hidas, M., McManus, M.A., Murphy, T., Newton, J., Roughan, M., Schaeffer, A., 2019. Coastal mooring observing networks and their data products: recommendations for the next decade. Front. Mar. Sci. 6, 180. https://doi.org/10.3389/fmars.2019.00180.

Banzon, V., Smith, T.M., Steele, M., Huang, B., Zhang, H.-M., 2020. Improved estimation of proxy sea surface temperature in the Arctic. J. Atmos. Ocean. Technol. 37, 341–349. https://doi.org/10.1175/jtech-d-19-0177.1.

Bell, M.J., Lefèbvre, M., le Traon, P.-Y., Smith, N., Wilmer-Becker, K., 2009. GODAE the global ocean data assimilation experiment. Oceanography 22, 14–21. https://doi.org/10.1016/j.marpolbul.2011.06.026.

Berry, D.I., Kent, E.C., 2011. Air–sea fluxes from ICOADS: the construction of a new gridded dataset with uncertainty estimates. Int. J. Climatol. 31 (7), 987–1001. https://doi.org/10.1002/JOC.2059.

Bigorre, S.P., Plueddemann, A.J., 2021. The annual cycle of air-sea fluxes in the Northwest Tropical Atlantic. Front. Mar. Sci. 7. https://doi.org/10.3389/fmars.2020.612842.

Bingham, F.M., 2019. Subfootprint variability of sea surface salinity observed during the SPURS-1 and SPURS-2 field campaigns. Remote Sens. 11, 2689. https://doi.org/10.3390/rs11222689.

Bojinski, S., Verstraete, M., Peterson, T.C., Richter, C., Simmons, A., Zemp, M., 2014. The concept of essential climate variables in support of climate research, applications, and policy. Bull. Am. Meteorol. Soc. 95, 1431–1443. https://doi.org/10.1175/BAMS-D-13-00047.1.

Bourassa, M.A., 2006. Satellite-based observations of surface turbulent stress during severe weather. In: Perrie, W. (Ed.), Atmosphere—Ocean Interactions. vol. 2. Wessex Institute of Technology Press, pp. 35–52.

Bourassa, M.A., Rodriguez, E., Gaston, R., 2010. NASA's ocean vector winds science team workshops. Bull. Am. Meteorol. Soc. 91. https://doi.org/10.1175/2010BAMS2880.1.

Bourassa, M.A., Vincent, D.G., Wood, W.L., 1999. A flux parameterization including the effects of capillary waves and sea state. J. Atmos. Sci. 56, 1123–1139.

Bourlès, B., Araujo, M., McPhaden, M.J., Brandt, P., Foltz, G.R., Lumpkin, R., Giordani, H., Hernandez, F., Lefevre, N., Nobre, P., Campos, E., Saravanan, R., Trotte-Duha, J., Dengler, M., Hahn, J., Hummels, R., Lubbecke, J.F., Rouault, M., Cotrim, L., Sutton, A., Jochum, M., Perez, R.C., 2019. PIRATA: a sustained observing system for tropical Atlantic climate research and forecasting. Earth Space Sci. 6, 577–616. https://doi.org/10.1029/2018EA000428.

Bourlès, B., Lumpkin, R., McPhaden, M.J., Hernandez, F., Nobre, P., Campos, E.J.D., Yu, L., Planton, S., Busalacchi, A., Moura, A.D., et al., 2008. The PIRATA program: history, accomplishments, and future directions. Bull. Am. Meteorol. Soc. 89, 1111–1125.

Castro, S.L., Wick, G.A., Emery, W.J., 2003. Further refinements to models for the bulk-skin sea surface temperature difference. J. Geophys. Res. Oceans 108 (C12), 3377. https://doi.org/10.1029/2002JC001641.

Centurioni, L., Horanyi, A., Cardinali, C., Charpentier, E., Lumpkin, R., 2017. A global observing system for measuring sea level atmospheric pressure: effects and impacts on numerical weather prediction. Bull. Am. Meteorol. Soc. 98 (2), 231–238. https://doi.org/10.1175/BAMS-D-15-00080.1.

Centurioni, L.R., Turton, J., Lumpkin, R., Braasch, L., Brassington, G., Chao, Y., Charpentier, E., Chen, Z., Corlett, G., Dohan, K., Donlon, C., Gallage, C., Hormann, V., Ignatov, A., Ingleby, B., Jensen, R., Kelly-Gerreyn, B.A., Koszalka, I.M., Lin, X., Lindstrom, E., Maximenko, N., Merchant, C.J., Minnett, P., O'Carroll, A., Paluszkiewicz, T., Poli, P., Poulain, P.-M., Reverdin, G., Sun, X., Swail, V., Thurston, S., Wu, L., Yu, L., Wang, B., Zhang, D., 2019. Global in situ observations of essential climate and ocean variables at the air–sea interface. Front. Mar. Sci. 6, 419. https://doi.org/10.3389/fmars.2019.00419.

Clark, D.K., Yarbrough, M.A., Feinholz, M., Flora, S., Broenkow, W., Kim, Y.S., Johnson, B.C., Brown, S.W., Yuen, M., Mueller, J.L., 2003. MOBY, a radiometric buoy for performance monitoring and vicarious calibration of satellite ocean color sensors: measurement and data analysis protocols. In: Mueller, J.L., Fargion, G.S., McClain, G.S. (Eds.), Ocean Optics Protocols for Satellite Ocean Color Sensor Validation, Revision 4, Volume VI: Special Topics in Ocean Optics Protocols and Appendices. NASA, Goddard Space Flight Center, Greenbelt, MD, pp. 138–170.

Clayson, C.A., Fairall, C.W., Curry, J.A., 1996. Evaluation of turbulent fluxes at the ocean surface using surface renewal theory. J. Geophys. Res. 101, 28503–28513.

Cornillon, P., Park, K.A., 2001. Warm core ring velocities inferred from NSCAT. Geophys. Res. Lett. 28 (4), 575–578. https://doi.org/10.1029/2000GL011487.

Cronin, M.F., Gentemann, C.L., Edson, J., Ueki, I., Bourassa, M., Brown, S., Clayson, C.A., Fairall, C.W., Farrar, J.T., Gille, S.T., Gulev, S., Josey, S.A., Kato, S., Katsumata, M., Kent, E., Krug, M., Minnett, P.J., Parfitt, R., Pinker, R.T., Stackhouse Jr., P.W., Swart, S., Tomita, H., Vandemark, D., Weller, R.A., Yoneyama, K., Yu, L., Zhang, D., 2019. Air-sea fluxes with a focus on heat and momentum. Front. Mar. Sci. 6, 430. https://doi.org/10.3389/fmars.2019.00430.

Cronin, M.F., Meinig, C., Sabine, C.L., 2005. Flux mooring for the North Pacific's western boundary current: Kuroshio Extension Observatory (KEO). In: Annual report on the state of the ocean and the ocean observing system for climate, fiscal year 2004, NOAA/OGP/Office of climate observation, Section 3.13a, pp. 142–146.

Domingues, R., Goni, G., Bringas, F., Lee, S.-K., Kim, H.-S., Halliwell, G., Dong, J., Morell, J., Pomales, L., 2015. Upper ocean response to Hurricane Gonzalo (2014): salinity effects revealed by sustained and targeted observations from underwater gliders. Geophys. Res. Lett. 42 (17), 7131–7138. https://doi.org/10.1002/2015GL065378.

Dong, S., Goni, G., Volkov, D., Lumpkin, R., Foltz, G.R., 2017. Near-surface salinity and temperature structure observed from dual-sensor drifters in the subtropical South Pacific. J. Geophys. Res. Oceans 122, 5952–5969. https://doi.org/10.1002/2017JC012894.

Donlon, C.J., Casey, K.S., Robinson, I.S., Gentemann, C.L., Reynolds, R.W., Barton, I., et al., 2009. The GODAE high-resolution sea surface temperature pilot project. Oceanography 22, 34–45. https://doi.org/10.5670/oceanog.2009.64.

Donlon, C.J., Minnett, P.J., Gentemann, C., Nightingale, T.J., Barton, I.J., Ward, B., et al., 2002. Toward improved validation of satellite sea surface skin temperature measurements for climate research. J. Clim. 15, 353–369. https://doi.org/10.1175/1520-0442.

Draper, D.W., Long, D.G., 2004. Evaluating the effect of rain on SeaWinds scatterometer measurements. J. Geophys. Res. 109, C02005. https://doi.org/10.1029/2002JC001741.

Drennan, W.M., Taylor, P.K., Yelland, M.J., 2005. Parameterizing the sea surface roughness. J. Phys. Oceanogr. 35, 835–848. https://doi.org/10.1175/JPO2704.1.

Drushka, K., Gille, S.T., Sprintall, J., 2014. The diurnal salinity cycle in the tropics. J. Geophys. Res. Oceans 119, 5874–5890. https://doi.org/10.1002/2014JC009924.

Durack, P.J., Gleckler, P.J., Purkey, S.G., Johnson, G.C., Lyman, J.M., Boyer, T.P., 2018. Ocean warming: from the surface to the deep in observations and models. Oceanography 31 (2), 41–51. https://doi.org/10.5670/oceanog.2018.227.

Edson, J.B., Jampana, V., Weller, R.A., Bigorre, S.P., Plueddemann, A.J., Fairall, C.W., Miller, S.D., Mahrt, L., Vickers, D., Hersbach, H., 2013. On the exchange of momentum over the open ocean. J. Phys. Oceanogr. 43 (8), 1589–1610.

Elipot, S., Lumpkin, R., 2008. Spectral description of oceanic near-surface variability. Geophys. Res. Lett. 35, L05605. https://doi.org/10.1029/2007GL032874.

Elipot, S., Lumpkin, R., Perez, R., Lilly, J., Early, J., Sykulski, A., 2016. A global surface drifter data set at hourly resolution. J. Geophys. Res. Oceans 121 (5), 2937–2966. https://doi.org/10.1002/2016JC011716.

Fairall, C.W., Bradley, E.F., Godfrey, J.S., Wick, G.A., Edson, J.B., Young, G.S., 1996a. Cool-skin and warm-layer effects on sea surface temperature. J. Geophys. Res. 101, 1295–1308.

Fairall, C.W., Bradley, E.F., Rogers, D.P., Edson, J.B., Young, G.S., 1996b. Bulk parameterizations of air-sea fluxes for topical ocean-global atmosphere coupled-ocean atmosphere response experiment. J. Geophys. Res. 101, 3747–3764.

Flament, P., Firing, J., Sawyer, M., Trefois, C., 1994. Amplitude and horizontal structure of a large diurnal sea surface warming event during the coastal ocean dynamics experiment. J. Phys. Oceanogr. 24, 124–139. https://doi.org/10.1175/1520-0485.

Foltz, G.R., Brandt, P., Richter, I., Rodríguez-Fonseca, B., Hernandez, F., Dengler, M., Rodrigues, R.R., Schmidt, J.O., Yu, L., Lefevre, N., Da Cunha, L.C., McPhaden, M.J., Araujo, M., Karstensen, J., Hahn, J., Martín-Rey, M., Patricola, C.M., Poli, P., Zuidema, P., Hummels, R., Perez, R.C., Hatje, V., Lübbecke, J.F., Polo, I., Lumpkin, R., Bourlès, B., Asuquo, F.E., Lehodey, P., Conchon, A., Chang, P., Dandin, P., Schmid, C., Sutton, A., Giordani,

H., Xue, Y., Illig, S., Losada, T., Grodsky, S.A., Gasparin, F., Lee, T., Mohino, E., Nobre, P., Wanninkhof, R., Keenlyside, N., Garcon, V., Sánchez-Gómez, E., Nnamchi, H.C., Drévillon, M., Storto, A., Remy, E., Lazar, A., Speich, S., Goes, M., Dorrington, T., Johns, W.E., Moum, J.M., Robinson, C., Perruche, C., de Souza, R.B., Gaye, A.T., López-Parages, J., Monerie, P.-A., Castellanos, P., Benson, N.U., Hounkonnou, M.N., Duhá, J.T., Laxenaire, R., Reul, N., 2019. The Tropical Atlantic observing system. Front. Mar. Sci. 6, 206. https://doi.org/10.3389/fmars.2019.00206.

Foltz, G.R., Schmid, C., Lumpkin, R., 2018. An enhanced PIRATA data set for tropical Atlantic ocean-atmosphere research. J. Clim. 31, 1499–1524. https://doi.org/10.1175/JCLI-D-16-0816.1.

Freitag, H.P., McCarty, M.E., Nosse, C., Lukas, R., McPhaden, M.J., Cronin, M.F., 1999. COARE Seacat data: calibrations and quality control procedures. Pac. Mar. Environ. Lab., National Oceanic and Atmospheric Administration, Seattle, Wash. NOAA Tech. Memo. ERL PMEL-115, 89 pp.

Freitag, H.P., Sawatzky, T.A., Ronnholm, K.B., McPhaden, M.J., 2005. Calibration procedures and instrumental accuracy estimates of next generation ATLAS water temperature and pressure measurements. NOAA/Pacific Marine Environmental Laboratory, Seattle, WA. NOAA Tech. Memo. OAR PMEL-128, NTIS: PB2008-101764, 22 pp.

Gelaro, R., McCarty, W., Suárez, M.J., Todling, R., Molod, A., Takacs, L., Randles, C.A., Darmenov, A., Bosilovich, M.G., Reichle, R., Wargan, K., Coy, L., Cullather, R., Draper, C., Akella, S., Buchard, V., Conaty, A., da Silva, A.M., Gu, W., Kim, G.-K., Koster, R., Lucchesi, R., Merkova, D., Nielsen, J.E., Partyka, G., Pawson, S., Putman, W., Rienecker, M., Schubert, S.D., Sienkiewicz, M., Zhao, B., 2017. The modern-era retrospective analysis for research and applications, version 2 (MERRA-2). J. Clim. 30, 5419–5454. https://doi.org/10.1175/jcli-d-16-0758.1.

Gentemann, C.L., Donlon, C.J., Stuart-Menteth, A., Wentz, F.J., 2003. Diurnal signals in satellite sea surface temperature measurements. Geophys. Res. Lett. 30, 1140. https://doi.org/10.1029/2002GL016291.

Gentemann, C.L., Minnett, P.J., LeBorgne, P., Merchant, C.J., 2008. Multi-satellite measurements of large diurnal warming events. Geophys. Res. Lett. 35, L22602. https://doi.org/10.1029/2008GL035730.

Gentemann, C.L., Minnett, P.J., Ward, B., 2009. Profiles of ocean surface heating (POSH): a new model of upper ocean diurnal warming. J. Geophys. Res. Oceans 114 (C7), C07017. https://doi.org/10.1029/2008JC004825.

Gulev, S., Jung, T., Ruprecht, E., 2007a. Estimation of the impact of sampling errors in the VOS observations on air–sea fluxes. Part I: uncertainties in climate means. J. Clim. 20 (2), 279–301.

Gulev, S., Jung, T., Ruprecht, E., 2007b. Estimation of the impact of sampling errors in the VOS observations on air–sea fluxes. Part II: impact on trends and interannual variability. J. Clim. 20 (2), 302–315.

Hersbach, H., Bell, B., Berrisford, P., et al., 2020. The ERA5 global reanalysis. Q. J. R. Meteorol. Soc. 146, 1999–2049. https://doi.org/10.1002/qj.3803.

Hood, E.M., Sabine, C.L., Sloyan, B.M. (Eds.), 2010. The GO-SHIP repeat hydrography manual: a collection of expert reports and guidelines. (IOCCP Report Number 14). ICPO Publication Series, 134. http://www.go-ship.org/HydroMan.html.

Hsu, S.A., 1974. A dynamic roughness equation and its application to wind stress determination at the air-sea interface. J. Phys. Oceanogr. 4, 116–120. https://doi.org/10.1175/1520-0485.

Kamphaus, R., Cronin, M., Sabine, C., Emerson, S., Meinig, C., Robert, M., 2008. New surface mooring at station papa monitors climate. PICES Press 16 (2), 26–27.

Kara, A.B., Wallcraft, A.J., Bourassa, M.A., 2008. Air-sea stability effects on the 10m winds over the global ocean: evaluations of air-sea flux algorithms. J. Geophys. Res. 113, C04009. https://doi.org/10.1029/2007JC004324.

Kearns, E.J., Hanafin, J.A., Evans, R.H., Minnett, P.J., Brown, O.B., 2000. An independent assessment of pathfinder AVHRR sea surface temperature accuracy using the marine-atmosphere emitted radiance interferometer (MAERI). Bull. Am. Meteorol. Soc. 81, 1525–1536. https://doi.org/10.1175/1520-0477.

Kelly, K.A., Dickinson, S., McPhaden, M.J., Johnson, G.C., 2001. Ocean currents evident in satellite wind data. Geophys. Res. Lett. 28, 2469–2472.

Kilpatrick, K.A., Podestá, G., Walsh, S., Williams, E., Halliwell, V., Szczodrak, M., et al., 2015. A decade of sea surface temperature from MODIS. Remote Sens. Environ. 165, 27–41. https://doi.org/10.1016/j.rse.2015.04.023.

Kilpatrick, K.A., Podestá, G.P., Evans, R.H., 2001. Overview of the NOAA/NASA pathfinder algorithm for sea surface temperature and associated matchup database. J. Geophys. Res. 106, 9179–9198.

Laurindo, L.C., Mariano, A., Lumpkin, R., 2017. An improved surface velocity climatology for the global ocean from drifter observations. Deep-Sea Res. I 124, 73–92. https://doi.org/10.1016/j.dsr.2017.04.009.

Le Hénaff, M., Domingues, R., Halliwell, G., Zhang, J.A., Kim, H.S., Aristizabal, M., et al., 2021. The role of the Gulf of Mexico ocean conditions in the intensification of Hurricane Michael (2018). J. Geophys. Res. Oceans 126, e2020JC016969. https://doi.org/10.1029/2020JC016969.

Le Menn, M., Poli, P., David, A., Sagot, J., Lucas, M., O'Carroll, A., Belbeoch, M., Herklotz, K., 2019. Development of surface drifting buoys for fiducial reference measurements of sea-surface temperature. Front. Mar. Sci. 6. https://doi.org/10.3389/fmars.2019.00578.

Legeckis, R., 1975. Application of synchronous meteorological satellite data to the study of time dependent sea surface temperature changes along the boundary of the Gulf stream. Geophys. Res. Lett. 2, 435–438. https://doi.org/10.1029/GL002i010p00435.

Legeckis, R., 1977. Long waves in the eastern equatorial Pacific Ocean: a view from a geostationary satellite. Science 197, 1179–1181. https://doi.org/10.1126/science.197.4309.1179.

Lindstrom, E., Bryan, F., Schmitt, R., 2015. SPURS: salinity processes in the upper-ocean regional study. Oceanography 28, 14.

Lindstrom, E.J., Edson, J.B., Schanze, J.J., Shcherbina, A.Y., 2019. SPURS-2: salinity processes in the upper-ocean regional study 2—the eastern equatorial pacific experiment. Oceanography 32, 15–19.

Liu, W.T., Katsaros, K.B., Businger, J.A., 1979. Bulk parameterization of air-sea exchanges of heat and water vapor including the molecular constraints at the interface. J. Atmos. Sci. 36, 1722–1735.

Liu, W.T., Tang, W., 1996. Equivalent neutral wind. In: JPL Publication 96-17. Jet Propulsion Laboratory, Pasadena. 16 pp.

Lumpkin, R., Centurioni, L., Perez, R.C., 2016b. Fulfilling observing system implementation requirements with the global drifter array. J. Atmos. Ocean. Technol. 33, 685–695. https://doi.org/10.1175/JTECH-D-15-0255.1.

Lumpkin, R., Özgökmen, T., Centurioni, L., 2016a. Advances in the applications of surface drifters. Annu. Rev. Mar. Sci. 9, 59–81. https://doi.org/10.1146/annurev-marine-010816-060641.

Lumpkin, R., Pazos, M., 2007. Chapter 2: Measuring surface currents with surface velocity program drifters: the instrument, its data, and some recent results. In: Griffa, A., Kirwan, A.D., Mariano, A., Özgökmen, T., Rossby, T. (Eds.), Lagrangian Analysis and Prediction of Coastal and Ocean Dynamics. Cambridge University Press.

Luo, B., Minnett, P.J., Nalli, N.R., 2021. Infrared satellite-derived sea surface skin temperature sensitivity to aerosol vertical distribution – field data analysis and model simulations. Remote Sens. Environ. 252, 112151. https://doi.org/10.1016/j.rse.2020.112151.

Mahadevan, A., Paluszkiewicz, T., Ravichandran, M., Sengupta, D., Tandon, A., 2016. Introduction to the special issue on the Bay of Bengal: from monsoons to mixing. Oceanography 29 (2), 14–17. https://doi.org/10.5670/oceanog.2016.34.

May, J., Bourassa, M.A., 2011. Quantifying variance due to temporal and spatial difference between ship and satellite winds. J. Geophys. Res. 116. https://doi.org/10.1029/2010JC006931.

McPhaden, M.J., Busalacchi, A.J., Cheney, R., Donguy, J.-R., Gage, K.S., Halpern, D., Ji, M., Julian, P., Meyers, G., Mitchum, G.T., et al., 1998. The tropical ocean–global atmosphere (TOGA) observing system: a decade of progress. J. Geophys. Res. 103, 14169–14240.

McPhaden, M.J., Meyers, G., Ando, K., Masumoto, Y., Murty, V.S.N., Ravichandran, M., Syamsudin, F., Vialard, J., Yu, L., Yu, W., 2009. RAMA: the research moored array for African–Asian–Australian monsoon analysis and prediction. Bull. Am. Meteor. Soc. 90, 459–480.

Meinig, C., Burger, E.F., Cohen, N., Cokelet, E.D., Cronin, M.F., Cross, J.N., de Halleux, S., Jenkins, R., Jessup, A.T., Mordy, C.W., Lawrence-Slavas, N., Sutton, A.J., Zhang, D., Zhang, C., 2019. Public private partnerships to advance regional ocean observing capabilities: a Saildrone and NOAA PMEL case study and future considerations to expand to global scale observing. Front. Mar. Sci. 6, 448. https://doi.org/10.3389/fmars.2019.00448.

Merchant, C.J., Harris, A.R., Maturi, E., MacCallum, S., 2005. Probabilistic physically based cloud screening of satellite infrared imagery for operational sea surface temperature retrieval. Q. J. R. Meteorol. Soc. 131, 2735–2755. https://doi.org/10.1256/qj.05.15.

Meyssignac, B., Boyer, T., Zhao, Z., Hakuba, M.Z., Landerer, F.W., Stammer, D., et al., 2019. Measuring global ocean heat content to estimate the Earth energy imbalance. Front. Mar. Sci. 6, 432. https://doi.org/10.3389/fmars.2019.00432.

Minnett, P., Alvera-Azcárate, A., Chin, T.M., Corlett, G.K., Gentemann, C.L., Karagali, I., Li, X., Marsouin, A., Marullo, S., Maturi, E., Santoleri, R., Saux Picart, S., Steele, M., Vazquez-Cuervo, J., 2019. Half a century of satellite remote sensing of sea-surface temperature. Remote Sens. Environ. 233. https://doi.org/10.1016/j.rse.2019.111366.

Moltmann, T., Turton, J., Zhang, H.-M., Nolan, G., Gouldman, C., Griesbauer, L., Willis, Z., Piniella, Á.M., Barrell, S., Andersson, E., Gallage, C., Charpentier, E., Belbeoch, M., Poli, P., Rea, A., Burger, E.F., Legler, D.M., Lumpkin, R., Meinig, C., O'Brien, K., Saha, K., Sutton, A., Zhang, D., Zhang, Y., 2019. A Global Ocean Observing System (GOOS), delivered through enhanced collaboration across regions, communities, and new technologies. Front. Mar. Sci. 6, 291. https://doi.org/10.3389/fmars.2019.00291.

NDBC, 2009. Handbook of Automated Data Quality Control Checks and Procedures. NDBC Technical Document 09–02, National Data Buoy Center, Stennis Space Center, Mississippi.

O'Carroll, A.G., Armstrong, E.M., Beggs, H., Bouali, M., Casey, K.S., Corlett, G.K., et al., 2019. Observational needs of sea surface temperature. Front. Mar. Sci. 6, 420. https://doi.org/10.3389/fmars.2019.00420.

Oost, W.A., Komen, G.J., Jacobs, C.M.J., Van Oort, C., 2002. New evidence for a relation between wind stress and wave age from measurements during ASGAMAGE. Bound.-Lay. Meteorol. 103, 409–438. https://doi.org/10.1023/A:1014913624535.

Patoux, J., Foster, R.C., 2012. Cross-validation of scatterometer measurements via sea-level pressure retrieval. IEEE Trans. Geosci. Remote Sens. 50 (7), 2507–2517. https://doi.org/10.1109/TGRS.2011.2172620.

Patoux, J., Foster, R.C., Brown, R.A., 2003. Global pressure fields from scatterometer winds. J. Appl. Meteorol. 42 (6), 813–826. https://doi.org/10.1175/1520-0450.

Pawlowicz, R., McDougall, T., Reistel, R., Tailleux, R., 2012. An historical perspective on the development of the thermodynamic equation of seawater—2010. Ocean Sci. 8 (2), 161–174. https://doi.org/10.5194/os-8-161-2012.

Perez, R.C., Foltz, G.R., Lumpkin, R., Schmid, C., 2019. Direct measurements of upper ocean horizontal velocity and vertical shear in the tropical North Atlantic Ocean at 4°N, 23°W. J. Geophys. Res. Oceans 124, 4133–4151. https://doi.org/10.1029/2019JC015064.

Plagge, A.M., Vandemark, D., Chapron, B., 2012. Examining the impact of surface currents on satellite scatterometer and altimeter ocean winds. J. Atmos. Ocean. Technol. 29 (12), 1776–1793. https://doi.org/10.1175/JTECH-D-12-00017.1.

Poli, P., 2018. Note on the impact of meteorological data from PIRATA moorings on global weather forecasts., https://doi.org/10.5281/zenodo.1164620.

Poli, P., Lucas, M., O'Carroll, A., Le Menn, M., David, A., Corlett, G.K., Blouch, P., Meldrum, D., Merchant, C.J., Belbeoch, M., 2019. The Copernicus surface velocity platform drifter with barometer and reference sensor for temperature (SVP-BRST): genesis, design, and initial results. Ocean Sci. 15, 199–214. https://doi.org/10.5194/os-15-199-2019.

Quinn, P.K., et al., 2021. Measurements from the RV Ronald H. Brown and related platforms as part of the Atlantic tradewind ocean-atmosphere mesoscale interaction campaign (ATOMIC). Earth Syst. Sci. Data 13, 1759–1790. https://doi.org/10.5194/essd-13-1759-2021.

Reul, N., Grodsky, S.A., Arias, M., Boutin, J., Catany, R., Chapron, B., et al., 2020. Sea surface salinity estimates from spaceborne L-band radiometers: an overview of the first decade of observation (2010-2019). Remote Sens. Environ. 242, 111769. https://doi.org/10.1016/j.rse.2020.111769.

Reverdin, G., Morisset, S., Boutin, J., Martin, N., 2012. Rain-induced variability of near sea-surface T and S from drifter data. J. Geophys. Res. 117, C02032. https://doi.org/10.1029/2011JC007549.

Reverdin, G., Olivier, L., Foltz, G.R., et al., 2021. Formation and evolution of a freshwater plume in the northwestern tropical Atlantic in February 2020. J. Geophys. Res. Oceans 126, e2020JC016981. https://doi.org/10.1029/2020JC016981.

Reynolds, R.W., Liu, C., Smith, T.M., Chelton, D.B., Schlax, M.G., Casey, K.S., 2007. Daily high-resolution-blended analyses for sea surface temperature. J. Clim. 20, 5473–5496.

Reynolds, R.W., Rayner, N.A., Smith, T.M., Stokes, D.C., Wang, W., 2002. An improved in situ and satellite SST analysis for climate. J. Clim. 15, 1609–1625.

Reynolds, R.W., Smith, T.M., 1994. Improved global sea surface temperature analysis using optimum interpolation. J. Clim. 7, 929–948.

Rhein, M., Rintoul, S.R., Aoki, S., Campos, E., Chambers, D., Feely, R.A., et al., 2014. Chapter 3: Observations: ocean. In: Stocker, T.F., et al. (Eds.), Climate Change 2013: The physical science basis. Contribution of Working Group I to the Fifth Assessment Report of the Intergovernmental Panel on Climate Change. Cambridge University Press, Cambridge, United Kingdom and New York, NY, USA, pp. 255–315.

Ross, D.B., Cardone, V.J., Overland, J., McPherson, R.D., Pierson Jr., W.J., Yu, T., 1985. Oceanic surface winds. Adv. Geophys. 27, 101–138.

Schlundt, M., Farrar, J.T., Bigorre, S.P., Plueddemann, A.J., Weller, R.A., 2020. Accuracy of wind observations from open-ocean buoys: correction for flow distortion. J. Atmos. Ocean. Technol. 37 (4), 687–703. https://doi.org/10.1175/JTECH-D-19-0132.1.

Slivinski, L.C., Compo, G.P., Sardeshmukh, P.D., Whitaker, J.S., McColl, C., Allan, R.J., Brohan, P., Yin, X., Smith, C.A., Spencer, L.J., Vose, R.S., Rohrer, M., Conroy, R.P., Schuster, D.C., Kennedy, J.J., Ashcroft, L., Brönnimann, S., Brunet, M., Camuffo, D., Cornes, R., Cram, T.A., Domínguez-Castro, F., Freeman, J.E., Gergis, J., Hawkins, E., Jones, P.D., Kubota, H., Lee, T.C., Lorrey, A.M., Luterbacher, J., Mock, C.J., Przybylak, R.K., Pudmenzky, C., Slonosky, V.C., Tinz, B., Trewin, B., Wang, X.L., Wilkinson, C., Wood, K., Wyszyński, P., 2021. An evaluation of the performance of the twentieth century reanalysis version 3. J. Clim. 34 (4), 1417–1438. https://doi.org/10.1175/JCLI-D-20-0505.1.

Smith, N., 2000. GODAE High Resolution Sea Surface Temperature Workshop Proceedings. European Commission Joint Research Centre, Space Applications Institute, Marine Environment Unit, Ispra, Italy. 90 pp.

Smith, N.R., Koblinsky, C.J., 2001. The ocean observing system for the 21st Century: a consensus statement. In: Koblinsky, C., Smith, N.R. (Eds.), Observing the Oceans in the 21st Century. Bureau of Meteorology, Melbourne.

Soloviev, A., Lukas, R., 1997. Observation of large diurnal warming events in the near-surface layer of the western equatorial warm pool. Deep Sea Res. 44, 1055–1076.

Stevens, B., et al., 2021. EUREC^{4}A. Earth Syst. Sci. Data 13, 4067–4119. https://doi.org/10.5194/essd-13-4067-2021.

Stramma, L., Cornillon, P., Weller, R.A., Price, J.F., Briscoe, M.G., 1986. Large diurnal sea surface temperature variability: satellite and in situ measurements. J. Phys. Oceanogr. 16, 345–358.

Stuart-Menteth, A.C., Robinson, I.S., Challenor, P.G., 2003. A global study of diurnal warming using satellite-derived sea surface temperature. J. Geophys. Res. 108, 3155. https://doi.org/10.1029/2002JC001534.

Stull, R.B., 1988. Measurement and Simulation. An Introduction to Boundary Layer Meteorology. Kluwer Academic Publishers, pp. 405–440.

Sybrandy, A.L., Niiler, P.P., Martin, C., Scuba, W., Charpentier, E., Meldrum, D.T., 2009. Global drifter programme barometer drifter design reference. Data Buoy Cooperation Panel Report No. 4, Revision 2.2, August 2009.

Taylor, P.K., Yelland, M.J., 2001. The dependence of sea surface roughness on the height and steepness of the waves. J. Phys. Oceanogr. 31, 572–590.

Thuillier, G., Hersé, M., Labs, D., Foujols, T., Peetermans, W., Gillotay, D., Simon, P.C., Mandel, H., 2003. The solar spectral irradiance from 200 to 2400 nm as measured by the SOLSPEC spectrometer from the Atlas and Eureca missions. Sol. Phys. 214, 1–22.

Vinogradova, N., Lee, T., Boutin, J., Drushka, K., Fournier, S., Sabia, R., et al., 2019. Satellite salinity observing system: recent discoveries and the way forward. Front. Mar. Sci. 6, 243. https://doi.org/10.3389/fmars.2019.00243.

Volkov, D., Dong, S., Foltz, G.R., Goni, G., Lumpkin, R., 2019. Observations of near-surface salinity and temperature structure with dual-sensor Lagrangian drifters during SPURS-2. Oceanography 32, 66–75. https://doi.org/10.5670/oceanog.2019.214.

Voss, K.J., McLean, S., Lewis, M.R., Johnson, C., Flora, S., Feinholz, M., Yarbrough, M., Trees, C.C., Twardowski, M.S., Clark, D.K., 2010. An example crossover experiment for testing new vicarious calibration techniques for satellite ocean color radiometry. J. Atmos. Ocean. Technol. 27, 1747–1759.

Wang, M., Franz, B.A., 2000. Comparing the ocean color measurements between MOS and SeaWiFS: a vicarious intercalibration approach for MOS. IEEE Trans. Geosci. Remote Sens. 38, 184–197.

Wang, Z., Stoffelen, A., Fois, F., Verhoef, A., Zhao, C., Lin, M., Chen, G., 2017. SST dependence of Ku- and C-band backscatter measurements. IEEE J. Sel. Top. Appl. Earth Obs. Rem. Sens. 10 (5), 2135–2145.

Wilkinson, M., Dumontier, M., Aalbersberg, I., et al., 2016. The FAIR guiding principles for scientific data management and stewardship. Sci. Data 3, 160018. https://doi.org/10.1038/sdata.2016.18.

Wong, A.P.S., et al., 2020. Argo data 1999–2019: two million temperature-salinity profiles and subsurface velocity observations from a global array of profiling floats. Front. Mar. Sci. 7, 700. https://doi.org/10.3389/fmars.2020.00700.

World Meteorological Organization, 1981. Review of Reference Height for and Averaging Time of Surface Wind Measurements at Sea (F.W. Dobson). Marine Meteorology and Related Oceanographic Activities Report No. 3, Geneva.

Wright, E., Bourassa, M.A., Stoffelen, A., Bidlot, J.-R., 2021. Characterizing buoy wind speed error in high winds and varying sea state with ASCAT and ERA5. Rem. Sens. 13 (22), 4558. https://doi.org/10.3390/rs13224558.

Wunsch, C., Stammer, D., 1997. Atmospheric loading and the oceanic "inverted barometer" effect. Rev. Geophys. 35, 79–107. https://doi.org/10.1029/96RG03037.

Xu, F., Ignatov, A., 2014. In-situ SST quality monitor. J. Atmos. Ocean. Technol. 31, 164–180. https://doi.org/10.1175/JTECH-D-13-00121.1.

Zhang, M., Reynolds, R.W., Lumpkin, R., Molinari, R., Arzayus, K., Johnson, M., Smith, T.M., 2009. An integrated global ocean observing system for sea surface temperature using satellites and in situ data: research to operations. Bull. Am. Meteorol. Soc. 90, 31–38. https://doi.org/10.1175/2008BAMS2577.1.

Chapter 6

Surface-based thermal infrared spectrometers

Peter J. Minnett[a], Robert O. Knuteson[b], and Jonathan Gero[b]
[a]*Department of Ocean Sciences, Rosenstiel School of Marine and Atmospheric Science, University of Miami, Miami, FL, United States,* [b]*Space Science and Engineering Center (SSEC), University of Wisconsin-Madison, Madison, WI, United States*

Chapter outline

1. Introduction 101
2. FTIR principles 103
3. Calibration 103
4. M-AERI 104
   4.1 ARM M-AERI 105
5. Deployments 105
6. Geophysical variables 106
   6.1 Sea surface emissivity 106
   6.2 SST$_{skin}$ 107
   6.3 Thermal skin layer profiles 109
   6.4 Near-surface temperatures and air-sea temperature differences 110
   6.5 Atmospheric profiles 110
7. Applications 112
   7.1 Validating satellite SST$_{skin}$ retrievals 112
   7.2 Atmospheric profiles 113
   7.3 Greenhouse gas heating of the ocean 114
8. Summary and conclusions 115
Acknowledgments 116
References 116

God made the bulk; surfaces were invented by the devil.

Wolfgang Pauli (1900–1958)

1. Introduction

For half a century, sea-surface temperature (SST) has been the subject of remote sensing by infrared radiometers on earth-observation satellites (Minnett et al., 2019). The early instruments were used to produce images of SST features, such as the Gulf Stream in the Atlantic Ocean (Legeckis, 1979), Tropical Instability Waves in the Equatorial Pacific Ocean (Legeckis, 1977), and fronts and eddies throughout the ocean (Legeckis, 1978). The problem with the quantification of the SST values is in finding a sufficiently accurate way to correct for the effects of the intervening atmosphere on the infrared emission from the sea surface as it propagates to the height of the satellites. Another problem is to confidently identify the presence of clouds in the images as these occlude the features on the sea surface beneath. Both problems can be addressed by taking simultaneous measurements at more than one wavelength: at two or more infrared wavelengths for the correction of atmospheric effects (McMillin, 1975) and for cloud identification, several wavelengths, including in the visible for the sunlit part of the orbits. As the capabilities of the satellite radiometers improved, along with the development of increasingly more effective algorithms for cloud identification and atmospheric correction, the need developed for assessing the accuracy of the satellite SST retrievals, a process often referred to as "validation." The growth of satellite remote sensing of SST coincided with the development and deployment of drifting and moored buoys, initially deployed for specific experiments but soon becoming required data sources for driving numerical weather prediction models. The buoys have subsurface thermometers (Lumpkin et al., 2016; Centurioni et al., 2019) that became critical to the development of algorithms to retrieve cloud-free SSTs from satellite data and to validate the derived SSTs. Drifting buoys became especially relied upon because of their numbers that eventually sampled very large areas of the oceans (Lumpkin et al., 2017) and for their ability to transmit their measurements in real-time via satellite telemetry.

Field Measurements for Passive Environmental Remote Sensing. https://doi.org/10.1016/B978-0-12-823953-7.00004-6
Copyright © 2023 Elsevier Inc. All rights reserved.

As the accuracy of the satellite-derived SSTs improved, it became apparent that comparisons with buoy measurements were less than ideal. Analyses revealed that the accuracy of the drifter temperature measurements was ~0.20–0.25 K (O'Carroll et al., 2008; Gentemann, 2014), which was a significant, perhaps dominant, contribution to the statistics of the differences between the satellite SSTs and the buoy measurements (Kilpatrick et al., 2015; Minnett et al., 2020) and another principal concern is that the presence of vertical temperature gradients between the depth of the subsurface thermometer and the ocean surface (Minnett, 2003). The issues with the accuracy of the thermometer are being addressed and now, at the time of writing, a significant proportion of drifters have thermometers with accuracies better than 0.1 K (Poli et al., 2019). The consequences of vertical temperature gradients are much more difficult to deal with as these are a ubiquitous characteristic of the ocean and have two distinct characteristics caused by separate physical processes. The first is the result of the heating of the upper ocean driven primarily by the solar radiation that penetrates the ocean surface and is absorbed in the upper several meters increasing the SST (Price et al., 1986; Fairall et al., 1996; Ward, 2006), and wind which mixes heat downward reducing the SST (Gentemann and Minnett, 2008; Gentemann et al., 2009). The dependence on solar heating produces a signal with a diurnal characteristic, and the process is generally referred to as "diurnal heating." The second is the result of heat flow between the ocean and atmosphere, and since the ocean surface is nearly everywhere and nearly always warmer than the atmospheric boundary layer, the heat flow is predominantly from ocean to atmosphere.

Over long length scales, the viscosity of water can be neglected in considering many ocean processes but on the aqueous side of the interface, over depths of a fraction of a millimeter, it cannot be ignored and this thin layer of water, known as the viscous sublayer, suppresses turbulence and thus hinders the vertical transport of heat between the water beneath and the interface (Soloviev and Lukas, 2014). The vertical transport of heat through the viscous sublayer is supported by thermal conductivity, and this requires a vertical temperature gradient. The thickness of the water through which this temperature gradient exists is called the thermal skin layer. The viscous sublayer and thermal skin layer co-exist on similar scales to the infrared electromagnetic (e-m) skin layer that determines the depth through which emission occurs (Fig. 1). In the infrared, the e-m skin depth is a fraction of a millimeter (Bertie and Lan, 1996), but microwave emission occurs from a somewhat thicker layer, 2–3 mm at frequencies of 6–10 GHz (Robinson, 2004), where SST retrievals are made (Wilheit and Chang, 1980). Thus, an SST determination based on the measurement of the emitted infrared radiation is characteristic of the temperature in the thermal skin layer, SST_{skin}. The SST derived from satellite measurements in the infrared is thus SST_{skin}.

The complications caused by diurnal heating and the thermal skin effect in the comparisons between satellite-derived SST_{skin} and surface-based measurements can be eliminated by determining the SST_{skin} using ship-based infrared radiometers. These fall into two classes, one with ambient temperature detectors and the spectral response function given by optical filters (Donlon et al., 2014) and the other being spectroradiometers which measure the spectrum of the sea-surface emission over a wide range of wavenumbers by using Fourier-transform infrared radiometers (FTIRs; Griffiths and de Haseth, 1986), first developed by Michelson and Morely (1887). These interferometers require the detectors to be cooled,

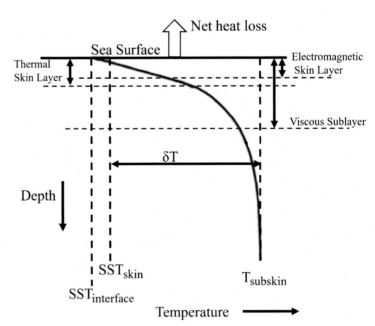

FIG. 1 A schematic representation of the temperature gradient beneath the ocean surface. The depths of the thermal and the infrared electromagnetic skin layers are indicated along with that of the viscous sublayer. The electromagnetic skin layer depth varies with wavenumber (Fig. 10), and the depths of the viscous sublayer and thermal skin layer become shallower with increasing wind. For winds $>\sim 4\,m\,s^{-1}$, the thermal skin layer is shallower than the electromagnetics skin layer depth at the wavenumbers where it is deepest. Beneath the viscous sublayer, the intensity of turbulence increases, and consequently the effectiveness of turbulent vertical heat transfer increases to that of the mixed layer beneath. $SST_{interface}$ is the temperature of the sea at the interface, on molecular scales, and SST_{skin}, the depth averaged temperature through the electromagnetic skin layer according to Beer's law, is the temperature derived from the measurements of infrared radiometers. $T_{subskin}$ is the temperature at the depth of the transition to the turbulent mixed layer.

typically to the temperatures of liquid N_2 (77 K). Both filter radiometers and FTIRs require internal calibration to provide the accuracy and stability of the measurements needed for the useful assessment of the accuracy of satellite-derived SST_{skin}.

In this chapter, we focus on the use of shipborne FTIRs.

2. FTIR principles

FTIR is an instrument approach to determine the infrared spectrum of a solid, liquid, or gas. Unlike a dispersive spectrometer that measures each wavelength separately, the FTIR spectrometer measures all wavelengths in the instrument passband simultaneously (Bell, 1972; Davis et al., 2001; Griffiths and de Haseth, 1986). This leads to the so-called "throughput or Jacquinot advantage" in signal to noise of the FTIR spectrometer over traditional spectrometers using slits to limit the bandpass. The bandwidth of the FTIR is limited only by the choice of materials for the optical elements/detectors and can span the electromagnetic spectrum from the far infrared to the ultraviolet. Perhaps the most famous FTIR spectrometer is the Michelson interferometer illustrated in Fig. 2 (Michelson and Morely, 1887). Coherent light enters from the left and is split into two beams using a birefringent crystal called the "beamsplitter" (e.g., ZnSe), which is oriented such that half the beam travels toward a fixed mirror and half travels toward a moveable mirror. When the two beams recombine each wavelength in the original light creates an interference pattern of light and dark fringes depending on the path difference taken by the two beams. The superposition of all the fringes from all the wavelengths as the moving mirror translates creates what is known as the interferogram. The Michelson interferometer has the remarkable characteristic that a mathematical Fourier transform of the measured interferogram generates a spectrum at a spectral resolution defined by the maximum optical path difference. Thus, the output spectrum resolution can be controlled by setting the length of the optical path of the moving mirror. While this principle was used scientifically since the late 1800s, it was not until the advent of digital computers to perform a fast Fourier transform and the invention of stable infrared lasers in the 1960s that practical field ready devices became possible (Shankland, 1964; Loewenstein, 1966).

3. Calibration

Unlike many applications of FTIR that emphasize the spectral fidelity of the measurements, the measurement of Earth infrared emission with high absolute accuracy also requires a sensor design that facilitates radiometric calibration for each spectral element (Taylor et al., 2020). Calibration is the procedure used to convert raw detector output after conversion to digital counts into physical engineering units. A typical calibration approach for single channel infrared radiometers is called a two-point calibration where the counts from a scene view are bounded by counts from high emissivity cold and hot targets at known temperatures (Workman, 2018). This approach requires a detector that has a linear response to incident radiation over the desired range. The extension of this approach to FTIR instruments at ambient temperatures originally encountered problems due to thermal emission of warm components within the instrument. These problems were solved by the use of complex arithmetic as described by Revercomb et al. (1988). Variations of the Revercomb calibration equation have been successfully used in the calibration of satellite, airborne, and surface-based sensors. In particular, the

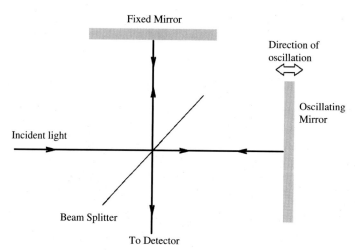

FIG. 2 A diagram of the Michelson Morley Interferometer that is the basis of the M-AERIs. Incoming light falls on a beam splitter that reflects half of the radiation to a plane mirror at a fixed position and transmits half to a plane mirror that oscillates. The two beams reflected from the two mirrors recombine at the beam splitter where a portion in reflected to a detector. When the path length difference between the two beams reflected by the plane mirrors is zero or an integer number of wavelengths of a component of the incoming light, the two beams combine constructively, the signal has a high value. When the path lengths differ by $n + ½$ wavelengths (where n is an integer, including zero), the two reflected beams combine destructively, and the detector signal is very small. The interferometer has converted spectral information in the incoming light into a temporal signal from the detector. The Fourier transform of the detector output over periods of many mirror oscillations produces the spectrum of in the incoming light beam. In the M-AERIs, corner cube reflectors are used in place of the plane mirrors, and both mirrors oscillate on a yoke so the reflectors move out of phase, so that when one is moving toward the beam splitter, the other is moving away.

surface-based atmospheric emitted radiance interferometer (AERI) was developed for the Department of Energy (DOE) Atmospheric Radiation Measurement (ARM) program (Stokes and Schwartz, 1994) to provide absolutely calibrated measurements of the downwelling atmospheric emission with better than 1% accuracy at cold sky temperatures (Knuteson et al., 2004a,b). The AERI achieves this accuracy using internal calibration involving measurement of the emission from two high emissivity (>0.995) cavity blackbodies at two temperatures (ambient air and ~60°C).

4. M-AERI

The Marine-AERI (M-AERI) was developed from the ARM AERI at the Space Science and Engineering Center (SSEC) at the University of Wisconsin-Madison, under contract to the Rosenstiel School of Marine and Atmospheric Science (RSMAS) at the University of Miami for the purpose of providing accurate SST_{skin} values for comparison with satellite retrievals (Minnett et al., 2001). The funding for the M-AERI development was provided by NASA. The FTIR, detector system, control and data acquisition software were taken from the AERI, with some software enhancements, but the fore-optics required modifications to permit a selectable range of view angles (Fig. 3) needed for the correction of reflected atmospheric emission (Smith et al., 1996; Minnett et al., 2001; Knuteson et al., 2004a,b; Minnett and Corlett, 2012b), which in turn required a repositioning of the black body calibration targets. The prototype M-AERI was field tested on a short cruise in the Gulf of Mexico in January 1995, the results of which were very successful (Smith et al., 1996), and which led to three units being built at SSEC and delivered to RSMAS for field deployment. In addition to redesigning the fore-optics, the M-AERIs were enclosed in a weather-proof enclosure to keep the instrument safe from precipitation and sea-spray when installed on ships. A rain gauge provided the signal to move the scan mirror to a safe position viewing the ambient temperature blackbody calibration target, turning its back to the weather. The original M-AERI used a liquid nitrogen Dewar to chill the detectors to operating temperatures, but this proved to be problematic for long duration research cruises starting in remote ports, and so the liquid refrigerant was replaced by a mechanical Stirling-Cycle cooler, which has been effective and reliable. The control and data acquisition electronics and computer were installed in a full-height 19″ electronics rack that was installed inside the ships in an air-conditioned environment and was connected to the M-AERI on the deck or instrument platform of the ships by a 96-conductor umbilical cable. Power was supplied from the ship's generators through a UPS system to guard against power interruptions.

After over a decade of deployments, the three M-AERIs were becoming difficult to repair and service as many components had become obsolete. A second-generation M-AERI was developed, again with NASA funding, by LRTech in Quebec, Canada. The three second-generation M-AERIs operated on the same principles as the original but are much smaller and easier to install on ships. The electronics are housed in a shoebox-sized container inside the environmental enclosure and connected to a laptop computer in the ship by a single cat-5 cable. A single third-generation M-AERI has been developed and deployed and this is yet smaller than its predecessors. The laptop computers are connected to the ships' LANs and the ships' satellite internet systems. The first-generation M-AERIs required the presence of operators on the ships. The second and third generations can operate autonomously with monitoring through satellite internet connections, thus facilitating much longer deployments, sometimes over a year with infrequent maintenance during port-calls, and much more cost-effective data acquisition.

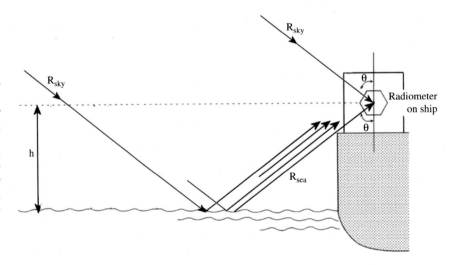

FIG. 3 Schematic of the measurement of the skin SST by a ship-based radiometer. The sea-viewing measurement (R_{sea}) at nadir angle θ includes not only emission from the sea surface that contains the information about the skin SST, but also a component of reflected sky radiance (R_{sky}) and of emission, both direct and reflected from the layer of atmosphere between the instrument height and the sea surface (h). (From Minnett, P. J., Corlett, G.K., 2012b. A pathway to generating Climate Data Records of sea-surface temperature from satellite measurements. Deep-Sea Res. II Top. Stud. Oceanogr. 77–80, 44–51. https://doi.org/10.1016/j.dsr2.2012.04.003, with permission.)

The internal calibration of the RSMAS M-AERIs is checked before and after each at-sea deployments using a laboratory water-bath calibration system (Fowler, 1995) designed at the National Institute of Science and Technology (NIST). The stabilized temperature of the water bath that surrounds a high-emissivity cone is monitored using a pair of thermometers with SI-traceable calibration with millikelvin accuracy. The performance of the M-AERIs and the laboratory calibrator is assessed periodically at workshops involving national metrology laboratories, NIST in the United States (Rice et al., 2004; Barton et al., 2004) and the National Physical Laboratory (NPL) in the United Kingdom (Theocharous et al., 2019). Propagating the accuracy estimates of the spectra through the algorithm to derive SST_{skin} leads to a distribution of the accuracy estimates for each deployment which has a modal value of ~40 mK.

To date, the RSMAS M-AERIs have been deployed on over 60 research cruises and on four cruise ships of the Royal Caribbean Group, including for 2072 days on *Explorer of the Seas*, starting in 2000, and once autonomous operations had become reliable, on *Celebrity Equinox*, *Allure of the Seas*, and *Adventure of the Seas*. Fig. 4 shows the layout of the components of a second-generation M-AERI on the *Adventure of the Sea*, and Fig. 5 is a photograph of the third-generation M-AERI on the NOAA S. *Ronald H. Brown*.

4.1 ARM M-AERI

ABB Inc. delivered the first of a new generation of M-AERI instruments in 2013, with ARM being the launch customer. This new instrument is based on the version-4 AERI design, with notable modifications including a front-end that allows viewing of elevation angles within the range ±45° of the horizon, and a marine-hardened TEC-cooled enclosure for the instrument back-end. The ARM M-AERI has been successfully deployed on three ship-going campaigns to date: ARM Cloud Aerosol Precipitation Experiment (ACAPEX) on the NOAA S. *Ronald H. Brown* (Fig. 5), Measurements of Aerosols, Radiation and CloUds over the Southern Ocean (MARCUS) on the RSV *Aurora Australis*, and the Multidisciplinary drifting Observatory for the Study of Arctic Climate (MOSAiC) on the PFS *Polarstern*. These have covered a large geographic region from the tropics, Southern Ocean and the Arctic Ocean, and the instrument has operated well in a variety of extremely challenging physical environments.

5. Deployments

The prototype SSEC M-AERI was first deployed during a short cruise in the northern Gulf of Mexico on the P/V *Pelican* in January 1995, when, for the first time, radiometric measurements of the SST_{skin} by an FTIR were derived, and when referenced to a subsurface temperature measured from a surface-following float these showed a physically reasonable behavior of the thermal skin layer (Smith et al., 1996). Following some improvements to ruggedize the prototype, an M-AERI was

FIG. 4 A second generation RSMAS M-AERI installed on an instrument platform built onto the side, behind the bridge, of the *Adventure of the Seas* of Royal Caribbean International. The top of the enclosure has been removed to reveal the components of the instrument.

106 PART | I Instrumentation

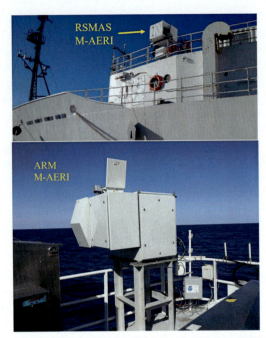

FIG. 5 The third generation RSMAS M-AERI (above) and the ARM M-AERI (below) installed on the NOAA S *Ronald H Brown*. Lower image courtesy of the U.S. Department of Energy of Atmospheric Radiation Measurement (ARM) user facility.

deployed on a month-long cruise of the NOAA S *Discoverer* in the Equatorial Pacific Ocean in 1996 (Post et al., 1997). The first two of three M-AERIs with Stirling Cycle coolers to chill the detectors were delivered to RSMAS in 1997 and were deployed side-by-side on the R/V *Roger Revelle* on a section from Hawai'i to New Zealand. The mean and standard deviations of the differences in the SST_{skin} retrievals was 5 mK ± 77 mK, $n = 890$ (Minnett et al., 2001). There then followed multiple deployments on research vessels that continue to the present. Fig. 6 shows the tracks of some of the research vessels colored by the SST_{skin} derived from the M-AERIs.

Through a collaboration between the Royal Caribbean Group and RSMAS, M-AERIs have been deployed on four cruise ships, beginning with *Explorer of the Seas* in 2000. Until the effects of the COVID-19 pandemic hit the cruise industry, calling a halt to cruises, three second-generation M-AERIs were deployed simultaneously and autonomously on three cruise vessels.

6. Geophysical variables

A great strength of the measurements of atmospheric and oceanic infrared emission spectra is the information conveyed in each, leading to the possibility of the retrieval of a range of geophysical variables. Additional benefit is gained by the fact that some are derived from different parts of the spectra and so are simultaneous determinations of more than a single variable.

6.1 Sea surface emissivity

The temperature dependence of the spectral emission from a surface is given by Planck's function, but the relationship is for an ideal emitter with unit emissivity, $\varepsilon(\lambda, \theta) = 1$, where λ is the wavelength of the emitted radiation and θ is the angle of emission measured to the normal of the surface at the point of emission. In reality, there are no ideal emitters and the emission is a fraction of that from the ideal, meaning that $1 > \varepsilon(\lambda, \theta) > 0$. In the thermal infrared, the emissivity of sea water is high, and as a result a measurement of the surface emission contains a small component of reflected sky emission (Fig. 7), where the reflectivity, r, is $r(\lambda, \theta) = 1 - \varepsilon(\lambda, \theta)$. To derive the spectrum of the surface emissivity requires the correction for the reflected sky emission and an accurate knowledge of $\varepsilon(\lambda, \theta)$.

The measured spectra provide a mechanism for deriving $\varepsilon(\lambda, \theta)$ as the sea view spectra include features that are the reflected emission from spectral lines of atmospheric gases (Smith et al., 1996; Hanafin and Minnett, 2005; Wong and Minnett, 2016b). The correct value of the emissivity, determined iteratively, reduces the reflected atmospheric emission

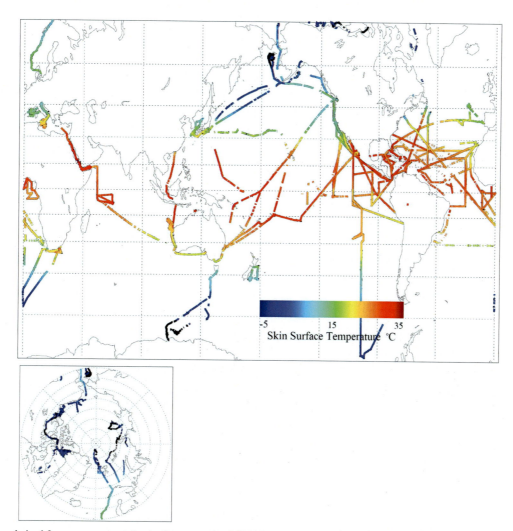

FIG. 6 SST$_{skin}$ derived from measurements by the first generation RSMAS M-AERIs deployed on research vessels, from 1997 to 2013.

lines to the noise level of the instrument (Fig. 8). The emission spectra from the M-AERI have been critical in supporting theoretical analyses of the wavelength dependence of the sea surface emissivity. The wind speed dependence of the surface emissivity determined empirically from the M-AERI data was found to be less pronounced than indicated in prior modeling studies (Hanafin and Minnett, 2005; Fig. 9), and this behavior was subsequently confirmed by more rigorous simulations (Nalli et al., 2008a,b; Masuda, 2006).

6.2 SST$_{skin}$

Given the option to select wavelengths from the M-AERI spectral measurements of the ocean and atmospheric emission, it is not necessary to use the same spectral intervals as the satellite radiometers as these are selected where the atmosphere is transmissive for given atmospheric conditions, typically in the 10–13 μm wavelength atmospheric "window." However, the relatively high transmission means that for ship-board radiometers making use of this window, the correction for reflected sky radiance is more uncertain in conditions of changing cloud amounts and properties given by the sky measurements. The M-AERI SST$_{skin}$ retrievals use measurements in the vicinity of $\lambda = 7.7$ μm (Minnett et al., 2001) where the atmospheric transmissivity is much lower, rendering the sky radiance correction less sensitive to changing cloud conditions. The emission depth of radiation at this wavelength is very close to that at $\lambda = 10$–13 μm so the SST$_{skin}$ derived from satellite-based and ship-based radiometers are essentially the same (see Fig. 10).

108 PART | I Instrumentation

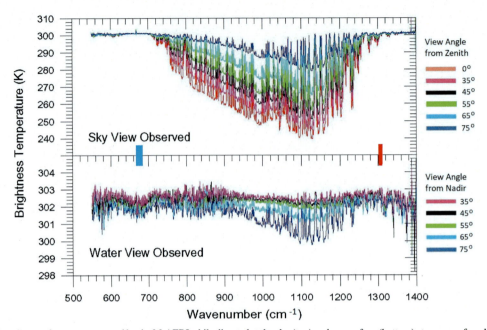

FIG. 7 Examples of parts of spectra measured by the M-AERI while directed at the sky (top) and sea surface (bottom) at a range of angles indicated by the colors. The spectra are represented as temperature, and those intervals where the sky temperatures are smallest indicate where the atmosphere is most transparent. The spikes in the atmospheric spectra are caused by emission lines. The small departure from unity in the emissivity of the sea surface results in reflection of the sky radiance in the sea-viewing measurement, including the atmospheric emission lines. The blue bar shows which spectral region is used to measure air temperature and the red bar skin sea-surface temperature. Note the change in temperature scales of the two panels. Most satellite imaging radiometers take two broadband spectral measurements in the 800–1000 cm^{-1} interval. These data were taken in the tropical western Pacific Ocean during the Combined Sensor Program Cruise in 1996 (Post et al. 1997). *(Based on Minnett, P.J., Knuteson, R.O., Best, F.A., Osborne, B.J., Hanafin, J.A., Brown, O.B., 2001. The Marine-Atmospheric Emitted Radiance Interferometer (M-AERI), a high-accuracy, sea-going infrared spectroradiometer. J. Atmos. Ocean. Technol. 18, 994–1013. © American Meteorological Society.)*

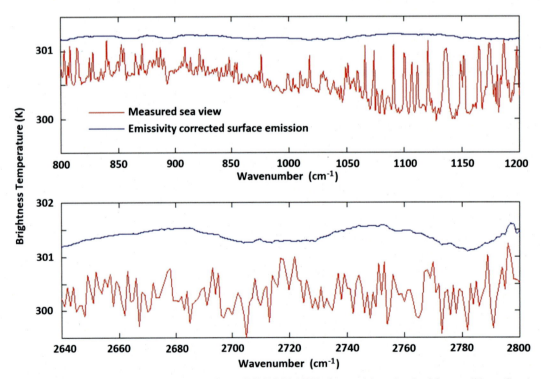

FIG. 8 Parts of a spectrum of the sea-viewing measurement from a RSMAS M-AERI taken at night under cloud-free conditions. *(Reprinted with permission from Wong, E.W., Minnett, P.J., 2016. Retrieval of the ocean skin temperature profiles from measurements of infrared hyperspectral radiometers – Part II: field data analysis. IEEE Trans. Geosci. Remote Sens. 54, 1891–1904. https://doi.org/10.1109/TGRS.2015.2501425. © 2016 IEEE.)*

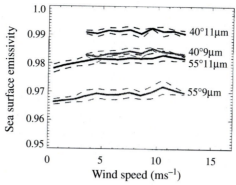

FIG. 9 Observed mean *(solid curves)* and standard deviation *(dashed curves)* of sea-surface emissivity in 1 m s^{-1} wind-speed bins for 9 and 11 μm at 40 and 55 degrees incidence angles. *(Reprinted with permission from Hanafin, J.A., Minnett, P.J., 2005. Infrared-emissivity measurements of a wind-roughened sea surface. Appl. Optics 44, 398–411. © The Optical Society.)*

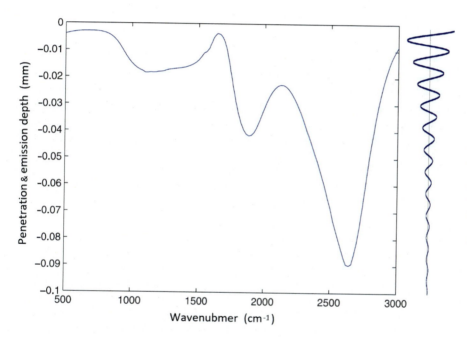

FIG. 10 Wavenumber dependence of the absorption/emission depth along the direction of propagation of infrared radiation beneath the air-sea interface, from data of Bertie and Lan (1996). The wave at right is a schematic indication of the amplitude of the absorbed electromagnetic wave, propagating down, or the emitted wave, propagating up. *(Reprinted with permission from Wong, E.W., Minnett, P.J., 2016. Retrieval of the ocean skin temperature profiles from measurements of infrared hyperspectral radiometers – Part II: field data analysis. IEEE Trans. Geosci. Remote Sens. 54, 1891–1904. https://doi.org/10.1109/TGRS.2015.2501425. © 2016 IEEE.)*

6.3 Thermal skin layer profiles

The derivation of temperature and humidity profiles through the atmosphere from multispectral and hyperspectral measurements from satellites of top-of-atmosphere infrared emission is quite a mature subject and has been the focus of research for several decades (Eyre, 1987; Bouillon et al., 2020; Smith and Barnet, 2020). A similar approach based on the spectrally varying emission depth in the infrared (Bertie and Lan, 1996; Fig. 10) can be adopted to derive the temperature profile through the thermal skin layer of the ocean from emission spectra measured from M-AERIs on ships. However, the governing equation (Wong and Minnett, 2016a) is highly nonlinear and ill-conditioned, meaning that the errors in the measurements are amplified possibly resulting in a meaningless solution. Introducing an additional criterion such that the solution is constrained can lead to a physically meaningful solution is obtained. The additional requirement here is that the retrieved profile is vertically monotonic in temperature, which is physically justifiable. As with all spectral inversion techniques, a reasonable first guess is required, which was provided by a complementary error function profile derived theoretically by Liu and Businger (1975). The algorithm derived by Wong and Minnett (2016a) using synthetic data was applied to at-sea measurements and produced realistic profiles (Wong and Minnett, 2016b), but since the part of the spectra close to $\lambda = 4$ μm is needed to provide information from the deepest emission depth (~90 μm) this technique can only be applied to nighttime measurements to avoid solar contamination.

6.4 Near-surface temperatures and air-sea temperature differences

An important benefit of using FTIRs to measure the oceanic and atmospheric emission is the ability to extract information on several geophysical variables at the same time from different parts of the measured spectra. A prime example is the determination of air temperature at the height of the instrument by selecting the measurements from emission with a path length of a few meters from CO_2 molecules (Minnett et al., 2005). Following a measurement sequence of an M-AERI installed on the *Explorer of the Seas* that included a horizontal measurement of atmospheric emission, it was found that averaging the parts of the spectra taken in the sea- and sky-viewing directions was sufficiently accurate, so that horizontal measurements are not required to determine the air temperature. When the M-AERI is mounted only a few meters above the sea surface, a correction for emission from the sea that contributes to the sea-viewing measurement is needed (Minnett et al., 2005). The radiometric measurement of air temperature has advantages over conventional approaches using contact thermometers in the air as these require heat to be exchanged between the air and the sensor, usually a very stable thermistor, and are prone to solar heating effects unless the thermometer is very carefully shielded (Berry et al., 2004; Berry and Kent, 2005). In tropical conditions, however, the radiometric approach shares a problem with conventional measurements as the air coming from directions that cause it to travel over the ship can be heated by the deck and superstructure before it is measured by a thermometer or before it enters the field of view of the spectrometer as a warm plume; in both cases, rejecting measurement taken with the relative wind coming from sectors occupied by the ship removes contaminated air temperature measurements (Berry and Kent, 2005; Minnett et al., 2005).

Given that both air temperature and SST_{skin} can be derived from different parts of the same spectrum, an accurate measurement of the air-sea temperature difference can be derived, avoiding many of the issues of the conventional method which is to form the difference between an air temperature measurement taken with a thermometer and a subsurface temperature measurement at a depth of a few meters. Thus, a temperature difference, often small, is derived from two thermometers with different calibration history in two fluids with very different thermal capacities resulting in a determination that sums the inaccuracies of each measurement. Even though a subsurface temperature can routinely be taken to high accuracy (<0.01 K inaccuracy), it is taken a depth below the interface, so it does not represent the temperature difference between the surface of the ocean and the near-surface atmosphere. Even in conditions when diurnal warming effects are negligible, the ubiquitous presence of the thermal skin layer introduces an error in the conventional measurement of the air-sea temperature difference; the temperature drop across the thermal skin layer can be a substantial part of the conventionally measured air-sea temperature difference (Kara et al., 2007). Variability in the air-sea temperature difference also plays a role in limiting the accuracy of satellite-derived SST_{skin} (May and Holyer, 1993).

Fig. 11 shows an example of histograms of air-sea temperature difference measured by M-AERIs, which is representative of those taken over a wide geographic range and different environmental conditions. Away from coasts and strong ocean frontal outcrops, there are generally very few instances where the air is warmer than the SST_{skin} (Fig. 12).

The air-sea temperature difference is critical variable in the calculation of the surface turbulent fluxes of heat, moisture, and momentum (Smith, 1988; Fairall et al., 2003), which are needed for many applications including weather forecasting and climate research. At the time of writing, there are no satellite instruments capable of deriving near-surface air temperature and air sea temperature differences from orbit, but it is foreseeable that new microwave hyperspectral radiometers will provide such measurements in the near future (Gentemann et al., 2020).

6.5 Atmospheric profiles

Ground-based thermal infrared spectrometers, such as the AERI, offer a powerful way to look at the thermodynamic structure and evolution of the lower atmosphere (Wulfmeyer et al., 2015). These instruments measure the emission from spectral regions that have varying optical depths, and thus are receiving signals from different heights in the atmosphere. These instruments measure in spectral bands associated with gases that can be assumed to be constant with altitude for temperature profiling, and also in spectral bands associated with emission by water vapor molecules to provide information on the vertical distribution of humidity. The high information content of AERI observations is particularly well suited for obtaining vertical profiles of temperature and water vapor in the boundary layer, and early efforts at retrieval development produced the clear-sky AERIprof retrieval (Smith et al., 1999; Feltz et al., 2003). Similarly, over the ocean, profiles of temperature and humidity of the marine atmosphere can be derived from the M-AERI spectra—see Section 7.2.1.

The Tropospheric Optimal Estimation Retrieval (TROPoe, formerly AERIoe; Turner and Löhnert, 2014; Turner and Blumberg, 2018) uses an optimal estimation-based physical retrieval algorithm to obtain profiles of temperature and water vapor, as well as cloud liquid water path and effective radius for a single liquid cloud layer, based on ground-based spectral radiance measurements. The algorithm produces retrievals under all-sky conditions (in the absence of precipitation) up to

Surface-based thermal infrared spectrometers **Chapter | 6** 111

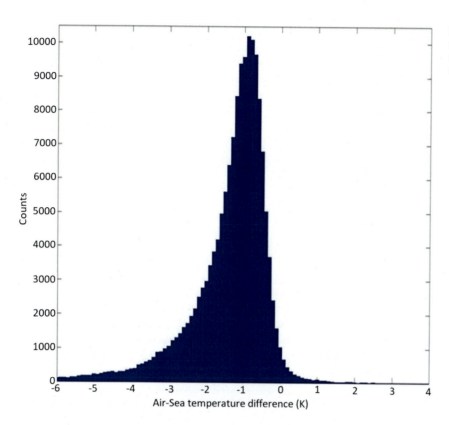

FIG. 11 Histogram of air-sea temperature difference measured radiometrically from the RCCL ship *Explorer of the Seas* in the Caribbean Sea and western North Atlantic from 2000 to 2006.

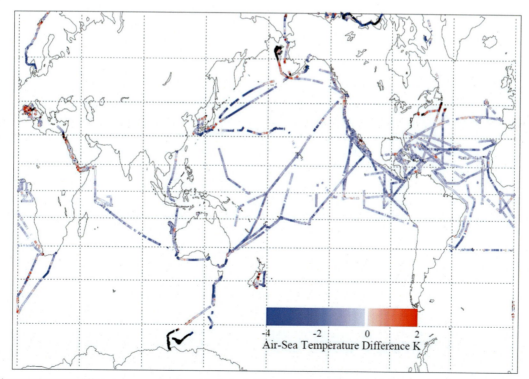

FIG. 12 Air-sea temperature difference measured radiometrically by the first generation RSMAS M-AERIs on research ships along the tracks shown in Fig. 6. For most of the oceans, the values are negative and positive occurrences are close to coasts or where there are strong ocean temperature fronts. Black indicates where the M-AERI was installed on icebreakers and the ship was in sea ice.

the cloud-base height. It provides a full error covariance matrix for each vertical profile and both the degrees of freedom for signal and the Shannon information content. Furthermore, the full error covariance matrix of the solution, the information content of the observations, and the true vertical resolution of the retrieved profiles are all natural outputs of the technique.

The algorithm has been rigorously validated in various climactic regimes (Arctic, mid-latitudes, tropics) through comparisons with radiosonde observations (Turner and Blumberg, 2018).

7. Applications

In this section, we present some examples of applications of M-AERI geophysical variables derived from M-AERI measurements; some are directly related to satellite data and others address pressing scientific problems.

7.1 Validating satellite SST$_{skin}$ retrievals

The primary motivation for developing the M-AERI was to provide ship-based measurements with which to assess the accuracy of the retrievals of SST$_{skin}$ from the MODerate-resolution Imaging Spectroradiometers on the Terra and Aqua satellites of NASA's Earth Observing System (Esaias et al., 1998) by comparisons of the satellite retrievals with those taken very close to the sea surface. As discussed earlier, the benefit is that this is a comparison of "like with like." The spatial resolution of the satellite measurements is typically 1 km^2 at the subsatellite point (Cracknell, 1997; Prata et al., 1990; Wolfe et al., 2002), growing toward the edges of the swath because of the geometrical spreading of the field of view; newer sensors have a higher spatial resolution at nadir, 0.5 km^2, with limited growth in pixel size across the swath (Schueler et al., 2013). Ideally, the M-AERI measurement should be in a cloud-free pixel in the satellite image, the position of which is well known. The inaccuracies in the pixel positions are typically much less than the pixel size (Wolfe et al., 2002, 2013), but not all pixels are cloud free. When the pixel containing the measurement is determined to be cloudy, then a nearby cloud-free pixel can be used. Studies have been made to determine the acceptable spatial separation between the satellite and surface-based measurements (Minnett, 1991; Corlett et al., 2014). Similarly, the M-AERIs may not be taking a sea-viewing measurement at the instant of the satellite measurements, and it is necessary to have an acceptable time interval between the two (Corlett et al., 2014). For the accuracy assessment of the SST$_{skin}$ values derived from many satellite radiometers, the intervals are taken as 10 km in space and 10 min in time. For M-AERI data taken from research ships on station, they are essentially point measurements, but for measurements taken from cruise ships traveling at ~22 km the integration interval of the M-AERI sea-viewing measurement results in a linear average along the ship's track. For the first-generation M-AERI, the integration time was 90 s resulting in an average along 1 km, comparable to the pixel size of many satellite imaging radiometers.

In addition to the assessment of accuracy, the comparisons are used to refine the cloud-screening and atmospheric correction algorithms to improve their performance, and this requires inclusion of many other variables that are capable of influencing the accuracy of the satellite retrievals, such as satellite zenith angle, solar zenith angle (to discriminate between daytime and nighttime measurements), wind speed and atmospheric water vapor content, are taken from a variety of sources to be included in the databases used for algorithm generation and refinement. These are frequently referred to as "Match-Up Data Bases" (MUDBs).

The M-AERI SST$_{skin}$ values have been used to validate the retrievals from the Advanced Very High Resolution Radiometer (AVHRR; Kearns et al., 2000; Vincent et al., 2008a,b), the MODerate-resolution Imaging Spectroradiometer (MODIS; Kilpatrick et al., 2015), the Advanced Along-Track Scanning Radiometer (AATSR; Corlett et al., 2006; Noyes et al., 2006), the Visible Infrared Imaging Radiometer Suite (VIIRS; Minnett et al., 2020), the Sea and Land Surface Temperature Radiometer (SLSTR; Luo et al., 2020a), and the Advanced Baseline Imager (ABI; Luo and Minnett, 2021), as well as contributing to many regional studies, including in the North Water Polynya in Baffin Bay (Vincent et al., 2008a,b), the Equatorial Indian Ocean (Vialard et al., 2009), and the Southern Ocean (Harvey et al., 2011; Minnett et al., 2011). Luo et al. (2020b) have used M-AERI data to assess the accuracy of the SST$_{skin}$ represented in reanalysis data, MERRA-2 (Gelaro et al., 2017) and ERA-Interim (Dee et al., 2011). The M-AERI data have also contributed to studies of the behavior of the thermal skin layer (Donlon et al., 2002; Minnett et al., 2011; Wong and Minnett, 2016b; Luo et al., 2022; Wong and Minnett, 2018) and diurnal heating of the ocean surface (Gentemann and Minnett, 2008; Gentemann et al., 2009; Luo et al., 2022).

7.1.1 Estimates of errors and uncertainties

The simplest way of representing the accuracy of the satellite SST$_{skin}$ retrievals by comparison with independent measurements is using gross statistics, as shown in Table 1 for MODIS and VIIRS on the Suomi National Polar-orbiting Partnership

TABLE 1 Global statistics for MODIS on *Terra* and *Aqua* and the S-NPP VIIRS nighttime SST$_{skin}$ retrievals relative to SST$_{skin}$ derived from measurements of the RSMAS M-AERIs. Temperature differences are in K.

Sensor	Mean	Median	Standard deviation	Robust standard deviation	Count
Terra MODIS	−0.058	−0.052	0.481	0.347	3069
Aqua MODIS	0.042	0.040	0.494	0.347	2070
S-NPP VIIRS	0.029	0.043	0.411	0.305	10,074

From Minnett, P.J., Kilpatrick, K.A., Podestá, G.P., Evans, R.H., Szczodrak, M.D., Izaguirre, M.A., Williams, E.J., Walsh, S., Reynolds, R.M., Bailey, S.W., Armstrong, E.M., Vazquez-Cuervo, J., 2020. Skin sea-surface temperature from VIIRS on Suomi-NPP—NASA continuity retrievals. Remote Sens. 12, 3369. https://doi.org/10.3390/rs12203369, by CC BY 4.0 license.

(S-NPP) satellite relative to M-AERI measurements. It is tempting to assign the differences revealed in the comparison to errors and uncertainties in the satellite SST$_{skin}$ retrievals, but there are contributions from other sources including the M-AERI and from spatial and temporal variations resulting from the mismatch in the fields of view of the satellite and ship-based radiometers, and the possible changes in the SST$_{skin}$ between the times of the two measurements (see Section 7.1 and Corlett et al., 2014).

The statistics serve a good purpose for assessing the gross performance of the SST$_{skin}$ retrieval algorithms, but they do not provide information about the conditions in which the performance is poorer than in others, and it is this information that is crucial to improving the algorithms. Following discussions in the Group for High Resolution Seas-Surface Temperature (GHRSST; Donlon et al., 2007), the concept of Sensor Specific Error Statistics (SSES) was promoted, that recommended that each SST retrieval, that is each pixel in an SST image, should be accompanied by an estimate of the mean, or bias, error, and the uncertainty of that pixel. Details of the SSESs are further discussed in Chapter 21.

7.1.2 Traceability to SI standards to generate CDRs

SST has been defined to be one of several Essential Climate Variables (Bojinski et al., 2014) that are indicators of a changing climate. Thus, a major motivation for deriving accurate SST$_{skin}$ retrievals from satellites is the need to produce global Climate Data Records (CDRs) that would fulfill the requirements of the research and climate monitoring communities (NRC, 2000, 2004; Fox et al., 2011). The accuracy and stability requirements for CDRs of SST are very stringent, being an absolute accuracy of ~0.1 K and a decadal stability of 0.04 mK (Ohring et al., 2005). These values are very difficult to achieve, and also very difficult to demonstrate whether they have been attained. The requirements of a CDR imply that SST$_{skin}$ fields be generated over multiple satellite missions, using different radiometer types. Each satellite radiometer is very well calibrated and characterized in the laboratory before launch, and once on orbit, by using internal calibration procedures, which are discussed by Minnett and Smith (2014), but the accuracies of the SST$_{skin}$ retrieval is limited by the effectiveness of the retrieval algorithms. Assessment of the retrieval accuracy using SI-traceable sensors, such as M-AERIs, permits the combination of SST$_{skin}$ fields from multiple satellite radiometers to generate multidecadal time series and to reduce the sampling errors introduced by clouds (Liu and Minnett, 2016; Liu et al., 2017).

The SI-traceable ship-board radiometers are the basis of generating SST$_{skin}$ CDRs, as discussed by Minnett and Corlett (2012a), but they are few compared to the global drifting buoy array—compare Fig. 6 and https://ships4sst.org/instruments/deployments with https://www.aoml.noaa.gov/phod/gdp/. Consequently, a combination of comparisons with radiometers and buoys is needed to provide the accuracy assessment of the satellite retrievals shown schematically in Fig. 13.

7.2 Atmospheric profiles

Thermodynamic profiles retrieved from AERI data have been used for many applications, including the investigation of cold fronts and drylines (Feltz et al., 1998; Turner et al., 2000), monitoring the evolution of convective indices and severe weather (Feltz and Mecikalski, 2002; Hu et al., 2019; Lewis et al., 2020), characterizing the differences between convective indices in tornadic and nontornadic storms (Wagner et al., 2008), studying the evolution of the boundary layer as atmospheric bores and other boundary propagate overhead (Bluestein et al., 2017; Toms et al., 2017; Grasmick et al., 2018; Haghi et al., 2019; Loveless et al., 2019), as input into a cumulus entrainment rate retrieval scheme (Wagner et al., 2013), studying nocturnal convection (Chipilski et al., 2020; Degelia et al., 2020).

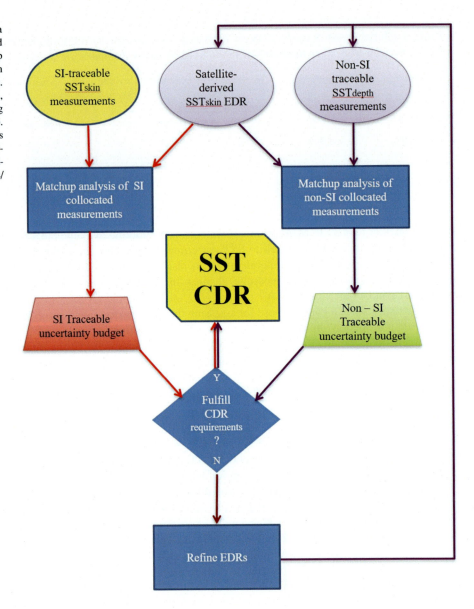

FIG. 13 Flow chart for generating a satellite-derived SST Climate Data Record (CDR) exploiting the SI-traceability of ship radiometers and also the data from the much larger number of drifting and moored buoys. EDR is Environmental Data Record, i.e., satellite retrievals of SST but not meeting the requirements of a CDR (NRC, 2000). The data flow is the result of discussions at a series of workshops hosted by the International Space Science Institute, Bern, Switzerland (https://www.issibern.ch/teams/satradio/index.html).

7.2.1 Saharan air layers

A particularly difficult situation for deriving atmospheric humidity and temperature profiles from hyperspectral measurements from the surface, from M-AERIs, and from space, using AIRS or IASI, is the Saharan Air Layer (SAL), a dry layer that originates as the name suggests in the Sahara Desert and is often associated with dust aerosols. Fig. 14 shows the SAL off W. Africa derived from M-AERI measurements taken from the NOAA S *Ronald H Brown* during an AEROSE (Aerosol and Ocean Science Expeditions; Nalli et al., 2011) cruise (Szczodrak et al., 2007). The red vertical lines are times of radiosonde launches from the ship, the data from which are used to initialize the retrievals. The tick marks indicate the times of M-AERI measurements of atmospheric emissions taken at zenith.

7.3 Greenhouse gas heating of the ocean

As discussed earlier, the infrared radiative exchanges between the ocean and atmosphere take place on the aqueous side in a very thin electromagnetic skin layer (Fig. 1), which is embedded in a viscous sublayer that inhibits the vertical transport of heat between the interface and the water beneath. Nevertheless, there is evidence that the ocean, throughout the water column, is heating in concert with the rate of increasing concentrations of greenhouse gases (Levitus et al., 2012). Using

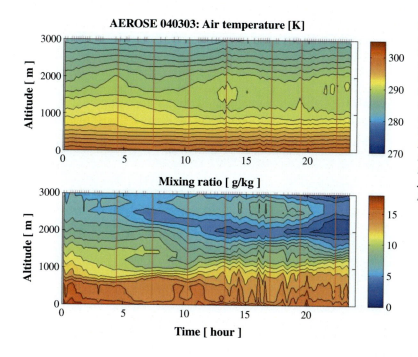

FIG. 14 Temperature and water vapor mixing ratio structure retrieved from zenith measurements taken by a first generation RSMAS M-AERI from March 3, 2004 during an AEROSE cruise on the NOAA S *Ronald H Brown*. First-guess profiles were constructed from radiosonde profiles launched from the ship at the times shown by red vertical lines. The cruise took place off W. Africa and the dry Saharan Air Layer is apparent at heights of 2–2.5 km. *(From Szczodrak, M., Minnett, P.J., Nalli, N.R., Feltz, W.F., 2007. Profiling the lower troposphere over the ocean with infrared hyperspectral measurements of the marine-atmosphere emitted radiance interferometer. J. Ocean. Atmos. Technol. 24, 390–402.* © *American Meteorological Society.)*

clouds as a source of increasing infrared radiation incident at the sea surface, Wong and Minnett (2018) investigated the response of the temperature drop across the thermal skin layer using profiles derived from M-AERI spectra, The presence of clouds can increase the incident infrared radiation by ~40 W m^{-2} in the tropics, and there is no evidence that this leads to a commensurate increase in the turbulent heat losses to the atmosphere, nor in the surface emission. Consequently, the increase in incident infrared radiation leads to a modification of the internal temperature gradient within the thermal skin layer resulting in the surface heat exchanges remaining more-or-less constant, but with the reduction of the heat flux drawn into the skin layer from below. Thus, a mechanism for greenhouse gas heating of the body of the ocean was demonstrated (Wong and Minnett, 2018; Fig. 15).

8. Summary and conclusions

FTIRs have been used on ships for a quarter century and have provided values of SST$_{skin}$ in a wide geographic range and in conditions that encompass the full gamut of atmospheric and oceanic variability for the validation of satellite-derived

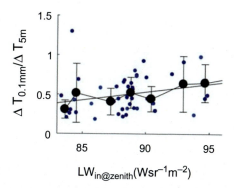

FIG. 15 The response of the ocean thermal skin layer to changes in incident infrared radiation, determined by integrating the spectra measured by a first-generation RSMAS M-AERI in the tropical pacific. Variations in the incident infrared radiation are caused by variable cloud conditions. The y-axis is the ratio of the temperature drop derived from the shallowest and deepest values in the temperature profiles in the electromagnetic skin layer derived from M-AERI spectra ($\Delta T_{0.1mm}$), to the temperature drop from the shallowest temperature to a $T_{subskin}$ measurement at a depth of 5 m (ΔT_{5m}); see Fig. 1. The data were taken at night during low wind speed conditions, <2 m s^{-1}. *(From Wong, E.W., Minnett, P.J., 2018. The response of the ocean thermal skin layer to variations in incident infrared radiation. J. Geophys. Res. Oceans 123, 19. https://doi.org/10.1002/2017JC013351.)*

SST_{skin}. Developments over many years have resulted in robust, accurate, stable instruments that can operate unattended for many months in the harsh environment of the marine atmospheric boundary layer. The data from the M-AERIs have determined the accuracy of the SST_{skin} retrievals from many satellite radiometers. The SI-traceable calibration of the M-AERIs permits the merging of SST_{skin} retrievals from many satellite instruments into a coherent data set that constitutes a Climate Data Record. In addition, the spectral information in the oceanic and atmospheric emission have supported many fundamental investigations, ranging from the sea-surface emissivity through diurnal heating to the response of the ocean to changing incident infrared radiation at the sea surface that has applications in the study of the ocean response to increasing levels of greenhouse gases.

The RSMAS M-AERI SST_{skin} and near-surface air temperature data from 2013 to 2020 are available from https://doi.org/10.17604/bswq-0119.

Acknowledgments

The development of the RSMAS M-AERIs was funded by NASA, and of the ARM M-AERI by the Department of Energy; these agencies also funded many deployments. Additional deployments of the RSMAS M-AERIs were supported by the NSF, DOE, NOAA, and the Ministry of Foreign Affairs, Italy. The Royal Caribbean Group is acknowledged for their longstanding support for the installation and operation of RSMAS M-AERIs on four of their cruise ships. The captains, officers, crews, and scientific colleagues of many research ships are thanked for their cooperation during many M-AERI deployments.

We would also like to acknowledge Werenfrid Wimmer for reviewing our draft and providing constructive feedback.

References

Barton, I.J., Minnett, P.J., Donlon, C.J., Hook, S.J., Jessup, A.T., Maillet, K.A., Nightingale, T.J., 2004. The Miami2001 infrared radiometer calibration and inter-comparison: 2. Ship comparisons. J. Atmos. Ocean. Technol. 21, 268–283.

Bell, R., 1972. Introductory Fourier Transform Spectroscopy. Elsevier, New York & London, p. 382. 0323152104, 0323152104.

Berry, D.I., Kent, E.C., 2005. The effect of instrument exposure on marine air temperatures: an assessment using VOSClim data. Int. J. Climatol. 25, 1007–1022.

Berry, D.I., Kent, E.C., Taylor, P.K., 2004. An analytical model of heating errors in marine air temperatures from ships. J. Atmos. Ocean. Technol. 21, 1198–1215.

Bertie, J.E., Lan, Z.D., 1996. Infrared intensities of liquids XX: the intensity of the OH stretching band revisited, and the best current values of the optical constants of H_2O (l) at 25°C between 15,000 and 1 cm^{-1}. Appl. Spectrosc. 50, 1047–1057.

Bluestein, H.B., Wienhoff, Z.B., Turner, D.D., Reif, D.W., Snyder, J.C., Thiem, K.J., Houser, J.B., 2017. A comparison of the finescale structures of a prefrontal wind-shift line and a strong cold front in the southern plains of the United States. Mon. Weather Rev. 145, 3307–3330. 1520-0493.

Bojinski, S., Verstraete, M., Peterson, T.C., Richter, C., Simmons, A., Zemp, M., 2014. The concept of essential climate variables in support of climate research, applications, and policy. Bull. Am. Meteorol. Soc. 95, 1431–1443. https://doi.org/10.1175/bams-d-13-00047.1.

Bouillon, M., Safieddine, S., Hadji-Lazaro, J., Whitburn, S., Clarisse, L., Doutriaux-Boucher, M., Coppens, D., August, T., Jacquette, E., Clerbaux, C., 2020. Ten-year assessment of IASI radiance and temperature. Remote Sens. 12, 2393. 2072-4292.

Centurioni, L.R., Turton, J., Lumpkin, R., Braasch, L., Brassington, G., Chao, Y., Charpentier, E., Chen, Z., Corlett, G., Dohan, K., Donlon, C., Gallage, C., Hormann, V., Ignatov, A., Ingleby, B., Jensen, R., Kelly-Gerreyn, B.A., Koszalka, I.M., Lin, X., Lindstrom, E., Maximenko, N., Merchant, C.J., Minnett, P., O'Carroll, A., Paluszkiewicz, T., Poli, P., Poulain, P.-M., Reverdin, G., Sun, X., Swail, V., Thurston, S., Wu, L., Yu, L., Wang, B., Zhang, D., 2019. Global in situ observations of essential climate and ocean variables at the air–sea interface. Front. Mar. Sci. 6. https://doi.org/10.3389/fmars.2019.00419.

Chipilski, H.G., Wang, X., Parsons, D.B., 2020. Impact of assimilating PECAN profilers on the prediction of bore-driven nocturnal convection: a multi-scale forecast evaluation for the 6 July 2015 case study. Mon. Weather Rev. 148, 1147–1175. 0027-0644.

Corlett, G.K., Barton, I.J., Donlon, C.J., Edwards, M.C., Good, S.A., Horrocks, L.A., Llewellyn-Jones, D.T., Merchant, C.J., Minnett, P.J., Nightingale, T.J., Noyes, E.J., O'Carroll, A.G., Remedios, J.J., Robinson, I.S., Saunders, R.W., Watts, J.G., 2006. The accuracy of SST retrievals from AATSR: an initial assessment through geophysical validation against in situ radiometers, buoys and other SST data sets. Adv. Space Res. 37, 764–769. https://doi.org/10.1016/j.asr.2005.09.037.

Corlett, G.K., Merchant, C.J., Minnett, P.J., Donlon, C.J., 2014. Assessment of long-term satellite derived sea surface temperature records. In: Zibordi, G., Donlon, C.J., Parr, A.C. (Eds.), Optical Radiometry for Ocean Climate Measurements. Experimental Methods in the Physical Sciences, vol. 47. Academic Press, pp. 639–677, https://doi.org/10.1016/B978-0-12-417011-7.00021-0. 1079-4042. Available from: http://www.sciencedirect.com/science/article/pii/B9780124170117000210.

Cracknell, A.P., 1997. The Advanced Very High Resolution Radiometer. CRC Press, Taylor and Francis, London, UK, p. 968. ISBN 9780748402090.

Davis, S.P., Abrams, M.C., Brault, J.W., 2001. Fourier Transform Spectrometry. Elsevier, p. 276. 0080506917, 0080506917.

Dee, D.P., Uppala, S.M., Simmons, A.J., Berrisford, P., Poli, P., Kobayashi, S., Andrae, U., Balmaseda, M.A., Balsamo, G., Bauer, P., Bechtold, P., Beljaars, A.C.M., van de Berg, L., Bidlot, J., Bormann, N., Delsol, C., Dragani, R., Fuentes, M., Geer, A.J., Haimberger, L., Healy, S.B., Hersbach, H., Hólm, E.V., Isaksen, L., Kållberg, P., Köhler, M., Matricardi, M., McNally, A.P., Monge-Sanz, B.M., Morcrette, J.J., Park, B.K., Peubey, C., de

Rosnay, P., Tavolato, C., Thépaut, J.N., Vitart, F., 2011. The ERA-Interim reanalysis: configuration and performance of the data assimilation system. Q. J. Roy. Meteorol. Soc. 137, 553–597. https://doi.org/10.1002/qj.828.

Degelia, S.K., Wang, X., Stensrud, D.J., Turner, D.D., 2020. Systematic evaluation of the impact of assimilating a network of ground-based remote sensing profilers for forecasts of nocturnal convection initiation during PECAN. Mon. Weather Rev. 148, 4703–4728. 0027-0644.

Donlon, C.J., Minnett, P.J., Gentemann, C., Nightingale, T.J., Barton, I.J., Ward, B., Murray, J., 2002. Toward improved validation of satellite sea surface skin temperature measurements for climate research. J. Climate 15, 353–369. https://doi.org/10.1175/1520-0442(2002)015<0353:TIVOSS>2.0.CO;2.

Donlon, C.J., Robinson, I., Casey, K.S., Vazquez-Cuervo, J., Armstrong, E., Arino, O., Gentemann, C., May, D., LeBorgne, P., Piollé, J., Barton, I., Beggs, H., Poulter, D.J.S., Merchant, C.J., Bingham, A., Heinz, S., Harris, A., Wick, G., Emery, B., Minnett, P., Evans, R., Llewellyn-Jones, D., Mutlow, C., Reynolds, R.W., Kawamura, H., Rayner, N., 2007. The global ocean data assimilation experiment high-resolution sea surface temperature pilot project. Bull. Am. Meteorol. Soc. 88, 1197–1213.

Donlon, C.J., Minnett, P.J., Jessup, A., Barton, I., Emery, W., Hook, S., Wimmer, W., Nightingale, T.J., Zappa, C., 2014. Ship-borne thermal infrared radiometer systems. In: Zibordi, G., Donlon, C.J., Parr, A.C. (Eds.), Optical Radiometry for Ocean Climate Measurements. Experimental Methods in the Physical Sciences, vol. 47. Academic Press, pp. 305–404, https://doi.org/10.1016/B978-0-12-417011-7.00011-8. 1079-4042. Available from: http://www.sciencedirect.com/science/article/pii/B9780124170117000118.

Esaias, W.E., Abbott, M.R., Barton, I., Brown, O.B., Campbell, J.W., Carder, K.L., Clark, D.K., Evans, R.H., Hoge, F.E., Gordon, H.R., Balch, W.M., Letelier, R., Minnett, P.J., 1998. An overview of MODIS capabilities for ocean science observations. IEEE Trans. Geosci. Remote Sens. 36, 1250–1265. https://doi.org/10.1109/36.701076.

Eyre, J.R., 1987. On systematic errors in satellite sounding products and their climatological mean values. Q. J. Roy. Meteorol. Soc. 113, 279–292. https://doi.org/10.1002/qj.49711347516.

Fairall, C., Bradley, E., Godfrey, J., Wick, G., Edson, J., Young, G., 1996. Cool-skin and warm-layer effects on sea surface temperature. J. Geophys. Res. 101, 1295–1308.

Fairall, C.W., Bradley, E.F., Hare, J.E., Grachev, A.A., Edson, J.B., 2003. Bulk parameterization of air-sea fluxes: updates and verification for the COARE algorithm. J. Climate 16, 571–591.

Feltz, W.F., Mecikalski, J.R., 2002. Monitoring high-temporal-resolution convective stability indices using the ground-based Atmospheric Emitted Radiance Interferometer (AERI) during the 3 May 1999 Oklahoma–Kansas tornado outbreak. Weather Forecast. 17, 445–455. 1520-0434.

Feltz, W.F., Smith, W.L., Knuteson, R.O., Revercomb, H.E., Woolf, H.M., Howell, H.B., 1998. Meteorological applications of temperature and water vapor retrievals from the ground-based atmospheric emitted radiance interferometer (AERI). J. Appl. Meteorol. 37, 857–875.

Feltz, W.F., Smith, W.L., Howell, H.B., Knuteson, R.O., Woolf, H., Revercomb, H.E., 2003. Near-continuous profiling of temperature, moisture, and atmospheric stability using the atmospheric emitted radiance interferometer (AERI). J. Appl. Meteorol. 42, 584–597.

Fowler, J.B., 1995. A third generation water bath based blackbody source. J. Res. Natl. Inst. Stand. Technol. 100, 591–599.

Fox, N., Kaiser-Weiss, A., Schmutz, W., Thome, K., Young, D., Wielicki, B., Winkler, R., Woolliams, E., 2011. Accurate radiometry from space: an essential tool for climate studies. Philos. Trans. R. Soc. A Math. Phys. Eng. Sci. 369, 4028–4063. https://doi.org/10.1098/rsta.2011.0246.

Gelaro, R., McCarty, W., Suárez, M.J., Todling, R., Molod, A., Takacs, L., Randles, C.A., Darmenov, A., Bosilovich, M.G., Reichle, R., Wargan, K., Coy, L., Cullather, R., Draper, C., Akella, S., Buchard, V., Conaty, A., da Silva, A.M., Gu, W., Kim, G.-K., Koster, R., Lucchesi, R., Merkova, D., Nielsen, J.E., Partyka, G., Pawson, S., Putman, W., Rienecker, M., Schubert, S.D., Sienkiewicz, M., Zhao, B., 2017. The modern-era retrospective analysis for research and applications, version 2 (MERRA-2). J. Climate 30, 5419–5454. https://doi.org/10.1175/jcli-d-16-0758.1.

Gentemann, C.L., 2014. Three way validation of MODIS and AMSR-E sea surface temperatures. J. Geophys. Res. Oceans 119, 2583–2598. https://doi.org/10.1002/2013JC009716.

Gentemann, C.L., Minnett, P.J., 2008. Radiometric measurements of ocean surface thermal variability. J. Geophys. Res. 113, C08017. https://doi.org/10.1029/2007JC004540.

Gentemann, C.L., Minnett, P.J., Ward, B., 2009. Profiles of ocean surface heating (POSH): a new model of upper ocean diurnal thermal variability. J. Geophys. Res. 114, C07017. https://doi.org/10.1029/2008JC004825.

Gentemann, C.L., Clayson, C.A., Brown, S., Lee, T., Parfitt, R., Farrar, J.T., Bourassa, M., Minnett, P.J., Seo, H., Gille, S.T., Zlotnicki, V., 2020. FluxSat: measuring the ocean–atmosphere turbulent exchange of heat and moisture from space. Remote Sens. (Basel) 12, 1796. https://doi.org/10.3390/rs12111796.

Grasmick, C., Geerts, B., Turner, D.D., Wang, Z., Weckwerth, T., 2018. The relation between nocturnal MCS evolution and its outflow boundaries in the stable boundary layer: an observational study of the 15 July 2015 MCS in PECAN. Mon. Weather Rev. 146, 3203–3226. 1520-0493.

Griffiths, P.R., de Haseth, J.A., 1986. Fourier Transform Infrared Spectrometry. John Wiley & Sons, New York. 0-471-09902-3, 0-471-09902-3.

Haghi, K.R., Geerts, B., Chipilski, H.G., Johnson, A., Degelia, S., Imy, D., Parsons, D.B., Adams-Selin, R.D., Turner, D.D., Wang, X., 2019. Bore-ing into nocturnal convection. Bull. Am. Meteorol. Soc. 100, 1103–1121. 0003-0007.

Hanafin, J.A., Minnett, P.J., 2005. Infrared-emissivity measurements of a wind-roughened sea surface. Appl. Optics 44, 398–411.

Harvey, M.J., Law, C.S., Smith, M.J., Hall, J.A., Abraham, E.R., Stevens, C.L., Hadfield, M.G., Ho, D.T., Ward, B., Archer, S.D., Cainey, J.M., Currie, K.I., Devries, D., Ellwood, M.J., Hill, P., Jones, G.B., Katz, D., Kuparinen, J., Macaskill, B., Main, W., Marriner, A., McGregor, J., McNeil, C., Minnett, P.J., Nodder, S.D., Peloquin, J., Pickmere, S., Pinkerton, M.H., Safi, K.A., Thompson, R., Walkington, M., Wright, S.W., Ziolkowski, L.A., 2011. The SOLAS air–sea gas exchange experiment (SAGE) 2004. Deep-Sea Res. II Top. Stud. Oceanogr. 58, 753–763. https://doi.org/10.1016/j.dsr2.2010.10.015.

Hu, J., Yussouf, N., Turner, D.D., Jones, T.A., Wang, X., 2019. Impact of ground-based remote sensing boundary layer observations on short-term probabilistic forecasts of a tornadic supercell event. Weather Forecast. 34, 1453–1476. 1520-0434.

Kara, A.B., Hurlburt, H.E., Loh, W.-Y., 2007. Which near–surface atmospheric variable drives air–sea temperature differences over the global ocean? J. Geophys. Res. 112, C05020. https://doi.org/10.1029/2006JC003833.

Kearns, E.J., Hanafin, J.A., Evans, R.H., Minnett, P.J., Brown, O.B., 2000. An independent assessment of Pathfinder AVHRR sea surface temperature accuracy using the marine-atmosphere emitted radiance interferometer (M-AERI). Bull. Am. Meteorol. Soc. 81, 1525–1536.

Kilpatrick, K.A., Podestá, G., Walsh, S., Williams, E., Halliwell, V., Szczodrak, M., Brown, O.B., Minnett, P.J., Evans, R., 2015. A decade of sea surface temperature from MODIS. Remote Sens. Environ. 165, 27–41. https://doi.org/10.1016/j.rse.2015.04.023.

Knuteson, R.O., Revercomb, H.E., Best, F.A., Ciganovich, N.C., Dedecker, R.G., Dirkx, T.P., Ellington, S.C., Feltz, W.F., Garcia, R.K., Howell, H.B., Smith, W.L., Short, J.F., Tobin, D.C., 2004a. Atmospheric emitted radiance interferometer. Part I: instrument design. J. Atmos. Oceanic Technol. 21, 1763–1776.

Knuteson, R.O., Revercomb, H.E., Best, F.A., Ciganovich, N.C., Dedecker, R.G., Dirkx, T.P., Ellington, S.C., Feltz, W.F., Garcia, R.K., Howell, H.B., Smith, W.L., Short, J.F., Tobin, D.C., 2004b. Atmospheric emitted radiance interferometer. Part II: instrument performance. J. Atmos. Ocean. Technol. 21, 1777–1789.

Legeckis, R., 1977. Long waves in the eastern equatorial Pacific Ocean: a view from a geostationary satellite. Science 197, 1179–1181. https://doi.org/10.1126/science.197.4309.1179.

Legeckis, R., 1978. A survey of worldwide sea surface temperature fronts detected by environmental satellites. J. Geophys. Res. Oceans 83, 4501–4522. https://doi.org/10.1029/JC083iC09p04501.

Legeckis, R.V., 1979. Satellite observations of the influence of bottom topography on the seaward deflection of the Gulf Stream off Charleston, South Carolina. J. Phys. Oceanogr. 9, 483–497. https://doi.org/10.1175/1520-0485(1979)009<0483:sootio>2.0.co;2.

Levitus, S., Antonov, J.I., Boyer, T.P., Baranova, O.K., Garcia, H.E., Locarnini, R.A., Mishonov, A.V., Reagan, J.R., Seidov, D., Yarosh, E.S., Zweng, M.M., 2012. World ocean heat content and thermosteric sea level change (0–2000 m), 1955–2010. Geophys. Res. Lett. 39, L10603. https://doi.org/10.1029/2012GL051106.

Lewis, W.E., Wagner, T.J., Otkin, J.A., Jones, T.A., 2020. Impact of AERI temperature and moisture retrievals on the simulation of a central plains severe convective weather event. Atmosphere 11, 729. 2073-4433.

Liu, W.T., Businger, J.A., 1975. Temperature profile in molecular sublayer near the interface of a fluid in turbulent motion. Geophys. Res. Lett. 2, 403–404.

Liu, Y., Minnett, P.J., 2016. Sampling errors in satellite-derived infrared sea-surface temperatures. Part I: global and regional MODIS fields. Remote Sens. Environ. 177, 48–64. https://doi.org/10.1016/j.rse.2016.02.026.

Liu, Y., Chin, T.M., Minnett, P.J., 2017. Sampling errors in satellite-derived infrared sea-surface temperatures. Part II: sensitivity and parameterization. Remote Sens. Environ. 198, 297–309. https://doi.org/10.1016/j.rse.2017.06.011.

Loewenstein, E.V., 1966. The history and current status of Fourier transform spectroscopy. Appl. Optics 5, 845–854. https://doi.org/10.1364/AO.5.000845.

Loveless, D.M., Wagner, T.J., Turner, D.D., Ackerman, S.A., Feltz, W.F., 2019. A composite perspective on bore passages during the PECAN campaign. Mon. Weather Rev. 147, 1395–1413. 0027-0644.

Lumpkin, R., Centurioni, L., Perez, R., 2016. Fulfilling observing system implementation requirements with the global drifter array. J. Atmos. Oceanic Tech. 33, 685–695. https://doi.org/10.1175/jtech-d-15-0255.1.

Lumpkin, R., Özgökmen, T., Centurioni, L., 2017. Advances in the application of surface drifters. Ann. Rev. Mar. Sci. 9, 59–81. 1941-1405.

Luo, B., Minnett, P.J., 2021. Skin sea surface temperatures from the GOES-16 ABI validated with those of the shipborne M-AERI. IEEE Trans. Geosci. Remote Sens., 1–12. https://doi.org/10.1109/TGRS.2021.3054895.

Luo, B., Minnett, P.J., Szczodrak, M., Akella, S., 2022. Regional and seasonal variability of the oceanic thermal skin effect. J. Geophys. Res. Oceans 127. https://doi.org/10.1029/2022JC018465. e2022JC018465.

Luo, B., Minnett, P.J., Szczodrak, M., Kilpatrick, K., Izaguirre, M., 2020a. Validation of sentinel-3A SLSTR derived sea-surface skin temperatures with those of the shipborne M-AERI. Remote Sens. Environ. 244, 111826. https://doi.org/10.1016/j.rse.2020.111826.

Luo, B., Minnett, P.J., Szczodrak, M., Nalli, N.R., Morris, V.R., 2020b. Accuracy assessment of MERRA-2 and ERA-Interim sea surface temperature, air temperature, and humidity profiles over the Atlantic Ocean using AEROSE measurements. J. Climate 33, 6889–6909. https://doi.org/10.1175/jcli-d-19-0955.1.

Masuda, K., 2006. Infrared sea surface emissivity including multiple reflection effect for isotropic Gaussian slope distribution model. Remote Sens. Environ. 103, 488–496. https://doi.org/10.1016/j.rse.2006.04.011.

May, D.A., Holyer, R.J., 1993. Sensitivity of satellite multichannel sea surface temperature retrievals to the air-sea temperature difference. J. Geophys. Res. Oceans 98, 12567–12577. https://doi.org/10.1029/93jc00913.

McMillin, L., 1975. Estimation of sea-surface temperatures from two infrared window measurements with different absorption. J. Geophys. Res. 80, 5113–5117.

Michelson, A.A., Morely, E.W., 1887. On the relative motion of the earth and the luminiferous ether. Am. J. Sci. 35, 333–345.

Minnett, P.J., 1991. Consequences of sea surface temperature variability on the validation and applications of satellite measurements. J. Geophys. Res. 96, 18475–18489.

Minnett, P.J., 2003. Radiometric measurements of the sea-surface skin temperature—the competing roles of the diurnal thermocline and the cool skin. Int. J. Remote Sens. 24, 5033–5047.

Minnett, P.J., Corlett, G.K., 2012. Generation of Climate Data Records of Sea-Surface Temperature From Current and Future Satellite Radiometers—Report of the First Workshop, March 26–30, 2012. (61 pp.). Bern, Switzerland. Available from: http://www.issibern.ch/teams/satradio/documents/ISSI_Sat_SST_CDR_Workshop1_FinalReport.pdf.

Minnett, P.J., Corlett, G.K., 2012b. A pathway to generating Climate Data Records of sea-surface temperature from satellite measurements. Deep-Sea Res. II Top. Stud. Oceanogr. 77–80, 44–51. https://doi.org/10.1016/j.dsr2.2012.04.003.

Minnett, P.J., Smith, D.L., 2014. Postlaunch calibration and stability: thermal infrared satellite radiometers. In: Zibordi, G., Donlon, C.J., Parr, A.C. (Eds.), Optical Radiometry for Ocean Climate Measurements. Experimental Methods in the Physical Sciences, vol. 47. Academic Press, pp. 201–243, https://doi.org/10.1016/B978-0-12-417011-7.00008-8. 1079-4042. Available from: http://www.sciencedirect.com/science/article/pii/B9780124170117000088.

Minnett, P.J., Knuteson, R.O., Best, F.A., Osborne, B.J., Hanafin, J.A., Brown, O.B., 2001. The Marine-Atmospheric Emitted Radiance Interferometer (M-AERI), a high-accuracy, sea-going infrared spectroradiometer. J. Atmos. Ocean. Technol. 18, 994–1013.

Minnett, P.J., Maillet, K.A., Hanafin, J.A., Osborne, B.J., 2005. Infrared interferometric measurements of the near surface air temperature over the oceans. J. Atmos. Ocean. Technol. 22, 1016–1029.

Minnett, P.J., Smith, M., Ward, B., 2011. Measurements of the oceanic thermal skin effect. Deep-Sea Res. II Top. Stud. Oceanogr. 58, 861–868. https://doi.org/10.1016/j.dsr2.2010.10.024.

Minnett, P.J., Alvera-Azcárate, A., Chin, T.M., Corlett, G.K., Gentemann, C.L., Karagali, I., Li, X., Marsouin, A., Marullo, S., Maturi, E., Santoleri, R., Saux Picart, S., Steele, M., Vazquez-Cuervo, J., 2019. Half a century of satellite remote sensing of sea-surface temperature. Remote Sens. Environ. 233, 111366. https://doi.org/10.1016/j.rse.2019.111366.

Minnett, P.J., Kilpatrick, K.A., Podestá, G.P., Evans, R.H., Szczodrak, M.D., Izaguirre, M.A., Williams, E.J., Walsh, S., Reynolds, R.M., Bailey, S.W., Armstrong, E.M., Vazquez-Cuervo, J., 2020. Skin sea-surface temperature from VIIRS on Suomi-NPP—NASA continuity retrievals. Remote Sens. (Basel) 12, 3369. https://doi.org/10.3390/rs12203369.

Nalli, N.R., Minnett, P.J., Maddy, E., McMillan, W.W., Goldberg, M.D., 2008a. Emissivity and reflection model for calculating unpolarized isotropic water surface-leaving radiance in the infrared. 2: validation using Fourier transform spectrometers. Appl. Optics 47, 4649–4671. https://doi.org/10.1364/AO.47.004649.

Nalli, N.R., Minnett, P.J., van Delst, P., 2008b. Emissivity and reflection model for calculating unpolarized isotropic water surface-leaving radiance in the infrared. I: theoretical development and calculations. Appl. Optics 47, 3701–3721. https://doi.org/10.1364/AO.47.003701.

Nalli, N.R., Joseph, E., Morris, V.R., Barnet, C.D., Wolf, W.W., Wolfe, D., Minnett, P.J., Szczodrak, M., Izaguirre, M.A., Lumpkin, R., Xie, H., Smirnov, A., King, T.S., Wei, J., 2011. Multiyear observations of the tropical Atlantic atmosphere: multidisciplinary applications of the NOAA aerosols and ocean science expeditions. Bull. Am. Meteorol. Soc. 92, 765–789. https://doi.org/10.1175/2011BAMS2997.1.

Noyes, E.J., Minnett, P.J., Remedios, J.J., Corlett, G.K., Good, S.A., Llewellyn-Jones, D.T., 2006. The accuracy of the AATSR sea surface temperatures in the Caribbean. Remote Sens. Environ. 101, 38–51.

NRC, 2000. Issues in the Integration of Research and Operational Satellite Systems for Climate Research: II. Implementation. National Academy of Sciences, Washington, DC, ISBN: 0-309-06994-7.

NRC, 2004. Climate Data Records From Environmental Satellites. National Academy of Sciences, Washington, DC, USA. 0-309-53080-6, 150 pp.

O'Carroll, A.G., Eyre, J.R., Saunders, R.W., 2008. Three-way error analysis between AATSR, AMSR-E, and in situ sea surface temperature observations. J. Atmos. Oceanic Technol. 25, 1197–1207.

Ohring, G., Wielicki, B., Spencer, R., Emery, B., Datla, R., 2005. Satellite instrument calibration for measuring global climate change: report of a workshop. Bull. Am. Meteorol. Soc. 86, 1303–1313.

Poli, P., Lucas, M., O'Carroll, A., Le Menn, M., David, A., Corlett, G.K., Blouch, P., Meldrum, D., Merchant, C.J., Belbeoch, M., 2019. The Copernicus surface velocity platform drifter with barometer and reference sensor for temperature (SVP-BRST): genesis, design, and initial results. Ocean Sci. 15, 199–214. 1812-0784.

Post, M.J., Fairall, C.W., White, A.B., Han, Y., Ecklund, W.L., Weickmann, K.M., Cooper, D.I., Minnett, P.J., Quinn, P.K., Sekelsky, S.M., McIntosh, R.E., Knuteson, R.O., 1997. The combined sensor program: an air-sea science mission in the central and western Pacific Ocean. Bull. Am. Meteorol. Soc. 78, 2797–2815.

Prata, A.J., Cechet, R.P., Barton, I.J., Llewellyn-Jones, D.T., 1990. The along-track scanning radiometer for ERS-1—scan geometry and data simulation. IEEE Trans. Geosci. Remote Sens. 28, 3–13.

Price, J.F., Weller, R.A., Pinkel, R., 1986. Diurnal cycling: observations and models of the upper ocean response to diurnal heating, cooling and wind mixing. J. Geophys. Res. 91, 8411–8427.

Revercomb, H.E., Buijs, H., Howell, H.B., LaPorte, D.D., Smith, W.L., Sromovsky, L.A., 1988. Radiometric calibration of IR Fourier transform spectrometers: solution to a problem with the High-Resolution Interferometer Sounder. Appl. Optics 27, 3210–3218.

Rice, J.P., Butler, J.J., Johnson, B.C., Minnett, P.J., Maillet, K.A., Nightingale, T.J., Hook, S.J., Abtahi, A., Donlon, C.J., Barton, I.J., 2004. The Miami2001 infrared radiometer calibration and intercomparison: 1. Laboratory characterization of blackbody targets. J. Atmos. Ocean. Technol. 21, 258–267.

Robinson, I.S., 2004. Measuring the Oceans from Space: The Principles and Methods of Satellite Oceanography. Springer Science & Business Media, p. 670. ISBN 978-3-540-42647-9.

Schueler, C.F., Lee, T.F., Miller, S.D., 2013. VIIRS constant spatial-resolution advantages. Int. J. Remote Sens. 34, 5761–5777. https://doi.org/10.1080/01431161.2013.796102.

Shankland, R.S., 1964. Michelson-Morley experiment. Am. J. Phys. 32, 16–35. 0002-9505.

Smith, S.D., 1988. Coefficients for sea surface wind stress, heat flux, and wind profiles as a function of wind speed and temperature. J. Geophys. Res. 93, 15467–15472.

Smith, N., Barnet, C.D., 2020. CLIMCAPS observing capability for temperature, moisture, and trace gases from AIRS/AMSU and CrIS/ATMS. Atmos. Meas. Tech. 13, 4437–4459. https://doi.org/10.5194/amt-13-4437-2020.

Smith, W.L., Knuteson, R.O., Revercomb, H.E., Feltz, W., Howell, H.B., Menzel, W.P., Nalli, N., Brown, O., Brown, J., Minnett, P., McKeown, W., 1996. Observations of the infrared radiative properties of the ocean—implications for the measurement of sea surface temperature via satellite remote sensing. Bull. Am. Meteorol. Soc. 77, 41–51.

Smith, W.L., Feltz, W.F., Knuteson, R.O., Revercomb, H.E., Woolf, H.M., Howell, H.B., 1999. The retrieval of planetary boundary layer structure using ground-based infrared spectral measurements. J. Atmos. Ocean. Technol. 16, 323–333.

Soloviev, A.V., Lukas, R., 2014. The Near-Surface Layer of the Ocean: Structure, Dynamics, and Applications, second ed. Springer, New York, ISBN: 978-94-007-7620-3, p. 552, https://doi.org/10.1007/978-94-007-7621-0.

Stokes, G.M., Schwartz, S.E., 1994. The atmospheric radiation measurement (ARM) program: programmatic background and design of the cloud and radiation test bed. Bull. Am. Meteorol. Soc. 75, 1201–1221.

Szczodrak, M., Minnett, P.J., Nalli, N.R., Feltz, W.F., 2007. Profiling the lower troposphere over the ocean with infrared hyperspectral measurements of the marine-atmosphere emitted radiance interferometer. J. Oceanic Atmos. Technol. 24, 390–402.

Taylor, J.K., Revercomb, H.E., Best, F.A., Tobin, D.C., Gero, P.J., 2020. The infrared absolute radiance interferometer (ARI) for CLARREO. Remote Sens. (Basel) 12, 1915. 2072-4292.

Theocharous, E., Fox, N.P., Barker-Snook, I., Niclòs, R., Santos, V.G., Minnett, P.J., Göttsche, F.M., Poutier, L., Morgan, N., Nightingale, T., Wimmer, W., Høyer, J., Zhang, K., Yang, M., Guan, L., Arbelo, M., Donlon, C.J., 2019. The 2016 CEOS infrared radiometer comparison: part II: laboratory comparison of radiation thermometers. J. Atmos. Ocean. Technol. 36, 1079–1092. https://doi.org/10.1175/jtech-d-18-0032.1.

Toms, B.A., Tomaszewski, J.M., Turner, D.D., Koch, S.E., 2017. Analysis of a lower-tropospheric gravity wave train using direct and remote sensing measurement systems. Mon. Weather Rev. 145, 2791–2812. 1520-0493.

Turner, D.D., Blumberg, W.G., 2018. Improvements to the AERIoe thermodynamic profile retrieval algorithm. IEEE J. Sel. Top. Appl. Earth Obs. Remote Sens. 12, 1339–1354. 1939-1404.

Turner, D.D., Löhnert, U., 2014. Information content and uncertainties in thermodynamic profiles and liquid cloud properties retrieved from the ground-based atmospheric emitted radiance interferometer (AERI). J. Appl. Meteorol. Climatol. 53, 752–771. https://doi.org/10.1175/jamc-d-13-0126.1.

Turner, D.D., Feltz, W.F., Ferrare, R.A., 2000. Continuous water profiles from operational ground-based active and passive remote sensors. Bull. Am. Meteorol. Soc. 81, 1301–1317.

Vialard, J., Duvel, J.P., McPhaden, M.J., Bouruet-Aubertot, P., Ward, B., Key, E., Bourras, D., Weller, R., Minnett, P., Weill, A., Cassou, C., Eymard, L., Fristedt, T., Basdevant, C., Dandonneau, Y., Duteil, O., Izumo, T., de Boyer Montégut, C., Masson, S., Marsac, F., Menkes, C., Kennan, S., 2009. Cirene: air-sea interactions in the Seychelles-Chagos thermocline ridge region. Bull. Am. Meteorol. Soc. 90, 45–61. https://doi.org/10.1175/2008BAMS2499.1.

Vincent, R.F., Marsden, R.F., Minnett, P.J., Buckley, J.R., 2008a. Arctic waters and marginal ice zones: part 2—an investigation of Arctic atmospheric infrared absorption for AVHRR sea surface temperature estimates. J. Geophys. Res. 113, C08044. https://doi.org/10.1029/2007JC004354.

Vincent, R.F., Marsden, R.F., Minnett, P.J., Creber, K.A.M., Buckley, J.R., 2008b. Arctic waters and marginal ice zones: a composite Arctic sea surface temperature algorithm using satellite thermal data. J. Geophys. Res. 113, C04021. https://doi.org/10.1029/2007JC004353.

Wagner, T.J., Feltz, W.F., Ackerman, S.A., 2008. The temporal evolution of convective indices in storm-producing environments. Weather Forecast. 23, 786–794. 1520-0434.

Wagner, T.J., Turner, D.D., Berg, L.K., Krueger, S.K., 2013. Ground-based remote retrievals of cumulus entrainment rates. J. Atmos. Ocean. Technol. 30, 1460–1471. 0739-0572.

Ward, B., 2006. Near-surface ocean temperature. J. Geophys. Res. 111, C02005. https://doi.org/10.1029/2004JC002689.

Wilheit, T.T., Chang, A.T.C., 1980. An algorithm for retrieval of ocean surface and atmospheric parameters from the observations of the scanning multi-channel microwave radiometer. Radio Sci. 15, 525–544.

Wolfe, R.E., Nishihama, M., Fleig, A.J., Kuyper, J.A., Roy, D.P., Storey, J.C., Patt, F.S., 2002. Achieving sub-pixel geolocation accuracy in support of MODIS land science. Remote Sens. Environ. 83, 31–49.

Wolfe, R.E., Lin, G., Nishihama, M., Tewari, K.P., Tilton, J.C., Isaacman, A.R., 2013. Suomi NPP VIIRS prelaunch and on-orbit geometric calibration and characterization. J. Geophys. Res. Atmos. 118, 2013JD020508. https://doi.org/10.1002/jgrd.50873.

Wong, E.W., Minnett, P.J., 2016a. Retrieval of the ocean skin temperature profiles from measurements of infrared hyperspectral radiometers—part I: derivation of an algorithm. IEEE Trans. Geosci. Remote Sens. 54, 1879–1890. https://doi.org/10.1109/TGRS.2015.2483746.

Wong, E.W., Minnett, P.J., 2016b. Retrieval of the ocean skin temperature profiles from measurements of infrared hyperspectral radiometers—part II: field data analysis. IEEE Trans. Geosci. Remote Sens. 54, 1891–1904. https://doi.org/10.1109/TGRS.2015.2501425.

Wong, E.W., Minnett, P.J., 2018. The response of the ocean thermal skin layer to variations in incident infrared radiation. J. Geophys. Res. Oceans 123, 19. https://doi.org/10.1002/2017JC013351.

Workman, J.J., 2018. A review of calibration transfer practices and instrument differences in spectroscopy. Appl. Spectrosc. 72, 340–365.

Wulfmeyer, V., Hardesty, R.M., Turner, D.D., Behrendt, A., Cadeddu, M.P., Di Girolamo, P., Schlüssel, P., Van Baelen, J., Zus, F., 2015. A review of the remote sensing of lower tropospheric thermodynamic profiles and its indispensable role for the understanding and the simulation of water and energy cycles. Rev. Geophys. 53, 819–895. https://doi.org/10.1002/2014RG000476.

Chapter 7

Sun photometers

Carlos Toledano[a], Victoria E. Cachorro[a], David Mateos[a], Roberto Román[a], Ramiro González[a], Alexander Smirnov[b,c], Julian Gröbner[d], Stelios Kazadzis[d], and Natalia Kouremeti[d]

[a]*Group of Atmospheric Optics, University of Valladolid, Valladolid, Spain,* [b]*NASA Goddard Space Flight Center, Greenbelt, MD, United States,* [c]*Science Systems and Applications, Inc., Lanham, MD, United States,* [d]*Physikalisch-Meteorologisches Observatorium Davos, World Radiation Center (PMOD/WRC), Davos, Switzerland*

Chapter outline

1. Introduction	121
2. Basics of Sun photometry	122
2.1 Theoretical background	122
2.2 Instrument specifications	124
3. Measurement networks and data products	125
3.1 Aerosol robotic network (AERONET)	125
3.2 Maritime aerosol network (MAN) as a component of AERONET	126
3.3 Global atmosphere watch: Precision filter radiometer network (GAW-PFR)	127
3.4 Brewer networks	129
3.5 Other networks	130
4. Calibration and uncertainty	131
5. Aerosol data interpretation	132
5.1 Aerosol types	132
5.2 Quality assurance and quality control	132
6. Conclusions	135
Acknowledgments	135
References	135

The three most important issues in Sun photometry are: calibration; calibration; and calibration.

(J. Michalsky, Univ. of Boulder).

1. Introduction

The measurement of the solar irradiance spectrum at the ground is a fundamental technique for the derivation of the atmospheric composition (Shaw, 2007). The scattering and absorption of the solar radiation by the atmosphere constituents produces extinction in the direct beam as well as the sky (diffuse) radiation. There exists a wealth of passive remote sensing techniques and applications based on the observation of the solar radiation, the moon or stars for nighttime, as well as the sky radiance, with a variety of target wavelength range and measurement strategies, including the classic global/direct/diffuse solar irradiance with pyranometers, differential optical absorption spectroscopy (DOAS) or Fourier-transform infrared spectroscopy (FTIR), to cite some. We need to focus on certain techniques in this chapter, but it is important to bear in mind that photometry in atmospheric research includes many other applications.

This chapter focuses on spectral Sun photometry and its applications mainly for aerosol properties' monitoring and its capability for calibration/validation of satellite sensors. We also provide some information about water vapor and ozone column derivation.

The extinction of the direct solar irradiance by the air molecules and various constituents (water vapor, ozone, aerosol particles, nitrogen dioxide, etc.) produces distinct attenuation of different parts of the solar spectrum. Thus, the spectral extinction is used to derive column-integrated properties of the aerosol particles, as well as to obtain the column abundances of absorbing species.

Long-term monitoring programs of NASA (the Aerosol Robotic Network, AERONET) and the World Meteorological Organization (Global Atmosphere Watch, GAW) were successfully created for this purpose, providing accurate and long-term observations based on robust instruments, procedures, and calibration techniques. These observations have been extensively used for aerosol characterization, satellite, and model validation and contribute to synergetic analyses with other atmospheric investigation techniques.

2. Basics of Sun photometry

2.1 Theoretical background

The extinction of the direct solar irradiance through the atmosphere is described by the radiative transfer equation. In the absence of emission in a certain air volume:

$$dE(\lambda) = -\kappa(\lambda) \cdot E(\lambda) \cdot dz \tag{1}$$

where $E(\lambda)$ is the irradiance at λ-wavelength, $\kappa(\lambda)$ is the extinction coefficient, and dz is the distance through the air volume. If we assume that $\kappa(\lambda)$ does not depend on z, the solution is the well-known Beer-Bouguer-Lambert law:

$$E(\lambda) = E_0(\lambda) \cdot e^{-\kappa(\lambda) \cdot z} \tag{2}$$

where E_0 is the initial irradiance entering the air volume. We can express the exponent in a more general way, assuming that $\kappa(\lambda)$ is a function of z, with the definition of the optical depth, which is the integral of the extinction along a certain path:

$$\tau(\lambda) = \int_{z_1}^{z_2} \kappa(\lambda, z) \cdot dz \tag{3}$$

Note that this expression is true for each wavelength, since the extinction κ (scattering plus absorption) strongly depends on wavelength. The scattering by air molecules (Rayleigh scattering) and aerosol particles (Mie scattering) is a smooth function of wavelength, whereas the absorption by atmospheric gases (H_2O, O_3, CO_2, NO_2, etc.) concentrates in absorption bands with abrupt changes as a function of wavelength.

In a single layer, plane-parallel atmosphere, the geometry of the direct solar irradiance measurement is illustrated in Fig. 1, where it can be seen that there is a geometrical relation between the optical depth in the measurement (slant) path and the optical depth in the vertical, which is denoted by air mass, m and depends on the solar zenith angle as $m = sec(\theta)$ in our plane-parallel approximation. More elaborated formulas exist for the air mass (Kasten and Young, 1989; Komhyr et al., 1989) that take into account the vertical distribution of the scatterer/absorber, the Earth curvature, or the refraction in the atmosphere. The solar zenith angle can be calculated following the formulation by Michalsky (1988).

Therefore, we can reformulate the Beer-Bouguer-Lambert law as follows:

$$E(\lambda) = E_0(\lambda) \cdot e^{-\tau(\lambda) \cdot m} \tag{4}$$

In this expression, the total optical depth (TOD) $\tau(\lambda)$ does not depend on the solar zenith angle and can be used to monitor the atmospheric extinction throughout the day. $E(\lambda)$ is the spectral irradiance at ground level and $E_0(\lambda)$ is the exo-atmospheric or extraterrestrial irradiance spectrum. TOD in cloudless conditions (i.e., Sun not obscured by clouds) is the result, for each wavelength, of the scattering and absorption contributions by the different atmospheric constituents:

$$\tau(\lambda) = \tau_R(\lambda) + \tau_a(\lambda) + \tau_{H_2O}(\lambda) + \tau_{O_3}(\lambda) + \cdots \tag{5}$$

where subscript R stands for Rayleigh scattering, a for aerosol scattering and absorption, H_2O and O_3 for water vapor and ozone absorptions, respectively, and we have assumed the same air mass for all components. The suspension points indicate that other species can be considered, depending on the wavelength range of interest. The Rayleigh optical depth can be obtained from different formulas (e.g., Bodhaine et al., 1999) and needs to be scaled to the actual pressure value, multiplying by P/P_0 (with $P_0 = 1013.25\,hPa$), because it depends on the actual air column above the photometer.

An example of the comparison between the direct solar irradiance before entering the atmosphere and at the ground is given in Fig. 2, where the absorption bands as well as the overall attenuation due to scattering (larger for shorter

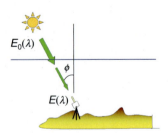

FIG. 1 Direct solar irradiance measurement geometry in a plane-parallel atmosphere.

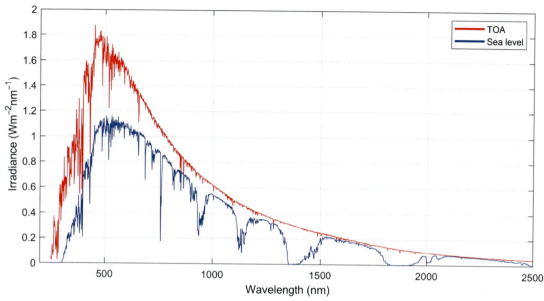

FIG. 2 Solar extraterrestrial spectrum and direct solar irradiance at the ground. Ground spectrum simulated with libRadtran version 1.0.4, US standard atmosphere, SZA = 32 degrees, O_3 = 300 DU, H_2O = 1 cm, Ånsgstrom parameters $\alpha = 1.1$, $\beta = 0.1$. Extraterrestrial spectrum from Kurucz (1994) (1 nm resolution).

wavelengths) can be seen. Depending on the intended application, the spectral measurements will concentrate on different parts of the solar spectrum. For instance, ozone retrievals are made by means of spectral irradiance measurements in the ultraviolet (UV) region. For aerosol research, spectral regions without strong gaseous absorption are chosen, while water vapor abundance can be retrieved in the corresponding absorption bands. The same kind of selection applies to the passive satellite sensors. In aerosol research, it is common to calculate the spectral slope of the aerosol optical depth (hereinafter AOD), following the empirical Ångström power law (Angström, 1930):

$$\tau_a(\lambda) = \beta \cdot \lambda^{-\alpha} \tag{6}$$

where α is the Ångström exponent (AE) and β equals to the AOD at 1 μm wavelength (λ). AE is a good indicative of the particle size predominance, being lower (close to 0) for coarse particles like mineral dust and higher (up to 2.5) for very fine particle aerosol, like urban pollution. The Ångström parameters (Cachorro et al., 1987) are retrieved with a linear fit of $\ln \tau_a$ versus $\ln \lambda$ (Fig. 3). Other authors propose second- or third-order polynomial fits that better capture the spectral variations of the AOD (see mineral dust case in Fig. 3) and allow the retrieval of microphysical parameters like fine-mode radius and the separation between fine- and coarse-mode AOD (O'Neill et al., 2003).

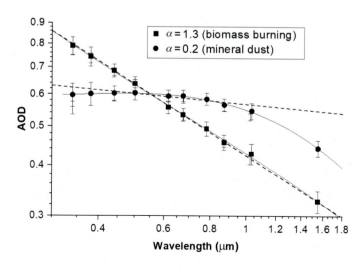

FIG. 3 Aerosol optical depth versus wavelength representation (log-log scale) for derivation of the Ångström exponent: fine-mode aerosol from biomass burning (*squares*); coarse mineral dust aerosol (*circles*). *Dashed line* represents the Ångström linear fit to the data in the 440–870 nm spectral range. *Solid line* is the second-order fit for the full spectral range.

If the other components are known or can be estimated (aerosol, Rayleigh), the $\tau_{H_2O}(\lambda)$ in a water vapor absorption band can be used to retrieve the precipitable water column. In bands of strong spectral variation of molecular absorption, the transmission is highly variable with wavelength and the Beer-Bouguer-Lambert law is no longer applicable for narrow-band filters. Therefore, the band and source-weighted water vapor transmittance (T_w) needs to be computed for the specific interference filter using radiative transfer modeling (Schmid et al., 2001). T_w can be modeled as

$$T_w = \exp\left[-a(u \cdot m)^b\right] \tag{7}$$

where u is the columnar water vapor, m is the water vapor air mass, and a and b are the instrument-specific constants, previously derived by a curve-fitting procedure of T_w over a range of slant path water vapor amounts (Ingold et al., 2000).

The approach to retrieve ozone amounts is somewhat different, because it is based on the differential absorption between two spectral channels, typically one band with strong and another one nearby with a weak absorption (Kerr, 2002). Following the Beer-Bouguer-Lambert law for two channels, i:

$$\ln E(\lambda_i) = \ln E_0(\lambda_i) - \tau_R(\lambda_i) \cdot m - C(\lambda_i) \cdot \Omega \cdot m_{O_3} \tag{8}$$

where Ω is the ozone amount, $C(\lambda_i)$ is the ozone absorption coefficient for the wavelength (band) λ_i, and m_{O_3} is the ozone air mass (Komhyr et al., 1989), which is slightly different to the air mass m due to the high elevation of the ozone layer in the atmosphere. From the difference between the above expression for channels 1 and 2 (Morys et al., 2001):

$$\Omega = \frac{\ln(E_0(\lambda_1)/E_0(\lambda_2)) - \ln(E(\lambda_1)/E(\lambda_2)) - [\tau_R(\lambda_1) - \tau_R(\lambda_2)] \cdot m}{[C(\lambda_1) - C(\lambda_2)] \cdot m_{O_3}} \tag{9}$$

Note that in this expression we have assumed that the aerosol has similar contribution in the two close wavelengths λ_1 and λ_2; therefore, it cancels out or has a minor effect (Basher and Thomas, 1979).

2.2 Instrument specifications

As mentioned earlier, the Beer-Bouguer-Lambert law is strictly valid only for monochromatic radiation; therefore, spectral measurements of the direct solar beam are needed. The necessary spectral resolution will depend on the application and the required uncertainty. In any case, the Sun photometers will generally have a fore optics that will define the aperture of the system and the field of view; they will also have an optical setup for wavelength discrimination; and finally, a detector or detector array will collect the energy. Each manufacturer provides its solution to this problem, although many elements are in common. For example, the Sun angular size (plan angle) is approximately 0.5 degrees in the sky; therefore, any fore optics must limit the field of view so that a minimum amount of diffuse radiation in the solar aureole enters the Sun photometer but avoiding the risk of vignetting the solar disk due to inaccurate pointing at the Sun direction.

About the spectral discrimination, there are several possible solutions. The first one is the use of a monochromator, based for example on a diffraction grating or prism that separates the incoming spectrum in different angular directions depending on wavelength. A detector array can be placed to collect the spectrum energy for all wavelengths simultaneously (spectrograph) or a single detector can be placed, and the grating be rotated to scan the different wavelengths onto the detector (spectroradiometer). Some instruments may have a double monochromator to reduce stray light and increase the spectral resolution. Another solution is the use of spectral filters, normally narrow-band interference filters with filter widths in the order of few nanometers (for the UV) up to tenths of nanometers (for the short-wave infrared). While monochromators allow for continuous spectral measurements (sometimes denoted as "hyperspectral"), the use of filters limits the spectral information to a certain number of narrow bands. The out-of-band blocking of this filters can be as high as optical density 6–7, which is needed for accurate UV measurements in the solar spectrum. Another approach is the use of interferometers, as it is the case of the Fourier transform infrared spectrometry (FTIR). This technique is described in Chapter 6.

As for the detectors, the choice depends upon the spectral range of interest. Silicon photodiodes are commonly used in the UV and visible (VIS) ranges, but they are no longer sensitive beyond 1 μm wavelength. In the UV, also photomultipliers are used. In the short-wave infrared (SWIR) range, from 0.7 up to 2.5 μm, other detector materials are needed. Indium-gallium-arsenide (InGaAs) sensors are the primary sensors used in SWIR imaging, for they are sensitive up to 2.5 μm wavelength and can cover the solar irradiance spectrum and overlap with silicon detectors in the region about 1 μm. This fact can be used to ensure the spectral measurement continuity as well as for quality assurance purposes, as it will be described in Section 5.2.

Some ground-based instruments can also measure sky radiances (this is the case for the AERONET radiometers, for example). The diffuse solar radiation is very useful to derive aerosol optical and microphysical properties by means of inversion methods (Dubovik and King, 2000; Dubovik et al., 2014; Nakajima et al., 1996; Sinyuk et al., 2020).

These complex algorithms include an aerosol model (Mie scattering for spherical particles, spheroid model for dust, etc.), radiative transfer calculations and a mathematical inversion algorithm (e.g., Multi-Term Least Squares Method). Polarization measurements can also be used to infer the degree of linear polarization of the sky light (Li et al., 2010). All these methods, generally based on the combination of AOD with spectral sky radiance at a set of scattering angles, will not be considered in this chapter. We will concentrate on the direct solar irradiance observation and its use to derive atmospheric constituents in the atmospheric column.

3. Measurement networks and data products

The large temporal and spatial variability of atmospheric constituents like water vapor or aerosol particles, makes it critical to monitor their properties on a global scale. The same principle applies to the satellite sensor validation, which needs to be accomplished under all possible conditions concerning geographical location, surface reflectance, cloudiness, pollution, humidity, aerosol type, etc. For this reason, large measurement networks are devoted to these tasks, with the aim of providing sufficient coverage and high-quality observations in a standardized manner.

3.1 Aerosol robotic network (AERONET)

The paradigm for this is the Aerosol Robotic Network (AERONET, Holben et al., 1998), a program launched by the NASA in the 1990s and focused on aerosol characterization and validation of satellite-based aerosol products. The main characteristic of this network is the standardization of instrument, calibration and data, as well as the public dissemination through their website (https://aeronet.gsfc.nasa.gov).

The radiometer used by AERONET is the Cimel Electronique CE318. This instrument has evolved over time. Three generations are currently used: analog models (starting 1992), digital (starting 2002), and "triple" 318-T (starting 2013) (Barreto et al., 2016; Toledano et al., 2018). The basic components and functioning have been maintained, though. It consists of a rotating filter wheel with nine narrow-band interference filters (Table 1) and two detectors, silicon for the ultraviolet and visible channels, and indium-gallium-arsenide for the short-wave infrared. The 1020 nm channel is observed with both detectors for quality control purposes (see Section 5.2). The field of view is 1.2 degrees, and a collimator tube is used to prevent from dust deposition on the front lenses and reduce stray light in the sky radiance measurements. The sensor head, equipped with a four-quadrant detector, is mounted on a two-axis robot equipped with step motor for accurate Sun tracking. The system is automatic, has low-power consumption, is able to transfer data through internet or using a satellite transmitter, and has proven to be very robust as it has been deployed in all kinds of environments: tropical, deserts, islands, polar, etc. The Cimel 318-T can also measure the moon direct irradiance to derive AOD during the night (Barreto et al., 2016).

TABLE 1 Cimel CE318 spectral channels: Nominal central wavelengths and filter widths at half maximum (FWHM).

Channel no.	Central wavelength (nominal) (nm)	FWHM (nm)
1	1020	10
2	1640 (InGaAs)	25
3	870	10
4	675	10
5	440	10
6	500	10
7	1020 (InGaAs)	10
8	936	10
9	380	4
10	340	2

All channels are measured with silicon detector except 1640 nm and the measurement at 1020 nm, which is made with both Si and InGaAs detectors.

AERONET provides long-term and near-real-time aerosol data that are used for aerosol monitoring, satellite and model calibration/validation purposes, and synergy with other instruments. The three pillars of the network are the standardization of instruments, calibration, and processing. The calibration is carried out at four calibration centers: Goddard Space Flight Center (GSFC, at Greenbelt, Maryland), NEON (National Ecological Observatory Network, at Longmont, Colorado), Laboratory of Atmospheric Optics (LOA, at Lille, France), and Group of Atmospheric Optics (GOA, at Valladolid, Spain) together with the Izaña Atmospheric Research Center (Tenerife, Spain). The calibration follows strict protocols and is designed to maintain high-quality data and homogeneity throughout the network. More details about the calibrations are given in Section 4.

The data processing is centralized at GSFC. Until now, three collections or processing versions have been developed: version 1 (Eck et al., 1999; Holben et al., 1998), version 2 (Smirnov et al., 2004), and the current version 3 (Giles et al., 2019). The corrections for the different wavelengths (gaseous absorptions, Rayleigh scattering), temperature characterization of all instruments, or cloud-screening procedures are key elements in the processing that have been improved over time. Another major improvement refers to the real-time quality control of the data: a number of automated checks are applied, based on the knowledge of the instrument characteristics and the physics of the measurement and magnitudes involved (Giles et al., 2019). These tests allow for near-real-time delivery of high-quality data for multiple validation and assimilation purposes. The network has largely expanded over 30 years. As shown in Fig. 4, the network greatly expanded from 1993 until 2021. The number of sites already exceeds 700. For comparison, up to 56 sites have more than 10 years of data (as for June 2021).

3.2 Maritime aerosol network (MAN) as a component of AERONET

Despite the large spatial coverage of AERONET, the oceans remain with poor coverage except for few remote island locations. Given the importance of the natural marine aerosol (Smirnov et al., 2002), and linked to AERONET, it emerged the

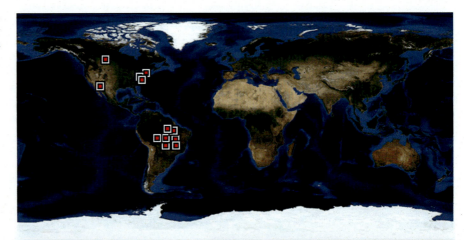

FIG. 4 Map of AERONET sites in 1993 (*top panel*) and 2021 (*bottom panel*). Only sites with more than 6 months of data are shown. *(Credit: NASA.)*

Maritime Aerosol Network (MAN, Smirnov et al., 2009). This initiative organizes AOD measurements on board ships (mainly research vessels although not only), based on the handheld Microtops II Sun photometer. These data are highly valuable since they provide observations over the ocean and validation points for satellite and aerosol transport models.

The Microtops II is a portable, handheld Sun photometer equipped with five spectral channels that are selected with interference filters. Two main filter configurations are 380, 440, 675, 870, and 936 nm or 440, 500, 675, 870, and 936 nm. The field of view is 2.5 degrees. The operator needs to manually point at the Sun and collect the measurements. A GPS attached to the main unit provides accurate time and geographical position to each measurement.

Both the calibration and the data processing are carried out within the AERONET network. The calibration is transferred from the reference Cimel radiometer at the GSFC facility. The processing of this shipborne data fostered the addition of specific features to the AERONET algorithms. In particular, the sites can have changing location (coordinates) and thus specific coordinates can be assigned to each single observation. The Microtops II direct Sun measurements allow calculating AOD, water vapor, and the Ångström parameters, following the AERONET direct Sun algorithm. For further information about MAN network, cruises (since 2004), and related publications, the reader can visit the website https://aeronet.gsfc.nasa.gov/new_web/maritime_aerosol_network.html. Fig. 5 shows a map with all cruises and corresponding AOD data, as for June 2021.

3.3 Global atmosphere watch: Precision filter radiometer network (GAW-PFR)

The Global Atmosphere Watch, Precision Filter Radiometer network (GAW-PFR, Wehrli, 2005), aims at monitoring the long-term changes in aerosol content and properties in the atmosphere. Therefore, it is composed by key sites, mainly located in remote environments around the globe, see Fig. 6, and employs very precise and stable instruments, to reduce at most the measurement uncertainty and be able to monitor trends at sites, where the aerosol content is generally very low. This is the case for high mountain stations like Mauna Loa in Hawai, Izaña in the Canary Islands or Jungfraujoch, Switzerland; or remote locations like Ny-Ålesund, Svalbard, Alice Springs, Australia, and others.

Moreover, GAW-PFR aims at providing intercomparison information between networks by overlapping at selected sites. The network is colocated with instruments of AERONET, SKYNET, CARSNET, the Australian Network, and other Sun photometers types at several sites. For instance, studies have been conducted for the comparison between GAW-PFR and AERONET (Cuevas et al., 2019; Kazadzis et al., 2018a, b), with an excellent agreement in spite of the technical differences, different (and independent) calibration, and data processing methodologies.

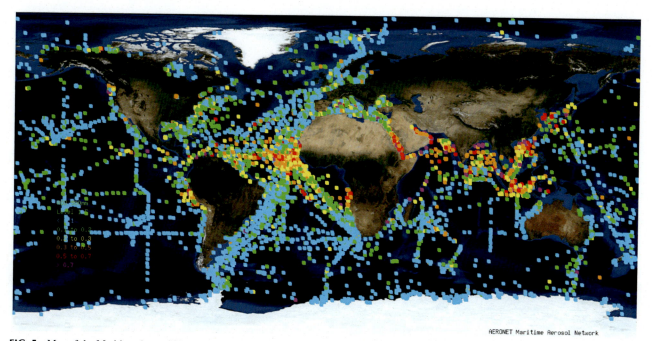

FIG. 5 Map of the Maritime Aerosol Network (MAN) cruises, as for June 2021. Color code indicates aerosol optical depth. Data are cloud screened. *(Credit: NASA.)*

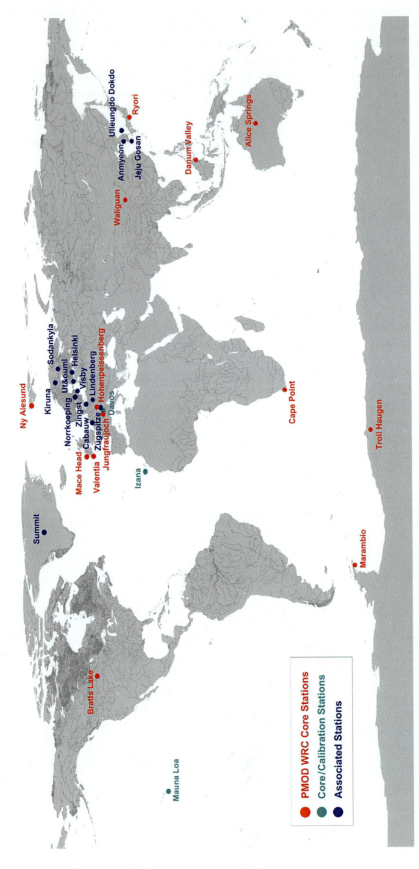

FIG. 6 Map of the Global Atmosphere Watch, Precision Filter Radiometer network for aerosol optical depth monitoring. *(Credit: PMOD-WRC.)*

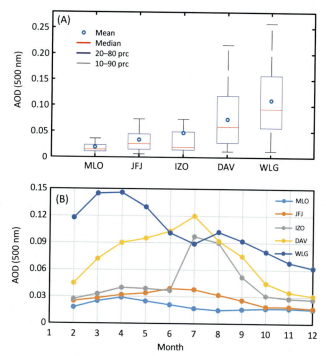

FIG. 7 (A) AOD (500 nm) mean and median for five high-altitude stations: (MLO 3.4 km, IZO 2.3 km, JFG 3.5 km, WLG 3.8 km, DAV 1.6 km). (B) Intraannual monthly mean AOD at 500 (*right*). *DAV*, Davos; *IZO*, Izaña; *JFG*, Jungfraujoch; *MLO*, Mauna Loa; *WLG*, Mount Walliguan.

The Precision Filter Radiometer (PFR, Wehrli, 2000) is the standard instrument of the World Meteorological Organization for AOD observations. It is designed for continuous and automated measurement of the direct solar irradiance at four narrow spectral channels (368, 412, 500, and 862 nm), with 5 nm FWHM. The full field of view is 2.5 degrees and it needs to be mounted on a solar tracker to follow the Sun. The PFR is designed for long-term operation: it is stabilized in temperature, filled with overpressure of dry nitrogen and the detectors are behind a shutter, to avoid exposure, except for the brief sampling periods.

The calibration of the network relies on a reference triad of instruments that are operated at the World Optical Depth Research Calibration Center (WORCC), established in 1996 at the World Radiation Center in Davos, Switzerland. Another two PFR instruments are frequently deployed at the high-altitude stations Mauna Loa (Hawaii) and Izaña (Canary islands) to perform Langley plot calibrations, and then are deployed at the WORCC in order to check and maintain the triad stability (Kazadzis et al., 2018b). The AOD climatology for several high-altitude stations is provided in Fig. 7. Such low AOD is needed to ensure the accuracy of the Langley plot calibration (see Section 4).

The algorithm for AOD processing is described in detail by Wehrli (2000) and WMO (2016). An automated cloud-screening scheme is also applied. Quality control is extensively applied to the data, in order to ensure maximum quality (Kazadzis et al., 2018b). Further information can be found at https://www.pmodwrc.ch/en/world-radiation-center-2/worcc/.

3.4 Brewer networks

The Brewer spectrophotometers are one of the main sources for measuring long-term ozone changes. This instrument was developed in the early 1980s with the aim of providing accurate solar irradiance spectrum in the ultraviolet region, between 300 and 320 nm (Kerr et al., 1981), from which the total ozone column can be derived. More than 200 instruments are currently installed around the world, many of them grouped into networks and coordination initiatives: the Network for the Detection of Atmospheric Composition Change (NDACC, https://www.ndaccdemo.org), the World Ozone and Ultraviolet Radiation Data Centre (WOUDC, https://woudc.org), the European Brewer Network (EUBREWNET, http://www.eubrewnet.org), and others. EUBREWNET was initiated by a European Union action (Rimmer et al., 2018), in order to improve the harmonization of observations, data processing, calibrations and operating procedures, and foster other products like solar UV irradiance and AOD or NO_2 from the Brewer instruments (Cede et al., 2006; López-Solano et al., 2018).

FIG. 8 Map of the SKYNET network. *(Credit: International SKYNET committee.)*

The basic measurement principle is to determine the total ozone column by comparing the direct solar measurements at wavelengths in the ultraviolet that are strongly and weakly absorbed by ozone. Specifically, four wavelengths between 310 and 320 nm are used to calculate total ozone. The "direct Sun" algorithm is the primary measurement mode of the Brewer, although it is also possible to employ zenith sky radiances under cloudy conditions (Fioletov et al., 2011).

Two main versions of the Brewer instrument are available, depending on whether they are equipped with single (MKII and MKIV) or double monochromator (MKIII), for improved stray light rejection, which is crucial in the UV region, where signals are weak and rapidly change with wavelength. The measurement frequency is not standardized but there are typically several direct Sun and zenith sky observations every hour during daytime. Brewers can also derive ozone columns using the Moon as light source (Kerr et al., 1990).

3.5 Other networks

Finally, it is necessary to mention that many other photometer networks exist worldwide, for example, SKYNET in Asia and Europe (Nakajima et al., 2020; Takamura et al., 2004), the Australian AOD network (Mitchell et al., 2017), the Chinese Aerosol Remote Sensing network (Che et al., 2015), National Oceanic and Atmospheric Administration surface radiation budget (SURFRAD) network (Augustine et al., 2008), to cite some.

SKYNET employs the Prede-POM photometer. The standard version has seven wavelengths (315, 400, 500, 675, 870, 940, and 1020 nm) for aerosol and water vapor retrieval. The POM-02 is an extended version, with UV wavelengths of 340 and 380 nm and shortwave infrared wavelengths of 1600 and 2200 nm. In this model, the 315 nm channel is used to retrieve ozone abundance. FWHM of the filters are <3 nm below 380 nm, 10 nm between 400 and 940 nm, and 20 nm for the infrared channels. The field of view is 1 degree. The instrument is equipped with a two-axis robot and is controlled by a PC. A modified version of POM-02 has been employed for lunar photometry (Uchiyama et al., 2019). A historical overview and references about data processing algorithms and calibration procedures is provided by Nakajima et al. (2020). Further information is provided on the website https://www.skynet-isdc.org. Most of the 50 SKYNET sites are in Asia, see Fig. 8. In Europe, the European SKYNET Radiometers network (http://www.euroskyrad.net, Campanelli et al., 2012) includes more than 15 sites.

4. Calibration and uncertainty

It is widely recognized that the key task in photometry is the instrument calibration. All the networks described earlier, make huge efforts in providing accurate calibration to their field instruments. The basic calibration process for spectral irradiance is to find a coefficient that converts the instrument output signals into physical units in Watts per square meter per unit wavelength. Standard lamps, with calibration certification by some metrology institution, can be used for this purpose. However, the determination of column amounts as described in Section 2, using the Beer-Bouger-Lambert law, does not strictly require such conversion into physical units. Actually, it is necessary to determine what the signal output of the instrument (at each spectral band) would be at the top of the atmosphere. The procedure for this calibration is the well-known Langley plot method (Shaw, 1983). A historical review on this method was provided by Shaw (2007).

The Langley plot determines the extraterrestrial constants (i.e., calibration coefficients) of the photometers by plotting the natural logarithm of the signals versus the air mass (Fig. 9). The data are fitted to a straight line, which intercept corresponds to air mass equal to zero, that is, signal outside the atmosphere. From the Beer-Bouguer-Lambert law:

$$\ln V(\lambda) = \ln V_0(\lambda) - \tau(\lambda) \cdot m \tag{10}$$

In this equation, we have replaced irradiances ("E") by raw signals ("V" for voltage, counts, etc.), to highlight the fact that there is no need to actually convert signals into physical units for the derivation of the optical depth.

The main requirement for the Langley plot method to effectively retrieve the extraterrestrial signal is that the atmospheric transmission does not change either temporarily or spatially (within about 50 km of the observer) during the acquisition of solar irradiance at a sufficient air mass range (or solar elevations). In practice, if the optical depth intends to be determined within ±0.01 uncertainty, the Langley plot needs to be performed at stations with both high altitude—located in the free troposphere—and also low latitude, to ensure that the Sun elevation rapidly changes at sunrise or sunset (Toledano et al., 2018) and that AOD does not significantly change during the Langley measurement period. Several alternatives have been developed (e.g., Cachorro et al., 2004; Campanelli et al., 2004; Forgan, 1994; Herman et al., 1981), in order to reduce the calibration error in case of atmospheric instability.

However, the classic Langley plot method is the one that is basically used in the photometer networks (Bais, 1997; Holben et al., 1998; Wehrli, 2005). The Mauna Loa (Hawaii) and Izaña (Canary Islands) stations are key sites for this purpose. The number of clear-sky and pristine days suitable for Langley plot calibrations is about 200 days/year (Toledano et al., 2018). The reference (or master) instruments are calibrated at these sites, and then used in intercalibration sites (at GSFC, Lille, Valladolid, or Davos) for side-to-side calibration of field instruments. Logistically, it is not feasible to transport several hundreds of instruments every year to the reference stations. Traveling standards are also used in some cases.

The uncertainty in the AOD can be derived with the classical derivation formula (JCGM, 2008):

$$\delta\tau_a = \sqrt{\left(\frac{1}{m}\cdot\frac{\delta V}{V}\right)^2 + \left(\frac{1}{m}\cdot\frac{\delta V_0}{V_0}\right)^2 + \left(\delta\tau_R \cdot \frac{P}{P_0}\right)^2 + \left(\delta\tau_g\right)^2 + \cdots} \tag{11}$$

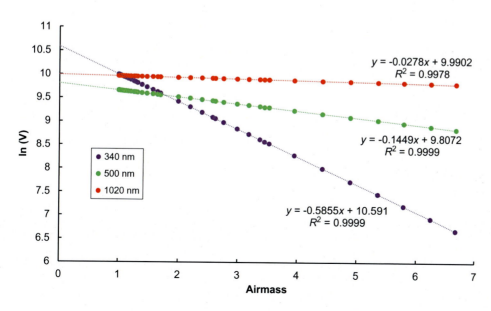

FIG. 9 Example of Langley plot for the determination of the top-of-the-atmosphere instrument signal.

Note that the signal (V) and calibration (V_0) errors produce an error in AOD that depends on the air mass (m), and therefore presents a symmetry with local noon (Cachorro et al., 2008). This is an important fact that is further discussed in the quality control part of Section 5.2.

Some of the uncertainty components are systematic, whereas others are not and therefore could be reduced by measurement averaging. Calibration errors are systematic. Other example of systematic error is the diffuse signal entering the field of view in the direct Sun measurements. This excessive signal produces systematically lower AOD than real, the difference depending on the field of view of the instrument as well as the actual aerosol amount and type (Cuevas et al., 2019). The finite bandwidth of the spectral measurements, incorrect out-of-band blocking, or instrument nonlinearity are other sources of systematic error (Reagan et al., 1986). On the other hand, electronic noise or inaccuracy in atmospheric pressure or gas corrections due to changing atmosphere fall in the category of random errors. Overall, the signal and calibration uncertainties have the largest impact. For instance, 1% error in calibration will result in 0.01 optical depth error for air mass equal to 1.

The overall uncertainty for AOD measurements is 0.01–0.02 for AERONET (larger for shorter wavelengths), although it is better for reference instruments (<0.002–0.005). That is the uncertainty range for GAW-PFR reference instruments too. This is achieved by several months of Langley plot calibrations at the Mauna Loa or Izaña sites. The microtops are calibrated by transfer from the AERONET reference Cimel Sun photometers. The estimated AOD uncertainty does not exceed 0.02, which is slightly higher than the uncertainty of AERONET field instruments (Giles et al., 2019; Kazadzis et al., 2018a; Smirnov et al., 2009).

In order to avoid instrument relocations, needed with the high mountain-based Langley calibration, an absolute calibration (instrument signal units [W m^{-2}]) and an absolute solar irradiance spectrum at the top of the atmosphere (e.g., Coddington et al., 2021; Gröbner et al., 2017; Thuillier et al., 2003) can be used. Such methods have been investigated in the past (e.g., Bais, 1997) and the recent results of the European-funded project Metrology of Aerosol Optical properties, showed very good agreement among the relative (Langley) and the absolute (SI traceable direct solar irradiance) method (Gröbner and Kouremeti, 2019; Kouremeti et al., 2022).

The water vapor derivation implies much larger uncertainty because water vapor changes rapidly in the atmosphere. Thus, the Langley plot determination of the extraterrestrial signal in the water vapor channels (e.g., 940 nm band) is subject to large errors. AERONET estimates the water vapor column uncertainty does not exceed 10% (Giles et al., 2019; Smirnov et al., 2004).

The uncertainty of the ozone column derived from the Brewer spectrophotometers is less than 1% (Parra-Rojas et al., 2021). Other gases like sulfur dioxide and nitrogen dioxide also absorb in the wavelength range used in the retrieval; the additional uncertainty will depend on the column amounts of those species. The scattering and absorption by the aerosol particles is minimized by the procedure of using ratios of measurements at close wavelengths (Gröbner et al., 2021).

5. Aerosol data interpretation

5.1 Aerosol types

The direct Sun observation with narrow-band photometers allows the retrieval of AOD at several wavelengths and the derivation of the Ångström parameters. An example of a dataset on several consecutive days is given in Fig. 10 (daytime only in this case).

The spectral variation of the AOD is related to the particle size predominance in the aerosol size distribution. By using the AE versus AOD plot, an aerosol-type classification scheme can be attempted. It allows the distinction of the amount and the size in a graphical way; thus it is possible to assign average values for various aerosol types (Holben et al., 2001; Toledano et al., 2007). This is shown in Fig. 11, in which the two branches showing higher AOD correspond to mineral dust and biomass burning or pollution. The overall plot with all types resembles a croissant. The typical values for each key aerosol type can be found in Holben et al. (2001).

5.2 Quality assurance and quality control

The quality assurance is the set of systematic and planned activities designed to ensure the quality of the final product, in this case, the AOD. The quality control focuses on defect identification, in our case, flagging or screening out data of insufficient quality. Both kinds of procedures are extensively developed in the operational networks (Giles et al., 2019; Kazadzis et al., 2018b).

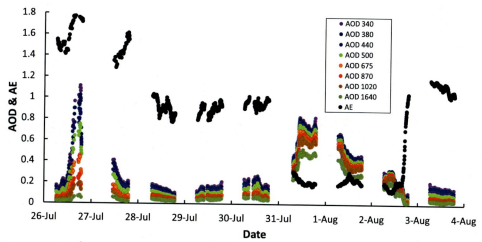

FIG. 10 Time series of AOD and Ångström exponent derived from an AERONET photometer during nine consecutive clear-sky days.

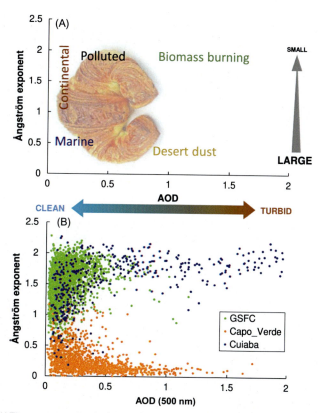

FIG. 11 Plot of Ångström exponent (AE) versus aerosol optical depth, for aerosol-type identification. (A) Schematic representation of the AE-AOD plot. (B) AERONET daily mean data at Cape Verde (predominant mineral dust), GSFC (urban), and Cuiaba (biomass burning).

The quality assurance includes the routine calibrations before and after each deployment period. Before deployment in the field, an extensive checkup list is applied to the instruments: the raw signals and derived AOD must fit that of the reference instrument within given thresholds, etc. The dark currents are routinely recorded to monitor the detector and electronics performance. In AERONET, the 1020 nm AOD observation is duplicated in the silicon and InGaAs channels, using different optical assembles. This instrumental design allows for monitoring possible obstructions in the optical path. In GAW-PFR, a four-quadrant detector is used to monitor the correct pointing at the Sun direction by the external solar tracker.

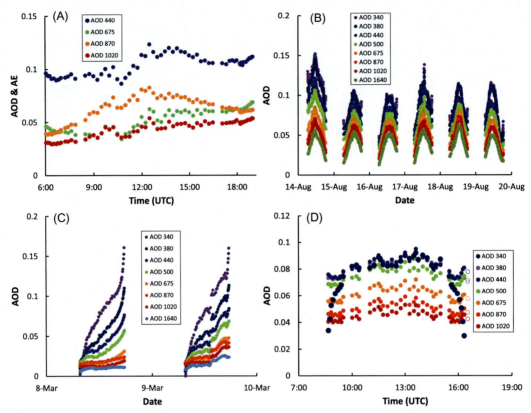

FIG. 12 (A) Crossing AOD channels denoting incorrect AOD; (B) diurnal cycle in AOD (all channels) due to obstruction; (C) asymmetric diurnal behavior in AOD due to shifted clock; and (D) bad filter blocking in 340 nm, resulting in too low AOD at large air mass (sunrise and sunset), that is, the "moustache" effect.

Other example of quality assurance is the design of the measurement protocol to allow for screening of cloud contamination. AERONET performs consecutive three direct Sun measurements within 1 min (triplets) and GAW-PFR collects data continuously, to evaluate variability according to certain thresholds. The cloud-screening algorithms are based on temporal variability checks, wavelength dependence of the AOD, or some other physical features that allow separating cloud and aerosol scenes (Giles et al., 2019; Harrison et al., 1994; Kazadzis et al., 2018b; Smirnov et al., 2000).

The quality control includes two main types of checks: instrument oriented and physical meaning oriented. The deep knowledge of the instrument performance is crucial for the first type. GAW-PFR and AERONET accumulate more than 20 years of experience with their standard instruments and have an extensive list of inspections to signals, dark current, instrumental status reports, internal temperature dependency, power supply voltages, etc. (Shaw, 1976). With respect to the physical testing, there are automated checks for several important characteristics of the AOD, based on a priori acceptable physical behavior of this parameter: negative values are not acceptable; AOD should generally decrease with wavelength, at least in the visible range; time series of different spectral channels should evolve in parallel or at least not cross each other (Fig. 12A).

Any systematic diurnal variation will very likely be related with either signal or calibration problem: filter degradation, obstruction in the optical path, incorrect extraterrestrial (calibration) coefficient, etc. The Beer-Bouger-Lambert law is extremely friendly for quality control, because of the 1/air mass dependence of AOD errors: AOD will show a diurnal cycle perfectly centered at solar noon in case calibration or signal errors occur. Depending on the optical design (filter wheel and common detector and fore optics for AERONET Cimel radiometers; one separate detector and optical path for each channel in the PFR or microtops; etc.) it is possible to find only one channel with diurnal cycle (pointing at a certain filter or detector) or all channels at a time (pointing at obstruction in the common path: dust, spider web, etc.). An example of these situations is provided in Fig. 12A and B.

If the instrument clock is shifted, the processing for AOD calculation will result in a diurnal cycle. In the case of fast clock, AOD shows too low in the morning and too high in the evening; and the opposite if the clock is slow. Shorter wavelengths are more affected than longer ones. An example is given in Fig. 12C.

Incorrect out-of-band blocking can be detected in several ways. During calibration, a side-by-side comparison with the reference instrument would show air mass dependence in the signal ratios. If Langley plot is attempted, the data would show separation from a straight line at large air masses, because the signal would be too high as light from other wavelengths leaks into the detector. Or the AOD can be checked for abnormal behavior consisting of a decrease at large air masses in some particular channel, the so-called "moustache" effect, meaning an excess of signal (incorrect blocking) in that particular filter (Fig. 12D).

Very important aspects in quality control are the automation and real-time delivery of the data flagging. This is critical for models that assimilate data in real time for improving forecasts. Recent efforts of AERONET in particular have been very successful in this aspect of operation, that is, quality-controlled data delivery in near-real time (Giles et al., 2019).

Finally, it is important to highlight the crucial role that is played by the site managers in the operation of the photometric networks. The routine checks and maintenance on site allow early detection and troubleshooting of instrumental issues, improving overall data quality and dataset continuity.

6. Conclusions

Sun photometry is a powerful and cost-effective technique in ground-based monitoring of the atmospheric composition. The atmospheric variables provided by Sun photometers depend on the measurement types and spectral ranges; the core products are the aerosol optical depth, precipitable water vapor, and ozone column. In order to provide global coverage of these atmospheric composition variables, satellite sensors and retrieval methods are used. The evaluation and improvement of such measurements and algorithms require accurate and homogeneous surface-based Sun photometer measurements.

Global operational networks carry out long-term observation programs for monitoring aerosol, ozone, UV radiation, etc., and provide invaluable data for satellite product validation, model development, and validation. The Aerosol Robotic Network (AERONET) program, the Global Atmosphere Watch-Precision Filter Radiometer network (GAW-PFR), SKYNET, and the Maritime Aerosol Network (MAN) are some of the main global networks devoted to aerosol monitoring. The Brewer spectrophotometer network is devoted to ozone column monitoring.

The data quality relies on strict and standardized calibration procedures, deep knowledge of the instruments, and systematic quality control of the data. The Langley plot is the preferred calibration method for the earlier-mentioned operational networks, as it provides uncertainties in the order of 1%–2% for the calibration coefficients (top-of-the-atmosphere instrument signal). The automation of the quality checks allows for real-time data provision, which is key in many applications. The detection of atmospheric composition long-term changes requires instrument stability and a continuous effort toward decreasing the measurement uncertainties.

Acknowledgments

The authors gratefully acknowledge the coordination work carried out by N. Nalli as well as the careful review made by P. Ciren.

References

Angström, A., 1930. On the atmospheric transmission of sun radiation II. Geogr. Ann. H11, 1301–1308.

Augustine, J.A., Hodges, G.B., Dutton, E.G., Michalsky, J.J., Cornwall, C.R., 2008. An aerosol optical depth climatology for NOAA's national surface radiation budget network (SURFRAD). J. Geophys. Res. Atmos. 113 (D11). https://doi.org/10.1029/2007JD009504.

Bais, A.F., 1997. Absolute spectral measurements of direct solar ultraviolet irradiance with a brewer spectrophotometer. Appl. Opt. 36 (21), 5199–5204. https://doi.org/10.1364/AO.36.005199.

Barreto, A., Cuevas, E., Granados-Muñoz, M.-J., Alados-Arboledas, L., Romero, P.M., Gröbner, J., Kouremeti, N., Almansa, A.F., Stone, T., Toledano, C., Román, R., Sorokin, M., Holben, B., Canini, M., Yela, M., 2016. The new sun-sky-lunar Cimel CE318-T multiband photometer—a comprehensive performance evaluation. Atmos. Meas. Tech. 9 (2), 631–654. https://doi.org/10.5194/amt-9-631-2016.

Basher, R.E., Thomas, R.W.L., 1979. Atmospheric aerosol effect on Dobson total ozone measurements: a simple approach. Appl. Opt. 18 (20), 3361–3362. https://doi.org/10.1364/AO.18.003361.

Bodhaine, B.A., Wood, N.B., Dutton, E.G., Slusser, J.R., 1999. On Rayleigh optical depth calculations. J. Atmos. Ocean. Technol. 16 (11), 1854–1861. https://doi.org/10.1175/1520-0426(1999)016<1854:ORODC>2.0.CO;2.

Cachorro, V.E., de Frutos, A.M., Casanova, J.L., 1987. Determination of the angstrom turbidity parameters. Appl. Opt. 26 (15), 3069–3076. https://doi.org/10.1364/AO.26.003069.

Cachorro, V.E., Romero, P.M., Toledano, C., Cuevas, E., de Frutos, A.M., 2004. The fictitious diurnal cycle of aerosol optical depth: a new approach for "in situ" calibration and correction of AOD data series. Geophys. Res. Lett. 31, L12106. https://doi.org/10.1029/2004GL019651.

Cachorro, V.E., Toledano, C., Sorribas, M., Berjon, A., de Frutos, A.M., Laulainen, N., 2008. An "in situ" calibration-correction procedure (KCICLO) based on AOD diurnal cycle: comparative results between AERONET and reprocessed (KCICLO method) AOD-alpha data series at El Arenosillo, Spain. J. Geophys. Res. 113, D02207. https://doi.org/10.1029/2007JD009001.

Campanelli, M., Nakajima, T., Olivieri, B., 2004. Determination of the solar calibration constant for a sun-sky radiometer: proposal of an in-situ procedure. Appl. Opt. 43 (3), 651–659. https://doi.org/10.1364/AO.43.000651.

Campanelli, M., Estellés, V., Smyth, T., Tomasi, C., Martínez-Lozano, J., Claxton, B., Muller, J.-P., Pappalardo, G., Pietruczuk, A., Shanklin, J., Colwell, S., Wrench, C., Lupi, A., Mazzola, M., Lanconelli, C., Vitale, V., Congeduti, F., Dionisi, D., Cardillo, F., Nakajima, T., 2012. Monitoring of Eyjafjallajoekull volcanic aerosol by the new European SkyRad users (ESR) sun-sky radiometer network. Atmos. Environ. 48, 33–45.

Cede, A., Herman, J., Richter, A., Krotkov, N., Burrows, J., 2006. Measurements of nitrogen dioxide total column amounts using a brewer double spectrophotometer in direct sun mode. J. Geophys. Res. Atmos. 111 (D5). https://doi.org/10.1029/2005JD006585.

Che, H., Zhang, X.-Y., Xia, X., Goloub, P., Holben, B., Zhao, H., Wang, Y., Zhang, X.-C., Wang, H., Blarel, L., Damiri, B., Zhang, R., Deng, X., Ma, Y., Wang, T., Geng, F., Qi, B., Zhu, J., Yu, J., Chen, Q., Shi, G., 2015. Ground-based aerosol climatology of China: aerosol optical depths from the China aerosol remote sensing network (CARSNET) 2002–2013. Atmos. Chem. Phys. 15 (13), 7619–7652. https://doi.org/10.5194/acp-15-7619-2015.

Coddington, O.M., Richard, E.C., Harber, D., Pilewskie, P., Woods, T.N., Chance, K., Liu, X., Sun, K., 2021. The TSIS-1 hybrid solar reference spectrum. Geophys. Res. Lett. 48 (12). https://doi.org/10.1029/2020GL091709. e2020GL091709.

Cuevas, E., Romero-Campos, P.M., Kouremeti, N., Kazadzis, S., Räisänen, P., García, R.D., Barreto, A., Guirado-Fuentes, C., Ramos, R., Toledano, C., Almansa, F., Gröbner, J., 2019. Aerosol optical depth comparison between GAW-PFR and AERONET-Cimel radiometers from long-term (2005–2015) 1 min synchronous measurements. Atmos. Meas. Tech. 12 (8), 4309–4337. https://doi.org/10.5194/amt-12-4309-2019.

Dubovik, O., King, M., 2000. A flexible inversion algorithm for retrieval of aerosol optical properties from sun and sky radiance measurements. J. Geophys. Res. 105 (D16), 20673–20696.

Dubovik, O., Lapyonok, T., Litvinov, P., Herman, M., Fuertes, D., Ducos, F., Lopatin, A., Chaikovsky, A., Torres, B., Derimian, Y., Huang, X., Aspetsberger, M., Federspiel, C., 2014. Grasp: a versatile algorithm for characterizing the atmosphere. SPIE Newsroom. https://doi.org/10.1117/2.1201408.005558.

Eck, T.F., Holben, B.N., Reid, J.S., Dubovik, O., Smirnov, A., O'Neill, N.T., Slutsker, I., Kinne, S., 1999. The wavelength dependence of the optical depth of biomass burning, urban and desert dust aerosols. J. Geophys. Res. 104, 31333–31350.

Fioletov, V.E., McLinden, C.A., McElroy, C.T., Savastiouk, V., 2011. New method for deriving total ozone from Brewer zenith sky observations. J. Geophys. Res. Atmos. 116 (D8). https://doi.org/10.1029/2010JD015399.

Forgan, B.W., 1994. General method for calibrating sun photometers. Appl. Opt. 33 (21), 4841–4850. https://doi.org/10.1364/AO.33.004841.

Giles, D.M., Sinyuk, A., Sorokin, M.G., Schafer, J.S., Smirnov, A., Slutsker, I., Eck, T.F., Holben, B.N., Lewis, J.R., Campbell, J.R., Welton, E.J., Korkin, S.V., Lyapustin, A.I., 2019. Advancements in the aerosol robotic network (AERONET) version 3 database—automated near-real-time quality control algorithm with improved cloud screening for Sun photometer aerosol optical depth (AOD) measurements. Atmos. Meas. Tech. 12 (1), 169–209. https://doi.org/10.5194/amt-12-169-2019.

Gröbner, J., Kouremeti, N., 2019. The precision solar spectroradiometer (PSR) for direct solar irradiance measurements. Sol. Energy 185, 199–210.

Gröbner, J., Kröger, I., Egli, L., Hülsen, G., Riechelmann, S., Sperfeld, P., 2017. The high-resolution extraterrestrial solar spectrum (QASUMEFTS) determined from ground-based solar irradiance measurements. Atmos. Meas. Tech. 10 (9), 3375–3383. https://doi.org/10.5194/amt-10-3375-2017.

Gröbner, J., Schill, H., Egli, L., Stübi, R., 2021. Consistency of total column ozone measurements between the Brewer and Dobson spectroradiometers of the LKO Arosa and PMOD/WRC Davos. Atmos. Meas. Tech. 14 (5), 3319–3331. https://doi.org/10.5194/amt-14-3319-2021.

Harrison, L., Michalsky, J.J., Berndt, J., 1994. Automated multifilter rotation shadow-band radiometer: an instrument for optical depth and radiation measurements. Appl. Opt. 33, 5118–5125.

Herman, B.M., Box, M.A., Reagan, J.A., Evans, C.M., 1981. Alternate approach to the analysis of solar photometer data. Appl. Opt. 20, 2925–2928.

Holben, B., Eck, T., Slutsker, I., Tanré, D., Buis, J., Setzer, A., Vermote, E., Reagan, J., Kaufman, Y., 1998. AERONET—a federated instrument network and data archive for aerosol characterization. Remote Sens. Environ. 66, 1–16.

Holben, B.N., Tanre, D., Smirnov, A., Eck, T.F., Slutsker, I., Abuhassan, N., Newcomb, W.W., Schafer, J., Chatenet, B., Lavenue, F., Kaufman, Y., Vande Castle, J., Setzer, A., Markham, B., Clark, D., Frouin, R., Halthore, R., Karnieli, A., O'Neill, N.T., Pietras, C., Pinker, R.T., Voss, K., Zibordi, G., 2001. An emerging ground-based aerosol climatology: aerosol optical depth from AERONET. J. Geophys. Res. 106, 12067–12097.

Ingold, T., Schmid, B., Mätzler, C., Demoulin, P., Kämpfer, N., 2000. Modeled and empirical approaches for retrieving columnar water vapor from solar transmittance measurements in the 0.72, 0.82, and 0.94 μm absorption bands. J. Geophys. Res. Atmos. 105 (D19), 24327–24343. https://doi.org/10.1029/2000JD900392.

JCGM, 2008. JCGM 100: Evaluation of Measurement Data-Guide to the Expression of Uncertainty in Measurement (GUM). Joint Committee for Guides in Metrology.

Kasten, F., Young, A.T., 1989. Revised optical air mass tables and approximation formula. Appl. Opt. 28, 4735–4738.

Kazadzis, S., Kouremeti, N., Diémoz, H., Gröbner, J., Forgan, B.W., Campanelli, M., Estellés, V., Lantz, K., Michalsky, J., Carlund, T., Cuevas, E., Toledano, C., Becker, R., Nyeki, S., Kosmopoulos, P.G., Tatsiankou, V., Vuilleumier, L., Denn, F.M., Ohkawara, N., Ijima, O., Goloub, P., Raptis, P.I., Milner, M., Behrens, K., Barreto, A., Martucci, G., Hall, E., Wendell, J., Fabbri, B.E., Wehrli, C., 2018a. Results from the Fourth WMO Filter Radiometer Comparison for aerosol optical depth measurements. Atmos. Chem. Phys. 18 (5), 3185–3201. https://doi.org/10.5194/acp-18-3185-2018.

Kazadzis, S., Kouremeti, N., Nyeki, S., Gröbner, J., Wehrli, C., 2018b. The World Optical Depth Research and Calibration Center (WORCC) quality assurance and quality control of GAW-PFR AOD measurements. Geosci. Instrum. Methods Data Syst. 7 (1), 39–53. https://doi.org/10.5194/gi-7-39-2018.

Kerr, J.B., 2002. New methodology for deriving total ozone and other atmospheric variables from Brewer spectrophotometer direct sun spectra. J. Geophys. Res. Atmos. 107 (D23). https://doi.org/10.1029/2001JD001227. ACH22-1–ACH22-17.

Kerr, J.B., McElroy, C.T., Olafson, R.A., 1981. Measurements of ozone with the Brewer ozone spectrophotometer. In: London, L. (Ed.), Proc. Quadrennial Int. Ozone Symp., 4–9 August 1980, Boulder, Colorado, vol. I. Int. Ozone Comm. (IAMAP), pp. 74–79.

Kerr, J.B., McElroy, C.T., Wardle, D.I., Dorokhov, V., 1990. Measurements of arctic total ozone during the polar winter. Atmos. Ocean 28 (4), 383–392. https://doi.org/10.1080/07055900.1990.9649384.

Komhyr, W.D., Grass, R.D., Leonard, R.K., 1989. Dobson spectrophotometer 83: a standard for total ozone measurements, 1962–1987. J. Geophys. Res. Atmos. 94 (D7), 9847–9861. https://doi.org/10.1029/JD094iD07p09847.

Kouremeti, N., Nevas, S., Kazadzis, S., Gröbner, J., Schneider, P., Schwind, K., 2022. SI-traceable solar irradiance measurements for aerosol optical depth retrieval. Metrologia 59, 044001. https://doi.org/10.1088/1681-7575/ac6cbb.

Kurucz, R.L., 1994. Synthetic infrared spectra. Symp. Int. Astron. Union 154, 523–531. https://doi.org/10.1017/S0074180900124805.

Li, Z., Blarel, L., Podvin, T., Goloub, P., Chen, L., 2010. Calibration of the degree of linear polarization measurement of polarized radiometer using solar light. Appl. Opt. 49 (8), 1249–1256. https://doi.org/10.1364/AO.49.001249.

López-Solano, J., Redondas, A., Carlund, T., Rodriguez-Franco, J.J., Diémoz, H., León-Luis, S.F., Hernández-Cruz, B., Guirado-Fuentes, C., Kouremeti, N., Gröbner, J., Kazadzis, S., Carreño, V., Berjón, A., Santana-Díaz, D., Rodríguez-Valido, M., De Bock, V., Moreta, J.R., Rimmer, J., Smedley, A.R. D., Boulkelia, L., Jepsen, N., Eriksen, P., Bais, A.F., Shirotov, V., Vilaplana, J.M., Wilson, K.M., Karppinen, T., 2018. Aerosol optical depth in the European Brewer network. Atmos. Chem. Phys. 18 (6), 3885–3902. https://doi.org/10.5194/acp-18-3885-2018.

Michalsky, J.J., 1988. The astronomical Almanac's algorithm for approximate solar position (1950–2050). Sol. Energy 40 (3), 227–235. https://doi.org/10.1016/0038-092X(88)90045-X.

Mitchell, R.M., Forgan, B.W., Campbell, S.K., 2017. The climatology of Australian aerosol. Atmos. Chem. Phys. 17 (8), 5131–5154. https://doi.org/10.5194/acp-17-5131-2017.

Morys, M., Mims III, F.M., Hagerup, S., Anderson, S.E., Baker, A., Kia, J., Walkup, T., 2001. Design, calibration, and performance of MICROTOPS II handheld ozone monitor and Sun photometer. J. Geophys. Res. Atmos. 106 (D13), 14573–14582. https://doi.org/10.1029/2001JD900103.

Nakajima, T., Tonna, G., Rao, R., Boi, P., Kaufman, Y., Holben, B., 1996. Use of sky brightness measurements from ground for remote sensing of particulate polydispersions. Appl. Opt. 35 (15), 2672–2686.

Nakajima, T., Campanelli, M., Che, H., Estellés, V., Irie, H., Kim, S.-W., Kim, J., Liu, D., Nishizawa, T., Pandithurai, G., Soni, V.K., Thana, B., Tugjsurn, N.-U., Aoki, K., Go, S., Hashimoto, M., Higurashi, A., Kazadzis, S., Khatri, P., Kouremeti, N., Kudo, R., Marenco, F., Momoi, M., Ningombam, S.S., Ryder, C.L., Uchiyama, A., Yamazaki, A., 2020. An overview of and issues with sky radiometer technology and SKYNET. Atmos. Meas. Tech. 13 (8), 4195–4218. https://doi.org/10.5194/amt-13-4195-2020.

O'Neill, N.T., Eck, T.F., Smirnov, A., Holben, B.N., Thulasiraman, S., 2003. Spectral discrimination of coarse and fine mode optical depth. J. Geophys. Res. 108 (D17), 4559. https://doi.org/10.1029/2002JD002975.

Parra-Rojas, F.C., Redondas, A., Berjón, A., López-Solano, J., 2021. A new data set for the brewer spectrophotometer uncertainty budget in the total ozone column measurements. In: vEGU21, the 23rd EGU General Assembly, held online 19–30 April, Id.EGU21-15316.

Reagan, J.A., Thomason, L.W., Herman, B.M., Palmer, J.M., 1986. Assessment of atmospheric limitations on the determination of the solar spectral constant from ground-based spectroradiometer measurements. IEEE Trans. Geosci. Remote Sens. GE-24 (2), 258–266. https://doi.org/10.1109/TGRS.1986.289645.

Rimmer, J.S., Redondas, A., Karppinen, T., 2018. EuBrewNet—a European Brewer network (COST Action ES1207), an overview. Atmos. Chem. Phys. 18 (14), 10347–10353. https://doi.org/10.5194/acp-18-10347-2018.

Schmid, B., Michalsky, J.J., Slater, D.W., Barnard, J.C., Halthore, R.N., Liljegren, J.C., Holben, B.N., Eck, T.F., Livingston, J.M., Russell, P.B., Ingold, T., Slutsker, I., 2001. Comparison of columnar water-vapor measurements from solar transmittance methods. Appl. Opt. 40 (12), 1886–1896. https://doi.org/10.1364/AO.40.001886.

Shaw, G.E., 1976. Error analysis of multi-wavelength sun photometry. Pure Appl. Geophys. 114, 1–14.

Shaw, G.E., 1983. Sun photometry. Bull. Am. Meteorol. Soc. 64, 4–10.

Shaw, G.E., 2007. Genesis of sun photometry. J. Appl. Remote Sens. 1 (1), 1–13. https://doi.org/10.1117/1.2712483.

Sinyuk, A., Holben, B.N., Eck, T.F., Giles, D.M., Slutsker, I., Korkin, S., Schafer, J.S., Smirnov, A., Sorokin, M., Lyapustin, A., 2020. The aeronet version 3 aerosol retrieval algorithm, associated uncertainties and comparisons to version 2. Atmos. Meas. Tech. 13 (6), 3375–3411. https://doi.org/10.5194/amt-13-3375-2020.

Smirnov, A., Holben, B.N., Eck, T.F., Dubovik, O., Slutsker, I., 2000. Cloud-screening and quality control algorithms for the AERONET database. Remote Sens. Environ. 73, 337–349.

Smirnov, A., Holben, B., Kaufman, Y., Dubovik, O., Eck, T., Slutsker, I., Pietras, C., Halthore, R., 2002. Optical properties of atmospheric aerosol in maritime environments. J. Atmos. Sci. 59, 501–523.

Smirnov, A., Holben, B., Lyapustin, A., Slutsker, I., Eck, T.F., 2004. AERONET processing algorithms refinement. In: AERONET Workshop, El Arenosillo, Spain, pp. 10–14.

Smirnov, A., Holben, B.N., Slutsker, I., Giles, D.M., McClain, C.R., Eck, T.F., Sakerin, S.M., Macke, A., Croot, P., Zibordi, G., Quinn, P.K., Sciare, J., Kinne, S., Harvey, M., Smyth, T.J., Piketh, S., Zielinski, T., Proshutinsky, A., Goes, J.I., Nelson, N.B., Larouche, P., Radionov, V.F., Goloub, P., Krishna Moorthy, K., Matarrese, R., Robertson, E.J., Jourdin, F., 2009. Maritime aerosol network as a component of aerosol robotic network. J. Geophys. Res. Atmos. 114 (D6). https://doi.org/10.1029/2008JD011257.

Takamura, T., Nakajima, T., SKYNET Community Group, 2004. Overview of SKYNET and its activities. Opt. Pura Apl. 37, 3303–3308.

Thuillier, G., Hersé, M., Labs, D., Foujols, T., Peetermans, W., Gillotay, D., Simon, P.C., Mandel, H., 2003. The solar spectral irradiance from 200 to 2400 nm as measured by the SOLSPEC spectrometer from the Atlas and Eureca missions. Sol. Phys. 214, 1–22.

Toledano, C., Cachorro, V.E., Berjon, A., de Frutos, A.M., Sorribas, M., de la Morena, B.A., Goloub, P., 2007. Aerosol optical depth and Angstrom exponent climatology at El Arenosillo AERONET site (Huelva, Spain). Q. J. R. Meteorol. Soc. 133, 795–807. 10.102/Qj.54.

Toledano, C., González, R., Fuertes, D., Cuevas, E., Eck, T.F., Kazadzis, S., Kouremeti, N., Gröbner, J., Goloub, P., Blarel, L., Román, R., Barreto, A., Berjón, A., Holben, B.N., Cachorro, V.E., 2018. Assessment of Sun photometer Langley calibration at the high-elevation sites Mauna Loa and Izaña. Atmos. Chem. Phys. 18 (19), 14555–14567.

Uchiyama, A., Shiobara, M., Kobayashi, H., Matsunaga, T., Yamazaki, A., Inei, K., Kawai, K., Watanabe, Y., 2019. Nocturnal aerosol optical depth measurements with modified skyradiometer POM-02 using the moon as a light source. Atmos. Meas. Tech. Discuss. 2019, 1–41. https://doi.org/10.5194/amt-2019-230.

Wehrli, C., 2000. Calibration of filter radiometers for determination of atmospheric optical depth. Metrologia 37, 419–422.

Wehrli, C., 2005. GAW-PFR: a network of aerosol optical depth observations with precision filter radiometers. In: WMO/GAW Experts Workshop on a Global Surface Based Network for Long Term Observations of Column Aerosol Optical Properties. GAW Report No. 162, WMO TD No. 1287.

WMO, 2016. WMO/GAW Aerosol Measurement Procedures, Guidelines and Recommendations, second ed. World Meteorological Organization, Scientific Advisory Group on Aerosol. WMO report 227.

Chapter 8

The AirCore atmospheric sampling system

Bianca C. Baier[a,b], Colm Sweeney[b], and Huilin Chen[c,d]

[a]*Cooperative Institute for Research in Environmental Sciences (CIRES), University of Colorado-Boulder, Boulder, CO, United States,* [b]*NOAA Global Monitoring Laboratory (GML), Boulder, CO, United States,* [c]*Centre for Isotope Research, University of Groningen, Groningen, The Netherlands,* [d]*Joint International Research Laboratory of Atmospheric and Earth System Sciences, School of Atmospheric Sciences, Nanjing University, Nanjing, China*

Chapter outline

1. Background — 139
   1.1 Atmospheric trace gases — 139
   1.2 Trace gas vertical profiling — 140
2. The AirCore atmospheric sampling system — 141
   2.1 Principles of operation — 141
   2.2 AirCore instrument package and accompanying flight string components — 142
   2.3 Profile retrieval and vertical resolution — 144
   2.4 AirCore sampler configurations and associated platforms — 145
   2.5 Retrieved profile uncertainty and reproducibility — 146
3. AirCore balloon-borne measurement sites — 149
4. AirCore evaluation of remote sensing retrievals — 149
   4.1 Calculation of AirCore total-column trace gas abundances — 150
   4.2 Case study: AirCore-calculated XCO_2 comparison to satellite XCO_2 — 150
   4.3 Case study: Linking remote sensing total column trace gas abundances to WMO standard scales using AirCore — 151
5. Conclusions and future outlook — 152
   5.1 Toward a global AirCore network — 152
   5.2 On the use of high-altitude uncrewed aircraft systems — 152
Acknowledgments — 153
References — 153

One accurate measurement is worth a thousand expert opinions.

Grace Hopper, American computer scientist

1. Background

1.1 Atmospheric trace gases

Accurately monitoring atmospheric trace gases affecting climate and air quality on a routine basis is critical for advancing our understanding of their sources, sinks, trends, and variability over space and time. The levels of atmospheric greenhouse gases and other trace gas pollutants (here, collectively termed "trace gases") have grown considerably since the pre-industrial era largely due to increases in anthropogenic emissions. Of the trace gases mentioned in this chapter, carbon dioxide (CO_2) and methane (CH_4) are the two most important greenhouse gases due to their relative impact on the Earth's radiative budget. Nitrous oxide (N_2O), in addition to other fluorinated trace gas species also, has significant global warming potential per unit of mass (IPCC, 2015). Other trace gases including halogenated species and chlorofluorocarbons are known for their significant ozone depleting potential and are internationally regulated by the Montreal Protocol and Kigali Amendment (UNEP, 1987).

Atmospheric inversions are tools routinely used to estimate trace gas sources and sinks by combing trace gas mole fraction observations with an atmospheric transport model. However, multiple sources of error and bias can arise in this approach, both in the model framework itself and in other model inputs such as meteorological transport, background levels of trace gases, and a priori flux estimates, and error structures therein. Direct measurements of atmospheric trace gases are therefore critical for constraining these models.

Global surface in situ observational networks routinely monitor the atmospheric abundances of climate- and air quality-relevant trace gases under the oversight of the World Meteorological Organization's Global Atmospheric Watch (WMO/

GAW, 2011). The WMO/GAW promotes the accurate detection of trace gas trends over space and time by requiring that participating measurement laboratories maintain traceability to standard calibration scales and by placing strict, interlaboratory compatibility requirements on global surface-based measurements. While surface-based measurements provide insight into long-term trends and regional spatial and temporal behavior of trace gases, vertical measurements spanning the atmospheric column can uniquely contribute to our understanding of larger scale variability of species. Measurements of trace gas vertical distributions provide a critical constraint for evaluating modeled vertical mixing (Stephens et al., 2007); for informing trace gas fluxes from distant and near-field sources (e.g., Bakwin et al., 2003; Crevoisier et al., 2010; Sweeney et al., 2015, 2022), and for serving as an evaluation tool for total-column trace gas retrievals from remote sensing instrumentation (Wunch et al., 2010; Inoue et al., 2013; Deeter et al., 2014; Hedelius et al., 2019; Zhou et al., 2019; Sha et al., 2020). Multispecies vertical distribution measurements that extend into the upper troposphere and middle to lower stratosphere uniquely allow for the determination of age of air in the stratosphere and for the evaluation of climate-driven feedbacks or trends in trace gases given the robust correlations between long-lived species in this region (Harries et al., 1996; Andrews et al., 2001a,b; Ray et al., 2002, 2014; Engel et al., 2017; Laube et al., 2020).

1.2 Trace gas vertical profiling

Throughout this chapter, we refer to the mole fraction distribution of atmospheric trace gases observed in the vertical dimension as *profiles*. The portion of the atmosphere measured, extending from the surface to the top of the atmosphere in the vertical dimension, is referred to as an atmospheric *column*. Profiling of some reactive trace gases throughout the majority of the atmospheric column (i.e., to ∼32,000 m above mean sea level (AMSL)) such as ozone or nitrogen dioxide can be performed in situ at high resolution using lightweight continuous sensors on small (less than 3000 g) unmanned balloons. Such instruments take advantage of well-known chemical reactions of these species with other reagents to produce signals proportional to the trace gas abundance in the atmosphere (see Chapters 1, 3, and 4). Other, more inert trace gases that absorb radiation in the infrared part of the electromagnetic spectrum, such as CO_2 and CH_4, typically require gas chromatography or spectroscopic measurement techniques and larger, more costly instrumentation. For these species, profiling is therefore most commonly performed aboard aircraft or large balloon platforms using, for example, a continuous analyzer (providing high-resolution measurements in the vertical dimension) or whole-air flasks (providing discrete samples in the vertical dimension). Routine aircraft sampling platforms typically reach altitudes of 4,000–12,000 m AMSL (Fig. 1, see also Chapter 13) with larger, more expensive aircraft needed to perform measurements at higher altitudes (i.e., up to 24,000 m AMSL). Until the past decade, large balloons (112,000–230,000 m^3 total volume) suspending gondolas instrumented with in situ analyzers were the only alternative to aircraft for continuous trace gas profiling at altitudes exceeding 24,000 m AMSL (Fig. 1; Daube et al., 2002; Moore et al., 2003; NRC, 2010).

The profiling techniques mentioned above involve the use of (a) high value, calibrated instrumentation to measure trace gas abundances throughout the atmosphere and (b) high-value platforms that are both operationally and logistically expensive relative to smaller, balloon-borne operations. Airspace regulations in many countries define an allowable weight carried on a balloon string payload, which limits trace gas sampling capabilities. For example, the U.S. Federal Aviation Administration's (FAA) Code of Federal Regulations for Aeronautics and Space (14 CFR Part 101.3; FAA, 1991) limits the payload weights of instrumented small, unmanned free balloons (e.g., weather balloons) to 5.4 kg (12 lbs). Though small, in situ trace gas sensors have been developed in the past decade that could be flown on a balloon string to high altitude for continuous measurements (e.g., Joly et al., 2020; Golston et al., 2017), they typically lack the precision and/or accuracy to satisfy WMO inter-laboratory compatibility goals for trace gas measurements (WMO/GAW, 2011).

The AirCore atmospheric sampling system (Tans, 2009; Karion et al., 2010), developed at the National Oceanic and Atmospheric Administration's (NOAA) Global Monitoring Laboratory (GML) and described in the remainder of this chapter provides a lightweight, low-cost alternative to passively sample more than 90% of the atmospheric column using a small balloon. With the AirCore, trace gas vertical profiles calibrated to WMO/GAW standards can be retrieved from altitude ceilings that extend well beyond those accessible from aircraft. These continuous profiles are then traceable to the WMO scales and compatible with other surface-based observing networks. To date, AirCore sampling provides one of the only methods for routine, low-cost monitoring of non-reactive trace gas abundances (i.e., greenhouse gases, chlorofluorocarbons, and sulfur-related species) from the stratosphere (∼30,000 m AMSL) to the Earth's surface. These data are used to evaluate atmospheric chemistry and global forward and inverse models of CO_2 and CH_4 (Jacobson et al., 2020; Peiro et al., 2021) and to quantify carbon cycle emissions and exchange throughout the atmospheric column (Hooghiem et al., 2020). AirCore profiles have also added critical multispecies information for evaluating trace gas trends and potential climate-driven dynamical changes in the stratosphere (Engel et al., 2017; Laube et al., 2020). Finally, and

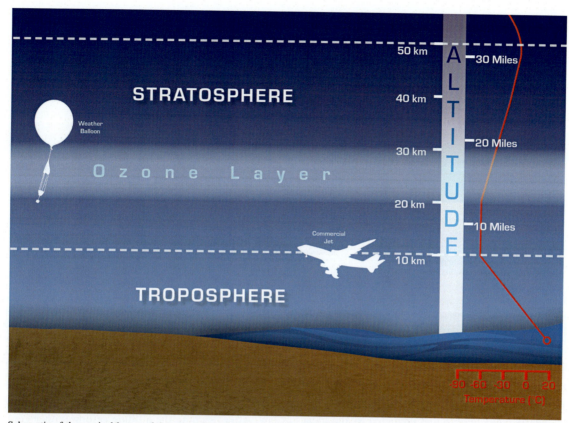

FIG. 1 Schematic of the vertical layers of the atmosphere from the surface through the stratosphere and associated temperature profile. The relative altitude ceilings of aircraft and balloons are indicated for each platform. *(Credit: Sydnee Macias, NOAA\GML.)*

related to this reference book, AirCore profiles are used to evaluate total-column trace gas abundances retrieved from remote sensing platforms (Hedelius et al., 2019; Roche et al., 2021; Nalli et al., 2020).

From here, Section 2 outlines AirCore sampling principles and operation on balloon-borne platforms, measurement of these samples, profile retrieval and evaluation of profile reproducibility and uncertainties, and different variants of the AirCore used to date. Next, Section 3 details current global AirCore measurement sites and institutions. Finally, Section 4 details the use of the AirCore for remote sensing retrieval evaluation, highlighting the utility of this sampling system for evaluation of NASA's Orbiting Carbon Observatory version 2 (OCO-2). We close with a discussion of next steps toward advancing this sampling system and observing capacity worldwide.

2. The AirCore atmospheric sampling system

2.1 Principles of operation

The AirCore atmospheric sampling system was developed and patented by Tans (2009) at the United States NOAA\GML. The name implies a "core" of the atmosphere analogous to an ice core that contains air bubbles frozen in ice buried deep in the glaciers of Greenland and Antarctica. The principle behind the AirCore is that air captured in a long tube is slow to mix with adjacent air because, in the absence of convective mixing, only molecular diffusion and Taylor dispersion (Karion et al., 2010; Taylor, 1953) can smooth the air sample. Relative to the length of a typical AirCore ($\sim$50–100 m), the length scale of molecular diffusion over the 4 h (the length of time that it typically takes to analyze the contents of the atmospheric sample) is small, and $\sim$0.5 m. The AirCore sampling technique is additionally simplified by the fact that it is flown with one end open and the other end closed and exploits ambient pressure to passively extract air from the tube (during ascent) and collect air (during descent).

Describing this operation in more detail, the AirCore tubing is previously filled with a known calibration gas called the "fill gas" and ascends on a balloon while open on one end and closed on the other. The tubing empties on ascent as ambient atmospheric pressure decreases. As the balloon nears its altitude ceiling, the pressure inside of the AirCore tubing is now

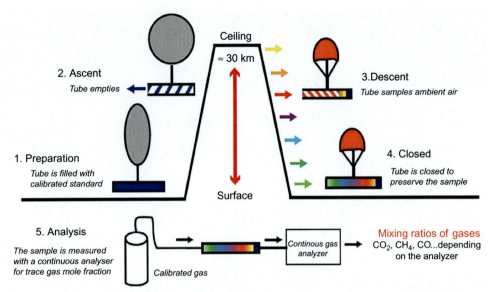

FIG. 2 Schematic of AirCore operations from preparation of flight through analysis procedures. The AirCore tubing coil is suspended from a balloon and ascends to ∼30,000 m AMSL before the balloon is cut away and the payload descends. As the AirCore descends, it passively collects an air sample throughout the vertical atmospheric column, denoted by different colors in the schematic. After flight, the air sample is measured on a continuous trace gas analyzer that has been calibrated to WMO trace gas standards (Membrive et al., 2017). *(Original credit: Membrive, O., Crevoisier, C., Sweeney, C., Danis, F., Hertzog, A., Engel, A., Bönisch, H., Picon, L., 2017. AirCore-HR: a high-resolution column sampling to enhance the vertical description of CH_4 and CO_2. Atmos. Meas. Tech. 19, distributed under CC-BY license.)*

relatively equilibrated with the ambient air pressure, and the tubing is almost empty (Fig. 2), containing the amount of fill gas that is nearly equivalent to that predicted by the ideal gas law. After reaching the altitude ceiling (either via balloon burst, or a pre-determined balloon cutaway), the AirCore quickly descends through the atmosphere on a parachute, capturing the surrounding air as its internal pressure equilibrates with ambient pressures from high altitude to the ground. The AirCore inlet is automatically closed upon landing, preserving the collected air sample. The AirCore is then either analyzed in the field or transported from its landing location to a laboratory for subsequent analysis on a calibrated, laboratory-grade trace gas analyzer. The AirCore thus preserves a continuous sample of atmospheric trace gas mole fractions along its descent path, ultimately providing a trace gas profile registered in altitude from the stratosphere to the Earth's surface.

2.2 AirCore instrument package and accompanying flight string components

A schematic of an AirCore tubing coil and its other components is shown in Fig. 3. The AirCore tubing is composed of stainless steel and treated with a chemically inert coating (SilcoTek 1000, Bellefonte, PA, United States) to prevent interactions at the surface of the tubing. Thus, any trace gas that is chemically stable in the atmosphere can theoretically be preserved within the AirCore. Air enters the AirCore through a small tubing section attached to the bottom inlet valve. This tubing section (#1, Fig. 3) is filled with coarse magnesium perchlorate ($Mg(ClO_4)_2$) granules for air sample drying upon collection. In addition to ambient pressure measurements, small temperature sensors (#5, Fig. 3) are placed in various locations throughout the tubing coil to account for the effect of temperature on the volume of air sampled during flight. A small data logging and control board placed within the AirCore foam housing structure (#3, Fig. 3) are used to record tubing coil temperature, ambient pressure, and redundant GPS location throughout the flight. This datalogger also records the differential pressure measured (#7, Fig. 3) between the closed end of the AirCore and the ambient environment for validating the modeled pressure distribution within the AirCore and subsequent altitude registration. Because the AirCore must be quickly recovered and returned to the laboratory for analysis of the air sample, shutoff valves (#2,6, Fig. 3) on the inlet and closed ends of the AirCore allow for the air sample to be preserved and precisely measured by a laboratory grade analyzer. Further details of the AirCore analysis procedure are described in Karion et al. (2010).

Unlike a typical weather balloon payload, AirCore samplers must be tracked continuously throughout flight in order to be recovered and reused. To ensure efficient recovery, telemetry components, and GPS devices (Fig. 4) are added to the balloon flight string with two to three times redundancy. The balloon string is threaded through a heated wire coil mechanism, which melts the tether string, cuts away the biodegradable latex balloon at a desired altitude (∼30,000 m AMSL),

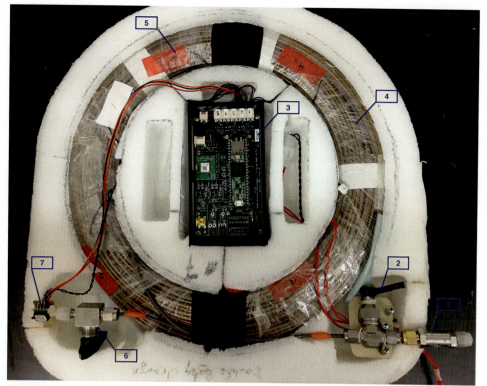

FIG. 3 Example of a 600 mL AirCore sampler and components made by NOAA's Global Monitoring Laboratory (GML). The AirCore is housed in structural foam for balloon-borne flight and parachute-based landing. Components of the AirCore are labeled as the following: (1) air sample drying tube and inlet for tubing coil; (2) bottom inlet valve of AirCore, which is closed immediately upon landing; (3) datalogger and control board; (4) coated tubing coil; (5) bead thermistors (labeled with numbered tape) placed throughout tubing coil for monitoring temperature distribution within the AirCore; (6) AirCore shutoff valve used on the closed ("top") end, alongside bottom inlet valve (2) to analyze the air sample after flight; (7) differential pressure sensor for measurement of pressure difference between the closed end of AirCore and ambient environment.

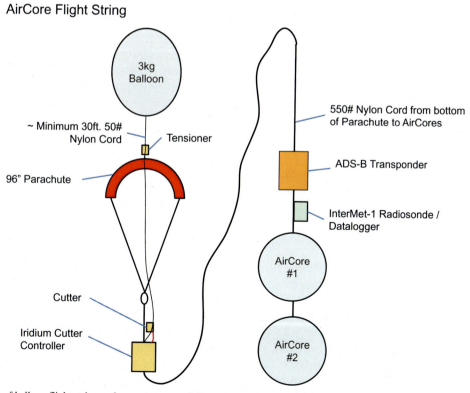

FIG. 4 Schematic of balloon flight string and components used for typical NOAA AirCore flights. The AirCore is suspended from a 3000 g balloon, equipped with telemetry components (ADS-B, Iridium) for tracking during flight and for quick recovery upon landing. A standard radiosonde (iMet-1, *InterMet Systems*) is flown on the same balloon string for altitude registration of the AirCore sample and for measurement of meteorological variables throughout the flight. Because NOAA AirCores are designed to be lightweight, dual AirCores are suspended from the same balloon string for evaluating repeatability between air samples measured and resulting trace gas profiles calculated.

and begins the parachute descent to the ground. Because meteorological variables are required for the altitude registration of an AirCore, a standard radiosonde is flown for simultaneous in situ measurement of temperature, relative humidity, pressure, and GPS position. The size of the balloons used for AirCore flights is small (∼1200–3000 g), with the AirCore housed in a lightweight foam structure (see Fig. 3) to protect the tubing coil and electronics from damage upon landing. As will be discussed later, the AirCore design can be versatile and sufficiently lightweight such that dual samplers can be suspended from the same balloon string.

2.3 Profile retrieval and vertical resolution

2.3.1 AirCore analysis and profile retrieval

The AirCore is analyzed by pushing the collected sample gas through a continuous trace gas analyzer at laboratory pressure. The analyzer reports trace gas mole fractions at a fixed frequency, leading to equal volume measurements of the sample that each represent a pressure "thickness" in the atmosphere. Each volume measurement is registered with altitude using a fluid dynamics model that calculates the moles of air sample entering the tubing upon descent—given measured ambient pressure and temperature—for retrieval of a full trace gas mole fraction profile (Karion et al., 2010; Tans, 2022). The modeled altitude registration of AirCore profiles of a specific dimension has been experimentally evaluated via a method by which a known trace gas is injected into the AirCore tubing at pre-determined altitudes, indicating good agreement within the troposphere, and larger uncertainties in altitude registration in the stratosphere due to assumptions in the AirCore tubing pressure equilibrium at low atmospheric pressures (Wagenhäuser et al., 2021). Differences between these two methods arise because the open end of the AirCore is not in equilibrium with the closed end of the tubing coil due to the length, width of tubing and the viscosity and velocity of the air moving through the tubing. To correct for this disequilibrium and its impact on the calculated total mass of air entering the AirCore, the pressure differential between the open and closed ends of the AirCore is also modeled (Tans, 2022). This *modeled* pressure differential can be directly compared to the *in situ measurement* of the pressure differential between ambient air and the closed end of the AirCore for evaluation (Fig. 3).

The species for which an AirCore profile is retrieved is dependent upon the trace gas analyzer used. To date, operational measurements of CO_2, CH_4, carbon monoxide (CO), and N_2O measurements have all been made successfully from the AirCore via continuous, wavelength-scanned cavity ring-down spectrometers (CRDS). The most commonly used CRDS analyzer used for AirCore analysis provides near-simultaneous measurement of CO_2, CH_4, CO, and water vapor (H_2O) at a frequency of approximately 0.5 Hz (Picarro, Inc. G2401-m, Chen et al., 2010, 2013; Crosson, 2008). While there are slight differences from analyzer to analyzer, measurement precisions at 0.5 Hz are typically less than 0.025 ppm, 2 ppb, and 5 ppb for CO_2, CH_4, and CO, respectively. Absolute trace gas mole fractions are derived from the calibration of the analyzer using gas standards traceable to current WMO scales.

Several research groups have expanded the measurement of AirCore samples to include analysis of halocarbons and other ozone-depleting substances by gas chromatography/mass spectrometry (Mrozek et al., 2016; Laube et al., 2020) and also extraction and graphitization of air samples for analysis by accelerator mass spectrometry (Paul et al., 2016).

2.3.2 Vertical resolution of profiles

Diffusion is one limiting factor in the vertical resolution of an AirCore trace gas profile, along with the cell volume used by the sample analyzer. Karion et al. (2010) describe the root mean square diffusion distance in the form of Eq. (1):

$$x_{\text{rms}} = \sqrt{2 D_{\text{eff}} t}, \tag{1}$$

where x_{rms} is the root mean square diffusion distance, and D_{eff} is an "effective" diffusivity constant that accounts for both molecular diffusion and Taylor dispersion during the storage time, t, of the AirCore sample prior to analysis. Though the diffusion distance is species-dependent, this quantity will be lower for molecules stored in a shorter, smaller-diameter tubing than for those stored in a longer, larger diameter tubing for the same time period. If we consider a 100 m long NOAA AirCore coil of constant inner diameter, a typical CO_2 molecule's root-mean-square diffusion distance is less than 0.5 m within the tubing coil (assuming a temperature of 20°C) over the 4-h period that it typically takes to retrieve the balloon payload and analyze the sample. In this circumstance, the air sample consists of approximately 100 independent trace gas measurements, assuming diffusion of molecules in both directions (i.e., $2x_{\text{rms}}$, Karion et al., 2010). AirCores developed over the past decade for routine, near total-column sampling have achieved an average vertical profile resolution better than 1000 m at an altitude of approximately 20,000 m AMSL and better than 2500 m at an altitude of 30,000 m AMSL

(Karion et al., 2010; Membrive et al., 2017; Engel et al., 2017). Because AirCore tubing dimensions can be augmented to achieve a desired vertical resolution throughout different layers of the atmosphere, we provide a detailed examination of the effect of varied AirCore tubing lengths and diameters on the profile vertical resolution in the next section.

2.4 AirCore sampler configurations and associated platforms

2.4.1 Balloon-borne, high-resolution passive sampler configurations

The first reported AirCore atmospheric sampler, described in Karion et al. (2010), was comprised of coated 304-grade stainless-steel tubing 152 m in length with an outer diameter of 0.64 cm (1/4 in.) and wall thickness of 0.025 cm, providing a total volume of approximately 4 L. The total weight of the original AirCore was roughly 6.7 kg, which required a U.S. FAA waiver to unmanned balloon flight regulations (FAA, 1991). The dimensions and total weight of the AirCore have evolved and been augmented over the past decade to increase its ease of deployment on balloon payloads given flight regulations, to establish comparability standards between AirCores flown, and to achieve various scientific objectives.

Impact of tubing dimensions on pressure drop across AirCore

Air moves through the long AirCore tubing coil in a laminar flow regime, and frictional forces cause a small buildup of a pressure gradient inside of the AirCore. In the stratosphere, slow equilibration of the AirCore internal pressure with ambient pressure due to this frictional effect, along with rapid descent rates ($\sim$50–70 m s^{-1}) in low-density air, result in the inefficient collection of sample air during early stages of the parachute payload descent. Smaller diameter tubing has a larger surface area to volume ratio and will exacerbate frictional effects in laminar flow more so than larger diameter tubing (Fig. 5). The result is that less air is collected at the highest altitudes achieved during flight for smaller-diameter tubing.

Impact of tubing dimensions on altitude resolution of retrieved profiles

Depending on scientific requirements for the vertical resolution of trace gas measurements throughout the atmosphere, the AirCore can be configured to use varied diameter and lengths of tubing throughout the tubing coil. Fig. 6 demonstrates the tradeoffs between sampling resolution for varied length and diameter tubing coils: a 600 mL AirCore composed of a single 300 m length of thin-walled 0.32 cm (1/8 in.) outside diameter (OD) tubing, versus a 2 L AirCore, with 80% of the bottommost (tropospheric) AirCore sample stored in 166 m of 0.64 cm OD (1/4 in.) tubing, and the topmost 20% (i.e., air collected at high altitude) in 234 m of smaller diameter, 0.32 cm (1/8 in.) OD tubing. With equal volumes needed for each

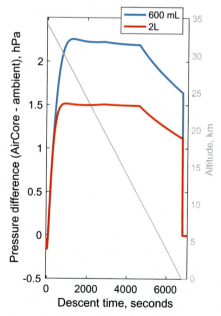

FIG. 5 Comparison of AirCore pressure differential across tubing coils with differing dimensions: a 600 mL AirCore, all 0.320 cm (1/8 in.) OD tubing, and a 2 L AirCore, 20% of which is 0.320 cm (1/8 in.) OD for the stratospheric AirCore portion, and 80% of which is larger diameter, 0.64 cm (1/4 in.) OD tubing on the bottom portion of the AirCore. The pressure differential is displayed as the difference between the closed end of the AirCore and ambient air pressure.

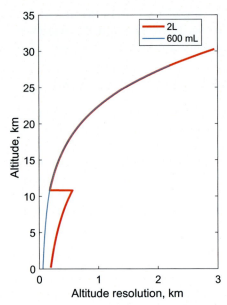

FIG. 6 Comparison of AirCore altitude resolution with differing dimensions: a 600 mL AirCore, all 0.320 cm (1/8″) OD tubing *(blue)*, and a 2 L AirCore, 20% of which is 0.320 cm (1/8″) OD and 80% of which is 0.64 cm (1/4″) OD *(red)*. The altitude resolution of each sampler configuration is calculated assuming that diffusion occurs within the AirCore for a 4 h sample residence time prior to analysis.

measurement, there will be more mixing between samples in the large-diameter tubing, where the volume to length ratio is larger and allows for more complete mixing by diffusion. Thus, the sampling resolution in the 2 L AirCore is lower for the larger diameter, tropospheric portion of the profile than that in the 600 mL, small diameter tropospheric tubing portion for the same 4 h storage time prior to analysis (Fig. 6). Other AirCore configurations for higher resolution sampling are detailed in Membrive et al. (2017) and Engel et al. (2017).

2.4.2 Balloon-borne dual-AirCore payload configurations

By modifying the AirCore described in Karion et al. (2010) to include small diameter, thin-walled tubing coils with a 600 mL total sample volume, two AirCores can be flown on a single balloon string under unmanned balloon regulations. This modification comes at the expense of less stratospheric air sampled in the AirCore, but allows for the evaluation of the repeatability of retrieved trace gas mole fraction profiles, and improves the quality control and assurance review for AirCore data. Fig. 7 shows the repeatability in NOAA/GML AirCore CO_2 mole fraction profiles observed as a function of pressure between paired, 600 mL AirCores that were flown together on the same balloon string. The mean absolute difference in CO_2 calculated for 100 hPa pressure bins throughout the atmospheric column is generally less than 0.2 ppm CO_2 (Fig. 7), while mean differences for CH_4 and CO mole fraction profiles are less than 3 parts per billion (ppb) and 2 ppb throughout the column, respectively.

2.4.3 Active AirCore sampling systems

The AirCore can also be configured to *actively* sample the atmosphere for spatially-resolved measurements of trace gases in a location of interest. Instead of relying on a pressure gradient between the AirCore and the atmosphere to passively sample the atmospheric column, the "Active AirCore" uses a small pump to compress or pull air into the tubing coil at a constant flow rate. This technique allows for sampling of trace gases in both horizontal and vertical directions along a flight path (Andersen et al., 2018). To date, the Active AirCore has been demonstrated on small unmanned aerial vehicles for the purposes of quantifying trace gas emissions from localized "hot spots," or areas with hypothesized large emissions sources (Andersen et al., 2021; Anderson et al., 2022; Vinković et al., 2022).

2.5 Retrieved profile uncertainty and reproducibility

Uncertainties in AirCore trace gas profiles are attributed to (a) air sampling artifacts and/or storage biases within the tubing coil, (b) instrument calibration, and (c) sample measurement precision and altitude registration uncertainties. Air sampling

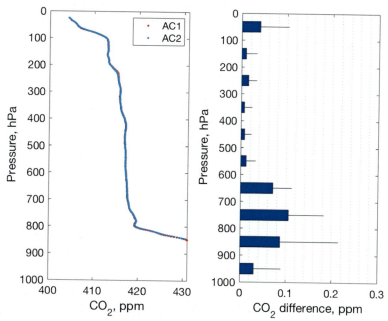

FIG. 7 CO_2 profiles retrieved from a March 8, 2021, NOAA AirCore flight with dual AirCores (labeled "AC1" and "AC2") flown together on the same payload string (left). The absolute CO_2 mole fraction average difference between profiles retrieved from dual AirCores flown on the same balloon string since 2012 is shown on the right ($N=40$ pairs, AirCore data v20201223 (Baier et al., 2021)). *Bars* indicate the average absolute difference for 100-hPa pressure bins, and error bars represent the standard error in the calculated profile differences for all available AirCore paired profiles in each pressure bin.

and storage biases due to surface contamination or tubing defects can be assessed in a laboratory setting by flowing a known mixture of air through the AirCore and assessing the mole fractions of trace gases at the output of the AirCore, or by storing a known mixture of air in the AirCore and assessing the mole fractions of trace gases within the tubing coil after a known storage time. Karion et al. (2010) indicated that these biases were less than 0.07 ppm and 0.7 ppb for CO_2 and CH_4, respectively. CRDS analyzers used to measure trace gases within AirCore samples provide high sampling precision to meet WMO compatibility requirements (see Section 2.3.1) and can be calibrated to WMO standard scales. Therefore, below we discuss the remaining uncertainties involved in the AirCore profile retrieval including evaluation of the accuracy of the trace gas profile mole fractions through comparisons of independent measurements, evaluation of the reproducibility between AirCore samplers measuring the same atmospheric column, and evaluation of the accuracy of the AirCore altitude registration.

2.5.1 Evaluation of AirCore profile measurement accuracy

Evaluation efforts regarding the accuracy of AirCore trace gas mole fraction profiles throughout the vertical column have been limited to date due to sparse independent, co-located calibrated measurements for comparison. Karion et al. (2010) compared the AirCore with continuous CRDS measurements and discrete flask samples aboard a common aircraft platform for evaluation of retrieved AirCore CO_2 and CH_4 dry mole fraction profiles up to approximately 9000 m AMSL. They found that AirCore differences from the continuous trace gas analyzer were unbiased and had a standard deviation between 0.2 and 0.3 ppm for CO_2 and about 5 ppb for CH_4. The absolute difference in pressure-weighted column averages was within 0.1 ppm and 2.2 ppb for CO_2 and CH_4, respectively. AirCore absolute mole fraction differences from 12 discrete whole-air flasks sampled throughout the vertical column were slightly higher, with CO_2 and CH_4 within 0.7 ppm CO_2 and 5 ppb CH_4.

AirCore profile evaluation has been conducted at higher altitudes using the lightweight stratospheric air sampler (LISA), designed to collect four multi-layer foil or Tedlar bag samples of volumes ranging from 180 to 800 mL at standard temperature and pressure (STP). LISA samples were later analyzed for CO_2, CH_4, and CO using a continuous, CRDS analyzer (Hooghiem et al., 2018). Integrated LISA sampler measurements, when registered in altitude by the number of moles of air sampled at each pressure level and the mean sampling pressure, agreed relatively well with the AirCore with a mean ($\pm 1\sigma$ standard deviation) of 0.6 ± 0.13 ppm for CO_2, -5.1 ± 13.1 ppb for CH_4, and -9.2 ± 5.2 ppb for CO (Hooghiem et al., 2018). Further, verification of the AirCore profile accuracy is needed. Large-balloon campaigns with gondolas instrumented with other, co-located, well-calibrated continuous trace gas sensors offer one pathway to such an evaluation.

2.5.2 Evaluation of AirCore profile reproducibility

The European Union-funded Readiness of ICOS (the Integrated Carbon Observation System) for Necessities of Integrated Global Observations (RINGO) project marked the first major comprehensive AirCore inter-comparison campaign among international institutions that have adopted this sampling technique. The first RINGO campaign took place in Sodankylä, Finland in June 2018 with 10 balloon flights and 26 vertical profiles retrieved from AirCore samplers (Chen et al., 2022, in preparation). Multiple AirCores were flown on the same balloon payload. Because air sample differences due to atmospheric variability will be small, inter-comparisons between trace gas profiles provide a valuable opportunity to evaluate the impacts of various AirCore dimensions, sampling techniques, and altitude registration methods on the AirCore profile retrieval.

During RINGO, the overall uncertainty in AirCore trace gas mole fraction profiles was assessed through comparison of the stratospheric portion of the profiles. This method was chosen because day-to-day variability in the trace gas mole fractions measured in AirCore stratospheric samples is relatively small and the age of air in this region is approximately 1–5 years (Engel et al., 2009), thus air sampled at these altitudes is somewhat disparate from that of lower altitudes. For CO_2, inter-comparisons were performed for profile altitudes between 20,000 and 25,000 m. It was found that the average difference in CO_2 measured in AirCores and retrieved independently by four different institutions was within 0.5 ppm (Fig. 8). CO profiles were compared between the altitude ranges of 13,000 and 17,000 m, indicating average differences generally less than 15 ppb (Fig. 8).

The reproducibility of the AirCore profile altitude registration for varying lengths and dimensions of AirCore tubing coils was assessed through comparisons of profiles retrieved independently by participating institutions during RINGO. Due to the rapid descent rate of the parachute payload at low atmospheric pressures, the stratospheric trace gas profile retrieval contains the largest source of error in the altitude registration. Eight pairs of independently-retrieved AirCore

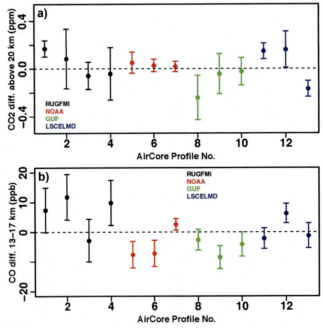

FIG. 8 RINGO comparison results of CO_2 (A) and CO (B) for selected parts of retrieved stratospheric profiles. Comparisons were made between balloon-borne AirCores flown and retrieved independently from four research groups: University of Groningen (RUG)/Finnish Meteorological Institute (FMI); NOAA/Global Monitoring Laboratory (NOAA); Goethe University Frankfurt (GUF); and Laboratoire des Sciences du Climat et de l'Environnement/Laboratoire des Sciences du Climat Laboratoire de Météorologie Dynamique (LSCELMD). *(From Chen, H., Hooghiem, J.J.D., Brownlow, R., Kivi, R., Heikkinen, P., Leuenberger, M., Nyfeler, P., Ramonet, M., Lopez, M., Engel, A., Wagenhaeuser, T., Laube, J., Baier, B., Sweeney, C., Danis, F., Crevoisier, C., 2022. Towards accurate vertical profile measurements of greenhouse gases using AirCore (in preparation).)*

profiles above 13,000m (~130hPa) were compared across pressure coordinates to estimate the pressure shift that maximizes the correlations of individual profile pairs. In this way, Chen et al. (2022, in preparation) estimated an average pressure altitude registration uncertainty of 0.5 hPa and a maximum absolute pressure shift of 4 hPa.

3. AirCore balloon-borne measurement sites

Since the AirCore technique was patented by Tans (2009), the AirCore sampling system has been adopted by several additional institutions: Netherlands (University of Groningen); France (the Laboratoire de Météorologie Dynamique); Germany (Forschungszentrum Jülich's Institute of Energy and Climate Research and Goethe University Frankfurt); and Finland (Finnish Meteorological Institute). Current and historical sampling locations are shown in Fig. 9.

Of these aforementioned institutions, several conduct routine AirCore measurement programs, as indicated in Table 1. Routine sampling within the NOAA AirCore program in the United States began in 2010 and has seen over one decade of routine launches and over 150 balloon-borne flights occurring on a near-monthly basis from Boulder, CO, United States. The French AirCore program (AirCore-FR) was initiated in 2013 in a collaboration by the Laboratoire de Météorologie Dynamique and Laboratoire des Sciences du Climat et de l'Environnement, with over 140 balloon-borne routine flights occurring since 2014 at sites in Traînou and Aire-sur-l'Adour, France. The Finnish Meteorological Institute in Sodankylä has been launching balloon-borne AirCores routinely in Sodankylä, Finland, across all seasons since 2013 with over 50 flights to date.

4. AirCore evaluation of remote sensing retrievals

Given its profiling capability, an important application for the AirCore is the evaluation of satellite and complementary ground-based total-column trace gas retrievals (typically denoted as X$[gas]$). While satellite retrievals provide unprecedented spatial and temporal information on greenhouse gas abundances, they cannot be directly calibrated and therefore require extensive evaluation (Ohring et al., 2005; Zhou et al., 2016) as outlined in Chapter 1. Ground-based remote sensing instrumentation (TCCON, Wunch et al., 2010, 2011; COCCON, Frey et al., 2015, 2019) have similar evaluation requirements. Furthermore, all of these remote sensing retrievals must be traceable to surface-based in situ networks (e.g., linked to WMO standard scales) in order to fill pertinent gaps in surface-based observing systems.

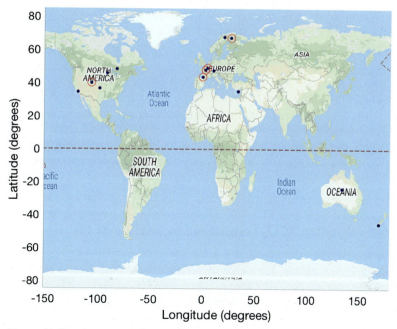

FIG. 9 Current and historical known AirCore measurement sites. Measurement locations with routine AirCore balloon-borne flights for long-term sites listed in Table 1 are circled in *red*. All other sites indicate shorter-term sampling or historical intensive campaign sites. *(Figure produced using Google Maps.)*

TABLE 1 AirCore long-term measurement site locations in France, United States, and Finland and corresponding institutions responsible for operations.

Long-term site	Aire-sur-l'Adour	Traînou[a]	Reims	Boulder	Sodankylä[a]
Latitude (degrees N)	43.71	47.97	49.24	39.991	67.416
Longitude (degrees E)	−0.25	2.09	4.07	−105.261	26.589
Institution(s)	Laboratoire de météorologie dynamique (LMD) Laboratoire des sciences du climat et de l'environnement (LSCE) Groupe de spectrométrie moléculaire et atmosphérique (GSMA) Observatoire de Physique du Globe de Clermont-Ferrand (OPGC) Centre National d'Etudes Spatiales (CNES)			National Oceanic and Atmospheric Administration (NOAA)	Finnish Meteorological Institute (FMI) University of Groningen (RUG)
Country	FRA			United States	FIN, NLD

[a]Denotes a Total Carbon Column Observing Network (TCCON) cooperative site.

4.1 Calculation of AirCore total-column trace gas abundances

The AirCore requires little interpolation or extrapolation of measurements to derive total-column trace gas abundances. This capability offers a greater potential for error or bias reduction in remote sensing total-column trace gas retrievals than measurements spanning a smaller extent of the atmospheric column.

Remote-sensing instrumentation is sensitive to only part of the atmosphere and to retrieve a total-column trace gas abundance, their raw information is convolved with a prior estimate of the atmospheric trace gas abundance that provides constraints on the part of the column to which the instrument is not sensitive. To compare an AirCore profile to a retrieved column, the measured profile must be similarly convolved with the retrieval's prior estimate using its averaging kernel via Eq. (2) from Rodgers and Connor (2003),

$$x_s = x_a + A(x_h - x_a), \qquad (2)$$

in which x_s describes a resulting averaging-kernel smoothed profile that is derived given an a priori remote sensing profile, x_a; the atmospheric "truth" state, x_h, from the AirCore; and the remote sensing averaging kernel, A. This convolution puts the retrieved and the AirCore-measured profiles on the same basis, so that their difference removes the influence of the a priori profile and involves only the part of the profile to which the satellite instrument is sensitive. The column mole fraction, $X[gas]$, for an AirCore trace gas species of interest can then be calculated by integrating x_s vertically along pressure coordinates.

4.2 Case study: AirCore-calculated XCO$_2$ comparison to satellite XCO$_2$

An example of the AirCore profile integration method is demonstrated through comparison to NASA's Orbiting Carbon Observatory version 2 (OCO-2) sun-synchronous near-IR satellite measuring total-column CO$_2$ (XCO$_2$) in Fig. 10 (Eldering et al., 2017). Here, the original AirCore CO$_2$ profile retrieved is extrapolated to the top of the atmosphere using the OCO-2 XCO$_2$ model-derived a priori profile and convolved with the OCO-2 averaging kernel and a priori profile using Eq. (2). To date, multiple remote sensing validation teams have utilized AirCore profiles in a similar manner for evaluation of total-column trace gas retrievals from both ground-based remote sensing instrumentation (e.g., TCCON) and satellites (e.g., MOPITT, TROPOMI, Suomi-NPP, NOAA-20; Hedelius et al., 2019; Nalli et al., 2020; Roche et al., 2021), indicating the value of the AirCore field measurements for passive remote sensing evaluation worldwide.

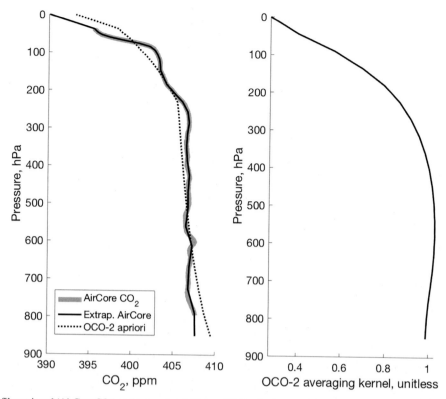

FIG. 10 AirCore profile retrieved (AirCore CO$_2$) and extrapolated (Extrap. AirCore) using an a priori profile for integration (OCO-2 apriori) with respect to NASA's Orbiting Carbon Observatory-2 (OCO-2) sensor averaging kernel, A (using Eq. 2), and retrieval of total-column CO$_2$ (XCO$_2$) (left). An example of the OCO-2 averaging kernel, A, is displayed between 0 and 1, with 1 indicating the greatest sensitivity throughout the vertical column for a particular time of year (right). Here, the AirCore CO$_2$, OCO-2 CO$_2$ a priori, and OCO-2 averaging kernel are all shown for northern hemisphere, mid-latitude winter as a case study for December 12, 2017, and for an AirCore mean profile location of (40.0253, −104.1086).

4.3 Case study: Linking remote sensing total column trace gas abundances to WMO standard scales using AirCore

Remotely sensed X[gas] retrievals have been linked to World Meteorological Organization (WMO) standard scales through comparisons to in situ measurements (e.g., Wunch et al., 2010, 2017). In doing so, the standard error from a best fit of the remote sensing X[gas] data to the in situ-derived X[gas] data—taking into account the uncertainties in both of these quantities—provides an estimate of the total uncertainty on the scaled retrievals. Using the NASA OCO-2 satellite's averaging kernel as an example, we demonstrate the potential for the AirCore to reduce scaled trace gas retrieval errors or biases due to its ability to provide a calibrated, retrieved profile for >90% of the atmospheric column.

We hypothetically estimate the total error budget in calculated XCO$_2$ from CO$_2$ vertical profiles collected on measurement platforms with varying altitude ceilings: the AirCore (altitude ceiling ∼ 30,000 m AMSL), a research jet (altitude ceiling of ∼ 12,000 m AMSL), and a light aircraft (altitude ceiling of ∼ 4000 m AMSL). Here we assume that all profiles were measured on a CRDS laboratory-grade analyzer. The error in each platform's CO$_2$ total column profile is estimated as a function of height for all three platforms (e.g., an error "profile"). For the AirCore, we sum in quadrature the 2σ precision on the CO$_2$ measurement and the mean difference in CO$_2$ calculated between dual AirCore profiles (Fig. 7) from the surface to the AirCore's altitude ceiling (typically, ∼ 30,000 m AMSL). The CO$_2$ error profile above the AirCore altitude ceiling to the top of the atmosphere is estimated using a conservative 2 ppm CO$_2$ uncertainty. For the two other aircraft platforms outlined in Table 2, error profiles are constructed similarly, with a 2σ CO$_2$ measurement precision below the altitude ceiling of the platform, and 2 ppm CO$_2$ uncertainty above that altitude. Error profiles for each platform are integrated along vertical pressure coordinates with respect to the OCO-2 averaging kernel (see Fig. 10 for NH mid-latitude winter). While the calculated platform XCO$_2$ error will vary based on the averaging kernel of the remote sensing instrument, time of year, and geographic location, this example highlights (a) that the contribution of the error profile above each platform's altitude ceiling is the largest source of error in the integrated column error budget and (b) that the AirCore provides the lowest estimated XCO$_2$ error of the three platforms. For remote sensing instrumentation with similar measurement sensitivities

TABLE 2 Integrated XCO$_2$ error with respect to the OCO-2 averaging kernel using CO$_2$ profiles collected from three measurement platforms: AirCore, a light aircraft, and a research jet.

Platform	Altitude ceiling (m AMSL)	Estimated XCO$_2$ error (ppm, 2σ)	Estimated average cost per profile ($USD)
Light aircraft	~4,000	1.25	1.5K
Jet	~12,000	0.35	10K
AirCore	~30,000	0.15	5K

Error estimates use the OCO-2 averaging kernel for the December 12, 2017, case shown in Fig. 10. Average estimated costs per profile are shown in the final column.

in the vertical dimension (i.e., similar averaging kernels), the AirCore therefore provides the largest potential for error and/or bias reduction in scaled trace gas retrievals at a significantly lower cost (Table 2).

5. Conclusions and future outlook

5.1 Toward a global AirCore network

Tans et al. (1996) outlined a vision for the global monitoring of atmospheric trace gases on tall towers, at surface sites, and aboard aircraft—all with carefully calibrated instrumentation for long-term monitoring of the global carbon cycle and other ozone-depleting substances. Other scientific agencies have outlined similar sampling programs (Ciais et al., 2014), yet in situ measurement networks today are still too sparse to reliably constrain regional trace gas budgets for establishing concrete carbon mitigation strategies. We discussed in Section 1 the unique capabilities and constraints that vertical profile measurements provide for advancing knowledge of these atmospheric trace gases.

The AirCore system currently provides a capability to monitor trace gases for the near-total atmospheric column at a low cost, and several countries have recognized the advantage of such a sampling system, adopting this technique. As indicated in Fig. 9, however, more work is needed to expand routine AirCore profiling to other regions of the world. Three-dimensional inverse models of trace gases and remote sensing platforms could benefit greatly from the added information that a global network of routine profiling sites would offer. A global AirCore network dedicated to routine sampling would provide added value for satellite and ground-based remote sensing evaluation if sites are co-located with ground-based trace gas remote sensing instrumentation such as those in the TCCON or COCCON. Long-term AirCore sampling sites in Sodankylä, Finland, and Traînou, France, are two examples of remote sensing evaluation sites with co-located, routine AirCore sampling. Additional sites are needed in traditionally under-observed (and climate-relevant) locations such as the tropics. Vertical sampling in the middle- to high-northern latitudes provides further knowledge of the composition of air residing within dynamical features such as the polar vortex and help for quantifying troposphere-stratosphere exchange of air.

Uniform operating procedures, and data delivery and archival requirements could be defined and adhered to at cooperating laboratories for defining robust sampling techniques, profile retrieval methods, and streamlined data reporting across groups. Additional work is needed to stringently evaluate AirCore data and profile uncertainties. For example, implementing the technique of dual AirCore sampling on a single balloon string has been shown to provide additional quality control and assurance of profiles for quick recognition of sampling or analysis biases. Routine inter-comparison studies between AirCore groups and other, independent, in situ measurement systems are also needed to further the AirCore technology, to fully understand uncertainties in this relatively new sampling system and to recognize potential artifacts that may evolve over time.

5.2 On the use of high-altitude uncrewed aircraft systems

AirCore launch locations and development of a global AirCore network are both limited to regions where the recovery of the AirCore and its accompanying scientific payload are both predictable (based on forecast meteorology) and feasible (based on ground-based obstacles), as there is no current technology to actively control the descent or landing of the parachute-based recovery system. Thus, trees, water, and local terrain provide significant challenges for today's AirCore launch and recovery operations. New, high-altitude, fixed-wing small uncrewed aircraft system (UAS) technology utilizing a weather balloon for ascent to high altitudes has been developed within the past decade. This technology provides one avenue for controlled recovery of the scientific payloads such as the AirCore (Craine, 2020; Kräuchi et al., 2016; Takahashi and Thomas, 2020). With a balloon-based ascent, the UAS can be operated at low cost relative to aircraft and platforms using large balloons. Because balloon drift distances can be in excess of 70,000 m downwind of the original launch location, such vehicles will require a sufficiently high glide ratio (horizontal distance divided by the change in altitude) and the ability to carry the weight of the AirCore tubing coil. In addition to the ability to land such a platform in a pre-determined recovery area, an added advantage of a UAS for AirCore recovery is the much-needed ability to control its descent rate to below $10-12\,\mathrm{m\,s^{-1}}$ in the stratosphere. One such platform has been designed specifically for use with the AirCore atmospheric sampling system and other, additional scientific payloads (Takahashi et al., 2020).

Despite regulatory hurdles to such operations, the use of UAS as a return vehicle shows great promise for future atmospheric trace gas sampling using the AirCore over both land and sea. As such, this technology provides a potential pathway for high-altitude AirCore sampling in remote areas such as the tropics and high latitudes where observations of trace gases are lacking and where climate impacts on atmospheric trace gas abundances are especially significant.

Acknowledgments

The authors thank both Dr. Andrew Jacobson and Dr. Anne Thompson for their generous time committed to reviewing this chapter. We thank our international AirCore network colleagues, whose collaboration over the past decade has led to significant contributions to the advancement and the use of this technology throughout the world, and we thank the satellite observing and evaluation communities, whose unrelenting efforts have made significant strides in improving our understanding of atmospheric greenhouse and trace gases.

References

Andersen, T., et al., 2018. A UAV-based active AirCore system for measurements of greenhouse gases. Atmos. Meas. Tech. 11 (5), 2683–2699. https://doi.org/10.5194/amt-11-2683-2018.

Andersen, T., Vinkovic, K., Vries, M.D., Kers, B., Necki, J., Swolkien, J., Roiger, A., Peters, W., Chen, H., 2021. Quantifying methane emissions from coal mining ventilation shafts using an unmanned aerial vehicle (UAV)-based active AirCore system. Atmos. Environ. X 12, 100135. ISSN 2590-1621 https://doi.org/10.1016/j.aeaoa.2021.100135.

Andrews, A.E., Boering, K.A., Wofsy, S.C., Daube, B.C., Jones, D.B., Alex, S., Loewenstein, M., Podolske, J.R., Strahan, S.E., 2001a. Empirical age spectra for the midlatitude lower stratosphere from in situ observations of CO_2: quantitative evidence for a subtropical "barrier" to horizontal transport. J. Geophys. Res.-Atmos. 106, 10257–10274. https://doi.org/10.1029/2000JD900703.

Anderson, T., de Vries, M., Necki, J., Swolkien, J., Menoud, M., Röckmann, T., Roiger, A., Fix, A., Peters, W., Chen, H., 2022. Local to regional methane emissions from the Upper Silesia Coal Basin (USCB) quantified using UAV-based atmospheric measurements. Atmos. Chem. Phys. https://doi.org/10.5194/acp-2021-1061.

Andrews, A.E., Boering, K.A., Daube, B.C., Wofsy, S.C., Loewenstein, M., Jost, H., Podolske, J.R., Webster, C.R., Herman, R.L., Scott, D.C., Flesch, G.J., 2001b. Mean ages of stratospheric air derived from in situ observations of CO_2, CH_4, and N_2O. J. Geophys. Res.-Atmos. 106 (D23), 32295–32314.

Baier, B., Sweeney, C., Tans, P., Newberger, T., Higgs, J., Wolter, S., NOAA Global Monitoring Laboratory, 2021. NOAA AirCore atmospheric sampling system profiles (Version 20201223) [Data set]. NOAA GML. https://doi.org/10.15138/6AV0-MY81.

Bakwin, P.S., Tans, P.P., Stephens, B.B., Wofsy, S.C., Gerbig, C., Grainger, A., 2003. Strategies for measurement of atmospheric column means of carbon dioxide from aircraft using discrete sampling. J. Geophys. Res.-Atmos., 108. https://doi.org/10.1029/2002JD003306.

Chen, H., Winderlich, J., Gerbig, C., Hoefer, A., Rella, C.W., Crosson, E.R., Van Pelt, A.D., Steinbach, J., Kolle, O., Beck, V., Daube, B.C., Gottlieb, E.W., Chow, V.Y., Santoni, G.W., Wofsy, S.C., 2010. High-accuracy continuous airborne measurements of greenhouse gases (CO_2 and CH_4) using the cavity ring-down spectroscopy (CRDS) technique. Atmos. Meas. Tech. 3, 375–386. https://doi.org/10.5194/amt-3-375-2010.

Chen, H., Hooghiem, J.J.D., Brownlow, R., Kivi, R., Heikkinen, P., Leuenberger, M., Nyfeler, P., Ramonet, M., Lopez, M., Engel, A., Wagenhaeuser, T., Laube, J., Baier, B., Sweeney, C., Danis, F., Crevoisier, C., 2022. Towards accurate vertical profile measurements of greenhouse gases using AirCore. (in preparation).

Chen, H., Karion, A., Rella, C.W., Winderlich, J., Gerbig, C., Filges, A., Newberger, T., Sweeney, C., Tans, P.P., 2013. Accurate measurements of carbon monoxide in humid air using the cavity ring-down spectroscopy (CRDS) technique. Atmos. Meas. Tech. 6, 1031–1040. https://doi.org/10.5194/amt-6-1031-2013.

Ciais, P., Dolman, A.J., Bombelli, A., Duren, R., Peregon, A., Rayner, P.J., Miller, C., Gobron, N., Kinderman, G., Marland, G., Gruber, N., Chevallier, F., Andres, R.J., Balsamo, G., Bopp, L., Bréon, F.-M., Broquet, G., Dargaville, R., Battin, T.J., Borges, A., Bovensmann, H., Buchwitz, M., Butler, J., Canadell, J.G., Cook, R.B., DeFries, R., Engelen, R., Gurney, K.R., Heinze, C., Heimann, M., Held, A., Henry, M., Law, B., Luyssaert, S., Miller, J., Moriyama, T., Moulin, C., Myneni, R.B., Nussli, C., Obersteiner, M., Ojima, D., Pan, Y., Paris, J.-D., Piao, S.L., Poulter, B., Plummer, S., Quegan, S., Raymond, P., Reichstein, M., Rivier, L., Sabine, C., Schimel, D., Tarasova, O., Valentini, R., Wang, R., van der Werf, G., Wickland, D., Williams, M., Zehner, C., 2014. Current systematic carbon-cycle observations and the need for implementing a policy-relevant carbon observing system. Biogeosciences 11, 3547–3602. https://doi.org/10.5194/bg-11-3547-2014.

Craine, N., 2020. Unmanned stratospheric glider for satellite calibration. EGU General Assembly Conference Abstracts, p. 9447.

Crevoisier, C., Sweeney, C., Gloor, M., Sarmiento, J.L., Tans, P.P., 2010. Regional US carbon sinks from three-dimensional atmospheric CO_2 sampling. PNAS 107, 18348–18353.

Crosson, E., 2008. A cavity ring-down analyzer for measuring atmospheric levels of methane, carbon dioxide, and water vapor. Appl. Phys. B 92, 403–408. https://doi.org/10.1007/s00340-008-3135-y.

Daube, B.C., Boering, K.A., Andrews, A.E., Wofsy, S.C., 2002. A high-precision fast-response airborne CO_2 analyzer for in situ sampling from the surface to the middle stratosphere. J. Atmos. Ocean. Technol. 19, 1532–1543. https://doi.org/10.1175/1520-0426(2002)019<1532:AHPFRA>2.0.CO;2.

Deeter, M.N., Martínez-Alonso, S., Edwards, D.P., Emmons, L.K., Gille, J.C., Worden, H.M., Sweeney, C., Pittman, J.V., Daube, B.C., Wofsy, S.C., 2014. The MOPITT Version 6 product: algorithm enhancements and validation. Atmos. Meas. Tech. 7, 3623–3632. https://doi.org/10.5194/amt-7-3623-2014.

Eldering, A., O'Dell, C.W., Wennberg, P.O., Crisp, D., Gunson, M.R., Viatte, C., Avis, C., Braverman, A., Castano, R., Chang, A., Chapsky, L., 2017. The Orbiting Carbon Observatory-2: first 18 months of science data products. Atmos. Meas. Tech. 10 (2), 549–563.

Engel, A., Möbius, T., Bönisch, H., Schmidt, U., Heinz, R., Levin, I., Atlas, E., Aoki, S., Nakazawa, T., Sugawara, S., Moore, F., Hurst, D., Elkins, J., Schauffler, S., Andrews, A., Boering, K., 2009. Age of stratospheric air unchanged within uncertainties over the past 30 years. Nat. Geosci. 2, 28–31. https://doi.org/10.1038/ngeo388.

Engel, A., Bönisch, H., Ullrich, M., Sitals, R., Membrive, O., Danis, F., Crevoisier, C., 2017. Mean age of stratospheric air derived from AirCore observations. Atmos. Chem. Phys. 17, 6825–6838. https://doi.org/10.5194/acp-17-6825-2017.

Federal Aviation Administration (FAA), 1991. Docket Number 1457, 29 FR 47, Jan 3, 1964. Available from: https://www.ecfr.gov/. (Accessed 4 June 2021).

Frey, M., Hase, F., Blumenstock, T., Groß, J., Kiel, M., Mengistu Tsidu, G., Schäfer, K., Sha, M.K., Orphal, J., 2015. Calibration and instrumental line shape characterization of a set of portable FTIR spectrometers for detecting greenhouse gas emissions. Atmos. Meas. Tech. 8, 3047–3057. https://doi.org/10.5194/amt-8-3047-2015.

Frey, M., Sha, M.K., Hase, F., Kiel, M., Blumenstock, T., Harig, R., Surawicz, G., Deutscher, N.M., Shiomi, K., Franklin, J.E., Bösch, H., Chen, J., Grutter, M., Ohyama, H., Sun, Y., Butz, A., Mengistu Tsidu, G., Ene, D., Wunch, D., Cao, Z., Garcia, O., Ramonet, M., Vogel, F., Orphal, J., 2019. Building the COllaborative Carbon Column Observing Network (COCCON): long-term stability and ensemble performance of the EM27/SUN Fourier transform spectrometer. Atmos. Meas. Tech. 12, 1513–1530. https://doi.org/10.5194/amt-12-1513-2019.

Golston, L.M., Tao, L., Brosy, C., Schäfer, K., Wolf, B., McSpiritt, J., Buchholz, B., Caulton, D.R., Pan, D., Zondlo, M.A., Yoel, D., Kunstmann, H., McGregor, M., 2017. Lightweight mid-infrared methane sensor for unmanned aerial systems. Appl. Phys. B Lasers Opt. 123, 170. https://doi.org/10.1007/s00340-017-6735-6.

Harries, J.E., Ruth, S., Russell III, J.M., 1996. On the distribution of mesospheric molecular hydrogen inferred from HALOE measurements of H_2O and CH_4. Geophys. Res. Lett. 23, 297–300. https://doi.org/10.1029/95GL03197.

Hedelius, J.K., He, T.-L., Jones, D.B.A., Baier, B.C., Buchholz, R.R., De Mazière, M., Deutscher, N.M., Dubey, M.K., Feist, D.G., Griffith, D.W.T., Hase, F., Iraci, L.T., Jeseck, P., Kiel, M., Kivi, R., Liu, C., Morino, I., Notholt, J., Oh, Y.-S., Ohyama, H., Pollard, D.F., Rettinger, M., Roche, S., Roehl, C.M., Schneider, M., Shiomi, K., Strong, K., Sussmann, R., Sweeney, C., Té, Y., Uchino, O., Velazco, V.A., Wang, W., Warneke, T., Wennberg, P.O., Worden, H.M., Wunch, D., 2019. Evaluation of MOPITT Version 7 joint TIR–NIR X_{CO} retrievals with TCCON. Atmos. Meas. Tech. 12, 5547–5572. https://doi.org/10.5194/amt-12-5547-2019.

Hooghiem, J.J.D., de Vries, M., Been, H.A., Heikkinen, P., Kivi, R., Chen, H., 2018. LISA: a lightweight stratospheric air sampler. Atmos. Meas. Tech. 11, 6785–6801. https://doi.org/10.5194/amt-11-6785-2018.

Hooghiem, J.J.D, Popa, M.E., Röckmann, T., Grooß, J-U., Tritscher, I., Müller, R., Kivi, R., Chen, H., 2020. Wildfire smoke in the lower stratosphere identified by in situ CO observations. Atmos. Chem. Phys. 20, 13985–14003. https://doi.org/10.5194/acp-20-13985-2020.

Inoue, M., Morino, I., Uchino, O., Miyamoto, Y., Yoshida, Y., Yokota, T., Machida, T., Sawa, Y., Matsueda, H., Sweeney, C., Tans, P.P., Andrews, A.E., Biraud, S.C., Tanaka, T., Kawakami, S., Patra, P.K., 2013. Validation of XCO_2 derived from SWIR spectra of GOSAT TANSO-FTS with aircraft measurement data. Atmos. Chem. Phys. 13, 9771–9788. https://doi.org/10.5194/acp-13-9771-2013.

International Panel on Climate Change (IPCC), 2015. Climate Change 2014: Synthesis Report. Contribution of Working Groups I, II and III to the Fifth Assessment Report of the Intergovernmental Panel on Climate Change [Core Writing Team, R.K. Pachauri and L.A. Meyer (eds.)]. IPCC, Geneva, Switzerland. 151 pp.

Jacobson, A.R., Schuldt, K.N., Miller, J.B., Oda, T., Tans, P., Andrews, A., Mund, J., Ott, L., Collatz, G.J., Aalto, T., Afshar, S., Aikin, K., Aoki, S., Apadula, F., Baier, B., Bergamaschi, P., Beyersdorf, A., Biraud, S.C., Bollenbacher, A., Bowling, D., Brailsford, G., Abshire, J.B., Chen, G., Chen, H., Chmura, L., et al., 2020. CarbonTracker CT2019B, 2020: Model Published by NOAA Global Monitoring Laboratory. https://gml.noaa.gov/ccgg/carbontracker/CT2019B/.

Joly, L., Coopmann, O., Guidard, V., Decarpenterie, T., Dumelié, N., Cousin, J., Burgalat, J., Chauvin, N., Albora, G., Maamary, R., Miftah El Khair, Z., Tzanos, D., Barrié, J., Moulin, É., Aressy, P., Belleudy, A., 2020. The development of the Atmospheric Measurements by Ultra-Light Spectrometer (AMULSE) greenhouse gas profiling system and application for satellite retrieval validation. Atmos. Meas. Tech. 13, 3099–3118. https://doi.org/10.5194/amt-13-3099-2020.

Karion, A., Sweeney, C., Tans, P., Newberger, T., 2010. AirCore: an innovative atmospheric sampling system. J. Atmos. Oceanic Tech. 27, 1839–1853. https://doi.org/10.1175/2010JTECHA1448.1.

Kräuchi, A., Philipona, R., Romanens, G., Hurst, D.F., Hall, E.G., Jordan, A.F., 2016. Controlled weather balloon ascents and descents for atmospheric research and climate monitoring. Atmos. Meas. Tech. 9, 929–938. https://doi.org/10.5194/amt-9-929-2016.

Laube, J.C., Elvidge, E.C.L., Adcock, K.E., Baier, B., Brenninkmeijer, C.A.M., Chen, H., Droste, E.S., Grooß, J.-U., Heikkinen, P., Hind, A.J., Kivi, R., Lojko, A., Montzka, S.A., Oram, D.E., Randall, S., Röckmann, T., Sturges, W.T., Sweeney, C., Thomas, M., Tuffnell, E., Ploeger, F., 2020. Investigating stratospheric changes between 2009 and 2018 with halogenated trace gas data from aircraft, AirCores, and a global model focusing on CFC-11. Atmos. Chem. Phys. 20, 9771–9782. https://doi.org/10.5194/acp-20-9771-2020.

Membrive, O., Crevoisier, C., Sweeney, C., Danis, F., Hertzog, A., Engel, A., Bönisch, H., Picon, L., 2017. AirCore-HR: a high-resolution column sampling to enhance the vertical description of CH_4 and CO_2. Atmos. Meas. Tech. 19, 2163–2181.

Moore, F.L., Elkins, J.W., Ray, E.A., Dutton, G.S., Dunn, R.E., Fahey, D.W., McLaughlin, R.J., Thompson, T.L., Romashkin, P.A., Hurst, D.F., Wamsley, P.R., 2003. Balloonborne in situ gas chromatograph for measurements in the troposphere and stratosphere. J. Geophys. Res.-Atmos. 108. https://doi.org/10.1029/2001JD000891.

Mrozek, D.J., van der Veen, C., Hofmann, M.E.G., Chen, H., Kivi, R., Heikkinen, P., Röckmann, T., 2016. Stratospheric air sub-sampler (SAS) and its application to analysis of $\Delta^{17}O(CO_2)$ from small air samples collected with an AirCore. Atmos. Meas. Tech. 9, 5607–5620. https://doi.org/10.5194/amt-9-5607-2016.

Nalli, N.R., Tan, C., Warner, J., Divakarla, M., Gambacorta, A., Wilson, M., Zhu, T., Wang, T., Wei, Z., Pryor, K., Kalluri, S., Zhou, L., Sweeney, C., Baier, B.C., McKain, K., Wunch, D., Deutscher, N.M., Hase, F., Iraci, L.T., Kivi, R., Morino, I., Notholt, J., Ohyama, H., Pollard, D.F., Té, Y., Velazco, V.A., Warneke, T., Sussmann, R., Rettinger, M., 2020. Validation of carbon trace gas profile retrievals from the NOAA-unique combined atmospheric processing system for the cross-track infrared sounder. Remote Sens. (Basel) 12, 3245. https://doi.org/10.3390/rs12193245.

National Research Council (NRC), 2010. Revitalizing NASA's Suborbital Program: Advancing Science, Driving Innovation, and Developing Workforce., https://doi.org/10.17226/12862.

Ohring, G., Wielicki, B., Spencer, R., Emery, B., Datla, R., 2005. Satellite instrument calibration for measuring global climate change: report of a workshop. Bull. Am. Meteorol. Soc. 86, 1303–1314. https://doi.org/10.1175/BAMS-86-9-1303.

Paul, D., Chen, H., Been, H.A., Kivi, R., Meijer, H.A.J., 2016. Radiocarbon analysis of stratospheric CO_2 retrieved from AirCore sampling. Atmos. Meas. Tech. 9, 4997–5006. https://doi.org/10.5194/amt-9-4997-2016.

Peiro, H., Crowell, S., Schuh, A., Baker, D.F., O'Dell, C., Jacobson, A.R., Chevallier, F., Liu, J., Eldering, A., Crisp, D., Deng, F., Weir, B., Basu, S., Johnson, M.S., Philip, S., Baker, I., 2021. Four years of global carbon cycle observed from OCO-2 version 9 and in situ data, and comparison to OCO-2 v7. Atmos. Chem. Phys. Discuss., 1–50. https://doi.org/10.5194/acp-2021-373.

Ray, E.A., Moore, F.L., Elkins, J.W., Hurst, D.F., Romashkin, P.A., Dutton, G.S., Fahey, D.W., 2002. Descent and mixing in the 1999–2000 northern polar vortex inferred from in situ tracer measurements. J. Geophys. Res.-Atmos. 107. https://doi.org/10.1029/2001JD000961. SOL 28-1-SOL 28-18.

Ray, E.A., Moore, F.L., Rosenlof, K.H., Davis, S.M., Sweeney, C., Tans, P., Wang, T., Elkins, J.W., Bönisch, H., Engel, A., Sugawara, S., Nakazawa, T., Aoki, S., 2014. Improving stratospheric transport trend analysis based on SF6 and CO_2 measurements. J. Geophys. Res.-Atmos. 119, 14110–14128. https://doi.org/10.1002/2014JD021802.

Roche, S., Strong, K., Wunch, D., Mendonca, J., Sweeney, C., Baier, B., Biraud, S.C., Laughner, J.L., Toon, G.C., Connor, B.J., 2021. Retrieval of atmospheric CO_2 vertical profiles from ground-based near-infrared spectra. Atmos. Meas. Tech. 14, 3087–3118. https://doi.org/10.5194/amt-14-3087-2021.

Rodgers, C.D., Connor, B.J., 2003. Intercomparison of remote sounding instruments. J. Geophys. Res.-Atmos. 108. https://doi.org/10.1029/2002JD002299.

Sha, M.K., De Mazière, M., Notholt, J., Blumenstock, T., Chen, H., Dehn, A., Griffith, D.W.T., Hase, F., Heikkinen, P., Hermans, C., Hoffmann, A., Huebner, M., Jones, N., Kivi, R., Langerock, B., Petri, C., Scolas, F., Tu, Q., Weidmann, D., 2020. Intercomparison of low- and high-resolution infrared spectrometers for ground-based solar remote sensing measurements of total column concentrations of CO_2, CH_4, and CO. Atmos. Meas. Tech. 13, 4791–4839. https://doi.org/10.5194/amt-13-4791-2020.

Stephens, B.B., et al., 2007. Weak northern and strong tropical land carbon uptake from vertical profiles of atmospheric CO_2. Science 316 (5832), 1732–1735. https://doi.org/10.1126/science.1137004.

Sweeney, C., Chatterjee, A., Wolter, S., McKain, K., Bogue, R., Newberger, T., Hu, L., Ott, L., Poulter, B., Schiferl, L., Weir, B., Zhang, Z., Miller, C.E., 2022. Using atmospheric trace gas vertical profiles to evaluate model fluxes: a case study of Arctic-CAP observations and GEOS simulations for the ABoVE domain. Atmos. Chem. Phys. 9, 6347–6364. https://doi.org/10.5194/acp-22-6347-2022.

Sweeney, C., Karion, A., Wolter, S., Newberger, T., Guenther, D., Higgs, J.A., Andrews, A.E., Lang, P.M., Neff, D., Dlugokencky, E., Miller, J.B., Montzka, S.A., Miller, B.R., Masarie, K.A., Biraud, S.C., Novelli, P.C., Crotwell, M., Crotwell, A.M., Thoning, K., Tans, P.P., 2015. Seasonal climatology of CO_2 across North America from aircraft measurements in the NOAA/ESRL Global Greenhouse Gas Reference Network. J. Geophys. Res.-Atmos. 120, 5155–5190. https://doi.org/10.1002/2014JD022591.

Takahashi, T.T., Thomas, P.R., 2020. The aerodynamic design of a high-altitude operational returning unmanned system for atmospheric science. In: AIAA AVIATION 2020 FORUM. p. 2658.

Tans, P.P., 2009. System and Method for Providing Vertical Profile Measurements of Atmospheric Gases. US7597014B2.

Tans, P.P., 2022. Fill dynamics and sample mixing in the AirCore. Atmos. Meas. Tech. 15, 1903–1916. https://doi.org/10.5194/amt-15-1903-2022.

Tans, P.P., Bakwin, P.S., Guenther, D.W., 1996. A feasible Global Carbon Cycle Observing System: a plan to decipher today's carbon cycle based on observations. Glob. Chang. Biol. 2, 309–318. https://doi.org/10.1111/j.1365-2486.1996.tb00082.x.

Taylor, G.I., 1953. Dispersion of soluble matter in solvent flowing slowly through a tube. Proc. R. Soc. Lond. A Math. Phys. Sci. 219 (1137), 186–203.

United Nations Environment Programme (UNEP), 1987. Montreal Protocol on Substances that Deplete the Ozone Layer. vol. 26 US Government Printing Office, Washington, DC, pp. 128–136.

Vinković, K., Andersen, T., de Vries, M., Kers, B., van Heuven, S., Peters, W., Hensen, A., van den Bulk, P., Chen, H., 2022. Evaluating the use of an Unmanned Aerial Vehicle (UAV)-based active AirCore system to quantify methane emissions from dairy cows. Sci. Total Environ. 831, 154898. https://doi.org/10.1016/j.scitotenv.2022.154898. In review.

Wagenhäuser, T., Engel, A., Sitals, R., 2021. Testing the altitude attribution and vertical resolution of AirCore measurements with a new spiking method. Atmos. Meas. Tech. 14, 3923–3934. https://doi.org/10.5194/amt-14-3923-2021.

WMO/GAW, 2011. 15th WMO/IAEA Meeting of Experts on Carbon Dioxyide, Other Greenhouse Gases and Related Tracers Measurement Techniques, WMO/TD. WMO, Geneva. GAW Report, 194.

Wunch, D., Toon, G.C., Wennberg, P.O., Wofsy, S.C., Stephens, B.B., Fischer, M.L., Uchino, O., Abshire, J.B., Bernath, P., Biraud, S.C., Blavier, J.-F.L., Boone, C., Bowman, K.P., Browell, E.V., Campos, T., Connor, B.J., Daube, B.C., Deutscher, N.M., Diao, M., Elkins, J.W., Gerbig, C., Gottlieb, E.W., Griffith, D.W.T., Hurst, D.F., Jiménez, R., Keppel-Aleks, G., Kort, E.A., Macatangay, R., Machida, T., Matsueda, H., Moore, F., Morino, I., Park, S., Robinson, J., Roehl, C.M., Sawa, Y., Sherlock, V., Sweeney, C., Tanaka, T., Zondlo, M.A., 2010. Calibration of the total carbon column observing network using aircraft profile data. Atmos. Meas. Tech. https://doi.org/10.5194/amt-3-1351-2010.

Wunch, D., Toon, G.C., Blavier, J.-F.L., Washenfelder, R.A., Notholt, J., Connor, B.J., Griffith, D.W.T., Sherlock, V., Wennberg, P.O., 2011. The total carbon column observing network. Philos. Trans. R. Soc. A Math. Phys. Eng. Sci. 369, 2087–2112. https://doi.org/10.1098/rsta.2010.0240.

Wunch, D., Wennberg, P.O., Osterman, G., Fisher, B., Naylor, B., Roehl, C.M., O'Dell, C., Mandrake, L., Viatte, C., Kiel, M., Griffith, D.W.T., Deutscher, N.M., Velazco, V.A., Notholt, J., Warneke, T., Petri, C., De Maziere, M., Sha, M.K., Sussmann, R., Rettinger, M., Pollard, D., Robinson, J., Morino, I., Uchino, O., Hase, F., Blumenstock, T., Feist, D.G., Arnold, S.G., Strong, K., Mendonca, J., Kivi, R., Heikkinen, P., Iraci, L., Podolske, J., Hillyard, P.W., Kawakami, S., Dubey, M.K., Parker, H.A., Sepulveda, E., García, O.E., Te, Y., Jeseck, P., Gunson, M.R., Crisp, D., Eldering, A., 2017. Comparisons of the Orbiting Carbon Observatory-2 (OCO-2) XCO2 measurements with TCCON. Atmos. Meas. Tech. 10, 2209–2238. https://doi.org/10.5194/amt-10-2209-2017.

Zhou, L., Divakarla, M., Liu, X., 2016. An overview of the joint polar satellite system (JPSS) science data product calibration and validation. Remote Sens. (Basel) 8, 139. https://doi.org/10.3390/rs8020139.

Zhou, M., Langerock, B., Sha, M.K., Kumps, N., Hermans, C., Petri, C., Warneke, T., Chen, H., Metzger, J.-M., Kivi, R., Heikkinen, P., Ramonet, M., De Mazière, M., 2019. Retrieval of atmospheric CH4 vertical information from ground-based FTS near-infrared spectra. Atmos. Meas. Tech. 12, 6125–6141. https://doi.org/10.5194/amt-12-6125-2019.

Part II

Intensive campaigns

Chapter 9

High-altitude aircraft radiometric calibration-validation campaigns

Joseph K. Taylor, David C. Tobin, Fred A. Best, Raymond K. Garcia, Michelle L. Loveless, Henry E. Revercomb, and William L. Smith, Sr.

Space Science and Engineering Center (SSEC), University of Wisconsin-Madison, Madison, WI, United States

Chapter outline

1. Introduction	159
2. The Scanning High-resolution Interferometer Sounder	160
3. S-HIS radiometric calibration, verification, and traceability	163
4. S-HIS high-altitude airborne calibration validation results	166
4.1 Calibration validation of the Cross-track Infrared Sounder	166
4.2 Calibration validation of the Advanced Baseline Imager	175
5. Summary	180
Acknowledgments	180
References	180

An experiment is a question which science poses to Nature, and a measurement is the recording of Nature's answer.

Maxwell Planck

1. Introduction

There is an established need for higher accuracy and more refined error characterization of radiance measurements from space for improved weather prediction and climate change monitoring (Ohring et al., 2007; Wielicki et al., 2013; Hewison et al., 2020). Climate change detection and attribution analyses, numerical weather prediction (NWP), as well as intercalibration applications require rigorous uncertainty analyses following established metrological principles. These analyses start with understanding the uncertainties associated with the spectral radiance observations and propagating these and other sources of uncertainty into radiance and geophysical product analyses. At the present time, in the absence of satellite infrared reference sensors with on-orbit traceability to absolute standards like those defined by the CLARREO program (Wielicki et al., 2013; Taylor et al., 2020), satellite under-flight comparisons with calibrated reference measurements obtained from high-altitude aircraft are uniquely qualified for assessing infrared satellite observations with high accuracy and reasonably direct traceability to absolute standards.

High-altitude airborne radiometric calibration validation typically involves collecting high-quality reference data from accurately calibrated airborne instruments that are temporally and spatially co-located with the satellite sensor measurements and then completing a detailed comparison between the satellite-based radiance measurements and the corresponding airborne sensor reference data. The detailed comparison between the satellite-based radiance measurements and those from a high-altitude aircraft must account for all known sources of error, including instrument noise and scene variations, as well as differences in the instrument altitudes, observation view angles, spatial footprints, and spectral response. For the calibration validation process to be accurate, traceable, and repeatable, it is most important that the reference data instrument be extremely well characterized and understood, carefully maintained, and accurately calibrated with strong traceability to absolute standards. The Scanning High-resolution Interferometer Sounder (S-HIS) (discussed more in Section 2) meets and exceeds these requirements and has proven to do so on multiple airborne platforms, each with significantly different instrument operating environments (Revercomb et al., 1998, 2004; Best and Revercomb, 2005; Revercomb and Best, 2005; Taylor et al., 2005; Tobin et al., 2006).

For the future, we emphasize the value of an on-orbit radiometric transfer standard for testing and improving the calibration accuracy and traceability of the operational infrared sounders (Tobin et al., 2016; Taylor et al., 2019a) as well as for setting a climate benchmark (Ohring et al., 2007; Wielicki et al., 2013; Taylor et al., 2020). Technology advancements at University of Wisconsin Space Science and Engineering Center (UW-SSEC) have led to the development of the Absolute Radiance Interferometer (ARI), an infrared spectrometer designed to serve as an on-orbit radiometric reference with ultra-high measurement accuracy (better than 0.1 K 3-σ brightness temperature at scene temperature) and on-orbit traceability to absolute standards (Taylor, 2014; Taylor et al., 2020). The ARI incorporates an On-orbit Absolute Radiance Standard (OARS), which utilizes miniature phase change cells and emissivity monitoring technologies to provide direct on-orbit verification and testing of the ARI accuracy with direct traceability to International Standards (SI) over a wide range of brightness temperatures (Best et al., 2010, 2012; Gero et al., 2012). An on-orbit asset such as ARI would not only provide the accuracy and traceability needed to optimize the measurement of the long-term changes of the Earth's atmosphere and surface but could also be used to inter-calibrate the international fleet of infrared (IR) hyperspectral sounders, such as the Atmospheric Infrared Sounder (AIRS), the Infrared Atmospheric Sounding Interferometer (IASI), and the Cross-track Infrared Sounder (CrIS), to similar measurement accuracy if flown in an orbit that frequently crosses sun-synchronous low Earth orbits. It is important to note that even if an on-orbit radiometric transfer standard is successfully implemented that there will still be a significant need for high-altitude airborne reference measurements which can be used to target specific scenes, provide higher spatial resolution than satellite sensors, and effectively bridge in-situ and satellite sensor measurements during field campaigns.

Climate benchmark measurements

Benchmark measurements in the context of long-term climate monitoring include the following characteristics (Goody, 2001; Ohring et al., 2007; Wielicki et al., 2013)
- Accuracy that extends over decades, or indefinitely
- Traceable to fundamental International System of Units (SI) standards and robust to gaps in the measurement record
- Sufficient information content and accuracy to determine decadal trends in essential climate change variables
- Time, space, and angle sampling sufficient to reduce aliasing bias error in global decadal change observations to well below predicted decadal climate change and below natural climate variability
- Observation strategy designed to reveal systematic errors through independent cross-checks, open inspection, and continuous interrogation

---

2. The Scanning High-resolution Interferometer Sounder

The Scanning High-resolution Interferometer Sounder (S-HIS) is a Fourier transform spectrometer (FTS) that measures thermal infrared radiation from 580 to 2850 cm^{-1} (3.5–17.3 μm) at 0.5 cm^{-1} spectral sampling resolution. High-spectral resolution offers inherent advantages for both radiometric and spectral calibration (Goody and Haskins, 1998). The S-HIS has a 0.100 rad angular field of view that provides a 2 km nadir footprint diameter from a 20 km observing altitude, and imaging of the scene is accomplished via cross-track scanning. Since 1998, the S-HIS has participated in 35 field campaigns on the NASA ER-2, DC-8, Proteus, WB-57, and Global Hawk airborne platforms. It has proven to be extremely dependable with high calibration accuracy and consistent performance on all platforms.

The S-HIS (Fig. 1) is an advanced version of the original High-resolution Interferometer Sounder (HIS) NASA ER-2 instrument (Revercomb et al., 1988a,b, 1989; Smith et al., 1987, 1990). The S-HIS was initially designed to fly on an unmanned aircraft vehicle (UAV) with limited payload capacity that drove it to be small, lightweight, and modular, with low power consumption. It was developed between 1996 and 1998, and refined beyond that, at the University of Wisconsin (UW) Space Science and Engineering Center (SSEC) with the combined support of the US DOE, NASA, and the then NPOESS Integrated Program Office. Its design and calibration techniques benefitted from the earlier experience with the HIS, as well as the ground-based Atmospheric Emitted Radiance Interferometer (AERI) (Knuteson et al., 2004a,b) developed for the DOE Atmospheric Radiation Measurement (ARM) program. The nadir-only spatial sampling of the original HIS was replaced with programmable cross-track coverage while maintaining similar sized footprints. A zenith (up-looking) view is available on the WB-57 and Proteus aircraft and enables further calibration verification analysis and upper atmosphere studies. The S-HIS is packaged in three modules (optics module, electronics module, and data storage computer) each mounted to the S-HIS instrument frame which is in turn mounted to the aircraft structural frame. Since the system is modular, the three enclosures may be mounted on a different structural frame if required. The flight calibration assembly is directly mounted to the front wall of the optics module and includes a scene selection mirror, two calibration

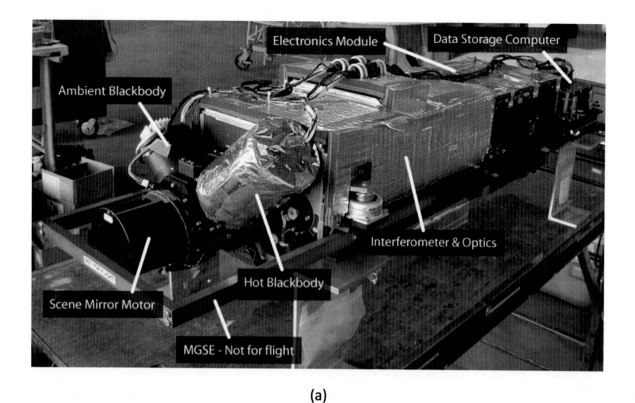

FIG. 1 (A) The Scanning High-resolution Interferometer Scanner (S-HIS) during Proteus integration at Scaled Composites (July 2004). (B) A diagram of the S-HIS optics module and flight calibrator assembly viewed from the nadir perspective. The IR beams between the beamsplitter and the Michelson mirror are not shown in the figure. *(Credit: Joe K. Taylor.)*

sources, and the scene mirror motor and encoder. The scene selection mirror allows the instrument to image the scene using cross-track scanning with contiguous cross-track coverage as broad as nadir ±45°. The scene selection mirror sequence is programmable and typically consists of multiple views of the Earth, a zenith view on compatible aircraft, and views of the two flight calibration sources, one at ambient and another controlled to a fixed temperature (~300 K in flight).

The Fourier transform spectrometer (FTS)

The design of most Fourier transform infrared (FTIR) spectrometers can be traced back to Michelson's original two-beam interferometer (Michelson, 1891, 1902). A Michelson interferometer separates the collimated observed source radiance into two beams via a beamsplitter. These beams are reflected back on themselves, one via a fixed mirror and the other via a moving mirror, and again made incident on the beamsplitter where they will be recombined or split again. This process retards the optical path length of one beam with respect to the other, with the optical path difference dependent on the position of the moving mirror. The recombination of the beams and measurement of the interference condition for a range of optical path differences creates an "interferogram," which contains the measurement information for all wavelengths within the passband. It can be shown that the interferogram is a measurement of the autocorrelation function of the optical field at the entrance of the instrument. Accordingly, the spectrum may be calculated from the interferogram using a simple Fourier transform.

For a historical review and perspective of FTS the reader is referred to Davis, Abrams, and Brault (Davis et al., 2001). Additionally, the theory and implementation of FTS instruments are well described by several texts (Bell, 2012; Davis et al., 2001; Griffiths and De Haseth, 2007) (cf. Chapter 6, Minnett et al.).

The S-HIS optical design is very efficient and provides useful signal-to-noise performance from a single interferogram dwell time of less than 0.5 s. The short interferogram collection time allows for contiguous scene coverage with a broad cross-track swath. As noted, the onboard reference blackbodies are also viewed via the scene mirror as part of each cross-track scan sequence, providing updated calibration information every 10–20 s. The S-HIS employs a customized commercial dynamically aligned plane-mirror interferometer (Bomem DA5). The Michelson mirror is voice coil driven and the original porch-swing mechanism has been replaced with a support mechanism that was designed and built at UW-SSEC and utilizes a linear bearing approach to minimize vibration induced tilt errors. A helium neon (HeNe) metrology laser is used to control optical delay sampling and provides very well known and highly stable spectral characteristics for the measurements. Any residual interferometer misalignments due to vibration induced velocity or tilt variations that are not fully compensated by the dynamic alignment system are measured at each interferogram sample, recorded, and corrected during data processing.

The spectral coverage is divided into three bands with separate detectors for each band (two photoconductive HgCdTe detectors and one InSb detector) to achieve the required signal to noise performance. The three detectors use a common field stop to ensure accurate spatial co-alignment between all three bands, and the fundamental measurement consists of an interferogram for each spectral band. The longwave band (580–1200 cm^{-1}, 8.33–17.3 μm) provides the primary information for temperature sounding as well as cloud phase and particle size. The midwave band (1030–1810 cm^{-1}, 5.5–9.7 μm) provides the primary water vapor sounding information and additional cloud property information. The shortwave band (1760–3000 cm^{-1}, 3.33–5.7 μm) provides information on cloud reflectance and augments sounding information. The detectors are cooled to 77 K using a Litton 0.6 W split-cycle Stirling cooler.

Scene- and calibration-view interferograms, along with calibration and engineering data, are stored to an onboard solid-state hard drive. For platforms without a high bandwidth downlink, the level 0 data (interferograms, calibration blackbody data, and engineering data such as instrument temperature, voltage, and current measurements) are downloaded from the instrument over an ethernet connection post-flight and processed to geolocated calibrated radiances (Level 1b) and temperature, water vapor, and trace gas retrievals (Level 2). Preliminary Level 1b and Level 2 products are typically available within a few hours of data download. This allows the data to be reviewed by the science team and initial conclusions to be made in a timely way for evaluating the success of experiment objectives as the field campaign progresses. Detailed instrument health and performance data are also processed each day to ensure that a healthy instrument is ready for the next flight. When a high bandwidth downlink is available on the aircraft, the Level 0 data are downlinked during the flight and the Level 1b and Level 2 products are processed using a real-time ground data processing system that can deliver atmospheric profiles, radiance data, and engineering status to mission support scientists in less than 1 min from the time of observation. This capability was developed and utilized for the Hurricane and Severe Storm Sentinel (HS3) mission on the NASA Global Hawk (Hoese et al., 2013; Taylor et al., 2015a). The Revercomb complex calibration method is used for radiometric calibration (Revercomb et al., 1988a), and operationally generated retrievals of temperature, water vapor, CO, CH$_4$, O$_3$, SO$_2$, N$_2$O profiles, total column CO$_2$, and surface temperature and emissivity are produced via the Dual Regression Retrieval with De-Aliasing (DRDA) algorithm (Smith et al., 2012; Weisz et al., 2013; Smith et al., 2021).

Past and current uses of the S-HIS measurements include retrievals of temperature and water vapor profiles (Smith et al., 2012; Weisz et al., 2013; Deslover et al., 2016), cloud radiative properties (Maestri and Holz, 2009; Verlinde et al., 2007), cloud top height (Holz et al., 2006; Smith et al., 2012; Weisz et al., 2013), surface emissivity and temperature (Knuteson et al., 2001, 2003; Tanamachi et al., 2001), and trace gases (Vinson et al., 2003). Furthermore, the radiances and retrieval products have been used for the evaluation of radiative transfer models (Liu, 2015; Liu et al., 2016; Newman et al., 2009); the characterization of the thermodynamic environment around hurricanes and tropical storms (Braun and Newman, 2012; Braun et al., 2016; Deslover et al., 2016; Munsell et al., 2018); the characterization of fire development, emission processes, plume evolution, and downwind impacts on air quality (King et al., 2003; McCourt et al., 2004; Roberts et al., 2020); and the topic of this chapter, satellite calibration validation (Kataoka et al., 2013; Moeller et al., 2003; Shephard et al., 2008; Taylor et al., 2015b; Tobin et al., 2006).

Generally, S-HIS is highly valuable within field programs where highly accurate radiance spectra and an extensive range of retrieval products are needed, with higher spatial resolution and better temporal overlap than can be provided by satellite sounding instruments, and more extensive coverage than what in situ observations can provide. This turns out to be a large fraction of field programs focused on climate and weather process studies.

3. S-HIS radiometric calibration, verification, and traceability

The S-HIS radiometric calibration, calibration verification, and traceability can be divided into four primary elements:

- Calibration of onboard blackbody references at the subsystem level.
- Pre- and post-deployment end-to-end calibration verification.
- Periodic end-to-end radiance evaluations under flight-like conditions with NIST transfer sensors.
- Instrument calibration during flight using the two onboard calibration blackbody references.

Calibration, uncertainty, and traceability

The vocabulary and uncertainty analysis methods described in this chapter follow the recommendations presented in the Joint Committee for Guides in Metrology (JCGM) Guide for Uncertainty in Measurement (GUM) (BIPM et al., 2008) and International Vocabulary of Metrology (known as the VIM) (BIPM et al., 2012) and the US National Institute of Standards and Technology (NIST) guidelines (Taylor and Kuyatt, 1994).

Calibration establishes the relationship between instrument output and the corresponding values realized by standards, producing results compatible with accepted units. Radiometric and spectral calibration define the relationship of the instrument output to the measured radiance and fix the ordinate (y) scale of the measurement to the radiance scale and the abscissa (x) scale to the optical frequency scale.

Measurement uncertainty is a required component of any complete measurement. Measurement uncertainty reflects the lack of exact knowledge of the value of the measurand. The corresponding state of knowledge is best described by means of a probability distribution over the set of possible values for the measurand.

Coverage factor is the number of standard deviations included in the reported uncertainty. For a normal distribution, $k=1$ (1-σ), $k=2$ (2-σ), and $k=3$ (3-σ) provide confidence levels of approximately 68%, 95%, and 99.7%, respectively.

Metrological traceability is defined as "the property of the result of a measurement or the value of a standard whereby it can be related to stated references, usually national or international standards, through an unbroken chain of comparisons all having stated uncertainties" (BIPM et al., 2012).

Calibration traceability to absolute standards provides an independence from unproven assumptions on stability. The primary method currently used to provide SI traceability for infrared radiances measured from on-orbit instruments is based on pre-flight instrument characterizations and calibrations, which cannot account for drift over very long periods on orbit, coupled with validation activities that help constrain the on-orbit performance.

Calibration of the onboard blackbody cavity references (thermistors and readout electronics) is typically performed on the order of every 5 years. The S-HIS blackbody thermistor readout electronics calibration is verified to within 5 mK using a series of six reference resistors, each calibrated to an absolute accuracy of better than 0.5 mK (3-σ) equivalent temperature using a Fluke 8508A DMM. The S-HIS onboard calibration blackbody thermistors are calibrated at 10 temperatures over the range from 213.5 to 333.15 K (−60°C to 60°C). These tests are done in a controlled isothermal environment using a NIST traceable temperature probe that is calibrated at Hart Scientific to an accuracy of 5 mK (3-σ). Following these tests, the onboard calibration blackbodies and readout electronics are integrated to the S-HIS instrument. Results from previous blackbody calibrations have shown no significant change in the key temperature ranges, with typical changes of less than 20 mK for the wide temperature range of the ambient blackbody, and less than 5 mK for the hot blackbody. These results have confirmed no significant change in blackbody thermometry.

Prior to and after each field campaign, end-to-end calibration verification is performed. End-to-end calibration verification is conducted using a variable temperature blackbody at the instrument's zenith viewport and an ice bath blackbody at the instrument's nadir view location. The S-HIS calibrated radiance measurements of the variable temperature blackbody at ambient (i.e., room temperature), 318.15 K (45°C), and 333.15 K (60°C) setpoints and the ice bath blackbody are compared to those calculated for the verification cavity blackbodies, based on the measured cavity temperature, knowledge of the cavity emissivity, and measurements of the background temperature. The variable temperature blackbody used for S-HIS calibration validation has its heritage rooted in the AERI instrument. These blackbodies have had their emissivity measured at NIST using three methods: the Complete Hemispherical Infrared Laser-based Reflectometer (CHILR); the Thermal Infrared Transfer Radiometer (TXR); and the Advanced Infrared Radiometry and Imaging Facility (AIRI) (Best et al., 2009). The ice bath blackbody is geometrically similar to the AERI blackbody, and its cavity is coated with the same paint. The pre- and post-mission end-to-end calibration verification tests show agreement within the established instrument 3-σ uncertainty and very little variation from test to test. A plot showing the historical record from 2007 through 2017 of pre- and post-mission end-to-end calibration verification test brightness temperature (cf. Chapter 1, Nalli and Kalluri) residuals are provided in Fig. 2. An atmospheric radiative transfer model is not used in the calculation of the verification body radiances, and the atmospheric absorption and emission lines are evident in the residuals shown in the figure. The S-HIS nonlinearity correction is optimized for the in-flight instrument temperature (and instrument thermal background), which is much colder than it is in the laboratory environment. The nonlinear response of the detectors has a small dependence on the instrument thermal background, and this results in a slightly less than optimal correction for measurements taken in the warmer laboratory environment. The impact on the end-to-end calibration verification results is evident as a small offset in the midwave band residual for the ice bath blackbody view. The results demonstrate excellent repeatability over the 10-year period, which is profoundly important for the purpose of environmental remote sensing and calibration validation of satellite sensors. Only the April 16, 2013, midwave band residual is slightly out of family for the ice bath blackbody measurement. The Stirling cooler was failing during the data collection for this test, and the S-HIS detector temperature increased to approximately 85 K during the calibration verification testing. The primary impact of the increased detector operating temperature was a change in midwave nonlinear response, which resulted in the small but evident offset to the midwave ice-bath blackbody residual.

Tests with the NIST Thermal Infrared Transfer Radiometer (TXR) (Rice and Johnson, 1998) solidly confirm the S-HIS calibration uncertainty estimates and provide direct NIST traceability. To verify the S-HIS calibration accuracy and provide direct NIST traceability of the S-HIS radiance observations, tests of the S-HIS and the NIST TXR were conducted using a laboratory thermal chamber to simulate flight temperatures for the S-HIS instrument. Two basic tests were conducted: a comparison of radiances measured by the S-HIS to those from the TXR, and the measurement of the reflectivity of a UW-SSEC blackbody by using the TXR as a stable detector (Best et al., 2007; Taylor et al., 2008). The radiance comparison involved the S-HIS and the TXR each observing a highly stable (and accurate) AERI blackbody over a wide range of temperatures (227–290 K). The test results showed mean agreement between the (1) predicted AERI blackbody radiance and the S-HIS NIST TXR Channel 2 equivalent spectral band of 60 ± 90 mK, (2) predicted AERI blackbody radiance and NIST TXR channel 2 (10 μm) of -22 mK, (3) NIST TXR channel 2 and the S-HIS band equivalent of less than 40 mK, and (4) predicted AERI BB radiance and the S-HIS NIST TXR channel 1 (5 μm) equivalent of 40 ± 85 mK.

Interferograms are collected for views of the onboard blackbody references (ambient blackbody and hot blackbody) as part of every cross-track scene select mirror scan sequence and are used for calibration of the S-HIS Earth scene measurements. The S-HIS ambient blackbody (ABB) runs at the ambient environmental temperature (typically between 218 and 245 K in flight and between 290 and 295 K in the laboratory), and the hot blackbody (HBB) is controlled to a flight setpoint near 300 K and a laboratory setpoint of 333 K.

For high spectral resolution infrared sensors, radiometric uncertainty (RU) characterizes the accuracy of the observed radiance spectra and the dependency of the combined uncertainty on relevant parameters such as scene temperature, wavelength, time, and viewing angle. Radiometric uncertainty represents an upper limit of the bias with respect to the true radiance and does not include effects such as detector noise that vary randomly from one spectrum to another. Example contributors to radiometric uncertainty include the uncertainties in the knowledge of the calibration blackbody temperature and emissivity, uncertainties in the degree of polarization and polarization angle of the sensor, and uncertainties in detector nonlinearity characterization. A measurement is not complete or traceable without uncertainty characterization. RU is required for any traceable application of the dataset and is particularly important for intercalibration studies and for radiometric calibration validation.

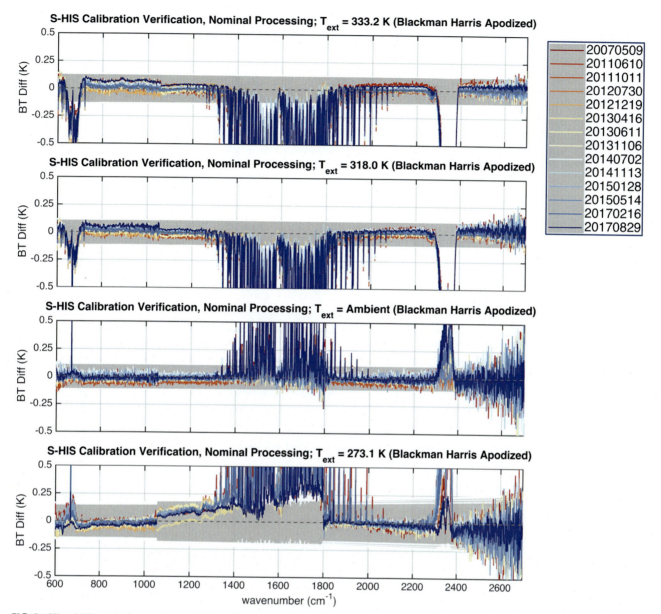

FIG. 2 Historical record of pre- and post-mission end-to-end calibration verification brightness temperature residuals from 2007 through 2017. The 3-σ combined uncertainty for the calibration verification is indicated by the *gray* shaded region on the plots.

The radiometric uncertainty of a given measurement can be determined via a differential analysis of the calibration equation and knowledge of the primary contributors to the uncertainty, as prescribed by the Guide to Uncertainty in Measurement (BIPM et al., 2008). For a wide range of scene temperatures, the calibration uncertainty (3-σ or $k=3$ coverage factor) estimate for S-HIS is less than 0.2 K (Best and Revercomb, 2005; Revercomb and Best, 2005). The primary contributors to the S-HIS radiometric uncertainty are the calibration reference temperature and emissivity as well as the nonlinearity correction of the nonlinear longwave and midwave detectors. The associated uncertainties are summarized in Table 1. The S-HIS absolute spectral calibration is determined by adjusting the effective metrology laser frequency to create optimal agreement with the positions of well-known spectral features present in clear-sky calculated spectra. Analyses of clear-sky spectra have been used to determine the S-HIS spectral calibration with an uncertainty of ±0.5 ppm (3-σ uncertainty in the mean) with no detectable changes with time (Tobin et al., 2003) and as such, the S-HIS spectral calibration is not a significant contributor to its overall radiometric uncertainty.

An example of the primary S-HIS radiometric uncertainty contributions and total RU for a clear-sky scene over ocean is shown in Fig. 3 and for a clear-sky scene over Greenland in Fig. 4. In these figures, the measured brightness temperature

TABLE 1 Primary contributors to the S-HIS radiometric uncertainty.

Calibration reference	Temperature uncertainty (3-σ)	Emissivity uncertainty (3-σ)
Ambient blackbody (ABB)	0.055 K	0.001
Hot blackbody (HBB)	0.055 K	0.001
Reflected radiance	5 K	
Detector nonlinearity		
Longwave band	10% of radiance difference between nominal calibration including nonlinearity correction and calibration without nonlinearity correction	
Midwave band		
Shortwave band	Linear, no nonlinearity correction needed	

spectra for each S-HIS spectral band are shown in the top panel, and radiometric uncertainty contributions with total radiometric uncertainty are provided in the middle panels. The bottom panels present scatter plots of the 3-σ radiometric uncertainty versus scene temperature, shaded by wavenumber. The gray vertical line indicates the ambient blackbody (ABB) temperature, and the black vertical line (∼300 K) indicates the S-HIS hot blackbody (HBB) temperature. For the warm, clear tropical scene, the 3-σ uncertainty is less than 0.2 K for all bands for scene brightness temperatures greater than 220 K. Uncertainties are minimized between the two onboard blackbody temperatures but grow rapidly when extrapolating to colder scenes. The impact of not having access to a cold space view for the cold calibration reference is apparent in these plots, evident by the relatively sharp increase in midwave and shortwave brightness temperature RU for scenes colder than the ABB temperature. It is also notable that the ambient blackbody temperature is roughly 10 K warmer in the example in Fig. 4 compared to Fig. 3, which results in higher radiometric uncertainty for scene temperatures below 240 K.

4. S-HIS high-altitude airborne calibration validation results

The ability to accurately validate infrared spectral radiances measured from space by direct comparison with airborne spectrometer radiances was first demonstrated using the S-HIS aircraft instrument flown under the AIRS sensor on the NASA Aqua spacecraft in 2002 (Revercomb et al., 2003; Tobin et al., 2004, 2006). Ensuing AIRS calibration validation underflights were completed in 2004 and 2006, providing successful comparisons over a wide range of scene conditions. Similar comprehensive and successful calibration validation efforts have also been conducted with S-HIS for the Moderate Resolution Imaging Spectroradiometer (MODIS) (Moeller et al., 2003), the Tropospheric Emission Spectrometer (TES) (Sarkissian et al., 2005; Shephard et al., 2008), IASI (Smith et al., 2008; Newman et al., 2012), the GOSAT Thermal and Near Infrared Sensor for Carbon Observation-Fourier Transform Spectrometer (TANSO-FTS) (Kataoka et al., 2013), CrIS (Taylor et al., 2013, 2015b, 2016, 2017), and the Advanced Baseline Imager (ABI) (Taylor et al., 2019b).

Examples of calibration validation of the low Earth orbit (LEO) Cross-track Infrared Sounder and geostationary Earth orbit (GEO) ABI using the S-HIS measurements are presented in the following sections. The results, and future calibration validation flights, are valuable for providing a periodic end-to-end assessment of the satellite sensor's calibration accuracy with low radiometric uncertainty and well understood traceability.

4.1 Calibration validation of the Cross-track Infrared Sounder

4.1.1 The Cross-track Infrared Sounder

The Cross-track Infrared Sounder (CrIS) (Han et al., 2013) is a high spectral resolution FTS onboard the Suomi National Polar-Orbiting Partnership (SNPP) and NOAA-20 satellites that provides operational observations of top-of-atmosphere thermal infrared radiance spectra for weather and climate applications. The CrIS optical system was designed to provide an optimum combination of optical performance and compact packaging (Kohrman and Luce, 2002; Stumpf and Overbeck, 2002). Its key subsystems include a step and settle two-axis scene selection module with image motion compensation capability, a full-aperture internal calibration source, a large-aperture Michelson interferometer, a three-element all reflective telescope, a cooled aft optics module, and a multiple-stage passive cooler. The CrIS interferometer uses a flat-mirror Michelson configuration equipped with a dynamic alignment system to minimize misalignments within the interferometer

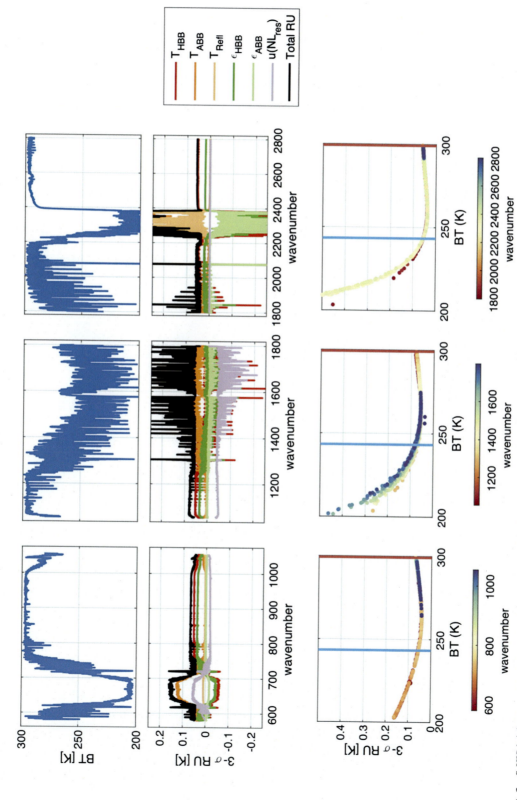

FIG. 3 S-HIS brightness temperature and radiometric uncertainty for a typical warm tropical scene over ocean (2017-04-13).

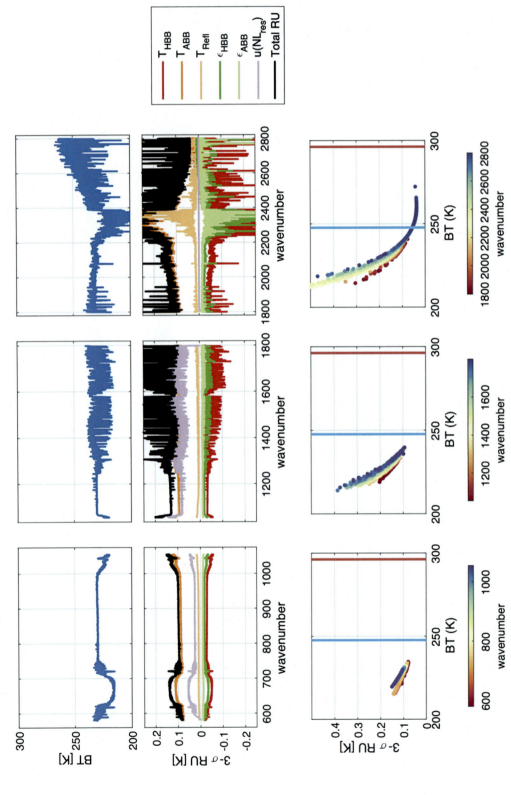

FIG. 4 S-HIS brightness temperature and radiometric uncertainty for a cold and dry scene over Greenland (2015-03-29).

and has a maximum optical path difference of ±0.8 cm. Each of the three spectral bands (longwave, midwave, and shortwave) uses 3 × 3 detector arrays, with an approximately 14 km near-nadir surface diameter (or "footprint") for each field of view (FOV).

Tobin et al. (2013) provides an excellent summary of the SNPP CrIS radiometric calibration uncertainty assessment that was based on prelaunch and on-orbit efforts to estimate calibration parameter uncertainties and also provides example results of SNPP postlaunch validation efforts to assess the predicted uncertainty. The CrIS prelaunch RU estimates computed for the thermal vacuum test environment were determined to be ∼0.2 K (3-σ) or less for blackbody scene temperatures above 250 K. The primary radiometric uncertainty contributors for CrIS have been identified as the In-flight Calibration Target (ICT) temperature and emissivity and reflected radiance terms, sensor polarization, and detector nonlinearity. Early analyses indicated that the impact of sensor polarization would have negligible impact on radiometric calibration, but further evaluation showed that when the polarized emission from the scene mirror is included in the analysis the effect becomes nonnegligible for cold scenes and a correction was added to the calibration algorithm. The CrIS RU originally presented by Tobin et al. (2013) was prior to this finding and did not include polarization as an RU contributor. The CrIS radiometric uncertainty assessments in this chapter include the uncertainty associated with the CrIS polarization correction, in addition to the uncertainty contributors originally summarized by Tobin et al. (2013).

4.1.2 The 2013 and 2015 SNPP airborne calibration validation campaigns

The first dedicated Suomi NPP airborne calibration validation campaign was conducted in May 2013 with a primary objective of providing detailed validation of CrIS radiance observations and meteorological products. During this calibration validation campaign, the NASA ER-2 aircraft instrument payload included the S-HIS, the NPOESS Atmospheric Sounder Testbed-Interferometer (NAST-I) (an FTIR sensor with similar measurement coverage to the S-HIS), the NPOESS Atmospheric Sounder Testbed-Microwave Spectrometer (NAST-M), the NASA MODIS/ASTER airborne simulator (MASTER), and the NASA JPL Airborne Visible/Infrared Imaging Spectrometer (AVIRIS) sensors. The campaign was based in Palmdale, California with 11 ER-2 under-flights of the Suomi NPP satellite, primarily targeting overpasses over the Pacific Ocean. Selection of the CrIS and S-HIS footprints included in each calibration validation comparison must take into consideration the spatial and temporal co-location of the two sets of observations, the spatial uniformity of the scene, and measurement noise reduction provided by co-adding individual fields of view for each sensor. With these considerations, the best conditions for radiance validation of CrIS with S-HIS for this campaign were encountered for the 2013-05-15, 2013-05-30, 2013-05-31, and 2013-06-01 flights. During each overpass, the ER-2 flew a straight and level flight leg at approximately 20 km altitude (50 hPa) along the suborbital track of SNPP. The flight path for all 11 flights, along with the VIIRS high-resolution image band I5 (I05) brightness temperature maps at the coincidence time for the four flights with the best conditions for radiance validation of CrIS with S-HIS are shown in Fig. 5.

The second Suomi NPP airborne calibration validation campaign was conducted in March 2015, with flights over Greenland with the S-HIS, NAST-I, NAST-M, and MASTER sensors integrated into the NASA ER-2 payload. The primary goals of this campaign were to assess the radiometric calibration and environmental product retrievals (e.g., temperature and moisture profiles) for polar conditions. The mission was conducted out of Keflavik, Iceland, and seven flights over the Greenland ice sheet were completed over 3 weeks with multiple under-flights of the Suomi-NPP, METOP-A, METOP-B, and Aqua satellites. While the weather conditions were challenging, and at times extreme, the high latitude created several overpass opportunities for each flight. The flights with the best conditions for radiance validation of CrIS with S-HIS during this campaign were on March 15, 2015 (two overpasses); March 28, 2015; and March 29, 2015 (two overpasses). The flight paths for all science flights are shown in Fig. 6, with each flight track overlaid on a VIIRS I05 band brightness temperature image for each flight.

Coincident data from infrared imagers, Visible Infrared Imaging Radiometer Suite (VIIRS) on the SNPP satellite and MASTER onboard the ER-2, were used to assess spatial uniformity of the scene within the S-HIS and CrIS footprints and optimize selection of the co-located CrIS and S-HIS footprints independently of the CrIS and S-HIS radiances. This was particularly important for the 2015 campaign, which in general presented more challenging scene conditions for calibration validation than that of the 2013 overpasses that were typically for clear-sky scenes over ocean.

An example of three co-location assessment plots for the March 29, 2015, SNPP overpass is provided in Fig. 7. Each of these plots uses VIIRS I05 data to assess the co-location of the CrIS and S-HIS fields of view. Fig. 7A illustrates the CrIS and S-HIS footprint outlines for the overpass region in gray, with the co-located footprints selected for intercomparison indicated by heavier weight black outlines, overlaid on the VIIRS I05 brightness temperature image. This allows quick visual assessment of scene temperature variability within the overpass region. Fig. 7B provides histograms of the number of VIIRS I05 pixels within the CrIS (red) and S-HIS (blue) co-located footprints versus the VIIRS I05 brightness temperature. The similar shape of each distribution affords confidence that the incomplete coverage of the CrIS footprints by the

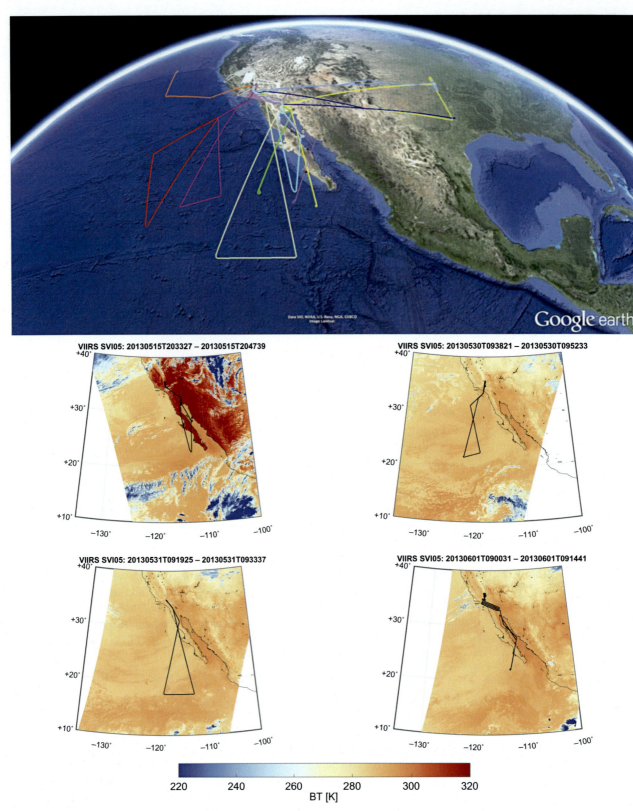

FIG. 5 The flight paths for the 11 ER-2 under-flights of the SNPP satellite conducted during the 2013 airborne calibration validation campaign are shown in the top panel. The lower four panels correspond to the flights with the best conditions for radiance validation of CrIS by S-HIS. In each of these panels, the ER-2 flight path is plotted in *black* over the VIIRS I05 observations at the ER-2 and SNPP coincidence. The VIIRS image is colored by the I05 band scene brightness temperature with the scale provided at the bottom of the figure.

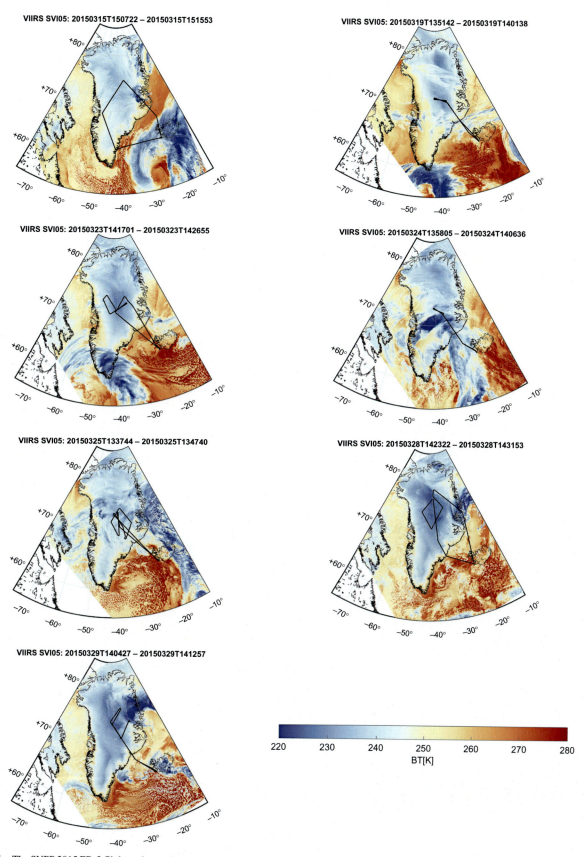

FIG. 6 The SNPP 2015 ER-2 flight paths are shown in *black* and are plotted over a VIIRS I05 band brightness temperature image for each flight with the brightness temperature color scale indicated at the bottom right of the figure.

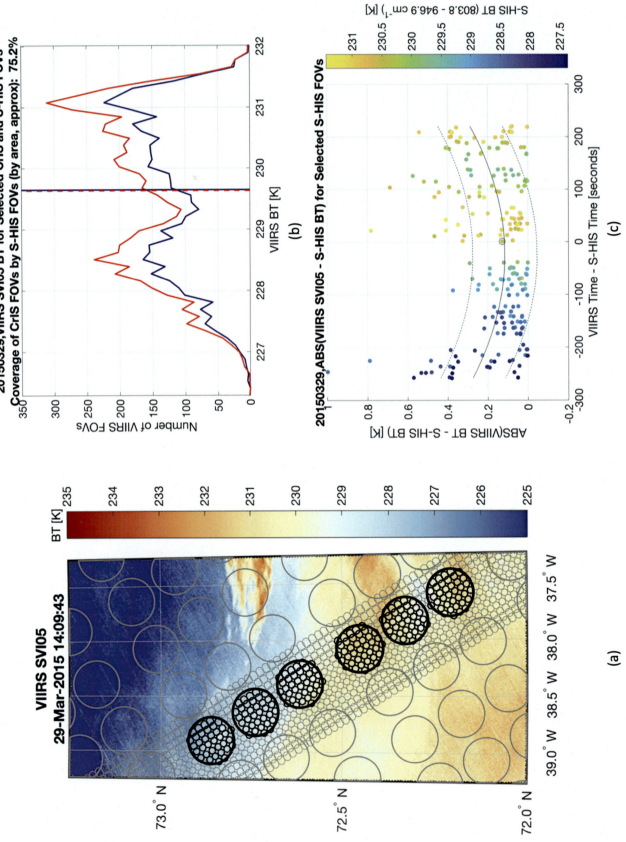

FIG. 7 An example of three co-location assessment plots for the March 29, 2015, SNPP overpass. (A) The CrIS and S-HIS footprint outlines for the overpass region are plotted in *gray*, with the co-located footprints selected for intercomparison indicated by heavier weight *black* outlines and plotted over the VIIRS I05 brightness temperature image. (B) Histograms of the VIIRS I05 pixels within the CrIS (*red*) and S-HIS (*blue*) co-located footprints versus the VIIRS I05 brightness temperature. (C) The magnitude (unsigned absolute value) of the difference between the co-located VIIRS and S-HIS observed brightness temperature versus the elapsed time between the observations.

S-HIS footprints (approximately 75.2% for this case) still accurately captures the spatial variation of the scene for the selected fields of view. Since each CrIS cross-track scan takes 8 s to complete, it will take roughly 16 s to collect two consecutive near-nadir 3 × 3 fields of regard. The ER-2 cruising speed is much slower than the satellite ground speed, and it takes the S-HIS onboard the ER-2 approximately 450 s to cover the same region. The elapsed time between the VIIRS I05 and S-HIS brightness temperature observations (Fig. 7C) can be used to verify that the selection of co-located CrIS and S-HIS footprints is optimized with respect to co-location uncertainties due to this temporal mismatch. The optimum temporal coincidence occurs at the minimum of a quadratic fit to the data.

The double observation minus calculation (DOMC) methodology (Tobin et al., 2006) has been used for the radiance calibration validation examples presented in this chapter. The S-HIS and the satellite sensor have different observation altitudes, footprint sizes, and spectral characteristics. To avoid co-location errors when creating comparisons of the satellite sensor and S-HIS observations, temporally and spatially coincident data collected under clear-sky and spatially uniform conditions are desirable. However, despite a careful selection of such conditions, scene variations within the larger CrIS footprints can be significant. To ensure that both CrIS and S-HIS are observing the same scene, it is necessary to use multiple S-HIS footprints collected over a range of view angles to provide as contiguous spatial coverage as possible of the larger CrIS footprints. In addition to the different observation altitudes and spectra characteristics, the view angle differences also need to be accounted for when comparing satellite sensor and S-HIS observations. The technique selected for doing this is to make use of calculations that include the actual spectral and spatial characteristics of each instrument. The radiance calculations for each instrument assume the same surface conditions, atmospheric state, and forward models. This results in systematic errors that are common to both sets of calculations and to first order removes the fundamental effects of altitude and view angle differences. Accordingly, the use of the calculated spectra allows the observed minus calculated difference for each instrument to be compared, avoiding the first-order effects of the altitude and view angle differences. The monochromatic calculations used for the DOMC-based calibration validation results presented in this chapter were completed using the Line-by-Line Radiative Transfer Model (LBLRTM) (Clough et al., 2005; Clough and Iacono, 1995; Clough et al., 1992) with analysis fields (pressure, temperature, water vapor, ozone, surface pressure, and temperature) from ERA-interim reanalysis (Dee et al., 2011), CO_2 and CH_4 from the NOAA ESRL Carbon Tracker profiles (Peters et al., 2007), the Combined ASTER MODIS Emissivity over Land (CAMEL) surface emissivity (Borbas et al., 2018; Feltz et al., 2018), and the Nalli ocean emissivity (Nalli et al., 2001, 2008a,b).

In the DOMC method, the spectral resolutions of both instruments are reduced to the lowest common spectral resolution for the two instruments and differences in instrument line shapes and spectral response functions (SRFs) are accounted for in both the observations and the calculations. This is a necessary step for an accurate comparison of two instruments. For the CrIS to S-HIS comparisons, matching the spectral resolution and line shapes for the two instruments is quite simple yet mathematically rigorous since they are both FTS instruments.

For the calibration validation to be complete and traceable, a rigorous uncertainty analysis must be included. An example of the 3-σ radiometric uncertainty associated with the DOMC calibration validation for the SNPP overpass during the June 1, 2013, flight is shown in Fig. 8. The primary contributors to the DOMC intercomparison method are the instrument (measurement) radiometric uncertainty, calculated radiance (radiative transfer model) uncertainty, and co-location uncertainty associated with spatial nonuniformity of the scene and spatial and temporal matching differences between the two instruments.

As noted by Tobin et al. (2006), the uncertainties in atmospheric state or forward model physics can result in errors in the calculated emission or absorption above the aircraft, which will have a different impact on the CrIS and S-HIS calculated radiances. Furthermore, a second order effect can also occur due to the larger S-HIS range of view angles needed to completely cover the near-nadir CrIS footprints, with errors in computed radiances amplified for large scan angles. To ensure that the uncertainty sufficiently accounts for these effects, 20% of the difference in the mean S-HIS and CrIS calculated radiances is used as an estimate of the 3-σ uncertainty contribution associated with the forward model radiances in the DOMC methodology. This can result in large uncertainties for spectral regions with sensitivity to atmosphere above the aircraft altitude, and it is reasonable to objectively exclude these regions from the comparison.

The uncertainty associated with spatial nonuniformity of the scene and spatial and temporal matching differences between the two instruments is calculated using the expressions for spatial and temporal mismatch uncertainty described for infrared sounder intercalibration via CLARREO (Revercomb et al., 2016; Tobin et al., 2016), noting the correction to the calculation of uncertainty due to spatial mismatch provided in Taylor et al. (2020). The time differences between the S-HIS and CrIS observations can result in scene changes between the observation times of the two instruments, particularly for scenes with cloud contamination or significant atmospheric variability. As described by Tobin et al. (2016), the temporal mismatch can effectively be treated as a spatial mismatch when completing the uncertainty analysis, by assuming simple advection of the scene with time. The temporal mismatch between the CrIS and S-HIS observations was estimated

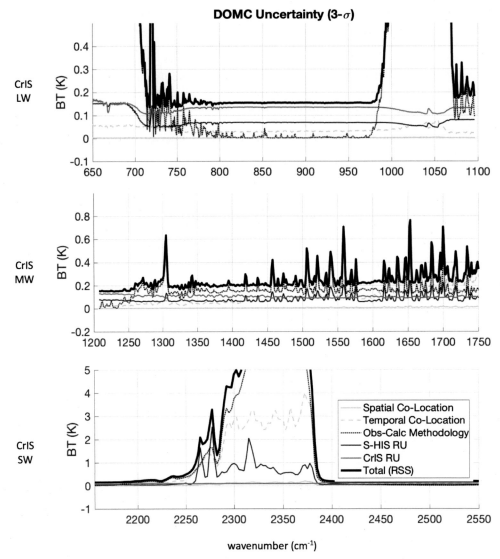

FIG. 8 The 3-σ radiometric uncertainty (expressed as brightness temperature) associated with the double observation minus calculation calibration validation, including the CrIS and S-HIS 3-σ radiometric uncertainty for flight and scene conditions for the SNPP overpass during the June 1, 2013, flight. All radiances have been reduced to the CrIS spectral resolution and Hamming apodization has been applied.

as the mean of the elapsed time between the CrIS and S-HIS observations and then combined with a wind speed of 20 m/s to calculate the effective displacement of the radiance field for determining the radiometric uncertainty due to temporal mismatch of the observations. The 2015 Greenland Summit Station radiosonde data were used to determine a conservative bound on wind speed.

The DOMC results for the flights with the best conditions for radiance validation of the CrIS sensor data record (SDR) calibrated radiances with the S-HIS are shown in Figs. 9 and 10, for the 2013 and 2015 SNPP calibration validation campaigns, respectively. Hamming apodization has been applied to both CrIS and S-HIS radiances (at CrIS resolution), and the noise in the S-HIS radiances has been further reduced via principal component noise filtering (Antonelli et al., 2004), which removes a large percentage of the spectrally random noise while retaining the spectral and radiometric fidelity of the spectra.

Due to ER-2 safety protocols and the runway configuration at the base of operations in Keflavik, Iceland, all the 2015 flights were conducted during daylight hours. The May 15 flight in 2013 was also conducted during the day. For daylight conditions, the upwelling shortwave radiance is subject to solar reflection and non-LTE (local thermodynamic equilibrium) which is not accounted for in the radiative transfer model and, as such, the shortwave results in the affected spectral regions for those flights are not included in the analysis. Even for the night flights, much of the CrIS shortwave spectral coverage is sensitive to significant CO_2 and N_2O absorption above the aircraft altitude. As expected, large differences and

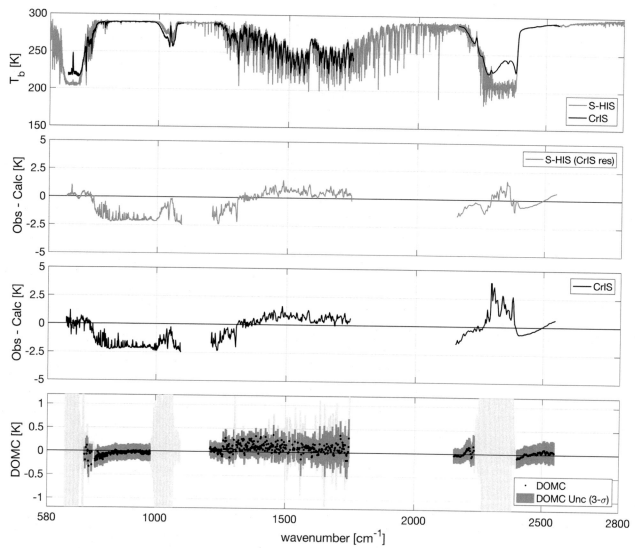

FIG. 9 Mean CrIS to S-HIS double observation minus calculation result for the four SNPP under-flights from 2013 with the optimal calibration validation conditions as discussed in the main text. Hamming apodization has been applied. The top panel shows the mean S-HIS and CrIS observed brightness temperature spectra at the native instrument spectral resolutions. The second and third panel show the mean difference between the observed and calculated brightness temperature spectra for S-HIS (reduced to CrIS resolution) and CrIS, respectively. The mean DOMC result and associated 3-σ uncertainty estimate is provided in the bottom panel. Spectral channels that are impacted by solar reflection and non-LTE are not included in the mean DOMC computation. The DOMC result is not shown for spectral channels with high sensitivity to the atmosphere above the aircraft, but the 3-σ DOMC uncertainty for these spectral regions is provided (*light gray* shaded regions).

intercomparison uncertainties are observed in spectral regions where the satellite sensor is sensitive to significant contributions from above the aircraft altitude. However, in other spectral regions, the comparison shows excellent agreement, with most residual differences less than 0.1 K and well within the total 3-σ radiometric uncertainty. These results illustrate the unique capability of satellite under-flight comparisons for periodic end-to-end assessment of the LEO hyperspectral IR sounder calibration with low uncertainty and well understood traceability.

4.2 Calibration validation of the Advanced Baseline Imager

4.2.1 The Advanced Baseline Imager

The Advanced Baseline Imager (ABI) on the GOES-R series of operational US geostationary weather satellites is a state-of-the-art 16-band imaging radiometer with spectral bands covering the visible, near-infrared (NIR), and infrared portions of the electromagnetic spectrum. Many attributes of ABI offer great improvements over the prior generation of GOES imagers, including more than three times the spectral information, four times the spatial resolution, and more than five

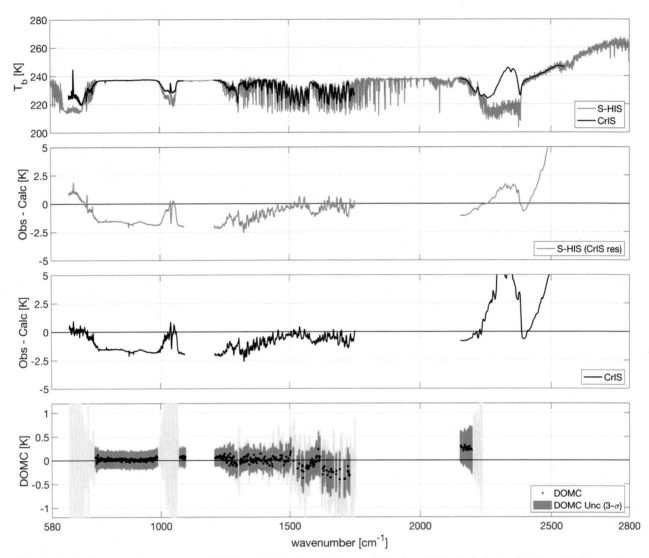

FIG. 10 As Fig. 9, except showing the mean DOMC result for the five SNPP under-flights from 2015 with the best calibration validation conditions. Spectral channels subject to solar reflection and non-LTE are not included in the DOMC computation, which leads to no DOMC result above 2240 cm^{-1} for this data set, since all the 2015 flights were conducted during daylight hours.

times faster temporal coverage. The calibrated and geolocated ABI information is used to produce a wide variety of weather and environmental data products and to monitor rapidly changing weather (Schmit et al., 2017). The nadir spatial resolution of the ABI infrared bands is 2 km, and the instrument is designed to permit a wide range of programmable image collection options.

4.2.2 The 2017 GOES-16 post launch test airborne calibration validation campaign

The GOES-16 post launch test (PLT) airborne calibration validation campaign was conducted March through May 2017. The primary goal of the campaign was to validate the ABI and Geostationary Lightning Mapper (GLM) product performance post-launch using independent measurements (Padula et al., 2017). The campaign was conducted out of Palmdale, California and Warner Robins Air Force Base, Georgia, with high altitude under-flights on the NASA ER-2 over a wide range of North American targets. During this calibration validation campaign, the NASA ER-2 payload consisted of the S-HIS and AVIRIS instruments, along with the Fly's Eye GLM Simulator (FEGS), the Lighting Instrument Package (LIP), the Cloud Physics Lidar (CPL), the ER-2 Doppler Radar (EXRAD), and the Geostationary Coastal and Air Pollution Event Airborne Simulator (GCAS) instruments. A total of 16 ER-2 flights were conducted during the campaign. The primary ABI thermal emissive band (TEB) calibration validation flight was conducted over the Gulf of Mexico on April 13, 2017, and the primary reflected solar band (RSB) calibration validation flights were conducted over the Sonoran Desert on March 23 and March 28, 2017. ABI North-South-Scan (NSS) data were collected in ABI Mode 25 (a special ABI scene scan sequence) for 90 min during all three ABI calibration validation flights, and the results of the ABI TEBs (channels 7–16) calibration validation with the S-HIS for the April 13 flight are presented in this section. North-South-Scan operation allows every detector element of each of the ABI channels to sequentially observe the same target location on the Earth. Each Mode 25 sequence takes 5 min to complete and includes 4 NSS sweeps for ABI channels 4–16, 5 NSS sweeps for ABI channels 1–3, with mesoscale data collection interleaved at 30 s intervals. The ABI NSS capability enabled thousands of spatially and temporally co-located ABI and S-HIS observations for each ABI infrared band in a 90-min observation period. The top panel of Fig. 11 shows the mean S-HIS spectrally resolved brightness temperature for the co-located S-HIS and ABB NSS observations during the April 13, 2017, flight. The ABI normalized SRFs are overlaid on the S-HIS mean brightness temperature spectra for reference. The lower panel of the figure illustrates the corresponding mean S-HIS 3-σ radiometric uncertainty for these observations, with the mean value over the passband for each ABI channel indicated by heavier weight horizontal bars.

The DOMC method was also used for intercomparison of the ABI and S-HIS measurements. To account for differences in the S-HIS and ABI spectral resolution and spectral response functions, the S-HIS observed radiances and the calculated monochromatic radiances were convolved with the ABI SRFs. Given the narrow line-shape of the S-HIS measurements compared to the much broader ABI SRFs it is not necessary to convolve the ABI observations with the S-HIS line-shape during the DOMC analysis as the operation would have negligible impact on the ABI measurements. To reduce comparison uncertainty associated with the significant differences in view geometry between the two instruments, the S-HIS and ABI observation co-location was also constrained by local zenith angle and azimuth angle, in addition to spatial and temporal match-up constraints. Co-location of the S-HIS and ABI measurements for the comparison was limited to ABI and S-HIS observations with a maximum difference in local zenith angle of 5°, maximum difference in azimuth angle of 5°, maximum difference in observation time of 60 s, and partial to full spatial overlap of the S-HIS and ABI geolocated footprints.

An example of a 3-panel ABI to S-HIS co-location summary plot, which was produced for every ABI NSS data file with a successful co-location with S-HIS observations, is shown in Fig. 12. In the NSS mode, there are roughly 2 s of ABI data

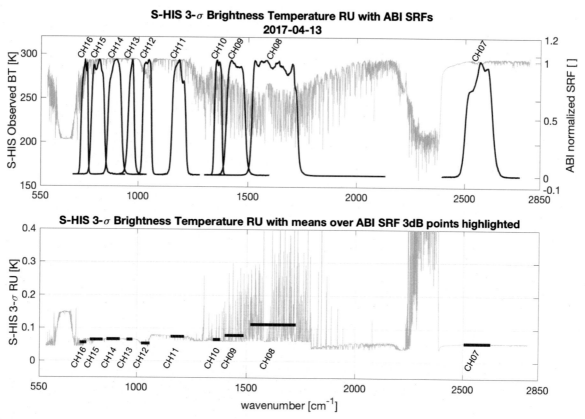

FIG. 11 The mean S-HIS spectral brightness temperature for the co-located S-HIS and ABI NSS observations during the April 13, 2017, flight are shown in the top panel, with the ABI normalized SRFs overlaid for reference. The ABI SRF is plotted on the right-hand scale of the plot, while the brightness temperature data are plotted on the left-hand scale. The corresponding mean S-HIS 3-σ radiometric uncertainty is provided in the bottom panel, with the mean value over the passband for each ABI channel indicated by heavier weight horizontal bars.

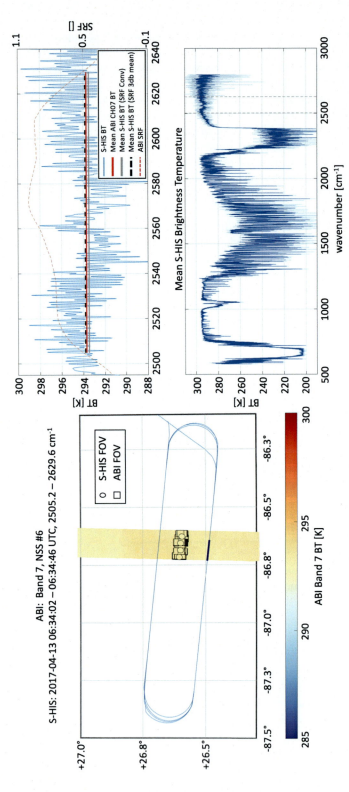

FIG. 12 An example 3-panel plot generated during the ABI to S-HIS co-location assessment process (ABI Band 7, north-south-scan swath position 6). The left panel maps the quantities being compared and the co-located footprints for a single NSS file. The ABI North-South swath observations are colored by the ABI brightness temperature and clearly demark a North to South scanline of the ABI Band 7 detectors. The co-located (elliptical) S-HIS and (rectangular) ABI footprints which meet all co-location requirements, including view angle, are outlined in *black*. The mean co-located ABI and S-HIS observations are shown in the top right panel, with the mean S-HIS brightness temperature spectrum in *blue*, mean ABI brightness temperature convolved with the ABI SRF in *gray*, mean ABI brightness temperature in *solid red*, and the ABI normalized SRF in *dashed red*. The ABI SRF is plotted on the scale on the right-hand side of the plot, while the brightness temperature data are plotted on the left-hand scale. The bottom right panel shows the S-HIS mean brightness temperature spectra for the co-located observations.

per file and during the 90-min data collection there were 1206 NSS files recorded for each ABI band. While there is one targeted North-South swath per ABI band in this mode, given the size of the ER-2 racetrack flight pattern, there were also secondary swath locations for the ABI IR bands where successful co-location with the S-HIS observations occurred. The left panel of Fig. 12 provides a map of the quantities being compared and the co-located footprints for a single NSS file. The ABI North-South swath observations are colored by the ABI brightness temperature, and the co-located (elliptical) S-HIS and (rectangular) ABI footprints are outlined in black. The mean co-located ABI and S-HIS observations are shown in the top right panel, with the mean S-HIS brightness temperature spectrum in blue, mean S-HIS brightness temperature convolved with the ABI SRF in gray, mean ABI brightness temperature in solid red, and the ABI normalized SRF in dashed red. The ABI SRF is plotted on the scale on the right-hand side of the plot, while the brightness temperature data are plotted on the left-hand scale. Finally, the bottom right panel shows the S-HIS mean brightness temperature spectra for the co-located observations.

The ABI to S-HIS DOMC result for all ABI NSS and S-HIS co-located observations from the April 13, 2017, ABI primary TEB calibration validation flight is shown in Fig. 13. The ABI radiometric uncertainty is not included in the 3-σ error bars, and the primary radiometric uncertainty contributors are the S-HIS radiometric uncertainty, spatial variability of the scene, and the uncertainty in the DOMC methodology (calculation uncertainty). Since the co-location conditions included a constraint on local zenith and azimuth viewing angles, the uncertainty contribution associated with the DOMC methodology for the S-HIS to ABI comparison was calculated as 10% of the difference in the mean S-HIS and ABI calculated radiances. Since the S-HIS and ABI footprints are approximately the same size and relatively tight temporal co-location constraints were enforced, the uncertainty due to spatial nonuniformity of the scene was calculated as the root sum of the squares of the standard deviations of the S-HIS and ABI observations, with each standard deviation reduced by a factor of the square root of the respective number of observations. For this case, the co-location is occurring for an extremely uniform clear-sky scene over warm ocean. Accordingly, the S-HIS radiometric uncertainty is the dominant contributor to the root-sum-squared (RSS) total uncertainty, except for ABI channels 9, 10, and 12. The dominant uncertainty contributor for channels 9 and 10 is spatial variability, and DOMC methodology uncertainty for channel 12. The ABI channel 12 measurement is sensitive to atmospheric ozone which is primarily above the ER-2 altitude, resulting in a much larger radiometric uncertainty for this channel. The ABI radiometric accuracy specification is 1 K (1-σ) at 300 K scene temperature and the DOMC results are all well within this specification, even without converting the residuals to an

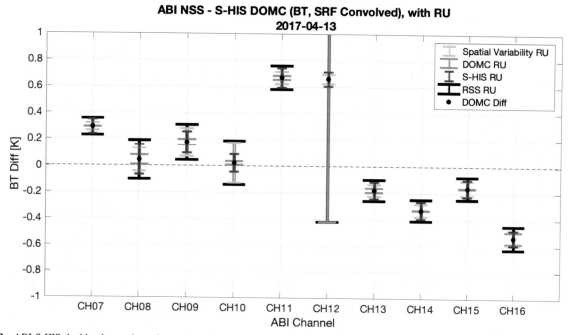

FIG. 13 ABI-S-HIS double observation minus calculation (DOMC) results for ABI North-South-Scan and S-HIS co-located footprints during the April 13, 2017, calibration validation flight. Error bars represent 3-σ DOMC uncertainty contributors and root-sum-squared (RSS) total. The results are well within the ABI radiometric accuracy specification (<1 K, 1-σ, at 300 K scene temperature). ABI spectral bands: 3.9 μm (CH07, Shortwave window), 6.2 μm (CH08, Upper-level water vapor), 6.9 μm (CH09, Midlevel water vapor), 7.3 μm (CH10, Lower-level water vapor), 8.4 μm (CH11, Cloud-top phase), 9.6 μm (CH12, Ozone), 10.3 μm (CH13, "Clean" longwave window), 11.2 μm (CH14, Longwave window), 12.3 μm (CH15, "Dirty" longwave window), 13.3 μm (CH16, CO_2 longwave).

effective residual for a 300 K scene, which would result in a band dependent reduction in all residuals by between 0.01 and 0.09 K for this case. A similar analysis was completed for the mesoscale data which was collected at 30 s intervals during the 90 min of data collection. The analysis produced comparable residuals to the NSS data assessment, which provides further confidence in the ABI NSS data processing.

5. Summary

In this chapter, the value of high-altitude aircraft field measurements for radiometric calibration validation of satellite infrared radiance measurements was presented along with details of the validation approach and analyses. An overview of the S-HIS instrument used in these field campaigns was provided, along with the results for the radiometric validation of the LEO CrIS and GEO ABI instruments using S-HIS measurements. In the absence of on-orbit SI traceable infrared reference sensors like those defined by the CLARREO program, satellite under-flight comparisons are uniquely capable of assessing infrared satellite observations with high accuracy and strong measurement traceability. Furthermore, if an on-orbit reference standard is established, there will continue to be a significant need for high-altitude airborne reference measurements like those described in this chapter as they can be used to target specific scenes and provide additional information on subpixel variability (e.g., clouds, water vapor, and surface emissivity) within the field of view of the satellite observations. Of course, for the calibration validation process to be accurate, traceable, and repeatable, it is most important that the reference data instrument be extremely well characterized and understood, carefully maintained, and accurately calibrated with traceability to absolute standards. It has been demonstrated that the S-HIS meets and surpasses these requirements and has proven to do so on multiple airborne platforms, each with significantly different instrument operating environments.

A detailed intercalibration assessment between the CrIS and S-HIS instruments for under flights from the 2013 and 2015 SNPP airborne calibration validation campaign were presented and between the ABI and S-HIS instruments for the primary ABI TEB calibration validation flight from the 2017 GOES-PLT campaign. The radiometric uncertainty contributions of the measurements, along with the radiometric uncertainty contribution associated with the comparison methodology (i.e., "mismatch errors" and forward model uncertainties), are a critical component of the intercalibration and have been included in the analysis and summary results. The CrIS to S-HIS comparisons show excellent agreement, with residual differences for much of the CrIS spectral coverage of less than 0.1 K, and well within the combined radiometric uncertainty estimates (except for spectral regions with significant sensitivity above the aircraft altitude). The S-HIS calibration validation of the broadband ABI instrument on GOES-16 produced intercomparison residuals between 0.04 and 0.67 K for the ABI infrared channels, well within the ABI 1-σ radiometric accuracy specification of 1 K at 300 K scene temperature. Furthermore, conversion of the residuals to their effective values for a 300 K scene results in a band-dependent reduction in all residuals by between 0.01 and 0.09 K for this case.

High-altitude airborne radiometric calibration provides a means for periodically verifying the absolute calibration of spacecraft instruments based on high accuracy radiometric field measurements, such as those obtained from the S-HIS, for which the calibration can be carefully maintained and tested to absolute standards on the ground. This capability is especially valuable for assuring the long-term consistency and accuracy of climate observations.

Acknowledgments

The authors are pleased to recognize the US DOE, NASA, NOAA, the NPOESS Integrated Program Office, and the University of Wisconsin (UW) Space Science and Engineering Center (SSEC) for their combined support of the development and refinement of the S-HIS instrument and more than two decades of field campaign deployments. We would also like to recognize and express our gratitude to the large team of individuals that have contributed to the success of the S-HIS program, and Nadia Smith (STC) and Nicholas Nalli (NOAA/NESDIS/STAR) for reviewing the chapter draft and providing constructive feedback.

References

Antonelli, P., Revercomb, H., Sromovsky, L., Smith, W., Knuteson, R., Tobin, D., Garcia, R., Howell, H., Huang, H.L., Best, F., 2004. A principal component noise filter for high spectral resolution infrared measurements. J. Geophys. Res.-Atmos. 109, 1–22.

Bell, R., 2012. Introductory Fourier Transform Spectroscopy. Elsevier.

Best, F.A., Revercomb, H.E., 2005. Calibration of the Scanning High-resolution Interferometer Sounder (S-HIS) infrared spectrometer: blackbody reference standards (Part 2). In: Conference on Characterization and Radiometric Calibration for Remote Sensing (CALCON), Logan, UT, USA.

Best, F.A., O'Connell, J., Rice, J.P., Tobin, D.C., Knuteson, R.O., Adler, D.P., Ciganovich, N.N., Dutcher, S., Taylor, J.K., Laporte, D.D., Ellington, S.D., Werner, M.W., Garcia, R.K., Revercomb, H.E., 2007. High accuracy infrared radiances for weather and climate, Part 1: NIST TXR validation of scanning HIS radiances and a UW-SSEC blackbody. In: Joint 2007 EUMETSAT Meteorological Satellite & 15th AMS Satellite Meteorology and Oceanography Conference, Amsterdam, The Netherlands.

Best, F.A., Knuteson, R.O., Revercomb, H.E., Tobin, D.C., Gero, P.J., Taylor, J.K., Rice, J.P., Hanssen, L., Mekhontsev, S., 2009. Measurements of the Atmospheric Emitted Radiance Interferometer (AERI) blackbody emissivity and radiance using multiple techniques. In: Conference on Characterization and Radiometric Calibration for Remote Sensing (CALCON), Logan, UT, USA.

Best, F.A., Adler, D.P., Pettersen, C., Revercomb, H.E., Perepezko, J.H., 2010. On-orbit absolute temperature calibration using multiple phase change materials: overview of recent technology advancements. In: Proceedings of SPIE (7857), Multispectral, Hyperspectral, and Ultraspectral Remote Sensing Technology, Techniques, and Applications. International Society for Optics and Photonics. 78570J-78570J-10.

Best, F.A., Adler, D.P., Pettersen, C., Revercomb, H.E., Gero, P.J., Taylor, J.K., Knuteson, R.O., Perepezko, J.H., 2012. On-orbit absolute radiance standard for the next generation of IR remote sensing instruments. In: Proceedings of SPIE (8527), Multispectral, Hyperspectral, and Ultraspectral Remote Sensing Technology, Techniques and Applications. International Society for Optics and Photonics. 85270N-85270N-10.

BIPM, IEC, IFCC, ILAC, IUPAC, IUPAP, ISO & OIML, 2008. Evaluation of Measurement Data—Guide to the Expression of Uncertainty in Measurement. 100 JCGM, pp. 1–116.

BIPM, IEC, IFCC, ILAC, IUPAC, IUPAP, ISO & OIML, 2012. The international Vocabulary of Metrology—Basic and General Concepts and Associated Terms (VIM), third ed. JCGM (Joint Committee for Guides in Metrology). JCGM 200: 2012.

Borbas, E.E., Hulley, G., Feltz, M., Knuteson, R., Hook, S., 2018. The combined ASTER MODIS emissivity over land (CAMEL) part 1: methodology and high spectral resolution application. Remote Sens. (Basel) 10, 643.

Braun, S., Newman, P., 2012. Hurricane and Severe Storm Sentinel (HS3): summary of the 2012 deployment. In: AGU Fall Meeting Abstracts.

Braun, S.A., Newman, P.A., Heymsfield, G.M., 2016. NASA's hurricane and severe storm sentinel (HS3) investigation. Bull. Am. Meteorol. Soc. 97, 2085–2102.

Clough, S.A., Iacono, M.J., 1995. Line-by-line calculation of atmospheric fluxes and cooling rates: 2. Application to carbon dioxide, ozone, methane, nitrous oxide and the halocarbons. J. Geophys. Res.-Atmos. 100, 16519–16535.

Clough, S.A., Iacono, M.J., Moncet, J.L., 1992. Line-by-line calculations of atmospheric fluxes and cooling rates: application to water vapor. J. Geophys. Res.-Atmos. 97, 15761–15785.

Clough, S., Shephard, M., Mlawer, E., Delamere, J., Iacono, M., Cady-Pereira, K., Boukabara, S., Brown, P., 2005. Atmospheric radiative transfer modeling: a summary of the AER codes. J. Quant. Spectrosc. Radiat. Transf. 91, 233–244.

Davis, S.P., Abrams, M.C., Brault, J.W., 2001. Fourier Transform Spectrometry. Elsevier.

Dee, D.P., Uppala, S.M., Simmons, A., Berrisford, P., Poli, P., Kobayashi, S., Andrae, U., Balmaseda, M., Balsamo, G., Bauer, D.P., 2011. The ERA-interim reanalysis: configuration and performance of the data assimilation system. Q. J. Roy. Meteorol. Soc. 137, 553–597.

Deslover, D.H., Taylor, J., Smith, W., Weisz, E., Revercomb, H., 2016. Analysis of S-HIS dual-regression retrievals during the HS3 field campaign. In: Optics and Photonics for Energy and the Environment. Optical Society of America, p. 33. Jw4A.

Feltz, M., Borbas, E., Knuteson, R., Hulley, G., Hook, S., 2018. The combined ASTER MODIS emissivity over land (CAMEL) part 2: uncertainty and validation. Remote Sens. (Basel) 10, 664.

Gero, P.J., Taylor, J.K., Best, F.A., Garcia, R.K., Revercomb, H.E., 2012. On-orbit absolute blackbody emissivity determination using the heated halo method. Metrologia 49, S1.

Goody, R., 2001. Report of the NOAA Science Advisory Board's Panel on Strategies for Climate Monitoring. NOAA Science Advisory Board, Silver Spring, MD. 26 pp.

Goody, R., Haskins, R., 1998. Calibration of radiances from space. J. Climate 11, 754–758.

Griffiths, P.R., De Haseth, J.A., 2007. Fourier Transform Infrared Spectrometry. John Wiley & Sons.

Han, Y., Revercomb, H., Cromp, M., Gu, D., Johnson, D., Mooney, D., Scott, D., Strow, L., Bingham, G., Borg, L., 2013. Suomi NPP CrIS measurements, sensor data record algorithm, calibration and validation activities, and record data quality. J. Geophys. Res.-Atmos. 118, 12734–12748.

Hewison, T.J., Doelling, D.R., Lukashin, C., Tobin, D., John, V.O., Joro, S., Bojkov, B., 2020. Extending the Global Space-Based Inter-Calibration System (GSICS) to Tie Satellite Radiances to an Absolute Scale. Remote Sens. (Basel) 12, 1782.

Hoese, D.J., Barnes, C.M., Best, F.A., Garcia, R.K., Gero, P.J., Knuteson, R.O., Revercomb, H.E., Sullivan, D.V., Taylor, J.K., Tobin, D.C., 2013. Real-time ground data processing for the Airborne Scanning High-Resolution Interferometer Sounder. In: Fourier Transform Spectroscopy. Optical Society of America. FM1D. 2.

Holz, R.E., Ackerman, S., Antonelli, P., Nagle, F., Knuteson, R.O., McGill, M., Hlavka, D.L., Hart, W.D., 2006. An improvement to the high-spectral-resolution CO_2-slicing cloud-top altitude retrieval. J. Atmos. Oceanic Tech. 23, 653–670.

Kataoka, F., Knuteson, R.O., Kuze, A., Suto, H., Shiomi, K., Harada, M., Garms, E.M., Roman, J.A., Tobin, D.C., Taylor, J.K., 2013. TIR spectral radiance calibration of the GOSAT satellite borne TANSO-FTS with the aircraft-based S-HIS and the ground-based S-AERI at the Railroad Valley desert playa. IEEE Trans. Geosci. Remote Sens. 52, 89–105.

King, M.D., Platnick, S., Moeller, C.C., Revercomb, H.E., Chu, D.A., 2003. Remote sensing of smoke, land, and clouds from the NASA ER-2 during SAFARI 2000. J. Geophys. Res.-Atmos. 108, 1–12.

Knuteson, R., Dedecker, R., Feltz, W., Osborne, B., Revercomb, H., Tobin, D., 2001. Progress towards a characterization of the infrared emissivity of the land surface in the vicinity of the ARM SGP Central Facility: surface (S-AERI) and airborne sensors (NAST-I/S-HIS). In: Proceedings of the Eleventh ARM Science Team Meeting.

Knuteson, R.O., Deslover, D.H., Larar, A.M., Osborne, B., Revercomb, H.E., Short, J.F., Smith, W.L., Tanamachi, R., 2003. Infrared land surface remote sensing using high spectral resolution observations. In: Multispectral and Hyperspectral Remote Sensing Instruments and Applications. International Society for Optics and Photonics, pp. 24–35.

Knuteson, R.O., Revercomb, H.E., Best, F.A., Ciganovich, N.N., Dedecker, R.G., Dirkx, T.P., Ellington, S.C., Feltz, W.F., Garcia, R.K., Howell, H.B., 2004a. Atmospheric emitted radiance interferometer. Part I: instrument design. J. Atmos. Ocean. Technol. 21, 1763–1776.

Knuteson, R.O., Revercomb, H.E., Best, F.A., Ciganovich, N.N., Dedecker, R.G., Dirkx, T.P., Ellington, S.C., Feltz, W.F., Garcia, R.K., Howell, H.B., 2004b. Atmospheric emitted radiance interferometer. Part II: Instrument performance. J. Atmos. Ocean. Technol. 21, 1777–1789.

Kohrman, R.J., Luce, S.D., 2002. Mechanical design of the crosstrack infrared sounder (CrIS). In: Infrared Spaceborne Remote Sensing IX. International Society for Optics and Photonics, pp. 445–455.

Liu, X., 2015. Advanced radiative models and retrieval algorithms for hyperspectral remote sensing data. In: Hyperspectral Imaging and Sounding of the Environment. Optical Society of America. HT2B. 1.

Liu, X., Yang, Q., Li, H., Jin, Z., Wu, W., Kizer, S., Zhou, D.K., Yang, P., 2016. Development of a fast and accurate PCRTM radiative transfer model in the solar spectral region. Appl. Optics 55, 8236–8247.

Maestri, T., Holz, R.E., 2009. Retrieval of cloud optical properties from multiple infrared hyperspectral measurements: a methodology based on a line-by-line multiple-scattering code. IEEE Trans. Geosci. Remote Sens. 47, 2413–2426.

McCourt, M., McMillan, W., Ackerman, S., Holz, R., Revercomb, H., Tobin, D., 2004. Using the "blue spike" to characterize biomass-burning sites during Southern African Regional Science Initiative (SAFARI) 2000. J. Geophys. Res. Atmos. 109, 1–8.

Michelson, A.A., 1891. Visibility of interference-fringes in the focus of a telescope. Publ. Astron. Soc. Pac. 3, 217–220.

Michelson, A.A., 1902. Light Waves and Their Uses. University of Chicago Press.

Moeller, C.C., Revercomb, H.E., Ackerman, S.A., Menzel, W.P., Knuteson, R.O., 2003. Evaluation of MODIS thermal IR band L1B radiances during SAFARI 2000. J. Geophys. Res.-Atmos. 108, 1–12.

Munsell, E.B., Zhang, F., Braun, S.A., Sippel, J.A., Didlake, A.C., 2018. The inner-core temperature structure of Hurricane Edouard (2014): observations and ensemble variability. Mon. Weather Rev. 146, 135–155.

Nalli, N.R., Smith, W.L., Huang, B., 2001. Quasi-specular model for calculating the reflection of atmospheric-emitted infrared radiation from a rough water surface. Appl. Optics 40, 1343–1353.

Nalli, N.R., Minnett, P.J., Maddy, E., McMillan, W.W., Goldberg, M.D., 2008a. Emissivity and reflection model for calculating unpolarized isotropic water surface-leaving radiance in the infrared. 2: Validation using Fourier transform spectrometers. Appl. Optics 47, 4649–4671.

Nalli, N.R., Minnett, P.J., Van Delst, P., 2008b. Emissivity and reflection model for calculating unpolarized isotropic water surface-leaving radiance in the infrared. I: Theoretical development and calculations. Appl. Optics 47, 3701–3721.

Newman, S.M., Knuteson, R.O., Zhou, D.K., Larar, A.M., Smith, W.L., Taylor, J.P., 2009. Radiative transfer validation study from the European Aqua Thermodynamic Experiment. Q. J. Roy. Meteorol. Soc. 135, 277–290.

Newman, S.M., Larar, A.M., Smith, W.L., Ptashnik, I.V., Jones, R.L., Mead, M.I., Revercomb, H., Tobin, D.C., Taylor, J.K., Taylor, J.P., 2012. The joint airborne IASI validation experiment: an evaluation of instrument and algorithms. J. Quant. Spectrosc. Radiat. Transf. 113, 1372–1390.

Ohring, G., Tansock, J., Emery, W., Butler, J., Flynn, L., Weng, F., Germain, K.S., Wielicki, B., Cao, C., Goldberg, M., 2007. Achieving Satellite Instrument Calibration for Climate Change. Wiley Online Library.

Padula, F., Goodman, S.J., Pearlman, A., Cao, C., 2017. Goes-R advanced baseline imager (ABI) and geostationary lightning mapper (GLM) calibration/validation from a field campaign perspective. In: 2017 IEEE International Geoscience and Remote Sensing Symposium (IGARSS). IEEE, pp. 301–304.

Peters, W., Jacobson, A.R., Sweeney, C., Andrews, A.E., Conway, T.J., Masarie, K., Miller, J.B., Bruhwiler, L.M., Pétron, G., Hirsch, A.I., 2007. An atmospheric perspective on North American carbon dioxide exchange: CarbonTracker. Proc. Natl. Acad. Sci. U. S. A. 104, 18925–18930.

Revercomb, H.E., Best, F.A., 2005. Calibration of the scanning high-resolution interferometer sounder (S-HIS) infrared spectrometer: overview (Part 1). In: Conference on Characterization and Radiometric Calibration for Remote Sensing (CALCON), Logan, UT, USA.

Revercomb, H.E., Buijs, H., Howell, H.B., Laporte, D.D., Smith, W.L., Sromovsky, L., 1988a. Radiometric calibration of IR Fourier transform spectrometers: solution to a problem with the High-Resolution Interferometer Sounder. Appl. Optics 27, 3210–3218.

Revercomb, H.E., Laporte, D.D., Smith, W.L., Buijs, H., Murcray, D.G., Murcrayr, F.J., Sromovsky, L.A., 1988b. High-altitude aircraft measurements of upwelling IR radiance: prelude to FTIR from geosynchronous satellite. Microchim. Acta 95, 439–444.

Revercomb, H.E., Buijs, H., Howell, H.B., Knuteson, R.O., Laporte, D.D., Smith Sr., W.L., Sromovsky, L.A., Woolf, H.W., 1989. Radiometric calibration of IR interferometers: experience from the High-resolution Interferometer Sounder (His) aircraft instrument. In: RSRM. vol. 87, pp. 89–102.

Revercomb, H.E., Walden, V.P., Tobin, D.C., Anderson, J., Best, F.A., Ciganovich, N.C., Dedecker, R.G., Dirkx, T., Ellington, S.C., Garcia, R.K., 1998. Recent results from two new aircraft-based Fourier transform interferometers: the scanning high-resolution interferometer sounder and the NPOESS atmospheric sounder testbed interferometer. In: 8th International Workshop on Atmospheric Science from Space Using Fourier Transform Spectrometry (ASSFTS), Toulouse, France, pp. 16–18.

Revercomb, H.E., Knuteson, R.O., Best, F.A., Tobin, D.C., Smith, W.L., Laporte, D.D., Ellington, S.D., Werner, M.W., Dedecker, R.G., Garcia, R.K., 2003. Scanning high-resolution interferometer sounder (S-HIS) aircraft instrument and validation of the Atmospheric InfraRed Sounder (AIRS). In: Optical Remote Sensing. Optical Society of America. JMA4.

Revercomb, H.E., Tobin, D.C., Knuteson, R.O., Best, F.A., Smith, W., Laporte, D., Ellington, S., Werner, M., Garcia, R., Ciganovich, N., 2004. Highly accurate FTIR observations from the scanning HIS aircraft instrument. Proc. SPIE, 41–53.

Revercomb, H., Best, F., Tobin, D., Knuteson, B., Smith, N., Smith, S.R., William, L., Weisz, E., 2016. Monitoring climate from space: a metrology perspective. In: Earth Observing Missions and Sensors: Development, Implementation, and Characterization IV. International Society for Optics and Photonics. 98810F.

Rice, J.P., Johnson, B.C., 1998. The NIST EOS thermal-infrared transfer radiometer. Metrologia 35, 505.

Roberts, J., Trainer, M., Murphy, D., Brown, S., Brewer, A., Gao, R.-S., Fahey, D., 2020. Fire Influence on Regional to Global Environments and Air Quality (FIREX-AQ). FIREX-QA, pp. 943–944. https://www. esrl. noaa. gov/csd/projects/firex-aq/whitepaper. pdf.

Sarkissian, E., Worden, H., Bowman, K., Fisher, B., Rider, D., Aumann, H., Apolinski, M., Debaca, R., Gluck, S., Madatyan, M., 2005. TES radiometric assessment. Eos. Trans. AGU 86.

Schmit, T.J., Griffith, P., Gunshor, M.M., Daniels, J.M., Goodman, S.J., Lebair, W.J., 2017. A closer look at the ABI on the GOES-R series. Bull. Am. Meteorol. Soc. 98, 681–698.

Shephard, M.W., Worden, H.M., Cady-Pereira, K.E., Lampel, M., Luo, M., Bowman, K.W., Sarkissian, E., Beer, R., Rider, D.M., Tobin, D.C., 2008. Tropospheric Emission Spectrometer nadir spectral radiance comparisons. J. Geophys. Res.-Atmos. 113, 1–12.

Smith Sr., W.L., Revercomb, H.E., Howell, H.B., Woolf, H.M., Laporte, D.D., 1987. The High Resolution Interferometer Sounder (HIS). American Meteorological Society, Boston, MA, pp. 271–281.

Smith Sr., W.L., Revercomb, H.E., Howell, H.B., Huang, H., Knuteson, R.O., Koenig, E.W., Laporte, D.D., Silverman, S., Sromovsky, L.A., Woolf, H.M., 1990. GHIS—the GOES high-resolution interferometer sounder. J. Appl. Meteorol. Climatol. 29, 1189–1204.

Smith Sr., W.L., Larar, A.M., Taylor, J.P., Revercomb, H., Kireev, S., Zhou, D.K., Liu, X., Tobin, D.C., Newman, S.M., Schluessel, P., 2008. Joint Airborne IASI Validation Experiment (JAIVEx)—an overview. In: Proc. Int. ATOVS Study Conf. XVI, Angra dos Reis, Brazil, CIMSS. University of Wisconsin-Madison, Angra dos Reis, Brazil, pp. 7–13. Citeseer.

Smith Sr., W.L., Weisz, E., Kireev, S.V., Zhou, D.K., Li, Z., Borbas, E.E., 2012. Dual-regression retrieval algorithm for real-time processing of satellite ultraspectral radiances. J. Appl. Meteorol. Climatol. 51, 1455–1476.

Smith, W.L., Revercomb, H.E., Weisz, E., Tobin, D.C., Knuteson, R.O., Taylor, J., Menzel, W.P., 2021. Hyperspectral satellite radiance atmospheric profile information content and its dependence on spectrometer technology. IEEE J. Sel. Top. Appl. Earth Obs. Remote Sens. 14, 4720–4736.

Stumpf, K.D., Overbeck, J.A., 2002. CrIS optical system design. In: Infrared Spaceborne Remote Sensing IX. International Society for Optics and Photonics, pp. 437–444.

Tanamachi, R.L., Knuteson, R., Ackerman, S., Walden, V., 2001. Comparison of IMG observations with scanning-HIS observations over the SAFARI 2000 region. In: AGU Fall Meeting Abstracts.

Taylor, J.K., 2014. Achieving 0.1 K Absolute Calibration Accuracy for High Spectral Resolution Infrared and far Infrared Climate Benchmark Measurements. Doctorat en génie électrique, Université Laval.

Taylor, B.N., Kuyatt, C.E., 1994. Guidelines for Evaluating and Expressing the Uncertainty of NIST Measurement Results. United States Department of Commerce Technology Administration, National Institute of Standards and Technology. NIST Technical Note 1297. U.S. Government Printing Office, Washington.

Taylor, J.K., Best, F.A., Ciganovich, N.N., Dutcher, S., Ellington, S.D., Garcia, R.K., Howell, H.B., Knuteson, R.O., Laporte, D.D., Nasiri, S., 2005. Performance of an infrared sounder on several airborne platforms: the scanning high resolution interferometer sounder (S-HIS). In: Earth Observing Systems X. International Society for Optics and Photonics, p. 588214.

Taylor, J.K., O'Connel, J., Rice, J.P., Revercomb, H.E., Best, F.A., Tobin, D.C., Knuteson, R.O., Adler, D.P., Ciganovich, N.N., Dutcher, S.T., 2008. SI traceable infrared radiance measurements and sources: NIST TXR Validation of S-HIS radiances and a UW-SSEC Blackbody. In: AGU Fall Meeting Abstracts. 0761.

Taylor, J.K., Revercomb, H., Best, F., Tobin, D., Knuteson, R., Gero, P., Garcia, R., Ciganovich, N., Laporte, D., Werner, M., 2013. Suomi NPP/JPSS Cross-track Infrared Sounder (CrIS): calibration validation with the aircraft based Scanning High-resolution Interferometer Sounder (S-HIS). In: 9th Annual Symposium on Future Operational Environmental Satellite Systems. American Meteorological Society, Austin, TX.

Taylor, J.K., Revercomb, H.E., Hoese, D.J., Garcia, R.K., Smith Sr., W.L., Weisz, E., Tobin, D.C., Best, F.A., Knuteson, R.O., Sullivan, D.V., 2015a. Real-time Data Processing and Visualization for the Airborne Scanning High-Resolution Interferometer Sounder (S-HIS). In: AGU Fall Meeting Abstracts.

Taylor, J.K., Tobin, D.C., Revercomb, H.E., Best, F.A., Garcia, R., Motteler, H., Goldberg, M., 2015b. Suomi NPP/JPSS cross-track infrared sounder (CrIS): calibration validation with the aircraft based Scanning High-resolution Interferometer Sounder (S-HIS). In: Fourier Transform Spectroscopy. Optical Society of America. FW1A. 3.

Taylor, J.K., Tobin, D.C., Revercomb, H.E., Best, F.A., Garcia, R.K., Merrelli, A., Goldberg, M., 2016. Calibration validation of the Cross-track Infrared Sounder (CrIS) with the aircraft based Scanning High-resolution Interferometer Sounder (S-HIS). In: 12th Annual Symposium on New Generation Operational Environmental Satellite Systems. American Meteorological Society, New Orleans.

Taylor, J.K., Tobin, D.C., Revercomb, H.E., Best, F.A., Garcia, R.K., Merrelli, A., Goldberg, M., 2017. Cold scene calibration validation of the Cross-track Infrared Sounder (CrIS) with the aircraft based scanning High-resolution Interferometer Sounder (S-HIS). In: 13th Annual Symposium on New Generation Operational Environmental Satellite Systems. American Meteorological Society, Seattle.

Taylor, J.K., Revercomb, H.E., Best, F.A., Knuteson, R.O., Tobin, D.C., Gero, P.J., Adler, D., Mulligan, M., 2019a. An on-orbit infrared intercalibration reference standard for decadal climate trending of the Earth. In: Sensors, Systems, and Next-Generation Satellites XXIII. International Society for Optics and Photonics, p. 1115116.

Taylor, J.K., Tobin, D.C., Revercomb, H.E., Best, F.A., Garcia, R.K., Knuteson, B., Feltz, M., Padula, F., Goodman, S., 2019b. Calibration validation of the GOES-16 Advanced Baseline Imager (ABI) with the high-altitude aircraft based Scanning High-resolution Interferometer Sounder (S-HIS). In: Fourier Transform Spectroscopy. Optical Society of America. FTu2B. 2.

Taylor, J.K., Revercomb, H.E., Best, F.A., Tobin, D.C., Gero, P.J., 2020. The infrared Absolute Radiance Interferometer (ARI) for CLARREO. Remote Sens. (Basel) 12, 1915.

Tobin, D., Revercomb, H., Knuteson, R., 2003. On-orbit spectral calibration of the geosynchronous imaging fourier transform spectrometer (GIFTS). In: Proceedings of CALCON.

Tobin, D.C., Revercomb, H.E., Moeller, C.C., Knuteson, R.O., Best, F.A., Smith, W.L., Van Delst, P., Laporte, D.D., Ellington, S.D., Werner, M.D., 2004. Validation of Atmospheric InfraRed Sounder (AIRS) spectral radiances with the Scanning High-resolution Interferometer Sounder (S-HIS) aircraft instrument. In: Remote Sensing of Clouds and the Atmosphere IX. International Society for Optics and Photonics, pp. 383–392.

Tobin, D.C., Revercomb, H.E., Knuteson, R.O., Best, F.A., Smith, W.L., Ciganovich, N.N., Dedecker, R.G., Dutcher, S., Ellington, S.D., Garcia, R.K., 2006. Radiometric and spectral validation of Atmospheric Infrared Sounder observations with the aircraft-based Scanning High-Resolution Interferometer Sounder. J. Geophys. Res.-Atmos. 111, 1–14.

Tobin, D., Revercomb, H., Knuteson, R., Taylor, J., Best, F., Borg, L., Deslover, D., Martin, G., Buijs, H., Esplin, M., 2013. Suomi-NPP CrIS radiometric calibration uncertainty. J. Geophys. Res.-Atmos. 118, 10589–10600.

Tobin, D., Holz, R., Nagle, F., Revercomb, H., 2016. Characterization of the Climate Absolute Radiance and Refractivity Observatory (CLARREO) ability to serve as an infrared satellite intercalibration reference. J. Geophys. Res.-Atmos. 121, 4258–4271.

Verlinde, J., Harrington, J.Y., Mcfarquhar, G.M., Yannuzzi, V., Avramov, A., Greenberg, S., Johnson, N., Zhang, G., Poellot, M., Mather, J.H., 2007. The mixed-phase Arctic cloud experiment. Bull. Am. Meteorol. Soc. 88, 205–222.

Vinson, K.H., Revercomb, H.E., Howell, H.B., Knuteson, R.O., Smith, W.L., 2003. Carbon monoxide and methane column retrieval from the scanning high resolution interferometer sounder (S-HIS) and the national polar-orbiting operational environmental satellite system airborne sounder testbed-interferometer (NAST-I). In: Optical Remote Sensing of the Atmosphere and Clouds III. International Society for Optics and Photonics, pp. 472–482.

Weisz, E., Smith Sr., W.L., Smith, N., 2013. Advances in simultaneous atmospheric profile and cloud parameter regression based retrieval from high-spectral resolution radiance measurements. J. Geophys. Res.-Atmos. 118, 6433–6443.

Wielicki, B.A., Young, D., Mlynczak, M., Thome, K., Leroy, S., Corliss, J., Anderson, J., Ao, C., Bantges, R., Best, F., 2013. Achieving climate change absolute accuracy in orbit. Bull. Am. Meteorol. Soc. 94, 1519–1539.

Chapter 10

Aircraft dropsonde campaigns

Holger Vömel[a] and Jason Dunion[b]
[a]National Center for Atmospheric Research, Boulder, CO, United States, [b]NOAA/AOML Hurricane Research Division (HRD), Miami, FL, United States

Chapter outline

1. Introduction	185	3.2 Surface winds	190
2. Measurement technology	186	3.3 Wind profiles	192
2.1 Dropsonde description	186	3.4 Clouds and precipitation	192
2.2 Launchers	187	4. Summary	193
2.3 Data processing and quality control	187	Acknowledgments	193
3. Cal val campaigns	189	References	193
3.1 Temperature and humidity	189		

In the fields of observation, chance favors only those minds, which are prepared.

Louis Pasteur

1. Introduction

Oceans cover two-thirds of our planet and give rise to the most severe storms and other weather systems. Atmospheric observations and in situ profiling over oceans are vital to understand the formation and dynamics of these storms (e.g., Rappaport et al., 2009). Meteorologists have deployed small instruments from aircraft, which descend through the atmosphere and radio observations back to the aircraft, for nearly 80 years. Forsythe (1946) called these instruments droppable radiosondes or the now more common term "dropsondes." Today, these instruments provide essential observations over oceans and sometimes over land for analysis, forecasting, and basic research ranging from tropical cyclones to polar exploration.

Historically, dropsondes only measured temperature, pressure, and humidity. Unlike radiosondes (see Chapter 2: Balloon-Borne Radiosondes), which could determine winds through optical or radio tracking, dropsondes started measuring winds only with the introduction of global radio-navigation systems in the 1970s. The first radio navigation system integrated into dropsondes and radiosondes was Omega, followed by LORAN-C. Modern dropsondes use Global Navigation Satellite Systems (GNSS), such as the American Global Positioning Satellites (GPS) to measure position and winds directly (Hock and Franklin, 1999).

Currently, there is a small number of manufacturers providing dropsonde systems for atmospheric research. The National Center for Atmospheric Research (NCAR) has developed the Airborne Vertical Atmospheric Profiling System (AVAPS) dropsonde that dates back more than 50 years. Vaisala (Finland) produces and markets these dropsondes in license from NCAR. These dropsonde systems are used in hurricane surveillance by the Hurricane Hunters of the US Air Force and the National Oceanic and Atmospheric Administration (NOAA), in research projects supported by NCAR, and by a number of different agencies in the U.S. and globally. Dropsonde systems are also produced by Meisei (Japan), Yankee Environmental Systems (United States), and Modem (France). Specialized dropsondes have also been developed by the Karlsruhe Institute of Technology (KIT, Germany) and several other companies.

Here we focus on NCAR's AVAPS dropsondes, produced and marketed by Vaisala, which constitute more than 95% of all dropsondes that have been deployed over the past 25 years. We explicitly refer to other suppliers where appropriate.

2. Measurement technology

2.1 Dropsonde description

Dropsondes perform a similar function and contain nearly the identical sensors as radiosondes (see Chapter 2: Balloon-Borne Radiosondes). Yet, there are some subtle, but important differences between these instruments. Operators launch dropsondes from fast moving aircraft; therefore, these instruments have to be very rugged and must consistently function through the launch process.

The two most recent AVAPS dropsonde models, which NCAR has developed, are the RD41 and NRD41, which replaced the previous RD94 and NRD94 models in 2019. Both share identical sensor packages and the same telemetry format but are different in size (see Table 1) to accommodate different aircraft configurations. The NRD41 dropsonde model (Fig. 1) uses a smaller form factor and can be launched from modified manual launchers, from automated launchers, and from long duration balloons. The larger RD41 is used mostly in operational missions conducted by NOAA and the Air Force Hurricane Hunters and by other research organizations.

Since dropsondes continuously measure the atmosphere during descent, their temperature and humidity sensors are installed at the bottom of the instrument, which provides relatively undisturbed air sampling. The dropsonde deploys a parachute after launch, timed either by a mechanical delay ribbon or by an electronic timer. A very short tether connects the parachute to the dropsonde, which all but eliminates any pendulum motion.

The overall measurement time for a deployed dropsonde ranges from a just few minutes to as long as 15 min when launched from the highest-flying aircraft currently capable of deploying dropsondes. The larger RD41 dropsonde also uses

TABLE 1 Current AVAPS dropsonde types developed at NCAR.

Dropsonde type	Diameter (cm)	Length (cm)	Weight (g)	Parachute size (cm × cm)	Launcher type
RD41	7.0	40.6	323	30 × 30	Manual
NRD41	4.5	30.5	169	20 × 20	Manual or automatic

FIG. 1 NRD41 dropsonde.

a larger parachute than the NRD41 such that the fall velocity of both sondes is very similar. Fall rates range from between about $22\,m\,s^{-1}$ at an altitude of 14 km to $10\,m\,s^{-1}$ near the surface. To achieve a similar vertical resolution as radiosondes, the data transmission rate for both dropsonde models has been increased to 2 Hz for pressure, temperature, and humidity and 4 Hz for winds.

The sensor modules of AVAPS dropsondes are based on the Vaisala RS41 radiosonde and have to be sufficiently fast to support this increase in data rate. The temperature sensors have a response time of less than 0.5 s. The humidity sensors have a response time of less than 0.3 s at 20°C, which, however, increases to 10 s at −40°C. Correction algorithms can recover some of the increase in response time, but combined with faster fall rates higher in the atmosphere, the vertical resolution of dropsonde humidity measurements decreases with altitude.

At launch, a dropsonde may experience large changes in environmental temperature and moisture as it exits the aircraft. This is particularly pronounced for high-flying aircraft, where the temperature can change from +20°C inside the cabin to −60°C outside. Equilibration to ambient conditions after launch can take up to 30 s or more and may lead to a loss of data of 500 m or more below the aircraft. For dropsonde releases made from lower flying aircraft, the equilibration is typically significantly faster.

The AVAPS dropsondes use Global Navigation Satellite System (GNSS) signals to determine winds and position. The AVAPS system initializes sondes prior to launch to allow faster acquisition of satellite signals after launch. The vertical resolution of wind measurements is higher compared to balloon measurements, since no pendulum motions have to be filtered out.

2.2 Launchers

Dropsondes can be launched through manual launch tubes (Fig. 2, left) or by automated dropsonde launchers (Fig. 2, right). Most dropsonde capable aircraft are equipped with manual launch tubes. Automated dropsonde launchers allow remote controlled operation from any position onboard the aircraft or even remote controlled operation from the ground, saving space that would otherwise be occupied by an onboard dropsonde operator.

Dropsondes have also been launched from Unmanned Aerial Systems (UAS, Wick et al., 2018) and from long duration balloons. Concordiasi in 2010 was the most recent of the long duration balloon projects and was led by the French Space Agency CNES. It was designed to study the atmosphere over Antarctica (Cohn et al., 2013).

2.3 Data processing and quality control

Fig. 3 shows a sample dropsonde profile obtained during the Organization of Tropical East Pacific Convection (OTREC) field campaign. Vömel et al. (2021) describe the quality control of this data set in detail and list typical and infrequent quality control issue that may occur in dropsonde observations.

Dropsondes send their data by telemetry back to the aircraft, where they are processed by the dropsonde system software in real time. The operating environments into which dropsondes are released are usually much more challenging than those

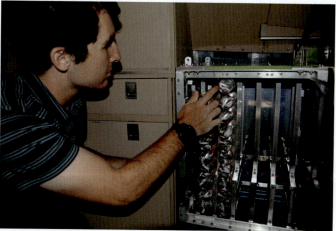

FIG. 2 (Left) Manual dropsonde launch onboard the NOAA WP-3D research aircraft. (Right) Automated dropsonde launcher onboard the NSF/NCAR GV research aircraft.

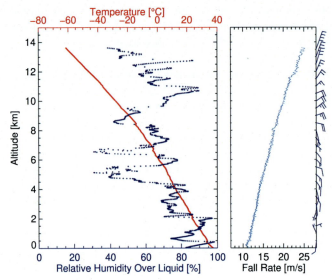

FIG. 3 Dropsonde profile obtained during OTREC south of Costa Rica on September 4, 2019. (Left) Ambient temperature *(red curve)* and humidity *(dark blue curve)*. (Right) Fall rate and horizontal wind profile.

for radiosondes. Meteorologists or scientists onboard the aircraft typically evaluate the data quality and check the data for any unusual instrument behavior. Occasionally, the automated detection of launch and the time of splashing in the water have to be adjusted.

The Atmospheric Sounding Processing Environment (ASPEN; Martin and Suhr, 2021) processing software typically removes several tenths to several hundreds of meters of data immediately following the sonde release due to the equilibration of the sonde to ambient environmental conditions. A small correction is applied to the temperature and humidity measurements due to the inherent lag of each sensor. Since dropsondes do not exactly follow the ambient winds, a small correction is also applied to the wind measurements (Hock and Franklin, 1999).

The expected uncertainty of atmospheric measurements using the current RD41 and NRD41 dropsonde models is summarized in Table 2. The uncertainty of the measurements of pressure, temperature, and humidity is based on the standard deviation of repeated calibrations (Vaisala, 2020). The uncertainty of the horizontal and meridional wind speed components is reported by the GNSS receiver and provided with the raw data. The number reported here is an estimate of the average uncertainty. The geopotential height is typically calculated from the surface upward and its uncertainty is mostly limited by the time between the last received dropsonde data frame and its splash into the water. The uncertainty of the GNSS height is larger than that of the geopotential height due to the challenges of measuring accurate GNSS heights from a falling object.

Due to the relatively higher dropsonde fall speed compared to radiosonde ascents and the different sensor orientation, a correction for solar radiation for the dropsonde temperature measurement is not required. Finally, unlike balloon-borne radiosonde observations, in almost all applications of dropsondes, sensor icing does not occur because the sondes fall from

TABLE 2 Expected performance of RD41 and NRD41 dropsonde measurements, based on repeated laboratory calibrations, and processing uncertainties for geopotential height and geometric GNSS height.

Parameter	Uncertainty
Pressure	0.4 hPa
Temperature	0.2 K
Relative humidity	2%
Zonal and meridional wind speed	0.5 m s^{-1}
Geopotential altitude	10 m
GNSS height	20 m

All uncertainties are given at a confidence level of $k=2$.

a colder to a warmer environment. Other AVAPS dropsonde data artifacts rarely occur and are generally detected and automatically removed by the software.

After processing on the aircraft, AVAPS dropsonde data are relayed via satellite to the ground, where additional processing may take place. Most operational profiles are then transmitted to the data centers connected to the Global Telecommunication System (GTS) of the World Meteorological Organization (WMO) and to dedicated forecasting centers, such as NOAA's National Hurricane Center (NHC) and the National Weather Service (NWS).

The data formats of AVAPS dropsondes are similar to those used for radiosondes. The FM-37 TEMPDROP format (WMO, 2019) contains only a limited number of vertical levels, but adds parameters, which provide important information to forecasters at the NHC. To satisfy the need for higher resolution observations in Numerical Weather Prediction (NWP) models, dropsonde observations were also migrated to the Binary Universal Form for the Representation of meteorological data (BUFR). NCAR and other centers also provide dropsonde data in NetCDF format for scientific research projects and dedicated satellite validation studies.

3. Cal val campaigns

Satellite observations, in particular those produced by nadir viewing sensors, can be validated with relatively fewer complications over oceans than over land. The ocean generally provides a radiatively well-defined surface, and topography and topographically induced atmospheric features are less pronounced away from land. On the other hand, reference observations over oceans require either ship- or aircraft-deployed instrumentation, or observations from small islands. Aircraft observations have the inherent benefit that they can be coordinated in time and space with satellite overpasses of interest. Furthermore, for in situ profiling, aircraft can release dropsondes over oceans with relatively fewer restrictions than over land. In particular, dropsondes released from high-flying research aircraft, such as the NSF/NCAR GV, the NOAA G-IV, the DLR HALO, the NASA ER-2, or the NASA WB-57 can measure a nearly complete tropospheric profile of the atmospheric state parameters. Here, we provide an overview of satellite validation and calibration campaigns for the atmospheric parameters measured by dropsondes.

Airborne assets provide an effective means for directly comparing observations with satellite radiances and, unlike fixed ground sites, can be repositioned anywhere (Larar et al., 2010). Airborne campaigns deploy remote sensing spectrometers to measure radiative spectra and in situ sensors, most notably dropsondes to characterize the atmospheric column contributing to the observed spectra. The higher the aircraft, the more accurate the measurement of the atmospheric column; however, instruments may also be distributed aboard different aircraft flying in coordination at different altitudes. Collocation is a significant issue comparing observations from radiosondes and satellites (Sun et al., 2010) as is using global radiosonde archives, which contain sondes from different manufacturers and with unique instrument characteristics. The ability to deploy dropsondes anywhere allows a significant reduction of collocation mismatches and targeting specific atmospheric features.

3.1 Temperature and humidity

Numerous validation campaigns have been conducted over the past 20 years to collect measurements of temperature and humidity in support of satellite validation and model evaluation. Zhou et al. (2007) developed retrieval algorithms for hyperspectral instruments, which determine cloud parameters and thermodynamic parameters based on spectral information. They applied these algorithms to observations from the National Polar-orbiting Operational Environmental Satellite System (NPOESS) Airborne Sounder Testbed Interferometer (NAST-I) during THe Observing Research and Predictability EXperiment (THORPEX) Atlantic Regional Campaign (ATReC). NAST-I flew aboard the NASA ER-2, which also carried the NASA Cloud Physics Lidar (CPL). Additionally, dropsondes were launched from the NOAA G-IV research aircraft, which flew in coordination with the NASA ER-2. Comparisons between the in situ soundings and retrievals from the remote sensing instruments demonstrated the capabilities of the retrievals under cloudy and partly cloudy conditions.

The European Aqua Thermodynamic Experiment (EAQUATE) campaign was designed to validate radiance measurements and derived products from the Atmospheric Infrared Sounder (AIRS) on NASA's Aqua satellite (Taylor et al., 2008). During underflights of the Aqua satellite, up to 14 dropsondes were launched from the British Aerospace (BAe) 146 aircraft, providing high-resolution in situ data. Similar to the approach by Zhou et al. (2007), these observations were used to evaluate the retrievals of the observations by NAST-I, which flew simultaneously on the Proteus aircraft in coordination with the BAe 146 flight track. The validated NAST-I retrievals were then used to evaluate AIRS retrievals over the same location. The high density of the collocated dropsondes within the NAST-I retrieval footprint captured the vertical and horizontal variability of water vapor in the atmosphere and were essential for a useful assessment of the satellite retrieval performance.

Validations of AIRS retrievals during the NASA African Monsoon Multidisciplinary Analyses (NAMMA) field campaign in 2006 show that the performance of AIRS in the Saharan Air Layer (SAL) and in cloudy regions is approaching the error limits of clear sky retrievals (Wu, 2009). Relative to dropsonde observations at standard pressure layers from 1000 to 300 hPa, AIRS shows root mean square (RMS) differences in temperature and relative humidity of 1.35°C and 13.8%, respectively. Pu and Zhang (2010) expanded this comparison and included dropsonde observations from the THORPEX-Pacific Asian Regional Campaign (T-PARC). As in the previous studies, they found good agreement between AIRS retrievals of temperature and dropsonde observations, but some bias in AIRS retrievals of relative humidity over the tropical oceans where tropical cyclones developed. In the two tropical cyclones they studied, correcting the biases produced significant improvements to the track forecasts.

During the Joint Airborne IASI Validation Experiment (JAIVEx) the NASA WB-57 carried NAST-I and the BAe 146, which flew in coordination, launched dropsondes (Zhou et al., 2009) using a similar approach employed by Taylor et al. (2008) for AIRS. The dropsondes and additional land-based radiosondes highlighted the horizontal variability, which was resolved by both IASI and AIRS. The same approach was later used in the evaluation of the Suomi National Polar-orbiting Partnership (NPP) cross-track infrared sounder (CrIS; Zhou et al., 2016), which is a similar hyperspectral infrared spectrometer as AIRS and IASI.

The Concordiasi field campaign launched long duration balloons over Antarctica from October through December 2010. These balloons deployed 635 dropsondes in the study region (Cohn et al., 2013), providing a unique data set to validate the performance of 10 different satellite data products from five different instruments (Wang et al., 2013). The NOAA PROducts Validation System (NPROVS) for routine monitoring of satellite-based data products (Reale et al., 2012) provided the collocation of dropsonde observations and satellite retrievals and generated a data set to investigate the performance of the satellite retrievals. Significant temperature biases relative to the dropsonde measurements existed in all satellite products except one. Boylan et al. (2015) showed that updates in the temperature retrieval algorithm by AIRS reduced this bias. On the other hand, the vertical smoothing of satellite observations limits their ability to detect shallow inversion layers above the surface, which are common over Antarctica and in Saharan Air Layer outbreaks over the tropical North Atlantic, can be challenging for remote sensing instruments (Boylan et al., 2016).

Korea's Geostationary Multi-Purpose Satellite-2A (GK2A) Advanced Meteorological Imager (AMI) uses nine infrared channels to retrieve vertical profiles of temperature and humidity. To be independent of radiosonde profiles, which are partly used to generate the a priori estimate, Kim et al. (2020) used dropsondes for the evaluation of the retrieval algorithms. They show a similar behavior of the retrieval algorithms for temperature over land as over ocean, but a better performance of the humidity retrieval over oceans than over land in cloud free observations and at cloud edges.

3.2 Surface winds

Near surface winds over the oceans can be detected by the amount of microwave energy emitted or reflected by the ocean surface, which is influenced by the wave structure and foam coverage and therefore by the surface winds. The algorithms used in space-based radiometers and scatterometers to derive ocean surface winds have been validated against surface winds from buoys (e.g., Goodberlet et al., 1990). Dropsondes became relevant in the validation of these retrievals with the implementation of high vertical resolution GPS winds. Dropsondes provide the vertical structure of the wind profile above and near the surface at any location where dropsondes can be released. Satellite surface winds are averages over a larger horizontal footprint, typically with 25 or 50 km resolution, whereas dropsonde and buoy observations are point observations. Therefore, comparisons require a larger number of coincident observations to average collocation and resolution differences.

Passive microwave imagers cannot retrieve accurate surface wind speeds in rainy conditions due to the interference of rain droplets, which, for example, affects the observations of atmospheric rivers. Dropsonde observations have been shown to significantly improve the characterization and prediction of these events (Ralph et al., 2004, 2005). Satellites with active scatterometers typically operate at C- and/or Ku-band and can detect surface winds even under moderate rainfall conditions. However, these wind retrievals require careful calibration of the algorithms based on in situ observations. Donnelly et al. (1999) conducted the first study to include GPS dropsondes into the set of in situ observations for the improvement of the algorithms for the NASA scatterometer (NSCAT) under clear atmospheric conditions. Chen (2007) studied retrievals by the Special Sensor Microwave Imager (SSM/I) and the Quick Scatterometer (QuikSCAT) winds and compared these to dropsonde winds. Biases between these instruments were relatively small and justified the assimilation of the satellite-derived winds into NWP models, where they significantly improved the forecast skill. Chou et al. (2010) found similar biases in tropical cyclone surveillance observations for QuikSCAT and Advanced Scatterometer (ASCAT) surface winds and concluded that these observations have utility for characterizing the size of tropical cyclone wind field.

As example, Fig. 4 shows the overpass of ASCAT on the Metop-B satellite over Hurricane Ida on August 28, 2021. The swath of surface winds that can be observed by this satellite instrument have a typical horizontal resolution of about 25 km. Black wind barbs are flagged and indicate possible rain contamination in the retrieval of surface winds. A NOAA Hurricane Hunter flight was probing this eye of the storm at the time of the ASCAT overpass and its flight track and the near-surface wind observations by dropsondes are shown as well. The movement of the storm during the research flight poses a challenge for the exact overpass timing of dropsondes.

The Cyclone Global Navigation Satellite System (CYGNSS, Ruf et al., 2016, 2019) makes use of the scattering of microwave signals by GNSS satellites. Their signals use a wavelength that is much less attenuated by rain than active scatterometers making observations even under precipitation feasible. On the other hand, the constellation of CYGNSS satellites generates a much smaller number of observations than active scatterometers, making validation more challenging. One system that can be used in the validation of CYGNSS retrievals is the Stepped Frequency Microwave Radiometer (SFMR) onboard the Air Force and NOAA Hurricane Hunter aircraft (Laughland and Grogan, 2020). To provide reliable observations, the SMFR retrievals in turn are characterized using dropsondes (Uhlhorn et al., 2007), which leads to a slight improvement in the evaluation of surface winds derived from CYGNSS.

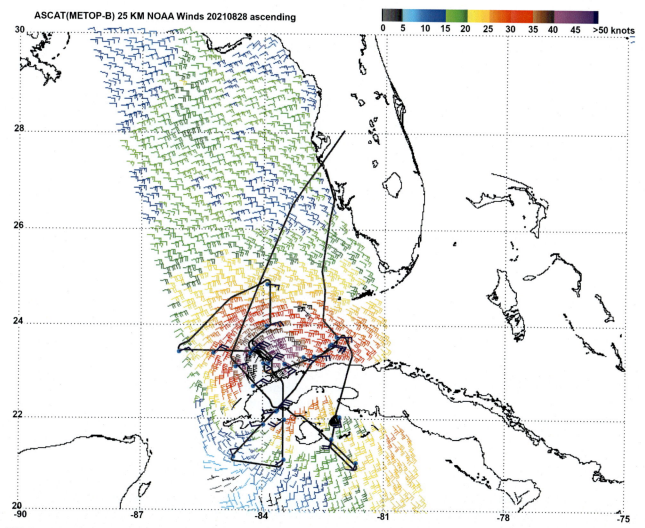

FIG. 4 ASCAT (METOP-B) satellite overpass showing ocean surface winds on August 28, 2021, at ~0224 UTC during Hurricane Ida, which is located near Western Cuba. Maximum sustained surface winds were 70 kt (36 m s^{-1}). The flight track of a NOAA Hurricane Hunter research flight is shown a *solid line* and dropsonde observations of near surface winds are shown as larger *dark blue* wind barbs. (*ASCAT image courtesy of NOAA/NESDIS Center for Satellite Applications and Research.*)

3.3 Wind profiles

Measuring the vertical structure of wind from space is more challenging than estimating surface winds. The earliest approach that was implemented in numerical analyses was the derivation of Atmospheric Motion Vectors (AMV) generated using visible, infrared, and water vapor satellite imagery to track the movement of cloud and water vapor features. The algorithms used to derive these wind proxies are routinely validated by numerical centers against radiosondes and other in situ and remote sensing observations (see Chapter 2: Balloon-Borne Radiosondes); however, this limits the validation to observations over land, where radiosonde observations are available. Over oceans, validation is more challenging due to the dearth of in situ or ground based remote sensing wind observations. In particular, near tropical disturbances with strong shear and anomalous flow, validation of satellite derived AMVs is challenging. Sears and Velden (2012) used dropsonde observations collected during the Pre-Depression Investigation of Cloud-Systems in the Tropics (PREDICT) in the tropical west Atlantic Ocean to characterize AMVs in tropical disturbances to better understand their error characteristics over the ocean. In general, the properties of AMVs over the ocean are comparable to those in global analyses. However, high cirrus and environments with strong vertical wind shear at times may lead to larger errors in the height assignment and degradation of their overall AMV quality.

Direct measurements of vertically resolved wind components from space were first achieved with the launch of the Atmospheric Laser Doppler Instrument (ALADIN) onboard the Aeolus satellite (Endemann et al., 2004) in late 2018. This instrument measures only the wind component along its horizontal line-of-sight, but still provides important input to NWP models.

Validation of space-based wind observations requires a detailed characterization of errors and a large enough statistics to be significant. Zhai et al. (2020) describe the ALADIN airborne demonstrator (A2D), which was developed at the German Aerospace Center (DLR) and deployed in several field campaigns prior and after the launch of Aeolus to validate the ALADIN measurement principle and wind retrievals. A series of dropsondes were launched during the North Atlantic Waveguide and Downstream Experiment (NAWDEX) and provided the reference to characterize response and errors of A2D, which in turn provide information about the expected performance of ALADIN. Witschas et al. (2020) compared early observations of winds by ALADIN with the DLR Doppler Wind Lidar (DWL). The errors of DWL had been characterized in a number of field campaigns prior to the launch of ALADIN using dropsondes and serve as reference for the validation of ALADIN by DWL. A similar approach to the validation of ALADIN was taken by Bedka et al. (2021). Dropsondes made by Yankee Environmental Systems, Inc. were used to validate and characterize the wind measurements of the Doppler Aerosol WiNd (DAWN) lidar onboard the NASA DC-8 research aircraft. These data showed generally good agreement between the airborne wind lidar and dropsonde winds, but some significant differences with the spaced based observations, which were attributed to early issues in the retrieval of winds by ALADIN. The water vapor observations of this Yankee Environmental Systems dropsondes did not have adequate sensitivity to detect vertical water vapor gradients.

3.4 Clouds and precipitation

Dropsondes are frequently used to characterize the atmospheric state in support of satellite observations, even if satellite measurements are not directly validated. The Global Precipitation Measurement (GPM) satellite, which was launched in 2014, extends the record of the Tropical Rainfall Measuring Mission (TRMM) satellite and expands the observations to higher latitudes not previously covered by TRMM. The Olympic Mountains Experiment (OLYMPEX) studied how accurately the instruments onboard GPM were able to measure different forms of precipitation over mountain regions (Houze et al., 2017). Numerous instruments were deployed on the Olympic Peninsula to observe frontal systems, while three different aircraft characterized the systems offshore and over land. Dropsondes measured profiles of temperature, humidity, and winds to best characterize these offshore weather systems and their associated moisture transport in NWP models.

Satellites are ideally suited to provide global cloud observations; however, in situ validation of these remotely sensed observations remains challenging. Dropsondes measure a complete profile of relative humidity from the aircraft flight level to the surface, from which an integrated water vapor (IWV) column can be calculated. Airborne microwave radiometers can simultaneously retrieve IWV, the Liquid Water Path (LWP), and the Rain Water Path (RWP) using the same wavelength bands (Jacob et al., 2019). IWV derived from dropsonde observations can be used to validate microwave radiometer-derived IWV to increase the confidence in retrievals of LWP and RWP, even though these latter two parameters are not directly measured by dropsondes. The fine spatial structure of clouds requires that the sonde drift and the cloud structure be considered in the validation.

4. Summary

Dropsondes provide in situ profiles of temperature, water vapor, and winds over the oceans and occasionally over land. Their main use is to provide targeted observations for NWP, in particular for tropical cyclones and other severe weather events. In addition, dropsonde measurements provide essential observations, for the direct and indirect validation of satellite observations. Dropsondes can be released almost anywhere over open oceans and can therefore be coordinated in time and space with satellite measurements. Since they are typically released from aircraft, dropsonde observations are also typically coincident with measurements from all other instruments on that aircraft platform. The high vertical resolution and high quality of pressure, temperature, relative humidity, and wind measurements of AVAPS dropsondes make them highly suitable for the direct validation of these observations by satellites.

Satellite validation over oceans has the advantage that the surface is radiatively well characterized, which allows for a better separation of the effects of surface versus atmospheric emissions on the satellite retrieval.

Dropsonde observations also provide important input data for NWP analyses, in particular over the oceans, where they may be used to anchor NWP fields. Using well-constrained NWP analyses significantly increases the amount of satellite observations that can be used in their validation. In addition, dropsonde observations are used to calibrate and characterize other aircraft remote sensors, which in turn are used to validate satellite observations and characterize their error characteristics. Traceability of the airborne remote sensors to an in situ reference is an essential element in building confidence in satellite observations.

Dropsondes provide critical observations that help improve model forecasts, support calibration and validation of satellite and airborne remote sensing data, and promote advancements in basic science understanding.

Acknowledgments

We would like to acknowledge Herman Smit and Ken Pryor (NOAA/NESDIS/STAR) for reviewing the draft and providing constructive feedback.

References

Bedka, K.M., Nehrir, A.R., Kavaya, M., Barton-Grimley, R., Beaubien, M., Carroll, B., Collins, J., Cooney, J., Emmitt, G.D., Greco, S., Kooi, S., Lee, T., Liu, Z., Rodier, S., Skofronick-Jackson, G., 2021. Airborne lidar observations of wind, water vapor, and aerosol profiles during the NASA Aeolus Cal/Val test flight campaign. Atmos. Meas. Tech. 14, 4305–4334. https://doi.org/10.5194/amt-14-4305-2021.

Boylan, P., Wang, J., Cohn, S.A., Fetzer, E., Maddy, E.S., Wong, S., 2015. Validation of AIRS version 6 temperature profiles and surface-based inversions over Antarctica using Concordiasi dropsonde data. J. Geophys. Res. 120, 992–1007. https://doi.org/10.1002/2014JD022551.

Boylan, P., Wang, J., Cohn, S.A., Hultberg, T., August, T., 2016. Identification and intercomparison of surface-based inversions over Antarctica from IASI, ERA-Interim, and Concordiasi dropsonde data. J. Geophys. Res. 121, 9089–9104. https://doi.org/10.1002/2015JD024724.

Chen, S., 2007. The impact of assimilating SSM/I and QuikSCAT satellite winds on Hurricane Isidore simulations. Mon. Weather Rev. 135 (2), 549–566. https://doi.org/10.1175/MWR3283.1.

Chou, K.-H., Wu, C.-C., Lin, P.-H., Majumdar, S., 2010. Validation of QuikSCAT wind vectors by dropwindsonde data from Dropwindsonde Observations for Typhoon Surveillance Near the Taiwan Region (DOTSTAR). J. Geophys. Res. 115, D02109. https://doi.org/10.1029/2009JD012131.

Cohn, S.A., et al., 2013. DRIFTSONDES providing in situ long-duration dropsonde observations over remote regions. Bull. Am. Meteorol. Soc. 94 (11), 1661–1674. https://doi.org/10.1175/bams-d-12-00075.1.

Donnelly, W.J., Carswell, J.R., McIntosh, R.E., Chang, P.S., Wilkerson, J., Marks, F., Black, P.G., 1999. Revised ocean backscatter models at C and Ku band under high-wind conditions. J. Geophys. Res. 104 (C5), 11485–11497. https://doi.org/10.1029/1998JC900030.

Endemann, M., Dubock, P., Ingmann, P., Wimmer, R., Morançais, D., Demuth, D., 2004. The ADM–AEOLUS mission—the first wind lidar in space. In: Pappalardo, G., Amodeo, A. (Eds.), Reviewed and Revised Papers Presented at the 22nd International Laser Radar Conference (ILRC 2004). vol. 2. ESTEC, pp. 953–956. ESA SP-561, 2004.

Forsythe, G.E., 1946. War-time developments in aircraft weather reconnaissance. Bull. Am. Meteorol. Soc. 27 (4), 160–163. https://doi.org/10.1175/1520-0477-27.4.160.

Goodberlet, M.A., Swift, C.T., Wilkerson, J.C., 1990. Ocean surface wind speed measurements of the special sensor microwave/imager (SSM/I). IEEE Trans. Geosci. Remote Sens. 28, 823–828. https://doi.org/10.1109/36.58969.

Hock, T.F., Franklin, J.L., 1999. The NCAR GPS Dropwindsonde. Bull. Am. Meteorol. Soc. 80 (3), 407–420. https://doi.org/10.1175/1520-0477(1999)080<0407:TNGD>2.0.CO;2.

Houze, R.A., McMurdie, L.A., Petersen, W.A., Schwaller, M.R., Baccus, W., Lundquist, J.D., Chandrasekar, V., 2017. The Olympic Mountains Experiment (OLYMPEX). Bull. Am. Meteorol. Soc. 98 (10), 2167–2188. https://doi.org/10.1175/BAMSD-16-0182.1.

Jacob, M., Ament, F., Gutleben, M., Konow, H., Mech, M., Wirth, M., Crewell, S., 2019. Investigating the liquid water path over the tropical Atlantic with synergistic airborne measurements. Atmos. Meas. Tech. 12, 3237–3254. https://doi.org/10.5194/amt-12-3237-2019.

Kim, T.M., Lee, S.J., Ahn, M.H., et al., 2020. Evaluation of atmospheric profile retrieval algorithm for GK2A satellite with dropsonde observations. Asia-Pacific J. Atmos. Sci. 56, 225–233. https://doi.org/10.1007/s13143-019-00154-5.

Larar, A.M., Smith, W.L., Zhou, D.K., Liu, X., Revercomb, H., Taylor, J.P., Newman, S.M., Schlüssel, P., 2010. IASI spectral radiance validation intercomparisons: case study assessment from the JAIVEx field campaign. Atmos. Chem. Phys. 10, 411–430. https://doi.org/10.5194/acp-10-411-2010.

Laughland, J., Grogan, P.T., 2020. Analyzing the Calibration and Validation Support Architecture for CYGNSS as a Design Problem, AIAA 2020-4057, ASCEND 2020, November 2020., https://doi.org/10.2514/6.2020-4057.

Martin, C., Suhr, I., 2021. NCAR/EOL Atmospheric Sounding Processing ENvironment (ASPEN) Software. Version 3.4.5. Available from: https://www.eol.ucar.edu/content/aspen. (Accessed 9 March 2021).

Pu, Z., Zhang, L., 2010. Validation of Atmospheric Infrared Sounder temperature and moisture profiles over tropical oceans and their impact on numerical simulations of tropical cyclones. J. Geophys. Res. 115, D24114. https://doi.org/10.1029/2010JD014258.

Ralph, F.M., Neiman, P.J., Wick, G.A., 2004. Satellite and CALJET aircraft observations of atmospheric rivers over the eastern North Pacific Ocean during the winter of 1997/98. Mon. Weather Rev. 132 (7), 1721–1745. https://doi.org/10.1175/1520-0493(2004)132<1721:SACAOO>2.0.CO;2.

Ralph, F.M., Neiman, P.J., Rotunno, R., 2005. Dropsonde observations in low-level jets over the Northeastern Pacific Ocean from CALJET-1998 and PACJET-2001: mean vertical-profile and atmospheric-river characteristics. Mon. Weather Rev. 133 (4), 889–910. https://doi.org/10.1175/MWR2896.1.

Rappaport, E.N., Franklin, J.L., Avila, L.A., Baig, S.R., Beven II, J.L., Blake, E.S., Burr, C.A., Jiing, J., Juckins, C.A., Knabb, R.D., Landsea, C.W., Mainelli, M., Mayfield, M., McAdie, C.J., Pasch, R.J., Sisko, C., Stewart, S.R., Tribble, A.N., 2009. Advances and challenges at the National Hurricane Center. Weather Forecast. 24 (2), 395–419. https://doi.org/10.1175/2008WAF2222128.1.

Reale, T., Sun, B., Tilley, F.H., Pettey, M., 2012. The NOAA Products Validation System (NPROVS). J. Atmos. Ocean. Technol. 29, 629–645. https://doi.org/10.1175/JTECH-D-11-00072.1.

Ruf, C.S., Atlas, R., Chang, P.S., Clarizia, M.P., Garrison, J.L., Gleason, S., Katzberg, S.J., Jelenak, Z., Johnson, J.T., Majumdar, S.J., O'Brien, A., Posselt, D.J., Ridley, A.J., Rose, R.J., Zavorotny, V.U., 2016. New ocean winds satellite mission to probe hurricanes and tropical convection. Bull. Am. Meteorol. Soc. 97 (3), 385–395. https://doi.org/10.1175/BAMS-D-14-00218.1.

Ruf, C.S., Gleason, S., McKague, D.S., 2019. Assessment of CYGNSS wind speed retrieval uncertainty. IEEE J. Sel. Top. Appl. Earth Obs. Remote Sens. 12 (1), 87–97. https://doi.org/10.1109/JSTARS.2018.2825948.

Sears, J., Velden, C.S., 2012. Validation of satellite-derived atmospheric motion vectors and analyses around tropical disturbances. J. Appl. Meteorol. Clim. 51 (10), 1823–1834. https://doi.org/10.1175/JAMC-D-12-024.1.

Sun, B., Reale, A., Seidel, D.J., Hunt, D.C., 2010. Comparing radiosonde and COSMIC atmospheric profile data to quantify differences among radiosonde types and the effects of imperfect collocation on comparison statistics. J. Geophys. Res. 115, D23104. https://doi.org/10.1029/2010JD014457.

Taylor, J.P., Smith, W.L., Cuomo, V., Larar, A.M., Zhou, D.K., Serio, C., Maestri, T., Rizzi, R., Newman, S., Antonelli, P., Mango, S., Di Girolamo, P., Esposito, F., Grieco, G., Summa, D., Restieri, R., Masiello, G., Romano, F., Pappalardo, G., Pavese, G., Mona, L., Amodeo, A., Pisani, G., 2008. EAQUATE: an international Experiment for hyperspectral atmospheric sounding validation. Bull. Am. Meteorol. Soc. 89 (2), 203–218. https://doi.org/10.1175/BAMS-89-2-203.

Uhlhorn, E.W., Black, P.G., Franklin, J.L., Goodberlet, M., Carswell, J., Goldstein, A.S., 2007. Hurricane surface wind measurements from an operational stepped frequency microwave radiometer. Mon. Weather Rev. 135 (9), 3070–3085. https://doi.org/10.1175/MWR3454.1.

Vaisala, 2020. Dropsonde RD41 Data Sheet. https://www.vaisala.com/sites/default/files/documents/RD41-Datasheet-B211706EN.pdf.

Vömel, H., Goodstein, M., Tudor, L., Witte, J., Fuchs-Stone, Ž., Sentić, S., Raymond, D., Martinez-Claros, J., Juračić, A., Maithel, V., Whitaker, J.W., 2021. High-resolution in situ observations of atmospheric thermodynamics using dropsondes during the Organization of Tropical East Pacific Convection (OTREC) field campaign. Earth Syst. Sci. Data 13, 1107–1117. https://doi.org/10.5194/essd-13-1107-2021.

Wang, J., Hock, T., Cohn, S.A., Martin, C., Potts, N., Reale, T., Sun, B., Tilley, F., 2013. Unprecedented upper-air dropsonde observations over Antarctica from the 2010 Concordiasi Experiment: validation of satellite-retrieved temperature profiles. Geophys. Res. Lett. 40, 1231–1236. https://doi.org/10.1002/grl.50246.

Wick, G., Hock, T.F., Neiman, P.J., Vömel, H., Black, M.L., Spackman, J.R., 2018. The NCAR/NOAA global hawk dropsonde system. J. Atmos. Oceanic Tech. 35, 1585–1604. https://doi.org/10.1175/JTECH-D-17-0225.1.

Witschas, B., Lemmerz, C., Geiß, A., Lux, O., Marksteiner, U., Rahm, S., Reitebuch, O., Weiler, F., 2020. First validation of Aeolus wind observations by airborne Doppler wind lidar measurements. Atmos. Meas. Tech. 13, 2381–2396. https://doi.org/10.5194/amt-13-2381-2020.

WMO, 2019. Manual on Codes—International Codes, Volume I.1, Annex II to the WMO Technical Regulations: part A—Alphanumeric Codes, WMO-No. 306, 2011 Edition updated in 2019. World Meteorological Organization, Geneva. https://library.wmo.int/index.php?lvl=notice_display&id=13617.

Wu, L., 2009. Comparison of atmospheric infrared sounder temperature and relative humidity profiles with NASA African Monsoon Multidisciplinary Analyses (NAMMA) dropsonde observations. J. Geophys. Res. 114, D19205. https://doi.org/10.1029/2009JD012083.

Zhai, X., Marksteiner, U., Weiler, F., Lemmerz, C., Lux, O., Witschas, B., Reitebuch, O., 2020. Rayleigh wind retrieval for the ALADIN airborne demonstrator of the Aeolus mission using simulated response calibration. Atmos. Meas. Tech. 13, 445–465. https://doi.org/10.5194/amt-13-445-2020.

Zhou, D.K., Smith Sr., W.L., Liu, X., Larar, A.M., Mango, S.A., Huang, H., 2007. Physically retrieving cloud and thermodynamic parameters from ultraspectral IR measurements. J. Atmos. Sci. 64 (3), 969–982. https://doi.org/10.1175/JAS3877.1.

Zhou, D.K., Smith, W.L., Larar, A.M., Liu, X., Taylor, J.P., Schlüssel, P., Strow, L.L., Mango, S.A., 2009. All weather IASI single field-of-view retrievals: case study—validation with JAIVEx data. Atmos. Chem. Phys. 9, 2241–2255. https://doi.org/10.5194/acp-9-2241-2009.

Zhou, D.K., Liu, X., Larar, A.M., Tian, J., Smith, W.L., Kizer, S.H., Wu, W., Liu, Q., Goldberg, M.D., 2016. First Suomi NPP Cal/Val campaign: intercomparison of satellite and aircraft sounding retrievals. IEEE J. Sel. Top. Appl. Earth Obs. Remote Sens. 9, 4037–4046. https://doi.org/10.1109/JSTARS.2016.2516765.

Chapter 11

Ship-based cal/val campaigns

Nicholas R. Nalli[a,b], Gregory R. Foltz[c], Jonathan Gero[d], Laura Gibson[b,e], Robert O. Knuteson[d], Rick Lumpkin[c], Peter J. Minnett[f], Vernon R. Morris[g], Michael Ondrusek[b], Renellys C. Perez[c], Menghua Wang[b], and Jianwei Wei[b,h]

[a]I.M. Systems Group, Inc., College Park, MD, United States, [b]NOAA/NESDIS Center for Satellite Applications and Research (STAR), College Park, MD, United States, [c]NOAA/Atlantic Oceanographic and Meteorological Laboratory, Miami, FL, United States, [d]CIMSS, University of Wisconsin-Madison, Madison, WI, United States, [e]NOAA/OMAO, Silver Spring, MD, United States, [f]Department of Ocean Sciences, Rosenstiel School of Marine and Atmospheric Science, University of Miami, Miami, FL, United States, [g]School of Mathematical and Natural Sciences, Arizona State University, Phoenix, AZ, United States, [h]Global Science and Technology, Inc., Greenbelt, MD, United States

Chapter outline

1. Introduction		195	5. Collaborative remote sensing campaigns	205
2. Research vessel/ship platforms		196	5.1 PNE/AEROSE	205
3. At-sea instrument and sensor deployments		198	5.2 CalWater/ACAPEX	207
3.1 Atmospheric deployments		198	5.3 SHEBA/MOSAiC	208
3.2 Oceanographic deployments		201	5.4 EUREC^{4}A/ATOMIC	209
4. Early infrared remote sensing campaigns		203	6. Ocean color campaigns	210
4.1 Ocean temperature interferometric survey (OTIS)		204	7. Summary	214
			Acknowledgments	215
4.2 Combined sensor program		205	References	215

I often say, 'rock the boat'. But before you rock it, do three things: Measure the freeboard on the boat, notice the state of the sea, and the distance to the shore. Only then rock the boat.

Prof. Verner E. Suomi

1. Introduction

The world's oceans asymmetrically cover approximately 71% of the Earth's total surface area. The Southern Hemisphere (SH) in particular is dominated by water, especially the latitude zone between 40°S and 65.5°S, which is almost exclusively ocean (namely the Southern Ocean), as shown in Fig. 1. Furthermore, the high latent heat of water coupled with their sheer volume (multiplying the surface area by their average depth, ≈3.7 km) translates to an extremely high bulk thermal inertia. For these reasons, the ability to acquire global observations of the world's oceans unquestionably constitutes one of the greatest benefits of satellite remote sensing. Data from operational environmental satellites are known, for example, to make their biggest impact on numerical weather prediction (NWP), as demonstrated by consistently greater impact over the SH (e.g., Le Marshall, 2006; Noh et al., 2020).

From the earlier considerations, in situ field measurements over the open ocean are critical for remote sensing applications generally speaking. These include calibration/validation (cal/val) (e.g., Hagan and Minnett, 2003; Nalli et al., 2006) and other applications of passive sensors sensitive to the thermal-infrared (IR), microwave (MW), and reflected-solar spectral regimes. IR/MW applications particularly include hyperspectral sounder systems (cf. Chapter 18) designed for atmospheric profiling along with narrowband imager systems designed for sensing clouds and the surface. Because the sea surface skin temperature, spectral emissivity, and bidirectional reflectance distribution function are more uniform and better understood (e.g., Nalli et al., 2006, 2008b; Smith et al., 1996) than land surfaces, marine-based measurements facilitate greater control over hyperspectral sounder cal/val closure experiments. Because of the small signals associated

FIG. 1 Earth fractional surface areas as a function of latitude zone: (*dotted line*) ocean and (*solid line*) total.

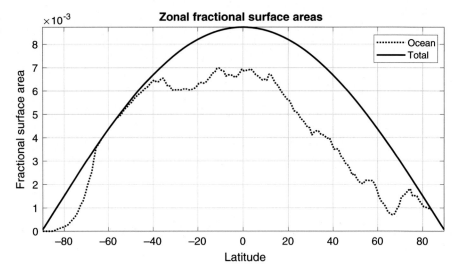

with reflected solar-spectrum radiances, vicarious calibration schemes (cf. Chapters 1, 5, and 15) are necessitated based on surface optical in situ measurements.

Although island sites and semipermanent platforms such as buoys allow routine periodic field measurements of surface parameters (cf. Chapter 5), buoy measurements are limited in observing capability, whereas island sites are location limited and can suffer from the influence of the land surface on the ambient atmosphere and surrounding ocean. Thus, comprehensive measurements in remote ocean basins require ship platforms known as research vessels (RVs) or oceanographic vessels, designed to provide the necessary deck, laboratory, computer, and mechanical resources for specific science objectives. Among other things, RV platforms can typically allow for the acquisition of vertical profiles of both oceanic and atmospheric parameters in remote locations. These are discussed in somewhat more detail in the forthcoming sections.

2. Research vessel/ship platforms

Ship-based intensive field campaigns (also variously referred to as *oceanographic research cruises* or *expeditions*) typically need to account for specific goals, materials, and platform needs well in advance, depending on the distance and duration. Advance planning and economization of equipment becomes increasingly more important the farther from shore a vessel is required to deploy. To these ends, RVs vary to a great degree in both design and function. Fig. 2 shows a three-dimensional (3D) schematic of a typical Global Class RV, modeled after the US NOAA Ship *Ronald H. Brown* (in service since 1997), with relevant nautical and scientific components annotated. This specific type of ship is designed for sustained voyages spanning well over 1 month and crossing entire oceans, and because these inevitably venture into oceanographic regions lacking phytoplankton (e.g., algae) or turbidity from coastal runoff, such vessels are also referred to as *blue water*.

Modern RVs are designed to facilitate scientific research, offering dedicated laboratories that are removed from the elements with various accessibilities and functions, as well as ample deck space for equipment and with access to the ocean surface, often including the stern or fantail (Fig. 2A). Other vessels (e.g., private ships and yachts) equipped with scientific instruments may also be exploited, although they are not explicitly designed for the science objectives of a field campaign.

Multiple considerations must be taken into account to justify the time and expense of RV operations, including especially the critical aspects of the scientific mission and requirements needed to meet those goals. Deck space and configuration is a critical consideration during project planning. Scientists must always prepare for setbacks, which may include the loss of days at sea due to weather, mechanical problems with the ship, personal emergency, etc. Generally speaking, no space aboard a boat or ship is free from the 3D axis motion (relative to the center of mass) caused by wave action, these being *roll*, *pitch*, and *yaw*, defined, respectively, as rotation about the long-horizontal, cross-horizontal, and vertical axes (not to mention *heave*, which involves linear vertical motion of the center of mass itself). Certain maritime regions may have additional restrictions or require permission for the collection of certain data during transit, for example, exclusive economic zones (EEZs). RVs and ships have specific guidelines for transporting chemicals and using/disposing of hazardous materials.

Onboard RVs, controlled-climate dry labs provide space for science workstations, and these are often outfitted with communication stations and processing computers, but many science parties also bring their own project-funded portable computers and software programmed to ingest and process data from field and lab instruments. Chemistry and/or biolab equipment will include fume hoods and safe cabinets for chemical and equipment stowage. Most RVs include their own permanent in situ

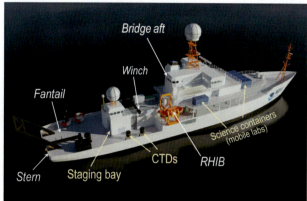

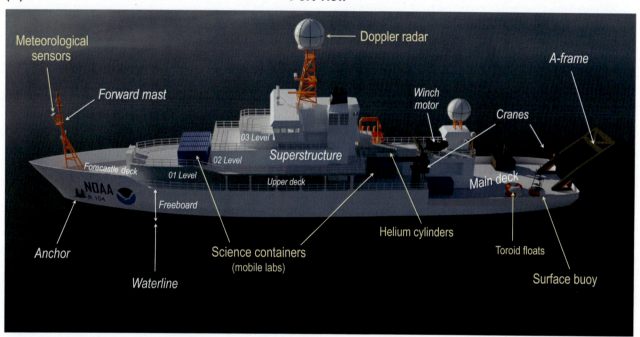

FIG. 2 Schematic of a global-class, blue-water RV (modeled after the NOAA Ship *Ronald H. Brown*): (A) starboard aft perspective view, (B) port forward perspective view, and (C) port broadside view. Nautical terms are labeled in *white italic font* and scientific terms labeled in *yellow font*. The RHIB (pronounced "rib") stands for "rigid-hull inflatable boat," which is used for ferrying passengers and for at-sea operations (e.g., buoy servicing or instrument special deployments). CTDs are oceanographic "conductivity-temperature-depth" instruments deployed on a rosette (a round metal frame that holds the CTD equipment during underwater deployment), deep beneath the surface via a cable and winch (Section 3.2.2). *(Three-dimensional (3D) ship rendering courtesy of Bertrand Dano (UM/CIMAS).)*

measurement systems for observation of both oceanographic and atmospheric parameters. These include flow-through water systems that continuously sample the near-surface water while in-transit or on station. Most RVs and ships of the World Meteorological Organization (WMO) Voluntary Observing Ships' (VOS) fleet also have a foremast, where sensors are mounted at a standard height (i.e., 10 m) for measuring near-surface meteorological variables (Fairall et al., 1997; Smith et al., 2018, 2019).

Cranes (Fig. 2C) are often needed for staging/destaging while docked in port or sometimes for deployment/recovery at sea. Oceanographic or hydrographic *winches* (Fig. 2A and C) are required for deploying oceanographic instrumentation during field campaigns (e.g., Carral et al., 2017; Talley et al., 2011). These may simply provide muscle to hoist or tow equipment, or they may also include advanced "smart" features, such as heave compensation (e.g., Carral et al., 2017). A winch with a conducting cable can communicate with an instrument, feed real-time video of a remotely operated vehicle (ROV) to the science party, or characterize the water column using conductivity-temperature-depth (CTD) sensors and associated electronics (Section 3.2.2). If a project has a need to sample water at depth for a specific parameter (salinity and temperature), a conducting winch can allow this capability.

Data acquired over the course of a cruise are typically archived as a function of Coordinated Universal Time (UTC), which is especially important when the vessel transits over multiple time zones. At any given time, an RV may have navigational, bathymetric, meteorological, and oceanographic data available to incorporate with specific "event" data, for example, a CTD cast or similar in situ deployment. Event-coincident data logging and data quality assurance are typical of modern RVs. For all events occurring during the campaign, frequent communication between the science party, crew, and vessel operators on the bridge is required. All hands must be aware when science and/or ship operations are underway for productivity as well as safety. Scientists may work with the RV's survey crew to collect calibration details and hardware requirements for onboard equipment to be incorporated into their research data stream. At the conclusion of a campaign, a cruise report is prepared by the Chief Scientist in collaboration with the participants; information in the cruise report can be very helpful for later interpretation of the data collected, as well as facilitating the next scheduled shipboard project.

RVs and ships are home to their crews, who must share their space with scientists as temporary onboard guests. The primary objective of the crew is the safe return to shore of all souls aboard. Scientists must thus allot time for manoverboard, fire, and abandon ship drills. An RV's specialized crew (survey technicians, deck crew, etc.) will have different levels of expertise depending on the vessel. The ship's bridge team, survey department, and deck crew coordinate navigating station locations and waypoints in conjunction with ship's speed as well as station operations; the engineering and steward departments ensure that the ship is functional, safe, and comfortable.

Although this chapter is focused on field campaign measurements from RVs, we also note here that useful data can also be obtained from instruments or systems on opportunistic commercial vessels, ships, and private yachts, especially if the equipment runs autonomously or requires only minor involvement of a crew member, such as periodic cleaning of optics or launching of expendable bathythermographs (XBTs), expendable CTDs (XCTDs), and/or drifting buoys. Such equipment has been mounted on cruise ships, freighters with transocean tracks, and ferries that spend significant periods away from the coast. When mounting equipment on cruise ships or ferries, it is usually necessary to install them where passengers do not have access. The following section will overview protocols for various shipboard operations with focus on satellite-related objectives. More specific campaign sections are found later in the chapter.

3. At-sea instrument and sensor deployments

3.1 Atmospheric deployments

Knowledge of the state of the atmosphere is important for understanding the performance of remote sensing algorithms to retrieve atmospheric and surface variables from measurements taken from satellites (Kilpatrick et al., 2015; Merchant and Leborgne, 2004; Merchant et al., 2009; Minnett, 1986; Minnett et al., 2019; Nalli et al., 2006, 2013b; Szczodrak et al., 2014).

3.1.1 Radiosonde and ozonesonde launches

A standard method of measuring the thermodynamic state of the atmosphere is by using *radiosondes*, sometimes simply called *sondes* for short and informally referred to as "weather balloons" by nonspecialists. Radiosondes are small, singleuse, lightweight instrument packages with sensors to measure pressure, temperature, and relative humidity, and telemetry electronics to transmit the measurements digitally to a receiver (cf. Chapter 2) on the ship. Extra sensors can be included to measure additional parameters, such as ozone and ionizing radiation; Global Positioning System (GPS) receivers can be included to measure the displacement of the radiosonde by winds. Most radiosonde launches are from weather stations on land, mainly in the Northern Hemisphere, with some from islands (Durre et al., 2018). This means sondes launched from ships over the open ocean are especially valuable for satellite remote sensing applications (e.g., Nalli et al., 2006, 2011), given that islands are sufficiently large to modify the marine atmosphere and have completely different surface radiative properties; similarly, measurements of radiosondes launched from coastal stations may not reflect marine conditions (Minnett and Key, 2007), and will be correlated with soundings taken inland.

A radiosonde system comprises the radiosonde itself, which is launched using a large latex balloon, along with an antenna and receiver, a laptop computer to log the data, and a device for calibrating the thermometer and hygrometer (cf. Chapter 2). Any offset in the pressure sensor is determined by comparison with the measurement of the ship's barometer, which is usually accurate especially if the ship belongs to the WMO VOS fleet (Smith et al., 2019). The data from the radiosonde are transmitted until the pressure measurement increases for a short period, indicating the balloon has burst and the balloon and sonde are falling. For the purposes of using the data in conjunction with measurements from a satellite, it is useful to coordinate the launch with the satellite overpass; in satellite cal/val parlance, such launches are typically referred to as *dedicated radiosondes* (e.g., Nalli et al., 2013b; Tobin et al., 2006).

Similar to standard radiosondes, *ozonesondes* are balloon-borne sensor packages consisting of a Teflon air pump and a rudimentary electrochemical cell that are interfaced with the radiosonde package (cf. Chapter 4). The small piston pumps bubbles of ambient air into the sensor cell usually containing a 1% potassium iodide solution. The reaction of ozone and iodide generates an electrical signal proportional to the amount of ozone. Ozonesondes launched for satellite overpasses are likewise referred to as *dedicated ozonesondes* (e.g., Nalli et al., 2018b).

Radiosondes are usually launched from an aft or upper deck (Fig. 2A and C), where there is some shelter from the wind when the ship is underway. Fig. 3 shows a dedicated radiosonde launch from an aft deck, in this case from the center of the fantail during fair weather conditions with low relative winds in the lee of the superstructure. At sea, helium has been typically used to inflate the balloons, as opposed to hydrogen, which is inflammable. However, helium cylinders are still considered hazardous because they are pressurized, and in recent years worldwide helium shortages have led to a reconsideration of hydrogen usage. The pressurized gas cylinders used for balloon inflation are usually secured somewhere outside on deck (Fig. 2C), with extra care obviously required for hydrogen; many ships have relocatable cylinder racks that can be fixed to a railing close to where the launches will take place. Given their exposed position, it is desirable to protect the reduction valve from the corrosive action of sea spray with a plastic bag or similar. The telemetry antenna is usually mounted aft and high (e.g., a railing of an aft upper deck), so there is a clear line of sight to the radiosonde as it ascends through the atmosphere. If the ship has a helicopter hangar, this is an ideal place to install the radiosonde receiver and other equipment. Details on other practical considerations and procedures associated with successful radiosonde and ozonesonde launches at sea may be found in Boxes 1 and 2, respectively.

3.1.2 Shipboard radiometers

There is well-founded recognition that skin sea surface temperature (SST_{skin} or T_s) derived from measurements of well-calibrated IR radiometers (Donlon et al., 2014b) (also see Chapters 6 and 21) mounted on ships are a valuable resource for

FIG. 3 Launch of a satellite-dedicated radiosonde from the fantail of the NOAA Ship *Ronald H. Brown* while underway, timed approximately 15–30 min prior to a forthcoming 13:30 SNPP LEO satellite overpass. Pictured is Howard University student Kafayat Olayinka with the radiosonde package (left hand) and a 200 g helium-filled balloon during a joint NOAA PIRATA Northeast Extension (PNE), Aerosols and Ocean Science Expedition (AEROSE) campaign (Section 5.1).

> **Box 1 Radiosonde launch procedures and safe practices at sea.**
>
> As mentioned earlier, *radiosondes* are usually launched from an aft or upper deck. If the launch time coincides with the ship being on station, then the launch can take place from an area in the lee of the superstructure. If there is sufficient space, inflating the balloon in the shelter of a hangar or *staging bay* is desirable (Fig. 2A). In the event of high relative winds, a reduction in ship speed and/or change in heading is sometimes necessary.
>
> Whenever possible, two people should be involved in the launch of a standard radiosonde from a ship, although single-handed launches are sometimes unavoidable with smaller science parties. Two-person teams facilitate the balloon operations on deck, but they also implicitly allow for *buddy-system* safe practices, especially at night. Most radiosonde manufacturers provide guidance on the steps needed to ensure a successful launch. Since a successful ascent takes approximately 1.5–2 h, preparations may commence approximately an hour or more before the overpass as this gives time to repeat the preparations if a problem is encountered, or if the balloon bursts prematurely.
>
> The first task in the preparations is to contact the Bridge to alert them of the intention to launch and ask permission to proceed before beginning the preparations. The radiosonde should only be removed from its protective packaging when the preparations start. Some radiosondes use standard alkaline or lithium batteries, while others use a wet cell that is activated by soaking for a few minutes in distilled water. Once the battery is connected, the radiosonde begins transmitting and a signal should be detected by the antenna and the data appear on the receiver display. The sonde should be placed with its sensors in the calibration unit. Once calibration is completed, the balloon inflation should begin. A narrow hose connects the low-pressure outlet of the reduction valve to an inflation nozzle inserted into the neck of the balloon, which can be secured by a special clip or attached using electrical tape. A weight, selected to match the radiosonde, may be attached to the nozzle so that at the point at which the balloon begins to lift the weight, it is adequately inflated so the gas flow can be halted. The neck of the balloon should be firmly gripped and sealed using string, electrical tape, or zip ties, to prevent gas from escaping. The radiosonde is then attached to the balloon using a dereeler about which the cord to the radiosonde is wound.
>
> After checking that the receiver is showing the data stream and it is being logged by the computer, the package is carefully brought to the launch site, avoiding allowing the balloon to touch anything that might cause it to burst. After release, the cord connecting the radiosonde to the balloon will unwind to full length in the first minute or so to ensure the radiosonde sensors are likely to be out of the turbulent wake of the balloon. Ideal ascent rates vary between 5 and 8 m s^{-1}, but lower rates provide better vertical resolution. Upon successful launch, the bridge is notified. When the sonde data transmission terminates, the logging program stops (usually automatically), and the data may then be examined.

> **Box 2 Ozonesonde launches at sea.**
>
> Because *ozonesondes* are merely instrument packages that are interfaced with standard radiosondes, much of the procedure is similar to that described in Box 1. However, deployments of ozonesondes require considerably more time than do radiosondes due to the advance preparation times of the electrochemical cell, including day of flight, along with increased balloon inflation deployment time and logistics. The latter consideration is due to the fact that a much larger balloon is required to carry the payload into the stratosphere (e.g., 1200–1400 g balloons versus a 200-g balloon for normal ship-based radiosondes).
>
> Because the larger balloon can be considerably more unwieldy under high wind conditions, a team of four to six scientists is usually required for successful control and deployment of the balloon; handling of the balloon requires latex gloves to avoid damage from skin oils. The ozonesondes can be launched from either position (as described earlier) but it is strongly preferred to launch them from the lower aft deck given the larger space. It is estimated that a minimum of approximately half an hour will be required for each ozonesonde deployment. For both ozonesonde and radiosonde deployments, it is important that there is sufficient space and clearance for maneuvering a helium- or hydrogen-filled balloon and launching from the fantail.

comparison with satellite-derived SST$_{skin}$, facilitating comparisons of SST$_{skin}$ to SST$_{skin}$ (Hagan and Minnett, 2003; Minnett and Corlett, 2012; Smith et al., 1996).

The radiometers are often mounted above the bridge or on the railing of a forward deck so that the instrument is directed at the surface ahead of the ship's bow wave to measure emission from the water surface undisturbed by the passage of the ship (Donlon et al., 2014a). If possible, the radiometers should be mounted on the windward side of the ship when on station. A clear view of the sky is needed to provide measurements to correct the sky emission reflected at the sea surface. Placing the instrument in a very exposed position carries risk of being contaminated by sea spray in rough conditions; the salt water can damage the fore optics and compromise the accuracy of the measurements. Consequently, a rain sensor is usually used to put the scan mirror into a safe position when rain or sea spray is encountered (Donlon et al., 2014b). Most instruments, which run autonomously and continuously in a noninterference mode with other ship activities, have a GPS receiver to provide latitude, longitude, and time for each measurement. In addition to operation on RVs, many IR radiometers have been installed on commercial vessels for long-duration deployment, including cruise ships of Royal Caribbean International and Cunard Line, ferries including those between the United Kingdom and Spain, and Denmark and Greenland, and on freighters that transit between Japan and the United States, and Hawaii and California.

Multifrequency microwave radiometers that measure atmospheric properties are mounted in a similar fashion to IR radiometers, but they can be in more sheltered positions as they do not measure sea surface emission. Nevertheless, they need a clear view of the sky from zenith to horizon to facilitate tipping calibration ("tip-cal") measurements. A tip-cal involves a series of measurements taken at different zenith angles to provide a calibration (Li et al., 2014).

3.1.3 In situ atmospheric surface measurements

As mentioned in Section 2, RVs have a foremast, where sensors measure near-surface meteorological variables (Fairall et al., 1997; Smith et al., 2018, 2019). The sensors must be well exposed to reduce the effects of the ship, such as acceleration of the wind by flow distortion (Popinet et al., 2004) and elevated air temperatures from heat-island effects when the structure of the ship has been heated by the sun (Goerss and Duchon, 1980; Minnett et al., 2005). Other sensors typically included are those to measure atmospheric humidity, and hemispheric incident shortwave radiation (solar flux) and longwave radiation (thermal IR flux primarily from the atmosphere). The shortwave measurements are especially prone to inaccuracies caused by motion of the ship as these change the apparent solar zenith angle; mounting the sensors on gimbals helps, as does averaging rapid samples over periods that are much longer than the roll and pitch periods of the ship. Accurate barometers are also part of the meteorological equipment, but these do not need to be mounted in well-exposed positions. As with all instruments used in scientific research, the sensors should be well calibrated, so some groups install their own instruments over which they have control of the calibration and the data logging.

For research expeditions focused on studying air-sea turbulent fluxes of heat, moisture, and gases, additional instruments may be mounted at the bow of the ship, often extending forward several meters on a boom to improve their exposure; this renders them vulnerable to damage in bad weather, so it is best to have the booms retractable to protect the sensors.

Trace gas and airborne particulate (or aerosol) monitoring can be conducted with a variety of miniaturized (usually electrochemically based) sensors, optical sensors (e.g., spectrometers), direct injection separation techniques (e.g., gas chromatographs), or mass spectral techniques. While the size and configuration of these instruments can vary significantly, many of these devices require that sampling inlets be deployed to draw ambient air into the measurement devices. These air intake lines should generally be situated as high off an uppermost deck or forward mast (e.g., the "jackstaff") as possible to avoid contamination by ship exhaust, external venting, and human activity onboard. It is critical that sample intakes be situated as far upwind of engine or galley exhaust (or any other source of pollutants, including smoking) as possible and under no circumstances adjacent to potential contaminants for the sake of sample integrity. Samplers, passive emission instruments, and open path instrumentation are typically rail or foremast mounted in such a way as to avoid heavy foot traffic and exhaust from any underway activity.

3.2 Oceanographic deployments

3.2.1 Moored buoys

Moored buoy operations are commonly conducted aboard research vessels that have sufficient deck space and the necessary equipment, such as a large A-frame and one or more cranes (Fig. 2C). The easiest and safest method for buoy recovery involves lifting the buoy onto the aft deck with the A-frame and then using a combination of winches and capstans to reel in the buoy's line (Fig. 4). The procedure takes several hours and requires close communication and coordination between the chief boatswain (pronounced, and often spelled, "bosun"), deck crew, mooring technicians, and bridge. When it is being reeled in, the mooring line has considerable tension on it, so one mistake can have severe consequences. Therefore, safety precautions and procedures are taken very seriously.

After the buoy is located, usually through a combination of GPS and radar, the mooring recovery begins by sending an acoustic signal to the mooring's release, which attaches the buoy's anchor to its line. Upon release, the surface buoy and line (with all instruments attached) are free to move, and the recovery can begin. Typically, the ship maneuvers alongside the buoy and then remains nearly stationary relative to it. One end of the rope is secured on deck and the other end is fed through the winch on the A-frame at the stern of the ship and around to the side of the ship where the buoy is located. The rope is then

FIG. 4 Recovery of a moored PIRATA buoy during an NOAA PNE/AEROSE (Section 5.1) campaign. Shown in the photo are the buoy's surface float and tower. The meteorological sensors, which were attached to the tower when deployed, have been removed. A thick rope attached to the tower through a pulley on the A-frame keeps the surface float suspended until the line of oceanographic sensors below the float can be secured to the deck. Smaller lines attached to the tower and A-frame keep the buoy secure. Note that the surface float is covered with dust aerosols (from the Sahara) on the top and marine life on the bottom.

hooked onto the buoy's tower and the ship moves slowly forward until the buoy is positioned at the stern. The A-frame is used to lift the buoy onto the ship's deck, and the mooring technicians remove the fragile meteorological sensors from the tower. The buoy is removed from its line and the load is transferred through the A-frame winch, around a deck winch, around a capstan, and onto a large wooden reel. As the capstan rotates, the line is reeled in onto the spool. The reeling is periodically stopped whenever there is an instrument on the line. The mooring technicians remove the instrument immediately after it emerges from water and the process continues until all instruments, mooring line, and the acoustic release have been recovered (cf. Fig. 4). Different stages of the buoy recovery require the ship to travel forward at different speeds in order to keep the proper tension on the line. For this reason, it is very important to maintain communication between the bosun, deck crew, and bridge.

Buoy deployment is normally done using one of the ship's cranes (Fig. 2). The buoy's line, with instruments attached, is first laid on deck. The crane then lifts up the buoy and carefully places it in water alongside the ship. The ship moves forward slowly until the buoy is directly at the stern. Several members of the deck crew and scientific party carefully place the instruments into water as the ship moves forward and the buoy line feeds out. The ship continues to move forward, while the rest of the line is fed out through the A-frame. The acoustic release and anchor are then attached and the line is secured to the ship's aft deck below the A-frame. The ship then tows the buoy to the preselected deployment site and the anchor is dropped. The exact deployment site is selected based on the ocean depth and flatness of bathymetry. For a taut-line buoy, if water is too deep, the buoy will be partially submerged, and if it is too shallow the buoy will have excess slack, resulting in too much lateral motion (cf. Chapter 5). After the buoy has been successfully deployed, the mooring technicians make sure the instruments on the buoy are working and transmitting data. If a surface sensor is not working, the technicians may take a small, rigid-hull inflatable boat (RHIB; Fig. 2A) to the buoy to perform repairs.

3.2.2 CTD casts

Conductivity-temperature-depth (CTD) casts are performed to measure various ocean parameters and obtain water samples at different depths. Normally, several sensors, and up to 24 sample bottles, are attached to the metal CTD frame, known as a *rosette*. The sensors transmit their data electronically in real time through the winch cable and into the ship's computer room, where the CTD console operator is monitoring them. The winch operator controls the speed at which the wire is let out and hence the descent/ascent speed of the rosette package. It is important to maintain constant communication between the ship's survey technicians, who lead the deployment and recovery of the rosette package on deck, along with the winch operator and the CTD computer console operator (the latter normally being a member of the scientific party). The console operator tells the winch operator when to stop the rosette package on its ascent so that a bottle can be closed to obtain a water sample (typically on the order of 10 L of water per depth). During CTD deployment and recovery, the survey technicians are in communication with the winch operator, bridge, and console operator to ensure that operations go smoothly. Salinity profiles are calculated from the conductivity and temperature measurements and calibrated using salinity values derived from the water samples, using an autosalinometer. Other ocean properties can be obtained from the water samples such as dissolved oxygen, nutrients, carbon, helium, tritium, and pH. Although oceanographic depth parameters are of limited direct use for satellite remote sensing, such data are all the more critical for ocean models, and it is often the case that remote sensing scientists must work alongside oceanographers within collaborative campaigns (cf. Section 5.1).

3.2.3 Expendable ocean profilers

Expendable bathythermographs (XBTs) are relatively inexpensive probes that can be deployed, while the vessel is underway. XBTs measure temperature to a depth of approximately 800 m with 0.5 m increments. XBTs can be deployed using a hand launcher, or loaded in deployment packages for automatically timed deployments from cargo vessels. In addition, expendable CTDs (XCTDs) have been developed that also measure conductivity, from which salinity and density can be derived.

3.2.4 Drifting buoys

Global Drifter Program (GDP) drifters and Argo floats (cf., Section 2.1 in Chapter 5) have best practices for deployment that balance ease of deployment with minimum chances of a failure upon deployment. GDP drifters can be deployed by untrained ship crew following simple deployment instructions. They are packaged in plastic shrink wrap to protect them from spray; this plastic is removed before deployment, but the paper tape securing the tether and drogue should always be left on to avoid potential injury when deploying in windy conditions. The activation magnet, held to the surface buoy by water-soluble tape, should be left on the buoy. This is because the buoy can reach extremely high temperatures while still on the deck of a ship, and removing it (activating the drifter) before seawater has cooled the buoy can significantly shorten its lifetime. The drifter is then dropped into the ocean from the stern, at the lowest possible deck (preferably less than 10 m above sea level, including heave), when the ship is traveling less than 20 knots. Deployment information can then be relayed to the GDP. Because of their larger size, Argo floats are more difficult to deploy. They must be activated before deployment, and are then lowered to the ocean surface from the ship. The easiest type of Argo float to deploy is lowered in a deployment box, which quickly disintegrates upon contact with water, and are released once they are at the surface. Other Argo floats, especially new deep Argo and biogeochemical Argo floats, require a more delicate deployment strategy. It is valuable to collect a CTD cast at the same location for calibration and validation purposes when Argo floats are being deployed.

Although field campaigns enable the deployment of instrumentation and collection of data, there is a growing use of autonomous systems to measure the ocean and near-surface atmosphere. One advantage is that they normally do not require a large research vessel (RV) and crew for deployment. For example, ocean gliders are routinely deployed and recovered with small boats, and saildrones are deployed and recovered from dockside. However, often the measurement capabilities are more limited compared to those that are possible from a large research ship during an intensive campaign. For more detailed information, Whitt et al. (2020) provide a comprehensive review of autonomous ocean observations and the vehicles that are most commonly used currently.

4. Early infrared remote sensing campaigns

The interaction between the Earth's oceans and atmosphere is widely recognized as a critical component of long-term climate change (e.g., Bryan et al., 1975; Manabe et al., 1975). In particular, the sea surface temperature (SST) is a major forcing parameter needed for that explaining the interannual variability of temperature in global climate models (GCMs) (e.g., Hansen et al., 1997). The ability to measure SST from ocean buoys using in situ subsurface temperature sensors has a long history (e.g., Soreide et al., 2001). However, the effective use of satellite thermal IR observations for the mapping of SST over the world's oceans is relatively more recent with several innovations as remote sensing technologies and calibration methods have matured (Minnett et al., 2019). Observations of the ocean emitted infrared radiance were demonstrated in two early field campaigns leading to the development of the Marine-Atmospheric Emitted Radiance Interferometer (MAERI) (Minnett et al., 2001). This section describes the challenges and lessons learned from two such early proof of concept campaigns, namely the 1995 Ocean Temperature Interferometric Survey (OTIS) and the 1996 Combined Sensor Program (CSP).

4.1 Ocean temperature interferometric survey (OTIS)

The late 1980s to mid-1990s saw the successful testing and proof of concept for IR Fourier transform spectrometers (FTS) possessing high calibration accuracy (Revercomb et al., 1988) onboard future low Earth orbit (LEO) satellite missions. Experimental instruments included the high-resolution interferometer sounder (HIS) (Smith et al., 1987) flown during high-altitude experimental aircraft campaigns onboard the NASA ER-2 (cf. Chapter 9). Remote sensing applications primarily focused on the improved capability of these instruments for obtaining passive soundings of temperature and water vapor from space with high vertical resolution (e.g., Smith, 1991; Smith et al., 1979), but consideration was also given to the possibility of high-accuracy measurement of surface parameters, especially sea surface skin temperatures (skin SSTs) (Nalli and Smith, 1998).

However, the question of skin SST retrievals was not as straightforward to answer given the "ground truth problem" (cf. Chapter 1, Section 5), namely the difficulty in obtaining a suitable correlative reference measurement. Unlike atmospheric sounding profiles, which could reasonably rely on concurrent radiosonde observations (although we know these also have their own issues with respect to the "ground truth problem" as discussed in Chapter 1, Section 5.2), in the case of SST there were only buoy measurements (moored and drifting). As discussed earlier in Chapter 5, buoys provide point measurements of the *bulk surface layer* on the order of 0.1–1 m in depth. The problem is, the electromagnetic skin temperature measured by a passive thermal-IR sensor is on the order of 10 µm depth, and the two temperatures are known to vary systemically from one another (cf. Chapter 6, Section 1, Fig. 1). Although there were commercially produced narrowband window radiometers available at the time (cf. Section 3.1.2), these could not account for the variable spectral emissivity (and bidirectional reflectance) of the surface, and their calibration was not necessarily traceable. It was therefore desirable to deploy a well-calibrated IR spectrometer that could measure the spectral radiance from the sea surface, but such a task would involve numerous practical issues associated with the deployment of expensive and delicate instrumentation in a harsh marine environment over open oceans.

Although these scientific and practical considerations posed a challenge, investigators at the University of Wisconsin-Madison and University of Miami proposed an experimental proof-of-concept campaign to deploy at-sea an FTS instrument, namely the Atmospheric Emitted Radiance Interferometer (AERI), onboard an RV. AERI instruments are uplooking, ground-based thermal-IR spectrometers which operate at land-based Department of Energy (DOE) Atmospheric Radiation Measurement (ARM) sites to provide well-calibrated measurements of the downwelling atmosphere with better than 1% accuracy at cold sky temperatures (Knuteson et al., 2004). AERI achieves this accuracy using high-emissivity ($\epsilon_v > 0.995$) cavity blackbodies at two temperatures (ambient and 333 K). AERI instruments were originally designed to invert downwelling atmospheric IR emission spectra to retrieve routine temperature and water vapor profiles of the planetary boundary layer at micrometeorological timescales. An AERI instrument was subsequently reconfigured for measuring IR spectra from both the sea surface (at multiple angles) and atmosphere (at zenith) for a 5-day cruise onboard the RV *Pelican* in the Gulf of Mexico in January 1995 (Smith et al., 1996), informally referred to as "OTIS" (named after one of the principal investigators).

It was then necessary to devise an algorithm to separate the thermodynamic skin temperature from the spectrum of IR emissivity within the measured AERI radiance spectra. From this approach, the skin SST was subsequently estimated with an absolute accuracy within 0.1 K, and the first ever field measurements of surface emissivity were obtained as a function of view angle and sea state. These field-measured radiative properties of the sea surface were found to be in good theoretical consistency with concurrent in situ measurements of the ocean bulk temperature. With the view of ocean emissivity at several angles and for a range of wind speeds, the data from this campaign were used for initially improving models of the ocean emissivity (Wu and Smith, 1997) and for the high-accuracy testing of IR satellite remote sensing of skin SST (Nalli and Smith, 1998; Smith et al., 1996).

4.2 Combined sensor program

Leveraging off the success of the Gulf of Mexico cruise (Smith et al., 1996), a new prototype, marine-based instrument based on the OTIS AERI was designed at the University of Wisconsin-Madison's Space Science and Engineering Center (SSEC) for routine deployment at sea; this new instrument was called the Marine Atmospheric Emitted Radiance Interferometer (MAERI) (Minnett et al., 2001) (cf. Chapter 6). The prototype MAERI would have its first at-sea demonstration during the much larger CSP from March 14 to April 13, 1996 (Post et al., 1997). The CSP campaign was a collaborative effort involving a 30-day survey of the Central and Tropical Western Pacific on the NOAA Ship *Discoverer*, from American Samoa to Manus Island, Papua New Guinea (where the TWP ARM site was located), then back to Hawaii. The campaign included a combination of in situ and remote sensors to study relationships between atmospheric and oceanic variables that affect radiative balance in the climatologically important *tropical warm pool* of the western Pacific Ocean, which include the highest SSTs measured on Earth. The DOE ARM Program used the mission to validate similar measurements made at the ARM site on Manus Island and to investigate the effect (if any) of landmasses on the marine environment. More details on the CSP campaign can be found in Post et al. (1997).

During CSP, in situ SST measurements were made at 10 cm and 5 m depths (via a special surface float and thermistor along with the standard ship intake, respectively) for studying the relationships between the bulk layer SST and the radiometric skin SST. During days with low winds, the heating of the upper ocean caused by the absorption of sunlight can result in vertical temperature gradients, called the *diurnal thermocline*, that tend to be eroded by turbulence the following night. Importantly, however, both day and night heat is constantly lost to the atmosphere through conductive, evaporative, and radiative energy transfer. This results in a "cool skin temperature" of approximately 0.1–0.2 K magnitude relative to the bulk water temperature below it. The evaporative heat loss to the atmosphere is a function of wind speed (Friehe and Schmitt, 1976; Minnett et al., 2019). Since IR satellite measurements observe the skin temperature and not the bulk temperature, the CSP observations provided experimental data to study the diurnal differences seen between satellite skin SSTs and ocean buoy SSTs at 1–2 m depth. The detector in the MAERI prototype required cooling that was provided via a liquid-nitrogen dewar that required periodic refilling performed by a four-person team assigned to 6-h shifts to attend to this and other science tasks underway.

The ultimate success of the OTIS and CSP cruises had far reaching implications for passive IR remote sensing in terms of improvements in the accuracies of forward models that are now often taken for granted today. It is worth mentioning here that the breakthroughs in IR remote sensing science afforded through these early campaigns were possible only through a certain degree of risk tolerance along with a healthy measure of thinking outside the box. When originally conceived, the idea of placing a piece of sophisticated scientific hardware worth hundreds of thousands of US dollars in a harsh environment at-sea onboard a (sometimes violently) moving platform was considered extremely risky and questionable. However, since then, the ship-based MAERI instruments have become a fundamental mainstay in the verification and improvement of global satellite-derived SST (e.g., Hagan and Minnett, 2003), sea, snow, and ice surface emissivity (e.g., Nalli et al., 2008a), and ocean heat flux parameterizations.

5. Collaborative remote sensing campaigns

As inferred earlier, ship-based field campaigns can be expensive, time consuming (in terms of operations and overhead), and assume a certain degree of risk. The experience of the 1996 CSP campaign suggested that remote sensing science objectives can be attained as a component of a broader overall marine science mission not otherwise related to remote sensing. Subsequently, many if not most ship-based remote sensing campaigns have been conducted as joint expeditions. Benefits to this collaborative approach include distribution of cost and risk, engagement of the broader science community, and cross-disciplinary science opportunities involving the application of remotely sensed satellite data (cf. Chapter 14).

This section highlights a handful of collaborative campaigns that were multidisciplinary in scope. Some of these joint expeditions were extended to accommodate the launching of dedicated radiosondes and ozonesondes (i.e., sondes dedicated to coincide with twice-per-day LEO satellite overpasses) to provide independent correlative measurements (intentionally not assimilated into NWP models) for validating sounder profile retrievals that subsequently formed a significant component of the overall NESDIS/JPSS validation program.

5.1 PNE/AEROSE

The multiyear NOAA Prediction and Research Moored Array in the Tropical Atlantic (PIRATA) (Bourlès et al., 2019) Northeast Extension (PNE) Aerosols and Ocean Science Expeditions (AEROSE) are a series of trans-Atlantic field experiments that have been carried out on an almost yearly basis since 2004 (Morris et al., 2006; Nalli et al., 2011). AEROSE was originally proposed by the Howard University NOAA Center for Atmospheric Sciences (NCAS) as a NOAA-collaborative effort for sampling the variation of Saharan dust outflows during their transport across the tropical Atlantic Ocean. Collaborating with NOAA/NESDIS, AEROSE was selected as the main science mission onboard the NOAA Ship *Ronald H. Brown* in March 2004 (Morris et al., 2006). However, multiyear opportunities have since transpired onboard the *Ronald H. Brown* and NATO RV *Alliance* that were made possible through a partnership with the NOAA PNE Project as part of their yearly buoy-servicing and deployment campaigns. AEROSE has since yielded an unprecedented collection of in situ measurements of the Saharan air layer (SAL) (Dunion and Velden, 2004; Nalli et al., 2005) and associated African dust outflows over the tropical Atlantic Ocean.

As a direct result of the NCAS-NESDIS collaboration, one of the emergent science objectives of AEROSE campaigns has been satellite remote sensing applications (Nalli et al., 2006, 2011). These have included satellite IR sounding validation of temperature, water vapor (Nalli et al., 2013b, 2018a; Xie et al., 2013), and ozone profiles (Nalli et al., 2013b, 2018b); the effect of dust aerosols on IR soundings (Maddy et al., 2012; Xie et al., 2013) and SST retrievals (Luo et al., 2020, 2021); systematic errors in cloud-cleared radiances (CCRs) (e.g., Nalli et al., 2013a); sea surface emissivity (Nalli et al., 2008a); study and validation of marine boundary layer profile retrievals from MAERI (Szczodrak et al., 2007); among others.

Ocean-based remote sensing campaigns are particularly valuable for passive IR and MW applications because the ocean surface is well characterized (Nalli et al., 2006, 2011) (in contrast to land/snow/ice), and oceans are also where satellite data have the greatest impact on numerical weather prediction (NWP) (Le Marshall et al., 2006). Perhaps equally important, the AEROSE data are truly independent in the sense that they are intentionally not assimilated, and they are far removed (on the order of 10^3 km), and thus "decoupled" meteorologically speaking, from land-based synoptic radiosonde sites.

5.1.1 Cross-sectional observations of the SAL

As mentioned earlier, the PIRATA buoy region and AEROSE domain are of interest in terms of synoptic and mesoscale meteorology, as well as air quality, including dust aerosol outflows, and anthropogenic biomass burning aerosols and trace gases, including tropospheric ozone. The periodic launching of radiosondes along linear cruise tracks are well suited for observing the space-time cross-sectional variation of atmospheric temperature and moisture profiles. Fig. 5 shows the locations of trans-Atlantic dedicated radiosondes launched during the 2019 PNE/AEROSE campaign collocated and coincident with NOAA-20 satellite overpasses. Fig. 6 shows the corresponding cross-sectional analyses of relative humidity (RH) for both the radiosondes (top plot) and passive IR/MW satellite profile retrievals (bottom plot) from the NOAA-Unique Combined Atmospheric Processing System (NUCAPS). Synoptic and mesoscale moisture features of interest evident in the

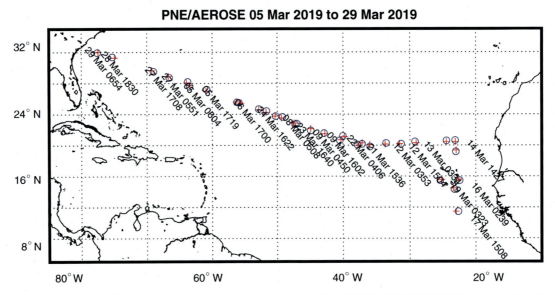

FIG. 5 Locations of trans-Atlantic dedicated radiosonde launches (*red crosses*) with collocated NOAA-20 NUCAPS fields of regard (*blue circles*) during the 2019 PNE/AEROSE campaign. The radiosondes were launched approximately 15–30 min prior to the NOAA-20 overpasses.

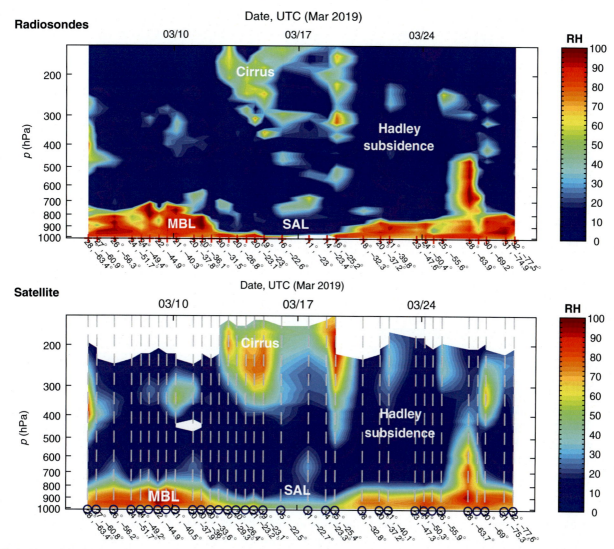

FIG. 6 Trans-Atlantic space-time cross sections of relative humidity (RH) profiles derived from (*top*) dedicated radiosondes launched during the 2019 PNE/AEROSE campaign and (*bottom*) collocated NOAA-20 NUCAPS passive IR/MW sounder retrievals (*dashed gray lines*). Launch lat/lon coordinates are indicated along the bottom axes. Meteorological features of interest are annotated, including the marine boundary layer (MBL) and Saharan air layer (SAL).

radiosonde data are also seen in the satellite retrievals, including the dry, stable SAL, cirrus cloud outflows, deep dry layers due to Hadley cell subsidence, and the moist marine boundary layer (MBL).

5.2 CalWater/ACAPEX

The CalWater (short for "California Water") 2015 campaign was a multifaceted, large-scale field campaign (consisting of land-, aircraft-, and ship-based components) aimed to improve our understanding and modeling of large-scale dynamics and cloud and precipitation processes associated with atmospheric rivers (ARs) and aerosol-cloud interactions that influence precipitation variability and extremes in the western United States (namely the Californian coast). The ship-based component of CalWater was provided by the DOE ARM Cloud Aerosol Precipitation Experiment (ACAPEX) onboard the *Ronald H. Brown*, which hosted ARM's second mobile facility (AMF2). The vessel was deployed off of the west coast of California between January 14 and February 12, 2015, during which time conditions under two ARs were sampled. Primary instruments in the AMF2 included a new DOE Marine MAERI (Gero et al., 2015), Aerosol Observing System, High Spectral Resolution Lidar, Ka-band Zenith Radar, Micropulse Lidar, W-Band (95 GHz) Cloud Radar, Microwave Radiometer, and well as frequent radiosonde and ozonesonde launches. This observational suite was used to quantify the moisture budget and cloud and precipitation processes associated with ARs and to characterize aerosols and aerosol-cloud-precipitation interactions associated with aerosols from long-range transport in the Pacific Ocean. The strong winds and high waves during AR conditions presented a challenging observing environment for the *Ronald H. Brown* and the AMF2 instruments. Nevertheless, ACAPEX produced valuable observations and scientific results on aerosol-cloud interactions, precipitation processes, and satellite validation.

ACAPEX was unique for contributing to the larger multiagency collaborative CalWater 2015 campaign (Ralph et al., 2016), which, as mentioned earlier, also included land and aircraft-based observations obtained by the US National Aeronautics and Space Administration (NASA), the National Science Foundation (NSF), NOAA, DOE, and the State of California. Four aircraft were used: the NASA ER-2, NOAA P-3, NOAA G-IV, and DOE G-1. In addition, ground-based observations were contributed by the California Department of Water Resources extreme precipitation network, and the NSF-sponsored aerosol and precipitation measurements at Bodega Bay. Coordination of all of these diverse observing assets operated by multiple agencies was a logistically complex endeavor. The combined CalWater campaign, augmented by satellite observations, provided unique opportunities to study aerosol-cloud interactions and precipitation processes observed in ARs (Leung, 2016).

Neiman et al. (2017) describe the structure and evolution of the long-lived AR that occurred from January 20–25, 2015 based on observations from the *Ronald H. Brown* during ACAPEX. Dedicated radiosondes and ozonesondes collected during ACAPEX were also used to validate passive IR/MW atmospheric profile retrievals under AR conditions, similar to that done during PNE/AEROSE campaigns under SAL conditions (Nalli et al., 2016). ACAPEX observations have informed the parameterizations of ice-nucleating particles, and characterized the spatial distribution, variability, and temperature dependence of ice nuclei in oceanic regions (Welti et al., 2020). Furthermore, the AR forecast process and tools that provided guidance for flight and ship position planning during ACAPEX are described by Cordeira et al. (2017).

5.3 SHEBA/MOSAiC

The hypothesis that the Earth's North Pole was covered by a permanent cap of sea ice was confirmed by the expedition of Norwegian explorer and scientist Fridtjof Nansen in 1893–96, onboard the *Fram*, a specially built wooden-hulled vessel. The need for in situ observations in the Arctic Ocean and in particular over the polar ice cap has continued into the twenty-first century. During the summer, the radiative balance is such that the surface is absorbing more energy than it is reflecting, enhancing sea ice melt, whereas during the winter the opposite is true. The transition between the seasons happens quite suddenly at the end of the summer season when the sun dips below the horizon.

Two large multidisciplinary scientific research campaigns that used icebreakers as a base of operations for year-long field radiation observations in a manner similar to the *Fram* are the 1997–98 Surface Heat Budget of the Arctic Ocean (SHEBA) (Uttal et al., 2002) and the 2019–20 Multidisciplinary drifting Observatory for the Study of Arctic Climate (MOSAiC) (Shupe et al., 2021). SHEBA was the first Arctic expedition onboard a modern icebreaker during the polar winter, although the Canadian icebreaker *Amundsen* also overwintered in 2006–07 as part of the International Polar Year (IPY).

The MOSAiC campaign was a successful multinational effort (including hundreds of researchers from 20 countries) to replicate the track of the *Fram* while recording radiation measurements upwelling from the surface and downwelling from the atmosphere (Shupe et al., 2021). Commencing in September 2019, the German icebreaker RV *Polarstern* set sail from Tromsø, Norway, to spend a year drifting through the Arctic Ocean while trapped in ice. The goal of the MOSAiC expedition was to take the closest look ever at the Arctic as an epicenter of global warming and to gain fundamental insights that are key to understand better global climate change. MOSAiC was led by the Alfred Wegener Institute, Helmholtz Centre for Polar and Marine Research (AWI); more details can be found from the organization website at https://mosaic-expedition.org/.

To support IR remote sensing applications, an AMF2 MAERI was deployed onboard the *Polarstern* during the MOSAiC campaign as shown in Fig. 7. A video camera (moscam) captured time-lapse images of the MAERI down view to monitor the conditions of the snow/ice surface. Three photos are shown in bottom panels of Fig. 7. The left, middle, and right photos are from March 1 before the sun appears above the horizon, March 14 showing a "lead" in the ice, and August 21 when melt ponds appear on the ice during the austral summer. The measurements from the MAERI combined with observations from many other sensors are being used to measure the change in heat released by the ocean into the atmosphere as the ice pack breaks up. Measurements of this type are difficult from space due to the large satellite footprints relative to the small-scale structure of the sea ice fractures. As with the IR sea surface emissivity studies mentioned earlier (Section 4), the MAERI data obtained from MOSAiC can provide insight into the spectral radiative properties of snow/ice surfaces, where the uncertainties are still a significant source of error within satellite data assimilation and retrieval systems.

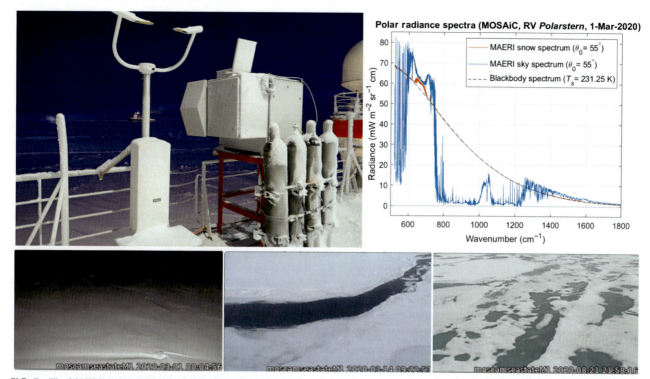

FIG. 7 The MAERI during the 2019–20 MOSAiC campaign: (*top left*) MAERI installed on the port railing of icebreaker RV *Polarstern*, (*top right*) example uplooking and downlooking observed IR radiance spectra (indicative of extreme cold temperatures), and (*bottom*) camera images of the MAERI downlooking view from port side of the *Polarstern*.

5.4 EUREC⁴A/ATOMIC

EUREC⁴A (Elucidating the Role of Cloud-Circulation Coupling in Climate), together with its UK cloud physics component (EUREC⁴A-UK), US ocean-atmosphere program of ATOMIC (Atlantic Tradewind Ocean-Atmosphere Mesoscale Interaction Campaign), and European EUREC⁴A-OA (Ocean-Atmosphere), was a large international field campaign that took place over 5 weeks during January–February 2020 (Quinn et al., 2021; Stevens et al., 2021). The main region of operation was the northwestern tropical Atlantic, extending eastward from Barbados to the Northwest Tropical Atlantic Station for air-sea flux measurement (NTAS) buoy at 15°N, 51°W, and southeastward to about 5°N. The original and primary objective was to gain a better understanding of the processes that affect clouds in trade-wind regions, including their interactions with the large-scale environment, and how shallow cumulus clouds may respond to global warming. Through ATOMIC, EUREC⁴A-OA, and EUREC⁴A-UK, the campaign evolved to address other important topics, such as the relative roles of micro- and macro-physical processes in rain formation; factors that influence the evolution of mesoscale eddies; the impact of ocean eddies, submesoscale fronts, and filaments on air-sea interaction and cloud formation; and validation of satellite retrievals and high-resolution numerical models.

The complexity of the processes investigated in EUREC⁴A and the wide range of temporal and spatial scales required a large array of aircraft, RVs, and autonomous vehicles acquiring a diverse set of measurements. There were four research aircraft and four global-class research ships, a cloud and aerosol observatory in Barbados, a large collection of oceanic and atmosphere autonomous systems and floating platforms (drones, ocean gliders, saildrones, wave gliders, surface drifting buoys, air-sea flux observing prototypes, Argo floats), together with remote sensing and experiments with high-resolution weather and climate models. Approximately 2500 radiosondes were launched from aircraft and ships, and more than 10,000 ocean profiles were obtained from ship-based CTDs, underway CTDs, vessel profilers, microstructure profilers, XBTs, XCTDs, Doppler current meter profilers, Argo floats, and ocean gliders. The most intensive atmospheric measurements from aircraft and ships were made in the area between Barbados and the NTAS mooring, while farther south ocean observations were carried out with the goals of understanding ocean eddy and freshwater plume formation and evolution (Reverdin et al., 2021) and air-sea interaction. Research, satellite retrieval algorithm validation, and numerical model evaluation based on EUREC⁴A is underway and will continue for years, with the goals of advancing knowledge of ocean-atmosphere and cloud processes, creating more accurate satellite retrievals, and improving numerical model predictions and projections.

6. Ocean color campaigns

In the field of ocean color remote sensing, it was recognized early on that shipboard measurements were critical in validating satellite remote sensing performance. The first satellite ocean color validation cruise took place after the 1978 launch of the first ocean color satellite, the Coastal Zone Color Scanner (CZCS). The CZCS lasted until 1986, and though the instrumentation was not as sophisticated as today, several validation cruises were conducted during this period (Clark, 1981). After 1986, there were no global ocean color sensors in orbit until the Japanese Ocean Color and Temperature Scanner (OCTS) and French Polarization and Directionality of the Earth's Reflectances (POLDER) onboard the Advanced Earth Observing Satellite (ADEOS), which operated during a short period from August 1996 to June 1997 (Wang et al., 2002). Routine global ocean color measurements started with the Sea-viewing Wide Field-of-view Sensor (SeaWiFS) in 1997 (McClain et al., 2004). In preparation for the SeaWiFS launch, NOAA, in collaboration with NASA, deployed the Marine Optical Buoy (MOBY) (Clark et al., 1997, see also Chapter 5) off Hawaii in the mid-1990s and started a series of Marine Optical Characterization Experiments (MOCE). The MOCE cruises included risk reduction experiments before the launch of SeaWiFS and initialization and validation experiments after the launch. Since SeaWiFS, there have been continuous ocean color radiometers on satellites along with near-continuous shipboard validation measurements.

Ocean color validation measurements are conducted during both dedicated cal/val expeditions and cruises of opportunities. Dedicated cal/val cruises allow investigators to plan for and target cloud-free areas necessary when relating in situ measurements to those of satellites. Such cruises often provide the investigators sufficient time and space to conduct duplicate or multiple casts of measurements of remote sensing reflectance (Rrs), chlorophyll-a (Chl-a) concentration, phytoplankton absorption coefficient, and other ocean property data, with a variety of instruments. The measurements collected allow better characterization of the water mass properties around the validation site, which helps us to understand better the matchups between the in situ and satellite measurements. However, dedicated cruises are costly and do not occur frequently. Many more validation measurements are conducted aboard cruises of opportunity similar to those discussed in Section 5. These allow collection of radiometric measurements for validation of satellite-derived ocean color products and for algorithm development. Similar to the PNE/AEROSE campaigns described earlier (Section 5.1), this cal/val "piggybacking" is typically mutually beneficial as the ocean color cal/val team members get the opportunity to conduct much-needed validation measurements at a few stations per day, and the bio-optical measurements often contribute to the cruise project. Table 1 lists dedicated and "piggyback" cruises, where the NOAA/NESDIS Ocean Color Cal/Val Team has conducted validation measurements since the launch of SeaWiFS. In all, over 1000 validation stations were sampled across the ocean.

TABLE 1 List of hyperspectral ocean color cal/val experiments and cruises.

Cruise name	Date	Region	Number of stations
MOCE Turbid-1 Moss Landing	October-91	West Coast	5
MOCE-3 Hawaiian Archipelago	October-94	Hawaii	18
MOCE Turbid-3 Mill Creek	July-95	Chesapeake Bay	10
MOCE Turbid-5 Mill Creek	September-95	Chesapeake Bay	15
MOCE-4 Oahu to Big Island	January-98	Hawaii	18
MOCE-5 Baja California	October-99	West Coast	20
MOCE-6 Lanai	April-00	Hawaii	5
MOCE-7 Lanai	November-00	Hawaii	15
MOCE-8 Lanai	February-01	Hawaii	9
MOCE Turbid-7	November-02	Chesapeake Bay	9
Turbid-8	May-03	Chesapeake Bay	4
Turbid-9	September-04	Chesapeake Bay	10
GEOCAPE	July-11	Chesapeake Bay	57
Bay 2011	2011	Chesapeake Bay	6
AOML	March-12	Florida	17
Oahu, HI	September-12	Hawaii	21
Bay 2012	December-12	Chesapeake Bay	2

TABLE 1 List of hyperspectral ocean color cal/val experiments and cruises—cont'd

Cruise name	Date	Region	Number of stations
CUNY Ches. Bay Exp.	August-13	Chesapeake Bay	42
GOM GEOCAPE	October-13	Gulf of Mexico	67
Bay 2013	2013	Chesapeake Bay	37
Puerto Rico 2014	May-14	Caribbean Sea	15
Hawaii/MOBY	April-14	Hawaii	12
VIIRS Cal/Val 2014	November-14	East Coast	23
Bay 2014	2014	Chesapeake Bay	27
Puerto Rico 2015	March-15	East Coast	15
East Coast Ocean Acidification	July-15	East Coast	74
VIIRS Cal/Val 2015	December-15	East Coast	27
Bay 2015	2015	Chesapeake Bay	7
KORUS OC	June-16	Korea	35
West Coast Ocean Acidification	June-16	West Coast	35
Fiji to Australia	July-16	South Pacific	24
CORAL PRISM	October-16	Australia	37
VIIRS Cal/Val 2016	October-16	East Coast	12
Bay 2016	2016	Chesapeake Bay	13
P18 South Pacific	February-17	South Pacific	45
JOEYS	March-17	Chesapeake Bay	12
Bay 2017	2017	Chesapeake Bay	15
AOML 1801	January-18	Florida	17
AOML 1803	March-18	Florida	12
VIIRS Cal/Val 2018	2018	Gulf of Mexico	22
ECOMON_1806	June-18	East Coast	21
ECOA_18_1	June-18	East Coast	30
ECOA_18_2	July-18	East Coast	20
RNG	August-18	East Coast	4
ECOMON_1808	August-18	East Coast	17
AOML_1810	November-18	Florida	21
ECOMON_1810	November-18	East Coast	5
Bay 2018	2018	Chesapeake Bay	10
Investigator	May-19	Australia	19
Whale Song	June-19	Australia	13
AOML_0819	August-19	East Coast	11
ECOMON_1908	August-19	East Coast	23
VIIRS Cal/Val 2019	September-19	East Coast	27
VIIRS Cal/Val 2021	March-21	Gulf of Mexico	33
		Total	1120

In terms of ocean color applications, *hyperspectral* refers to spectral resolution of less than 1 nm over 380–900 nm (cf. Section 8.2 of Chapter 5).

The basic principle of the ocean color satellites is to use the sunlight propagating into the upper water column of the ocean and detect the backscattered light. The dependence on the sun implies a series of restrictions to the operation of the ocean color satellites. First, it is difficult to derive reliable ocean color products from satellites when the solar zenith angle, $\theta_\odot$, is large (i.e., $\theta_\odot > 70$ degrees) (Mikelsons et al., 2020). Second, it is a complex or formidable task to make utilizable ocean color measurements through clouds, smoke, or desert dust. Third, it is challenging to separate the sun glint contribution from the ocean color signals. Fourth, it is not easy to retrieve high-quality ocean color products in the vicinity of land and clouds, due to cloud shadow, stray-light, and adjacent effects (Bulgarelli et al., 2014; Jiang and Wang, 2013). One must take into account these essential facts associated with satellite observations when planning the field validation cruises. Weather and cloudiness can limit the utility of a validation measurement. In addition, inter- and intrapixel variability of water can make it difficult to relate the in situ measurement to the satellite signal (i.e., mismatch error; cf. Chapter 1, Section 5.1). Researchers encounter many different conditions when utilizing ocean remote sensing data in their work, and therefore, validation measurements in a variety of conditions are necessary to understand the uncertainties in all these conditions.

The NOAA ocean color team has been conducting dedicated annual VIIRS ocean color cal/val campaigns aboard NOAA ships since 2014 (Ondrusek et al., 2015, 2016, 2017, 2019, 2021). These projects have covered many different regions and seasons in an attempt to characterize ocean color properties under varying conditions (Fig. 8). The Joint Polar Satellite System (JPSS) Ocean Color Cal/Val Team members participate in these cruises to conduct validation measurements for VIIRS and other ocean color satellites utilized operationally by NOAA. These validation measurements include both radiances and derived ocean color products. In addition, the JPSS Ocean Color Cal/Val Team members conduct intercomparisons of the measurements to understand better the uncertainties of the validation protocols and processing. Team members conduct profiling, floating, and above-water measurements of water-leaving radiances and derived products using identical and different instrumentation and methods that are used routinely by investigators.

During these cal/val cruises, in situ observations for satellite ocean color product validation and algorithm development are measured following the NASA ocean optics protocols for satellite ocean color sensor and algorithm validation (Mueller et al., 2003). In addition to the radiometric measurements described earlier, inherent optical properties (IOPs) such as absorption and backscattering coefficients, bio-optical properties including high-performance liquid chromatography (HPLC) and fluorometric phytoplankton pigments, and supporting ancillary data such as geolocation, clouds, and physical conditions are also documented. Water samples for the detection and analyses of various components are sampled from flow-through systems and by profilers. The measurements from multiple such instrumentations onboard these campaign provide the in situ correlative data for satellite ocean color validation (cf. Chapter 20) and for algorithm development and uncertainty evaluations. Analyzing the reproducibility and uncertainty between team members, for instance, allows understanding of the uncertainties in individual measurements. This is important for the understanding of the limitations of in situ

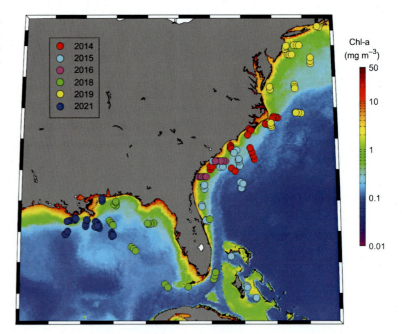

FIG. 8 Sampling stations of six NOAA Annual Ocean Color Cal/Val campaigns. The background is a VIIRS Chl-a composite image from NOAA CoastWatch Data Portal (https://coastwatch.noaa.gov, monthly 9-km product, October 2012).

databases such as the SeaWiFS Bio-optical Archive and Storage System (SeaBASS), NASA bio-Optical Marine Algorithm Dataset (NOMAD) (Werdell and Bailey, 2005), or NOAA's ocean color in situ database.

Examples of the intercomparison between different measures of water-leaving radiances from a station from four NOAA campaigns are given in Fig. 9. As stated earlier, there are generally three categories of measurements: in-water measurements (e.g., profilers and floaters), above-water measurements (handheld and mounted), and on-water measurements. The profilers were all deployed simultaneously off of the stern of the ship. These include multiple Seabird/Satlantic Hyperspectral Profilers (HyperPro) and a Biospherical Compact Optical Profiling System (COPS). A Satlantic MicroPro was utilized only during the November 2014 cruise (Ondrusek et al., 2015). The deployments of the profilers are spread out across the stern of the ship, and the instruments are profiled at a far enough distance from the ship to avoid ship shadowing (Shang et al., 2017). A description of the deployment can be found in Section 3.2 of Chapter 20 in-water section. Ideally, the sun is positioned off the port quarter during the measurement. This gives the correct angle for the above-water measurement to be conducted off the bow, where the measurements need to have an azimuth angle relative to the sun between 90 and 135 degrees. This allows the simultaneous measurements of the profilers and above-water instruments. The above-water measurement procedures are described in Section 3.1 of Chapter 20. The Analytical Spectral Device, Inc. Handheld II (ASD), the Spectra Vista GER1500 (GER), and the Spectral Evolution Field Spectroradiometers (SEV) are all handheld above-water radiometers, and the Seabird Hyperspectral Surface Acquisition System (HyperSAS) is a mounted radiometer. Floaters are deployed directly after the profilers. The Hyperspectral Tethered Spectral Radiometer Buoy (HTSRB) is an in-water floater that measures near-surface light field, while the radiometer with skylight-blocking apparatus (SBA) directly

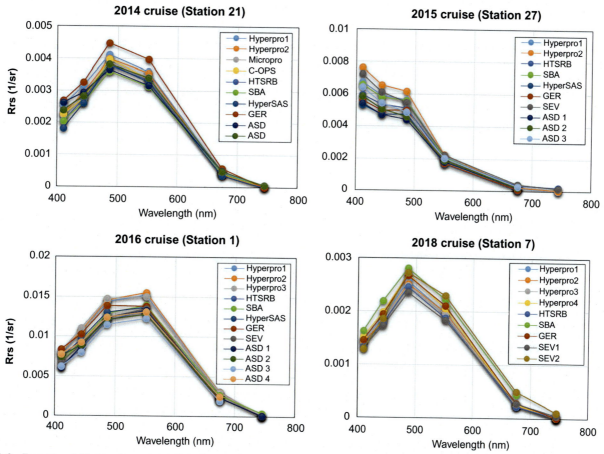

FIG. 9 Examples of ship-based, remote sensing reflectance intercomparisons from each of the first four NOAA cal/val campaigns. The 2014, 2015, and 2016 cruises took place aboard the NOAA Ship *Nancy Foster* off the east coast of the United States, and the 2018 cruise took place on the NOAA Ship *Okeanos Explorer* in the Gulf of Mexico. All plots are hyperspectral ship-based measurements spectrally weighted to the VIIRS bandpasses. Note that in plot legend Hyperpro is the Satlantic Hyperpro II, HTSRB is the Satlantic Hyperspectral Tethered Spectral Radiometer Buoy, SBA is the Satlantic Skylight-Blocking Apparatus, HyperSAS is the Seabird Hyperspectral Surface Acquisition System, GER is the Spectra Vista GER1500, SEV is the Spectral Evolution Field Spectroradiometer, and ASD is the Analytical Spectral Device, Inc. Handheld II. The number after the instrument name is replicate of the same instrument operated by different groups.

TABLE 2 Average of all-station instrument absolute percent difference (%) relative to the average of all instruments.

Band	410	443	486	551	675	Avg 410-551
All	9.76	7.68	6.66	7.9	22.75	8
Profiler	6.18	5.66	4.98	5.61	13.91	5.61
Floater	9.24	7.6	6.37	6.78	30.11	7.5
Above	14.01	10.01	8.19	11.18	26.38	10.85
Hyperpro	6.45	5.82	5.18	5.42	14.07	5.72
SBA	10.1	8.48	8.6	8.41	21.99	8.9
HTSRB	8.72	7.12	5.03	5.75	35	6.66
Hypersas	14.18	8.23	6.51	8.57	26.54	9.37
GER	9.49	6.77	5.89	8.06	22.43	7.55
Spec Ev	17.32	12.64	10.65	14.62	29.82	13.81
NOAA hyp	5.31	5.1	4.77	5.57	13.11	5.19

"Profiler" refers to HyperPros and C-OPS; "Floaters" are HTSRB and SBA combined; "Above" is the HyperSAS, GER, Spectral Evolution, and ASD combined. All instruments are hyperspectral except C-OPs.

measures the water-leaving radiance (Wei et al., 2021), which is an on-water instrument. Interested readers are referred to Chapter 20 for a detailed description of optical radiometry.

One way to analyze the measurement uncertainty among the different methods is to look at the mean absolute percentage difference. In Table 2, the instrument-specific absolute percentage difference of the water-leaving radiance measurements relative to the average of all-instrument measurements is calculated. Results show that the profiler usually has the closest measurements to the all-instrument mean values followed by floaters, then by above-water measurements. Note that other factors, such as the individual number of different instrument types and the environmental conditions occurring to each deployment, should also be considered to understand fully these uncertainties.

7. Summary

The vast global surface area and thermal inertia of the world's oceans renders them a critical component of the Earth's weather and climate system, but their lack of accessibility to humans also renders them extremely difficult to observe and measure in their integral whole using traditional in situ observing methods. For these reasons, satellite remote sensing plays a critical role in observation of the global oceans, complementing data from in situ systems such as buoys. Field measurements in support of ocean remote sensing include intensive campaigns that must be conducted from ships, especially in remote locations, where such data are otherwise sparse or nonexistent.

In this chapter, we discussed the importance of ship-based campaigns for satellite remote sensing. Such campaigns are typically designed and equipped to establish, acquire, and maintain critical in situ data over open-ocean locations. A basic introduction to ship platforms, specifically oceanographic RVs, was presented, along with the various instrumentation, data, techniques, and strategies needed for performing successful ship-based campaigns.

Included in this chapter was the discussion the logistical issues and challenges associated with ship-based campaigns and instrument deployments, including both atmospheric (e.g., radiosondes, ozonesondes, radiometers, drones) and oceanographic measurements (e.g., buoys, CTDs, XBTs). These practical considerations necessary for conducting successful field experiments at sea may be seen as an important element within the broader scientific method (i.e., experimental design), but they are often taken for granted or otherwise not discussed within subsequent technical papers and/or reports, which typically focus on the science results. Because of the unique challenges posed by the remoteness from land and the sometimes harsh marine environment, ship-based campaigns inevitably incur a certain degree of risk.

Finally, attention was given to historical and contemporary campaigns of note, along with some of the more prominent passive satellite applications, including emission-based IR/MW (e.g., atmospheric soundings and SST) and reflectance-based solar-spectrum (e.g., ocean color) sensor systems. These campaigns are noteworthy for field data contributions that led to notable advances in the science of passive environmental remote sensing.

Acknowledgments

The authors acknowledge the support of the JPSS Program Office and Proving Ground and Risk Reduction (PGRR) initiatives, the NOAA/NESDIS/STAR Satellite Meteorology and Climatology Division (SMCD), countless students and volunteers, and the crew and officers of RVs we have had the pleasure of sailing aboard. We also express our sincere appreciation to Bertrand Dano (UM/CIMAS) for graciously providing us with numerous 3D ship renderings (three of which being used in Fig. 2), and to Joe Taylor, Hank Revercomb, and Michelle Loveless (UW/SSEC) for reviewing the manuscript and providing constructive feedback. The scientific results and conclusions, as well as any views or opinions expressed herein, are those of the authors and do not necessarily reflect the views of NOAA or the Department of Commerce.

References

Bourlès, B., Araujo, M., McPhaden, M.J., Brandt, P., Foltz, G.R., Lumpkin, R., et al., 2019. PIRATA: a sustained observing system for tropical Atlantic climate research and forecasting. Earth Space Sci. 6. https://doi.org/10.1029/2018EA000428.

Bryan, K., Manabe, S., Pacanowski, R., 1975. A global ocean-atmosphere climate model. Part II. The oceanic circulation. J. Phys. Ocean. 5 (1), 30–46.

Bulgarelli, B., Kiselev, V., Zibordi, G., 2014. Simulation and analysis of adjacency effects in coastal waters: a case study. Appl. Opt. 53, 1523–1545.

Carral, L., de Lara Rey, J., Alvarez-Feal, J., Carral Couce, J., 2017. Winch control gear for CTD sampling with a system to compensate vertical motion heave when manoeuvring in rough seas. Ocean Eng. 135, 246–257. https://doi.org/10.1016/j.oceaneng.2017.03.005.

Clark, D.K., 1981. Phytoplankton pigment algorithms for the Nimbus-7 CZCS. In: Gower, J.F.R. (Ed.), Oceanography From Space. Marine Science. Springer, Boston, MA, https://doi.org/10.1007/978-1-4613-3315-9_28.

Clark, D.K., Gordon, H.R., Voss, K.J., Ge, Y., Broenkow, W., Trees, C., 1997. Validation of atmospheric correction over the ocean. J. Geophys. Res. 102, 17209–17217.

Cordeira, J., Ralph, F., Martin, A., Gaggini, N., Spackman, J., Neiman, P., Rutz, J., Pierce, R., 2017. Forecasting atmospheric rivers during CalWater 2015. Bull. Am. Meteorol. Soc. 98, 3. https://doi.org/10.1175/BAMS-D-15-00245.1.

Donlon, C.J., Minnett, P.J., Fox, N., Wimmer, W., 2014a. Strategies for the laboratory and field deployment of ship-borne fiducial reference thermal infrared radiometers in support of satellite-derived sea surface temperature climate data records. In: Optical Radiometry for Ocean Climate Measurements, Experimental Methods in the Physical Sciences, vol. 47. Academic Press, pp. 557–603, https://doi.org/10.1016/B978-0-12-417011-7.00018-0.

Donlon, C.J., et al., 2014b. Ship-borne thermal infrared radiometer systems. In: Optical Radiometry for Ocean Climate Measurements, Experimental Methods in the Physical Sciences, vol. 47. Academic Press, pp. 305–404, https://doi.org/10.1016/B978-0-12-417011-7.00011-8.

Dunion, J.P., Velden, C.S., 2004. The impact of the Saharan air layer on Atlantic tropical cyclone activity. Bull. Am. Meteorol. Soc. 85 (3), 353–385.

Durre, I., Yin, X., Vose, R.S., Applequist, S., Arnfield, J., 2018. Enhancing the data coverage in the integrated global radiosonde archive. J. Atmos. Ocean. Technol. 35 (9), 1753–1770. https://doi.org/10.1175/JTECH-D-17-0223.1.

Fairall, C., White, A., Edson, J., Hare, J., 1997. Integrated shipboard measurements of the marine boundary layer. J. Atmos. Ocean. Technol. 14, 338–359.

Friehe, C.A., Schmitt, K.F., 1976. Parameterization of air-sea interface fluxes of sensible heat and moisture by the bulk aerodynamic formulas. J. Phys. Oceanogr. 6 (6), 801–809.

Gero, P.J., et al., 2015. A new marine atmospheric emitted radiance interferometer for shipboard atmospheric and oceanic observations. In: Fourier Transform Spectroscopy and Hyperspectral Imaging and Sounding of the Environment, Optical Society of America, JM1A.2, https://doi.org/10.1364/FTS.2015.JM1A.2.

Goerss, J.S., Duchon, C.E., 1980. Effect of ship heating on dry-bulb temperature measurements in GATE. J. Phys. Oceanogr. 10, 478–479.

Hagan, D.E., Minnett, P.J., 2003. AIRS radiance validation over ocean from sea surface temperature measurements. IEEE Trans. Geosci. Remote Sens. 41 (2), 432–441.

Hansen, J., et al., 1997. Forcings and chaos in interannual to decadal climate change. J. Geophys. Res. 102 (D22), 25679–25720.

Jiang, L., Wang, M., 2013. Identification of pixels with stray light and cloud shadow contaminations in the satellite ocean color data processing. Appl. Opt. 52, 6757–6770.

Kilpatrick, K.A., et al., 2015. A decade of sea surface temperature from MODIS. Remote Sens. Environ. 165, 27–41.

Knuteson, R.O., et al., 2004. Atmospheric emitted radiance interferometer. Part I: instrument design. J. Atmos. Ocean. Technol. 21 (12), 1763–1776.

Le Marshall, J., et al., 2006. Improving global analysis and forecasting with AIRS. Bull. Am. Meteorol. Soc. 87 (7), 891–894.

Leung, L.R., 2016. Arm Cloud-Aerosol-Precipitation Experiment (ACAPEX) Field Campaign Report. Tech. Rep. US Department of Energy, DOE/SC-ARM-16-012.

Li, J., Guo, L., Lin, L., Zhao, Y., Cheng, X., 2014. A new method of tipping calibration for ground-based microwave radiometer in cloudy atmosphere. IEEE Trans. Geosci. Remote Sens. 52, 5506–5513.

Luo, B., Minnett, P.J., Szczodrak, M., Nalli, N.R., Morris, V.R., 2020. Accuracy assessment of MERRA-2 and ERA-interim sea surface temperature, air temperature, and humidity profiles over the Atlantic Ocean using AEROSE measurements. J. Clim. 33 (16), 6889–6909. https://doi.org/10.1175/JCLI-D-19-0955.1.

Luo, B., Minnett, P.J., Zuidema, P., Nalli, N.R., Akella, S., 2021. Saharan dust effects on North Atlantic sea-surface skin temperatures. J. Geophys. Res. Oceans 126. https://doi.org/10.1029/2021JC017282. e2021JC017282.

Maddy, E.S., et al., 2012. On the effect of dust aerosols on AIRS and IASI operational level 2 products. Geophys. Res. Lett. 39, L10809. https://doi.org/10.1029/2012GL052070.

Manabe, S., Bryan, K., Spelman, M., 1975. A global ocean-atmosphere climate model. Part I. The atmospheric circulation. J. Phys. Ocean. 5 (1), 3–29.

McClain, C.R., Feldman, G.C., Hooker, S.B., 2004. An overview of the seawifs project and strategies for producing a climate research quality global ocean bio-optical time series. Deep Sea Res. II Top. Stud. Oceanogr. 51, 5–42.

Merchant, C.J., Leborgne, P., 2004. Retrieval of sea surface temperature from space, based on modeling of infrared radiative transfer: capabilities and limitations. J. Atmos. Ocean. Technol. 21, 1734–1746.

Merchant, C.J., Harris, A.R., Maturi, E., Embury, O., MacCallum, S.N., Mittaz, J., Old, C.P., 2009. Sea surface temperature estimation from the geostationary operational environmental satellite-12 (GOES-12). J. Atmos. Ocean. Technol. 26, 570–581. https://doi.org/10.1175/2008JTECHO596.1.

Mikelsons, K., Wang, M., Jiang, L., 2020. Statistical evaluation of satellite ocean color data retrievals. Remote Sens. Environ. 237, 111601. https://doi.org/10.1016/j.rse.2019.111601.

Minnett, P.J., 1986. A numerical study of the effects of anomalous North Atlantic atmospheric conditions on the infrared measurement of sea-surface temperature from space. J. Geophys. Res. 91, 8509–8521.

Minnett, P.J., Corlett, G.K., 2012. A pathway to generating climate data records of sea-surface temperature from satellite measurements. Deep Sea Res. Part II Top. Stud. Oceanogr. 77–80, 44–51.

Minnett, P.J., Key, E.L., 2007. Meteorology and atmosphere–surface coupling in and around leads and polynyas. In: Polynyas: Windows to the World, Elsevier, The Netherlands, pp. 127–161.

Minnett, P.J., Knuteson, R.O., Best, F.A., Osborne, B.J., Hanafin, J.A., Brown, O.B., 2001. The marine-atmospheric emitted radiance interferometer (M-AERI): a high-accuracy, sea-going infrared spectroradiometer. J. Atmos. Ocean. Technol. 18, 994–1013.

Minnett, P.J., Maillet, K.A., Hanafin, J.A., Osborne, B.J., 2005. Infrared interferometric measurements of the near surface air temperature over the oceans. J. Atmos. Ocean. Technol. 22, 1019–1032.

Minnett, P.J., et al., 2019. Half a century of satellite remote sensing of sea-surface temperature. Remote Sens. Environ. 233, 111366. https://doi.org/10.1016/j.rse.2019.111366.

Morris, V., et al., 2006. Measuring trans-Atlantic aerosol transport from Africa. Eos Trans. AGU 87 (50), 565–571.

Mueller, J.L., Fargion, G.S., McClain, C.R., 2003. Data requirements for ocean color algorithms and validation (Chapter 3). In: Ocean Optics Protocols for Satellite Ocean Color Sensor Validation, vol. I: Introduction, Background and Conventions, NASA Goddard Space Flight Center, Greenbelt, MD, pp. 31–40.

Nalli, N.R., Smith, W.L., 1998. Improved remote sensing of sea surface skin temperature using a physical retrieval method. J. Geophys. Res. 103 (C5), 10527–10542.

Nalli, N.R., et al., 2005. Profile observations of the Saharan air layer during AEROSE 2004. Geophys. Res. Lett. 32, L05815. https://doi.org/10.1029/2004GL022028.

Nalli, N.R., et al., 2006. Ship-based measurements for infrared sensor validation during Aerosol and Ocean Science Expedition 2004. J. Geophys. Res. 111, D09S04. https://doi.org/10.1029/2005JD006385.

Nalli, N.R., Minnett, P.J., Maddy, E., McMillan, W.W., Goldberg, M.D., 2008a. Emissivity and reflection model for calculating unpolarized isotropic water surface leaving radiance in the infrared. 2: Validation using Fourier transform spectrometers. Appl. Opt. 47 (25), 4649–4671.

Nalli, N.R., Minnett, P.J., van Delst, P., 2008b. Emissivity and reflection model for calculating unpolarized isotropic water surface leaving radiance in the infrared. 1: Theoretical development and calculations. Appl. Opt. 47 (21), 3701–3721.

Nalli, N.R., et al., 2011. Multi-year observations of the tropical Atlantic atmosphere: multidisciplinary applications of the NOAA Aerosols and Ocean Science Expeditions (AEROSE). Bull. Am. Meteorol. Soc. 92, 765–789. https://doi.org/10.1175/2011BAMS2997.1.

Nalli, N.R., Barnet, C.D., Gambacorta, A., Maddy, E.S., Xie, H., King, T.S., Joseph, E., Morris, V.R., 2013a. On the angular effect of residual clouds and aerosols in clear-sky infrared window radiance observations 2. Satellite experimental analyses. J. Geophys. Res. Atmos. 118, 1–16. https://doi.org/10.1029/2012JD018260.

Nalli, N.R., et al., 2013b. Validation of satellite sounder environmental data records: application to the cross-track infrared microwave sounder suite. J. Geophys. Res. Atmos. 118, 13628–13643. https://doi.org/10.1002/2013JD020436.

Nalli, N.R., et al., 2016. Satellite sounder observations of contrasting tropospheric moisture transport regimes: Saharan air layers, Hadley cells, and atmospheric rivers. J. Hydrometeor. 17 (12), 2997–3006. https://doi.org/10.1175/JHM-D-16-0163.1.

Nalli, N.R., et al., 2018a. Validation of atmospheric profile retrievals from the SNPP NOAA-unique combined atmospheric processing system. Part 1: temperature and moisture. IEEE Trans. Geosci. Remote Sens. 56 (1), 180–190. https://doi.org/10.1109/TGRS.2017.2744558.

Nalli, N.R., et al., 2018b. Validation of atmospheric profile retrievals from the SNPP NOAA-unique combined atmospheric processing system. Part 2: Ozone. IEEE Trans. Geosci. Remote Sens. 56 (1), 598–607. https://doi.org/10.1109/TGRS.2017.2762600.

Neiman, P.J., et al., 2017. An analysis of coordinated observations from NOAA's Ronald Brown ship and G-IV aircraft in a landfalling atmospheric river over the North Pacific during CalWater-2015. Mon. Weather Rev. 145, 3647–3669. https://doi.org/10.1175/MWR-D-17-0055.1.

Noh, Y.-C., Lim, A.H.N., Huang, H.-L., Goldberg, M.D., 2020. Global forecast impact of low data latency infrared and microwave sounders observations from polar orbiting satellites. Remote Sens. 12 (14), 2193. https://doi.org/10.3390/rs12142193.

Ondrusek, M., et al., 2015. Report for dedicated JPSS VIIRS ocean color calibration/validation cruise. NOAA Technical Report NESDIS 146., https://doi.org/10.7289/V52B8W0Z. NOAA/NESDIS, Silver Spring, MD.

Ondrusek, M., et al., 2016. Report for dedicated JPSS VIIRS ocean color calibration/validation cruise, December 2015. NOAA Technical Report NESDIS 148., https://doi.org/10.7289/V5/TR-NESDIS-148. NOAA/NESDIS, Silver Spring, MD.

Ondrusek, M., et al., 2017. Report for dedicated JPSS VIIRS ocean color calibration/validation cruise, October 2016. NOAA Technical Report NESDIS 151., https://doi.org/10.7289/V5/TR-NESDIS-151. NOAA/NESDIS, Silver Spring, MD.

Ondrusek, M., et al., 2019. Report for dedicated JPSS VIIRS ocean color calibration/validation cruise, May 2018. NOAA Technical Report NESDIS 152., https://doi.org/10.25923/scyb-qf42. NOAA/NESDIS, Silver Spring, MD.

Ondrusek, M., et al., 2021. Report for dedicated JPSS VIIRS ocean color calibration/validation cruise, September 2019. NOAA Technical Report NESDIS 154., https://doi.org/10.25923/p9de-yw97. NOAA/NESDIS, Silver Spring, MD.

Popinet, S., Smith, M., Stevens, C., 2004. Experimental and numerical study of the turbulence characteristics of airflow around a research vessel. J. Atmos. Ocean. Technol. 21, 1575–1589.

Post, M.J., et al., 1997. The combined sensor program: an air-sea science mission in the central and Western Pacific Ocean. Bull. Am. Meteorol. Soc. 78 (12), 2797–2815.

Quinn, P.K., et al., 2021. Measurements from the RV Ronald H. Brown and related platforms as part of the Atlantic tradewind ocean-atmosphere mesoscale interaction campaign (ATOMIC). Earth Syst. Sci. Data 13 (4), 1759–1790. https://doi.org/10.5194/essd-13-1759-2021.

Ralph, F.M., Prather, K.A., Cayan, D., Spackman, J.R., DeMott, P., Dettinger, M., Fairall, C., Leung, R., Rosenfeld, D., Rutledge, S., Waliser, D., White, A.B., Cordeira, J., Martin, A., Helly, J., Intrieri, J., 2016. CalWater field studies designed to quantify the roles of atmospheric rivers and aerosols in modulating U.S. West Coast precipitation in a changing climate. Bull. Am. Meteorol. Soc. 97 (7), 1209–1228.

Revercomb, H.E., Buijs, H., Howell, H.B., LaPorte, D.D., Smith, W.L., Stromovsky, L.A., 1988. Radiometric calibration of IR Fourier transform spectrometers: solution to a problem with the high-resolution interferometer sounder. Appl. Opt. 27 (15), 3210–3218.

Reverdin, G., et al., 2021. Formation and evolution of a freshwater plume in the northwestern tropical Atlantic in February 2020. J. Geophys. Res. Oceans 126 (4). https://doi.org/10.1029/2020JC016981. e2020JC016981.

Shang, Z., Lee, Z.P., Dong, Q., Wei, J., 2017. Self-shading associated with a skylight-blocked approach system for the measurement of water-leaving radiance and its correction. Appl. Opt. 56, 7033–7040. https://doi.org/10.1364/AO.56.007033.

Shupe, M., et al., 2021. Multidisciplinary drifting observatory for the study of arctic climate (mosaic) (field campaign report)., https://doi.org/10.2172/1787856.

Smith, W.L., 1991. Atmospheric soundings from satellites—false expectation or the key to improved weather prediction? Q. J. R. Meteorol. Soc. 117 (498), 267–297.

Smith, W.L., Howell, H.B., Woolf, H.M., 1979. The use of interferometric radiance measurements for sounding the atmosphere. J. Atmos. Sci. 36, 566–575.

Smith, W.L., Revercomb, H.E., Howell, H.B., Woolf, H.M., LaPorte, D.D., 1987. The high resolution interferometer sounder (HIS). Atmospheric radiation: progress and prospects. In: Liou, K.-N., Xiuji, Z. (Eds.), Beijing International Radiation Symposium, American Meteorological Society, Boston, MA, pp. 271–281.

Smith, W.L., et al., 1996. Observations of the infrared properties of the ocean: implications for the measurement of sea surface temperature via satellite remote sensing. Bull. Am. Meteorol. Soc. 77, 41–51.

Smith, S.R., Briggs, K., Bourassa, M.A., Elya, J., Paver, C.R., 2018. Shipboard automated meteorological and oceanographic system data archive: 2005–2017. Geosci. Data J. 5, 73–86.

Smith, S.R., et al., 2019. Ship-based contributions to global ocean, weather, and climate observing systems. Front. Mar. Sci. 6. https://doi.org/10.3389/fmars.2019.00434.

Soreide, N.N., Woody, C.E., Holt, S.M., 2001. Overview of ocean based buoys and drifters: present applications and future needs. In: MTS/IEEE Oceans 2001. An Ocean Odyssey. Conference Proceedings, IEEE (Cat. No. 01CH37295), vol. 4, pp. 2470–2472.

Stevens, B., et al., 2021. EUREC^{4}A. In: Earth System Science Data Discussions, 2021, pp. 1–78, https://doi.org/10.5194/essd-2021-18. vol.

Szczodrak, M., Minnett, P.J., Nalli, N.R., Feltz, W.F., 2007. Profiling the lower troposphere over the ocean with infrared hyperspectral measurements of the marine-atmosphere emitted radiance interferometer. J. Atmos. Ocean. Technol. 34 (3), 390–402.

Szczodrak, M., Minnett, P.J., Evans, R.H., 2014. The effects of anomalous atmospheres on the accuracy of infrared sea-surface temperature retrievals: dry air layer intrusions over the tropical ocean. Remote Sens. Environ. 140, 450–465.

Talley, L.D., Pickard, G.L., Emery, W.J., Swift, J.H., 2011. Descriptive Physical Oceanography: An Introduction, sixth ed. Academic Press, Boston, MA, https://doi.org/10.1016/B978-0-7506-4552-2.10033-2.

Tobin, D.C., et al., 2006. Atmospheric radiation measurement site atmospheric state best estimates for atmospheric infrared sounder temperature and water vapor retrieval validation. J. Geophys. Res. 111, D09S14. https://doi.org/10.1029/2005JD006103.

Uttal, T., et al., 2002. Surface heat budget of the Arctic Ocean. Bull. Am. Meteorol. Soc. 83 (2), 255–276. https://doi.org/10.1175/1520-0477(2002)0832.3.CO;2.

Wang, M., Isaacman, A., Franz, B.A., McClain, C.R., 2002. Ocean color optical property data derived from the Japanese ocean color and temperature scanner and the French polarization and directionality of the earth's reflectances: a comparison study. Appl. Opt. 41, 974–990.

Wei, J., Wang, M., Lee, Z.P., Ondrusek, M., Zhang, S., Ladner, S., 2021. Experimental analysis of the measurement precision in spectral water-leaving radiance in different water types. Opt. Exp. 29 (2), 2780–2797. https://doi.org/10.1364/OE.413784.

Welti, A., et al., 2020. Ship-based measurements of ice nuclei concentrations over the Arctic, Atlantic, Pacific and Southern oceans. Atmos. Chem. Phys. 20 (23). https://doi.org/10.5194/acp-20-15191-2020.

Werdell, P.J., Bailey, S.W., 2005. An improved in-situ bio-optical data set for ocean color algorithm development and satellite data product validation. Rem. Sens. Environ. 98, 122–140. https://doi.org/10.1016/j.rse.2005.07.001.

Whitt, C., et al., 2020. Future vision for autonomous ocean observations. Front. Mar. Sci. 7, 697. https://doi.org/10.3389/fmars.2020.00697.

Wu, X., Smith, W.L., 1997. Emissivity of rough sea surface for 8–13 μm: modeling and validation. Appl. Opt. 36, 1–11.

Xie, H., et al., 2013. Integration and ocean-based prelaunch validation of GOES-R advanced baseline imager legacy atmospheric products. J. Atmos. Ocean. Technol. 30 (8), 1743–1756. https://doi.org/10.1175/JTECH-D-12-00120.1.

Chapter 12

Land-based cal/val campaigns

Lori A. Borg[a,b], Ruud J. Dirksen[c], and Robert O. Knuteson[a,b]

[a]Space Science and Engineering Center (SSEC), University of Wisconsin-Madison, Madison, WI, United States, [b]CIMSS, University of Wisconsin-Madison, Madison, WI, United States, [c]GRUAN Lead Centre, Deutscher Wetterdienst, Meteorologisches Observatorium Lindenberg, Lindenberg, Germany

Chapter outline

1. Introduction	219	3.3 RS92-RS41 radiosonde intercomparisons	228
2. DOE ARM campaigns	220	3.4 WMO radiosonde intercomparisons	228
2.1 ARM Water Vapor Intensive Operating Periods	220	4. Links to other land-based campaigns	230
2.2 AIRS Cal/Val at ARM	221	5. Summary	231
2.3 JPSS CrIS Cal/Val at ARM	222	Acknowledgment	231
3. GRUAN	226	References	231
3.1 GRUAN data products	227	Further reading	233
3.2 GRUAN sites	227		

One of the things we said from the start is we are not voting on this. We put a real emphasis on absolute understanding of the calibrations rather than an inter-comparison.

Hank Revercomb, University of Wisconsin-Madison

1. Introduction

The world's land mass covers only 30% of the Earth's total surface area. However in 2009, a World Bank economic development report indicated that 95% of the world's population is concentrated in just 10% of this land area. In a changing climate, the need for accurate remote sensing over land areas gains greater importance as mankind experiences extremes in weather whether it be severe storms, flood, drought, or fire. However, this need for improved weather information is as old as mankind itself.

The documented history of weather observations and forecasting goes back more than 2000 years. Around 340 B.C., the Greek philosopher Aristotle wrote *Meteorologica*, a philosophical treatise that included theories about the formation of rain, clouds, hail, wind, thunder, lightning, and hurricanes (Aristotle 340 BC, 1923). Aristotle in the introduction of his treatise wrote, "It [*Meteorologica*] studies also all the affections we may call common to air and water, and the kinds and parts of the earth and the affections of its parts. These throw light on the causes of winds and earthquakes and all the consequences the motions of these kinds and parts involve. Of these things some puzzle us, while others admit of explanation in some degree. Further, the inquiry is concerned with the falling of thunderbolts and with whirlwinds and fire-winds, and further, the recurrent affections produced in these same bodies by concretion." Aristotle made some remarkably astute inferences concerning the region between the Earth and the planets which we now call the atmosphere and his text was considered to be the authority on weather theory for almost 2000 years.

The development of scientific instrumentation during the Middle Ages led to the first organized measurements of weather at the Earth's surface. Under the auspices of the Royal Society, Robert Hooke, who experimented in a variety of scientific subjects, published in 1667 a set of instructions and examples called "A Method for Making a History of the Weather" (Hooke, 1667). During the early days of the United States, the founding fathers were particularly interested in practical scientific instruments. Benjamin Franklin experimented with the static electric charge difference between cloud and ground using kites to demonstrate the scientific explanation for lightning. Jefferson made regular observations at Monticello from 1772 to 1778 of pressure and temperature and participated in taking the first known simultaneous weather observations in America. George

Washington also made regular observations; the last weather entry in his diary was made the day before he died (Washington, 2007). During the early and mid-1800s, weather observation networks began to grow and expand across the United States. Although most basic meteorological instruments had existed for over 100 years, it was the telegraph that was largely responsible for the advancement of operational meteorology during the 19th century. With the advent of the telegraph, weather observations from distant points could be rapidly collected, plotted, and analyzed at one location. This organized weather data collection led to the development of National Weather Service bureau's in many countries around the world.

Despite the importance of the land surface for forestry, agriculture, transportation, and human habitation, the ability of space-based remote sensing over land to accurately measure properties of the surface and atmosphere poses special challenges. In particular, passive remote sensing from space requires some knowledge of the land surface reflectivity and emissivity properties in order to interpret observations in the microwave, thermal infrared, and shortwave infrared spectral bands. In contrast to the ocean, the land surface is highly variable on a variety of spatial and temporal scales. Not only is the land composed of many different soil and mineral types but the fraction of vegetation that covers most land surfaces has a strong seasonal dependence. Moreover, unlike the ocean, the lack of thermal conductivity of land surfaces leads to large diurnal variations of the surface skin. These surface temperature variations drive a dynamic boundary layer over land that extends from 1 to 3 km above the surface and is highly dependent on local topography. The planetary boundary layer (PBL) is one of the distinctive features of the atmosphere and offers a significant challenge for remote sensing from space (Stull, 1988). Satellites from the NASA Earth Observing System (EOS) have been used to improve our understanding of both natural and man-made effects on land surface temperature and surface emission properties (Hulley et al., 2019; Loveless et al., 2021). All of the above considerations lead scientists to emphasize the importance of uncertainties in the interpretation of satellite soundings over land.

This chapter presents a partial summary of the state of the art in land-based field measurements for the satellite sensors deployed at the end of the 20th century and the beginning of the 21st. This time period coincides with the recognition of the importance of carbon dioxide emission to the future warming of the planet's atmosphere. The satellite sensors used for weather monitoring then also become important indicators of shifts in the Earth's climate. By necessity we are somewhat selective in the topics included in this chapter, however, we believe that they are representative of activities in the scientific community. The main topic of discussion is the uncertainties of in situ and remote sensing instruments used in land-based field campaigns and their ability to vertically profile the atmosphere. We highlight some field campaigns that leverage the instrumentation at sites of the Department of Energy Atmospheric Radiation Measurement (ARM) program. The work that began in the 1990s at measurement sites in Oklahoma, Alaska, and in the Tropical Western Pacific continues to this day and is augmented by numerous international locations as part of the GCOS Reference Upper Air Network (GRUAN).

2. DOE ARM campaigns

The Department of Energy (DOE) Atmospheric Radiation Measurement (ARM) program has operated three permanent sites and two mobile facilities on a continuous basis since the mid-1990s (Stokes and Schwartz, 1994; Mather and Voyles, 2013; Miller et al., 2016; Ackerman et al., 2016; Cress and Sisterson, 2016). The original recommended ordered set of primary locales for long-term occupancy as ARM climate observation sites was as follows (Patrinos, 1991):

1. Southern U.S. Great Plains (SGP)
2. Tropical Western Pacific Ocean (TWP)
3. Eastern North Pacific Ocean or Eastern North Atlantic Ocean (ENA)
4. North Slope of Alaska (NSA)
5. Gulf Stream off Eastern North America, extending eastward

In addition to routine data collection from facility instruments, these ARM facilities were made available to the scientific community to support land-band field campaigns for a wide variety of atmospheric studies. In the view of an ARM program founder (Stokes, 2016), one of the key contributions of ARM was to the measurement of water vapor, which is described in Turner et al. (2016).

2.1 ARM Water Vapor Intensive Operating Periods

In the 1990s, the DOE ARM program sought to constrain the errors in downwelling infrared flux calculations to less than $1\,W\,m^{-2}$, corresponding to an approximate absolute accuracy of 2% in the total column water vapor amount. The main focus of the ARM Water Vapor Intensive Operating Periods (WVIOP) was to understand and resolve differences between various instruments considered to be candidates for absolute measurements of water vapor. In all cases, this requires a well calibrated and stable instrument, and a sound, proven way of deriving water vapor from the measurements. Initial quandaries regarding this absolute issue going into the 1997 IOP were as follows:

- 5% differences in total precipitable water vapor (PWV) common, range of 15%
- microwave instruments (ARM, NOAA's Environmental Technology Laboratory (ETL)) differ by 5%–10%
- ground-based GPS much drier than ARM microwave, slightly drier than ETL
- Solar instruments differ by 5%–10% also.

After intensive investigation and calibration refinement the 1997 WVIOP obtained the results shown in Fig. 1. This result represents an agreement better than 5% for measurements made with diverse calibration traceability to standards (Revercomb et al., 2003). While this is a considerable improvement, it falls short of the requirement of 2% in the water vapor column needed to constrain global climate models. This uncertainty also applies to the validation of satellite derived water vapor profiles over land where the uncertainty of the sensor used in validation should be significantly less than the satellite product.

2.2 AIRS Cal/Val at ARM

The Atmospheric Infrared Sounder (AIRS) on the NASA Aqua satellite was the first of a new generation of advanced satellite-based atmospheric sounders with the capability of obtaining high-vertical resolution profiles of temperature and water vapor. The high-accuracy retrieval goals of AIRS (e.g., 1 K RMS in 1 km layers below 100 mbar for air temperature, 10% RMS in 2 km layers below 100 mbar for water vapor concentration), combined with the large temporal and spatial variability of the atmosphere and difficulties in making accurate measurements of the atmospheric state, necessitate careful and detailed validation using well-characterized ground-based sites. As part of ongoing AIRS Science Team efforts and a collaborative effort between the NASA Earth Observing System (EOS) project and the Department of Energy Atmospheric Radiation Measurement (ARM) program, data from various ARM and other observations are used to create "best estimates" of the atmospheric column at the Aqua overpass times (Tobin et al., 2006a). The resulting validation data set was an ensemble of temperature and water vapor profiles created from radiosondes launched at the approximate Aqua overpass times, interpolated to the exact overpass time using time continuous ground-based profiles, adjusted to account for spatial gradients within the Advanced Microwave Sounding Unit (AMSU) footprints, and supplemented with limited cloud

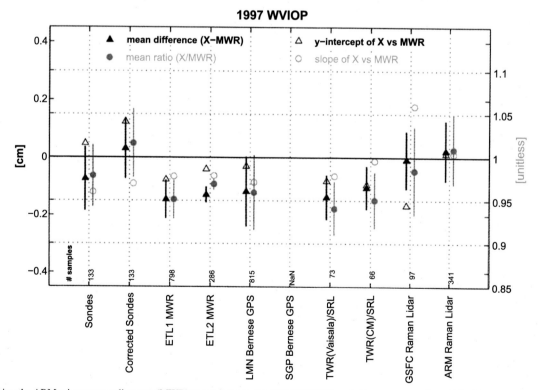

FIG. 1 Using the ARM microwave radiometer (MWR) as a reference, the 1997 WVIOP found all sensors agreed to within 5% despite a wide range of calibration methods. *(From Revercomb, H.E., Turner, D.D., Tobin, D.C., Knuteson, R.O., Feltz W.F., Barnard, J., Bösenberg, J., Clough, S., Cook, D., Ferrare, R., Goldsmith, J., Gutman, S., Halthore, R., Lesht, B., Liljegren, J., Linné, H., Michalsky, J., Morris, V., Porch, W., Richardson, S., Schmid, B., Splitt, M., Hove, T.V., Westwater, E., Whiteman, D., 2003. The ARM Program's water vapor intensive observation periods: overview, initial accomplishments, and future challenges. Bull. Am. Meteor. Soc. 84, 217–236. https://doi.org/10.1175/BAMS-84-2-217, Figure 11.)*

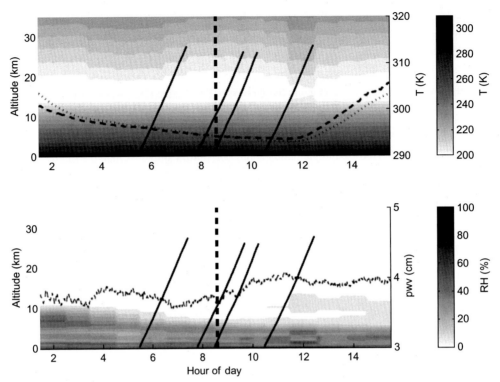

FIG. 2 Time series of ARM temperature and water vapor for July 25, 2002. The *vertical dashed line* indicates the time of the Aqua overpass and the *slanted lines* show the altitude of the radiosondes versus time for this day. *(From Tobin, D.C., Revercomb, H.E., Knuteson, R.O., Lesht, B.M., Strow, L.L., Hannon, S.E., Feltz, W.F., Moy, L.A., Fetzer, E.J., Cress, T.S., 2006. Atmospheric Radiation Measurement site atmospheric state best estimates for Atmospheric Infrared Sounder temperature and water vapor retrieval validation. J. Geophys. Res. Atmos. 111(D9), Figure 2.)*

observations. Estimates of the spectral surface infrared emissivity and local skin temperatures were also constructed. Relying on the developed ARM infrastructure and previous and ongoing characterization studies of the ARM measurements, the data set provides a good combination of statistics and accuracy which is essential for assessment of the advanced sounder products. Combined with the collocated AIRS observations, the products were used to study observed minus calculated AIRS spectra, aimed at evaluation of the AIRS forward radiative transfer model, AIRS observed radiances, and temperature and water vapor profile retrievals. The AIRS retrievals over the tropical ocean were found to have very good accuracy for both temperature and water vapor, with RMS errors approaching the theoretical expectation for clear sky conditions, while retrievals over a midlatitude land site have poorer performance. The results demonstrated the importance of using specialized "truth" sites for accurate assessment of the advanced sounder performance and motivated the continued refinement of the AIRS science team retrieval algorithm, particularly for retrievals over land.

One of the key innovations from the AIRS validation field campaign in 2002 was the launch of two dedicated radiosondes at the ARM sites, the first 1 h before the satellite overpass and the second at the time of the overpass, using two simultaneously operating radiosonde receivers. By using two radiosonde launches, the data can be interpolated to the time of the satellite overpass at each altitude. This greatly improved the representation of temporal change of the atmosphere over the ARM site and removed much of the problem of comparing point in situ observations over the course of a 90 min balloon ascent with the nearly instantaneous observation of the entire vertical profile from the satellite sensor. Fig. 2 illustrates the "dedicated" radiosonde launch concept from Tobin et al. (2006a), Figure 2. In a similar manner, GOES sounding data were used to characterize the spatial uniformity of the region around the ARM SGP site to account for spatial sampling gradients. The surface emissivity of the ARM SGP site region was also measured using a ground-based interferometer and NASA aircraft flights (Tobin et al., 2006b).

2.3 JPSS CrIS Cal/Val at ARM

The Cross-Track Infrared Sounder (CrIS) is an advanced sounder currently flying on two Joint Polar Satellite System (JPSS) Low Earth Orbit (LEO) satellite platforms Suomi National Polar-orbiting Partnership (SNPP) and NOAA-20 with the ability to obtain high-vertical resolution profiles of temperature, water vapor, and other trace gases. The high-accuracy retrieval goals of CrIS (e.g., 0.8–1.2 degree Kelvin RMS for temperature and 0.02–1.2 g/kg for water vapor in broad atmospheric layers and under

FIG. 3 ARM Sites at Eastern North Atlantic (ENA), North Slope Alaska (NSA) at Barrow, Southern Great Plains (SGP), and former ARM site at Tropical West Pacific (TWP) at Manus shown counter-clockwise from top-right. ARM site images are courtesy of the U.S. Department of Energy Atmospheric Radiation Measurement (ARM) user facility.

clear-sky to party-cloudy conditions (Nalli et al., 2018)) requires accurate and ongoing validation data from ground-based sites. Following efforts by Tobin et al. (2006a) to validate AIRS retrievals using the ARM sites, a similar collaborative effort began in July 2012 between the JPSS Project and the ARM program to target SNPP overpasses with radiosonde launches from the NSA, SGP, and TWP sites. The ENA site was added as a validation site in February 2015 approximately 9 months after the closure of the TWP site. These sites are shown in Fig. 3. This effort to target SNPP overpasses ran for approximately 5½ years then pivoted to targeting NOAA-20 overpasses in February of 2018, shortly after NOAA-20 launch. This validation effort was still underway in 2021, when this book was written and it is expected to continue through the JPSS-2 mission and beyond.

During this JPSS validation campaign, an effort is made to target a satellite overpass at each site approximately every 4 days, alternating between daytime and nighttime overpasses. Launch conditions are met if the satellite view angle to the site is less than 30 degrees, there is not steady precipitation, and cloud cover is less than 50% (75% at NSA). Over time, a statistically significant data set is built up that can be used to assess day/night, seasonal, and site-dependent variations. Radiosonde measurements made from launches at the ARM sites have been routinely collected in the NOAA Products Validation System for assessing the SNPP and NOAA-20 satellite sounding products (see Chapter 16).

Satellite overpasses are targeted with either a single radiosonde, which is launched 15 min prior to overpass or with radiosondes bracketing the overpass time, as shown in Fig. 4. With this bracketing strategy, two-balloons are used. The first balloon is launched 45 min prior to overpass, and the 2nd balloon is launched 5-min prior to overpass. The data from these two radiosondes are then interpolated to the time of the overpass at each altitude, which greatly improves the knowledge of the atmosphere at the overpass time. A best estimate (BE) of the atmospheric state is then created using the microwave radiometer to scale the relative humidity and water vapor mixing ratios, the same approach used by Tobin et al. (2006a). The satellite synchronized radiosonde data minimizes the radiosonde-satellite collocation mismatch, allowing for a better understanding of their consistency with satellite measurements (Calbet et al., 2017; Sun et al., 2021).

An optional balloon configuration can also be used in which two radiosondes are flown on the same balloon, also shown in Fig. 4. This configuration, also referred to as a twin sounding, is used to assess radiosonde differences and was used at the ENA, NSA, and SGP sites to assess differences between the Vaisala RS92 and RS41 radiosonde models as part of the GCOS

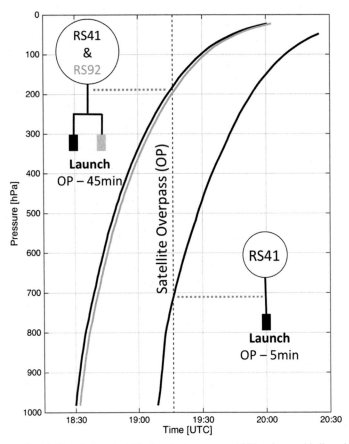

FIG. 4 Two-balloon launch strategy. First balloon is launched 45 min prior to overpass (OP) and second balloon is launched 5 min prior to OP. First balloon is shown with the optional configuration including two different radiosonde models, the Vaisala RS41 and RS92 on the same balloon.

Reference Upper Air Network (GRUAN) intercomparison campaign during the time period when the RS92 radiosonde was being phased out of production and replaced with the newer RS41 model. While analysis of the RS92 and RS41 radiosonde models is still ongoing, this type of launch configuration has allowed for robust comparisons of these two radiosonde models (Jensen et al., 2016; Sun et al., 2019, 2021) and has ensured the long-term homogeneity of the JPSS radiosonde measurement series.

While the ARM validation sites are limited in number, the radiosonde profiles consist of highly accurate measurements of a wide range of climatic conditions. A sample of these radiosonde profiles is shown in Fig. 5 for SNPP-targeted overpasses. ENA is an island site on Graciosa in the Azores. NSA is a coastal site in Alaska with cold and dry conditions. This site is also the most difficult retrieval site because of the low contrast of the clouds and the variable surface, which can include ocean, sea ice, tundra, etc. SGP is a continental site in Oklahoma with wide fluctuations in temperature and water vapor along with a complex surface emission (Knuteson et al., 2001). TWP was a tropical island ARM site with high water vapor amounts and low variability in the atmosphere. This site ceased operations in the Summer of 2014.

The 2-balloon launch strategy, as described previously, was used routinely at the NSA and SGP ARM sites to target SNPP and can be used to estimate the short-term variability in the temperature and water vapor profiles. This variability estimate is simply the differences between individual microwave-scaled radiosonde-pairs and is shown in Fig. 6 for NSA (top-row) and SGP (bottom-row). Individual radiosonde-pair differences are shown in dark gray, and the mean and RMS differences are shown in white dashed and solid lines, respectively. The variability in temperature that occurs within approximately 40 min (the time between radiosonde launches) is on the order of 3/4 K. The water vapor RMS percent differences range from 5% to 30%. Given the high-accuracy retrieval goals of the CrIS instrument, this short-term temporal variability is not negligible and shows how important it is to try to minimize the time between the radiosonde and the satellite measurements.

While validation of the CrIS instruments at the ARM sites is ongoing, measurements from this field campaign have been used to assess not only the maturity of the retrieval algorithm (Nalli et al., 2018) but also to assess the retrievals over many

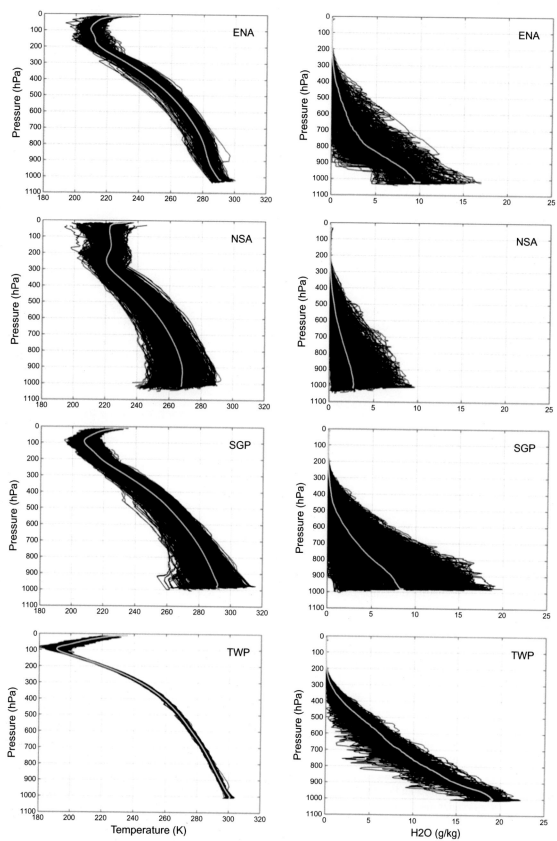

FIG. 5 Sample of radiosonde temperature (left-column) and water vapor (right-column) profiles from ENA (top-row), NSA (2nd-row), SGP (3rd-row), and TWP (bottom-row) ARM sites which targeted SNPP overpasses between July 2012 and January 2018. TWP launches ceased in May 2014 and began at ENA in February 2015. Mean profiles shown in *white*.

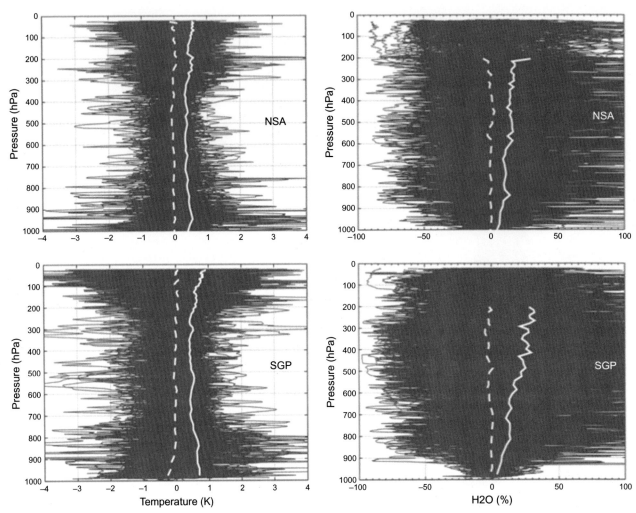

FIG. 6 Temporal variability in temperature (left-column) and water vapor (right-column) at NSA (top-row) and SGP (bottom-row) for SNPP targeted overpasses between July 2012 and January 2018. Individual differences between microwave-scaled radiosonde pairs are shown in *dark gray*. Mean and RMS differences are shown with *white-dashed and white-solid lines*, respectively.

years through various algorithm updates. This is a difficult process, which requires accurate and on-going validation data. The ARM sites are uniquely qualified for this purpose.

3. GRUAN

One of the main goals of the GCOS Reference Upper Air Network (GRUAN) is to perform reference observations of profiles of atmospheric temperature and humidity for the purpose of monitoring climate change and satellite validation (Bodeker et al., 2016; GCOS, 2013). In this respect, essential criteria for establishing a reference observation are measurement traceability, the correction of all known errors and biases, and the availability of measurement uncertainties. For more than 70 years, radiosondes have been providing in situ profiles of temperature, humidity, and pressure at unmatched vertical resolution, and with currently approximately 800 radiosoundings performed each day, radiosounding constitutes a vast source of data which can be used for, e.g., climate monitoring. However, data products from commercial radiosondes often rely on black-box or proprietary algorithms, which are not disclosed to the scientific user. Furthermore, inevitable changes in the hardware and/or the data processing cause discontinuities or inhomogeneities in the long-term time series. Both factors render these data unsuitable as reference observations. In the case of black-box software, the data processing is not traceable, which simply put means that it is unclear what has been done to the data and that, although manufacturers as a rule are diligent to produce reliable data, one cannot trust the data. The introduction of discontinuities or inhomogeneities by system changes means that long-term data records contain instrument artifacts which hampers trend

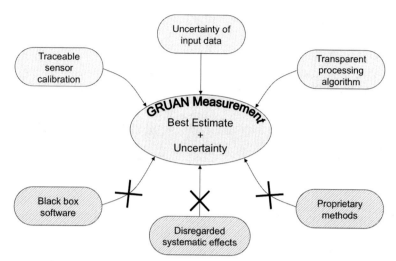

FIG. 7 Schematic representation of requirements for establishing a reference-quality data product. Reference data must be traceable to an accepted standard, include an estimate of the measurement uncertainty, and the data processing must be transparent and well-documented. The hatched boxes on the bottom denote components that are not reconcilable with reference data. The consequence of applying black box software is that one does not know what has been done with the data. Disregarded systematic effects basically means that the data are wrong and the application of proprietary methods means that the data cannot be reprocessed by a third party if the need arises.

analysis. These inhomogeneities are caused by differences in quality of the manufacturer-processed data, and in the ability of the processing software to correct for measurement errors. Incidentally, this implicitly supports the first reservation about trusting black-box processed data (Fig. 7).

3.1 GRUAN data products

The lack of availability of reference observations of water vapor in the upper atmosphere was recognized in IPCC Assessment Report 4, and this contributed to the foundation of GRUAN in 2008, to fulfill the need for reference observations in the free and upper atmosphere (IPCC, 2014). GRUAN's approach to establish reference data quality is by developing correction algorithms that build on an extensive characterization of the measurement errors. This characterization is performed in dedicated setups under laboratory conditions, and the resulting corrections are validated by comparing with other measurements. Examples of the most important sources of error are the solar heating of the temperature sensor, and the time lag of the humidity sensor at low temperatures. To ensure traceability, the GRUAN data processing algorithms are well-documented, the resulting measurement data are traceable to SI standards, and it includes an estimate of the measurement uncertainty.

Currently, GRUAN data products (GDPs) are available for the Vaisala RS92 (Dirksen et al., 2014), RS41 (von Rohden et al., 2021), and Meisei RS-11G (Kizu et al., 2018; Kobayashi et al., 2019) radiosondes. Data products for additional radiosonde models, including frostpoint hygrometers to measure stratospheric water vapor, as well as for other measurement techniques such as GNSS-derived precipitable water vapor column (GNSS-PW) (Ning et al., 2016), lidar and MWR are in various stages of development. The uncertainty estimate of GRUAN processed radiosonde data depends on altitude, and as a rule of thumb it scales with the magnitude of the correction applied. In the case of temperature profiles, the uncertainty increases from 0.1 K at the surface, where the calibration uncertainty dominates, to around 1 K at 30 km altitude for daytime measurements due to the radiation correction. In comparison, for other measurement techniques, such as GNSS-PW, the uncertainty in the retrieved water vapor column is approximately $0.3 \, kg/m^2$, which is considerably smaller than the uncertainty for the water vapor column derived from radiosonde data (around $3 \, kg/m^2$).

3.2 GRUAN sites

The map in Fig. 8 shows an evident scarcity of GRUAN sites on the African and South American continent, as well as over the Pacific. Efforts are made to promote the establishment of GRUAN sites in those regions, but the main reason for the lack of sites in these regions is a financial one: participation in GRUAN is a voluntary effort that is funded by the national weather service or by another government institute. The establishment of a continued long-term measurement program, which is the prerequisite for a GRUAN site, requires secure and stable funding that can not always be provided by National Meteorological Services (NMHs) that are faced with limited or uncertain budgets. Expendables such as radiosondes make up a relevant part of the operational costs for a site, with the result that financial concerns often play a role in the choice of the operational radiosonde. The selection of the operational radiosonde is fully at the discretion of the site or its governing

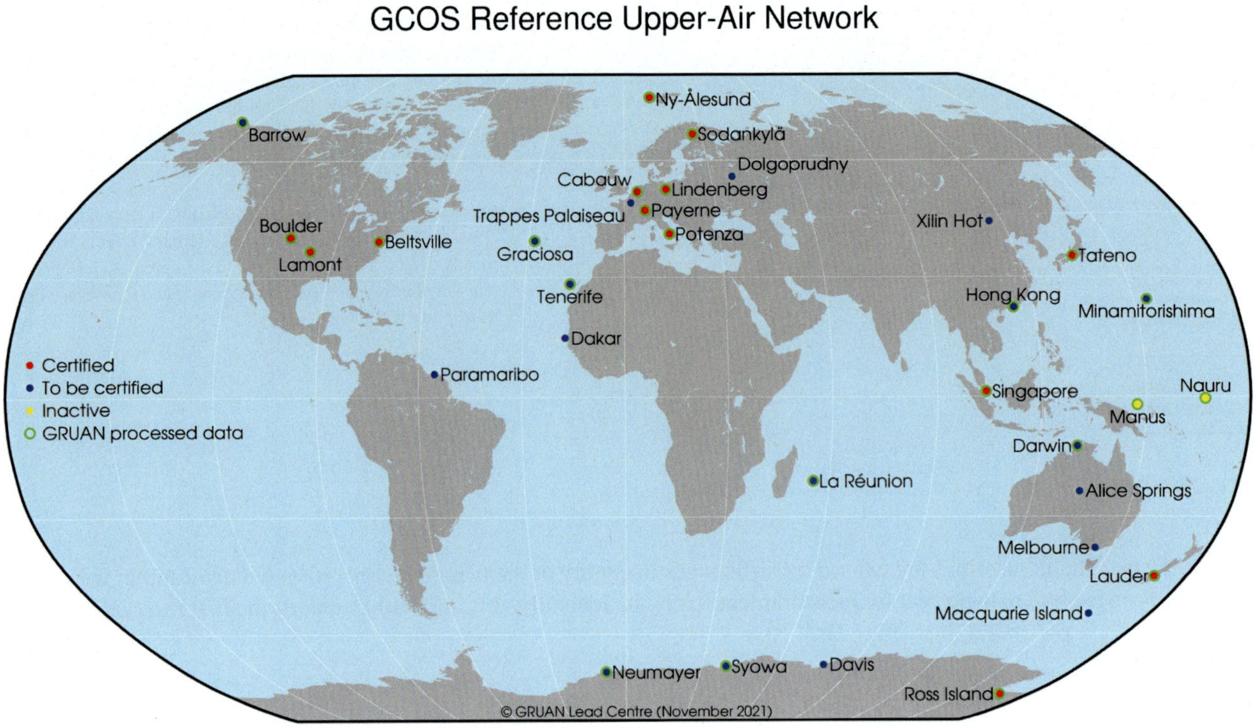

FIG. 8 Map showing the location of GRUAN sites (in July 2021). The color of the *dots* denotes the status of the site, where the classification certified means that the site's measurement program and change management procedures have been assessed and approved by the Working Group on GRUAN. For the former sites Manus and Nauru, the GRUAN measurement programs were terminated in the course of 2013/14.

body, and GRUAN as a World Meteorological Organization (WMO) subsidiary promotes competition in the marketplace and explicitly refrains from expressing a preference for, or from recommending a, specific manufacturer of radiosonde model. Nevertheless, the radiosondes that are employed at GRUAN sites have to fulfill certain requirements regarding measurement quality, so that a GDP of sufficient quality can be generated. In practical terms, this means that the radiosonde must be able to reliably measure humidity profiles up to the tropopause.

3.3 RS92-RS41 radiosonde intercomparisons

Comparison of measurement systems is an inseparable part of GRUAN. Be it through deliberate measurement redundancy or by dedicated campaigns, the principle remains that observing the same measurand with two (or more) independent measurement systems is essential to detect biases between the respective instruments and will provide information on the measurement uncertainties. It is more or less mandatory to perform comparisons in case of changes to the measurement system, in order to prevent aforementioned inhomogeneities in data records. A large-scale intercomparison program was performed following the transition from Vaisala RS92 to RS41 as the operational radiosonde at various GRUAN sites. This included regular RS92-RS41 twin soundings for an extended period. The exact length of this intercomparison effort varies among the sites and depends, e.g., on the sites resources. Further details on the management of this change by GRUAN are discussed by Dirksen et al. (2020). The analysis of the twin soundings from Lindenberg shows that daytime difference between the GDPs for both radiosondes is well below 0.1 K up to 25 km altitude. Above 25 km the GRUAN RS92 is consistently warmer than the RS41, and the difference gradually increases to approximately 0.3 K at 35 km. For nighttime measurements, the temperature difference between the GDPs is, not unexpectedly, negligible up to 30 km altitude, above this altitude it gradually approaches 0.1 K (Fig. 9).

3.4 WMO radiosonde intercomparisons

WMO radiosonde intercomparison campaigns are organized at regular intervals to get an overview of the performance of the various operational radiosounding systems that are currently available. Such campaigns provide important information to NMHs, who partially base the selection of an operational radiosounding system for their network on

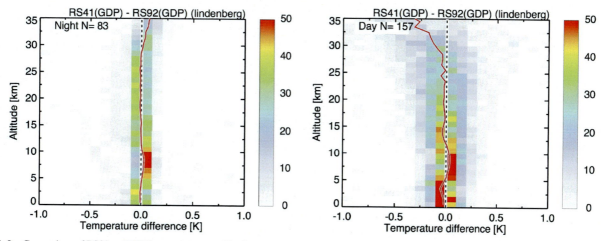

FIG. 9 Comparison of RS92 and RS41 temperature profiles from twin soundings performed between 2016 and 2021 at Lindenberg observatory. Shown are the temperature differences $T_{RS41} - T_{RS92}$ for GRUAN-processed data. The *colors* represent the number of data points in $0.1\,K \times 1\,km$ bins, and the *red trace* represents the median of the difference. Left: the result for 83 nighttime soundings, right: the result for 157 daytime soundings.

the outcome of these campaigns. Furthermore, these campaigns improve the quality and cost-effectiveness of upper air observing systems by providing recommendations on system performances, improvements of instruments and methods of observation, suitable working references to WMO Members and instrument manufacturers. Over time, the number of participating radiosonde models has grown from 5 during the first campaign in 1984/85 to 11 during the last campaign in Yangjiang in 2010 (Nash et al., 2011). With the increasing number of participants, the complexity of the logistics involved for the upcoming Upper Air Instrument Intercomparison (UAII2022) campaign, that is co-hosted by DWD and MeteoSwiss and will take place at Lindenberg observatory in 2022, the number of participating radiosonde manufacturers is limited to 12 (Table 1).

During the intercomparison, a balloon carries an extended rig with the participating radiosondes suspended with a string, allowing them to swing and rotate freely. Doing so ensures that the radiosondes all probe the same air mass under similar measurement conditions as during a regular radiosounding. When the number of participants exceeds the rig's capacity, additional soundings with separate rigs are performed, and a transfer to correlate the individual rigs is needed. This transfer is usually derived from the average of selected radiosondes. In the previous campaigns, the data were compared to a reference that was constructed from the mean of all profiles of an ascent. This gives a good indication of how a specific radiosonde model performs compared to the ensemble, but because of the lack of a reference it does not provide information on the real biases of the systems. During UAII2022, an assessment of the real biases will be achieved by comparing to a reference provided by GRUAN GDPs. These GDPs will also be used as the transfer between the rigs (Fig. 10).

TABLE 1 Year, location, and number of participants in past WMO radiosonde intercomparisons campaigns.

Year	Location	Participants
1984	Bracknell	5
1985	Wallops	5
1989	Dzhambul	5
1993	Tsukuba	7
1995–97	Moscow & Wallops	8
2001	Alcantara	17
2005	Mauritius	7
2010	Yangjiang	11

FIG. 10 Screenshot of a camera recording of a test flight with a rig carrying 10 radiosondes, that will be employed for the 2022 radiosonde intercomparison campaign.

4. Links to other land-based campaigns

This chapter is only able to briefly touch on a few of the numerous land-based field campaigns that continue to be conducted across the globe. We provide here a list of hyperlinks to experiments conducted on every continent to give a flavor of the diversity of field experiments, most of which provide opportunities for students to participate.

Africa
Sub-saharan RADAGAST
https://www.arm.gov/research/campaigns/amf2006radagast
South Africa SAFARI
https://daac.ornl.gov/cgi-bin/dataset_lister.pl?p=18

Antarctica
West Antarctic AWARE campaign
https://www.arm.gov/research/campaigns/amf2015aware
DOME-C Polar-AERI AIRS validation campaign
see Walden et al. (2006)
https://agupubs.onlinelibrary.wiley.com/doi/full/10.1029/2005JD006357

Arctic
ARM North Slope of Alaska Site
https://www.arm.gov/capabilities/observatories/nsa

Asia
China AEROSOL-INDIRECT
https://www.arm.gov/research/campaigns/amf2008aerosolindirect

Australia
NASA SMAP Campaign
https://earthobservatory.nasa.gov/blogs/fromthefield/2015/05/07/smap-campaign-collects-data-from-soil-sky-and-space/

Europe
Spain Grassland PBL
https://link.springer.com/article/10.1023/A:1002609509707
Northern Europe BALTEX
https://link.springer.com/article/10.1023/A:1006567924820

Great Plains
Kansas/Oklahoma, United States PECAN
https://www.arm.gov/research/campaigns/sgp2015aspecan

Greenland
International Summit Station
https://geo-summit.org/summit-station

India
Ganges Valley Aerosol Experiment (GVAX)
https://www.arm.gov/research/campaigns/amf2011gvax

Mountain
Colorado, U.S. SAIL
https://sail.lbl.gov/

South America
Argentina RELAMPAGO/CACTI
https://sites.google.com/illinois.edu/relampago/home
https://www.arm.gov/research/campaigns/amf2018cacti

5. Summary

Land based field campaigns have an advantage over other types of field measurements in that it is relatively easy to bring together traditional measurements, e.g., in situ meteorological observations, with advanced research instruments that require experts to operate and interpret the data. Historically land based remote sensing campaigns at sites with routine in situ observations have been used to obtain unique combined datasets that can be used for a number of purposes. In this chapter we have emphasized a limited selection of observations that have proven particularly useful for validation of satellite temperature and moisture sounding products. The common theme of these observations is to provide a well characterized traceability path from the atmospheric measurements back to international standards (SI). This emphasis on calibration and validation has proven to be important in order to advance the state of atmospheric remote sensing science to achieve the accuracies required for monitoring small changes in the Earth's climate over long periods of time.

Acknowledgment

We would like to acknowledge Bomin Sun (NOAA/NESDIS/STAR) for reviewing the chapter draft and providing constructive feedback.

References

Ackerman, T.P., Cress, T.S., Ferrell, W.R., Mather, J.H., Turner, D.D., 2016. The programmatic maturation of the ARM Program. Meteorol. Monogr. 57, 3-1.

Aristotle (340 BC), 1923. Meteorologica. Oxford Clarendon Press translation. https://archive.org/details/workstranslatedi03arisuoft.

Bodeker, G.E., Bojinski, S., Cimini, D., Dirksen, R.J., Haeffelin, M., Hannigan, J.W., Hurst, D.F., Leblanc, T., Madonna, F., Maturilli, M., Mikalsen, A.C., Philipona, R., Reale, T., Seidel, D.J., Tan, D.G.H., Thorne, P.W., Vömel, H., Wang, J., 2016. Reference upper-air observations for climate: from concept to reality. Bull. Am. Meteorol. Soc. 97, 123–135. https://doi.org/10.1175/bams-d-14-00072.1.

Calbet, X., Peinado-Galan, N., Rípodas, P., Trent, T., Dirksen, R., Sommer, M., 2017. Consistency between GRUAN sondes, LBLRTM and IASI. Atmos. Meas. Tech. 10, 2323–2335. https://doi.org/10.5194/amt-10-2323-2017.

Cress, T.S., Sisterson, D.L., 2016. Deploying the ARM sites and supporting infrastructure. Meteorol. Monogr. 57, 5-1.

Dirksen, R.J., Sommer, M., Immler, F.J., Hurst, D.F., Kivi, R., Vömel, H., 2014. Reference quality upper-air measurements: GRUAN data processing for the Vaisala RS92 radiosonde. Atmos. Meas. Tech. 7 (12), 4463–4490. https://doi.org/10.5194/amt-7-4463-2014.

Dirksen, R.J., Bodeker, G.E., Thorne, P.W., Merlone, A., Reale, T., Wang, J., Hurst, D.F., Demoz, B.B., Gardiner, T.D., Ingleby, B., Sommer, M., von Rohden, C., Leblanc, T., 2020. Managing the transition from Vaisala RS92 to RS41 radiosondes within the Global Climate Observing System Reference Upper-Air Network (GRUAN): a progress report. Geosci. Instrum. Methods Data Syst. 9 (2), 337–355. https://doi.org/10.5194/gi-9-337-2020. https://gi.copernicus.org/articles/9/337/2020/.

GCOS, 2013. The GCOS Reference Upper Air Network (GRUAN) Manual. Technical Report 170, WMO/WIGOS Technical Report 2013-02 https://library.wmo.int/index.php?lvl=notice_display&id=15181#.XyF0pDXgqUl.

Hooke, R., 1667. A method for making a history of the weather. In: The History of the Royal Society of London. For the Improving of Natural Knowledge. The Royal Society Publishing, London, pp. 173–179.

Hulley, G., et al., 2019. New ECOSTRESS and MODIS land surface temperature data reveal fine-scale heat vulnerability in cities: a case study for Los Angeles County, California. Remote Sens. (Basel) 11 (18), 2136.

IPCC, 2014. Climate change 2014: synthesis report. In: Core Writing Team, Pachauri, R.K., Meyer, M.A. (Eds.), Contribution of Working Groups I, II and III to the Fifth Assessment Report of the Intergovernmental Panel on Climate Change. IPCC, Geneva, Switzerland, p. 151.

Jensen, M.P., Holdridge, D.J., Survo, P., Lehtinen, R., Baxter, S., Toto, T., Johnson, K.L., 2016. Comparison of Vaisala radiosondes RS41 and RS92 at the ARM Southern Great Plains site. Atmos. Meas. Tech. 9, 3115–3129. https://doi.org/10.5194/amt-9-3115-2016.

Kizu, N., Sugidachi, T., Kobayashi, E., Hoshino, S., Shimizu, K., Maeda, R., Fujiwara, M., 2018. GRUAN Technical Document 5—Technical characteristics and GRUAN data processing for the Meisei RS-11G and iMS-100 radiosondes. Technical report, GRUAN Lead Centre, Lindenberg. https://www.gruan.org/documentation/gruan/td/gruan-td-5. (version 1.0).

Knuteson, R.O., Osborne, B., Short, J., Revercomb, H.E., Tobin, D.C., Nasiri, S.L., 2001. Progress towards a characteristic of the infrared emissivity of the land surface in the vicinity of the ARM SGP central facility: surface (S-AERI) and airborne sensor (NAST-I/S-HIS). In: Carrothers, D.A. (Ed.), Proceedings of the Eleventh Atmospheric Radiation Measurement (ARM) Science Team Meeting. U.S. Department of Energy, Richland, WA.

Kobayashi, E., Hoshino, S., Iwabuchi, M., Sugidachi, T., Shimizu, K., Fujiwara, M., 2019. Comparison of the GRUAN data products for Meisei RS-11G and Vaisala RS92-SGP radiosondes at Tateno (36.06N, 140.13E), Japan. Atmos. Meas. Tech. 12 (6), 3039–3065. https://doi.org/10.5194/amt-12-3039-2019. https://www.atmos-meas-tech.net/12/3039/2019/.

Loveless, M., et al., 2021. Climatology of the combined ASTER MODIS emissivity over land (CAMEL) version 2. Remote Sens. (Basel) 13 (1), 111.

Mather, J.H., Voyles, J.W., 2013. The ARM Climate Research Facility: a review of structure and capabilities. Bull. Am. Meteorol. Soc. 94 (3), 377–392.

Miller, M.A., Nitschke, K., Ackerman, T.P., Ferrell, W.R., Hickmon, N., Ivey, M., 2016. The ARM mobile facilities. Meteorol. Monogr. 57, 9-1.

Nalli, N.R., Gambacorta, A., Liu, Q., Barnet, C., Tan, C., Iturbide-Sanchez, F., Reale, T., Sun, B., Wilson, M., Borg, L., Morris, V., 2018. Validation of atmospheric profile retrievals from the SNPP NOAA-unique combined atmospheric processing system. Part 1: temperature and moisture. IEEE Trans. Geosci. Remote Sens. 56 (1). https://doi.org/10.1109/TGRS.2017.2744558.

Nash, J., Oakley, T., Vömel, H., Wei, L., 2011. WMO Intercomparison of High Quality Radiosonde Systems Yangjiang, China, 12 July—3 August 2010. Technical report, WMO. https://www.wmo.int/pages/prog/www/IMOP/publications/IOM-107%_Yangjiang.pdf. WMO/TD-No. 1580, Instruments And Observing Methods Report No. 107.

Ning, T., Wang, J., Elgered, G., Dick, G., Wickert, J., Bradke, M., Sommer, M., Querel, R., Smale, D., 2016. The uncertainty of the atmospheric integrated water vapour estimated from GNSS observations. Atmos. Meas. Tech. 9 (1), 79–92. https://doi.org/10.5194/amt-9-79-2016. http://www.atmos-meas-tech.net/9/79/2016/.

Patrinos, A., 1991. Identification, Recommendation, and Justification of Potential Locales for ARM Sites Executive Summary (No. DOE/ER-0494T). DOE Office of Science Atmospheric Radiation Measurement (ARM) Program (United States).

Revercomb, H.E., Turner, D.D., Tobin, D.C., Knuteson, R.O., Feltz, W.F., Barnard, J., Bösenberg, J., Clough, S., Cook, D., Ferrare, R., Goldsmith, J., Gutman, S., Halthore, R., Lesht, B., Liljegren, J., Linné, H., Michalsky, J., Morris, V., Porch, W., Richardson, S., Schmid, B., Splitt, M., Hove, T.V., Westwater, E., Whiteman, D., 2003. The ARM Program's water vapor intensive observation periods: overview, initial accomplishments, and future challenges. Bull. Am. Meteorol. Soc. 84, 217–236. https://doi.org/10.1175/BAMS-84-2-217.

Stokes, G.M., 2016. Original ARM concept and launch. Meteorol. Monogr. 57, 2-1.

Stokes, G.M., Schwartz, S.E., 1994. The Atmospheric Radiation Measurement (ARM) Program: programmatic background and design of the cloud and radiation test bed. Bull. Am. Meteorol. Soc. 75 (7), 1201–1222.

Stull, R.B., 1988. Boundary layer clouds. In: An Introduction to Boundary Layer Meteorology. Springer, Dordrecht, pp. 545–585.

Sun, B., Reale, T., Schroeder, S., Pettey, M., Smith, R., 2019. On the accuracy of Vaisala RS41 versus RS92 upper-air temperature observations. J. Atmos. Oceanic Technol. 36, 635–653. https://doi.org/10.1175/JTECH-D-18-0081.1.

Sun, B., Calbet, X., Reale, A., Schroeder, S., Bali, M., Smith, R., Pettey, M., 2021. Accuracy of Vaisala RS41 and RS92 upper tropospheric humidity compared to satellite hyperspectral infrared measurements. Remote Sens. (Basel) 13, 173. https://doi.org/10.3390/rs13020173.

Tobin, D.C., Revercomb, H.E., Knuteson, R.O., Lesht, B.M., Strow, L.L., Hannon, S.E., Feltz, W.F., Moy, L.A., Fetzer, E.J., Cress, T.S., 2006a. Atmospheric Radiation Measurement site atmospheric state best estimates for Atmospheric Infrared Sounder temperature and water vapor retrieval validation. J. Geophys. Res. 111, D09S14. https://doi.org/10.1029/2005JD006103.

Tobin, D.C., Revercomb, H.E., Knuteson, R.O., Best, F.A., Smith, W.L., Ciganovich, N.N., Dedecker, R.G., Dutcher, S., Ellington, S.D., Garcia, R.K., Howell, H.B., LaPorte, D.D., Mango, S.A., Pagano, T.S., Taylor, J.K., van Delst, P.Vinson, Werner, K.H., 2006b. Radiometric and spectral validation of Atmospheric Infrared Sounder observations with theaircraft-based Scanning High-Resolution Interferometer Sounder. J. Geophys. Res. 111, D09S02. https://doi.org/10.1029/2005JD006094.

Turner, D.D., Mlawer, E.J., Revercomb, H.E., 2016. Water vapor observations in the ARM program. The atmospheric radiation measurement program: the first 20 years. Meteorol. Monogr. 57. https://doi.org/10.1175/AMSMONOGRAPHS-D-15-0025.1. Amer. Meteor. Soc.

von Rohden, C., Sommer, M., Naebert, T., Motuz, V., Dirksen, R.J., 2021. Laboratory characterisation of the radiation temperature error of radiosondes and its application to the GRUAN data processing for the Vaisala RS41. Atmos. Meas. Tech. Discuss., 1–36. https://doi.org/10.5194/amt-2021-187.

Walden, V.P., Roth, W.L., Stone, R.S., Halter, B., 2006. Radiometric validation of the atmospheric infrared sounder over the Antarctic plateau. J. Geophys. Res. Atmos. 111 (D9), D09S03. https://doi.org/10.1029/2005JD006357.

Washington, G., 2007. The Papers of George Washington [Electronic Resource]: Digital Edition/Theodore J. Crackel, Editor in Chief. University Press of Virginia, Charlottesville, Va. 1732–1799 https://catalog.loc.gov/vwebv/holdingsInfo?bibId=15463891.

Further reading

ARM. https://arm.gov.
GRUAN. https://www.gruan.org.
IPCC. https://www.ipcc.ch/report/ar5/syr/.
JPSS Cal/Val. https://www.arm.gov/research/campaigns/sgp2017rdosnppjpss.
NWS. https://www.weather.gov/upperair/net-info.
WVIOP. https://arm.gov/research/campaigns/sgp2000fallwv.

Chapter 13

Aircraft vertical profile measurements for evaluation of satellite retrievals of long-lived trace gases

Eric A. Kort[a] and Kathryn McKain[b,c]

[a]*University of Michigan, Ann Arbor, MI, United States,* [b]*Cooperative Institute for Research in Environmental Sciences, University of Colorado, Boulder, CO, United States,* [c]*NOAA Global Monitoring Laboratory (GML), Boulder, CO, United States*

Chapter outline

1. Introduction — 235
2. "Global" intensive profiling campaigns — 236
   2.1 HIPPO: HIaper Pole-to-Pole Observations — 236
   2.2 ATom: Atmospheric Tomography Mission — 239
3. Sustained observations — 240
   3.1 Global Greenhouse Gas Reference Network's Aircraft Network — 240
   3.2 Deeter et al. (2019): Example of Role 1 — 241
   3.3 TCCON: Total Carbon Column Observing Network—Example of Role 3 — 241
4. Aggregating aircraft observations from regional intensives — 243
   4.1 Chevallier et al. (2019): Example of (2) — 243
5. Summary — 243
Acknowledgments — 243
References — 244

When you can measure what you are speaking about, and express it in numbers, you know something about it; but when you cannot measure it, when you cannot express it in numbers, your knowledge is of a meagre and unsatisfactory kind.

Lord Kelvin

1. Introduction

Atmospheric trace gas abundances can be highly variable throughout the atmosphere, and this variability is driven by a combination of emissions or production of the gas, removal or chemical destruction, and atmospheric transport and mixing. Measurements of atmospheric trace gases capture spatial and temporal variability, thus supporting studies to better understand controls on their abundances as well their impacts on climate and air quality. Space-based observations can provide insights on local to global-scale trends and processes, informing on important questions such as the magnitude and changes in emissions of gases such as carbon dioxide (CO_2) and methane (CH_4) (e.g., Liu et al., 2021; Wu et al., 2020; Zhang et al., 2020, 2021). To maximize the value of space-based observations of trace gases, which cannot be directly calibrated, they must be linked to robust calibration scales consistent with in situ measurements, and observed variability must be driven by true variability in atmospheric abundances.

Calibration, validation, and evaluation of remote sensing retrievals of trace gases are based on multiple pillars. Radiometric calibration provides evaluation of the radiometric signal observed from space (e.g., Bruegge et al., 2019). Validation of retrieved trace gas columns is often founded on comparison with the Total Carbon Column Observation Network (TCCON), where ground-based remote-sensing stations around the globe provide long-term evaluation and calibration correction of retrieved concentrations from space with calibrated measurements (Wunch et al., 2011). Evaluation of remote sensing retrievals can take many forms, including comparing with balloon-based profiles as highlighted in an earlier chapter (Baier et al., 2022). This chapter will focus on intensive programs to obtain vertical profile measurements of trace gases from aircraft. Such programs most often takes the form of flight campaigns that are designed to achieve a set of science objectives, which may include targeted sampling to specifically evaluate satellite-based observations. Routine aircraft vertical profile observations, such as from the network run by NOAA (Sweeney et al., 2015) and described in detail later in this chapter, can also support satellite-based retrieval evaluation. Aircraft in situ vertical profile measurements fulfill the following unique and complementary roles for evaluating remote-sensing measurements to ensure they produce robust and consistent products.

1. Direct evaluation of space-based trace gas estimates via comparison of coincident observations.

   This application of airborne measurements is perhaps the most obvious for supporting satellite remote sensing. When in situ vertical profiles are sampled where space-based sensors are observing, direct comparisons can be made between retrieved trace gas columns and direct in situ observations. This can be done for specific periods of time and on regional to global scales to support evaluations of the utility and limitations of satellite retrievals and to identify potential artifacts.

   When considering long-lived trace gases, measurement requirements are strict. For example, to be able to use space-based observations of CO_2 to improve understanding of natural and anthropogenic fluxes, they must meet stringent accuracy requirements. Estimates require biases less than 0.3% of global background abundances, corresponding to less than 0.5–1 ppm (Miller et al., 2007). This requirement is an extraordinary challenge for remote sensing, and to ensure such requirements are met, intensive airborne profiling campaigns are essential to evaluate whether such requirements are met.

2. Independent observations to evaluate surface fluxes derived from space-based observations.

   In this case, a regional or global modeling system will optimize surface fluxes to best match satellite retrievals of a trace gas, typically through an assimilation or inversion system. Intensive airborne campaigns provide high-resolution data that can be used to evaluate optimized model results inferred from satellite observations. By comparing airborne data with simulations driven by the optimized fluxes, one can evaluate if these optimized fluxes improve, worsen, or have negligible impact on the model's representation of the atmosphere and identify errors in inferred fluxes that may arise due to unrecognized systematic errors in remote sensing data.

3. Link in situ observations and remote sensing retrievals on one unified calibration scale.

   To improve compatibility between in situ and remote-sensing systems, observations, and retrievals must be on the same absolute calibration scale. In situ vertical profile measurements are needed to place remote sensing trace gas measurements onto established calibration scales, linking the tiered earth observing system.

   In this chapter, we highlight aircraft profiling efforts through both shorter-term, campaign-style intensives and long-term network measurements, identifying examples of how these three roles have been fulfilled. We include near-global scale campaigns and ongoing network observations, as well as aggregations of regional intensive campaigns, and focus primarily on two longer lived trace gases in the atmosphere, CO_2 and CH_4.

   Most aircraft campaigns focus on specific regions over short periods of time (weeks to months). Typically, regional campaigns are designed with a targeted set of scientific objectives in mind: for example, to better understand CO_2 air-sea gas exchange in the Southern Ocean (Stephens et al., 2018). Vertical profiling of the aircraft (i.e., flying in a pattern that samples from near the surface to higher altitudes, often limited by the aircraft's ceiling) can be necessary for the science objectives of the specific campaign, or be conducted specifically to support and evaluate remote sensing retrievals. Some airborne campaigns conduct extensive spatial sampling, but without deep (~15 km) vertical profiles, so their value for evaluating satellite remote sensing is more limited, as satellite retrievals typically have sensitivity through much of the troposphere. In this chapter, we focus on campaigns with extensive vertical profiling.

2. "Global" intensive profiling campaigns

2.1 HIPPO: HIaper Pole-to-Pole Observations

The HIPPO program was designed to provide finely vertically and horizontally resolved in situ observations of a large suite of trace gases spanning from high northern (~85N) to high southern latitudes (~67S) (Wofsy et al., 2011). Five deployments with the National Science Foundation (NSF) Gulfstream V aircraft took place covering all seasons between 2009 and 2011, sampling in vertical profiles from just above the ocean surface to nearly 15 km above sea level, mostly over the Pacific Ocean, with profiles occurring every 2–3 degrees of latitude (Fig. 1).

The aircraft payload primarily focused on long-lived atmospheric trace gases, with redundant in situ measurements of gases such as CO_2 and CH_4, as well as flask sampling. All in situ trace gases were calibrated and traced to the relevant World Meteorological Organization (WMO) calibration scale, linking the observed in situ trace gases to global standards.

This series of campaigns provided a first set of observations of many trace gases with such dense vertical and global coverage. This type of sampling provides unique value to satellite remote sensing evaluations—one common set of

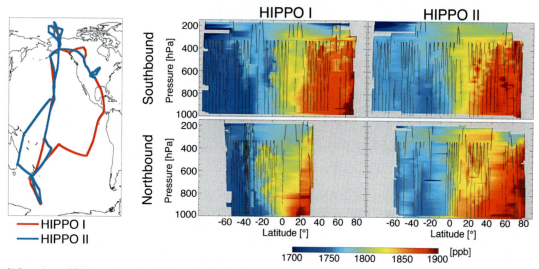

FIG. 1 (Left) Locations of flight tracks and vertical profiles for the first and second HIPPO deployment. (Right) Altitude-latitude cross-section of flight tracks and interpolated CH_4 observations. Note the extensive vertical profiling of the aircraft, which facilitates evaluation of satellite observations that span the globe. *(From Wecht, K.J., et al., 2012. Validation of TES methane with HIPPO aircraft observations: implications for inverse modeling of methane sources. Atmos. Chem. Phys. 12, 1823–1832. https://doi.org/10.5194/acp-12-1823-2012.)*

instruments sampled over a large span of the globe (with very different solar zenith angle, height of tropopause, etc.), providing a consistent and traceable truth, against which to challenge and evaluate space-based remote sensing retrievals. Since space-based remote sensing retrievals of trace gases detect total column abundances (commonly denoted as XGAS, such as XCO_2), deep vertical profiles (from near the surface up to ~15 km) are essential for evaluation, as trace gas abundance and variability occurring anywhere in the vertical column can impact the space-based total column retrieval (depending on the specific averaging kernel for that retrieval).

Below we give three examples of how HIPPO observations have been used to evaluate satellite remote sensing measurements of trace gases.

2.1.1 Wecht et al. (2012): Example of Role 1

Space-based remote sensing measurements of CH_4, with their global coverage, are often analyzed to find optimal estimates of emissions, thus improving our understanding of CH_4 emissions, resolved in space and by emissions sector. To ensure such an analysis is producing optimized emissions, and not aliasing satellite retrieval biases into fluxes, evaluation of the satellite product is critical. In Wecht et al. (2012), the authors evaluated two different retrievals of CH_4 from the Tropospheric Emissions Spectrometer (TES) and assessed the utility of the TES retrievals (V004, V005) for inverse modeling to determine optimized CH_4 emissions. Essential to this study were the 151 vertical profiles of CH_4 observations collected on the first two HIPPO deployments that span from high northern to high southern latitudes, enabling evaluation of TES retrievals across hemispheres and a large latitude range (Fig. 2).

This comparison illustrated a mean bias of 65.8 ppb (3.7%) and a residual standard deviation of 43.8 ppb (2.4%) in TES retrievals of CH_4. No clear latitudinal biases emerged, though larger residual noise was found at higher northern latitudes, coincident with when profiles occurred over land, where larger concentration variations are expected (Fig. 2). This type of latitudinal assessment is one of the unique values of a near-global aircraft profiling campaign, and the lack of obvious latitudinal dependencies suggests promise for optimizing emissions without aliasing latitudinally dependent systematic errors into emissions estimates. However, the residual noise in this case was of sufficient size, that propagating the findings of this evaluation into a model study found that the V004 TES retrieval has limited value for improving our understanding of surface fluxes, whereas V005, with improved precisions (though still in development at the time of publication) appeared to be more promising. This example from Wecht et al. (2012) demonstrates one example of the value of near-global vertical profiles for assessment of satellite retrievals and their utility in global inverse modeling.

2.1.2 Frankenberg et al. (2016): Example of Role 1

As discussed earlier, measurements of CO_2 must meet extremely strict requirements to be sufficient for improving our understanding of flux processes. In Frankenberg et al. (2016), HIPPO aircraft data were used to both evaluate three different

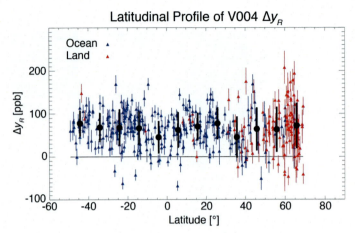

FIG. 2 Differences by latitude between the TES V004 product and CH$_4$ measured during HIPPO-1 and HIPPO-2. Each symbol represents an individual HIPPO profile, and the *vertical bars* show theoretical standard deviations (s.d.) in the TES retrievals. *Black circles* and *vertical bars* are means and s.d. binned by 10 degree latitude bins. No clear latitudinal biases emerge in this comparison. *(Adapted from Wecht, K.J., et al., 2012. Validation of TES methane with HIPPO aircraft observations: implications for inverse modeling of methane sources. Atmos. Chem. Phys. 12, 1823–1832. https://doi.org/10.5194/acp-12-1823-2012.)*

space-based remote sensing products (Greenhouse gases Observing SATellite (GOSAT), Tropospheric Emission Spectrometer (TES), and the Atmospheric InfRared Sounder (AIRS)) as well as two different models (CarbonTracker (Peters et al., 2007 with updates documented at http://carbontracker.noaa.gov) and MACC (Monitoring Atmospheric Composition and Climate, http://www.copernicus-atmosphere.eu/)) that assimilated in situ data to produce optimal surface fluxes.

The broad latitudinal coverage of HIPPO enabled a unique evaluation of both the space-based retrievals and model products. Overall comparison statistics between HIPPO observations and space-based retrievals as well as model estimates are shown in Table 1.

Both model simulations showed excellent overall agreement with HIPPO observations, exhibiting very high r^2 values, with the largest disagreements occurring at high latitudes, possibly attributable to vertical transport errors. The three satellite comparisons are significantly different, with GOSAT agreeing with in situ observations most closely, TES largely agreeing within instrument retrieval noise estimates, but AIRS exhibiting large and latitudinally dependent biases that would need to be addressed to use AIRS-retrieved CO$_2$ to infer fluxes.

Such a comparison led to clear guidance as to which satellite could be used for a given scientific purpose and where future efforts should be focused to meet scientific needs. AIRS' large biases and larger disagreement with HIPPO data compared with global models indicates that improvements in this product would be a prerequisite for use to improve understanding of surface fluxes beyond what can be learned from in situ data. TES appears to perform as designed, but is unlikely to improve evaluation of global simulations. GOSAT shows excellent performance relative to in situ data, suggesting that its use in a global inversion framework could provide improved flux estimates.

TABLE 1 Summary of all HIPPO comparisons.

	No. profiles	r^2	Slope	μ (ppm)	σ (ppm)	σ_{CT}	σ_{MACC}
GOSAT	94	0.85	0.99	−0.06	0.45	0.42	0.36
TES	135	0.75	1.45	0.34	1.13	0.36	0.3
AIRS	200	0.37	0.66	1.11	1.46	0.63	0.47
CT2013B	676	0.93	0.95	0.10	0.51	N/A	N/A
MACC	674	0.95	1.00	0.06	0.43	N/A	N/A

No. of profiles shows how many HIPPO profiles were used for the comparison. Correlation coefficients, fitted slope, mean difference μ, and standard deviation σ of the difference compared to HIPPO of all comparisons are computed using measurements normalized by the respective campaign average. For comparison, σ of model-HIPPO for the satellite co-locations and respective sensitivity are provided as well.
From Frankenberg, C., et al., 2016. Using airborne HIAPER Pole-to-Pole Observations (HIPPO) to evaluate model and remote sensing estimates of atmospheric carbon dioxide. Atmos. Chem. Phys. 16, 7867–7878. https://doi.org/10.5194/acp-16-7867-2016.

2.1.3 Deng et al. (2015): Example of Role 2

In some cases, aircraft in situ data may be used to correct model simulations or optimize satellite data, with implications for resulting surface flux estimates. Deng et al. (2015) considered a potential problem in model representation of the upper tropopause/lower stratosphere (UTLS) region of the atmosphere. Deep vertical profiles with in situ measurements across the UTLS, as provided by HIPPO, were central to conducting such an evaluation. Deng et al. used the HIPPO observations to correct GEOS-Chem simulations that had been optimized with space-based retrievals of CO_2 (Fig. 3).

An overestimate of CO_2 in the UTLS was identified and corrected, and the impact on flux estimates was evaluated. Such differences can lead to extraordinarily large shifts in carbon budgets. Correction of the Arctic UTLS mismatch shown in Fig. 3, led to changes of fluxes in North America, Europe, and tropical Asia of 19%, 13%, and 49%. A similar mismatch identified in the tropical UTLS, when corrected, led to reduction of emissions in South America by 77%. Identification of such problems and addressing them are critical steps needed to provide robust evaluation of surface fluxes of CO_2 from space-based observations, and global-scale intensive aircraft profile observations play an essential role in this task.

2.2 ATom: Atmospheric Tomography Mission

The Atmospheric Tomography Mission (ATom) was a second aircraft vertical profiling campaign with nearly pole-to-pole latitudinal and multi-seasonal coverage similar to HIPPO, but with flights over the Atlantic and Pacific Ocean basins and a larger suite of atmospheric composition measurements as enabled by the NASA DC8 research aircraft (Thompson et al. 2021). ATom took place during 2016–18 and consisted of four global circuits, one to cover each season, with vertical profiles spanning 0.15–13 km MSL in altitude (Fig. 4). Several hundred separate measurements of trace gas, aerosols, and radical species were made on ATom. For this chapter, we focus only on the measurements of the long-lived trace gases CO_2, CH_4, and carbon monoxide (CO) during ATom and their application for evaluating satellite remote sensing retrievals of the same gases.

2.2.1 Nalli et al. (2020): Example of Role 1

Nalli et al. leverage the ATom measurements to evaluate vertical profile measurements of CO_2, CH_4 and CO made from the Cross-track Infrared Sounder (CrIS) instrument aboard the NOAA-20 and Suomi National Polar-orbiting Partnership (SNPP) satellites. Co-location criteria for the comparison were ≤100 km distance and ≤1.5 h time difference. This comparison with ATom enabled the quantitative conclusion that CrIS retrievals perform with accuracy of 5% and precision of 15% for CO, and accuracies of 1% for CH_4 and CO_2, and precision of 1.5% for CH_4 and 0.5% for CO_2. This indicates that both retrievals are performing well within their Level 1 design requirements except for CH_4 precision, which is exceeding

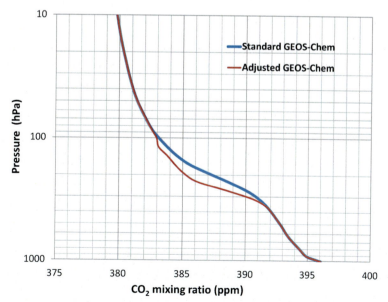

FIG. 3 The mean profile of CO_2 in the Arctic before *(in blue)* and after *(in red)* the adjustment in CO_2 in the UTLS based on the HIPPO-3 CO_2/O_3 correlations. *(From Deng, F., et al., 2015. Sensitivity analysis of the potential impact of discrepancies in stratosphere–troposphere exchange on inferred sources and sinks of CO_2. Atmos. Chem. Phys. 15, 11773–11788. www.atmos-chem-phys.net/15/11773/2015/.)*

240 PART | II Intensive campaigns

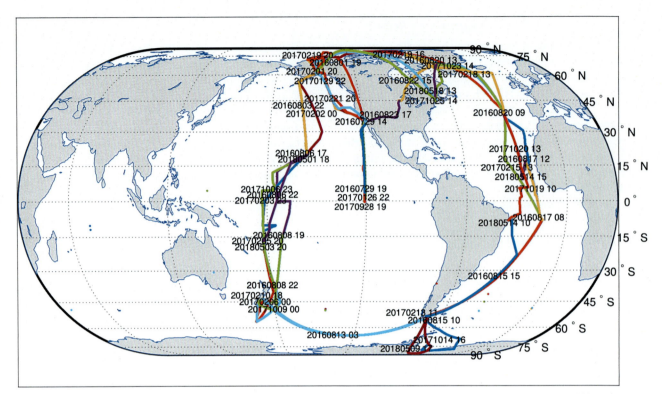

FIG. 4 Flight tracks from ATom, labeled and colored by date, illustrating the zonal and meridional coverage. *(From Nalli, N.R., et al., 2020. Validation of carbon trace gas profile retrievals from the NOAA-unique combined atmospheric processing system for the cross-track infrared sounder. Remote Sens. 12, 3245. https://doi.org/10.3390/rs12193245.)*

the stated requirement of 1%. This work highlights the fact that these global aircraft campaigns are the single most valued evaluation tool for SNPP and NOAA-20 and that the maturity of the satellite products rely heavily on how they compare with global campaign data sets like ATom.

2.2.2 Martinez-Alonzo et al. (2020): Example of Role 1

TROPOspheric Monitoring Instrument (TROPOMI) retrieves CO from solar-reflected radiances, including from over the ocean in the presence of clouds, by approximating the partial column below the cloud top using model-based reference profiles. In situ CO vertical profiles from ATom-4 ($N=103$) were used in a comparison with TROPOMI CO (Martinez-Alonzo et al., 2020) to evaluate this approach. Excellent agreement was found between ATom and TROPOMI CO (Fig. 5), well below the design requirements and with no apparent latitudinal bias, demonstrating the success of the approach for retrieving CO in the near-infrared from over open ocean.

The several examples given above demonstrate that global aircraft profiling campaigns like HIPPO and ATom yield some of the most useful data sets for direct evaluation of satellite retrievals on trace gas measurements using in situ measurements. HIPPO and ATom offer the opportunity to investigate potential latitudinal biases (and meridional in the case of ATom) with sample sizes of several hundred individual profiles. However, HIPPO and ATom represent snapshots in time and do not offer the opportunity to investigate long-term drift or interannual variability in satellite trace gas retrievals. Campaigns of the scale like HIPPO and ATom are challenging to fund and manage, but their value added is unique and future campaigns of this scale would provide substantial value for global satellite and model evaluation.

3. Sustained observations

3.1 Global Greenhouse Gas Reference Network's Aircraft Network

The NOAA Global Monitoring Laboratory (GML) operates a long-running program for collecting vertical profile measurements of trace gases from a network of sites mostly in North America (Sweeney et al., 2015). For this program, NOAA contracts with private pilots to collect flask air samples from a light aircraft approximately twice per month. This frequency

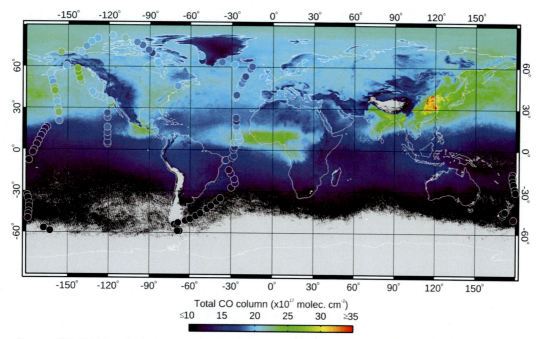

FIG. 5 Map of averaged TROPOMI total CO column values acquired between April 24 and May 21, 2018, the duration of the ATom-4 campaign. *Circles* show ATom-4 profiles spatially and temporally colocated with single TROPOMI retrievals; *circles* are color-coded according to their retrieval-simulated (smoothed) ATom total CO column value. There is good agreement between the two data sets, despite differences in the time span and footprint size each of them represents. *(From Martinez-Alonzo, S., et al., 2020. 1.5 years of TROPOMI CO measurements: comparisons to MOPITT and ATom. Atmos. Meas. Tech. 13, 4841–4864. https://doi.org/0.5194/amt-13-4841-2020.)*

has varied over time with different funding levels and based on site-related operational circumstances. The longest running site in the Aircraft Network has been running since 1992, with most other sites beginning in the early to mid-2000s. In the Aircraft Network, flask air samples are analyzed for a large suite of ~55 trace gases, including the major greenhouse gases, hydrocarbons, and halocarbons. Since the program utilizes light aircraft, maximum altitudes are ~7–8 km asl at most sites, which is lower than what can be achieved with larger dedicated research aircraft, such as those used for ATom and HIPPO. Whereas campaigns like HIPPO and ATom provide vertical profiles with uniquely comprehensive spatial coverage, data from the NOAA GML Aircraft Network provide unique temporal coverage of vertical profiles over long time periods.

3.2 Deeter et al. (2019): Example of Role 1

How does long-time series vertical profile information help in evaluating satellite retrievals? Deeter et al. (2019) consider MOPITT (Measurements of Pollution in the Troposphere) retrievals of CO made in the thermal and near infrared from the NASA Terra satellite. MOPITT observations have continued since 2000, meaning nearly 20 years of measurements were available at the time of this work. This long time series can support analysis on trends in CO emissions and atmospheric burdens, even small trends that manifest over longer time frames, an important science question. However, this type of analysis requires robust understanding of any drift that may be present in the satellite product. In situ vertical profiles of CO from the ongoing NOAA Aircraft Network enables just such an analysis (Fig. 6).

Significant drifts in MOPITT CO retrievals appear at both high and low altitudes, with near surface retrieals drifting downward while the high altitude retrievals drift in the opposite direction, upward. This competing drift largely canceled in the total column, fortuitously. This evaluation clarifies what type of change inferred from MOPITT can be robustly attributed to observations versus what may reflect instrumental drift. This analysis also determined that while the v7 version of the MOPITT retrieval exhibited drift exceeding 0.5%/year at high altitudes, the v8 version of the product maintained drift below 0.1%/year for all altitudes.

3.3 TCCON: Total Carbon Column Observing Network—Example of Role 3

While not an aircraft profiling network, TCCON links calibrated aircraft profile observations with satellite observations, providing evaluation and traceability to surface-based calibration scales. TCCON is a network of ground stations equipped

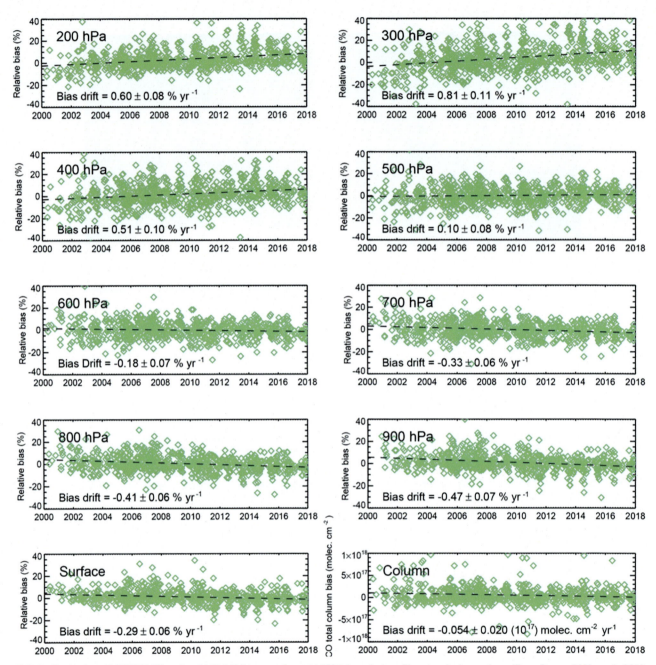

FIG. 6 Evaluation of MOPITT CO retrieval (v7) drift in comparison with NOAA vertical profile network observations. *(From Deeter, M.N., et al., 2019. Radiance-based retrieval bias mitigation for the MOPITT instrument: the version 8 product. Atmos. Meas. Tech. 12, 4561–4580. https://doi.org/10.5194/amt-12-4561-2019.)*

with high-resolution Fourier Transform Spectrometers (FTS) which view the sun at a range of short-wave infrared wavelengths (Wunch et al., 2011). With these remote sensing observations, total column abundances of trace gases such as CO_2 can be retrieved continuously when the sun is visible in a similar manner to space-based remote sensing. As such, TCCON forms the basis for ground-based evaluation of remote sensing retrievals of trace gases for a multitude of satellites currently in orbit. As the TCCON FTS retrievals cannot be directly calibrated, aircraft profiling conducted over TCCON sites using calibrated in situ measurements provide an evaluation, calibration and estimate of overall uncertainty in these retrievals. Then, with the TCCON observation now tied to the in situ calibration scales, bias corrections for satellite data can be derived based on their comparison to TCCON (O'Dell et al., 2018). These aircraft profiles at TCCON sites have been conducted as part of both global and regional intensive campaigns, as outlined in Wunch et al. (2011).

For our interest in this chapter, it is the aircraft profiles conducted over these TCCON sites that provide the foundational calibration that ultimately ties space-based remote sensing retrievals to ground-based observing networks.

4. Aggregating aircraft observations from regional intensives

Aircraft campaigns often provide dense data sets for their regions and time periods of study, but are rarely used alone in evaluations of satellite retrievals of CO_2 or CH_4. Instead, aircraft observations from a collection of intensive regional campaigns and sustained networks are typically considered in aggregate. In particular, these aircraft data are often used as independent data to evaluate models optimized with satellite observations (Role 2). By aggregating multiple distinct campaigns and network sites, model evaluation can be conducted across much of the globe and over a range of time periods. This aggregation of aircraft data sets is facilitated by the ObsPack (Observation Package, https://gml.noaa.gov/ccgg/obspack/) data products, which include data from a combination of global campaigns, network observations, regional campaigns, and measurements made on commercial airliners.

4.1 Chevallier et al. (2019): Example of (2)

Chevallier et al. (2019) consider how to objectively assess fluxes of CO_2 optimized to match satellite observations and how current inversions perform. The approach they outline relies on aircraft observations to be withheld from the inversion and used after optimization to evaluate model performance. By leveraging data from many airborne campaigns in the ObsPack dataset, an independent check across much of the globe and over much of the time frame of the analysis is provided. Evaluation of simulations optimized with satellite observations (GOSAT or OCO-2) or in situ observing networks can be conducted across campaigns. If performance against the withheld data set is significantly different between using satellite-based observations or in situ observations to drive the inverse analysis, it would suggest that observational product better captures real CO_2 variability. Chevallier et al. (2019) find comparable performance, suggesting that some satellite-based inversions can now provide results comparable in their performance as inversions driven by surface measurements, marking an advance in the potential for space-based remote sensing observations to improve understanding of the carbon cycle.

5. Summary

Aircraft vertical profile measurements play a unique and critical role for producing space-based observations of long-lived trace gases in the atmosphere by providing an independent data set for: (1) evaluation of space-based retrievals, (2) evaluation of models optimized to satellite data, and (3) linking space-based remote sensing observations to calibration scales. While all are important roles, aircraft observations increasingly are being utilized for Role 2, where the high-quality, vertically resolved, and independent nature of aircraft vertical profile observations make them ideal for evaluating models of optimized surface fluxes. Notably, aircraft observations that have nearly global coverage (HIPPO, ATom) or continuous temporal coverage (NOAA aircraft network) provide particular value for evaluating spaced-based observations. Regional intensive campaigns are used for evaluation, but typically in aggregate to provide the spatial and/or temporal coverage desired for more robust evaluation. Space-based remote sensing observations of long-lived trace gases such as CO_2 and CH_4 are improving in quality and resolution, and more satellites and measurements are planned. As this space-based fleet expands, it will be essential to continue and likely expand aircraft observations to ensure the satellite-fleet's observational veracity. Global-scale intensive campaigns have only been conducted twice (e.g., HIPPO and ATom), and without plans to continue this type of spatial coverage in the future, large spatial and temporal gaps may emerge in future evaluation of space-based remote sensing.

Acknowledgments

The authors would like to thank Bianca Baier and Colm Sweeney for feedback to improve this chapter. The authors would also like to thank the entire airborne and satellite trace-gas community, whose dedicated work continues to advance our understanding of important atmospheric constituents.

References

Baier, B., et al., 2022. The AirCore atmospheric sampling system. In: Nalli, N.R. (Ed.), Field Measurements for Passive Environmental Remote Sensing. Elsevier.

Bruegge, C.J., et al., 2019. Vicarious calibration of Orbiting Carbon Observatory-2. IEEE Trans. Geosci. Remote Sens. 57 (7), 5135–5145. https://doi.org/10.1109/TGRS.2019.2897068.

Chevallier, F., et al., 2019. Objective evaluation of surface- and satellite-driven carbon dioxide atmospheric inversions. Atmos. Chem. Phys. 19, 14233–14251. https://doi.org/10.5194/acp-19-14233-2019.

Deeter, M.N., et al., 2019. Radiance-based retrieval bias mitigation for the MOPITT instrument: the version 8 product. Atmos. Meas. Tech. 12, 4561–4580. https://doi.org/10.5194/amt-12-4561-2019.

Deng, F., et al., 2015. Sensitivity analysis of the potential impact of discrepancies in stratosphere–troposphere exchange on inferred sources and sinks of CO_2. Atmos. Chem. Phys. 15, 11773–11788. www.atmos-chem-phys.net/15/11773/2015/.

Frankenberg, C., et al., 2016. Using airborne HIAPER Pole-to-Pole Observations (HIPPO) to evaluate model and remote sensing estimates of atmospheric carbon dioxide. Atmos. Chem. Phys. 16, 7867–7878. https://doi.org/10.5194/acp-16-7867-2016.

Liu, J., et al., 2021. Carbon Monitoring System Flux Net Biosphere Exchange 2020 (CMS-Flux NBE 2020). Earth Syst. Sci. Data 13, 299–330. https://doi.org/10.5194/essd-13-299-2021.

Martinez-Alonzo, S., et al., 2020. 1.5 years of TROPOMI CO measurements: comparisons to MOPITT and ATom. Atmos. Meas. Tech. 13, 4841–4864. https://doi.org/10.5194/amt-13-4841-2020.

Miller, C.E., et al., 2007. Precision requirements for space-based XCO2 data. JGR-Atmos. https://doi.org/10.1029/2006JD007659.

Nalli, N.R., et al., 2020. Validation of carbon trace gas profile retrievals from the NOAA-unique combined atmospheric processing system for the cross-track infrared sounder. Remote Sens. (Basel) 12, 3245. https://doi.org/10.3390/rs12193245.

O'Dell, C.W., et al., 2018. Improved retrievals of carbon dioxide from Orbiting Carbon Observatory-2 with the version 8 ACOS algorithm. Atmos. Meas. Tech. 11, 6539–6576. https://doi.org/10.5194/amt-11-6539-2018.

Peters, W., et al., 2007. An atmospheric perspective on North American carbon dioxide exchange: CarbonTracker. PNAS 104 (48), 18925–18930. https://doi.org/10.1073/pnas.0708986104.

Stephens, B.B., et al., 2018. The O_2/N_2 ratio and CO_2 Airborne Southern Ocean (ORCAS) study. Bull. Am. Meteorol. Soc. https://doi.org/10.1175/BAMS-D-16-0206.1.

Sweeney, C., et al., 2015. Seasonal climatology of CO_2 across North America from aircraft measurements in the NOAA/ESRL Global Greenhouse Gas Reference Network. J. Geophys. Res. Atmos. 120, 5155–5190. https://doi.org/10.1002/2014JD022591.

Thompson, C.R., et al., 2021. The NASA Atmospheric Tomography (ATom) mission: imaging the chemistry of the global atmosphere. Bull. Am. Meteorol. Soc. https://doi.org/10.1175/BAMS-D-20-0315.1.

Wecht, K.J., et al., 2012. Validation of TES methane with HIPPO aircraft observations: implications for inverse modeling of methane sources. Atmos. Chem. Phys. 12, 1823–1832. https://doi.org/10.5194/acp-12-1823-2012.

Wofsy, S.C., et al., 2011. HIAPER Pole-to-Pole Observations (HIPPO): fine-grained, global-scale measurements of climatically important atmospheric gases and aerosols. Phil. Trans. R. Soc. A 369, 2073–2086. https://doi.org/10.1098/rsta.2010.0313.

Wu, D., et al., 2020. Space-based quantification of per capita CO_2 emissions from cities. Environ. Res. Lett. 15 (3). https://doi.org/10.1088/1748-9326/ab68eb.

Wunch, D., et al., 2011. The Total Carbon Column Observing Network. Phil. Trans. R. Soc. A 369, 2087–2112. https://doi.org/10.1098/rsta.2010.0240.

Zhang, Y., et al., 2020. Quantifying methane emissions from the largest oil-producing basin in the United States from space. Sci. Adv. 6 (17), eaaz5120. https://doi.org/10.1126/sciadv.aaz5120.

Zhang, Y., et al., 2021. Attribution of the accelerating increase in atmospheric methane during 2010–2018 by inverse analysis of GOSAT observations. Atmos. Chem. Phys. 21, 3643–3666. https://doi.org/10.5194/acp-21-3643-2021.

Chapter 14

Campaign situational awareness from operational satellite sounding retrievals

Rebekah Esmaili and Christopher D. Barnet
Science and Technology Corp. (STC), Columbia, MD, United States

Chapter outline

1. Introduction	245	4. Intensity Forecast EXperiment (IFEX)	251
2. Data sets and methods	247	5. Fire Influence on Regional to Global Environments and Air Quality (FIREX-AQ)	254
2.1 Satellite soundings	247		
2.2 Radiosondes	248	6. Summary	256
2.3 RAP model	249	Acknowledgments	257
2.4 Aircraft	249	References	257
3. Hazardous Weather Testbed	249	Further reading	260

The cross pollination of disciplines is fundamental to truly revolutionary advances in our culture.

Neil deGrasse Tyson

1. Introduction

Validation-focused campaigns are conventionally used to improve satellites instruments and products, such as through the calibration and validation efforts described in Chapters 9–13. This is often the focus for newly launched satellites or recently developed algorithms. However, mature satellite retrievals can also contribute observations to scientific field campaigns and, in turn, bridge scientific research with real-world environmental hazard monitoring. The latter has been the case with retrievals from the U.S. National Oceanic and Atmospheric Administration (NOAA) Unique Combined Atmospheric Processing System (NUCAPS). NUCAPS is a satellite sounding algorithm with a lineage dating to 2002 with the launch of the AIRS instrument on the Aqua satellite (Susskind et al., 2003, 2011). NUCAPS is an operational (i.e., routine and automated) satellite algorithm that retrieves global profiles of temperature, water vapor, ozone, carbon monoxide, and other trace gases from satellite radiance observations (Barnet et al., 2021). Because data are available in real time, NUCAPS has played a contributing role in scientific field campaigns such as the El Nino Rapid Response (ENRR; Dole et al., 2018), Intensity Forecasting EXperiment (IFEX; Rogers et al., 2006, 2013), and Fire Influence on Regional to Global Environments and Air Quality (FIREX-AQ; Junghenn Noyes et al., 2020). NUCAPS can enhance what we refer to as *situational awareness*, or "the perception of the elements in the environment within a volume of time and space, the comprehension of their meaning and the projection of their status in the near future" (Endsley, 1995). Scientific field campaigns typically involve expensive instrumentation and platforms (e.g., aircraft or ship-based measurements), so campaigns can optimize their data using situational awareness derived from operational soundings.

In addition to the profiles themselves, the NUCAPS spatial coverage is superior to surface observations alone. NUCAPS retrieval footprints range from 50km (at nadir) to 150km (at the scan edge) across 2200km wide swaths from pole to pole. In one day, a single satellite can collect up to 324,000 NUCAPS profiles. Presently, there are six satellites (Suomi NPP, NOAA-20, MetOp-A/-B/-C, and Aqua) with infrared (IR) and microwave (MW) sounders that NUCAPS can produce retrievals from. If fully synergized, NUCAPS can generate 1.6 million profiles globally per day. While sounding profiles have lower vertical resolution (~1–1.5km) than high-resolution radiosonde profiles (~5m), their greater spatial availability can complement the 1400 global radiosondes each day.

While impressive, the volume of satellite sounding data alone does not guarantee improvement in situational awareness for environmental hazard monitoring. Three sequential components of situational awareness are (1) accessing information, (2) interpreting information, and (3) anticipating future states (Endsley, 1995). To begin this chain, scientific field campaign organizers and participants must have access to the NUCAPS within their study timeframe. Within satellite remote sensing parlance, *data latency* is defined as the time it takes for the satellite to make a remote observation, transfer the observation data to Earth, process the data into a usable form, and provide that data to the forecaster and other stakeholders. If too much time elapses to complete data processing and delivery, then the value of that data to situational awareness is diminished. Data delivery is more than the receipt of data by the end user, but also requires that the data are easily visualized using the campaign's standard tools. This last step is most successful when NUCAPS developers collaborate directly with stakeholders. The term NUCAPS developers refers to the large community of scientists, programmers, and subject matter experts who improve the NUCAPS algorithm and its' applications. Stakeholders refers to scientists and forecasters who use the NUCAPS for real-time hazard monitoring or scientific field campaign decision making.

The NUCAPS algorithm was designed to support real-time operational weather monitoring. Convective storms can form and intensify rapidly, so their prediction requires short-term forecasts of less than 2 h, which is also called *nowcasting* (Browning and Collier, 1989). NUCAPS from NOAA-20 is delivered in less than 60 min to forecasters via conventional display tools, such as the Advanced Weather Interactive Processing System II (AWIPS-II). Because NUCAPS meets the less than 2 h nowcasting latency requirement, NUCAPS was reviewed favorably in forecaster evaluations (Wheeler et al., 2018; Smith et al., 2018, 2019; Esmaili et al., 2020) when used alongside traditional severe weather situational awareness and forecasting data sets (Chapter 23, Pryor). Like operational weather forecasting, campaigns require rapid data delivery and display, alongside tailored, application-specific products. Since operational NUCAPS meets nowcasting latency requirements so it can also enhance situational awareness to scientific field campaigns. As was the case with operational forecasting, NUCAPS' utility to campaign organizers benefits direct collaboration. We discuss this synergistic process in Sections 3–5. For the remainder of this section, we focus on how data is accessed, the first component of situational awareness, by NOAA National Weather Service (NWS) stakeholders.

NUCAPS is presently delivered to NWS forecasters operationally using the Cross-track Infrared Sounder (CrIS) and Advanced Technology Microwave Sounder (ATMS) instruments (JPSS/STAR, 2021), which are instruments on the Suomi NPP (launched in 2011) and NOAA-20 satellites (launched in 2017). From September 2014 to March 2019, NUCAPS was delivered to NWS forecasters from Suomi NPP through the satellite broadcast network (SBN), which is the operational NOAA pathway. During this time, NUCAPS was successfully utilized for monitoring large-scale and long-duration events, such as *cold air aloft*, a concern for aviation forecasters (Weaver et al., 2019). An example of a smaller-scale and short-duration event is the preconvective environment, which is in part characterized by a state of increased instability with elevated low-level moisture prior to convective storm initiation in a region. Satellite sounding products were found to be useful for calculating atmospheric stability parameters (Weisz et al., 2015; Iturbide-Sanchez et al., 2018; Bloch et al., 2019), which can be used to monitor the preconvective environment. While NUCAPS had skill for this application, the data latency of NUCAPS-Suomi NPP was up to 240 min through the SBN, which did not meet the two-hour nowcasting window. Thus, early research efforts were dedicated to faster data delivery pathways to forecasters.

NUCAPS-Suomi NPP was delivered more quickly using the community satellite processing package (CSPP) which processes data from the satellite's direct broadcast antenna. CSPP processes data from a network of satellite data receivers across North America and the Pacific Ocean. The data receivers can download a subset of instrument data as a satellite passes overhead, so forecasters could receive data products in less than 60 min. Forecasters found this low latency, real-time NUCAPS more valuable for monitoring convection during the Hazardous Weather Testbed (HWT; Wheeler et al., 2018; Smith et al., 2018; Smith et al., 2019; Esmaili et al., 2020). In March 2019, NUCAPS-Suomi NPP was removed from NWS operations and replaced with NUCAPS-NOAA-20. The operational latency for NUCAPS-NOAA-20 was significantly lower than NUCAPS-Suomi NPP primarily because (1) CrIS and ATMS measurements from NOAA-20 have two downlink stations (Svalbard, Norway, and McMurdo, Antarctica) versus the one from Suomi NPP (Svalbard) and (2) the order of data download was changed from a "last in, first out" to "first in, first out." The data download order is significant because on the ascending (i.e., local afternoon) overpass, the data from the Northern hemisphere was received "first" over Svalbard will pass over the Northern Hemisphere, thereby reducing the latency. The reduced latency allowed NUCAPS-NOAA-20 to be ingested in AWIPS-II (Section 4 in Chapter 18) via SBN in a timely manner for situational awareness over the Northern Hemisphere. The rest of this chapter will focus on the other two phases of situational awareness, which are (1) how campaign participants can interpret soundings from NUCAPS and (2) anticipate future states from NUCAPS.

The work highlighted in this chapter was performed through the Joint Polar Satellite Systems (JPSS) Proving Ground and Risk Reduction Program (PGRR). The JPSS-PGRR was established to improve user applications of NOAA operational satellites by funding interaction between researchers, technical experts, forecasters, and scientific field campaign organizers. Interacting with scientific field campaigns is productive and significantly cheaper than organizing an independent field campaign. For example, NSF-funded scientific field campaigns in 2015 had a median cost of $1.5 million, which primarily stemmed from deploying aircraft, supporting facilities, and campaign personnel funding (Avallone and Baeuerle, 2017). The JPSS-PGRR costs for the collaboration include the funding the developers' time or through dropsonde purchases, which is typically less than 5% of the cost of a typical field campaign.

2. Data sets and methods

2.1 Satellite soundings

NUCAPS is designed to retrieve atmospheric profiles of temperature, water vapor, and trace gases from IR and MW sounders. Operational satellite sounders currently include CrIS/ATMS on Suomi NPP and NOAA-20; the Infrared Atmospheric Sounding Interferometer (IASI) and Advanced Microwave Sounding Units (AMSU) on the MetOp series satellites; and the Atmospheric Infrared Sounder (AIRS) and AMSU on Aqua. NUCAPS is based on version 5 of the AIRS Science Team Algorithm (Susskind et al., 2003, 2011). The major steps in the NUCAPS algorithm are outlined in Fig. 1, and we will describe relevant steps in context of this chapter here. Detailed descriptions of NUCAPS exist elsewhere (Gambacorta and Barnet, 2013; Sun et al., 2017; Nalli et al., 2018; Barnet et al., 2021), and IR sounders are described in Chapter 18.

Like other inversion problems in remote sensing, NUCAPS must solve a mathematically ill-posed problem that has a finite number of radiances to retrieve temperature and trace gases as a function of pressure. NUCAPS employs a Bayesian optimal estimation technique (Rodgers, 2004), which requires a priori *measurements* to constrain the solution. Because of its orientation toward operational forecasting, NUCAPS does not use formal a priori measurements for temperature, water vapor, skin temperature, or emissivity, but instead iterates off a *first guess* derived from a linear regression algorithm. The regression algorithm predictor is based on the principal component (PC) scores from IR and MW sounder radiances (Goldberg et al., 2003). The predictand is the relevant atmospheric state variables from European Centre for Medium-Range Weather Forecasts (ECMWF). Note that while NUCAPS contains model statistics for the focus days, NUCAPS only uses spectral radiances in the retrieval so it is model independent. Where NUCAPS has skill in the atmospheric column

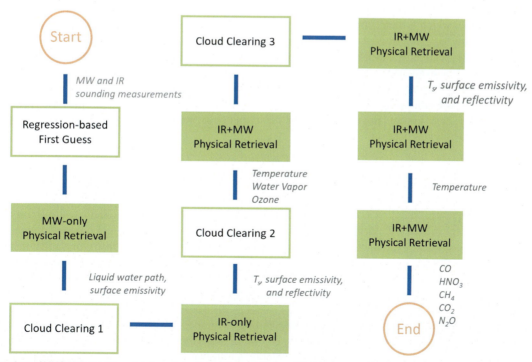

FIG. 1 Flowchart of NUCAPS algorithm showing relevant steps that are discussed in this chapter. All parameters retrieved in the physical retrieval steps are sequential to increase the solution stability, yield, and processing speed. Each physical retrieval step calculates error diagnostics to improve retrieval skill and assessment.

for a given variable, the first guess is adjusted using an atmospheric radiative transfer model and the observed radiances; this process is referred to as a *physical retrieval*.

NUCAPS first performs a MW-only physical retrieval (Rosenkranz, 2001, 2006), which measures several state variables, including temperature and water vapor, which are saved to the Environmental Data Record (EDR) file. However, only the retrieved liquid water path and surface emissivity are used downstream in NUCAPS; the MW-only retrievals themselves are output as the retrieval under "cloudy" overcast conditions when the IR algorithm fails to reach convergence.

Clouds make it difficult and computationally expensive to model atmospheric scattering, so only cloud-cleared IR radiances are passed into the IR + MW physical retrieval (Strow et al., 2003; Susskind et al., 2003). Cloud clearing is a technique that removes the radiative effects of clouds by combining radiance observations from multiple sounder fields of view to estimate clear column radiances. The first cloud clearing method removed a single cloud layer from two adjacent sounder fields of view (Smith, 1968). A later method used additional fields of view to clear multiple cloud layers (Chahine, 1977). NUCAPS employs the latter approach on a 3 × 3 field of view array that can clear up to four complex cloud formations over a ∼50 km footprint with realistic error estimates (Susskind et al., 2003). Scientific field campaigns provide opportunities to evaluate NUCAPS soundings with complex cloud layers for specific applications.

NUCAPS retrieves temperature, water vapor, ozone, and other trace gases sequentially. Sequential retrievals increase the solution stability, yield, and processing speed (Smith and Barnet, 2019). This is because NUCAPS (i) uses previously retrieved variables as the first guess for later retrieval steps and (ii) propagates the error of previously retrieved variables into all following retrieval steps. The latter is especially important because it allows NUCAPS to computes the information content of each variable (Susskind et al., 2003) to maximize the skill in scenes with difficult cloud conditions. While sounders can have thousands of channels, each variable is retrieved using a smaller and unique set of IR and MW channels. The selected channels are most sensitive to the retrieved variable while less sensitive or insensitive to all other variables (Gambacorta and Barnet, 2013). The final retrieval returns each profile on the vertical radiative transfer grid (Strow et al., 2003), which is a fixed set of 100 pressure levels for temperature or 100 layers for trace gases. NUCAPS also retrieves other parameters, of which cloud top fraction and cloud top pressure are the most relevant to this chapter.

Several steps are repeated in NUCAPS to further stabilize the final retrieval. Temperature and the surface parameters are improved during the second pass in part because (ii) the first guess uses the retrieved values of water vapor and ozone instead of the regression and (ii) the error estimates from earlier steps better characterize the information content. The improved surface parameter and temperature retrievals are then used to improve quality control metrics and downstream trace gas retrievals. Cloud clearing is performed multiple times to refine the clear radiance estimates.

The NUCAPS EDR archive (i.e., the operational version) is available on the Comprehensive Large Array-data Stewardship System (CLASS), which is an electronic collection of NOAA data sets from numerous meteorological satellites. NUCAPS-Suomi NPP and NUCAPS-NOAA-20 were approved for public release on CLASS beginning September 5, 2014, and October 24, 2019, respectively. Developers regularly improve NUCAPS and install upgrades into the operational version. The changes are first evaluated in an offline version of NUCAPS before installing them in the operational version. For instance, the offline version of NUCAPS (i) replaces the surface parameter regression with the CAMEL database (Hook, 2017) for land surface emissivity, (ii) uses a more robust first guess regression with additional focus days for temperature and water vapor, and (iii) contains the averaging kernel matrices, which are diagnostic parameters that are not currently saved to the operational data set.

2.2 Radiosondes

Ground reference data sets (also sometimes called *ground truth* or *baseline data*) are used to characterize errors in satellite soundings. Radiosondes (cf. Chapter 2, Vömel and Ingleby) are a reference data sets that are compared with NUCAPS (Fetzer et al., 2003; Reale et al., 2012; Sun et al., 2017; Nalli et al., 2013). Radiosondes are instruments that measure temperature and water vapor vertically through the atmosphere either by being attached to a weather balloon, or in the case of a dropsonde, upon release from an aircraft. NWS forecast offices launch balloons with radiosondes twice daily at 00 and 12 UTC. An archive of radiosonde data can be accessed from the Earth System Research Laboratory (ESRL) database (https://ruc.noaa.gov/raobs/).

There are differences between radiosondes and satellite soundings. For one, radiosondes are a point, or more accurately, a narrow path observation through the vertical atmospheric column. Furthermore, balloons can take up to 45 min to reach the top of the atmosphere. This contrasts with NUCAPS, which is a retrieval profile integrated over a footprint area (50–150 km). NUCAPS is also essentially an instantaneous measurement, meaning that the boundary layer and the top of the atmosphere are measured simultaneously. Radiosondes can take measurements below and within clouds, whereas

NUCAPS retrieves the atmosphere around clouds. While there should be general agreement between these two observations, some differences will arise, which is sometimes referred to as *mismatch error* (cf. Chapter 1, Nalli and Kalluri).

2.3 RAP model

Like satellite soundings, model data are gridded and thus more representative of a retrieval footprint than a radiosonde. The Rapid Refresh (RAP) model is an hourly model that covers the North American domain (Benjamin et al., 2016). The RAP is designed for short-range forecasting and decision making for severe weather and aviation. The RAP has a 13 km grid and 51 vertical pressure levels, so while not convection-resolving (~3 km or less), it is high resolution and within the drift of radiosondes. Bulk statistics versus radiosonde data show that analysis from the RAP has a < 2% RMS error in relative humidity through most of the atmospheric column under 400 hPa and 0.1–0.2 Kelvin (Benjamin et al., 2016). Most importantly, forecasters regularly use and trust the RAP model.

2.4 Aircraft

The Lockheed ER-2 aircraft is a high altitude, flying laboratory operated by the National Aeronautics and Space Administration (NASA). In particular, the ER-2 is equipped with two IR interferometers that produce 40 km swaths, which are the Scanning High-resolution Interferometer Sounder (S-HIS; 3.3–18 μm; Tobin et al., 2006) and NPOESS Atmospheric Sounder Testbed-Interferometer (NAST-I; 3.5–16 μm; Smith Sr. et al., 2005). Note that while this chapter focuses on NUCAPS retrievals, it is important to recognize how scanning interferometers provide high spatial resolution data for the evaluation of situational awareness for nonoperational retrieval products used by the sounding research community. For instance, the dual regression method (Smith et al., 2012; Weisz et al., 2015) processed sounding data from the S-HIS on uninhabited aircraft during a scientific field campaign (DeSlover et al., 2016) and using the S-HIS and NAST-I for FIREX-AQ campaign (Taylor et al., 2021). These studies in turn help identify general strengths and weaknesses of sounding retrievals for situational awareness. For additional information on the S-HIS and NAST-I instruments, the reader is referred to Chapter 9 (Taylor et al.). The case study discussed in this chapter was collected during the FIREX-AQ aircraft campaign and campaign data are available from: https://www-air.larc.nasa.gov/cgi-bin/ArcView/firexaq.

IFEX releases dropsondes from both WP-3D and NOAA Gulfstream IV-SP (G-VI) aircraft. In this chapter, we primarily discuss the Gulfstream IV flights because they can release dropsondes at a higher altitude than the WP-3D flights. These higher sondes yield more observations for evaluation in the middle atmosphere, where NUCAPS was expected to provide the most value for situational awareness. Given that near surface processes are also important for tropical cyclone dynamics, a future tropical cyclone campaign may provide opportunities to evaluate the lower atmosphere. For additional information on dropsonde use in scientific field campaigns, the reader is referred to Chapter 10 (Vömel and Dunion). Dropsondes used in this chapter were collected by the IFEX campaign and are available from: https://seb.noaa.gov/pub/flight/ASPEN_Data/.

3. Hazardous Weather Testbed

The Hazardous Weather Testbed (HWT) is an annual experiment where NWS forecasters monitor for severe weather using new data products and tools, including NUCAPS. The experiment takes place in Norman, Oklahoma but emulates the conditions of an NWS forecast office. Forecasters investigate if the new products provide additional value to live weather forecasts across the United States. Satellite product developers are also in the room to help forecasters interpret data. From a developer's perspective, the HWT is like a validation campaign because forecasters verify the accuracy and utility of new products and tools for meteorologically relevant cases. By design, the HWT fosters close collaboration between the research and operational communities.

NUCAPS temperature and dew point soundings have been evaluated in the HWT since 2015. Initially, forecasters wanted to verify the accuracy of NUCAPS and often performed intercomparisons with more traditional data sources, such as radiosondes and convective model data. In subsequent HWTs, developers enhanced NUCAPS data delivery, display, and training and that allowed forecasters to integrate NUCAPS into their analyses and identify interesting use cases. In the most recent HWTs, forecasters showed interest in the potential of using NUCAPS to capture small-scale features in the vertical profile.

In NUCAPS, forecasters documented observed *capping inversions*, which occur when layers of warm air overly cooler layers, thereby increasing atmospheric static stability. Inversions can inhibit convection either temporarily or entirely by preventing upward motion in preconvective environments. The vertical depth of an inversion can vary greatly across the globe, but a daytime inversion over midwestern US can have typical values from 800 to 1200 m. Forecasters have asked developers if NUCAPS can retrieve capping inversions. The answer requires some understanding of the NUCAPS algorithm. The physical retrieval step cannot capture fine features that are shallower than the theoretical vertical resolution allowed by the spectral radiances, which is ~1–1.5 km (Iturbide-Sanchez et al., 2018; Maddy and Barnet, 2008). However, the NUCAPS first guess can potentially see fine features as small as 250–500 m in scenes with low cloud contamination. So, it is possible for NUCAPS to capture capping inversions but evaluation requires using internal diagnostics. It is also necessary to evaluate preconvective cases because capping inversions are common but not all inhibit storm development.

Interactions between forecasters and developers in the HWT helped identify specific cases where preconvective capping inversions contribute to situational awareness. Furthermore, forecasters verify the presence of the preconvective capping inversion using evidence from other in situ, satellite, and model data sets. Combining multiple data sources can diagnose whether features are signal or noise. The forecaster perspective can augment traditional validation techniques (Chapter 16; Sun et al.). Developers can validate inversions if a radiosonde is available within 30 min and 50 km of the retrieval. This timing is possible using dedicated radiosondes but not probable because there are relatively few launch sites and the location may not have a risk of severe weather. Forecaster case identification is also important because traditional validation approaches generate statistics from large samples to ensure that NUCAPS profiles meet uncertainty requirements of 0.5 K for temperature and 10% for water vapor (Nalli et al., 2018). While important, bulk statistics do not evaluate NUCAPS' skill for specific applications like preconvective capping inversions. Forecasters' detailed narrative can corroborate the existence of relevant capping inversions for analysis and validation. Fig. 2 shows an example of a targeted case study from the HWT.

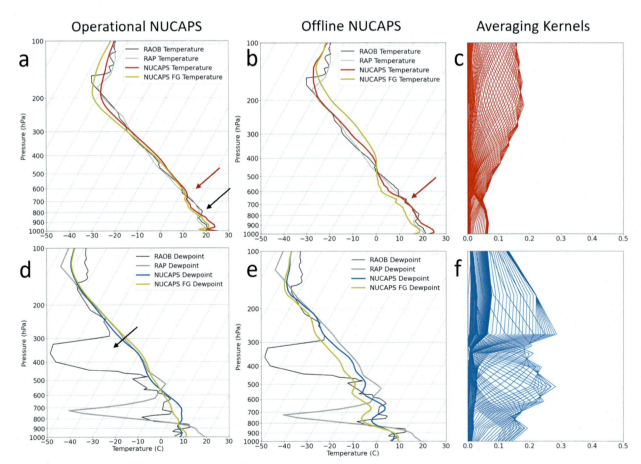

FIG. 2 Comparison of the first guess *(yellow)*, final retrieval *(red or blue)*, RAP model *(gray)*, and radiosonde observations *(black)* for temperature and dew point temperature from operational NUCAPS (A and D, respectively) and offline NUCAPS (B and E, respectively) from NOAA-20. Averaging kernel matrices (AKMs) are shown for (C) temperature and (D) water vapor. Profiles shown are from April 25, 2019, near 30.48°N, 86.33°W. *Arrows* indicate features of interest, which are two capping inversions in (A), one capping layer inversion in (B), and a dry layer in (D).

Fig. 2A shows a capping inversion captured by the operational NUCAPS version over the Florida panhandle (30.48°N, 86.33°W) on April 25, 2019. The feature is indicated by the top red arrow on the Skew-T Log-P diagram, roughly between 625 and 650 hPa. A radiosonde from the nearby Tallahassee, FL upper air site also shows an inversion, albeit one that is lower in the atmosphere (black arrow). The 18 UTC RAP is also shown (gray line) for reference. The forecaster indicated that convection was delayed on that given day, and the capping inversion may be responsible. Thus, the operational version of NUCAPS was able to augment the model and in situ observations to strengthen the forecaster's analysis of the event.

As mentioned previously, the offline science version of NUCAPS incorporates recent algorithm upgrades and internal diagnostics (Fig. 2B). The offline NUCAPS more closely matches the temperature profile of the radiosonde the capping inversion was much smaller. This indicates that in this case, the capping inversion present in operational NUCAPS may have been an artifact in the first guess.

The offline NUCAPS has internal diagnostics to measure retrieval skill such as the averaging kernel matrix (AKM), which quantifies the amount of information the measurement contributes to the first guess. AKM values range from 0 (no information content and the final retrieval replicates the first guess) to 1 (100% information content and the final retrieval replaces the first guess). More typically, AKM values fall somewhere in between and varies by scene (Smith and Barnet, 2020). Fig. 2C shows the AKMs for each of the 100 pressure levels in the offline NUCAPS, which are represented individually by lines. The information content of the temperature profile is lower in between 600 and 700 hPa (~0.05) as compared to upper-level retrievals, such as between 200 and 100 hPa (closer to 0.2). While there is greater information content in mid-levels, the retrieval has skill in the lower atmosphere, but forecasters are encouraged to compare surface soundings with other data sets.

This example illustrates why the physical step is critical in retrieval algorithms for operational applications. Recall that the fine vertical stricture in the retrieval comes from the first guess (yellow line, Fig. 2A, B, D and E), not from the physical retrieval. The physical retrieval acts to smoothly adjust the first guess into a more realistic representation of the atmosphere. Without the physical retrieval, the first guess is unrealistically cooler throughout the column when compared with the radiosonde; the profile is unrepresentative and would not provide any value to operational forecasting. Furthermore, detailed diagnostics like AKMs and some quality control metrics are generated in the physical retrieval step. The embedded information content in turn helps developers characterize why NUCAPS did or did not match reference observations.

While the capping layer nearly vanishes in offline NUCAPS, the dew point temperature more closely matches the radiosonde value in the lower atmosphere (Fig. 2E) than operational NUCAPS (Fig. 2D). In fact, the surface dew point is almost a near match in the offline version, which is important for calculating atmospheric stability. In contrast, the 18 UTC RAP has a significantly warmer dew point suggestion greater instability.

Between 300 and 400 hPa, there is a dry layer (black arrow) that is present in the radiosonde that is not captured in either the operational and offline NUCAPS (Fig. 2D and E). Examining the AKM (Fig. 2F) shows that there is comparatively less skill between 300 and 400 hPa. The dry layer is also absent in the 18 UTC RAP, so it is possible that the dry pocket is localized and may not be captured in NUCAPS or the RAP because they are volume observations.

This section highlights the importance of the research-to-operations feedback loop (Jedlovec, 2013), where targeted improvements in the operational product can improve forecasting and nowcasting. We showed how forecasters identify significant instances of capping inversions in NUCAPS. Presently, vertical diagnostics are performed offline by developers. In the future, developers could create a new quality control metric that is derived from the AKM to show where in the profile NUCAPS has skill. By working together, developers and forecasters can better understand the strengths and weaknesses of satellite soundings for severe weather. Many of the lessons learned in the HWT are transferrable to other scientific field campaigns.

4. Intensity Forecast EXperiment (IFEX)

Scientific field campaigns are organized to study atmospheric processes in depth. Collaborating with a campaign can evaluate if NUCAPS provides value for a specific application prior to utilization in a testbed or operations. NUCAPS developers collaborated with IFEX (Chapter 10, Vömel and Dunion) to determine if NUCAPS can enhance tropical cyclone situational awareness. The HWT helped NUCAPS developers understand forecaster needs for severe weather, but tropical cyclones involve atmospheric processes with different monitoring requirements. For instance, convection has a strong diurnal cycle, with peak formation typically in the afternoon (Liu and Zipser, 2008). For severe weather applications, soundings from the 1:30 pm local time equatorial overpass will have the most value to this forecasting community. In contrast, tropical cyclones continuously evolve during both day and night. Thus, soundings from the morning and afternoon overpasses can be useful for monitoring the tropics (Duran et al., 2021). Additionally, surface temperature and moisture retrievals (e.g., below 850 hPa) are heavily utilized for preconvective situational awareness, but mid-level retrievals (850–500 hPa) may have value for monitoring tropical cyclones.

Mid-level moisture (~500 hPa) plays an important role in tropical cyclone formation and intensification and satellites are routinely used for monitoring (DeMaria et al., 2001). The presence of moist air intensifies tropical cyclones, whereas dry air entrainment can enhance downdrafts and increase atmospheric stability, thereby suppressing the development of tropical cyclones. From July to October, satellite and radiosonde atmospheric profiles in the North Atlantic and Caribbean experience three distinct atmospheric states. These states are the moist tropical environment (66%), the Saharan Air Layer (SAL; 20%), and low- to mid-level dry air masses (14%) (Dunion, 2011). SAL conditions are characterized as a warm, dry layer aloft that originates over North Africa. The presence of dust within the SAL can inhibit development of tropical cyclones (Dunion and Velden, 2004; Strong et al., 2018). While NUCAPS has skill in convective situational awareness (Section 3), a collaboration with tropical cyclone scientists can help determine if NUCAPS adds value to the existing observations used to monitor tropical cyclones (Esmaili et al., 2022).

When a tropical cyclone is likely to make landfall, NOAA WP-3D, NOAA G-IV, and U.S. Air Force WC-130 hurricane hunter aircraft are dispatched and deploy dropsondes to sample temperature, moisture, wind speed, and wind shear in the tropical cyclone and surrounding environment (cf. Chapter 10, Vömel and Dunion. Forecasters use these measurements to determine the intensity of the storm. The dropsonde observations are also assimilated by forecast models. While valuable, flights are carefully planned because they are costly and put the crews at risk. Furthermore, flights are not possible if the tropical cyclone is beyond the flight range for an aircraft taking off from the Lakeland, Florida deployment site. For example, the flight range is ~7000 km for the G-IV. The flight timing prioritizes model data assimilation requirements and thus are infrequently collocated in time and space with satellite overpasses. For general validation purposes, the dropsondes launch point should be within 6 h and 150 km of the retrieval central position (Sun et al., 2017; Chapter 16, Sun et al.). However, the atmosphere evolves rapidly so applications like tropical cyclones require an even closer collocation of within 30 min and 25 km (Nalli et al., 2018; Chapter 18). A typical hurricane hunter flight will release dropsondes spaced over 6 h, so even with more relaxed criteria, many of these sondes do not meet the temporal criteria for collocation. For example, during a flight over Hurricane Dorian on 28 August 2019 less than a third of the 30 released dropsondes were within 2 h of the NOAA-20 (17:12 UTC) and Suomi NPP (18:54 UTC) overpasses.

The JPSS program provided dropsondes for IFEX Gulfstream-IV research flights to ensure that they were well-timed with satellite overpasses. To synergize NUCAPS collocation with IFEX campaign objectives, we suggested the organizers identify what tropical cyclones and conditions they wanted to target during the flights. This allowed the campaign organizers to identify features that are of importance to them, which we in turn could analyze both during the experiment and offline. This approach enabled the NUCAPS developers to learn more about the application and campaign organizers to learn more about NUCAPS. Through the collaboration, there were six successful flights in the 2018, 2019, and 2021 hurricane seasons, with additional flights planned. The 2018 research flight targeted a SAL environment (Barnet et al., 2019). Four flights in 2019 captured the tropical cyclone environment, one targeting Hurricane Jerry and three over Hurricane Lorenzo. Two satellites were operational in 2019, Suomi NPP and NOAA-20, which made it possible to study much change occurred between overpasses. A flight in 2021 targeted Hurricane Larry, which was very long-lived and historic as the first category 1 hurricane to make landfall in Newfoundland. Fig. 3 provides a visual of Hurricane Jerry on September 18, 2019, a day before its peak wind intensity of 46 m/s. Hurricane Jerry was weakened by strong winds at 250 hPa and dry mid-level air.

Fig. 3A shows the flight path around Hurricane Jerry using the 16 UTC local overpass of Suomi NPP. The blue line shows the flight path, while the dots represent where dropsondes were released and the color represents NUCAPS data quality. The dot colors use a scheme derived for NWS operations (Esmaili, 2019). Green dots represent footprints that passed the IR+MW physical retrievals. Yellow dots represent footprints that passed the MW-only physical retrieval but did not pass the IR+MW physical retrieval. These profiles likely have significant cloud contamination. Red dots represent footprints that likely have precipitation, so neither the MW-only nor IR+MW physical retrieval will pass. The Geostationary Operational Environmental Satellite 16 (GOES-16) ABI water vapor band (6.19 μm) is shown underneath in greyscale to help visually identify the location of the hurricane. The dropsondes were well-timed with Suomi NPP and NOAA-20 overpasses. Dropsonde #17 was within 1 min and 36 km of the NUCAPS-Suomi NPP footprint center. Dropsonde #24 was the closest in time to the NOAA-20 overpass, which was within 6 min and 7 km of the footprint center.

In addition to well-timed dropsondes with satellite overpasses, IFEX provided an opportunity to test the feasibility of operationally running and delivering NUCAPS for tropical cyclone applications. As a result, we set up a suite of tools to provide real-time NUCAPS from direct broadcast data from the Miami and Puerto Rico antenna sites. Under this configuration, Skew-T comparisons were made within 20 min of receiving the dropsonde data, which occurred mid-flight. Furthermore, a web-based version of Gridded NUCAPS (Berndt et al., 2020) was also provided to the campaign as part of a suite of satellite-based tools. So, the campaign was an opportunity demonstrate the latency of data delivery.

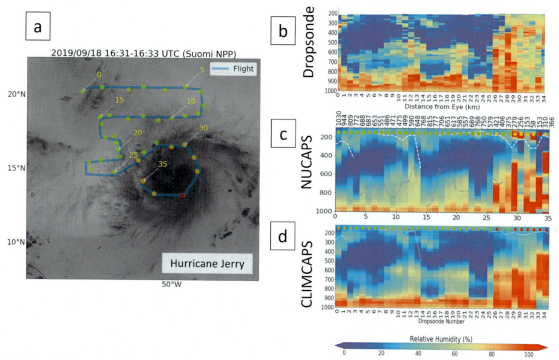

FIG. 3 (A) Flight path *(blue straight lines)* and quality flag of NUCAPS-Suomi NPP footprint center collocated with the dropsonde *(dots)* over Hurricane Jerry (darker features in the GOES-16 6.19 μm band) on September 18, 2019. (B) Profiles from dropsondes, (C) NUCAPS IR+MW retrievals, and (D) CLIMCAPS IR+MW retrievals, with pressure (hPa) on the y-axis and dropsonde number on the x-axis. CLIMCAPS (Smith and Barnet, 2019) is a sibling algorithm to NUCAPS that uses MERRA-II as the first guess.

Fig. 3B shows the cross section of relative humidity from the dropsondes across the flight path and Fig. 3C IR+MW retrievals from NUCAPS from the closest footprint. Smaller numbered sondes were further away from the hurricane center (up to 1030 km), while the larger numbered sondes were points closer to the hurricane center (as close as 58 km). The area farther from the hurricane represents the environment that surrounds Hurricane Jerry. In operations, the tropical cyclone environment is closely monitored for conditions that lead to intensification. As you can see from the lighter regions of Fig. 3A, the environment surrounding Hurricane Jerry was relatively dry. Thus, if the storm entrained this dry air, it could inhibit development. NUCAPS shows agreement with the dropsonde capturing dry air aloft. This was in part because NUCAPS has skill capturing mid-level dry conditions because these scenes have minimal cloud contamination and no precipitation. Over the hurricane, NUCAPS did not pass quality control so these profiles did not match the radiosonde. This is expected, as these scenes have heavy precipitation. There are documented cases where NUCAPS was able to make a retrieval in a hurricane eye when it was both large and clear. Temperature and moisture profiles inside the eye are useful for developers evaluating the quality control but have limited value for forecasting. Instead, forecasters can determine the eye position and shape using satellite imagers, MW sounders, ground radar, or by hurricane hunter flights (Landsea and Franklin, 2013).

A question that arose during IFEX was how close can NUCAPS have successful retrievals to a tropical cyclone, which is measured by the distance in kilometers from the storm center. Fig. 3B shows the distance across the top axes. The closest passing retrieval (dropsonde #28) was within 375 km of the eye. However, this retrieval does not agree with the dropsonde because there is significant cloud contamination, which is visible in the GOES-16 water vapor band in Fig. 3A. By visually comparing Fig. 3A and B, dropsonde #25 reasonably matches the radiosonde and is closer to 579 km from the footprint to the center of the hurricane. If IFEX collaborators find value in closer observations, there are options to test in future scientific field campaigns. For one, only the combined IR+MW physical retrieval is delivered to forecasters. Severe weather forecasts are very sensitive to the boundary layer and the derived stability parameters, so they only want data with a high vertical resolution. Tropical forecasters are different applications, such as characterizing larger air masses like those identified by Dunion (2011). Tropical forecasters may find value in the MW-only retrievals which have a much coarser vertical resolution.

Another approach is to use a model-based first guess. Severe weather forecasters requested model-independence, hence the NUCAPS first guess is based on a regression and not dynamic model data. However, at times, the models may provide a better initial estimate of the atmosphere. For instance, the RAP temperature profile in Fig. 2A and B more closely the radiosonde than the first guess. Tropical forecasters may prefer a blended approach, which could use a forecast model (e.g., using GFS for global coverage) as the first guess for the NUCAPS physical retrieval to smoothly adjust. To illustrate these possibilities, Fig. 3D shows Community Long-Term Infrared Microwave Combined Atmospheric Product System (CLIMCAPS; Smith and Barnet, 2019), a sibling algorithm to NUCAPS that uses MERRA-II as the first guess. Other distinctions are discussed in Smith and Barnet (2019). In this example, CLIMCAPS provides a better estimate of the dry layer structure than NUCAPS. Note that this comparison is only illustrative because CLIMCAPS is not a real-time product. Instead, CLIMCAPS is optimized for trace gas retrievals and is designed to be a long-term climate record. However, a future version of NUCAPS could incorporate a model-based first guess (e.g., using GFS) for evaluation with operational NUCAPS during future research flights.

The collaboration between NUCAPS developers and IFEX made timing the dropsondes with the satellite overpasses more cost effective. Using JPSS-funded dropsondes, a single coordinated flight produced a data set that was useful for IFEX goals, satellite product validation, and tailoring NUCAPS for tropical cyclone situational awareness. NUCAPS agrees with dropsondes released near Hurricane Jerry which shows that is has value for examining dry air layers in tropical cyclone environment. Retrievals over the tropical cyclone are possible using the data collected from the flights and new algorithm configurations. Additionally, NUCAPS was delivered to campaign organizers on time scales where it could provide situational awareness. With additional cycles of collaboration and refinement in the campaign, NUCAPS may have value in operational tropical cyclone situational awareness.

5. Fire Influence on Regional to Global Environments and Air Quality (FIREX-AQ)

Thus far, we have focused on temperature and water vapor retrievals from NUCAPS. These products are delivered via the SBN so that operational forecasters can visualize them within AWIPS-II alongside other data sources. NUCAPS also successfully retrieves trace gases, which are CO, CO_2, CH_4, HNO_3, O_3, N_2O, and SO_2. Except for O_3, trace gas products are not available operationally to forecasters because they are not often used to issue severe weather watches and warnings to the public. However, the trace gas retrievals are necessary for NUCAPS to provide high-quality temperature and water vapor retrievals. Additionally, they are useful for situational awareness of air quality and regional smoke transport. The joint NASA-NOAA scientific field campaign, FIREX-AQ (Junghenn Noyes et al., 2020), provided an opportunity to evaluate NUCAPS for these applications (cf. Chapter 13, Kort and McKain). FIREX-AQ targeted medium to small-scale fires between July and August 2019. As with the tropical cyclone flights in Section 4, ER-2 flights were well timed with satellite overpasses as part of the collaboration.

While aircraft instruments such as the S-HIS and the NAST-I are highly sensitive to trace gases, satellite retrievals like NUCAPS can provide sampling of long-range smoke transport and composition outside of the flight path. To support FIREX-AQ, developers created a web-based monitoring tool that displayed NUCAPS trace gas and cloud retrievals that were gridded on to maps.

Air quality forecasters often combine models and surface observations with satellite products, such as aerosol optical depth (AOD) (cf. Chapter 22, Ciren and Kondragunta), which can detect long-range smoke transport (Huff et al., 2021). There are several ways NUCAPS can work in tandem with AOD or provide estimates when AOD is unavailable. First, NUCAPS can show the level and composition of specific trace gases while AOD can show the concentration of all particles. Second, AOD uses visible and near-IR bands along with a cloud mask to identify where there are clouds. If clouds are present, the algorithm does not make a retrieval (NOAA/STAR, 2018; Sawyer et al., 2020). This contrasts with NUCAPS, which performs cloud clearing (Chahine, 1977; Susskind et al., 2003) and can make retrievals around clouds in partly cloudy scenes. Third, because AOD uses visible bands, AOD cannot be retrieved at night whereas NUCAPS products are available day and night. A useful wildfire case that allowed for data synergy was identified through the collaboration with FIREX-AQ.

Fig. 4A shows a Suomi NPP visible infrared imaging radiometer suite (VIIRS) true color image of the Williams Flat fire near Spokane in Washington State. The ER-2 aircraft flew over a wildfire that took place on 6 August 2019 at roughly 47.98°N, 118.62°W. While the Williams Flat fire was relatively small, AOD retrievals (Hsu et al., 2019) in Fig. 4B show that there was medium-range smoke transport. The scale of this fire is a challenging target for NUCAPS, which at nadir, has a 50 km footprint. This case provided an opportunity to test if NUCAPS trace gases, derived quantities, and diagnostics provide situational awareness for smaller fires.

Fig. 5A and B show the total precipitable water and temperature at 850 hPa, respectively, using the Gridded NUCAPS methodology (Berndt et al., 2020). Missing values within the swath did not pass the IR+MW physical retrieval step. These two figures, respectively, show that conditions were relatively dry and warm, which can increase the potential for wildfires.

Campaign situational awareness from satellite data **Chapter | 14** 255

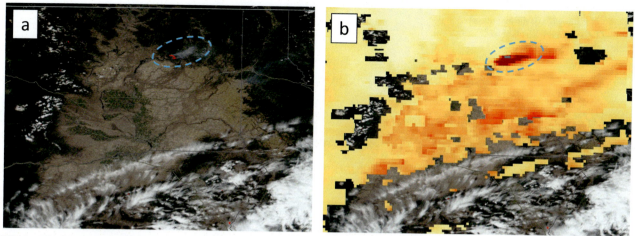

FIG. 4 Williams flat fire (6 August 2019) in (A) true color image from Suomi NPP and (B) the same true color image overlaid with AOD. Williams flat fire is indicated by the dashed circle. *(Source: NASA/Worldview (public domain).)*

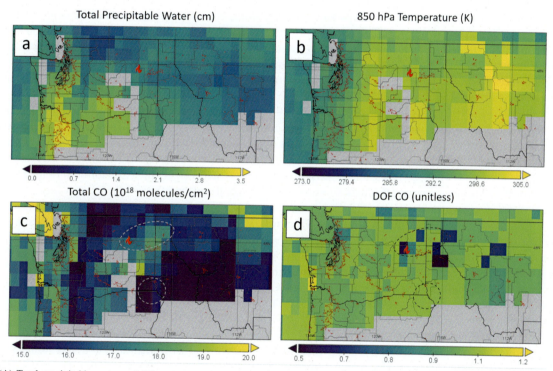

FIG. 5 (A) Total precipitable water and (B) temperature at 850 hPa, (C) total column CO, and (D) the degrees of freedom of CO measured from NUCAPS-NOAA-20 in the region of the Williams flat fire (6 August 2019).

Fig. 5C shows the total column CO, which is expressed as the sum of column density (molecules per cm^2). We will focus on two regions for the remainder of this section, which are (1) the area upstream of the fire (indicated by the red icon in Fig. 5C) and (2) a background region farther away from urban centers.

It is interesting that the CO concentration is not significantly elevated by this fire. This is in part because there are numerous sources of CO, including urban emissions. Given the 1–2-month lifetime of CO in the atmosphere it is possible that there is long-range transport of CO across the pacific. NUCAPS cannot distinguish the source of emissions. This fire is located upstream of Spokane so using a smaller inspection radius minimizes some of these effects.

FIREX-AQ organizers requested to see the NUCAPS CO degrees of freedom in real time. The degrees of freedom for each footprint are computed from the trace of the AKM and provide an estimate of the maximum number of independent pieces of information that NUCAPS can use for retrieving profiles of a given variable. Typical degrees of freedom in the

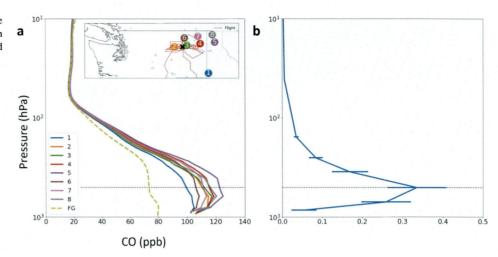

FIG. 6 (A) CO profiles around the Williams Flat fire and (B) the mean AKM for these profiles. The dashed profile is the first guess (FG).

midlatitudes for temperature is 6.4, water vapor is 3.85, ozone is 1.64, and carbon monoxide is 0.84 (Maddy and Barnet, 2008). Unlike AKMs, which would require campaign organizers to inspect each footprint profile, the degrees of freedom can be easily gridded and plotted on a map. This allowed organizers to diagnose the retrieval quality quickly and provided a greater level of detail than the red, yellow, and red quality flags that were developed for weather forecasting. In Fig. 5D, we show that for two regions of interest (indicated by dashed ovals) the degrees of freedom are between 0.6 and 1.0, thus showing there is reasonable CO retrieval skill. The degrees of freedom are likely lower near the wildfire because the smoke can cause cloud clearing to fail.

Fig. 6A shows eight NUCAPS CO profiles near the wildfire and in the background region, where the locations are shown on the inset map. Unlike the temperature and moisture profiles, NUCAPS uses a formal a priori measurement for the CO trace gas retrieval (Warner et al., 2010) and not a first guess based on a regression retrieval. In this case, the CO a priori measurement is a monthly CO climatology from MOPPITT, one each for the Northern and Southern Hemisphere. Thus, this region has a background level of CO (profile 1, in blue) that is higher than the Northern Hemisphere mean (dashed yellow line). It is interesting that the highest values of CO are not found directly over the fire (profiles 2, 3, and 6) as AOD and the visual imagery would indicate (Fig. 4). Instead, the highest CO values are further downstream in profiles 4, 5, and 8. By inspecting the mean AKM for these profiles (Fig. 6B), we can see that the peak sensitivity occurs at 500 hPa. Given that CO has roughly one degree of freedom (Fig. 5D), NUCAPS only has information about CO at one independent level. So, it can be assumed that the CO emissions are below 500 hPa near the fire, but become detectable by NUCAPS as the CO is transported over larger distances and rises farther into the atmosphere. NUCAPS cloud products were used to determine smoke plume height and location when combined with NUCAPS CO retrievals for a California wildfire (Smith et al., 2019). In this case, the plume was much smaller than the footprint (less than 10% cloud cover), so these fields did not provide situational awareness. The scale of this fire was also too small for the JPSS cloud fraction product.

FIREX-AQ captured high-resolution measurements of small wildfires using on-board sounders. The flights were well timed with satellite overpasses, so NUCAPS is comparable to the aircraft instrument retrievals. NUCAPS swaths extend up to 2200 km beyond the flight path to provide regional situational awareness of smoke transport. Through the collaboration, developers could evaluate NUCAPS trace gas and cloud retrievals' sensitivity to smaller fires that were selected by campaign organizers. At campaign organizers request, NUCAPS developers delivered diagnostic information such as the degrees of freedom of a variable, in real time. Like with IFEX campaign, collaboration with FIREX-AQ helped better understand NUCAPS capabilities for providing situational awareness for a targeted application.

6. Summary

NUCAPS retrievals provide situational awareness to scientific field campaigns when used alongside conventional assets such as radiosondes and high-resolution forecast models. Collaboration with scientific field campaigns like IFEX and FIREX-AQ allows developers to tailor NUCAPS retrievals of temperature, water vapor, and trace gases for specific applications. Scientific field campaigns can be considered an intermediate evaluation of NUCAPS usefulness before delivery to testbeds or operations.

The HWT opened a communication channel between developers and forecasters which improved NUCAPS for situational awareness and also trained forecasters on NUCAPS subtleties and skill (Smith et al., 2018; Esmaili et al., 2020). During the HWT, forecasters posed questions to NUCAPS developers, such as if NUCAPS can detect capping layers. To answer this question, developers can use the first guess and internal diagnostics like AKMs, which measures the observing capability of a scene. Forecasters identified preconvective capping inversion case using NUCAPS, model, and in situ data. The capping inversion was present in the operational NUCAPS first guess but not in offline NUCAPS, so it may have been an artifact. However, upgrades installed in the offline NUCAPS showed greater agreement to the radiosonde. As a result of this collaboration, NUCAPS developers have an interesting and operationally relevant case to study NUCAPS' signal to noise ratio and a target for improvement.

NUCAPS developers participated in two scientific field campaigns to determine when and where NUCAPS can provide situational awareness for tropical cyclones and for wildfire smoke transport. During a flight over Hurricane Jerry, NUCAPS showed agreement with dropsondes when capturing mid-level water vapor in the tropical cyclone environment. This assessment was only possible because the dropsondes were well timed with the Suomi NPP and NOAA-20 overpasses. Some future directions that resulted from this collaboration include using MW-only retrievals or a model-based first guess (as opposed to the current model-independent first guess) to improve NUCAPS over the tropical cyclone. FIREX-AQ organizers showed interest in using temperature and water vapor from Gridded NUCAPS for situational awareness of wildfire risk and smoke transpose. NUCAPS CO retrievals were useful for monitoring smoke transport once it leaves the boundary layer and reaches ~500 hPa, where NUCAPS has the most skill. FIREX-AQ organizers were also interested in diagnostic fields like the degrees of freedom to assess retrieval quality.

In closing, developers, scientists, and forecasters often work independently of each other. This chapter illustrates how synergizing these communities can improve situational awareness by determining (1) when and where satellite products have value to a campaign and (2) how to tailor novel satellite products for a specific application. A third benefit of open communication is increased awareness and understanding of a satellite product, which is especially important for operations. A reason for successful adoption of novel technologies is stakeholder awareness (Rogers, 2003), which is increased through scientific field campaign collaboration. Through their interaction, developers simultaneously improve NUCAPS while also informing campaign participants about the strengths and limitations of the retrievals. Thus, once tailored updates are available operationally worldwide, the community is already familiar with the product and know their feedback is embedded in the development. This chapter highlights how NUCAPS' provided situational awareness for NOAA-specific applications, but stakeholder-tailored improvements are also gaining traction for NASA applications, such as through CLIMCAPS. Similarly, other satellite products may benefit from direct collaboration with campaigns to facilitate their successful transition to operations.

Acknowledgments

The authors acknowledge the JPSS Proving Ground and Risk Reduction (PGRR) program for their support of the work described in this chapter. We would like to recognize the two reviewers for this chapter, Nadia Smith and Nicholas Nalli, for their insightful reviews. Their feedback helped improve the conceptual aspects of this chapter and tie the content into the rest of the book. We would also like to express appreciation to the large number of individuals who collaborated with the NUCAPS development team in the HWT, IFEX, and FIREX-AQ.

References

Avallone, L.M., Baeuerle, B., 2017. A 20-year history of NSF-supported atmospheric science field campaigns: statistics and demographics. Bull. Am. Meteorol. Soc. 98, 1333–1339. https://doi.org/10.1175/BAMS-D-15-00222.1.

Barnet, C.D., Smith, N., Folmer, M.J., Dunion, J.P., Zawislak, J., Layns, A., Esmaili, R., Goldberg, M.D., 2019. The use of temperature and water vapor profiles for tropical cyclone weather applications: recent activities within the NOAA/JPSS proving ground sounding initiative. In: 15th Annual Symposium on New Generation Operational Environmental Satellite Systems. Presented at the 99th American Meteorological Society Annual Meeting, Phoenix, AZ.

Barnet, C.D., Divakarla, M., Gambacorta, A., Iturbide-Sanchez, F., Nalli, N.R., Pryor, K., Tan, C., Wang, T., Warner, J., Zhang, K., Zhu, T., 2021. The NOAA Unique Combined Atmospheric Processing System (NUCAPS) Algorithm Theoretical Basis Document (v3.1), NOAA/NESDIS/STAR Joint Polar Satellite System, College Park, MD, USA. https://www.star.nesdis.noaa.gov/jpss/documents/ATBD/ATBD_NUCAPS_v3.1.pdf.

Benjamin, S.G., Weygandt, S.S., Brown, J.M., Hu, M., Alexander, C.R., Smirnova, T.G., Olson, J.B., James, E.P., Dowell, D.C., Grell, G.A., Lin, H., Peckham, S.E., Smith, T.L., Moninger, W.R., Kenyon, J.S., Manikin, G.S., 2016. A North American hourly assimilation and model forecast cycle: the rapid refresh. Mon. Weather Rev. 144, 1669–1694. https://doi.org/10.1175/MWR-D-15-0242.1.

Berndt, E., Smith, N., Burks, J., White, K., Esmaili, R., Kuciauskas, A., Duran, E., Allen, R., LaFontaine, F., Szkodzinski, J., 2020. Gridded satellite sounding retrievals in operational weather forecasting: product description and emerging applications. Remote Sens. (Basel) 12, 3311. https://doi.org/10.3390/rs12203311.

Bloch, C., Knuteson, R.O., Gambacorta, A., Nalli, N.R., Gartzke, J., Zhou, L., 2019. Near-real-time surface-based CAPE from merged hyperspectral IR satellite sounder and surface meteorological station data. J. Appl. Meteorol. Climatol. 58, 1613–1632. https://doi.org/10.1175/JAMC-D-18-0155.1.

Browning, K.A., Collier, C.G., 1989. Nowcasting of precipitation systems. Rev. Geophys. 27, 345–370. https://doi.org/10.1029/RG027i003p00345.

Chahine, M.T., 1977. Remote sounding of cloudy atmospheres. II. Multiple cloud formations. J. Atmos. Sci. 34, 744–757. https://doi.org/10.1175/1520-0469(1977)034<0744:RSOCAI>2.0.CO;2.

DeMaria, M., Knaff, J.A., Connell, B.H., 2001. A tropical cyclone genesis parameter for the tropical Atlantic. Weather Forecast. 16, 219–233. https://doi.org/10.1175/1520-0434(2001)016<0219:ATCGPF>2.0.CO;2.

DeSlover, D.H., Taylor, J.K., Smith, W.L., Weisz, E., Revercomb, H.E., 2016. Analysis of S-HIS dual-regression retrievals during the HS3 field campaign. In: Light, Energy and the Environment. Presented at the Fourier Transform Spectroscopy, OSA, Leipzig, p. JW4A.33, https://doi.org/10.1364/FTS.2016.JW4A.33.

Dole, R.M., Spackman, J.R., Newman, M., Compo, G.P., Smith, C.A., Hartten, L.M., Barsugli, J.J., Webb, R.S., Hoerling, M.P., Cifelli, R., Wolter, K., Barnet, C.D., Gehne, M., Gelaro, R., Kiladis, G.N., Abbott, S., Akish, E., Albers, J., Brown, J.M., Cox, C.J., Darby, L., de Boer, G., DeLuisi, B., Dias, J., Dunion, J., Eischeid, J., Fairall, C., Gambacorta, A., Gorton, B.K., Hoell, A., Intrieri, J., Jackson, D., Johnston, P.E., Lataitis, R., Mahoney, K.M., McCaffrey, K., McColl, H.A., Mueller, M.J., Murray, D., Neiman, P.J., Otto, W., Persson, O., Quan, X.-W., Rangwala, I., Ray, A.J., Reynolds, D., Dellaripa, E.R., Rosenlof, K., Sakaeda, N., Sardeshmukh, P.D., Slivinski, L.C., Smith, L., Solomon, A., Swales, D., Tulich, S., White, A., Wick, G., Winterkorn, M.G., Wolfe, D.E., Zamora, R., 2018. Advancing science and services during the 2015/16 El Niño: the NOAA El Niño rapid response field campaign. Bull. Am. Meteorol. Soc. 99, 975–1001. https://doi.org/10.1175/BAMS-D-16-0219.1.

Dunion, J.P., 2011. Rewriting the climatology of the tropical North Atlantic and Caribbean Sea atmosphere. J. Climate 24, 893–908. https://doi.org/10.1175/2010JCLI3496.1.

Dunion, J.P., Velden, C.S., 2004. The impact of the Saharan air layer on Atlantic tropical cyclone activity. Bull. Am. Meteorol. Soc. 85, 353–366. https://doi.org/10.1175/BAMS-85-3-353.

Duran, E.L., Berndt, E.B., Duran, P., 2021. Observation of the tropical cyclone diurnal cycle using hyperspectral infrared satellite sounding retrievals. Mon. Weather Rev. 149, 3671–3690. https://doi.org/10.1175/MWR-D-20-0415.1.

Endsley, M., 1995. Toward a theory of situation awareness in dynamic systems. Hum. Factors 37 (1), 32–64. https://doi.org/10.1518/001872095779049543.

Esmaili, R.B., 2019. NUCAPS Quality Flag Quick Guide [WWW Document]. https://weather.msfc.nasa.gov/nucaps/qg/NUCAPS-QF-quick-guide.pdf. (Accessed 11 March 2020).

Esmaili, R.B., Smith, N., Berndt, E.B., Dostalek, J.F., Kahn, B.H., White, K., Barnet, C.D., Sjoberg, W., Goldberg, M., 2020. Adapting satellite soundings for operational forecasting within the hazardous weather testbed. Remote Sens. (Basel) 12, 886. https://doi.org/10.3390/rs12050886.

Esmaili, R., Barnet, C., Dunion, J., Folmer, M., Zawislak, J., 2022. Evaluating satellite sounders for monitoring the tropical cyclone environment in operational forecasting. Remote Sens. 14 (13). https://doi.org/10.3390/rs14133189.

Fetzer, E., et al., 2003. AIRS/AMSU/HSB validation. IEEE Trans. Geosci. Remote Sens. 41 (2), 418–431. https://doi.org/10.1109/TGRS.2002.808293.

Gambacorta, A., Barnet, C.D., 2013. Methodology and information content of the NOAA NESDIS operational channel selection for the cross-track infrared sounder (CrIS). IEEE Trans. Geosci. Remote Sens. 51 (6), 3207–3216. https://doi.org/10.1109/TGRS.2012.2220369.

Goldberg, M.D., Qu, Y., McMillin, L., Wolf, W., Zhou, L., Divakarla, M., 2003. AIRS near-real-time products and algorithms in support of operational numerical weather prediction. IEEE Trans. Geosci. Remote Sens. 41 (2). https://doi.org/10.1109/TGRS.2002.808307.

Hook, S., 2017. Combined ASTER and MODIS Emissivity Database Over Land (CAMEL) Emissivity Monthly Global 0.05Deg V002., https://doi.org/10.5067/MEASURES/LSTE/CAM5K30EM.002.

Hsu, C., VIIRS Level 2 Deep Blue Aerosol (NRT), 2019. NASA. 10.5067/VIIRS/AERDB_L2_VIIRS_SNPP_NRT.011 [dataset].

Huff, A.K., Kondragunta, S., Zhang, H., Laszlo, I., Zhou, M., Caicedo, V., Delgado, R., Levy, R., 2021. Tracking smoke from a prescribed fire and its impacts on local air quality using temporally resolved GOES-16 ABI aerosol optical depth (AOD). J. Atmos. Oceanic Tech. 38, 963–976. https://doi.org/10.1175/JTECH-D-20-0162.1.

Iturbide-Sanchez, F., da Silva, S.R.S., Liu, Q., Pryor, K.L., Pettey, M.E., Nalli, N.R., 2018. Toward the operational weather forecasting application of atmospheric stability products derived from NUCAPS CrIS/ATMS soundings. IEEE Trans. Geosci. Remote Sens. 56, 4522–4545. https://doi.org/10.1109/TGRS.2018.2824829.

Jedlovec, G., 2013. Transitioning research satellite data to the operational weather community: the SPoRT paradigm [organization profiles]. IEEE Geosci. Remote Sens. Mag. 1, 62–66. https://doi.org/10.1109/MGRS.2013.2244704.

JPSS/STAR, 2021. Algorithm Maturity Matrix [WWW Document]. Algorithm Maturity Matrix. https://www.star.nesdis.noaa.gov/star/index.php. (Accessed 7 August 2021).

Junghenn Noyes, K.T., Kahn, R.A., Limbacher, J.A., Li, Z., Fenn, M.A., Giles, D.M., et al., 2020. Wildfire smoke particle properties and evolution, from space-based multi-angle imaging II: the Williams flats fire during the FIREX-AQ campaign. Remote Sens. (Basel) 12 (22), 3823. https://doi.org/10.3390/rs12223823.

Landsea, C.W., Franklin, J.L., 2013. Atlantic hurricane database uncertainty and presentation of a new database format. Mon. Weather Rev. 141, 3576–3592. https://doi.org/10.1175/MWR-D-12-00254.1.

Maddy, E.S., Barnet, C.D., 2008. Vertical resolution estimates in version 5 of AIRS operational retrievals. IEEE Trans. Geosci. Remote Sens. 46, 2375–2384. https://doi.org/10.1109/TGRS.2008.917498.

Nalli, N.R., Barnet, C.D., Reale, A., Tobin, D., Gambacorta, A., Maddy, E.S., Joseph, E., Sun, B., Borg, L., Mollner, A., Morris, V.R., Divakarla, M., Liu, X., Minnett, P.J., Knuteson, R.O., King, T.S., Wolf, W.W., 2013. Validation of satellite sounder environmental data records: application to the cross-track infrared microwave sounder suite. J. Geophys. Res. Atmos. 118, 13628–13643. https://doi.org/10.1002/2013JD020436.

Nalli, N.R., Gambacorta, A., Liu, Q., Barnet, C.D., Tan, C., Iturbide-Sanchez, F., Reale, T., Sun, B., Wilson, M., Borg, L., Morris, V.R., 2018. Validation of atmospheric profile retrievals from the SNPP NOAA-unique combined atmospheric processing system. Part 1: temperature and moisture. IEEE Trans. Geosci. Remote Sens. 56, 180–190. https://doi.org/10.1109/TGRS.2017.2744558.

NOAA/STAR, 2018. GOES-R Advanced Baseline Imager (ABI) Algorithm Theoretical Basis Document For Suspended Matter/Aerosol Optical Depth and Aerosol Size Parameter.

Reale, T., Sun, B., Tilley, F.H., Pettey, M., 2012. The NOAA products validation system (NPROVS). J. Atmos. Oceanic Tech. 29 (5), 629–645. https://doi.org/10.1175/JTECH-D-11-00072.1.

Rogers, E.M., 2003. Diffusion of Innovations, fifth ed. Free Press, New York.

Rogers, R., Aberson, S., Black, M., Black, P., Cione, J., Dodge, P., Dunion, J., Gamache, J., Kaplan, J., Powell, M., Shay, N., Surgi, N., Uhlhorn, E., 2006. The intensity forecasting experiment: a NOAA multiyear field program for improving tropical cyclone intensity forecasts. Bull. Am. Meteorol. Soc. 87, 1523–1538. https://doi.org/10.1175/BAMS-87-11-1523.

Rogers, R., Aberson, S., Aksoy, A., Annane, B., Black, M., Cione, J., Dorst, N., Dunion, J., Gamache, J., Goldenberg, S., Gopalakrishnan, S., Kaplan, J., Klotz, B., Lorsolo, S., Marks, F., Murillo, S., Powell, M., Reasor, P., Sellwood, K., Uhlhorn, E., Vukicevic, T., Zhang, J., Zhang, X., 2013. NOAA'S hurricane intensity forecasting experiment: a progress report. Bull. Am. Meteorol. Soc. 94, 859–882. https://doi.org/10.1175/BAMS-D-12-00089.1.

Rosenkranz, P.W., 2001. Retrieval of temperature and moisture profiles from AMSU-A and AMSU-B measurements. IEEE Trans. Geosci. Remote Sens. 39, 2429–2435. https://doi.org/10.1109/36.964979.

Rosenkranz, P.W., 2006. Cloud liquid-water profile retrieval algorithm and validation. J. Geophys. Res. Atmos. 111. https://doi.org/10.1029/2005JD005832.

Sawyer, V., Levy, R.C., Mattoo, S., Cureton, G., Shi, Y., Remer, L.A., 2020. Continuing the MODIS dark target aerosol time series with VIIRS. Remote Sens. (Basel) 12, 308. https://doi.org/10.3390/rs12020308.

Smith, W.L., 1968. An improved method for calculating tropospheric temperature and moisture from satellite radiometer measurements. Mon. Weather Rev. 96, 387–396.

Smith, N., Barnet, C.D., 2019. Uncertainty characterization and propagation in the community long-term infrared microwave combined atmospheric product system (CLIMCAPS). Remote Sens. (Basel) 11, 1227. https://doi.org/10.3390/rs11101227.

Smith, N., Barnet, C.D., 2020. CLIMCAPS Observing Capability for Temperature, Moisture and Trace Gases from AIRS/AMSU and CrIS/ATMS. Atmospheric Measurement Techniques Discussions, pp. 1–33, https://doi.org/10.5194/amt-2020-71.

Smith, W.L., Weisz, E., Kireev, S.V., Zhou, D.K., Li, Z., Borbas, E.E., 2012. Dual-regression retrieval algorithm for real-time processing of satellite ultraspectral radiances. J. Appl. Meteorol. Climatol. 51, 1455–1476. https://doi.org/10.1175/JAMC-D-11-0173.1.

Smith Sr., W.L., Zhou, D.K., Larar, A.M., Mango, S.A., Howell, H.B., Knuteson, R.O., Revercomb, H.E., Smith Jr., W.L., 2005. The NPOESS airborne sounding Testbed interferometer—remotely sensed surface and atmospheric conditions during CLAMS. J. Atmos. Sci. 62 (4), 1118–1134. https://doi.org/10.1175/JAS3384.1.

Smith, N., Barnet, C.D., Berndt, E., Goldberg, M., 2019. Why operational meteorologists need more satellite soundings. In: Presented at the 99th American Meteorological Society Annual Meeting, AMS, Phoenix, AZ.

Smith, N., Shontz, K., Barnet, C.D., 2018. What is a satellite measurement? Communicating abstract satellite science concepts to the world. In: Presented at the 98th American Meteorological Society Annual Meeting, AMS, Austin, TX.

Strong, J.D.O., Vecchi, G.A., Ginoux, P., 2018. The climatological effect of Saharan dust on global tropical cyclones in a fully coupled GCM. J. Geophys. Res. Atmos. 123, 5538–5559. https://doi.org/10.1029/2017JD027808.

Strow, L.L., Hannon, S.E., De Souza-Machado, S., Motteler, H.E., Tobin, D., 2003. An overview of the AIRS radiative transfer model. IEEE Trans. Geosci. Remote Sens. 41, 303–313. https://doi.org/10.1109/TGRS.2002.808244.

Sun, B., Reale, A., Tilley, F.H., Pettey, M.E., Nalli, N.R., Barnet, C.D., 2017. Assessment of NUCAPS S-NPP CrIS/ATMS sounding products using reference and conventional radiosonde observations. IEEE J. Sel. Top. Appl. Earth Obs. Remote Sens. 10, 2499–2509. https://doi.org/10.1109/JSTARS.2017.2670504.

Susskind, J., Barnet, C.D., Blaisdell, J.M., 2003. Retrieval of atmospheric and surface parameters from AIRS/AMSU/HSB data in the presence of clouds. IEEE Trans. Geosci. Remote Sens. 41 (2), 390–409. https://doi.org/10.1109/TGRS.2002.808236.

Susskind, J., Blaisdell, J.M., Iredell, L., Keita, F., 2011. Improved temperature sounding and quality control methodology using AIRS/AMSU data: the AIRS science team version 5 retrieval algorithm. IEEE Trans. Geosci. Remote Sens. 49 (3), 883–907. https://doi.org/10.1109/TGRS.2010.2070508.

Taylor, J.K., Tobin, D.C., Revercomb, H.E., Best, F.A., Garcia, R.K., Smith, W., Weisz, E., Pierce, R.B., Kalashnikova, O., Frost, G., Goldberg, M., 2021. A summary of scanning high-resolution interferometer sounder (S-HIS) observations during the FIREX-AQ ER-2 campaign. In: Buckley, S., Vanier, F., Shi, S., Walker, K., Coddington, I., Paine, S., Westberg, J. (Eds.), OSA Optical Sensors and Sensing Congress 2021 (AIS, FTS, HISE, SENSORS, ES). OSA Technical Digest (Optical Society of America). paper FTh2G.2.

Tobin, D.C., et al., 2006. Radiometric and spectral validation of atmospheric infrared sounder observations with the aircraft-based scanning high-resolution interferometer sounder. J. Geophys. Res. 111, D09S02. https://doi.org/10.1029/2005JD006094.

Warner, J.X., Wei, Z., Strow, L.L., Barnet, C.D., Sparling, L.C., Diskin, G., Sachse, G., 2010. Improved agreement of AIRS tropospheric carbon monoxide products with other EOS sensors using optimal estimation retrievals. Atmos. Chem. Phys. 10, 9521–9533. https://doi.org/10.5194/acp-10-9521-2010.

Weaver, G.M., Smith, N., Berndt, E.B., White, K.D., Dostalek, J.F., Zavodsky, B.T., 2019. Addressing the cold air aloft aviation challenge with satellite sounding observations. J. Oper. Meteorol., 138–152. https://doi.org/10.15191/nwajom.2019.0710.

Weisz, E., Smith, N., Smith Sr., W.L., 2015. The use of hyperspectral sounding information to monitor atmospheric tendencies leading to severe local storms. Earth Space Sci. 2, 369–377. https://doi.org/10.1002/2015EA000122.

Wheeler, A., Smith, N., Gambacorta, A., Barnet, C.D., 2018. Evaluation of NUCAPS products in AWIPS-II: results from the 2017 hazardous weather testbed. In: 14th Annual Symposium on New Generation Operational Environmental Satellite Systems; 98th Amer. Meteor. Soc. Annual Meeting, 8 Jan 2018, Austin, TX. Extended abstract available online: https://ams.confex.com/ams/98Annual/webprogram/Manuscript/Paper337401/AWheeler-etal_AMS2018_NUCAPS_preconvective_HWT_cases.pdf.

Further reading

Divakarla, M.G., Barnet, C.D., Goldberg, M.D., McMillin, L.M., Maddy, E., Wolf, W., Zhou, L., Liu, X., 2006. Validation of atmospheric infrared sounder temperature and water vapor retrievals with matched radiosonde measurements and forecasts. J. Geophys. Res. Atmos. 111. https://doi.org/10.1029/2005JD006116.

Feltz, M.L., Borg, L., Knuteson, R.O., Tobin, D., Revercomb, H., Gambacorta, A., 2017. Assessment of NOAA NUCAPS upper air temperature profiles using COSMIC GPS radio occultation and ARM radiosondes. J. Geophys. Res. Atmos. 122, 9130–9153. https://doi.org/10.1002/2017JD026504.

Iturbide-Sanchez, F., Liu, Q., Gambacorta, A., Barnet, C., Nalli, N.R., Tan, C., Santos da Silva, S.R., 2017. Using averaging kernels to study the vertical resolution of nucaps temperature and water vapor. In: 2017 IEEE International Geoscience and Remote Sensing Symposium (IGARSS). Presented at the 2017 IEEE International Geoscience and Remote Sensing Symposium (IGARSS), pp. 33–35, https://doi.org/10.1109/IGARSS.2017.8126886.

Knuteson, R., Antonelli, P., Best, F., Dutcher, S., Garcia, R., Howell, B., LaPorte, D., Revercomb, H., Taylor, J., Tobin, D., 2005. Scanning high-resolution interferometer sounder (S-HIS) observations during ADRIEX/EAQUATE. In: Fourier Transform Spectroscopy/Hyperspectral Imaging and Sounding of the Environment. Presented at the Hyperspectral Imaging and Sounding of the Environment, OSA, Alexandria, Virginia, p. HWD1, https://doi.org/10.1364/HISE.2005.HWD1.

Kuciauskas, A., Reale, A., Esmaili, R., Sun, B., Nalli, N., Morris, V., 2022. Investigating NUCAPS skill in profiling Saharan dust for near-real time forecasting. Remote Sens. Submitted for publication.

Kuciauskas, A., Esmaili, R., Reale, A., Nalli, N., 2020. Using NUCAPS to observe the thermodynamic structure of strong Saharan Air Layer outbreaks about its source within the deserts of Northeast Africa. In: 16th Annual Symposium on New Generation Operational Environmental Satellite Systems. Presented at the American Meteorological Society Meeting, Boston, MA.

Liu, C., Zipser, E.J., 2008. Diurnal cycles of precipitation, clouds, and lightning in the tropics from 9 years of TRMM observations. Geophys. Res. Lett. 35. https://doi.org/10.1029/2007GL032437.

Maddy, E.S., Barnet, C.D., Gambacorta, A., 2009. A computationally efficient retrieval algorithm for hyperspectral sounders incorporating a priori information. IEEE Geosci. Remote Sens. Lett. 6, 802–806. https://doi.org/10.1109/LGRS.2009.2025780.

Nalli, N.R., Joseph, E., Morris, V.R., Barnet, C.D., Wolf, W.W., Wolfe, D., Minnett, P.J., Szczodrak, M., Izaguirre, M.A., Lumpkin, R., Xie, H., Smirnov, A., King, T.S., Wei, J., 2011. Multiyear observations of the tropical Atlantic atmosphere: multidisciplinary applications of the NOAA aerosols and ocean science expeditions. Bull. Am. Meteorol. Soc. 92 (6), 765–789. https://doi.org/10.1175/2011BAMS2997.1.

Rodgers, C.D., 2004. Inverse methods for atmospheric sounding: theory and practice. In: Series on Atmospheric Oceanic and Planetary Physics, Reprinted edition. World Scientific, Singapore.

Smith, N., Barnet, C., 2018. How Forecasters Use NUCAPS (JPSS soundings) in Operational Decision-Making and Anticipated Benefits from Geostationary Orbit., https://doi.org/10.13140/RG.2.2.36476.44165.

Smith, N., White, K.D., Berndt, E.B., Zavodsky, B.T., Wheeler, A., Bowlan, M.A., Barnet, C.D., 2018b. NUCAPS in AWIPS: rethinking information compression and distribution for fast decision making. In: Presented at the 98th American Meteorological Society Annual Meeting, AMS, Austin, TX.

Smith, N., Esmaili, R., Barnet, C., Frost, G., McKeen, S., Trainer, M., Francoeur, C., 2020. Monitoring atmospheric composition and long-range smoke transport with NUCAPS. In: Satellite Soundings in Field Campaigns and Operations. Presented at the 100th American Meteorological Society Annual Meeting, AMS.

Smith, N., Shontz, K., Barnet, C.D., 2018a. What is a satellite measurement? Communicating abstract satellite science concepts to the world. In: Presented at the 98th American Meteorological Society Annual Meeting, AMS, Austin, TX.

# Part III

Satellite applications

Chapter 15

On-orbit VIIRS sensor calibration and validation in reflective solar bands (RSB)

Taeyoung Choi[a], Changyong Cao[b], Slawomir Blonski[a], Xi Shao[c], Wenhui Wang[c], and Sirish Uprety[c]

[a]*Global Science & Technology, Inc, Greenbelt, MD, United States,* [b]*NOAA/NESDIS Center for Satellite Applications and Research (STAR), College Park, MD, United States,* [c]*Cooperative Institute for Satellite and Earth System Studies (CISESS), Earth System Science Interdisciplinary Center, University of Maryland, College Park, MD, United States*

Chapter outline

1. Introduction	263	5. On-orbit calibration	272
2. The NOAA-20 VIIRS sensor	264	5.1 On-orbit SD F-factors	272
3. Overview of the prelaunch calibration	266	5.2 On-orbit lunar F-factor trends	274
4. On-orbit radiometric calibration for reflective solar band	267	5.3 Daily DCC trending	274
4.1 Primary VIIRS RSB calibration coefficient (SD H- and F-factors)	267	5.4 Simultaneous nadir overpass (SNO) trending validation	276
4.2 Scheduled lunar calibration collection	268	5.5 Decisions on operational calibration coefficients for VIIRS SDR data production	277
4.3 Lunar F-factor calculation algorithm	270	6. Conclusions	277
4.4 The DCC technique for VIIRS	271	Acknowledgments	278
4.5 Simultaneous nadir overpass (SNO) trend derivation algorithm	272	Disclaimer	278
		References	278

Keep your face to the sun and you will never see the shadows.

Helen Keller

1. Introduction

Ensuring accurate satellite on-orbit radiometer calibration has become an important endeavor within environmental satellite remote sensing because long-term climate change detection and accurate numerical weather prediction (NWP) models depend upon reliable radiometric data. Accordingly, reliable climate data records (CDRs) should include measurement time series of sufficient length, consistency, and continuity that will allow determinations of climate variability and change (N. R. Council, 2004). CDRs require that the remotely sensed data collected from satellite instruments ensure the long-term radiometric stability enabling detection of climate change. To verify the long-term stability, field measurements and vicarious calibration have been often performed that provide ground truth data to evaluate accurate performance of the on-orbit sensors at the time of satellite overpass (Uprety et al., 2013a; Chander et al., 2013a,b,c). However, vicarious calibration is very sensitive and subject to atmospheric conditions, procedures of data collection, and processing methodologies. In addition, field measurements are not cost effective in terms of number of collections and repeatability because of varying environments and human introduced uncertainties during the measurements. To circumvent these issues, satellite radiometers use on-board calibrators such as Solar Diffuser (SD) and Solar Diffuser Stability Monitor (SDSM) to track radiometer responsivity degradations over the lifetime. The SD became a standard for the Low Earth Orbit (LEO) sensors such as Terra and Aqua Moderate Resolution Imaging Spectroradiometer (MODIS), Landsat Operational Land Imager (OLI), Multi-angle Imaging Spectro-Radiometer (MISR), Suomi National Polar-orbiting Partnership (SNPP)/NOAA-20

VIIRS. The SD is also used for geostationary (GEO) imagers for the Advanced Baseline Imager (ABI) on GOES 16 and 17 satellites and Advanced Himawari Imager (AHI) on Himawari 8 satellite. The imaging radiometers with SD (or two SDs) as an on-orbit calibration source frequently view the on-orbit SD illuminations at the time of proper solar illumination geometry. With a proper angular correction, the SD observations provide the long-term radiometer changes compared to the prelaunch SD Bidirectional Reflectance Distribution Function (BRDF) (Choi and Cao, 2019). For reflective solar bands (RSBs), the accuracy of on-orbit calibration is mostly dependent on the understanding of the SD degradation over time. The degradation-corrected, time dependent SD BRDF provides radiometer response changes over time. In the case of VIIRS (Sun et al., 2018), this degradation is quantified and represented using an F-factor, which is defined as the radiance ratio between the theoretically expected radiance from the Sun and the observed SD radiance. The F-Factor can be interpreted as the inverse of the radiometer gain, and it reflects the radiometer responsivity changes over time, separated by optoelectrical components, such as band, detectors, Half Angle Mirror (HAM) sides, and gain state.

To independently validate the on-orbit radiometer calibration, moon-based vicarious calibration is also used separately for comparisons due to the unparalleled stability of the lunar surface reflectance (Sun et al., 2018; Xiong et al., 2016). The lunar calibration results are used offline to correct the long-term SD calibration trends, especially early orbits of SNPP VIIRS in the RSBs (Sun and Wang, 2016). Science teams from different agencies such as the NASA VIIRS team and NOAA Ocean Color and SDR teams perform independent calibration and have published slightly different long-term trend correction of the SD F-factors (Choi et al., 2018), but all have agreed that the SD F-factors need to be corrected by the lunar calibration. The calibration differences between the SD and lunar methods are also closely monitored for the NOAA-20 VIIRS on-orbit calibration (Choi et al., 2019a).

As a part of postlaunch product vicarious validation, Pseudo Invariant Calibration Site (PICS) observations are routinely used to detect and evaluate long-term calibration drifts of radiometers (Mishra et al., 2014; Helder et al., 2013). Similarly, deep convective cloud (DCC) observations are also used as a spectrally stable target in the visible and near infrared band long-term calibration (Wang and Cao, 2016, 2020a). Another methodology called simultaneous nadir overpasses (SNOs) for satellite pairs from LEO sensors is also frequently used to detect relative long-term calibration biases between two sensors with similar spectral bands (Uprety et al., 2013a).

In this chapter, detailed procedures and decision-making steps of the on-orbit radiometric calibration of NOAA-20 VIIRS are presented. A brief explanation of the VIIRS sensor is given in Section 2. Before launch, VIIRS went through extensive ground tests to characterize the sensor performance as summarized in Section 3 detailing the prelaunch calibration. In Section 4, the radiometric calibration coefficient (F-factor), is explained in detail along with the SD degradation estimation from the SDSM observations. In addition to the SD based calibration, lunar calibration is also introduced in Section 4. Besides lunar F-factors, the long-term DCC and SNO trends providing the best on-orbit calibration for producing stable and consistent NOAA-20 VIIRS products are also discussed in Section 5. Finally, the radiometric performance of NOAA-20 VIIRS is determined and summarized in Section 6.

2. The NOAA-20 VIIRS sensor

The NOAA-20 VIIRS was launched on November 18, 2017, as an advanced Earth observing, polar-orbiting sensor in a sun-synchronous orbit with a nominal equator crossing time at 13:25 in ascending node. With a nominal altitude of 829 km and a large scan angle of ±56°, VIIRS can observe the entire Earth in a day for both short- and long-term environmental monitoring and forecasting. Similar to VIIRS on Suomi National Polar-orbiting Partnership (SNPP), the NOAA-20 VIIRS has 14 RSBs, 7 Thermal Emissive Bands (TEBs) and one Day Night Band (DNB) covering a spectral range of 0.41–12.5 µm, with two spatial resolutions of 375 m for imaging (I) bands and 750 m for moderate resolution (M) bands. Details of center wavelength, band width, driving EDR, and pixel size at the nadir observation are listed in Table 1.

To provide accurate radiometric calibration, the VIIRS has on-board calibrators (OBCs) such as Solar Diffuser (SD), Solar Diffuser Stability Monitor (SDSM), Space View (SV), and Blackbody (BB), as shown in Fig. 1.

Among these OBCs, RSB calibration uses observations from SD, SDSM, and SV, whereas BB observations are used for TEB calibration as a thermal reference. There have been no major issues for TEB calibration because of the superior stability of the BB temperature throughout the lifetime of NOAA-20 VIIRS operation (Wang and Cao, 2021). On the other hand, accurate radiometer responsivity change estimations in the RSBs are challenging because of the time and wavelength-dependent surface reflectance degradations of SD, which is the primary calibration reference. The degradation of SD reflectance is caused by the exposure to the solar UV and energetic particle radiation especially for the shorter wavelength (0.4–0.6 µm) bands (Shao et al., 2019). The success of on-orbit calibration is mainly dependent on the accurate estimation of the SD degradation (or called H-factor (Baker et al., 2011)), which is measured by the SDSM. Under the assumption that the SD is an isotropic surface, the SDSM is a rationing radiometer measuring the surface reflectance ratio

TABLE 1 VIIRS SDR characteristics in the RSBs (Cao et al., 2014).

Band	Center wavelength (nm)	Band width (nm)	Driving EDR	Gain states	Nadir pixel size (m)
M1	411	19.8	Ocean color	High/low	750
M2	444	14.3	Ocean color	High/low	750
M3	486	19.0	Ocean color	High/low	750
M4	551	20.9	Ocean color	High/low	750
M5	672	20.0	Ocean color	High/low	750
M6	745	14.6	Atmos. correct.	Single	750
M7	862	38.7	Ocean color	High/low	750
M8	1238	27.1	Cloud particle	Single	750
M9	1375	15.0	Cirrus/cloud	Single	750
M10	1602	58.7	Snow fraction	Single	750
M11	2257	46.7	Clouds	Single	750
I1	639	77.5	Imagery EDR	Single	350
I2	862	39.4	NDVI	Single	350
I3	1602	57.2	Snow map	Single	350

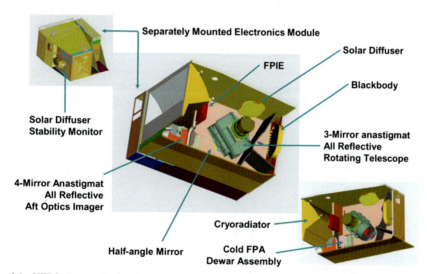

FIG. 1 A cut-out view of the VIIRS opto-mechanical module (Baker et al., 2011).

between the SD reflected signals from the SD screen and direct sun illumination reading through the SDSM sun screen, as shown in Fig. 2.

Once the on-orbit H-factor is established, the on-orbit RSB radiometric calibration coefficients (called F-factors (Baker et al., 2011)) are derived from the SD responses to the instantaneous incoming solar illumination through the rotating telescope assembly (RTA) as the primary fore-optics of VIIRS.

As an alternative approach, VIIRS can view the moon at a specific phase angle of −51° on a monthly basis through the SV port. Instead of using SD as a reference, lunar F-factors can be derived by using a lunar irradiance model, such as the Global Space-based Inter-Calibration System (GSICS) implementation of the Robotic Lunar Observatory (ROLO) model. The long-term trend of lunar F-factors is used to validate the daily SD F-factors as an alternative on-orbit radiometric calibration.

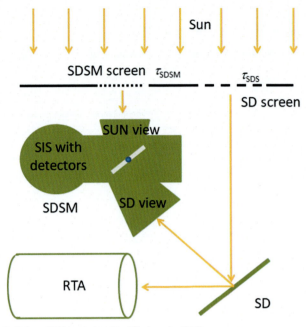

FIG. 2 A simplified diagram of SD, SDSM, and RTA relationship (Choi et al., 2020).

After applying the H and F-factors, the VIIRS SDRs achieved "beta provisional" status on February 1, 2018, and validated maturity status on February 19, 2018. After the "beta provisional" status, the NOAA-20 VIIRS SDR products were made available to the public through NOAA Comprehensive Large Array-Data Stewardship System (CLASS) distribution website. As a part of postlaunch tests, time-series results of DCCs, and simultaneous nadir overpasses (SNOs) were compared for validating the true VIIRS radiometer gain changes.

In this chapter, primary daily on-orbit NOAA-20 VIIRS RSB calibration coefficients (as derived by NOAA VIIRS team) are compared with the long-term, lunar-based calibration coefficients. Over the last 3 years, the SD RSB F-factors showed very stable responses with small degrees of annual oscillation patterns, except for the short wavelength bands from M1 to M4. These bands showed gradual SD F-factor decreases from 1.5% to 0.5% levels, but these decreasing trends were not observed in the lunar F-factors. The decreasing trends in the F-factor indicate that the radiometer gains are increasing in the short wavelength bands, which is not the usual case. To decide the true radiometer gain (or 1/F-factor) changes, the long-term DCC and SNO trending results were used for verification as a part of postlaunch tests (PLTs). Finally, the NOAA VIIRS team decided not to change the constant SD F-factors since April 2018 because the monthly lunar F-factors, daily DCC trends, and SNO results have not indicated any significant signs of radiometer gain changes of NOAA-20 VIIRS RSB bands. The details of the decision-making processes are summarized in the following sections.

3. Overview of the prelaunch calibration

Before launch of each JPSS satellite, VIIRS sensors undergo extensive testing both as stand-alone instruments and after integration with the spacecraft (Wolfe et al., 2013; Oudrari et al., 2015, 2016, 2018; McIntire et al., 2019). The tests are conducted under various environmental conditions (ambient and thermal vacuum) to characterize instrument performance and verify compliance with engineering performance requirements. In order to derive the ground-system processing parameters to be applied during on-orbit operations, a large sample of calibration and characterization data must be acquired under conditions similar to those encountered on orbit. Instrument characterization includes spectral response, spatial (detector) response, response versus scan angle, radiometric response (calibration), including electronic and optical cross-talk, noise, stability and uniformity of the response, band-to-band registration, polarization sensitivity, near-field response, stray light, and pointing knowledge. The measurements are completed at three thermal vacuum plateaus, namely cold, nominal, and hot, to cover the range of the expected on-orbit conditions. During the test program, two important decisions are made that define how the instrument is operated on orbit: (1) set point temperature for the cooled focal plane detectors is selected and (2) the primary and redundant sides of the instrument electronics are identified.

VIIRS RSB prelaunch calibration is measured at each thermal vacuum plateau for each gain state, detector, HAM side, and the electronics side (Oudrari et al., 2015, 2016, 2018). Two light sources are used: a 100-cm diameter spherical integrating source (SIS100) and a very bright three-mirror collimator (TMC) SIS. Additionally, a space view source is used to measure dark offsets needed to determine the background-corrected detector response. TMC SIS is used only to calibrate the low gain for bands M1–M3, as the SIS100 is not bright enough to cover the full dynamic range of these bands. Only SIS100 is traceable to the National Institute of Standards and Technology (NIST) calibrated light sources and has a radiance monitor to correct for signal fluctuations and drifts. Simultaneous SIS100 measurements are used to cross-calibrate TMC SIS over the radiance region of overlap between the two sources. Even with the radiance monitor, SIS100 is not sufficiently stable to directly calibrate VIIRS RSB radiometric response. Instead, an attenuator approach is used to derive the prelaunch calibration coefficients (the c coefficients in Eq. 3). The radiance measurements are performed with and without an attenuator screen inserted into the optical path between the light source and VIIRS, while the time between attenuator in and out is short to minimize effects of source instability on the measurements. The attenuator method allows only to calculate ratios of the calibration coefficients versus the linear term (c_1), and the linear term is still calculated directly from the calibration source select radiance values. The linear term dominates the calibration equation and is assumed to be sufficiently corrected on orbit with the F-factors.

VIIRS RSB RVS (response vs scan angle) is measured in a separate test in the ambient conditions (Oudrari et al., 2015). The test characterizes the relative reflectance of the HAM as a function of the light's angle of incidence (AOI). The measurements are conducted for a number of the AOI values, and an RVS function is estimated by fitting a quadratic polynomial of AOI to the data (Oudrari et al., 2018; McIntire et al., 2019; Lessel and McClain, 2007).

Before integration with the VIIRS instrument, elements of the onboard calibrator system are characterized as well, including the bidirectional reflectance of the solar diffuser at two view angles (RTA and SDSM) (Lessel and McClain, 2007; Murgai et al., 2014; Klein et al., 2017; Murgai and Klein, 2019), solar diffuser's solar attenuation screen transmission (Klein et al., 2019; Valencia et al., 2020), SDSM sun-view attenuation screen transmission (Murgai et al., 2015), and SDSM spectral response (Elgas and Murgai, 2020). These parameters are measured for a range of the illumination directions that are expected to occur on orbit. These data are then utilized in deriving processing parameters for the ground software.

4. On-orbit radiometric calibration for reflective solar band

4.1 Primary VIIRS RSB calibration coefficient (SD H- and F-factors)

The initial on-orbit NOAA-20 VIIRS radiometric calibration was tied to the prelaunch calibration through the SD BRDF measurements. Its degradation is measured by the Solar Diffuser Stability Monitor (SDSM), as shown in Fig. 2. Near the Earth's terminator from night to day, SD and SDSM have 14–15 opportunities per day to receive proper solar illumination of the diffuser (lasting for about 1 min). As shown in Fig. 2, the attenuated sunlight through the SD and SDSM screens illuminates both the SD surface and the SDSM Spherical Integrating Sphere (SIS) simultaneously. The SD observations through the Rotating Telescope Assembly (RTA) are used for radiometric calibration coefficient calculations (called F-factors) which correct the on-orbit changes of the sensor responsivity changes.

The wavelength and time-dependent SD degradation (called H-factor) are further modeled by the new Surface Roughness-induced Rayleigh Scattering (SRRs) model (Shao et al., 2016, 2019) and measured by the ratio between the Digital Count (DC) of the SD and Sun views with the proper screen effect corrections. It is defined in Eq. (1) at the eight SDSM detector center wavelengths, as shown in Table 2.

$$H = \frac{dc_{SD} \cdot \tau_{SDSM}}{dc_{sun} \cdot \cos(\theta_{SD}) \cdot \tau_{sds} \cdot BRDF_{SDSM} \cdot \Omega} \quad (1)$$

where dc_{SD} is the digital count (DC) from SD view, τ_{SDSM} is the SDSM sun screen transmittance function, dc_{sun} is the digital count (DC) of sun view, θ_{SD} is the solar incident angle to the SD screen, $\tau_{SD} BRDFD_{SDSM}$ is the combined function of SD screen transmittance and SD BRDF function to the SDSM view port, and Ω is the solid angle of the SDSM to SD view

TABLE 2 VIIRS SDSM detector center wavelength (CW).

Detector	1	2	3	4	5	6	7	8
CW (nm)	411.5	448.0	489.5	549.5	674.0	744.5	868.0	912.0

port. From Eq. (1), one can see the H-factor is defined as the ratio between the DC from the SD and the DC from the Sun with the screen correction factors including the solid angle of the SDSM to SD view (Ω), SDSM BRDF, and cosine of the incident angle to the SD screen (θ_{SD}). The SDSM was operated in every orbit initially since the NOAA-20 VIIRS activation on November 28, 2017, and it was reduced to every other orbit on December 14, 2017. When the functionality of the SDSM was verified, it was further reduced to once per day on January 5, 2018. To reduce the number of motor movements, the SDSM operation was reduced to once per week starting from February 28, 2019.

Regardless of the SDSM operations, VIIRS can make SD observations in every orbit through RTA as shown in Fig. 2. In each SD observation, the SD F-factor can be calculated 14–15 times per day. The SD F-factor is a ratio between the predicted solar radiance from the prelaunch measurements and the observed SD radiance on-orbit. The F-factor is calculated by following Eq. (2).

$$F_{SD} = \frac{\cos(\theta_{inc}) \cdot E_{sun} \cdot \tau_{SDS} \cdot BRDF_{SD_{RTA}} \frac{H(t)}{H_0} \cdot RVS_{SD}}{c_0 + c_1 \cdot dn_{SD} + C_2 \cdot dn_{SD}^2} \quad (2)$$

In Eq. (2), θ_{inc} is the incident solar illumination to the SD surface, E_{sun} is the solar irradiance with the earth-sun distance correction, $\tau_{SDS}BRDF_{SD\_RTA}$ is a combination of SD BRDF to the RTA view angle and the SD transmittance function, $H(t)/H_0$ is the normalized time-dependent H-factor, and RVS_{SD} is the response-versus-scan (RVS) angle to the Half-Angle-Mirror (HAM) surface. In the numerator part of Eq. (2), the observed lunar radiance was calculated by the offset removed DN that is indicated by the lower case from the SD (dn_{SD}) with the c-coefficients with the quadratic equation. From Eqs. (1) and (2), it should be noted that any changes in the H-factor will affect F-factors. In turn, the Earth view radiance is calculated from following Eq. (3).

$$L_{EV} = \frac{F \cdot (c_0 + c_1 \cdot dn_{EV} + c_2 \cdot dn_{EV}^2)}{RVS_{EV}} \quad (3)$$

The Earth view radiance is a function of F-factor, bias, and temperature effect corrected EV DN, and the corresponding RVS at the time of dn_{EV} was observed. Detailed explanation of the H- and F-factors is described in the VIIRS Algorithm Theoretical Basis Document (ATBD) (Baker et al., 2011).

4.2 Scheduled lunar calibration collection

The VIIRS can view the moon through the SV port at the scan angles from −66.1° to −62.25° before the Earth View scan as shown in Figs. 1 and 3. From the angle range, the VIIRS observes the moon with 48 frames (or pixels) of data with the Moderate (M) band resolution. The monthly scheduled lunar calibration collections are planned in advance about a month

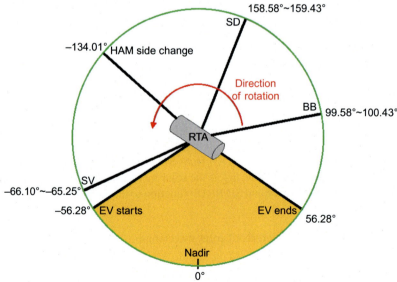

FIG. 3 RTA scan angle and on-board calibrators (Choi et al., 2015).

FIG. 4 NOAA-20 VIIRS scheduled lunar collection on 2018-5-25 in band M7. The center of the moon is located in aggregation zone 3 (Choi et al., 2019a).

except during June to September when the moon dips below the Earth limb by the geolocation relationship among Sun, Earth, and satellite (Patt et al., 2005). During each scheduled lunar collection, a spacecraft roll maneuver with a specific roll angle and EV sector rotation are applied together to place the moon in aggregation zone 3 of the EV frame as shown in Fig. 4. By applying the sector rotation, the EV pixels are co-registered compensating the offsets in different bands. For NOAA-20 VIIRS, the moon appears in the nonaggregation zone 3, whereas the moon is imaged at the center of the EV sector for the SNPP VIIRS case (Choi et al., 2019a). The spacecraft roll maneuver is calculated and applied on each lunar collection with a roll angle limit between 0° and −14°. During the scheduled lunar collection, the SV data are not valid for determining the basis level of detectors because of the sector rotation. The detector offsets need to be determined by the DN responses near the moon where the VIIRS views the deep space. Fig. 5 shows all the RSB bands for the scheduled lunar collection on May 25, 2018.

Table 3 shows an example of set of NOAA-20 scheduled lunar collections for purpose of illustration. The negative phase angle indicates the waxing lunar phases. The first lunar collection on December 29, 2017, placed the moon in the 2-sample aggregation zone whereas all other collections imaged the moon in the no-aggregation zone. In addition, the thermal and SWIR band responses were not available with the first collection because the cryo-radiator cooler was not yet functioning. As shown in Table 3, the lunar phase angles were intentionally selected around −51° to avoid additional source of uncertainty from the GIRO irradiance model.

Besides radiometric calibration, the scheduled lunar collections can be used to measure the band-to-band registration (BBR) from the selected middle scans. The initial BBR results suggested that the NOAA-20 VIIRS BBR have remained stable within ±0.2-pixel range in the scan direction and within ±0.05-pixel range in the track direction (Choi et al., 2019b).

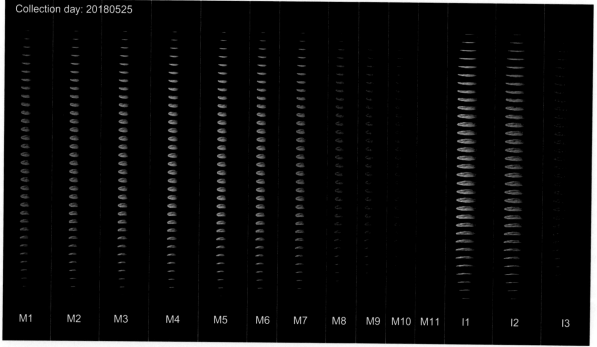

FIG. 5 NOAA-20 VIIRS scheduled lunar collection on 2018-05-25 in all RSB.

TABLE 3 NOAA-20 VIIRS scheduled lunar collection.

Date	UTC time (UTC)	Moon phase angle
2017-12-29[a]	10:03:56	−50.58
2018-1-27	19:22:49	−51.34
2018-2-26	04:47:03	−51.13
2018-3-27	12:32:59	−51.16
2018-4-25	20:21:36	−50.98
2018-5-25	05:53:34	−50.31
2018-6-23[b]	13:43:07	−51.42
2018-11-19	01:54:45	−50.99
2018-12-18	17:56:39	−51.32
2019-1-17	09:59:05	−50.81
2019-2-15	22:44:41	−50.84
2019-3-17[b]	08:11:05	−51.19
2019-4-15	15:59:10	−51.02
2019-5-14	22:07:57	−50.91
2019-6-13[b]	4:15:15	−50.87
2019-11-7[b]	23:37:01	−51.00
2019-12-7	19:03:36	−51.01
2020-1-6	16:08:16	−50.10
2020-2-5	08:14:31	−51.04
2020-3-5[b]	22:44:31	−51.27
2020-4-4[b]	09:4:39	−51.23
2020-5-3	17:44:39	−51.11
2020-6-1	23:55:09	−50.82
2020-11-25	15:06:14	−51.36
2020-12-25	12:09:45	−50.70

[a]Moon center was located in the 2-sample aggregation zone.
[b]No lunar roll maneuver was performed.

4.3 Lunar F-factor calculation algorithm

As an alternative source of calibration, moon observation can be used for radiometric calibration of VIIRS instead of using the primary on-orbit calibration source of SD. In the SD F-factor calculation in Eq. (2), the reference radiance is based on the prelaunch measured SD BRDF with the on-orbit reflectance change correction from the H-factor. Instead of using SD BRDF, the Global Space-based Inter-Calibration System (GSICS) Implementation of Robotic Lunar Observatory(ROLO) (GIRO) model is used for reference lunar irradiance accounting for geometric distance effect, lunar phase, and liberation (Kieffer and Stone, 2005). The original ROLO model was developed from 1000+ lunar observations of a ground-based telescope covering 250–2450 nm with 32 wavelengths over 10 years. The ROLO was implemented by the European Organization for the Exploitation of Meteorological Satellites (EUMETSAT) as a standard lunar calibration tool for GSICS community collaborating with Japan Aerospace Exploration Agency (JAXA), Centre National d'Etudes Spatiales (CNES), USGS, and NASA.

Once a scheduled lunar collection is performed correctly, the offset-removed VIIRS detector responses are calculated with the lower case "dn" as shown in Eq. (4). The B, D, H, and S represent band, detector, HAM side, and scan number. The offset level of the scan is calculated from the deep space DN responses on both sides of the moon (50 frames on each side) because the SV response are not valid due to the sector rotation

$$dn(B, D, H, S) = DN_{\text{moon}}(B, D, H, S) - \overline{DN_{\text{space}}(B, D, H)} \tag{4}$$

where DN_{moon} is the DN of the scheduled lunar collection and DN_{space} is the DN of the dark space near the moon. The radiance of lunar pixels is converted to radiance by applying the initial SD F-factor (F_0), c-coefficients and SV RVS in Eq. (5).

$$L_{\text{Pixel}}(B, D, H, S) = \frac{F_0\left[\sum_{i=0}^{3} c_i(B, D, H, S) dn(B, D, H, S)^i\right]}{RVS_{\text{SV}}(B, D)} \tag{5}$$

For NOAA-20 VIIRS SWIR band calibration, the third order of polynomials is used. In addition, the "c0" coefficients in Eq. (5) are set to zero in the RSB calibration. When all the lunar radiance values are calculated in the lunar image, the lunar irradiance is derived to match up with the unit of GIRO irradiance output. The final lunar F-factors are calculated by

$$F_{\text{lunar}}(B) = \frac{E_{\text{GIRO}}(B)}{E_{\text{observed}}(B)} = \frac{E_{\text{GIRO}}(B)}{\sum_{\text{pixel}} \frac{L_{\text{pixel}}(B, D, H, S)}{N} \frac{\pi R_{\text{moon}}^2}{D^2} \frac{1 + \cos e(\theta)}{2}} \tag{6}$$

where E is the irradiance, R is the lunar diameter, D is the distance between the moon and satellite, and $\square$ is the phase angle of the moon. In the denominator of Eq. (6), only effective pixels are considered.

4.4 The DCC technique for VIIRS

DCCs are stable targets with nearly Lambertian (i.e., diffuse) reflectance that have been widely used for the on-orbit calibration stability monitoring of satellite radiometers in the solar reflective spectrum (Doelling et al., 2013; Hu et al., 2004; Wang and Cao, 2015, 2016, 2020b). DCCs are extremely cold clouds above which the absorption due to water vapor and other gases is minimal in the VIS/NIR spectrum. They are abundant over the intertropical convergence zone (ITCZ, latitude within ±25°) and can be simply identified using a single longwave infrared (LWIR) channel centered at ~11 μm brightness temperature. VIIRS DCC pixels are identified based on the following criteria (Wang and Cao, 2015, 2016, 2020b):

(1) M15 (10.729 μm) BT (TB11) ≤ 205 K;
(2) Standard deviation of TB11 of the subject pixel and its eight adjacent pixels ≤ 1 K;
(3) Standard deviation of RSB reflectance of the subject pixel and its eight adjacent pixels ≤ 3%;
(4) Solar zenith angle ≤ 40°;
(5) View zenith angle ≤ 35° (to avoid the bow-tie effect in VIIRS SDRs).

The DCC technique generally consists of the following steps: (1) collecting satellite data over an area of interest; (2) identifying DCC pixels using the above describe criteria; (3) correcting for the anisotropic effect in DCC reflectance; (4) calculating DCC probability distribution functions (PDF); and (5) generating and analyzing DCC mean and mode time series (Wang and Cao, 2016). It is a statistical-based vicarious calibration method; therefore sufficient DCC samples need to be collected to ensure robust statistical analysis results. Both monthly and daily DCC time series have been used for VIIRS RSB on-orbit radiometric calibration stability and bias monitoring. The daily DCC method is similar to the monthly DCC method expect that the mode and mean of DCC reflectance are calculated on a daily basis. For the monthly DCC method, a subset of the ITCZ (such as a region defined by 25° S to 25° N and 150° W to 60° W) can be used to reduce the volume of data required to be processed. For the daily DCC method, DCCS over the entire ITCA need to be used to maximize the sample size for more reliable results. The DCC technique can be applied to all VIIRS RSBs, except M6 which saturated over DCCs.

Though DCCs have nearly Lambertian behavior, the anisotropic effect still exists in the DCC TOA reflectance and Angular Distribution Models (ADM) were developed to account for the effect (Doelling et al., 2013). The anisotropic effects in the VIIRS VIS/NIR bands (M1–M5, M7, and I1 and I2) are corrected using a 0.65 μm angular distribution model

(ADM) developed by Hu et al. (HU2004) (Hu et al., 2004). The HU2004 ADM performs well in bands with center wavelength closer to 0.65 µm, such as VIIRS bands M5 and M7. However, it is not applied to the VIIRS SWIR bands (M8–M11 and I3) due to the fact that the anisotropic effects in the bands are different. The seasonal variations were observed in the DCC time series for SWIR bands (without the anisotropic effect correction) as well as in the shorter wavelength VIS/NIR bands (after the anisotropic effect correction). A DCC seasonal cycle climatology was developed using 4 years of reprocessed SNPP VIIRS SDRs (Cao et al., 2021) and used to reduce the seasonal cycles in the daily and monthly DCC time series for all bands (Wang and Cao, 2020b).

4.5 Simultaneous nadir overpass (SNO) trend derivation algorithm

Independent validation of the VIIRS radiometric performance is critical in order to ensure that the radiometric accuracy and stability of the instrument is well within the specification.

Simultaneous nadir overpass (SNO) is a widely used technique for intercalibration and validation of satellite sensors. The methodology provides an excellent approach to assess the radiometric consistency between satellite sensors. When two or more satellites orbit the earth at different altitudes, SNOs periodically occur (Cao et al., 2004). The SNO technique provides a unique opportunity to compare multiple satellite instruments at their orbital intersection, with a small time difference between the instruments' observation. Comparison of simultaneous measurements between two or more instruments at their orbital intersection with almost identical viewing conditions makes this approach extremely suitable for reducing the uncertainties in intercomparison, associated with atmospheric variability, and BRDF. To use VIIRS data for time-series studies such as climate change, VIIRS calibration should be free of any artifacts and radiometrically consistent irrespective of the satellite platform.

This chapter uses SNO technique for intercomparing NOAA-20 and SNPP VIIRS. Although NOAA-20 and SNPP do not have direct SNOs between them, both have SNOs with Aqua MODIS and thus a double-differencing approach can be used for VIIRS intercomparison (Uprety et al., 2018). In addition to polar SNOs, SNOs between SNPP and Aqua satellites can be extended at low latitudes (SNOx), although with a larger time difference between the instruments' observations. This helps to compare NOAA-20 and SNPP VIIRS measurements at low latitudes that cover larger dynamic range (Uprety et al., 2013b).

For each SNO event, VIIRS reflectance product is collected from the NOAA CLASS (http://www.class.ncdc.noaa.gov), and MODIS level 1b (L1B) data are collected from NASA SIPS. Collocation is done by mapping the VIIRS reflectance data to the MODIS latitude/longitude grid. After mapping VIIRS data into MODIS grid for each SNO event, clear sky ROIs are extracted from both images near nadir in the overlapping region. Average reflectance is computed for both VIIRS and MODIS ROIs. The bias is calculated for each SNO event as, radiometric bias = (VIIRS-MODIS)*100%/MODIS. This process is repeated for all SNO events, resulting in a bias time series of SNPP/NOAA-20 VIIRS relative to MODIS. The difference in the VIIRS bias trends suggests the radiometric consistency between NOAA-20 and SNPP VIIRS. Spectral differences between the two VIIRS instruments are analyzed using spectral band adjustment factors (SBAF) based on Sciamachy (https://satcorps.larc.nasa.gov/cgi-bin/site/showdoc?mnemonic=SBAF). This approach of double differencing the SNO-based bias time series can be effectively used for future VIIRS sensors as well to evaluate the postlaunch radiometric performance.

5. On-orbit calibration

5.1 On-orbit SD F-factors

The SD degradation is calculated by the ratio of SD and Sun response with viewing effect corrections according to Eq. (1) as shown in Fig. 6. Because of the increased surface roughness that caused Rayleigh scattering, the largest SD reflectance degradation (approximately 25% at 1200 days after launch) is found at the shortest wavelength with SDSM detector 1 at 411.5 nm, whereas very small degradation of approximately 0.5% is found with detector 8 at 912 nm around the 1200 days after launch.

Once the H-factor is determined, the on-orbit SD F-factors can be derived from Eq. (2) as shown in Fig. 7. The SD degradations at the VIIRS center wavelengths are linearly interpolated by using the center wavelengths of the SDSM detectors. The inverse of the SD F-factors is equivalent to the radiometric response (or gain) changes over time. The starting point of NOAA-20 VIIRS SD F-factors is lower than unity which indicated that the observed SD radiances in the short wavelength of RSB bands (M1~M7, I1 and I2) are higher than the calculated radiance from the numerator part of Eq. (2). In addition, the spectrally matching bands such as I2 and M7 showed different patterns of SD F-factor responses.

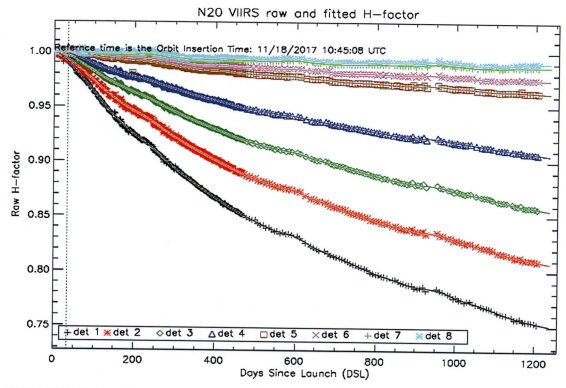

FIG. 6 NOAA-20 VIIRS H-factor after the τ_{SDSM} and 2 years of on-orbit SDSM updates.

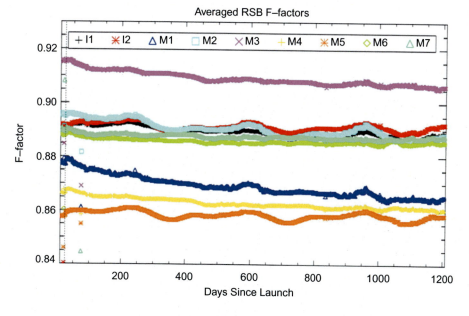

FIG. 7 Offline version of the NOAA-20 VIIRS SD F-factors as of April 12, 2021.

For the SNPP VIIRS case, most of the SD F-factors in the RSB bands started near unity except band M1 (Choi and Cao, 2019; Choi et al., 2018). This indicates that there could be possible prelaunch calibration error when the c-coefficients were calculated.

The short wavelength bands (M1∼M4) showed time-dependent degradation (or increase of gain) trends up to 1.5% level in band M1 (*blue triangle* in Fig. 7). On the other hand, the SWIR band SD F-factors show flat trends within 0.2% annual oscillations levels.

5.2 On-orbit lunar F-factor trends

To validate the long-term trends in the SD F-factors, the SD and lunar F-factors are properly normalized in Fig. 8. The lunar F-factors, symbols, are normalized on the SD F-factors at the 2nd lunar collection point in January 2018. This normalization process is focused to see whether there are significant differences between the SD and lunar F-factor in the long-term trends. Fig. 8 shows the detector averaged SD F-factors with solid lines and the lunar F-factors with symbols that are normalized to the 2nd lunar collection.

There are large annual gaps in the lunar collection around 300, 600, and 1000 days since launch (DSL) in Fig. 8. These are the no lunar collection months from June to October since the moon goes below the earth limb from the view of NOAA-20 VIIRS. In VIS/NIR bands, the long-term trends between the moon and SD F-factors are quite similar. But the SD F-factors are gradually deviating from the lunar F-factors in the short wavelength bands (M1~M4). The largest deviation of 1.5% approximately is found in the shortest wavelength of VIIRS in band M1. The degree of deviations is reduced to 1% level in band M2 and M3 and it becomes 0.5 in band M4. Even though there are annual oscillation patterns, the lunar F-factors indicate that there are no significant detector degradation in the all the RSB bands especially in the short wavelength bands (M1~M4).

There are some anomalous lunar F-factors near 500, 700, and 840 DSL that are caused by the no roll maneuver collections. Without the roll maneuver, the location of the moon is closer to the earth limb that introduced different levels in the offset estimation on both sides of the moon because of scattering light near the earth. This increased instability of the lunar F-factors for the no-roll maneuver lunar F-factors.

Fig. 9 shows the SD and lunar F-factor comparisons in the longer wavelength side of the RSB bands. The two F-factors show very stable response over 3 years of operation within 1% of annual oscillation levels.

5.3 Daily DCC trending

NOAA-20 VIIRS RSB on-orbit calibration has been monitored using the daily DCC method (Wang and Cao, 2020b). Monthly DCC method, as well as other vicarious methods, cannot provide reliable results in a timely manner during the VIIRS postlaunch test (PLT, or intensive cal/val) period, when frequent on-orbit calibration updates are made to transfer calibration from prelaunch to postlaunch and to further refine on-orbit calibration. Daily DCC time series were introduced in early 2018 to monitor NOAA-20 VIIRS RSB calibration and support the timely decision making of on-orbit calibration updates. The degradation trends derived using the daily and weekly DCC time series are consistent with that from the monthly DCC time series. Moreover, short-term calibration anomaly that lasts much shorter than a month, therefore, usually overlooked by monthly DCC time series can be successfully detected.

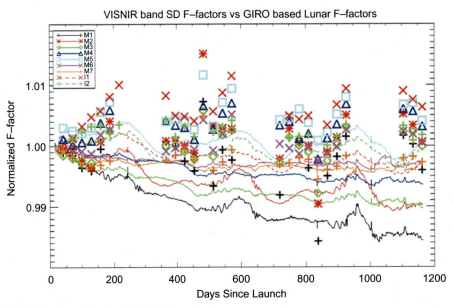

FIG. 8 NOAA-20 VIIRS SD and lunar F-factors in VIS/NIR bands.

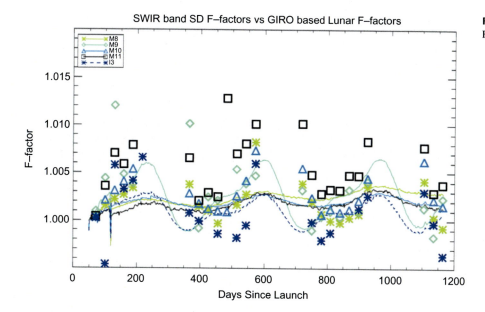

FIG. 9 NOAA-20 VIIRS SD and lunar F-factors in SWIR bands.

Fig. 10 shows NOAA-20 daily DCC time series for VIS/NIR bands (M1–M5, M7, and I1–I2) from the beginning of the mission to April 16, 2021. DCC trends and 95% confidence intervals (CI) were calculated using data after April 27, 2018, after the latest on-orbit calibration update. NOAA-20 VIIRS RSBs have been calibrated using constant calibration factors after this date. All statistics were estimated after the HU2004 ADM and residual annual cycle corrections. NOAA-20 VIIRS VIS/NIR bands have been very stable, with trends within ±0.08%/year for all bands. Bands M2 and M4 show relatively larger trends, but the trends are comparable to the uncertainty. DCC trending results indicate that there is no large change in the NOAA-20 NIR responsivity after more than 3 years on-orbit, different from SNPP that exhibits the significant in the NIR spectrum during its early mission (Cao et al., 2014). The results derived using the daily DCC method are generally consistent with those from lunar calibration (see Section 5.2).

Fig. 11 show the daily DCC time series for NOAA-20 SWIR bands (M8–M11 and I3). Similar to the VIS/NIR bands, DCC trends and 95% confidence intervals (CI) were calculated using data after April 27, 2018. Constant F-factors have also been used for SWIR bands on-orbit calibration since this date. Daily DCC trends indicate the on-orbit NOAA-20 SWIR responsivities changes are also small. Small upward trends were observed in all NOAA-20 SWIR bands. The trend in M8, M9, and M11 are small and similar to the uncertainty levels. Relative larger upward trends were observed in M10 and I3 (~0.3%/year).

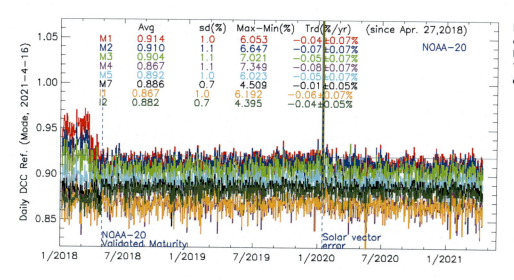

FIG. 10 Daily DCC time series of the NOAA-20 VIIRS VIS/NIR bands (M1–M5, M7, and I1–I2). VIIRS SDRs from the NOAA operational processing were used.

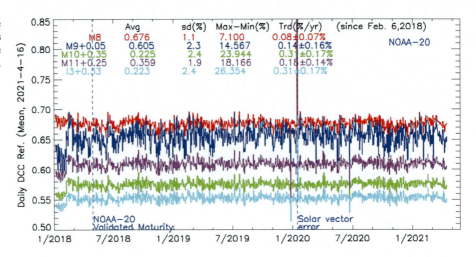

FIG. 11 Daily DCC time series of the NOAA-20 VIIRS SWIR bands (M8–M11 and I3). VIIRS SDRs from the NOAA operational processing were used.

5.4 Simultaneous nadir overpass (SNO) trending validation

SNO-based VIIRS comparison suggests that all the reflective solar bands of NOAA-20 VIIRS are biased lower than S-NPP. Other than bands M5 and M7, the rest of solar bands indicate bias ranging from 2% to 3% with the uncertainty (1-sigma) on the order 1%. M5 and M7 suggest larger biases, approximately 4% or more mainly because SNPP VIIRS absolute calibration is overestimated by about 2% (Uprety et al., 2018; Uprety and Cao, 2015). Fig. 12 shows the bias time series for M2 using SNO over polar region and SNOx over Saharan desert, both indicating the radiometric consistency between the two VIIRS sensors ranging from 2% to 3%. Similar to M bands, the I bands also suggest consistently lower responsivity for NOAA-20 compared to SNPP. The temporal trends for bias are nearly stable over time. NOAA-20 I1 indicates about −2% bias, whereas I2 indicate about −4% bias relative to SNPP. M7 and I2 RSRs match very well and thus suggests similar bias on the order of −4%. Although I1 and M5 spectral range matches well, the bias levels are different because SNPP I1

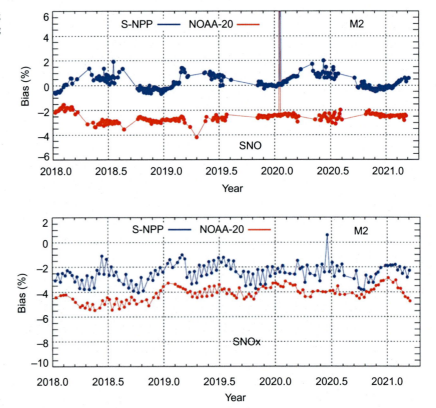

FIG. 12 VIIRS bias time series for M2 bands; (top) using SNO over polar region; (bottom) using SNOx over Saharan desert.

band indicates no overestimation in calibration. After accounting for −2% bias in NOAA-20 VIIRS for all M bands and +2% calibration overestimation in S-NPP VIIRS M5, M7, and I2, the two VIIRS instruments agree very well to within 1%. Uncertainties in bias estimation mainly come from individual sensor calibration uncertainties, BRDF, residual cloud contamination, and uncertainty in SBAF due to lack of in situ hyperspectral measurements over the intercomparison sites. The radiometric consistency between the two sensors can change over time and thus the bias time series needs to be monitored and analyzed frequently over the mission life.

5.5 Decisions on operational calibration coefficients for VIIRS SDR data production

As shown in Fig. 8, the primary calibration source, the SD F-factors, showed gradual degradations over time in the short wavelength bands in M1–M4. On the contrary, the monthly lunar F-factors have been very stable over the 3 years of operations, even though there were some unstable lunar F-factors that are associated with no-roll lunar maneuvers. Whenever there is no-roll lunar collection, the lunar F-factors become unstable because the moon appears closer to the Earth limb which introduced unstable detector bias-level estimation from the deep space observations. The SWIR bands show consistent calibration results between SD and lunar F-factors that have been very stable since launch within the 1% level as shown in Fig. 8. For the VIIRS SDR production, a decision should be made between the two different results from SD and lunar F-factors. For the best quality of VIIRS SDR products, DCC and SNO/SNOx results are used to determine the correct long-term radiometric calibration. With the DCC trends, the VIS/NIR bands showed very stable responses within ±0.08% per year including the short wavelength bands M1∼M3. In addition to DCC, the long-term SNO trends also showed stable results as shown in Table 4. These near real-time DCC and SNO trends are available at the NOAA Calibration Center (NCC) website (https://ncc.nesdis.noaa.gov/NOAA-20/VSTS.php). With all the possible sources of calibration, the historical operational F-factors are shown in Fig. 13. There are initial changes of F-factors because of the initial H-factor instability as discussed previously and with F-factor updates. The F-factors are set to a stable level since April of 2018, even though NOAA VIIRS SDR team has performed monthly comparisons with SD, lunar, DCC, and SNO trends to provide the best quality of VIIRS SDR products to the end users.

6. Conclusions

In order to meet the radiometric stability requirement for the CDR, the on-orbit RSB radiometric calibration of NOAA-20 VIIRS is described and demonstrated using a primary on-orbit calibration source called SD. The on-orbit SD degradation (H-factor) trends are stabilized after the SDSM Sun transmittance LUT updates from the yaw maneuver-derived results. The updated SD F-factors are compared to the secondary source of calibration using the moon. The lunar calibration has been used from the experience with the heritage instruments such as Terra/Aqua MODIS and SeaWiFS instruments. Over the 3 years of operations, the on-orbit SD calibration results are validated by the monthly lunar F-factors within 1% annual

TABLE 4 NOAA-20 VIIRS bias (relative to SNPP) and long-term stability derived using extended SNOs over Saharan desert.

Bands	N20 VIIRS bias relative to SNPP (SNOx)	NOAA-20 VIIRS stability
M1	−3.18%±1.3%	−0.14%±0.28%
M2	−1.67%±0.94%	0.37%±0.37%
M3	−2.25%±0.56%	0.35%±0.14%
M4	−1.37%±0.55%	0.36%±0.23%
M5	−4.86%±0.93%	−0.29%±0.09%
I1	–	–
M7	−3.85%±1.07%	−0.37%±0.33%
I2	–	–
M8	−3.22%±0.58%	0.11%±0.27%
M10	−2.22%±0.63%	−0.06%±0.15%

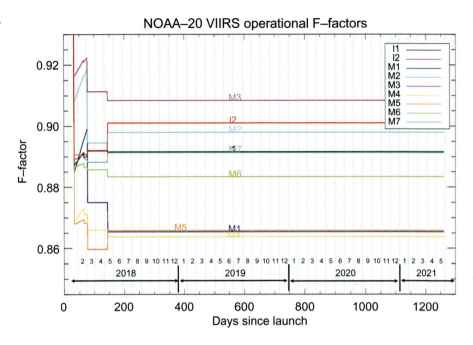

FIG. 13 NOAA-20 VIIRS operational F-factors for VIIRS SDR production.

oscillation level except the short wavelength bands from M1 to M4. These bands show consistent and linear SD F-factor drops approximately −1.5% to −0.5% deviations from the lunar F-factors. To validate these differences, the long-term trends of field measurements such as lunar, DCC, and SNO observations are also incorporated in current NOAA-20 VIIRS on-orbit calibration as two independent calibration sources. Even though there are larger uncertainty levels compared to the SD and lunar F-factors, the long-term DCC and SNO trends do not show any signs of gain changes especially in the short wavelength bands. Considering all the possible on-orbit calibration sources such as SD, moon, DCC, and SNO trends, the operational F-factors were kept in the constant level since April 2018 for the NOAA-20 VIIRS SDR products meeting the radiometric quality requirements.

In this chapter, detailed descriptions of the NOAA-20 VIIRS RSB calibration are provided from launch to on-orbit as an example. For the best quality of SDR products, the VIIRS calibration team has monitored and utilized all the possible on-orbit calibration sources from the primary on-orbit calibrators of SD and SDSM including the field measurements of moon, DCC, and SNO with weekly and monthly discussions and decisions whenever lunar calibration occurred. On top of these on-orbit calibration activities, VIIRS calibration team reprocessed the mission-long SNPP VIIRS SDR products and is preparing a set of reprocessed NOAA-20 VIIRS SDR products to meet end users' needs.

Acknowledgments

We would like to acknowledge Quanhua (Mark) Liu and Nicholas Nalli for reviewing the draft and providing constructive feedback.

Disclaimer

The scientific results and conclusions, as well as any views or opinions expressed herein, are those of the author(s) and do not necessarily reflect those of NOAA or the Department of Commerce.

References

Baker, N., Kilcoyne, H., 2011. In: Joint Polar Satellite System (JPSS) Ground Project (Ed.), Joint Polar Satellite System (JPSS) VIIRS Radiometric Calibration Algorithm Theoretical Basis Document (ATBD). NOAA and NASA.

Cao, C., Weinreb, M., Xu, H., 2004. Predicting simultaneous nadir overpasses among polar-orbiting meteorological satellites for the intersatellite calibration of radiometers. J. Atmos. Ocean. Technol. 21 (4), 537–542.

Cao, C., De Luccia, F.J., Xiong, X., Wolfe, R., Weng, F., 2014. Early on-orbit performance of the visible infrared imaging radiometer suite onboard the Suomi National Polar-Orbiting Partnership (S-NPP) satellite. IEEE Trans. Geosci. Remote Sens. 52 (2), 1142–1156.

Cao, C., et al., 2021. Mission-long recalibrated science quality Suomi NPP VIIRS radiometric dataset using advanced algorithms for time series studies. Remote Sens. (Basel) 13 (6), 1075.

Chander, G., et al., 2013a. Applications of spectral band adjustment factors (SBAF) for cross-calibration. IEEE Trans. Geosci. Remote Sens. 51 (3), 1267–1281.

Chander, G., Hewison, T.J., Fox, N., Wu, X., Xiong, X., Blackwell, W.J., 2013b. Overview of intercalibration of satellite instruments. IEEE Trans. Geosci. Remote Sens. 51 (3), 1056–1080.

Chander, G., Angal, A., Choi, T., Xiong, X., 2013c. Radiometric cross-calibration of EO-1 ALI with L7 ETM+ and Terra MODIS sensors using near-simultaneous desert observations. IEEE J. Sel. Top. Appl. Earth Obs. Remote Sens. 6 (2), 386–399.

Choi, T., Cao, C., 2019. S-NPP VIIRS on-orbit calibration coefficient improvements with yaw maneuver reanalysis. IEEE Trans. Geosci. Remote Sens. 57 (10), 7460–7465.

Choi, T., Shao, X., Cao, C., Weng, F., 2015. Radiometric stability monitoring of the Suomi NPP Visible Infrared Imaging Radiometer Suite (VIIRS) reflective solar bands using the moon. Remote Sens. (Basel) 8 (1), 15.

Choi, T., Shao, X., Cao, C., 2018. On-orbit radiometric calibration of Suomi NPP VIIRS reflective solar bands using the Moon and solar diffuser. Appl. Optics 57 (32), 9533–9542.

Choi, T., Shao, X., Blonski, S., Wang, W., Uprety, S., Cao, C., 2019a. NOAA-20 VIIRS initial on-orbit radiometric calibration using scheduled lunar observations. In: Presented at the Earth Observing Systems XXIV., https://doi.org/10.1117/12.2528459.

Choi, T., Shao, X., Cao, C., 2019b. NOAA-20 Visible Infrared Imaging Radiometer Suite (VIIRS) on-orbit band-to-band registration estimation for Reflective Solar Band (RSB) using scheduled lunar collections. In: Presented at the IGARSS, Yokohama, Japan.

Choi, T., Shao, X., Blonski, S., Cao, C., 2020. On-orbit NOAA-20 VIIRS solar diffuser bidirectional reflectance distribution function and screen transmittance characterization using yaw manoeuvres and regular on-orbit SDSM data. Int. J. Remote Sens. 41 (17), 6503–6526. 2020/09/01.

Doelling, D.R., Morstad, D., Scarino, B.R., Bhatt, R., Gopalan, A., 2013. The characterization of deep convective clouds as an invariant calibration target and as a visible calibration technique. IEEE Trans. Geosci. Remote Sens. 51 (3), 1147–1159.

Elgas, J., Murgai, V., 2020. JPSS-3,-4 VIIRS solar diffuser stability monitor relative spectral response. In: Presented at the Earth Observing Systems XXV., https://doi.org/10.1117/12.2568767.

Helder, D., et al., 2013. Absolute radiometric calibration of Landsat using a pseudo invariant calibration site. IEEE Trans. Geosci. Remote Sens. 51 (3), 1360–1369.

Hu, Y., Wielicki, B.A., Ping, Y., Stackhouse Jr., P.W., Lin, B., Young, D.F., 2004. Application of deep convective cloud albedo observation to satellite-based study of the terrestrial atmosphere: monitoring the stability of spaceborne measurements and assessing absorption anomaly. IEEE Trans. Geosci. Remote Sens. 42 (11), 2594–2599.

Kieffer, H.H., Stone, T., 2005. The spectral irradiance of the moon. Astron. J. 129 (6), 2887–2901.

Klein, S.N., Johnson, L.H., Murgai, V., 2017. Spectralon solar diffuser BRDF variation for NPP, JPSS J1 and J2. In: Presented at the Earth Observing Systems XXII., https://doi.org/10.1117/12.2275908.

Klein, S.N., Murgai, V., Johnson, L.H., 2019. Solar attenuation screen transmittance, modulation, and albedo for JPSS J2. In: Presented at the Earth Observing Systems XXIV., https://doi.org/10.1117/12.2530398.

Lessel, K., McClain, S., 2007. Low uncertainty measurements of bidirectional reflectance factor on the NPOESS/VIIRS solar diffuser. In: Presented at the Earth Observing Systems XII., https://doi.org/10.1117/12.753243.

McIntire, J., Moyer, D., Oudrari, H., Xiong, X., 2019. Pre-launch radiometric characterization of JPSS-2 VIIRS thermal emissive bands. Remote Sens. (Basel) 11 (6), 732.

Mishra, N., Helder, D., Angal, A., Choi, J., Xiong, X., 2014. Absolute calibration of optical satellite sensors using Libya 4 pseudo invariant calibration site. Remote Sens. (Basel) 6 (2), 1327–1346.

Murgai, V., Klein, S.N., 2019. Spectralon Solar Diffuser BRDF extrapolation to 2.25 microns for JPSS J1, J2, and J3. In: Presented at the Earth Observing Systems XXIV., https://doi.org/10.1117/12.2530399.

Murgai, V., Johnson, L., Moskun, E.M., 2014. BRDF characterization of solar diffuser for JPSS J1 using PASCAL. In: Presented at the Earth Observing Systems XIX., https://doi.org/10.1117/12.2063153.

Murgai, V., Yu, K., Nelson, N., McCarthy, J., 2015. JPSS-1 VIIRS solar diffuser stability monitor response versus sun angle of incidence. In: Presented at the Earth Observing Systems XX., https://doi.org/10.1117/12.2189328.

N. R. Council, 2004. Climate Data Records from Environmental Satellites: Interim Report. The National Academies Press, Washington, DC, p. 150.

Oudrari, H., et al., 2015. Prelaunch radiometric characterization and calibration of the S-NPP VIIRS sensor. IEEE Trans. Geosci. Remote Sens. 53 (4), 2195–2210.

Oudrari, H., et al., 2016. JPSS-1 VIIRS radiometric characterization and calibration based on pre-launch testing. Remote Sens. (Basel) 8 (1), 41.

Oudrari, H., et al., 2018. An overall assessment of JPSS-2 VIIRS radiometric performance based on pre-launch testing. Remote Sens. (Basel) 10 (12), 1921.

Patt, F.S., Butler, J.J., Eplee, R.E., Barnes, R.A., Meister, G., Butler, J.J., 2005. Use of the moon as a calibration reference for NPP VIIRS. In: Presented at the Earth Observing Systems X., https://doi.org/10.1117/12.618370.

Shao, X., Cao, C., Liu, T.-C., 2016. Spectral dependent degradation of the solar diffuser on Suomi-NPP VIIRS due to surface roughness-induced rayleigh scattering. Remote Sens. (Basel) 8 (3), 254.

Shao, X., Liu, T.-C., Xiong, X., Cao, C., Choi, T., Angal, A., 2019. Surface roughness-induced spectral degradation of multi-spaceborne solar diffusers due to space radiation exposure. IEEE Trans. Geosci. Remote Sens., 8658–8671.

Sun, J., Wang, M., 2016. VIIRS reflective solar bands calibration progress and its impact on ocean color products. Remote Sens. (Basel) 8 (3), 194.

Sun, J., Chu, M., Wang, M., 2018. On-orbit characterization of the VIIRS solar diffuser and attenuation screens for NOAA-20 using yaw measurements. Appl. Optics 57 (22), 6605–6619.

Uprety, S., Cao, C., 2015. Suomi NPP VIIRS reflective solar band on-orbit radiometric stability and accuracy assessment using desert and Antarctica Dome C sites. Remote Sens. Environ. 166, 106–115.

Uprety, S., Cao, C., Xiong, X., Blonski, S., Wu, A., Shao, X., 2013a. Radiometric intercomparison between Suomi-NPPVIIRS and Aqua MODIS reflective solar bands using simultaneous nadir overpass in the low latitudes. J. Atmos. Oceanic Tech. 30 (12), 2720–2736.

Uprety, S., Cao, C., Xiong, X., Blonski, S., Wu, A., Shao, X., 2013b. Radiometric intercomparison between Suomi-NPP VIIRS and Aqua MODIS reflective solar bands using simultaneous nadir overpass in the low latitudes. J. Atmos. Ocean. Technol. 30 (12), 2720–2736.

Uprety, S., Cao, C., Blonski, S., Shao, X., 2018. Assessing the NOAA-20 and S-NPP VIIRS radiometric consistency. In: Presented at the Earth Observing Missions and Sensors: Development, Implementation, and Characterization V., https://doi.org/10.1117/12.2324464.

Valencia, A., Klein, S., Murgai, V., 2020. Solar attenuation screen transmittance, modulation, and albedo for JPSS J3 and J4. In: Presented at the Earth Observing Systems XXV., https://doi.org/10.1117/12.2568562.

Wang, W., Cao, C., 2015. DCC radiometric sensitivity to spatial resolution, cluster size, and LWIR calibration bias sased on VIIRS observations. J. Atmos. Ocean. Technol. 32 (1), 48–60. 2015/01/01.

Wang, W., Cao, C., 2016. Monitoring the NOAA operational VIIRS RSB and DNB calibration stability using monthly and semi-monthly deep convective clouds time series. Remote Sens. (Basel) 8 (1), 32.

Wang, W., Cao, C., 2020a. Evaluation of NOAA-20 VIIRS reflective solar bands early on-orbit performance using daily deep convective clouds. IEEE J. Sel. Top. Appl. Earth Obs. Remote Sens. 13, 1.

Wang, W., Cao, C., 2020b. Evaluation of NOAA-20 VIIRS reflective solar bands early on-orbit performance using daily deep convective clouds recent improvements. IEEE J. Sel. Top. Appl. Earth Obs. Remote Sens. 13, 3975–3985.

Wang, W., Cao, C., 2021. NOAA-20 and S-NPP VIIRS thermal emissive bands on-orbit calibration algorithm update and long-term performance intercomparison. Remote Sens. (Basel) 13 (3), 448.

Wolfe, R.E., Lin, G., Nishihama, M., Tewari, K.P., Tilton, J.C., Isaacman, A.R., 2013. Suomi NPP VIIRS prelaunch and on-orbit geometric calibration and characterization. J. Geophys. Res. Atmos. 118 (20), 11508–11521.

Xiong, X., Sun, J., Fulbright, J., Wang, Z., Butler, J.J., 2016. Lunar calibration and performance for S-NPP VIIRS reflective solar bands. IEEE Trans. Geosci. Remote Sens. 54 (2), 1052–1061.

Chapter 16

The NOAA sounding Products Validation System (NPROVS)

Bomin Sun[a,b], Anthony Reale[b], Michael Pettey[a,b], Ryan Smith[a,b], and Charlie Brown[a,b]
[a]*I.M. Systems Group, Inc., College Park, MD, United States,* [b]*NOAA/NESDIS Center for Satellite Applications and Research (STAR), College Park, MD, United States*

Chapter outline

1. Introduction — 281
2. Data sources — 282
   2.1 Radiosonde data — 282
3. Collocation compiling — 287
4. Vertical statistics computation — 288
5. Graphic applications — 289
   5.1 Orbital Display System — 289
   5.2 Profile Display — 291
   5.3 NPROVS Archive Summary (NARCS) Statistics — 293
6. Summary — 294
Acknowledgments — 295
References — 295

Truth comes out of error more easily than out of confusion.

Francis Bacon

1. Introduction

Satellite sounders can produce three-dimensional retrievals of temperature, water vapor, and other trace gases and are used routinely for monitoring severe weather, flight conditions, and air quality (Chapter 18). These applications are possible because of significant validation have been conducted. Validation ensures that data meets the quality requirements for newly launched sounding instruments and is statistically consistent with other products with known accuracy. For mature data sets, validation allows scientists and developers to make sure upgrades improve. In terms of applications, rigorous validation can help communicate the strengths and weaknesses of a data set to data users.

The validation of satellite sounding products using collocated global conventional radiosonde (Chapter 2) has been a baseline component of operational satellite programs at NOAA since the deployment of the TIROS Operational Vertical Sounder (TOVS) operational soundings (Smith and Woolf, 1976) in 1979. Now, 40+ years later, the Center of Satellite Application and Research (STAR) at NOAA NESDIS supports multiple programs and scientific approaches for deriving atmospheric sounding products from a variety of radiometric sensor configurations onboard NOAA, EUMETSAT, and NASA EOS polar satellites and NOAA GOES satellites. Recently, Global navigation satellite system (GNSS) radio occultation (RO) constellations (Anthes et al., 2008) have been added with the capability of all weather and high vertical resolution atmospheric sounding.

However, each program has specific goals regarding sounding assessment, so there are many approaches for validation. For example, there are numerous strategies for selecting appropriate reference data sets, performing quality control, and selecting criteria for comparison. Furthermore, there is no guaranteed consistency in the type of statistical analysis performed or a common interpretation the results. Such differences can skew a given set of results and associated interproduct comparisons between scientific programs and across different sounding products. Recognizing this, the NOAA Joint Polar Satellite System (JPSS) program created (in 2006) a requirement for a centralized, radiosonde, numerical weather prediction (NWP), and multiple satellite product collocations enabling routine comparisons among legacy, existing and newly deployed satellite product suites.

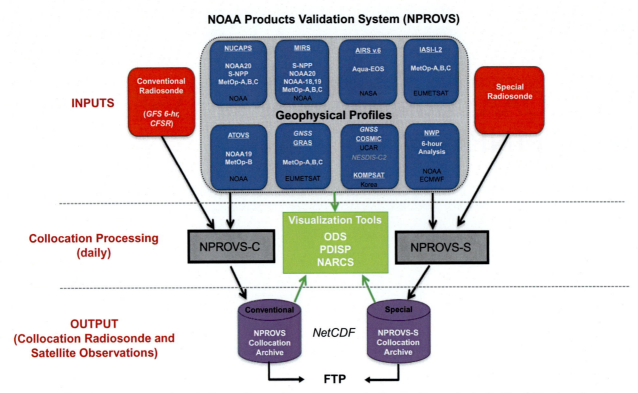

FIG. 1 NPROVS data management schematic diagram for compiling collocations of radiosondes/dropsondes *(red)* with satellite atmospheric temperature and water vapor profiles from over 20 independent satellite and NWP product suites *(blue)* and archives of collocations anchored to conventional and special radiosondes *(purple)*. Graphical Applications *(green)* provide tools for analyzing product suites and collocations.

Two years later in 2008, the NOAA Products Validation System (NPROVS) was deployed at STAR to meet these requirements with a primary objective of providing benchmark sounding product assessments in accordance with the JPSS Cal/Val program (Chapter 15). NPROVS assessment for hyperspectral IR and MW sounding products began in January 2012 in support of SNPP and currently supports NOAA-20 sounding product. NPROVS also includes product suites focused on atmospheric sounding from other satellite platforms and instruments, providing a robust system for interproduct comparisons.

This chapter serves as an introduction to the NPROVS, as summarized in the schematic diagram (Fig. 1). NPROVS provides a centralized capability for compiling collocations of radiosondes/dropsondes (red) with various satellite and NWP atmospheric temperature and water vapor sounding profiles from over 20 independently operated product suites (blue). A single closest sounding from each product suite is collocated to each radiosonde and archived (purple). Graphical applications tools are available to analyze all input and collocation data (green).

Section 2 introduces radiosondes, satellite, and NWP products collected in NPROVS. Preprocessing (prior to collocation) of radiosonde observations including sets of conventional and special (which include GCOS GRUAN and JPSS-funded dedicated radiosondes) is also described. Section 3 addresses the methodology used for collocating the satellite and NWP data with radiosonde data. Section 4 describes procedures to compute the vertical statistics focused on the mean bias, root-mean-square (RMS) error, and/or the standard deviation (SD) error of the products against radiosonde (or user designated) baseline data. Section 5 demonstrates the graphic monitoring and analysis capability that allow users to conveniently conduct products assessment and understand products performance under a variety of sampling constraints. Section 6 provides a summary.

2. Data sources

2.1 Radiosonde data

2.1.1 Conventional

Conventional radiosondes are those launched in the global operational upper air network of the World Meteorological Organization (WMO). These radiosondes are distributed via Global Communications System (GTS), which enables rapid data distribution and are utilized in NWP data assimilation and forecasting at major NWP centers across the world.

NPROVS collocates all satellite and NWP product systems with conventional radiosondes. The method of how to collocate different data sources with radiosondes are described in Section 3. Radiosondes used in NPROVS are accessed from the post data assimilation of NOAA Global Forecast System (GFS) 6-h PREPBUFR files (available online at http://www.emc.ncep.noaa.gov/mmb/data_processing/prepbufr). These files contain the original WMO radiosonde report data (at mandatory and significant levels) and include for each report level: quality control (QC) marks assigned during NWP assimilation and the GFS 6-h forecast (Initialization) 4D-interpolated during NWP assimilation. Those data are retained in a binary file, which we refer to as the unified radiosonde file and is available for use during satellite products assessment. The unified file also includes internal tests and profile characterizations that are performed when creating the unified file, including boundary layer meteorological features such as surface inversion, super adiabatic, supersaturation, and moisture vertical structure, tropopause level, and atmospheric stability. Information of radiosonde profile data completeness, the top-most level where the report is available, and the flag indicating day or night when the radiosonde is launched is also included in the file.

Details on the above and other tests can be found in Reale et al. (2012) and on the NPROVS website (https://www.star.nesdis.noaa.gov/smcd/opdb/nprovs/). These tests and characterizations allow the science team to subset collocation samples for analysis. For example, one can perform boundary layer assessments using the inversion flag to identify collocations for radiosondes with surface temperature inversions. These supplement the more basic radiosonde sampling options, for example, terrain, region, day/night, vertical extent, and radiosonde instrument type.

There are dozens of radiosonde instrument types in the WMO network. Measurement quality and accuracy can vary with sonde type (Sun et al., 2010). Radiosonde instrumentation type changes periodically as sensor technology advances. For example, the Vaisala RS41 has recently (since 2018) emerged as a dominant instrument type in the WMO and GRUAN (Dirksen et al., 2020) networks. RS41 shows improvements compared to its predecessor RS92 for both upper air temperature and humidity (Sun et al., 2019a, 2021). NPROVS graphical applications (Section 5.2) include collocation sorting by radiosonde instrument type. Fig. 2 shows typical global distributions of all radiosondes (top) and the Vaisala RS41 only (bottom). Users can decide whether it is better to use a technology advanced sensor at the expense of reduced global distribution. All conventional radiosonde data are those processed at the site using manufacturer software.

2.1.2 Special sondes

NPROVS special collocates all satellite and NWP product systems with special radiosondes, the ones from the GCOS GRUAN and the ones synchronized with satellite overpasses.

The GRUAN (Bodeker et al., 2016) is an international observing network, designed to meet climate requirements and to fill a major void in the current global observing system by providing reference observations. GRUAN is envisaged as a network of 30–40 sites; currently data from 25 sites are stored in NPROVS special. The long-term homogeneity of GRUAN data time series makes them ideally suited for providing a reference standard for space-based measurements. The GRUAN radiosondes are processed by GRUAN software (www.GRUAN.org, Dirksen et al., 2020) and provide fully characterized measurements and uncertainty estimates for each individual observation.

In addition to GRUAN radiosondes, NPROVS special also include satellite synchronized observations from the JPSS funded dedicated radiosonde program currently managed among NOAA STAR, the U.S. Department of Energy's (DOE) Atmospheric Radiation Measurement (ARM) network and the Space Science and Engineering Center (SSEC) at the University of Wisconsin at Madison. The objective of the dedicated launch programs is to assist in the Cal/Val of temperature and moisture soundings currently from the Cross-track Infrared Sounder (CrIS)/Advanced Technology Microwave Sounder (ATMS) instruments onboard the SNPP and more recently the NOAA-20 satellites (and occasionally MetOp). JPSS dedicated radiosondes are also processed into reference observations by GRUAN, providing the unique data source for robust satellite product assessment.

More details about the land and ship-based radiosonde launch programs can be found in Chapters 11 and 12. Here we summarize specific campaigns (with their data stored in NPROVS special) that have contributed to the Cal/Val of satellite hyperspectral sounding products developed at STAR (e.g., Sun et al., 2017a,b).

- **DOE-ARM sites**: Southern Great Plains in Oklahoma, United States, North Slope Alaska in Alaska, United States, and Eastern North Atlantic on Graciosa Island (Tobin et al., 2006; Nalli et al., 2013, and also see Chapter 12).
- **Radiosonde Intercomparison and Validation (RIVAL)** partnership between GRUAN/ARM/JPSS and managed by STAR and University of Madison at Wisconsin. Dual radiosonde launches (Vaisala RS41 and RS92 attached on one balloon) are launched within 40 min of NOAA-20 overpasses (Chapter 12). In selected cases, a dual launch is followed by a single launch about 30 min later, referred to as a sequential launch.
- **The trans-Atlantic AERerosols and Ocean Science Expeditions (AEROSE)** campaigns (Nalli, 2011; Nalli et al., 2018, and also see Chapter 11). These campaigns (stored in NPROVS beginning 2013) sample atmospheric soundings

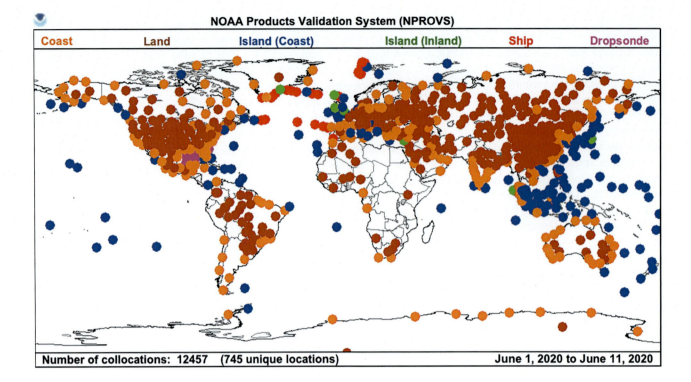

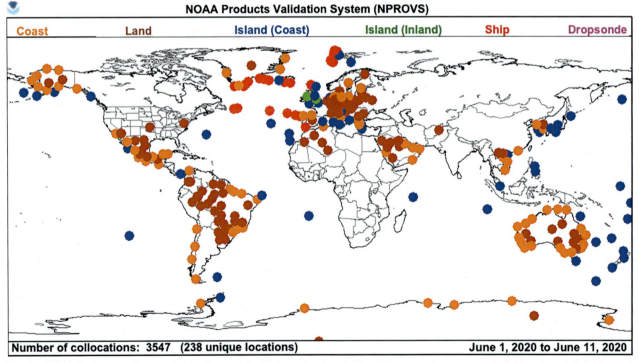

FIG. 2 Global distributions of conventional radiosondes for 10-day period June 1–11, 2020. Observations are color-coded based on radiosonde terrain with *brown* (land), *orange* (coast), *blue* (island), *green* (island inland), and *red* (ship). The upper panel is all radiosondes (12,457), and the lower panel is Vaisala RS41 only (3547).

in the Atlantic basin focused on the environments conducive to Saharan Aerosol Layer (SAL) dust events. These include JPSS funded dedicated radiosonde and subsequent GRUAN reference processing for selected campaigns.
- **The CalWater/ARM Cloud Aerosol Precipitation Experiment (ACAPEX)** over the eastern North Pacific from January to February 2015 (Nalli et al., 2016). Observations include those targeted with polar satellite overpass

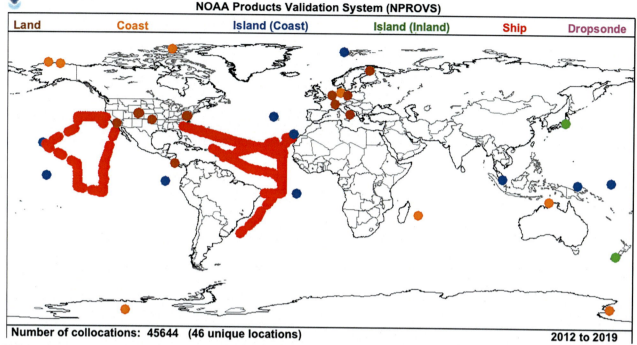

FIG. 3 Global distribution of NPROVS special (GRUAN, JPSS dedicated, field campaigns, etc.) radiosondes. Color-coded based on radiosonde terrain with *brown* (land), *orange* (coast), *blue* (island), *green* (island inland), and *red* (ship).

(SNPP and MetOp) in the vicinity of atmospheric rivers to better understand the physical processes and satellite products performance associated with these extreme precipitation events (impacting the U.S. West Coast).
- **The El Nino Rapid Response Experiment (ENRR)** spanning the central and eastern tropical Pacific during February–March 2016 (Nalli et al., 2016) with objectives similar to CalWater and including targeted observations.
- **US National Weather Service (NWS) Sterling Field Support Center (SFSC)** providing multiple radiosonde instrument data sets, some targeted with JPSS satellites.
- **DOE-ARM mobile sites** including the sites located in Barrow, Alaska, and Antarctica.

The NPROVS special collocation data set also contain naturally synchronized observations not targeted in any specific dedicated campaign; these span NOAA, EUMETSAT, NASA, and GNSS satellites and can provide 25+% increase in targeted samples, optimal for assessment (Fig. 3).

The special radiosondes listed in this section come with a high vertical density, with measurements taken once per second. These data come from a variety of data sources contingent on respective availability. Each source typically processes their data differently, including over time. As a result, the NPROVS team must accommodate multiple formats and file types over time. Once all the necessary variables have been read from the high vertical density file, the profile is converted to a predefined set of 100-layer effective pressures using the procedures consistent with those adopted by the NUCAPS team (Nalli et al., 2016; see also Chapter 18). The 100-layer radiosonde profiles at effective pressure are ultimately stored in the unified data file and the basis for all product suite assessments described in this chapter.

The objective using both conventional and special radiosondes is to leverage their respective strengths, namely large samples and high accuracy, facilitating the robust monitoring and assessment of any satellite product suite that is ingested.

2.1.3 Satellites

The following satellites and associated level-2 atmospheric sounding product suites are routinely ingested in NPROVS.

(a) The NOAA Unique Combined Atmospheric Processing System (NUCAPS)

NUCAPS is the operational retrieval system for soundings of atmospheric temperature, water vapor, and trace gas Environmental Data Record (EDR) products (Gambacorta et al., 2012; Gambacorta 2013; also see Chapter 18 and https://star.nesdis.noaa.gov/jpss/soundings.php for more details). The algorithm is adopted from the legacy NASA AIRS sounding retrieval system, which will be summarized in Section 2.1.3(e). The retrieval approach includes procedures of radiance calibration and adjustment, microwave retrieval, cloud clearing, and iterative IR physical retrieval. A fast eigenvector regression retrieval is used to compute initial estimates of temperature and water vapor profiles for computing cloud-cleared radiances. A second fast eigenvector regression using the cloud-cleared radiance is then used to compute the first-guess profile and iterative IR+MW retrieval. The IR+MW and MW-only retrievals are produced at 101 boundary pressure levels for temperature and 100 effective pressure levels for water vapor mixing ratio (see Section 4). Those retrievals are produced at every field-of-regard (FOR), consisting of 3×3 CrIS field-of-views (FOVs) for SNPP and NOAA-20, or 2×2 IASI FOVs for MetOp satellites. The FOR horizontal resolution varies from approximately 50 km at the nadir view to 150 km at the limb. The atmospheric sounding product EDRs are generated from the following instruments and satellites: (i) CrIS/ATMS on NOAA-20, (ii) CrIS/ATMS on SNPP, (iii) IASI/Advanced Microwave Sounding Unit (AMSU) on MetOP-C, (iv) IASI/AMSU on MetOp-B, and (v) IASI/AMSU on MetOp-A.

(b) The NOAA Microwave Integrated Retrieval System (MIRS)

The MiRS algorithm is a one-dimensional variational inversion scheme (1DVAR) that employs the Community Radiative Transfer Model (CRTM) as the forward and adjoint operator and aims at minimizing a cost function similar to the NWP variational assimilation. An important characteristic relates to solving simultaneously for the surface and the atmospheric parameters including hydrometeors, making it single, all weather approach. Retrievals are produced at 100 effective pressure layers for temperature and water vapor mixing ratio (more details see Boukabara et al., 2011, and Chapter 17), and the retrievals are produced at every ATMS (for SNPP and NOAA-20) or MHS FOV (for other polar satellites). The MW sounding product EDRs are generated from the following instruments and satellites: (i) ATMS on NOAA-20, (ii) ATMS SNPP, (iii) AMSU/Microwave Humidity Sounder (MHS) on NOAA-19, (iv) AMSU/MHS NOAA-18, (v) AMSU/MHS on MetOp-C, (vi) AMSU/MHS on MetOp-B, and (vii) AMSU/MHS on MetOp-A.

(c) EUMETAT Infrared Atmospheric Sounding Interferometer (IASI) Level-2

Retrievals are derived from the hyperspectral infrared atmospheric sounding interferometer (IASI) instrument on MetOp satellites (August et al., 2012). The EUMETAT IASI operational processing system allows retrievals to be generated under cloudy conditions by combined use of IASI and collocated advanced microwave sounding unit (AMSU) and microwave humidity sounder (MHS) measurements. For version 6 and newer versions after September 2014, the all-sky retrievals are first generated using piecewise regression methods across the infrared and microwave channel data. Those retrievals are then used as the first guess to generate the IR-based physical retrievals using an optimal estimation method (OEM). OEM is only attempted for clear sky only as identified using strict cloud screening procedures. Retrievals are produced at 101 boundary pressure levels for temperature and water vapor mixing ratio. And the physical retrievals are produced at IASI FOVs that are considered to be clear. See August et al. (2016) and Hultberg et al. (2016) for details of the IASI retrieval processing system and also eumetsat.int/iasi. Sounding products are generated from the following instruments and satellites: (i) IASI/AMSU/MHS on MetOp-C, (ii) IASI/AMSU/MHS on MetOp-B, and (iii) IASI/AMSU/MHS on MetOp-A.

(d) The NOAA Advanced TIROS Observational Vertical Sounder (ATOVS)

ATOVS is a NOAA legacy sounding retrieval system adopted from the original TOVS. The products are derived from the 20 channel High-resolution Infrared Radiation Sounder (HIRS), 15 channel AMSU-A and 5-channel AMS-B. The HIRS and AMSU measurements (interpolated to HIRS) are combined to create three classes of atmospheric soundings, namely clear, cloud-cleared, and MW-only retrievals; MW retrievals also use the higher peaking HIRS (channels 1–3). The first guess for retrieval is based on a library search/eigenvector technique using recent (up to 50 days) collocations from a radiosonde and satellite collocation data set. The final retrieval utilizes a minimum variance simultaneous solution. Soundings are retrieved at the 40 TOVS levels with the HIRS and AMSU measurements limb adjusted to nadir view prior to retrieval. The product is available for subset of clear, partly cloudy (cloud-cleared) and MW-only retrieval, respectively. Details can be found in Reale et al. (1994, 2008) and Reale (2001, 2002). The satellite products ingested in NPROVS now include the following instruments and satellites: (i) HIRS/AMSU on NOAA-19 and (ii) HIRS/AMSU on MetOp-B.

(e) The NASA-Earth Observing System (EOS) Aqua Advanced InfraRed Sounder (AIRS)

The AIRS was the first in a new generation of high-spectral resolution infrared sounder instruments flown aboard Aqua research mission. The AIRS processing system (APS) suite has been put into operation at National Aeronautics and Space Administration (NASA) Jet Propulsion Laboratory, NOAA/NESDIS Office of Research and Applications, and at Goddard Space Flight Center (GSFC) Distributed Active Archive Center (DAAC) for dissemination of AIRS data products to user communities. The APS suite includes calibration, microwave retrieval, cloud clearing, initial IR+MW retrieval, and a final physical retrieval (Susskind et al., 2003). The AIRS system was adopted at NESDIS to produce the original NUCAPS product for SNPP and NOAA-20 and MetOp satellites with temperature and water vapor mixing ratio products. Products are still generated from AIRS, but AMSU has degraded and no longer used after September 2016; AIRS has far exceeded its expected lifetime.

(f) GNSS RO profiles

GNSS-RO geophysical profiles are generated using the phase delay of radio waves transmitted by Global Positioning System (GPS) satellites as they pass through the Earth's atmosphere. High-resolution vertical profiles of the atmospheric structure are derived covering all weather conditions (Kursinski et al., 1997; Hajj et al., 2002). The atmospheric temperatures and moisture soundings, referred to as wet profiles, are retrieved with the One-Dimensional Variational method, which optimally combines a RO bending angle profile with a priori NWP model information to estimate temperature and moisture profiles. In addition, a dry temperature is also derived directly from GNSS RO measurements, with an average accuracy <0.1 K in atmospheric regions where water vapor is negligible. RO profiles have fine vertical resolution (0.5 km in the low troposphere and 1.5 km in the middle atmosphere) by satellite retrieval standards, they have a wide horizontal resolution, ranging from ~160 km at the low troposphere to 320 km in the upper atmosphere (i.e., ~250 km, defined by the distance traversed by the radio path as it enters and exits a layer (Kursinski et al., 1997)).

The high accuracy of GNSS RO dry temperature is of value in assessing the quality of temperature measurements of other observing systems in the upper troposphere and lower stratosphere, serving as a candidate reference temperature measurement. The high-density profiles, including bending angle and refractivity, are thinned to approximating 300 levels prior to integration into NPROVS and are available from the following sources: (i) University Corporation for Atmospheric Research (UCAR, http://www.cosmic.ucar.edu/cdaac) including Constellation Observing System for Meteorology Ionosphere and Climate (COSMIC-1 and COSMIC-2) and Korean Multi-Purpose Satellite-5 (KOMPSAT-5), and (ii) EUMETSAT Radio Occultation Meteorology (ROM) Satellite Application Facility (SAF) GNSS Receiver for Atmospheric Sounding data (GRAS; www.romsaf.org) using GPS receivers on MetOp-A, -B, and -C satellites.

2.1.4 Numerical weather prediction

In addition to satellite and radiosonde observations, NPROVS ingests and stores several NWP model profiles due to their complete global coverage. These include (i) the NCEP 6-h GFS forecast (64 vertical levels and 38-km horizontal resolution), which are 4D interpolated (J. Wollen, 2018, personal communication) to each mandatory and significant radiosonde level; (ii) gridded ECMWF (ECMWF, 2018) operational analyses, which are available at 0000, 0600, 1200, and 1800 UTC, with 91 vertical pressure levels thinned from the 137 model sigma levels and horizontal resolution of 0.258 (Eresmaa and McNally, 2014); and (iii) NOAA Climate Forecast System Reanalysis (CFSR) climate forecast (64 vertical levels and 38-km horizontal resolution), which use a 4D interpolation scheme similar to GFS (J. Wollen, 2018, personal communication; Saha et al., 2010).

2.1.5 Handling file and format differences across products

The original satellite and NWP product suite integrated into NPROVS has data format differences. For example, the NUCAPS and MiRS products are in netCDF, EUMETSAT IASI products are in native EPS, AIRS product is in HDF-EOS, and NWP products are in GRIB. To facilitate efficient collocation compiling, the different formats are converted and reconstructed into the same data format and structure before collocation compiling is conducted. Note, we do not change data values and all data appear as received during this preprocessing. But at the time when we compute vertical statistics for assessment, all the products of interest need to be converted into 100 effective layers if they are not at (see Section 4).

3. Collocation compiling

A truly accurate assessment requires perfectly synchronized and collocated between observations. However, in practice, this is highly improbable so scientists must develop a screening strategy using reasonable time and distance constraints to identify which observations are candidates for a match. NPROVS addresses this by providing a complete but efficient collocation data set from which user-analysts can apply constraints and create subsamples to meet specific assessment requirements.

The collocation strategy used in NPROVS identifies the single closest satellite profile from each suite among all candidates that are within 6 h and 150 km (250 km for GNSS RO) of a given radiosonde; radiosondes must extend at least 5 km to be considered. This essentially assures at least one collocation per satellite per radiosonde and facilitates a broad range of subsampling options including temporal and spatial impact studies. The strategy is the same for all satellite and NWP product suites and does not consider any ancillary information, such as terrain, QC flag, and cloudiness. The time and distance differences between radiosonde and satellite are computed based on the radiosonde launch time and location and the satellite profile time and location on the Earth' surface. Even though 150 km is used for the maximum collocation distance mismatch, 92% of collocations of radiosonde-satellite radiance derived products are within 50 km (Sun et al., 2017a). For GNSS profiles that are less dense than from polar satellites the distance window is expanded to 250 km at the 100 hPa level, with the radiosonde site location compared to GNSS location at 100 hPa (Sun et al., 2010, 2013, 2019b).

Satellite observations are essentially instantaneous, but it can take up to 2 h for a radiosonde attaching to a balloon to reach the top of the atmosphere. To account for this sampling time difference, 30 min is added to each radiosonde launch time, synchronizing the collocation to the vicinity of 300 hPa (Seidel et al., 2011). Furthermore, in cases of multiple candidate satellite orbits, adding 30 min can better assure that the radiosonde is in the air at overpass. Dropsondes, which are radiosondes released from aircraft, only take roughly 20 min to reach the surface (Chapter 10). Thus, 30 min is not added for dropsondes collocated with satellites and NWP products.

Typically, satellite profiles from a single satellite overpass (orbit) are candidates for collocation and the closest in distance is the default selection. However, in mid-latitude and polar regions particularly, multiple candidate orbits can occur. A more generic approach was needed to, respectively, weight the time and distance differences of each candidate to determine the single closest profile.

The approach to generically select a single closest profile from each product suite is based on a closeness parameter C, defined as

$$C = \text{Time difference} \times F + \text{Distance difference}$$

where F is a penalty factor (km/h). Setting F to zero would result in the closest observation in distance being selected regardless of time, and increasing F favors more the closest observation in time. The goal was to set a pragmatic value of F applicable among all satellites; $F = 30$ km/h is selected.

Another peculiarity regarding collocation compilation is that although most conventional and special radiosondes at a given site are typically available once or twice daily (00 Z and 12 Z), there are cases (<10%) when radiosondes at a given site can launch within 6 h (or less) or when two (or more) radiosondes at approximately the same time can be within 50-km (typical for dropsondes). In these cases, the same satellite sounding can be collocated with different radiosondes, these are allowed (less than 2% of collocations) and although adding complication for statistics but they can be valuable in case studies.

There is one ambiguity in collocation compiling. Collocation is conducted every day for radiosonde data at synoptic time 00, 06, 12, and 18 Z for the base-day but also include radiosondes from 18 Z of the previous day and 00 and 06 Z of the next day to try to obtain the collocations that are closer in time. This requires a step to remove the redundant collocations from the neighboring 2 days (i.e., same radiosondes are in two collocations with different soundings from a same satellite product) by selecting the collocation with a smaller C value. This is referred to as redundancy cleaning.

Ultimately, NPROVS creates daily collocation files containing individual collocation records. All information for a given radiosonde and the collocated product suites files is stored on the daily unified files. The daily files are in binary format for internal use and converted to netCDF format for outside user access.

The collocation procedure is summarized as the following steps:

(a) consider only radiosondes that pass the vertical extent QC,
(b) add 30-min to radiosonde launch time (except when collocating GNSS),
(c) identify candidate products within ±6 h and 150 km (250 km for GNSS) of a radiosonde,
(d) use C formula to select single, closest profile per product suite, and
(e) output collocations to daily files for internal/external users.

4. Vertical statistics computation

The sounding products in NPROVS collocation files, including radiosondes, satellite-radiance derived IR + MW, or MW-only geophysical profiles (e.g., NUCAPS, MiRS, and AIRS), NWP model outputs, and GNSS geophysical profiles, have different numbers of vertical levels from surface to the top of the atmosphere. For example, the special radiosondes and satellite radiance derived sounding profiles are at 100 levels. Conventional radiosondes have 50–150 vertical levels. ECMWF analysis has 91 vertical levels, while GFS has 46 levels. RO profiles have ~300 vertical levels for atmospheric observations of dry temperature (after the original profiles are thinned when ingested in NPROVS).

Ultimately, consistent with NPROVS primary objective to assess JPSS NUCAPS sounding products, all the sounding product suites are linearly interpolated (in the logarithmic pressure coordinate) to the 100 effective vertical layers from the surface to the top of the atmosphere (Nalli et al., 2013) for assessment, if the product profiles are originally not at 100 effective layers. The 100 effective layers are the vertical grids that University of Maryland at Baltimore County (UMBC) rapid transmittance algorithm (RTA), the Stand-alone AIRS Radiative Transfer Algorithm (SARTA) uses. In this context of the assessment, satellite sounding profiles serve or are considered as a type of radiosonde profile in the NWS Advanced Weather Interactive Processing System (AWIPS) forecasting application. Forecasters care about detailed structures of the satellite profile particularly within the atmospheric boundary layer, allowing them to prognose the threat of severe weather using, for example, Convective Available Potential Energy (CAPE) derived from satellite sounding profiles.

The 100 effective pressures represent a sub 500 m vertical density in the troposphere. Given the typical IR and MW sounder vertical sensitivities, the derived satellite sounding products (including NUCAPS) have a much lower vertical resolution, typically on the order of 2.0–2.5 km for IR sounder retrievals (Maddy and Barnet, 2008; Iturbide-Sanchez et al., 2017). To suppress small-scale atmospheric structures that can be captured by radiosonde profiles (typically void to radiance sensors) and minimize the impact of different vertical resolutions on the satellite data assessment, an approach typical of satellite hyperspectral sounding retrieval validation is also adopted by NPROVS (Susskind et al., 2003; Tobin et al., 2006). Namely, the satellite-minus-radiosonde vertical statistics (i.e., mean bias, RMS error, or SD error) are computed at 30 coarse layers (~1-km coarse layer) for temperature and 20 coarse layers (~2-km coarse layer) for water vapor. A detailed explanation of the differences between 101 RTA boundary pressure levels, 100 effective pressure layers and 30/20 coarse layers and how to convert effective layers to coarse layers is described in Nalli et al. (2013) and Smith and Barnet, (2020) (https://docserver.gesdisc.eosdis.nasa.gov/public/project/Sounder/CLIMCAPS_V2_L2_science_guides.pdf). Note, the statistics for water vapor are in mixing ratio (g/kg) percent differences (i.e., 100 satellite-minus-radiosonde divided by radiosonde) weighted by radiosonde mixing ratio layer amount. Statistics computed at the coarse layers are used to assess if the products meet the mission specifics (Sun et al., 2017; Nalli et al., 2018).

Statistics at the coarse layers are typically used to assess the NUCAPS sounding products in order to determine if they meet global requirements. These procedures include steps to assure that the representation of the statistics is consistent with global terrain distribution, namely that 70% of the earth is covered by sea. However, current NOAA NWS users of NUCAPS are more concerned with data over land (i.e., the contiguous United States, CONUS) using the 100-layer data. Subsequently, statistics at the 100 layers are needed including additional focus over land. NPROVS is designed to provide a range of assessments that can address NUCAPS mission specific global requirements as well as performance on regional (CONUS) and local scales in support of users.

Horizontal resolution varies among the radiosonde, NWP and satellite product suites as indicated in Section 2. Further research is needed to understand the impact of those differences on data comparisons discussed in the chapter.

5. Graphic applications

There are three graphical display and analysis applications, Orbital Display System (ODS), Profile Display (PDISP) and NPROVS Archive Summary (NARCS), which support NPROVS assessments. All three are written in Java language that allows platform independence so they run on any Java supported computer, such as Windows, Mac OS, and Linux servers. These applications support product assessments ranging from long-term global performance time series to periodic regional statistical assessments and deep-dive case studies. For more details, see the respective Quick Start and User Guides for each application available on the NPROVS website: https://www.star.nesdis.noaa.gov/smcd/opdb/nprovs/nprovs_documentation.php.

5.1 Orbital Display System

ODS shows horizontal gridded images of the Earth's atmosphere for each satellite product and NWP forecast suite. Gridded fields can be displayed for temperature or water vapor mixing ratio at any pressure level and for selected ancillary parameters (i.e., QC flag, first guess, etc.) over a broad range of color scales and default settings. User option for zooming and differencing selected gridded fields provide a useful tool for intercomparing horizontal fields from the various product suites integrated in NPROVS. Options include multiple panel display and a variety of global projection options to best portray a given assessment objective. Additionally, ODS can display vertical cross-sections of the atmosphere as well as individual vertical profiles (Skew-T, etc.) from any selected location within a given panel or sets of panels.

Fig. 4 shows an example of using ODS to portray the coverage of satellite observations in the vicinity of the special ARM/GRUAN radiosonde site (labeled as "R" on the maps) at Lamont, Oklahoma, a site of interest for NWS users of

290 PART | III Satellite applications

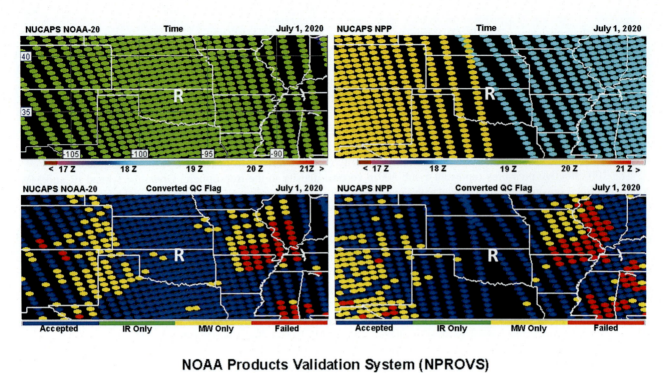

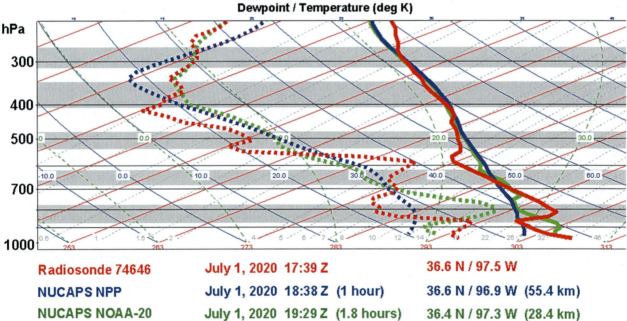

FIG. 4 The upper panels show examples of consecutive satellite overpass times for NOAA-20 (upper left) and SNPP (upper right) and corresponding NUCAPS QC flags indicating successful IR+MW *(blue)*, MW-only *(yellow)* and both failed *(red)* soundings in the vicinity of the SGP ARM/GRUAN site ("R"). The lower panel shows collocated radiosonde *(red)* and NUCAPS soundings from SNPP *(blue)* at NOAA-20 *(green)* at SGP compiled by NPROVS at the time of the upper panels.

NUCAPS. The top two panels show overpass times 17Z (purple) to 21Z (red) for NOAA-20 (left) and SNPP (right). As can be seen the radiosonde ("R") lies in between two orbits. The lower panels are the NUCAPS QC flag indicating successful IR+MW soundings (blue), successful MW-only soundings (yellow), and soundings for which the IR+MW and MW-only both failed (blue); NWS users only consider successful IR+MW soundings. The collocated NUCAPS soundings from SGP processed by NPROVS is shown on the right. Typically, NOAA-20 is 50min later than SNPP for most locations but exceptions can occur as indicated in the upper showing SNPP prior to NOAA-20 on the eastern side but later than NOAA-20 on the west side of each panel. The NPROVS collocation selected the sounding from the earlier SNPP orbit as it was closer in time to the radiosonde, affirming the C parameter approach used to select the single closest sounding when two (or more) candidate orbits are available (see Section 3). It can also be seen that both the SNPP (green) and NOAA-20 (blue) operational temperature profiles overall agree but diverge near the surface where sensor (CrIS/ATMS) sensitivity is weak.

5.2 Profile Display

PDISP is the centerpiece NPROVS graphical application that displays collocated radiosonde and satellite vertical profiles (Skew-T, etc.) and provides the calculation and display of vertical accuracy statistics and scatter plots for user selected product suites and regions. For example, users can view global locations where selected product suite collocations are mapped and users can subselect collocations for more detailed assessment including deep dive. A key function of PDISP is the calculation of vertical statistics. When comparing two or more suites, the user can easily identify a common set of collocations containing all the product suites under a given set of constraints (i.e., time window, QC flag, terrain, etc.). Comparisons of multiple products with the same ground truth is called an enterprise assessment. Enterprise assessments help increase confidence the observed statistical differences among the product suite are very likely to be real.

PDISP allows users to generate statistics using specific custom sets of collocations and numerous options to display the statistics in a variety of ways. The vertical statistic generation begins after a user has subselected the collocations, at which time the user is provided with additional options prior to the actual computations, including

(a) baseline profile (default is the radiosonde),
(b) selection of vertical and horizontal axis range,
(c) weighting options in the context of the global distribution and moisture, and
(d) the vertical pressure layers for computing statistics, for example 100 effective layers or 30 course layers (see Section 4).

The global distribution option provides three primary distribution options. The first or default method treats every collocation the same in terms of weighting in statistics computation. The second method is a grid method which sorts collocations into 10-degree by 10-degree bins and then computes statistics among each grid. A final method sorts the collocations into land (and coast) and sea sets and then applies a weighting of 0.7 to sea collocations and 0.3 to land collocations; this option is most suitable for assessments against the JPSS requirements for soundings. Once an option is selected, the program then begins calculating the statistics for every chosen profile. The program calculates running sums of the differences between the profile and the baseline while looping through all of the available collocations. The running sums are computed at every pressure. During the step the time differences, distance differences, maximum negative differences, and maximum positive differences are also calculated. Running sums are maintained for temperature, water vapor mixing ratio, relative humidity, and total precipitable water. After the running sums are computed for every collocation, the calculation of the statistics is performed. The calculated statistics are the baseline mean, profile mean, bias, SD, RMS, and R-squared. Properties of the matches are also computed, such as the average distance, the average time difference, and average of the absolute value of the time difference between matches.

After the statistics are calculated, the values are grouped and returned to the part of Profile Display that displays the statistics. At this point, the user has many options for displaying the statistics, including choices between showing temperature, water vapor mixing ratio fraction, or relative humidity. Users can also choose between displaying the statistics on a graph, table, or in a scatter plot.

Fig. 5 shows an example of vertical statistics for a 10-day period during June 2021. The product suites compared are the NUCAPS IR+MW and MiRS (microwave only) for NOAA-20, AIRS v6 IR-only, and GNSS COSMIC-2 wet profiles. The time window for all collocations is set to ±3h and the distance window is 150km from the radiosonde launch. The NUCAPS and AIRS are constrained to successful IR-based retrievals, MiRS for rain rates equal to zero and COSMIC-2 which passed QC. The assessment is enterprise meaning that the radiosonde baseline is the same for all suites. The pressure scale (y-axis) is set from the surface to 5hPa for temperature and to 200hPa for water vapor mixing ratio fraction covering the 100 effective pressures; the x-axis range is −1.5 to 3K for temperature and −25% to 75% for water vapor mixing ratio fraction. Bias and standard deviation statistics are plotted for all terrain and radiosonde instrument types using the default global distribution option (treat every collocation the same). The upper panel shows the global distribution after constraints are applied (compare to upper panel of Fig. 2), and the lower panels show the vertical statistics for temperature (left) and water vapor mixing ratio fraction (right).

The statistics among the four product suites appear overall consistent with temperature bias within 1K and standard deviation less than 2K in the middle troposphere. Water vapor mixing ratio fractions also show good consistency and

292 PART | III Satellite applications

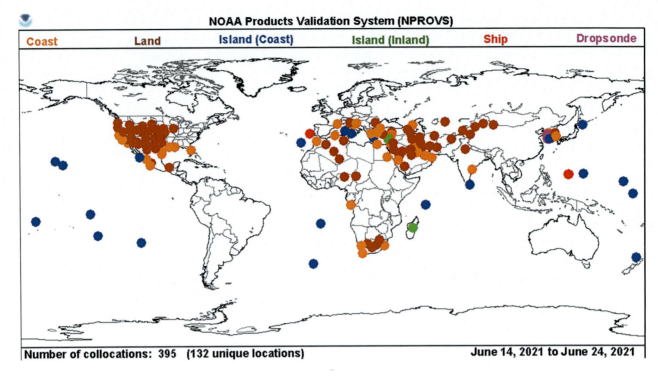

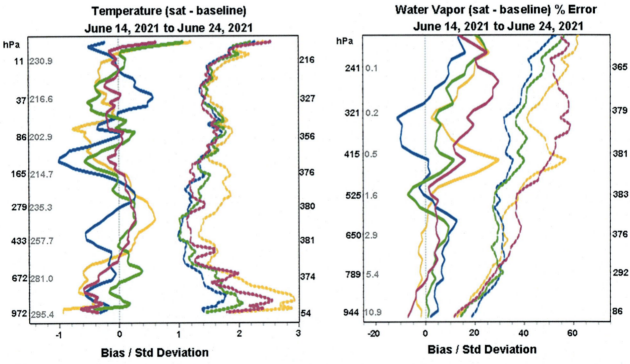

FIG. 5 Global distribution of enterprise sample of collocated COSMIC-2, NUCAPS NOAA-20, MiRS NOAA-20, and AIRS (top panel) and corresponding vertical statistics of temperature (left) and water vapor mixing ratio fraction (right) mean *(solid)* and standard deviation *(dash)* versus conventional radiosonde; 10-day period, June 2021.

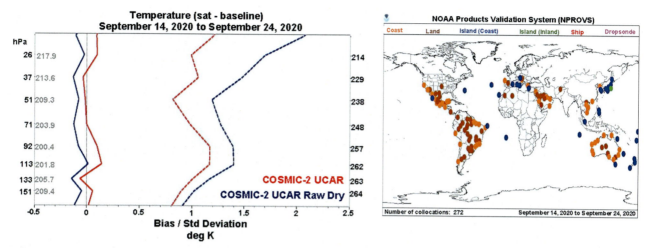

FIG. 6 Mean difference and standard deviation difference of UCAR COSMIC-2 wet temperature *(red)* and dry temperature *(blue)* vs Vaisala RS41 (left) based on global collocations (2 h/150 km) of September 4–14, 2020 (right). *Solid lines* are for mean bias and *dotted lines* are for standard deviation (left).

can be converted to the amount by multiplying by the mean water vapor mixing ratio listed along the inside left y-axis. Although the assessment is enterprise, the time difference between the satellite data can be up to 6 h for the COSMIC-2, whereas the three polar satellite systems are at similar (identical for MiRS and NUCAPS) overpass times. It is also noted that the global distributions are cutoff at about 50 latitude since the COSMIC-2 do not sense at high latitudes. The enterprise sample for these systems is on the order of 350 out of an original sample of 12,000.

Fig. 6 shows an example of using high-accuracy satellite data to assess radiosonde data. The left plot indicates the UCAR COSMIC-2 wet temperature matches well with RS41 data in the upper tropospheric and lower stratosphere and that could be because the wet temperature is impacted by the GFS background that assimilates the radiosonde data. A slightly negative difference of RO dry temperature minus RS41 suggests a residual warm bias (~0.1 K) may remain in RS41 (Sun et al., 2019).

5.3 NPROVS Archive Summary (NARCS) Statistics

NARCS displays long-term, vertical time series of statistical results over the lifetime of a given satellite and associated product suite. Individual time series are provided for each product suite for temperature and water vapor mixing ratio fraction for bias, RMS and SD, respectively. All time series are baselined against the radiosondes and routinely segregated into global, sea, and land terrain categories. NARCS time series are provided separately for conventional and special radiosondes. Separate time series are also shown for 100 vs 30/20 layer profiles (see Section 4) and for collocation time windows ±6 h and ±2 h, respectively. All-time series are precomputed but the NARCS application contains user options to select atmospheric pressure and profile data range focusing on specific time periods adding to the usefulness of this utility. Options also include the display of daily, weekly, or monthly averaged statistics for a given product suite users also.

Fig. 7 shows two examples of NARCS covering the history of NUCAPS soundings from SNPP compared to legacy ATOVS soundings beginning 2008. Shown are daily averages for RMS error at the 100 effective pressure levels for collocations within ±6 h. The transition from winter (blue) to summer (yellow) seasons is tracked along the horizontal axis with white indicating spring and gray fall (northern hemisphere). The color scale range is shown on the right (1–4 K).

Fig. 7 demonstrates the overall improved RMS (less yellow) in the troposphere and tropopause region for the NUCAPS versus ATOVS (less yellow). The improvement of NUCAPS over ATOVS is also demonstrated in the coarse layer statistics (see Chapter 18). NUCAPS is also improved in the lower troposphere but appears to retain the seasonal pattern of higher RMS during winter originally observed in ATOVS. Investigations show that the RMS increase at the surface, although related to more frequent occurrences of surface temperature inversions, is a general feature also observed in noninversion cases.

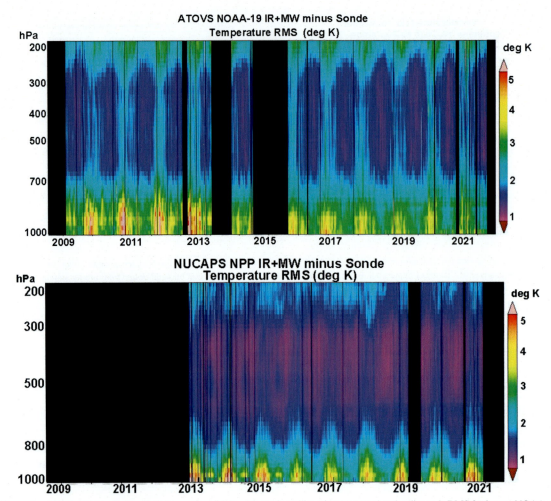

FIG. 7 Examples of NARCS historical vertical time series of daily averaged satellite minus conventional radiosonde RMS for legacy NOAA-19 ATOVS (upper) and SNPP NUCAPS IR+MW soundings at the 100 effective pressures from the surface to 200 hPa RMS (right axis) with difference color scale (0–5 K) shown on the left. Collocations within the 6-h window are used.

6. Summary

In this chapter, we describe centralized access and compilation of collocated radiosonde and satellite and NWP product suites and associated methods to assess these observations with a focus on the computation of vertical profile statistics. NPROVS sounding product monitoring capability, a unique capability focused on the intercomparison of internationally available satellite derived geophysical profiles, is demonstrated.

Since operated at NOAA STAR in 2008, NPROVS has contributed to the improvement of sounding product algorithms particularly the NUCAPS algorithm, through routine monitoring and trouble-shooting problem areas. Some of these are illustrated in the text. NPROVS has also been used in assessing NUCAPS products maturity review to ensure that mission requirements were met prior to releases the data to the public. NPROVS is open ended, ready to meet expansions to new satellite, and data product-scientific algorithms while maintaining and eventually retiring legacy product suites.

In addition to satellite product monitoring, NPROVS collocation data accumulated since 2008 have emerged as a source of "condensed" data set for products cross checking and validation and has also been utilized in research areas including (i) quantifying the uncertainty of time and distance mismatches on the bias and standard deviation of the product being validated (Sun et al., 2010, 2017a), (ii) assessing if satellite products meet mission requirements by taking advantage of the large radiosonde sample from conventional network and high-quality dedicated/GRUAN data (Sun et al., 2017a, b), (iii) quantifying and correcting radiation-induced temperature biases in conventional radiosonde types by using GNSS RO dry temperature as the target data (Sun et al., 2010, 2013), (iv) assessing the consistency of GRUAN observations with satellite measurements in the Global Space-based Inter-Calibration System context (Sun et al., 2019a), and (v) quantifying the measurement improvement of newly emerging Vaisala RS41 over Vaisala RS92 in support of their transition in GRUAN (Sun et al., 2019b, 2021).

Acknowledgments

The authors express our appreciation to Rebekah Esmaili for reviewing this chapter and providing constructive feedback. The scientific results and conclusions, as well as any views or opinions expressed herein, are those of the authors and do not necessarily reflect the views of NOAA or the Department of Commerce.

References

Anthes, R.A., et al., 2008. The COSMIC/FORMOSAT-3 mission: early results. Bull. Am. Meteorol. Soc. 89, 313–333.

August, T., Klaes, D., Schlüssel, P., Hultberg, T., Crapeau, M., Arriaga, A., O'Carroll, A., Coppens, D., Munro, R., Calbet, X., 2012. IASI on Metop-A: operational level 2 retrievals after five years in orbit. J. Quant. Spectrosc. Radiat. Transf. 113 (11), 1340–1371.

August, T., et al., 2016. EUMETSAT operational IASI L2 version 6. In: Presented at NASA AIRS Science Team Meeting, Greenbelt, Maryland. https://airs.jpl.nasa.gov/system/.../files/141_August-EUMETSAT-IASI-L2-V6.pdf.

Bodeker, G.E., et al., 2016. Reference upper-air observations for climate: from concept to reality. Bull. Am. Meteorol. Soc. 97, 123–135.

Boukabara, S.A., Garrett, K., Chen, W.C., Iturbide-Sanchez, F., Grassotti, C., Kongoli, C., Chen, R.Y., Liu, Q.H., Yan, B.H., Weng, F.Z., Ferrar, R., Kleespies, T.J., Meng, H., 2011. MiRS: an all-weather 1DVAR satellite data assimilation and retrieval system. IEEE Trans. Geosci. Remote Sens. 49 (9), 3249–3272. https://doi.org/10.1109/tgrs.2011.2158438.

Dirksen, R., et al., 2020. Managing the transition from Vaisala RS92 to RS41 radiosondes within the Global Climate Observing System Reference Upper-Air Network (GRUAN): a progress report. Geosci. Instrum. Methods Data Syst. 9, 337–355. https://doi.org/10.5194/gi-9-337-2020.

ECMWF, 2018. IFS documentation CY45r1. ECMWF Report, 103 pp.

Eresmaa, R., McNally, A.P., 2014. Diverse profile datasets from the ECMWF 137-level short-range forecasts. NWP Satellite Application Facilities Report. NWPSAF-EC-TR-017, 12 pp https://nwpsaf.eu/oldsite/reports/nwpsaf-ec-tr-017.pdf.

Gambacorta, A., 2013. The NOAA Unique CrIS/ATMS Processing System (NUCAPS). Algorithm Theoretical Basis Doc, Version 1.0. NOAA, p. 73. [Online]. Available from: http://www.ospo.noaa.gov/Products/atmosphere/soundings/nucaps/docs/NUCAPS_ATBD_20130821.pdf.

Gambacorta, A., Barnet, C., Wolf, W., Goldberg, M., King, T., Nalli, N., Maddy, E., Xiong, X., Divakarla, M., 2012. The NOAA Unique CrIS/ATMS Processing System (NUCAPS): first light retrieval results. In: Proceedings of ITSC-XVIII International TOVS Working Group (ITWG), Toulouse, France.

Hajj, G.A., Kursinski, E.R., Romans, L.J., Bertiger, W.I., Leroy, S.S., 2002. A technical description of atmospheric sounding by GPS occultation. J. Atmos. Sol. Terr. Phys. 64, 451–469.

Hultberg, T., Crapeau, M., August, T., 2016. IASI L2 Version 6: status, content, and assessment. In: Presented at GEWEX Water Vapor Assessment (G-VAP). http://gewex-vap.org/wp-content/uploads/2016/01/GVAP_WS5_04_August_IASI_L2_v6_GVAP_v3.pdf.

Iturbide-Sanchez, F., et al., 2017. Using averaging kernels to study the vertical resolution of NUCAPS temperature and water vapor. In: 2017 IEEE International Geoscience and Remote Sensing Symposium (IGARSS), pp. 33–35, https://doi.org/10.1109/IGARSS.2017.8126886.

Kursinski, E.R., Hajj, G.A., Schofield, J.T., Linfield, R.P., Hardy, K.R., 1997. Observing earth's atmosphere with radio occultation measurements using the Global Positioning System. J. Geophys. Res. 102 (D19), 23429–23465.

Maddy, E.S., Barnet, C., 2008. Vertical resolution estimates in version 5 of AIRS operational retrievals. IEEE Trans. Geosci. Remote Sens. 46 (8), 2375–2384.

Nalli, N.R., et al., 2011. Multi-year observations of the tropical Atlantic atmosphere: multidisciplinary applications of the NOAA Aerosols and Ocean Science Expeditions (AEROSE). Bull. Am. Meteorol. Soc. 92, 765–789. https://doi.org/10.1175/2011BAMS2997.1.

Nalli, N.R., et al., 2013. Validation of satellite sounder environmental data records: application to the Cross-track Infrared Microwave Sounder Suite. J. Geophys. Res. Atmos. 118, 1–16. https://doi.org/10.1002/2013JD020436.

Nalli, N.R., et al., 2016. Satellite sounder observations of contrasting tropospheric moisture transport regimes: Saharan air layers, Hadley cells, and atmospheric rivers. J. Hydrometeorol. 107, 2997–3006. https://doi.org/10.1175/JHM-D-16-0163.1.

Nalli, N.R., et al., 2018. Validation of atmospheric profile retrievals from the SNPP NOAA-Unique Combined Atmospheric Processing System. Part 1: temperature and moisture. IEEE Trans. Geosci. Remote Sens. 56 (1), 180–190. https://doi.org/10.1109/TGRS.2017.2744558.

Nalli, N.R., Gambacorta, A., Liu, Q., Barnet, C.D., Tan, C., Iturbide-Sanchez, F., Reale, A., Sun, B., Wilson, M., Borg, L., Morris, V.R., 2018. Validation of atmospheric retrievals from the SNPP NOAA-Unique Combined Atmospheric Processing System. Part 1: temperature and moisture. IEEE Trans. Geosci. Remote Sens. 56, 180–190.

Reale, A., 2001. NOAA operational sounding products from Advanced-TOVS polar orbiting environmental satellites. NOAA Technical Report, NESDIS 102, 61 pp.

Reale, A., 2002. NOAA operational sounding products for Advanced-TOVS. NOAA Technical Report, NESDIS 107, 29 pp.

Reale, A., Chalfant, M.W., Wagoner, R.V., Gardiner, T.J., 1994. TOVS operational sounding upgrades: 1990–1992. NOAA Technical Report, NESDIS 52, 57 pp.

Reale, A., Tilley, F., Ferguson, M., Allegrino, A., 2008. NOAA operational sounding products for ATOVS. Int. J. Remote Sens. 29, 4615–4651.

Reale, A., Tilley, F., Pettey, M., 2012. The NOAA products validation system (NPROVS). J. Atmos. Ocean. Technol. 29 (5), 629–645. https://doi.org/10.1175/jtech-d-11-00072.1.

Saha, S., et al., 2010. The NCEP climate forecast system reanalysis. Bull. Am. Meteorol. Soc. 91, 1015–1057. https://doi.org/10.1175/2010BAMS3001.1.

Seidel, D.J., Sun, B., Pettey, M., Reale, A., 2011. Global radiosonde balloon drift statistics. J. Geophys. Res. 116, D07102. https://doi.org/10.1029/2010JD014891.

Smith, N., Barnet, C.D., 2020. CLIMCAPS observing capability for temperature, moisture, and trace gases from AIRS/AMSU and CrIS/ATMS. Atmos. Meas. Tech. 13 (8), 4437–4459. https://doi.org/10.1109/TGRS.2002.808244.

Smith, W.L., Woolf, H.M., 1976. The use of eigenvectors of statistical co-variance matrices for interpreting satellite sounding radiometer observations. J. Atmos. Sci. 33, 1127–1140.

Sun, B., Reale, A., Seidel, D.J., Hunt, D.C., 2010. Comparing radiosonde and COSMIC atmospheric profile data to quantify differences among radiosonde types and the effects of imperfect collocations on comparison statistics. J. Geophys. Res. 115. https://doi.org/10.1029/2010JD014457.

Sun, B., Reale, A., Schroeder, S.R., Seidel, D.J., Ballish, B.A., 2013. Toward improved corrections for radiation induced biases in radiosonde temperature observations. J. Geophys. Res. Atmos. 118, 4231–4243. https://doi.org/10.1002/jgrd.50369.

Sun, B., Reale, A., Tilley, F.H., Pettey, M., Nalli, N.R., Barnet, C.D., 2017a. Assessment of NUCAPS S-NPP CrIS/ATMS sounding products using reference and conventional radiosonde observations. IEEE J. Sel. Top. Appl. Earth Obs. Remote Sens. 10, 1–18.

Sun, B., Reale, A., Pettey, M., Smith, R., Nalli, N.R., Zhou, L., 2017b. Leveraging the strength of dedicated, GRUAN and conventional radiosondes for EUMETSAT IASI atmospheric sounding assessment. In: Proceedings for the 2017 EUMETSAT Meteorological Satellite Conference, 2–6 October 2017, Rome, Italy, pp. 1–8.

Sun, B., Calbet, X., Reale, A., Bali, M., 2019a. GSICS use of GRUAN humidity observations in the context of satellite sensor assessment. GSICS Newsletter Summer 2019 Issue 13. https://doi.org/10.25923/63j6-sb72.

Sun, B., Reale, A., Schroeder, S., Pettey, M., Smith, R., 2019b. On the accuracy of Vaisala RS41 versus RS92 upper-air temperature observations. J. Atmos. Ocean. Technol. 36, 635–653.

Sun, B., Calbet, X., Reale, A., Schroeder, S., Bali, M., Smith, R., Pettey, M., 2021. Accuracy of Vaisala RS41 and RS92 upper tropospheric humidity compared to hyperspectral satellite infrared measurements. Remote Sens. (Basel) 13, 173. https://doi.org/10.3390/rs13020173.

Susskind, J., Barnet, C.D., Blaisdell, J., 2003. Retrieval of atmospheric and surface parameters from AIRS/AMSU/HSB data under cloudy conditions. IEEE Trans. Geosci. Remote Sens. 41 (2), 390–409.

Tobin, D.C., et al., 2006. Atmospheric radiation measurement site atmospheric state best estimates for Atmospheric Infrared Sounder temperature and water vapor validation. J. Geophys. Res. 111, D09S14. https://doi.org/10.1029/2005JD006103.

Chapter 17

Satellite microwave sounder product retrieval validation

Christopher Grassotti[a], Quanhua Liu[b], Yong-Keun Lee[a], Shuyan Liu[c], and Yan Zhou[a]
[a]*Cooperative Institute for Satellite and Earth System Studies (CISESS), Earth System Science Interdisciplinary Center, University of Maryland, College Park, MD, United States,* [b]*NOAA/NESDIS Center for Satellite Applications and Research (STAR), College Park, MD, United States,* [c]*Cooperative Institute for Research in the Atmosphere (CIRA), Colorado State University, Fort Collins, CO, United States*

Chapter outline

1. Overview	297	4.3 Water vapor profile validation	308
2. Microwave sounder data	297	4.4 Total precipitable water validation	311
3. Microwave integrated retrieval system (MiRS)	299	5. Summary	315
4. MiRS product validation	301	Acknowledgments	316
4.1 Validation background	301	References	316
4.2 Temperature profile validation	304	Further reading	317

Look at the moon in the sky, not the one in the lake.

Jalaluddin Rumi

1. Overview

Validation of data products from microwave sounder retrieval systems is a key element of the meteorological satellite product development enterprise that encompasses satellite mission and instrument planning, development, deployment, pre- and post-launch sensor calibration, retrieval algorithm development and tuning, operational and/or research implementation, delivery of retrieval products to users, and validation itself. This process is not unidirectional; user experience and feedback, as well as validation results themselves can inform the developers of the retrieval algorithms on the strengths and weaknesses of the retrieval data and the algorithms that produce them, which can lead to improvements in the retrieval systems. Additionally, variations over time in retrieval biases can indicate potential changes in the calibration of input sensor data that require attention. In this chapter, we briefly review the primary space-based microwave sounder systems currently or recently deployed, and their most salient characteristics and applications. Then, by way of example, we focus on a typical example of an operational microwave retrieval algorithm (i.e., MiRS), describing the algorithm's approach, its main features, and finally review results from validation of its sounding-related retrieval products. It should also be noted that another extremely important application of microwave sounder data is its use within global data assimilation systems that provide the initial conditions needed for operational numerical weather prediction models, which are run at various forecasting centers around the world. However, this application is beyond the scope of this chapter and will not be discussed here.

2. Microwave sounder data

Generally, most passive microwave (PMW) instruments designed for atmospheric sounding applications share common spectral and spatial sampling characteristics. Spectral similarity exists because they exploit the same atmospheric absorption features at microwave and millimeter wavelengths that allow for vertical temperature and water vapor profiling, most commonly the oxygen absorption line near 60 GHz and the water vapor absorption line near 183 GHz (Weng, 2018).

Additional channels may also be present that allow for estimation of surface characteristics (e.g., emissivity and cryospheric variables), or of clouds and precipitation. Space-based deployment of passive microwave instruments is mainly limited to polar orbiting satellites and this, along with instrument and antenna design constraints, largely determine their spatial sampling characteristics. Currently operating sensors include the legacy cross-track scanning instruments the Advanced Microwave Sounding Unit (AMSU) and Microwave Humidity Sounder (MHS), flown aboard the older generation National Oceanic and Atmospheric Administration (NOAA) Polar Operational Environmental Satellite (POES) series, as well as the European Meteorological Operational (MetOp) series (Klaes et al., 2007). Additional sounding and surface sensing capability has been provided by the conically scanning Special Sensor Microwave Imager/Sounder (SSMIS) instruments flown aboard the Defense Meteorological Satellite Program (DMSP) series satellites (Kunkee et al., 2006). More recently, under the framework of the Joint Polar Satellite System (JPSS) program (Goldberg et al., 2013), the NOAA/NASA-developed cross-track Advanced Technology Microwave Sounder (ATMS) instrument has been deployed aboard the Suomi-National Polar orbiting Partnership (S-NPP) and NOAA-20 satellites, since 2011 and 2017, respectively. Three additional ATMS instruments are planned for launch aboard JPSS satellites in the 2022–2031 timeframe. The instrument specifications and vertical weighting functions for the ATMS instrument are shown below in Table 1 and Fig. 1, respectively. The weighting functions are the derivative of the vertical transmittance at each channel with respect to pressure or height. The relative smoothness of the weighting functions implies that the inherent vertical resolution of the measurements with respect to temperature and water vapor is relatively low and that retrieved profiles

TABLE 1 ATMS instrument characteristics.

Channel number	Center frequency (GHz)	Band width (GHz)	Sensitivity (NEDT) (K)	Accuracy (K)	Beam width (degrees)
1	23.8	0.27	0.7	1.0	5.2
2	31.4	0.18	0.8	1.0	5.2
3	50.3	0.18	0.9	0.75	2.2
4	51.76	0.4	0.7	0.75	2.2
5	52.8	0.4	0.7	0.75	2.2
6	53.596±0.115	0.17	0.7	0.75	2.2
7	54.4	0.4	0.7	0.75	2.2
8	54.94	0.4	0.7	0.75	2.2
9	55.5	0.33	0.7	0.75	2.2
10	57.2903	0.33	0.75	0.75	2.2
11	57.2903±0.115	0.078	1.2	0.75	2.2
12	57.2903	0.036	1.2	0.75	2.2
13	57.2903±0.322	0.016	1.5	0.75	2.2
14	57.2903±0.322±0.010	0.008	2.4	0.75	2.2
15	57.2903±0.322±0.004	0.003	3.6	0.75	2.2
16	87–91(88.20)	2.0	0.5	1.0	2.2
17	164–167	3.0	0.6	1.0	1.1
18	183.31±7	2.0	0.8	1.0	1.1
19	183.31±4.5	2.0	0.8	1.0	1.1
20	183.31±3	1.0	0.8	1.0	1.1
21	183.31±1.8	1.0	0.8	1.0	1.1
22	183.31±1.0	0.5	0.9	2.0	1.1

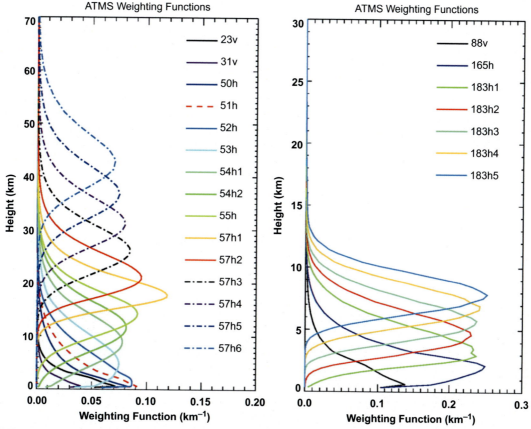

FIG. 1 ATMS weighting functions for channels 1–15 (*left*), and 16–22 (*right*), based on a US Standard Atmosphere. The weighting functions provide a synthesis of how the atmospheric state at each vertical layer (primarily through absorption/emission) contributes to the observed upwelling radiance at the top of the atmosphere for each channel. The water vapor channel weighting functions can shift upward (or downward) in atmospheres with more (or less) water vapor.

from a well-constrained retrieval system should also be relatively smooth if the information content of the measurements is correctly used. The upcoming European Polar System—Second Generation (EPS-SG) plans for a series of three cross-track Microwave Sounder (MWS) instruments (among others) to be deployed in the 2022–2045 timeframe (Schlüssel and Kayal, 2017). The oxygen absorption feature at 118 GHz, rather than at 60 GHz can also be exploited for temperature sounding, as with the radiometer onboard the Micro-sized Microwave Atmospheric Satellite (MicroMAS-2) to be flown in the planned Time-Resolved Observations of Precipitation structure and storm Intensity with a Constellation of Smallsats (TROPICS) constellation mission (Blackwell et al., 2018), as well as the MicroWave Humidity Sounder-2 (MWHS-2) instrument onboard the Chinese FY-3C satellite (Li and Qin, 2016). In the case of the TROPICS mission, which will be constellation of SmallSats, the choice is driven mainly by limitations on antenna size imposed by lower weight and power requirements.

3. Microwave integrated retrieval system (MiRS)

The Microwave Integrated Retrieval System (https://www.star.nesdis.noaa.gov/mirs), a passive microwave retrieval algorithm, has been running operationally at NOAA since 2007 routinely processing PMW measurements from a range of low earth orbit satellites. Compared to visible and infrared radiation, microwaves have a longer wavelength, and thus can penetrate through the atmosphere more effectively. This feature allows microwave observations under almost all weather conditions including in cloudy and rainy atmospheres. MiRS follows a one-dimensional variational (1DVAR) methodology (Boukabara et al., 2011, 2013). The inversion is an iterative physical algorithm in which the fundamental physical attributes affecting the microwave observations are retrieved physically, including the profiles of atmospheric temperature, water vapor, nonprecipitating cloud, hydrometeors, as well as surface emissivity and skin temperature (Iturbide-Sanchez et al., 2011; Liu et al., 2017, 2020; Boukabara et al., 2018; Grassotti et al., 2020). The Joint Center for Satellite Data Assimilation (JCSDA) Community Radiative Transfer Model (CRTM) (Han et al., 2006; Ding et al., 2011) is used as the forward and Jacobian operator to simulate the radiances at each iteration prior to fitting the measurements to within the combined instrument and forward model noise level, along with satisfying additional a priori constraints. An additional means of regularizing the solution includes using a limited number of empirical orthogonal functions (EOFs) as basis function representation for the profile variables (i.e., temperature, water vapor, clouds, and hydrometeors), and surface emissivity spectrum. Essentially, the algorithm seeks to minimize a cost function which in effect balances a fit of the simulated to observed radiances with the departure of the solution from an a priori climatology. Prior to the 1DVAR retrieval itself, preprocessing of the satellite-observed radiances is performed, which consists of two steps. First, depending on the particular instrument and scan geometry spatial resampling and/or footprint matching is performed so that, to the extent possible, all channel measurements are valid for the same area on the Earth's surface. Second, because the CRTM physical model is used as part of the 1DVAR inversion, a bias correction is added to the original radiometric measurements. This is needed to account for the systematic differences between observed and simulated radiances that are known to exist, even when knowledge of the Earth-atmosphere system provided to the model is accurate. The bias correction itself is currently static and computed off-line using a large set of collocated satellite measurements and corresponding atmospheric profiles obtained from global numerical weather analysis fields. After the core parameters of the state vector are retrieved in the 1DVAR step, an additional post-processing is performed to retrieve derived parameters based on inputs from the core 1DVAR retrieval. The postprocessing products are either the result of vertically integrating vertical profiles from the core retrieval (e.g., total precipitable water (TPW) and integrated cloud liquid water (CLW)), or are determined by predetermined relationships between the core retrieval and the derived products (e.g., snow water equivalent (SWE), sea ice concentration (SIC), and precipitation or rain rate (RR)). Snowfall rate (SFR) is a separate algorithm that uses primarily the higher frequency channel measurements sensitive to ice crystal scattering. A schematic of the MiRS processing components and data flow is shown in Fig. 2. MiRS retrieval products are also listed in Table 2 where, for each product, the processing step in which it is produced is indicated.

Given the wide range of retrieval products produced by the MiRS retrieval system from the combination of sounding and surface-sensitive channels, it is beyond the scope of this chapter to detail validation activities and results for all of them. Instead, we focus on the primary vertical sounding products of interest, temperature, and water vapor and the vertically integrated water vapor or total precipitable water.

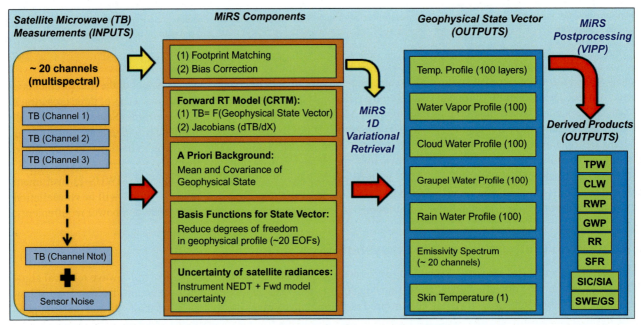

FIG. 2 Schematic of MiRS processing components and data flow showing MiRS core retrieval and post-processing components. Core products are retrieved simultaneously as part of the state vector. Post-processing products are derived through vertical integration or through predefined relationships (i.e., look-up tables or analytic functions) between state vector variables and the derived products. Note: abbreviations are defined as follows: *TB*, brightness temperature; *dTB/dX*, Jacobian rate of change of brightness temperature with respect to a change in the geophysical state vector X; *EOF*, empirical orthogonal function, the basis functions used to represent the geophysical profile in the retrieval; *NEDT*, noise equivalent delta temperature; *TPW*, total precipitable water; *CLW*, cloud liquid water; *RWP*, rain water path; *GWP*, graupel water path; *RR*, rain rate; *SFR*, snowfall rate; *SIC/SIA*, sea ice concentration/sea ice age; *SWE/GS*, snow water equivalent/snow grain size.

TABLE 2 List of MiRS retrieval products and the processing step in which they are produced.

Retrieval product	Core 1DVAR	Post-processed
Temperature profile	x	
Water vapor profile	x	
Cloud water profile	x	
Rain water profile	x	
Graupel water profile	x	
Surface skin temperature	x	
Surface emissivity	x	
Total precipitable water		x
Total cloud liquid water path		x
Total rain water path		x
Total graupel water path		x
Sea ice concentration and age		x
Snow water equivalent and grain size		x
Precipitation rate		x
Snowfall rate		x

Validation results for those products highlighted in bold are presented in this chapter.

4. MiRS product validation

In this section, we discuss validation of PMW sounder retrieval products. This discussion is first motivated by an overview of the data sets most widely used as references in a validation activity. Their corresponding advantages and disadvantages are cited as well as their primary characteristics, particularly in terms of temporal and spatial coverage and resolution. The importance of characterizing the uncertainties in the reference data themselves is also highlighted. Then we present a series of validation results for, in turn, vertical temperature profiles, vertical water vapor profiles, and finally, total vertically integrated water vapor or total precipitable water.

4.1 Validation background

MiRS vertical profiles are retrieved on a fixed set of 100 pressure layers. While typical PMW sounders only have on the order of 20 channels, the use of EOFs as vertical basis functions regularizes the solution and allows for the specification of the retrieval on the high vertical resolution grid. Fig. 3 shows an example of a typical retrieved temperature and water vapor profile in a subtropical ocean region. For comparison, a profile from the temporally and spatially collocated operational European Centre for Medium-Range Weather Forecasts (ECMWF) analysis is also shown. The ECMWF profile was vertically interpolated from 91 layers to the 100-layer vertical grid used in MiRS. The lack of small-scale vertical variations in the retrieved profiles is a due both to the smoothness of the ATMS weighting functions and to the use of a relatively small number of EOF basis functions, which limit the degrees of freedom permitted in the profile. There is good qualitative agreement with the ECMWF profile for both temperature and water vapor.

In the context of temperature and water vapor soundings, a validation exercise may be thought of as having two possible goals, with the purpose determining the scale and scope of validation. In the first instance, one may be interested in considering the detailed mechanisms and behavior of the retrieval process—for example, in a particular case study of a meteorological event such as a tropical cyclone or severe weather outbreak. In this case, the goal is not necessarily to obtain a large number of samples in order to generate robust validation statistics, but to better understand the mechanisms of the retrieval, and its behavior relative to the meteorological conditions and the available information content of the radiometric measurements (Rodgers, 2000). Here the focus would be on individual profile retrievals with, to the extent possible,

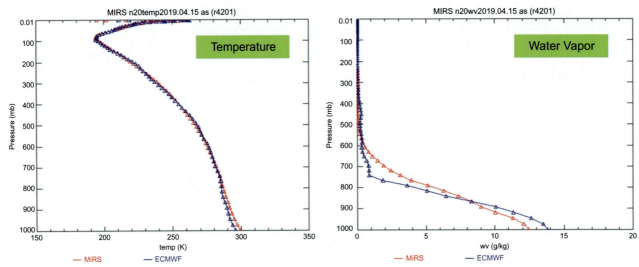

FIG. 3 Example of MiRS and collocated ECMWF temperature *(left)* and water vapor *(right)* profiles. Profiles were retrieved from NOAA-20 ATMS measurements on April 15, 2019, at 20.2 N latitude and 170.2 W longitude. MiRS retrievals are retrieved on the 100 pressure layers indicated, and ECMWF analysis profiles are interpolated from 91 original layers to the 100 layers.

comparisons to in situ measurements from radiosondes, dropsondes, ground-based upward looking remote sensing systems, other space-based observing systems, or perhaps analysis/forecast fields from a high-resolution NWP model.

In the second instance (and the one that is emphasized in this chapter), the goal is to obtain reliable statistics of retrieval performance on regional and global scales. In this case, it is desirable to have as large a sample size as possible so as to reduce the uncertainty in the performance estimates themselves. For this type of validation, the use of both global radiosonde data sets as well as operational NWP analysis fields are often used. These may be supplemented with other more specialized data sets, such as ground-based GPS measurements. GNSS-RO profiles of temperature and moisture also have a role to play here; particularly for validation studies on somewhat longer timescales (i.e., monthly or longer), and for larger geographic regions, GNSS-RO measurements can be an extremely valuable reference with accuracy on par with radiosondes.

When considering reference data sets for validation activities, a key point is that there is no perfect reference measurement. All data sets, whether in situ or remotely sensed will have uncertainties, and these uncertainties should be considered when judging the degree of agreement or disagreement between the PMW retrieval and the reference. Additional complexity is introduced because (1) uncertainties in reference data may be themselves a function of location, vertical layer, and atmospheric conditions and (2) each reference data set has its own particular temporal and spatial sampling characteristics. The latter issue can lead to the phenomenon of representativeness error in which the PMW and reference data are not referring to the same 4-dimensional view of the atmosphere. This implies that some amount of disagreement between the PMW sounding and the reference data are acceptable (and perhaps desirable).

Conventional radiosonde measurements of temperature and water vapor have typically been considered a primary source of validation reference data. The GRUAN radiosonde network (Dirksen et al., 2014) contains high-quality measurements designed for calibration and validation activities and is noteworthy due to the inclusion of uncertainty estimates. Certain observation campaigns have dedicated radiosonde launches specifically timed to coincide with JPSS satellite overpasses (Nalli et al., 2013).

Radiosonde data sets and NWP analyses have their respective advantages and disadvantages. Radiosonde profiles of temperature and water vapor have the advantage of being direct in situ measurements. The global network of twice daily coordinated radiosonde launches traditionally served as a backbone of the global observing system prior to the widespread availability of satellite-based observations. Radiosonde observations continue to provide a strong constraint on global meteorological analyses, especially over regions with adequate sampling such as northern hemisphere continental regions. The error characteristics of the various sondes typically deployed have been estimated and the profiles are generally considered to be of high quality. Most recently, Sun et al. (2019) and Sun et al. (2021) have analyzed the uncertainties in currently deployed temperature and water vapor sondes in the upper troposphere and lower stratosphere. The two most commonly deployed sondes are the Vaisala RS41 and RS92, with the former gradually replacing the RS92. Generally speaking for temperature measurements, radiation induced warm biases are seen in the upper tropospheric and lower stratospheric measurements, while for humidity, sensor time-lag and warm-temperature induced dry bias are significant sources of uncertainty. The manufacturer cites uncertainties of temperature and water vapor (relative humidity) in the troposphere of 0.3 K and 4%, respectively (Vaisala, 2017). Representativeness error must also be considered when comparing satellite soundings to radiosondes as the sonde is effectively a point measurement with high vertical resolution and the satellite sounding is an area and volume-averaged measurement with relatively low vertical resolution.

From a regional and global validation perspective, radiosonde measurements have several disadvantages. The first is the heterogeneous geographic distribution—with the vast majority of profiles obtained over populated land areas, and with a greater concentration in more developed countries (e.g., US, Europe, Southeast Asia, Australia, and portions of S. America). Sparsely populated land areas and oceans are poorly sampled. Therefore, to the extent that retrieval performance itself is spatially nonuniform, validation statistics will reflect the geographic distribution of the radiosonde launches, which may differ from true global or regional performance. Fig. 4 shows a typical global distribution of radiosondes for a 10-day period in January 2018. A second limitation from a satellite validation perspective is that with most radiosondes launched at the 00 and 12 UTC times, temporal and spatial collocation with satellite overpasses is rarely perfect. Most collocation procedures use time and space windows to increase the number of valid matchups. But increasing sample sizes by way of larger space and/or time windows will normally result in lower agreement with the reference data. Finally, as noted, the radiosonde can be considered a point measurement (albeit with a small height-dependent geolocation due to balloon drift), while the satellite soundings are a spatial average of conditions within the measurement field of view, which is on the order of 25–100 km in size, depending on the sensor viewing geometry.

Objective meteorological analyses produced by operational global data assimilation systems have the advantage of global coverage, specification on a regular horizontal and vertical grid, high spatial resolution relative to PMW measurements, and generally high accuracy as they assimilate a wide variety of conventional and satellite observations. The use of the physical constraints provided by the forecast model generally yields analyses that are spatially, temporally, and physically self-consistent. These characteristics mean that, regardless of geographic location, an analysis profile can always be found that is both spatially and temporally collocated to the satellite sounding. This allows for more fine-grained stratification of performance assessments by region, season, surface type, and meteorological conditions. It also allows for a more complete 3 and 4-dimensional view of soundings that can identify spatio-temporal artifacts in the retrieved fields—something not possible when aggregate statistics are used. There are also potential disadvantages of using NWP analyses for validation. In data sparse regions, the analyses will likely have higher uncertainties. Particularities of the input observation quality control procedures, assimilation system approaches, and embedded forecast models can result in systematic differences between various operational analyses. These differences can affect retrieval validation statistics. The fact that most data assimilation systems directly assimilate many of the temperature and water vapor sounder channel measurements from PMW instruments means that there is not complete independence of the analysis values and the satellite soundings. In practice, an approach that uses both radiosondes and numerical model analyses, as well as other observing systems (e.g., ground based GPS, etc.) is usually recommended.

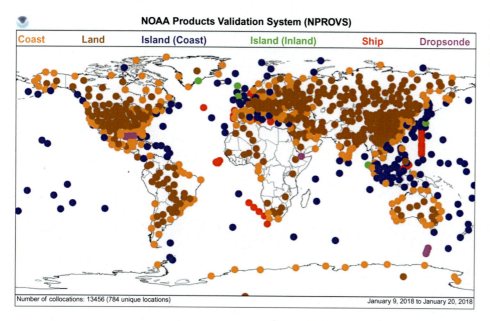

FIG. 4 Distribution of radiosonde and dropsonde observations for the period January 9–20, 2018. *(Data obtained from the NOAA Products Validation System (NPROVS).)*

With respect to ECMWF analyses, Hersbach et al. (2020) provided a comprehensive update on the ERA5 reanalysis system, including some discussion of uncertainty estimates. Generally speaking, uncertainty of the atmospheric temperature analyses ranged from 0.1 to 0.5 K, depending on vertical pressure level. Water vapor specific humidity was estimated to have an uncertainty of 4%–15% depending on vertical level, with higher values at 500 and 300 hPa. In this summary, we utilize the near real-time ECMWF analyses as a reference rather than ERA5, so uncertainties are expected to be somewhat larger. With respect to GDAS analyses, there are no published statistics on analysis uncertainty. However, given other studies (e.g., Hoffman et al., 2018) that show ECMWF forecast model skill to be slightly, but consistently higher than the NOAA/NCEP forecasts, we may infer that uncertainty of the GDAS analyses would be of the same order, but somewhat higher than that seen for ECMWF. Cases where validation results for the same time period show different results when using ECMWF versus GDAS analyses would be indicative of differences in the reference data uncertainty. For example, differences in mean global biases between Figs. 10 and 11 below would have to be related to differences in the reference data. In these types of cases, use of a third independent reference data set can help determine which reference data would be more accurate. Conversely, when comparison to two different (and ideally independent) reference data show similar bias characteristics, confidence is increased in the reliability of the reference data and in the estimated bias.

Global Navigation Satellite System Radio Occultation (GNSS-RO) data can also play an important role as they have very high accuracy down to 0.1 K and vertical resolution as fine as 100 m above the planetary boundary layer (Anthes, 2011; Ho et al., 2020). GNSS-RO data coverage is on the order of ~4000 profiles/day (in the case of COSMIC-2) with a distinct latitudinal dependence. For example, for a single polar-orbiting PMW sounder, one can typically obtain about 75–200 collocations/day near the equator, with a decrease to roughly 30–100 collocations/day between 40 degrees and 50 degrees latitude. While the locations of GNSS-RO measurements change each day, which represent a challenge if targeting validation in a specific synoptic condition or geographic location, these high-quality measurements, that are independent of PMW retrievals should be considered competitive with the global radiosonde observing system. Additionally, the global network of ground-based Global Positioning System (GPS) receivers (e.g., SuomiNet) produces accurate estimates of TPW (Ware et al., 2000); stations are placed at inland, coastal, and island locations (see Fig. 5). Vaquero-Martinez and Anton (2021) recently reviewed validation results from ground-based GPS water vapor measurements obtained from multiple independent studies over more than 20 years and found that the uncertainty range of TPW was between 0.5 and 4 mm with biases generally less than ±1 mm, depending on a number of factors including reference data, geographic location, and time of day.

Fig. 6 shows an example of global NOAA-20/ATMS channel 6 and 20 measurements for one full day along with corresponding MIRS retrievals and ECMWF analyses of temperature and water vapor at 500 hPa. Channel 6 (53.6 GHz) is a mid to upper tropospheric temperature sounding channel, and channel 20 (183.31 ± 3 GHz) is a mid-tropospheric water vapor sounding channel. Good qualitative agreement is seen between the MiRS sounding products and the analyses.

4.2 Temperature profile validation

Fig. 7 shows S-NPP and NOAA-20 temperature sounding validation results from global radiosonde collocations during 2019. The collocation criteria were a search radius of 250 km and ±6 h. The S-NPP satellite is situated one-half orbit (~50 min) behind NOAA-20 with an early afternoon local crossing time. When conducting the global matchups with

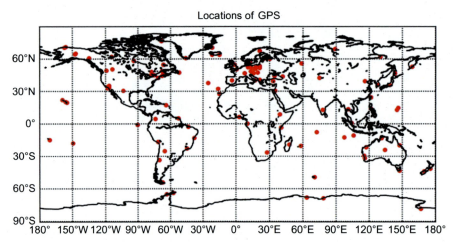

FIG. 5 Global distribution of ground-based GPS measurements of TPW.

Satellite microwave sounder product retrieval validation **Chapter | 17** 305

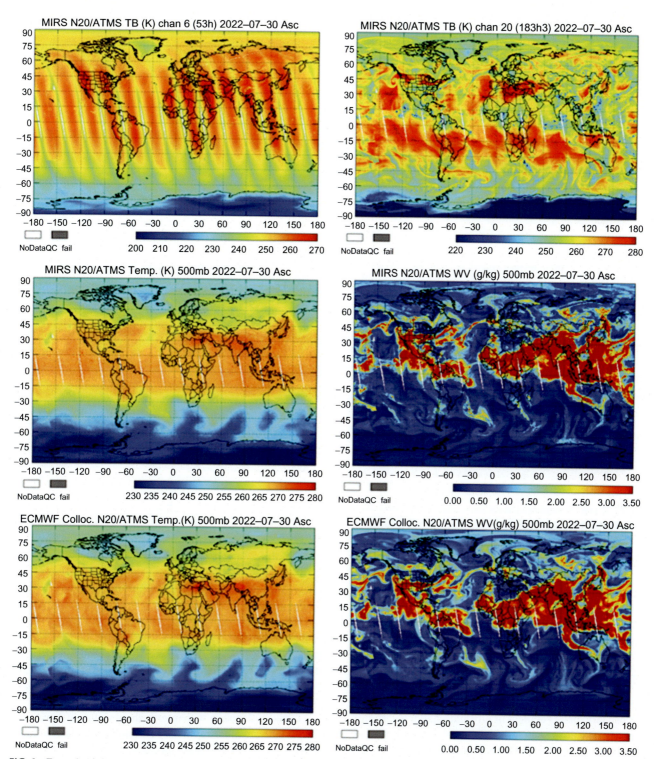

FIG. 6 Example of observed global NOAA-20/ATMS brightness temperatures *(top row)*, MiRS retrieved temperature and water vapor *(middle row)*, and operational ECMWF analysis *(bottom row)*. ATMS brightness temperatures are from temperature sounding channel 6 (53.6 GHz) and water vapor sounding channel 20 (183.31 ± 3 GHz). The temperature and water vapor maps are for the vertical pressure layer at 500 hPa valid on 30 July 2022.

radiosonde measurements, the collocation criteria is a typical consideration. Reducing the space and time window can reduce the component of the retrieval "error" related to collocation differences, but this will be achieved at the expense of smaller sample sizes. In the vertical, radiosonde measurements are interpolated to the 100 pressure layers used in the MiRS retrieval profiles.

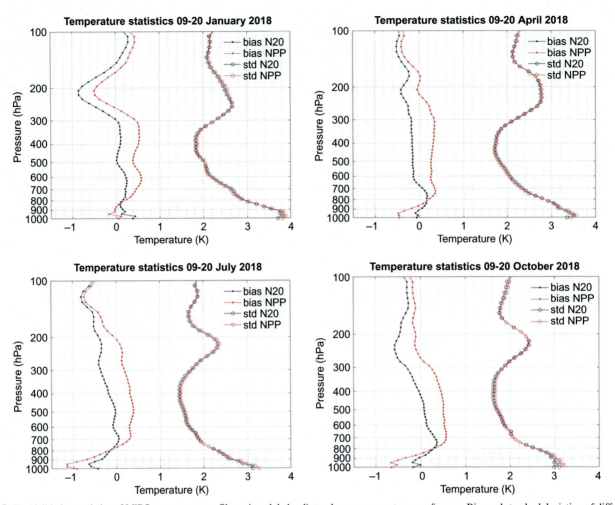

FIG. 7 Validation statistics of MiRS temperature profiles using global radiosonde measurements as a reference. Bias and standard deviation of differences are presented for four 10-day periods in 2018: 9–20 January *(top left)*, 9–20 April *(top right)*, 9–20 July *(bottom left)*, and 9–20 October *(bottom right)*. Results from both S-NPP *(red)* and NOAA-20 *(black)* ATMS retrievals are shown.

Results are shown from four 10-day periods representing performance through one annual cycle. Several features are notable. First, while the S-NPP and NOAA-20 error standard deviation is nearly identical, the retrieval bias is somewhat different, with the NOAA-20 soundings generally between 0 and 0.5 K colder than S-NPP, depending on vertical layer. Only the lowest layer in the tropopause shows NOAA-20 retrievals to be up to 0.5 K warmer than S-NPP. Ideally, radiometric bias corrections can reduce and eliminate systematic differences between soundings from different satellites, but in practice, depending on the nature of calibration-related differences in measured radiances, not all retrieval differences can be entirely removed. Also, calibration changes over time can result in changes in retrieval biases if the radiometric biases are not updated. The other notable feature of the retrieval performance is the seasonal dependence. There are larger negative tropospheric biases evident in July and smaller and slightly positive biases in January, with intermediate values in April and October. The error standard deviation is also temporally variable with largest values in January and smallest values in July and October. For the systematic error (bias), the skill of the retrieval a priori background climatology may have a seasonal component. The behavior of the retrieval random error (standard deviation) is likely a reflection of the higher density of radiosondes in the northern hemisphere and shows the seasonal changes in tropospheric climate, with higher variability in northern hemisphere winter.

Fig. 8 shows an example of MiRS global NOAA-20 and S-NPP 500 hPa temperature retrievals and corresponding differences with ECMWF analyses. The maps show considerable structure in the retrieval differences with generally positive differences in the tropics and southern hemisphere and slightly smaller and negative differences in the northern hemisphere extratropics. In the northern hemisphere, there is an indication of slightly larger positive differences over land relative to

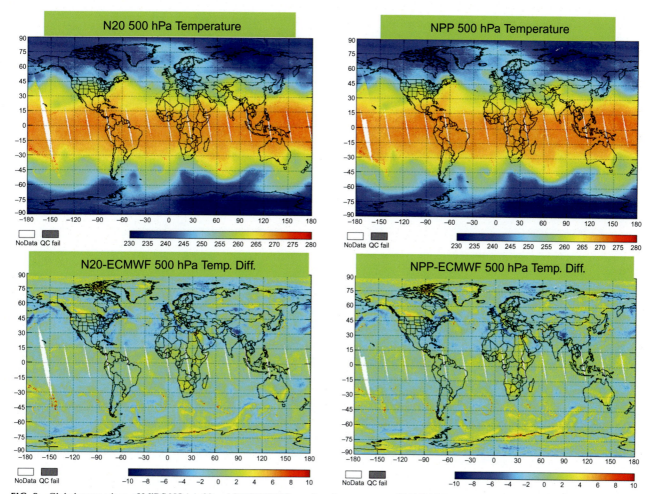

FIG. 8 Global comparison of MiRS NOAA-20 and S-NPP 500 hPa retrieved temperature with ECMWF analysis on April 15, 2019. *Top row* shows the MiRS retrieval, and *bottom row* shows the difference with ECMWF analysis.

ocean surfaces. Retrievals over southern hemisphere oceans seem to show slightly higher differences in the vicinity of frontal regions, indicating possible impacts of higher water vapor and/or clouds on the temperature soundings.

For a more quantitative view, Fig. 9 shows scatterplots of MiRS NOAA-20 clear-sky, over-ocean temperature at two pressure layers compared with ECMWF. These results from a single day of global retrievals show biases all below 1 K and error standard deviations less than 2 K. It should be noted that the global statistics based on collocations with either ECMWF or NOAA Global Data Assimilation System (GDAS) analyses are disproportionally weighted toward performance in the higher latitudes. This is because the collocations are made for every available satellite retrieval, and the increased overlap of polar-orbiting satellite swaths means that locations become increasingly oversampled as one moves poleward. To obtain performance estimates based on equal area representation would require a cosine of latitude weighting of the individual collocations, or a stratification of statistics by latitude zone.

Another view can be seen in Figs. 10 and 11, which are time series of global retrieval performance using ECMWF and GDAS analyses as references, respectively. The time period represented is more than one year (November 2017—May 2019), and retrieval performance at four pressure layers is shown for both land and ocean surfaces. While results from the ECMWF and GDAS appear largely consistent, one notable feature is that the biases with respect to GDAS are roughly 1 K lower (less positive or more negative), depending on vertical layer and time of year. This highlights the importance of using multiple validation references, when possible, as results may be somewhat dependent on the reference data. A small seasonal dependence of the bias is seen with respect to both references with somewhat lower (absolute) values during northern hemisphere summer. Likewise, the standard deviation generally shows lower values during this period as well, which reflects the lower tropospheric variability of the northern hemisphere warm season. Retrieval random errors are similarly higher over land than ocean, especially in the lower troposphere, owing to the higher variability over continental regions.

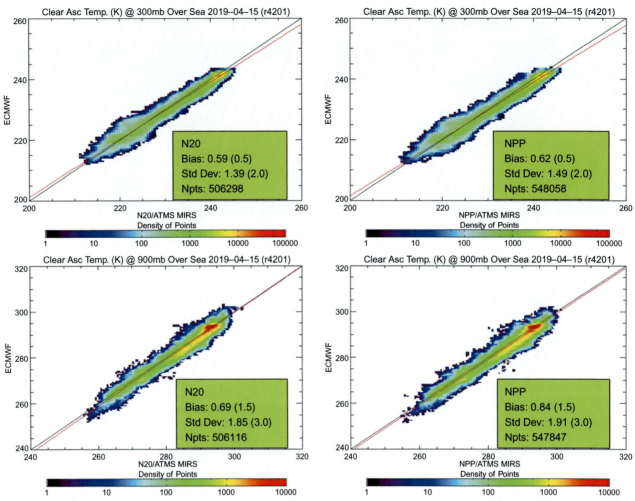

FIG. 9 Global scatterplot comparison of MiRS NOAA-20 *(left)* and S-NPP *(right)* retrieved temperature at 300 and 900 hPa with ECMWF analyses on April 15, 2019. Comparison is for clear scenes only. Statistics in parentheses are the JPSS program requirements for retrieval accuracy (bias) and precision (standard deviation).

4.3 Water vapor profile validation

Similar to temperature profile validation, a large proportion of water profile assessment and validation is typically done using comparisons with radiosondes, and with objective analyses from operational assimilation and forecasting systems. Fig. 12 shows results based on S-NPP and NOAA-20 collocations with global radiosondes, again stratified into four 10-day periods in different seasons. The bias and error standard deviation analyzed are the percentage of the reference (radiosonde) layer mixing ratio. This normalization is one way of accounting for the very large decrease with height in water vapor mixing ratio usually seen in the atmosphere. While the error standard deviation appears quite stable across seasons and similar between satellites (showing a general increase with height up to about 400 hPa), the bias does exhibit differences in different seasons and to a certain extent between satellites. The overall vertical structure is similar with relative minima near the surface and between 400 and 500 hPa. However, at higher levels, there are some differences between satellites with NOAA-20 retrievals showing smaller differences with the radiosondes than S-NPP. Another factor that can impact performance estimates is the quality of the sonde measurements themselves. For instance, the apparent dry bias at very high levels above 300 hPa may in fact be an artifact of a moist bias in the radiosonde measurements. As noted in Section 4.2, radiometric bias corrections can also impact the retrievals, particularly the biases seen with respect to reference data, as in the MiRS system the bias is applied statically and affects all measured radiances in the same way, regardless of location.

Use of global analyses from operational data assimilation systems allows assessment and validation of the spatial structure of sounding variables and their errors—something not possible with point-based profile measurements such as radiosondes. Fig. 13 shows an example of retrieved 700 hPa water vapor mixing ratio from both NOAA-20 and

Satellite microwave sounder product retrieval validation **Chapter | 17** 309

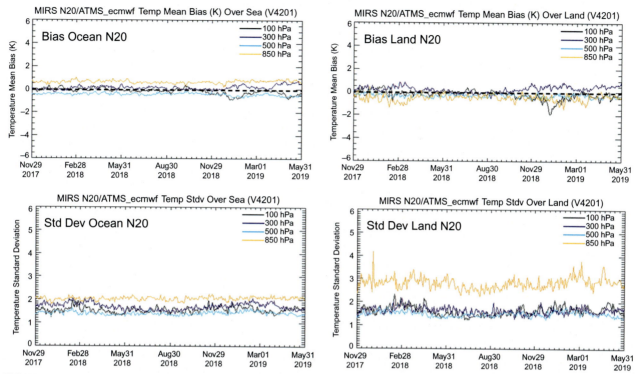

FIG. 10 Time series of MiRS NOAA-20 ATMS temperature validation statistics at selected pressure layers (850, 500, 300, and 100 hPa) with respect to ECMWF analyses for the period November 29, 2017—May 31, 2019. Bias *(top)* and standard deviation *(bottom)* are shown separately for ocean *(left)* and land *(right)*.

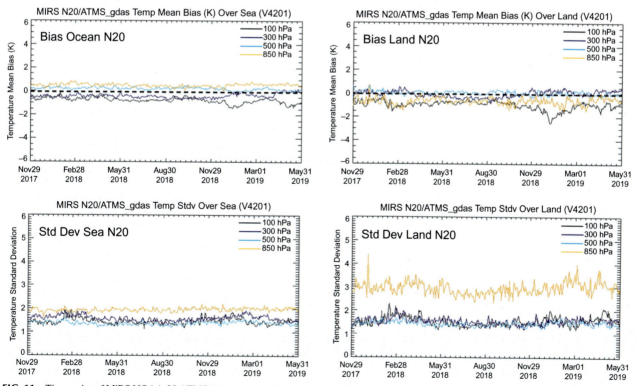

FIG. 11 Time series of MiRS NOAA-20 ATMS temperature validation statistics at selected pressure layers (850, 500, 300, and 100 hPa) with respect to GDAS analyses for the period November 29, 2017—May 31, 2019. Bias *(top)* and standard deviation *(bottom)* are shown separately for ocean *(left)* and land *(right)*.

310 PART | III Satellite applications

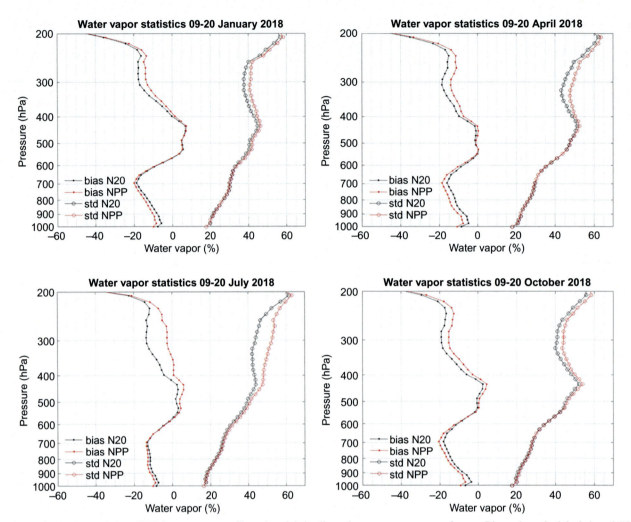

FIG. 12 Validation statistics of MiRS water vapor profiles using global radiosonde measurements as a reference. Bias and standard deviation of differences (as percentage of radiosonde layer mixing ratio) are presented for four 10-day periods in 2018: 9–20 January *(top left)*, 9–20 April *(top right)*, 9–20 July *(bottom left)*, and 9–20 October *(bottom right)*. Results from both S-NPP *(red)* and NOAA-20 *(black)* ATMS retrievals are shown.

S-NPP, along with difference maps using the ECMWF analysis as a reference. The difference maps can be quite useful in characterizing the retrieval behavior as a function of location and synoptic situation. For example, it is apparent that in the tropics there is a tendency for underestimation at 700 hPa with respect to ECMWF over tropical land surfaces. Over tropical oceans, the differences appear to be more evenly distributed between positive and negative values. In mid-latitudes, the differences seem to be correlated with synoptic features such as fronts and cyclones. If differences related to uncertainties in the reference data can be accounted for, these features may provide valuable information to algorithm developers, highlighting areas for improvement in the retrieval system.

Time series of water vapor retrieval performance allow users and developers to get a further understanding of the seasonal and interannual behavior of the soundings. Figs. 14 and 15 show time series of bias and error standard deviation with respect to ECMWF and GDAS analyses, respectively, during the period November 2017—May 2019. While the time-series patterns are quite similar with respect to both references, as with the temperature series in Section 4.2, there are some systematic differences in both the bias and error standard deviation, which highlight the importance of using multiple references to obtain a more complete view. Over land, biases appear at most levels to be slightly larger (more positive) with respect to GDAS than ECMWF analyses, implying that GDAS analyses over these regions are somewhat drier than those of ECMWF. The error standard deviations over ocean at 500 and 300 hPa are somewhat lower for the GDAS analyses than when ECMWF is used as a reference. That is to say that the random component of the retrieval error is smaller vs. GDAS than ECMWF analyses. Regardless of reference, the error standard deviations trend lower during Northern Hemisphere summer, likely a reflection of lower amounts of atmospheric variability during the warm season.

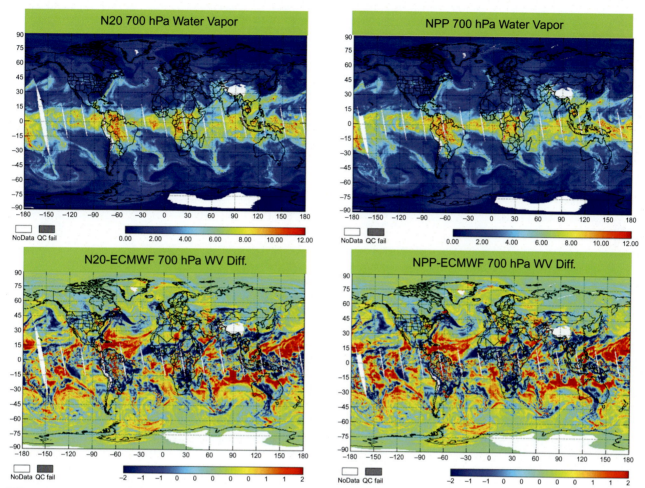

FIG. 13 Global comparison of MiRS NOAA-20 and S-NPP 700 hPa retrieved water vapor mixing ratio with ECMWF analysis on April 15, 2019. *Top row* shows the MiRS retrieval and *bottom row* shows the difference with ECMWF analysis.

4.4 Total precipitable water validation

The vertically integrated column water vapor, based on the retrieved water vapor profile, is often referred to as total precipitable water (TPW). This product is frequently used by operational forecasters and research scientists to gain situational awareness or quantify the intensity of possible significant weather events such as atmospheric rivers and associated heavy precipitation. Therefore, validation of this sounder-derived product is commonly performed in addition to, or instead of, water vapor profile validation.

Fig. 16 shows an example of a full day of MiRS NOAA-20 global TPW retrievals, as well as analyses from GDAS and ECMWF. Additionally, the corresponding difference maps with GDAS and ECMWF analyses are shown. As with the previous comparisons, the overall patterns of differences are quite similar. The retrievals generally capture the large-scale features seen in the TPW analyses of both GDAS and ECMWF. However, it can be seen that differences with ECMWF are systematically higher, by approximately 1–2 mm. This is true for both land and ocean surfaces, with the largest differences focused in the tropics where higher TPW values are typically found.

Using global gridded analyses as reference data allows more precise stratification of validation results, as shown in Figs. 17 and 18. Fig. 17 shows NOAA-20 TPW retrieval performance relative to GDAS analyses over ocean and sea ice, while Fig. 18 shows TPW results for land and snow-covered surfaces. Results are stratified by atmospheric conditions of clear and cloudy. For comparison, validation statistics also include in parentheses JPSS program requirements for bias and standard deviation. The results show that for all surface types the random error of TPW retrievals (standard deviation) increases under cloudy conditions.

Finally, as noted above, ground-based GPS measurements of TPW have been shown to be of high quality, and can be used as a reference for validation of PMW retrievals. Fig. 4 shows the global distribution of the SuomiNet observation

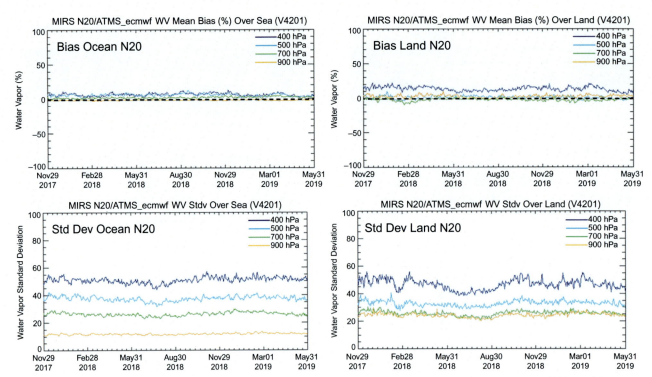

FIG. 14 Time series of MiRS NOAA-20 ATMS water vapor validation statistics at selected pressure layers (900, 700, 500, and 400 hPa) with respect to ECMWF analyses for the period November 29, 2017–May 31, 2019. Bias *(top)* and standard deviation *(bottom)* are shown separately for ocean *(left)* and land *(right)* and are expressed in percentage of the respective ECMWF layer mixing ratio.

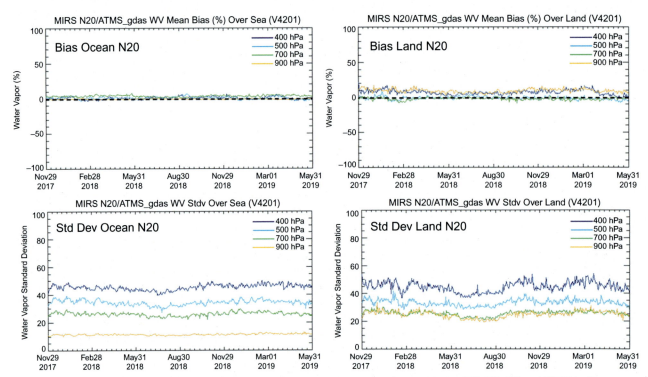

FIG. 15 Time series of MiRS NOAA-20 ATMS water vapor validation statistics at selected pressure layers (900, 700, 500, and 400 hPa) with respect to GDAS analyses for the period 29 November 2017–31 May 2019. Bias *(top)* and standard deviation *(bottom)* are shown separately for ocean *(left)* and land *(right)* and are expressed in percentage of the respective GDAS layer mixing ratio.

Satellite microwave sounder product retrieval validation **Chapter | 17** 313

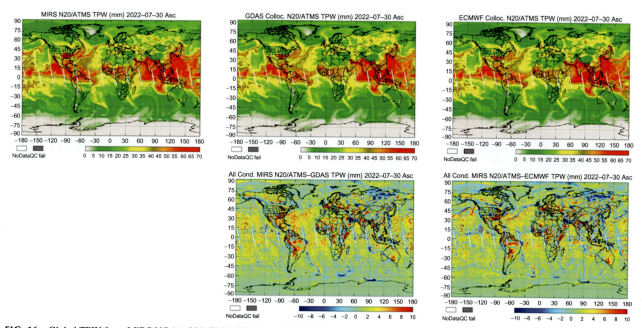

FIG. 16 Global TPW from MiRS NOAA-20/ATMS, GDAS, and ECMWF analyses *(top row)*, along with corresponding retrieval-analysis differences *(bottom row)*. Example is from 30 July 2022.

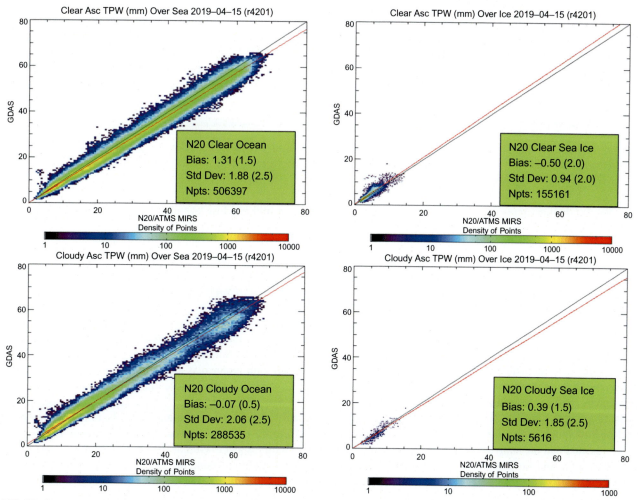

FIG. 17 Scatterplots of MiRS NOAA-20/ATMS TPW vs. GDAS analysis on 15 April 2019 for scenes classified as ocean *(left)* or sea ice *(right)*. Results are stratified into clear *(top)* and cloudy *(bottom)*. Statistics shown in parentheses are the JPSS program performance requirements for accuracy (bias) and precision (standard deviation).

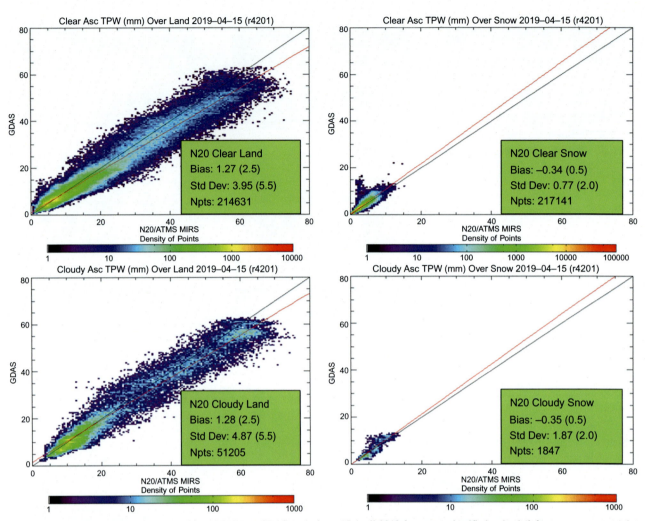

FIG. 18 Scatterplots of MiRS NOAA-20/ATMS TPW vs. GDAS analysis on 15 April 2019 for scenes classified as land *(left)* or snow cover *(right)*. Results are stratified into clear *(top)* and cloudy *(bottom)*. Statistics shown in parentheses are the JPSS program performance requirements for accuracy (bias) and precision (standard deviation).

network. While most reporting stations are in continental locations at mid and low latitudes, a small number are located along coasts and on islands, which may allow for characterization of oceanic retrievals if enough data are collected.

Fig. 19 shows results from a validation study of MiRS NOAA-20 TPW retrievals during 2019 (Lee et al., 2022). Both the mean (bias) and standard deviation of differences with respect to ground-based GPS TPW measurements were computed based on collocations, which were aggregated each month (green and cyan curves). Collocation with the GPS data used a time window of ±30 min and a maximum distance of 30 km. Since the GPS data are generally available every 30 min, this means in practice that matchups were usually no more than 15 min from the nearest in time GPS measurement. For comparison, ECMWF 0.25-degree analyses were also matched directly to the time and location of the GPS measurements (black curves). The results show that the MiRS retrieval bias is generally twice as large as that of the ECMWF analyzed values (both are always positive), and that the satellite estimated TPW error standard deviation is approximately 1.5 to 2 mm larger than the analysis. A distinct seasonal dependence is seen with larger bias and standard deviation occurring during the May to September period. As seen in Fig. 18, the GPS measurements are preferentially located in the northern hemisphere, and so the statistics reflect the seasonal increase in average TPW during the warm season at these locations, indicating that the uncertainty of the TPW estimates is proportional to the TPW magnitude itself. As noted above, it is unclear how much of the differences seen are due to the retrieval and how much are related to the reference itself. It is to be expected that the analysis values will have smaller differences with respect to GPS as they benefit from a wide range of conventional and satellite observations and are constrained in space and time using the physics of the forecast model. Another reason that the satellite retrievals will exhibit higher random uncertainty than the ECMWF analysis is related to the

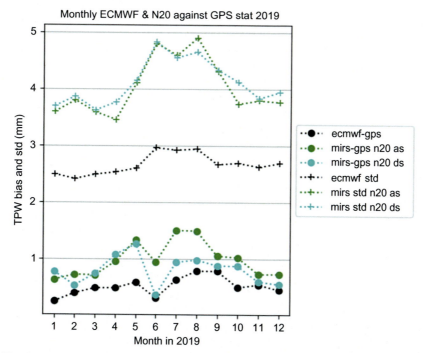

FIG. 19 Time series in 2019 of monthly TPW bias *(filled circles)* and difference standard deviation *(plus symbols)* of MiRS NOAA-20 retrieved TPW with respect to ground-based GPS TPW *(green and cyan curves)* and ECMWF analyses TPW *(black curves)*. MiRS statistics are separated by ascending and descending orbits. Sample sizes for the satellite collocations are approximately 10,000 per month and about 65,000 per month for the direct ECMWF to GPS collocations. *(Original figure is from Lee, Y.-K., Grassotti, C., Liu, Q., Liu, S., Zhou, Y., 2022. In depth evaluation of MiRS TPW from S-NPP and NOAA-20 ATMS. Earth Space Sci. 9 (2). https://doi.org/10.1029/2021EA002042.)*

respective spatial resolutions and water vapor spatial scales of variability. The PMW retrievals have an effective resolution of roughly 25 to 100 km (depending on viewing geometry) compared with the 0.25-degree resolution of the ECMWF analysis. The lower spatial resolution of PMW retrievals will always be associated with a representativeness error when compared with higher resolution measurements of parameters such as water vapor, which are characterized by significant variability on smaller subpixel scales.

5. Summary

We have summarized the primary issues and approaches to validation of PMW sounding products, defined here as profiles of temperature and water vapor, as well as total vertically integrated water vapor, or total precipitable water. The reference data sets most widely used include in situ profiles obtained from radiosonde measurements, global objective analyses from major operational weather forecast centers, such as ECMWF and GDAS analyses, and regular ground-based remote sensing measurements, such as GPS-based estimates (e.g., SuomiNet). Other satellite-based measurements may also play a role, such as GNSS-RO profiles of temperature and moisture, which have been shown to be of high quality. While these latter measurements were not considered here, they may be quite useful in validating and assessing retrieval performance on seasonal to interannual timescales, and for detecting possible time-dependent changes in performance. As more GNSS-RO satellite missions become operational, the issues related to spatial coverage will diminish, and we expect that these measurements will become more attractive as a reference for a range of PMW sounder validation activities. Advantages and disadvantages of the various validation references were discussed, along with the importance of considering uncertainties in the reference data themselves. A major conclusion is that an approach that considers multiple reference data sets is preferred, as it can provide higher levels of confidence in the sounding performance estimates. Additionally, multiple references can help quantify how much of the difference between the satellite soundings is true retrieval error, how much is due to uncertainties and biases in the reference data, and, importantly, what proportion is due to time and space collocation differences. Similarly, while difficult to estimate, representativeness error should also be considered, particularly in comparisons between areal average and point measurements (e.g., radiosondes).

Acknowledgments

This work was supported by NOAA under grant NA19NES4320002 to the Cooperative Institute for Satellite and Earth System Studies (CISESS) at the University of Maryland/Earth System Science Interdisciplinary Center (ESSIC). The scientific results and conclusions, as well as any views or opinions expressed herein, are those of the author(s) and do not necessarily reflect those of NOAA or the Department of Commerce. We also thank Dr. Robbie Iacovazzi (NESDIS/STAR) for reviewing the manuscript and providing very helpful comments that improved the overall quality.

References

Anthes, R.A., 2011. Exploring Earth's atmosphere with radio occultation: contributions to weather, climate and space weather. Atmos. Meas. Tech. 4, 1077–1103. https://doi.org/10.5194/amt-4-1077-2011.

Blackwell, W.J., Braun, S., Bennartz, R., et al., 2018. An overview of the TROPICS NASA earth venture mission. Q. J. Roy. Meteorol. Soc. 144 (Suppl. 1), 16–26. https://doi.org/10.1002/qj.3290.

Boukabara, S.A., Garrett, K., Chen, W., Iturbide-Sanchez, F., Grassotti, C., Kongoli, C., Chen, R., Liu, Q., Yan, B., Weng, F., Ferraro, R., Kleespies, T.J., Meng, H., 2011. MiRS: an all-weather 1DVAR satellite data assimilation and retrieval system. IEEE Trans. Geosci. Remote Sens. 49 (9), 3249–3272. https://doi.org/10.1109/tgrs.2011.2158438.

Boukabara, S.A., Garrett, K., Grassotti, C., 2018. Dynamic inversion of global surface microwave emissivity using a 1DVAR approach. Remote Sens. (Basel) 10 (5), 679–696. https://doi.org/10.3390/rs10050679.

Boukabara, S.A., Garrett, K., Grassotti, C., Iturbide-Sanchez, F., Chen, W., Jiang, Z., Clough, S.A., Zhan, X., Liang, P., Liu, Q., Islam, T., Zubko, V., Mims, A., 2013. A physical approach for a simultaneous retrieval of sounding, surface, hydrometeor, and cryospheric parameters from S-NPP/ATMS. J. Geophys. Res. Atmos. 118 (22), 12600–12619. https://doi.org/10.1002/2013jd020448.

Ding, S., Yang, P., Weng, F., Liu, Q., Han, Y., Van Delst, P., Li, J., Baum, B., 2011. Validation of the community radiative transfer model. J. Quant. Spectrosc. Radiat. Transf. 112 (6), 1050–1064. https://doi.org/10.1016/j.jqsrt.2010.11.009.

Dirksen, R.J., Sommer, M., Immler, F.J., Hurst, D.F., Kivi, R., Vömel, H., 2014. Reference quality upper-air measurements: GRUAN data processing for the Vaisala RS92 radiosonde. Atmos. Meas. Tech. 7, 4463–4490. https://doi.org/10.5194/amt-7-4463-2014.

Goldberg, M.D., Kilcoyne, H., Cikanek, H., Mehta, A., 2013. Joint polar satellite system: the United States next generation civilian polar-orbiting environmental satellite system. J. Geophys. Res. Atmos. 118, 13,463–13,475. https://doi.org/10.1002/2013JD020389.

Grassotti, C., Liu, S., Liu, Q., Boukabara, S.-A., Garrett, K., Iturbide-Sanchez, F., Honeyager, R., 2020. Precipitation estimation from the microwave integrated retrieval system (MiRS). In: Satellite Precipitation Measurement. vol. 1, pp. 153–168, https://doi.org/10.1007/978-3-030-24568-9_9.

Han, Y., Van Delst, P., Liu, Q., Weng, F., Yan, B., Treadon, R., Derber, J., 2006. Community Radiative Transfer Model (CRTM),—Version 1. NOAA Technical Report 122, 33 pp.

Hersbach, H., Bell, B., Berrisford, P., et al., 2020. The ERA5 global reanalysis. Q. J. Roy. Meteorol. Soc. 146, 1999–2049. https://doi.org/10.1002/qj.3803.

Ho, S., et al., 2020. The COSMIC/FORMOSAT-3 radio occultation mission after 12 years: accomplishments, remaining challenges, and potential impacts of COSMIC-2. Bull. Am. Meteorol. Soc. 101 (7), E1107–E1136. https://doi.org/10.1175/BAMS-D-18-0290.1.

Hoffman, R.N., Kumar, V.K., Boukabara, S., Ide, K., Yang, F., Atlas, R., 2018. Progress in forecast skill at three leading global operational NWP centers during 2015–17 as seen in summary assessment metrics (SAMs). Weather Forecast. 33 (6), 1661–1679. https://doi.org/10.1175/WAF-D-18-0117.1.

Iturbide-Sanchez, F., Boukabara, S.-A., Chen, R., Garrett, K., Grassotti, C., Chen, W., Weng, F., 2011. Assessment of a variational inversion system for rainfall rate over land and water surfaces. IEEE Trans. Geosci. Remote Sens. 49 (9), 3311–3333. https://doi.org/10.1109/TGRS.2011.2119375.

Klaes, K.D., Cohen, M., Buhler, Y., Schlüssel, P., Munro, R., Luntama, J., von Engeln, A., Clérigh, E.Ó., Bonekamp, H., Ackermann, J., Schmetz, J., 2007. An introduction to the EUMETSAT polar system. Bull. Am. Meteorol. Soc. 88 (7), 1085–1096. https://doi.org/10.1175/BAMS-88-7-1085.

Kunkee, D., Boucher, D., Poe, G., Swadley, S., 2006. Evaluation of the defense meteorological satellite program (DMSP) special sensor microwave imager sounder (SSMIS). In: 2006 IEEE International Symposium on Geoscience and Remote Sensing, pp. 101–104, https://doi.org/10.1109/IGARSS.2006.31.

Lee, Y.-K., Grassotti, C., Liu, Q., Liu, S., Zhou, Y., 2022. In depth evaluation of MiRS TPW from S-NPP and NOAA-20 ATMS. Earth Space Sci. 9 (2). https://doi.org/10.1029/2021EA002042.

Li, J.Z., Qin, G.L., 2016. A new generation of Chinese FY-3C microwave sounding measurements and the initial assessments of its observations. Int. J. Remote Sens. 37, 4035–4058.

Liu, S., Grassotti, C., Chen, J., Liu, Q., 2017. GPM products from the microwave-integrated retrieval system. IEEE J. Sel. Top. Appl. Earth Obs. Remote Sens. 10 (6), 2565–2574. https://doi.org/10.1109/JSTARS.2017.2716356.

Liu, S., Grassotti, C., Liu, Q., Lee, Y.-K., Honeyager, R., Zhou, Y., Fang, M., 2020. The NOAA microwave integrated retrieval system (MiRS): validation of precipitation from multiple polar orbiting satellites. IEEE J. Sel. Top. Appl. Earth Obs. Remote Sens. 13 (6), 3019–3031. https://doi.org/10.1109/JSTARS.2020.3000348.

Nalli, N.R., et al., 2013. Validation of satellite sounder environmental data records: application to the cross-track infrared microwave sounder suite. J. Geophys. Res. Atmos. 118, 13,628–13,643. https://doi.org/10.1002/2013JD020436.

Rodgers, C.D., 2000. Inverse Methods for Atmospheric Sounding: Theory and Practice. World Scientific Publishing Company, Incorporated, London, https://doi.org/10.1142/3171.

Schlüssel, P., Kayal, G., 2017. Introduction to the next generation EUMETSAT polar system (EPS-SG) observation missions. In: Proc. SPIE 10423, Sensors, Systems, and Next-Generation Satellites XXI, 104230G, 29 September 2017., https://doi.org/10.1117/12.2282181.

Sun, B., Calbet, X., Reale, A., Schroeder, S., Bali, M., Smith, R., Pettey, M., 2021. Accuracy of Vaisala RS41 and RS92 upper tropospheric humidity compared to satellite hyperspectral infrared measurements. Remote Sens. (Basel) 13, 173. https://doi.org/10.3390/rs13020173.

Sun, B., Reale, T., Schroeder, S., Pettey, M., Smith, R., 2019. On the accuracy of Vaisala RS41 versus RS92 upper-air temperature observations. J. Atmos. Oceanic Tech. 36 (4), 635–653. https://doi.org/10.1175/JTECH-D-18-0081.1.

Vaisala, 2017. Vaisala RS41-SG Technical Data Specifications Sheet. https://www.vaisala.com/sites/default/files/documents/WEA-MET-RS41-Datasheet-B211321EN.pdf. Retrieved from Web on 5 October 2021.

Vaquero-Martinez, J., Anton, M., 2021. Review on the role of GNSS meteorology in monitoring water vapor for atmospheric physics. Remote Sens. (Basel) 13, 2287. https://doi.org/10.3390/rs13122287.

Ware, R.H., et al., 2000. Suominet: a real-time national GPS network for atmospheric research and education. Bull. Am. Meteorol. Soc. 81, 677–694.

Weng, F., 2018. Passive Microwave Remote Sensing of the Earth for Meteorological Applications. Wiley-VCH, Weinheim, Germany.

Further reading

Kim, E., Lyu, C.-H.J., Anderson, K., Leslie, R.V., Blackwell, W.J., 2014. S-NPP ATMS instrument prelaunch and on-orbit performance evaluation. J. Geophys. Res. Atmos. 119, 5653–5670. https://doi.org/10.1002/2013JD020483.

Weng, F., Zou, X., Zou, X., Sun, N., Yang, H., Tian, M., Blackwell, W.J., Wang, X., Lin, L., Anderson, K., 2013. Calibration of Suomi national polar-orbiting partnership advanced technology microwave sounder. J. Geophys. Res. Atmos. 118, 187–200. https://doi.org/10.1002/jgrd.50840.

Chapter 18

Considerations for thermal satellite infrared sounder applications

Nicholas R. Nalli[a,b], Nadia Smith[c], Rebekah Esmaili[c], Bomin Sun[a,b], Anthony Reale[b], and Christopher D. Barnet[c]

[a]*I.M. Systems Group, Inc., College Park, MD, United States,* [b]*NOAA/NESDIS Center for Satellite Applications and Research (STAR), College Park, MD, United States,* [c]*Science and Technology Corp. (STC), Columbia, MD, United States*

Chapter outline

1. Satellite thermal infrared remote sounding	319	4. Applications	329
2. Algorithm development	322	5. Summary and outlook	332
3. Algorithm validation	323	Acknowledgments	332
3.1 Objectives	323	References	332
3.2 Strategies and methodologies	323		
3.3 Validation highlights using field data	327		

Because much of our planet is ocean, satellite sensing of the atmospheric state would appear to be crucial…

Prof. William L. Smith

1. Satellite thermal infrared remote sounding

In satellite remote sensing parlance, the terminology *satellite sounder* is a descriptive term referring to any instrument designed to *sound* (i.e., measure the profile of) the atmosphere from space (analogous to radiosondes; cf. Chapter 2), the original terminology harkening back to traditional ocean sounding methods (see Box 1). However, unlike balloon-borne in situ soundings from radiosondes passive thermal infrared (IR) and microwave (MW) atmospheric sounding involves obtaining indirect estimates of vertical profiles of state parameters from spectral radiance measurements and inversion of the radiative transfer equation (RTE). State parameters of interest include temperature and water vapor as a function of pressure, p, that is, $T(p)$ and $Q(p)$, respectively. Radiosondes directly measure temperature and relative humidity (RH) as a function of p, these being essential for global numerical weather prediction (NWP). Passive sensors designed for atmospheric sounding have sensitivity to upwelling Earth-emitted radiance in multiple spectral channels, each possessing varying degrees of interaction with the intervening atmosphere. For a cloud-free atmosphere, the thermal IR Earth emission spectrum (i.e., wavenumbers 600–3000 cm^{-1}) is particularly well suited for this purpose, given the relatively high spectral information content. Given spectral IR observations, algorithms are derived to "retrieve" the parameters of interest (e.g., T and Q) by solving the RTE.

The first operational satellite passive sounders consisted of multispectral filter radiometers featuring on the order of 20 narrowband channels with weighting functions (cf. Chapter 1) peaking at various pressures throughout the troposphere (e.g., Smith, 1991). These early sounding systems included the Advanced TIROS Operational Vertical Sounder (ATOVS), consisting of the High-Resolution Infrared Radiation Sounder (HIRS), flown on low-Earth orbit (LEO) satellites (Li et al., 2000), namely the NOAA Polar Operational Environmental Satellites (POES) and EUMETSAT Metop series, as well as geostationary Earth orbit (GEO) satellites (e.g., the NOAA Geostationary Operational Environmental Satellite Sounder). These IR sounders initially showed promise for NWP, but ultimately their spectral resolution was too coarse for obtaining sufficient vertical resolution throughout the atmospheric column (Smith, 1991). Thus, narrowband IR sounders have largely been superseded by the newer generation of *hyperspectral* sounders (Smith et al., 2009), beginning in 2002 with the US NASA Atmospheric IR Sounder (AIRS) onboard *Aqua* (Chahine et al., 2006), followed by the EUMETSAT Infrared

> **BOX 1 On the origin of the term "soundings."**
>
> In atmospheric science, a *sounding* is a vertical profile measurement of the atmosphere (i.e., multiple measurements taken vertically through some portion of the column). The measurements in question typically include pressure, temperature, and moisture, but they can also include trace gases and winds. Radiosondes get their name from the fact that they *sound* the atmosphere using in situ sensors mounted to a balloon, then radioing the measured data back to a ground station (cf. Chapter 2). Likewise, atmospheric vertical profile measurements obtained from satellite passive instruments designed for this purpose are referred to as "soundings," and the instruments themselves, while fundamentally being spectrometers or radiometers, are thus referred to as *sounders*.
>
> At first glance, one might reasonably presume that these terminologies originated from sonar *echo-sounding* of the ocean, especially given that sonar is an earlier remote sensing technology based on sound waves. Interestingly enough, however, this turns out to be a misconception. In actuality, the term "sounding" dates well back before sonar to antiquity, when seafarers would use what were called "sounding weights" (basically weights attached to a length of line that would be lowered into the water) to determine the depth and bottom topography of a body of water (Oleson, 2008). One example of this premodern usage can be found in the New Testament Book of Acts (c. CE 62):
>
> *On the fourteenth night we were still being driven across the Adriatic Sea, when about midnight the sailors sensed they were approaching land. They took soundings and found that the water was a hundred and twenty feet deep. A short time later they took soundings again and found it was ninety feet deep.*
>
> Acts 27:27–28
>
> The original terminology apparently refers to the geographical term *sound*, namely an ocean inlet or bay used for marine access (e.g., Puget Sound). Thus, it would appear to be only a happy coincidence that this same terminology would be perfectly suitable when sonar was later invented and used to sound water bodies. Nevertheless, the adaptation of the "sounding" terminology for satellite remote sensing is undoubtedly better associated with the sonar analogue than it is with "sounding weights."

Atmospheric Sounding Interferometer (IASI) onboard Metop satellites, and most recently the US Joint Polar Satellite System (JPSS) Cross-track Infrared Sounder (CrIS) onboard the Suomi National Polar-orbiting Partnership (SNPP) and NOAA-20 satellites. Unlike the earlier operational sounders, advanced IR sounders obtain well-calibrated, high-resolution spectral measurements, with thousands of channels across the IR spectrum, AIRS being a *grating spectrometer*, and IASI, and CrIS being *Fourier transform spectrometers* (FTS) (e.g., Revercomb et al., 1988). Example brightness temperature spectra from AIRS, IASI and CrIS (obtained by taking the inverse Planck function of the radiances; cf. Chapter 1) are illustrated in Fig. 1. For reference Fig. 1 also includes the bandpasses of narrowband radiometer systems, including HIRS along with VIS/IR imagers MODIS, AVHRR, and VIIRS. Of the three sounder systems, only IASI (middle plot) provides spectrally continuous coverage of the thermal IR spectrum.

The nearly global, twice-daily radiance observations obtained from these advanced sounders have had a high positive impact on NWP via direct assimilation of the sensor data records (SDRs) (i.e., Level 1B/1C calibrated spectral radiances) within variational analysis schemes (Le Marshall et al., 2008; Noh et al., 2020). However, inversion schemes continue to be used to retrieve environmental data records (EDRs) (i.e., Level 2 products) within operational retrieval algorithms, for example, the NOAA-Unique Combined Atmospheric Processing System (NUCAPS) (Barnet et al., 2021; Susskind et al., 2003) and the Community Long-Term Infrared Microwave Combined Atmospheric Product System (CLIMCAPS) (Smith and Barnet, 2019). To address the problem of global cloud cover, the IR data are typically supplemented with passive MW data obtained from a collocated MW sounder, for example, the Advanced Technology Microwave Sounder (ATMS) onboard SNPP and the JPSS series. The combined hyperspectral IR + MW data are used within a cloud-clearing algorithm to derive clear-column radiance spectra in the presence of clouds (Susskind et al., 2003). NUCAPS then inverts the cloud-cleared IR radiances (via optimal estimation) (Smith and Barnet, 2019; Susskind et al., 2003) using different channel sets (Gambacorta and Barnet, 2013) sequentially to retrieve atmospheric profiles of temperature (T), water vapor (H_2O), ozone (O_3), carbon monoxide (CO) (e.g., Gambacorta et al., 2014; Warner et al., 2013), methane (CH_4) (e.g., Xiong et al., 2013), and carbon dioxide (CO_2) (e.g., Maddy et al., 2008). Fig. 2 shows a schematic of the JPSS CrIS/ATMS satellite sounding system used for retrieving these profile EDRs under global clear to partly-cloudy conditions; note that similar setups are used for both the NASA AIRS/AMSU and the Metop IASI/AMSU systems.

While atmospheric vertical temperature and moisture profile EDRs retrieved by NUCAPS are operationally used by NOAA National Weather Service (NOAA/NWS) weather forecast offices (WFOs) within the Advanced Weather Interactive Processing System (AWIPS), validation of the retrieved EDRs also indirectly facilitates the routine monitoring of the SDRs themselves (e.g., the NOAA integrated calibration/validation system, https://www.star.nesdis.noaa.gov/icvs/index.php).

Considerations for thermal satellite infrared sounder applications **Chapter | 18** 321

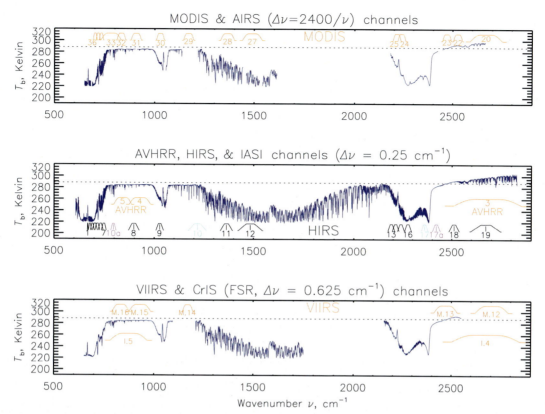

FIG. 1 Thermal infrared (IR) Earth emission spectra (*blue lines*) rendered as brightness temperatures (K) via the inverse Planck function: (*top*) NASA Aqua AIRS, (*middle*) EUMETSAT Metop IASI, and (*bottom*) JPSS CrIS. Bandpasses for narrowband imagers are shown in all three plots as *orange-red*, and the HIRS sounder is shown with different colors along the bottom axis of the center plot.

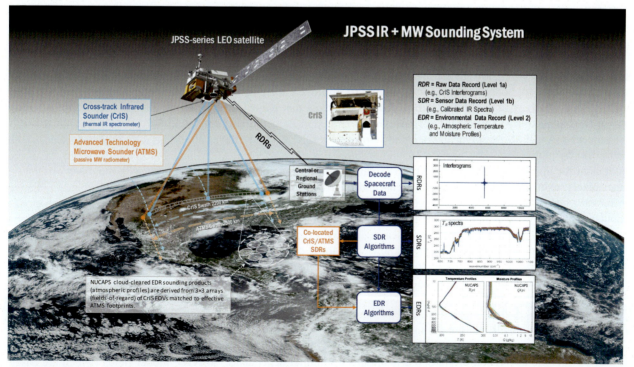

FIG. 2 Combined IR + MW satellite sounding system schematic. "Whisk broom" scans of the Earth's surface (cf. Chapter 1) allow collocation of the IR and MW fields of view (FOVs) such that the MW data can provide information below clouds, which facilitates "cloud clearing" of the IR spectra. In the case of an FTS instrument (e.g., the JPSS CrIS and Metop IASI), the "raw data records" are the interferograms obtained from the Michelson interferometer (cf. Chapters 6 and 9). These are processed into the "sensor data records" (i.e., geolocated radiances) that are used within inversion schemes to produce the "environmental data records" (i.e., retrievals).

2. Algorithm development

NUCAPS is based on the AIRS Science Team Version 5 algorithm (Strow et al., 2003; Susskind et al., 2003, 2011) operational at NOAA for CrIS/ATMS on Suomi-NPP and NOAA-20. Through an agreement with EUMETSAT, NOAA also runs NUCAPS on IASI/AMSU measurements from Metop-A, -B, and -C. At NOAA, an algorithm is considered "operational" when it runs continually on satellite measurements as they are received and transmitted via ground stations. Being operational, NUCAPS soundings are globally available from multiple satellite platforms throughout the diurnal cycle for every instrument field of view (FOV). This not only makes it relevant to a wide range of end users but also imposes algorithm design constraints because an operational algorithm needs to fulfill mission requirements (e.g., Nalli et al., 2013; Zhou et al., 2019). In this section, we discuss how NOAA requirements have informed the design and development of the NUCAPS algorithm.

The NUCAPS algorithm is a series of mathematical methods and sequential steps developed to solve an inverse problem that is ill-composed and under determined. In practice, this means that a given radiance measurement cannot be inverted into a unique set of atmospheric variables about the true state at the time of measurement. Instead, being an under determined problem, there are multiple solutions to radiance inversion that each approximate to a possible state. NUCAPS retrieves a set of atmospheric state variables from a given set of spectral radiance measurements in a manner that satisfies NOAA requirements, including that soundings be available globally, day and night, in clear and cloudy conditions with a latency that is low enough to meet NWS forecaster data needs. Retrieval algorithms, therefore, have purpose in their design and are typically developed to satisfy target applications. Nevertheless, although there is no single "best" way to perform a retrieval—inversion algorithms can be tailored for different applications based on the same radiance measurements—the solutions should still approximate the true state and fall within the range of possible solutions for a given set of measurements.

The amount of information IR measurements contain about the atmospheric state depends on the ambient conditions at the time of measurement (Smith and Barnet, 2020). A satellite IR measurement made during a cloudy Polar winter night over a snow-covered area has lower information content compared to a measurement made on a clear day over the Tropical ocean. At its core, NUCAPS is an iterative, optimal estimation (OE) (Rodgers, 2000), *physical retrieval algorithm* (cf. Chapter 1). To achieve daily global coverage and retrieve soundings from spectral measurements with varying information content (e.g., Maddy et al., 2009), NUCAPS deviates from the traditional OE approach (which typically solves for multiple parameters simultaneously) in a number of important ways. First, NUCAPS achieves stability in the signal-to-noise ratio (SNR) of its retrievals by quantifying and propagating all known sources of uncertainty at a target retrieval scene, including uncertainty caused by clouds. Second, NUCAPS adopted the AIRS science team method that separates signal from noise in the IR measurements using singular value decomposition (Smith and Barnet, 2019, 2020; Susskind et al., 2003). This ensures that the retrievals minimize their dependence on background assumptions of information content and dynamically adjusts the amount of the measurement to believe at run-time, maximizing it when measurement SNR is high, and minimizing it when SNR is very low. In this manner, the retrieval will depend more on either the first-guess or a priori estimate only where it needs to, both vertically and spatially. To improve global stability, NUCAPS retrieves its set of sounding variables sequentially, one at a time, from subsets of channels that are selected based on their sensitivity to the target variable and insensitivity to all other variables (Gambacorta and Barnet, 2013).

To meet NOAA/JPSS requirements for cloudy atmospheres, NUCAPS retrieves atmospheric soundings in scenes with cloud cover as high as 90% by using a technique known as *cloud-clearing* (Chahine, 1977; Smith, 1968). With the requirement for low latency soundings, NUCAPS cannot afford to colocate each IR footprint with hundreds of satellite imager pixels, gain an estimate of cloud optical- and microphysical properties, and then run complex cloudy radiative transfer calculations multiple times for each retrieval variable. Not only are radiative transfer calculations through clouds computationally expensive, but they are also inconsistent in their accuracy of cloudy atmospheres across the globe (Aumann et al., 2018). The NUCAPS cloud-clearing removes the radiative effects of clouds from each IR measurement without the requirement for scene-dependent knowledge about cloud properties or their impact on spectral information content. Cloud-clearing, instead, uses the spatial information content from each 3×3 cluster of IR measurements to derive a single cloud-free IR spectrum. While cloud-clearing imposes a courser spatial resolution on the radiance measurements, it allows NUCAPS to maintain low latency, stable retrievals in complex cloudy scenes, and achieve a global retrieval yield of 75% on most days. A cloud-cleared NUCAPS sounding can be interpreted as a representation of the clear atmospheric state surrounding the clouds, but not through them.

With NWS forecasters as primary end users, NUCAPS algorithm development goals evolve with their data needs. The current operational version of NUCAPS maintains a minimal dependence on NWP model fields (which are based on both satellite and field measurements) because forecasters require model-independent observations for verification

(Smith et al., 2018a, b; Wheeler et al., 2018). NUCAPS achieves "model-independence" by using linear principal component analysis-based regression retrievals (Goldberg et al., 2003), an early form of machine learning, as the first-guess estimates in its OE inversion. As more forecasters voice their need for satellite soundings that more effectively depict where and how satellite measurements deviate from NWP models, the regression retrieval first guess can easily be replaced with model estimates. Any such algorithm change can be implemented if forecasters reach consensus and make a formal request. In this way, NUCAPS is a user-driven product. Characterization of the NUCAPS algorithm global profile EDR uncertainties is attained via validation with respect to independent in situ field measurements, discussed more below.

3. Algorithm validation

Validation of satellite sounder observations, both the SDRs (Level 1 radiances) and EDRs (Level 2 retrievals), involves "the process of ascribing uncertainties...through comparison with correlative observations" (Fetzer et al., 2003). As with any assessment exercise, the validation task is typically performed with respect to some sort of established criteria that serves as the metric for success.

3.1 Objectives

IR sounder accuracy specifications were first established for the AIRS/AMSU/HSB system for both the radiances (Level 1B) and retrieved (Level 2) products. These were later refined by the JPSS Cal/Val Program, which has more specifically defined Performance Requirements for the IR sounder profile EDRs on broad atmospheric layers (i.e., dividing the troposphere and lower stratosphere into three to five layers), including T, Q, and O_3, along with carbon trace gases CO, CH_4, and CO_2. It is worth noting here that program requirements are defined based on the reasonable expectations of the system as determined in simulations and from experience (e.g., from the legacy ATOVS system), which were based in part on comparisons with field measurements. It should go without saying that it makes little sense to define requirements that would be unattainable in practice.

As mentioned above, the JPSS program requirements are defined for global, nonprecipitating cases on three to five atmospheric "broad-layers" that are computed as an average of 1–5 km vertical "coarse-layers" for temperature and 2 km for moisture. "Partly Cloudy" conditions are defined by successful cloud-clearing and IR retrieval, whereas "Cloudy" conditions are defined by cases where cloud clearing was not successful, thereby resulting in a MW-only algorithm solution as the final product. Tables defining these metrics by which the EDR products are considered to have reached validated maturity and have met requirements may be found in studied by Nalli et al. (2018a, b, 2020). The requirements for carbon trace gases are defined for total column quantities. These broad-layer and total column requirements are primarily due to the fact that passive thermal IR sounders (e.g., CrIS, AIRS, IASI) have limited vertical sensitivity, with overlapping weighting functions (cf. Chapter 1) spanning finite layers of the atmosphere.

A comment is warranted on the utility and limitations of establishing "global requirements." By defining requirements "globally," we are then charged with specifying the uncertainty characteristics of the system for the global population of environmental cases, which is in principle the basic task of validation. The validation then provides users with a single, simple-to-understand metric on the expected performance of the system on average. However, in practice, IR sounding products often have their greatest utility for conditions falling outside the expected distribution (e.g., extreme weather). Thus, while it remains necessary to provide quantitative estimates of global uncertainty through the process of formal validation, it is also important to assess the performance of the system under various specific atmospheric conditions of interest to users. Assessing the performance under a wide range of conditions also provides diagnostic information to improve the retrieval algorithms under those conditions.

3.2 Strategies and methodologies

As mentioned earlier, sounder validation involves the process of empirically assigning uncertainties to a given measurement system. This essentially involves performing quantitative statistical analyses versus well-established, independent correlative measurements (specifically, in the case of passive sounders, atmospheric profiles of the parameters of interest) that are considered to be accurate representations of the true state. Statistical analyses versus reference datasets enable the determination of systematic (bias), random (variability), and total combined errors.

On a practical level, operational validation is an ongoing process that continues throughout the lifetime of a satellite instrument, as the sensor performance changes, and algorithms are refined and upgraded over time, as in the case of the JPSS cal/val program (Zhou et al., 2016). Furthermore, while it may be conceptually simple in principle, global characterization of a satellite system is not necessarily a straightforward task given the limitations inherent in the available established reference data, including field measurements. To contend with these real-world considerations, "hierarchal" approaches have been devised based roughly on the relative strengths of the datasets used (Barnet, 2009; Nalli et al., 2013). Techniques/datasets at the beginning/bottom of the hierarchy are those typically used at the early stages of EDR validation (i.e., early in the satellite mission), with those at the top usually employed at later maturity stages.

3.2.1 Temperature, moisture, and ozone hierarchy

The hierarchal methodology for validation of satellite sounder retrieved atmospheric temperature (T), moisture (Q), and ozone (O_3) profile EDRs was derived from years of experience with the NASA AIRS sounder (Barnet, 2009; Fetzer et al., 2003; Nalli et al., 2013), and may be summarized as follows.

1. **NWP model global comparisons.** NWP model analysis/forecast outputs (e.g., from ECMWF and/or NCEP) allow for large, truly global samples acquired from "Focus Days"; they are useful for sanity checks, bias tuning, and regression, but are not fully independent reference data because satellite data (in addition to field data) are assimilated. But more importantly, models have become essential for performing "double-differences" (i.e., taking the difference between two differences; e.g., to compare errors from two different satellite platforms) and can serve as a "transfer-standard" for analyzing retrievals from different platforms.
2. **Satellite EDR intercomparisons.** As with NWP models, intercomparisons with other satellite sounder EDRs (e.g., CrIS, IASI, AIRS, COSMIC) allow for global samples acquired from Focus Days; however, other sensors may have similar error characteristics, and a rigorous assessment would need to account for the averaging kernels (vertical sensitivities) of both systems (e.g., Rodgers and Connor, 2003). For satellites in different LEO (polar) orbits, the technique of simultaneous nadir-overpasses (SNO) can facilitate collocations, but these are primarily limited to polar regions.
3. **Radiosonde matchup assessments.** Measurements from operational radiosonde and ozonesonde networks (Chapters 2 and 3) can be collocated with satellite overpasses within broad space-time matchup windows to allow characterization of the satellite EDRs versus in situ field measurements with representation of global zones and large samples within a couple months' acquisition period (e.g., Divakarla et al., 2006, 2008; Nalli et al., 2018b). This technique has been routinely implemented as part of the NOAA Products Validation System (NPROVS) (Reale et al., 2012; Sun et al., 2017), which also facilitates long-term monitoring and is discussed in detail in Chapter 16. However, limitations include skewed distribution toward Northern Hemisphere continents as well as mismatch errors arising from differing overpass and launch times (Sun et al., 2010).
4. **Dedicated/reference radiosonde assessments.** Dedicated radiosondes and ozonesondes are allocated for the explicit purpose of satellite validation, these generally being state-of-the-art instrument packages that are launched synchronized with satellite overpasses, thereby minimizing measurement and mismatch uncertainties. Traceable reference measurements (e.g., frost-point hygrometer, FPH, and GRUAN-corrected sondes) are traceable to a standard and typically including uncertainty estimates. While these datasets provide an optimal reference measurement with error characterization, they are limited by relatively small sample sizes and sparse geographic coverage. Radiosonde, frost-point hygrometers, and ozonesondes are discussed in detail Chapters 2, 3, and 4, respectively.
5. **Intensive field campaign assessments/dissections.** Intensive field campaigns (e.g., closure experiments) for IR sounder $T/H_2O/O_3$ validation include dedicated sondes and ancillary datasets (e.g., lidar, MAERI, sun photometer); ideally they also include an aircraft IR sounder (cf. Chapter 9). The combination of these datasets enables detailed performance assessment, geophysical state specification (closure), as well as detailed case studies. Examples of traditional funded cal/val intensive field campaigns include European Aqua Thermodynamic Experiment (EAQUATE) (Taylor et al., 2008), Joint Airborne IASI Validation Experiment (JAIVEX) (Newman et al., 2012), and the AIRS Water Vapor Experiment-Ground (AWEX-G) (Miloshevich et al., 2006). More recently, multidisciplinary opportunistic campaigns have been leveraged for the remote sensing cal/val objective, including Aerosols and Ocean Science Expeditions (AEROSE) (Morris et al., 2006; Nalli et al., 2006, 2011), the CalWater (Ralph et al., 2016), ARM Cloud Aerosol and Precipitation Experiment (ACAPEX) (Neiman et al., 2017), and the El Niño Rapid Response (ENRR) (Dole et al., 2018). Detailed discussions of such intensive campaigns may be found in Chapters 9–14.

3.2.2 Trace gas hierarchy

Although ozone is a trace gas, its retrieval is facilitated by a prominent absorption band between 990 and 1080 cm^{-1} and ozonesondes are a mature in situ field measurement technology with established global networks (cf. Chapter 4). Ozone data are routinely assimilated into NWP models, and JPSS requirements for O_3 are defined on coarse layers similar to temperature and moisture. For these reasons, IR ozone profile validation was accomplished under the same validation hierarchy as T and Q.

This is not the case, however, for other important trace gases aside from ozone, especially carbon greenhouse gases CO, CH_4, CO_2, along with NH_3, SO_2, among others. Thus, an analogous validation hierarchy was subsequently devised as follows (Nalli et al., 2020).

1. **Global dynamical-chemistry model comparisons.** Trace gas retrievals from IR sounders may be compared against global chemistry models for global Focus Days, and can serve as a "transfer-standard" for double-difference comparisons (as also mentioned earlier). Models suitable for trace gases include the NOAA CarbonTracker (Jacobson et al., 2020) and the Copernicus Atmosphere Monitoring Service (CAMS) (Inness et al., 2019).
2. **Satellite EDR intercomparisons.** As with O_3, intercomparisons can be performed for global Focus Days against EDRs retrieved from other satellite sensors, especially for air quality and carbon missions, including from the Tropospheric Monitoring Instrument (TROPOMI) onboard the Copernicus Sentinel-5 Precursor, the NASA Orbiting Carbon Observatory (OCO-2), Greenhouse gases Observing Satellite (GOSAT), and the Aura Microwave Limb Sounder (MLS). The JPSS requirements to validate total column quantities (as opposed to profiles) helps mitigate the need for applying averaging kernels in performing such intercomparisons, given the varying vertical sensitivities of different satellite instruments.
3. **Surface-based network matchup assessments.** Ground-based observations obtained periodically from surface-based observing networks provide independent reference field measurements with a global distribution reasonably representing global latitude zones roughly analogous to ozonesondes. A notable example is the Total Carbon Column Observing Network (TCCON) (Wunch et al., 2011), consisting of uplooking solar-spectrum FTS instruments that retrieve total column trace gases. A newer source of in situ data are vertical profiles obtained from the balloon-borne AirCore sampling system (Karion et al., 2010; Membrive et al., 2017), highlighted in Chapter 8. Limitations in these datasets include the time latencies needed for acquiring reasonable collocation sample sizes, the need for unit conversions involving other parameters, and different sensitivities to atmospheric layers. These data provide valuable independent "spot checks" at collocated stations, supplementing large datasets available from the model and satellite comparisons indicated earlier.
4. **Aircraft campaign data assessments.** At the pinnacle of the hierarchy for trace gases are aircraft campaigns that provide episodic, but comprehensive sets of in situ vertical profile field data from ascents and descents of dedicated aircraft flying over specified regions. Aircraft campaigns allow for detailed performance assessment over regions of interest. Examples of atmospheric composition campaigns suitable for IR sounder validation include the Atmospheric Tomography (ATom) mission (Wofsy et al., 2018) and, previously, the HIAPER Pole-to-Pole Observations (HIPPO) (Wofsy, 2011) campaigns (covered in detail in Chapter 13).

Both NUCAPS and CLIMCAPS perform retrievals of trace gas concentrations of carbon monoxide (CO) and methane (CH_4) (as well as H_2O) in *layer abundances* (molecules cm^{-2}), whereas carbon dioxide (CO_2) is treated differently and is retrieved in mixing ratios (PPM). Thus, for CO and CH_4, the column assessments are performed for total column quantities simply by integrating the retrieved layer abundances (discussed more below), whereas for CO_2 the assessments are performed for simply column averages; for more details, the reader is referred to Nalli et al. (2020).

3.2.3 Reduction of high-resolution field data to correlative layers

Comparisons between high vertical resolution in situ field measurements (e.g., dedicated radiosonde, ozonesonde, or AirCore profiles) with coarser-resolution satellite physical retrievals (EDRs) require that the field measurements first be reduced to effective layer quantities consistent with the forward radiative transfer model (RTM), sometimes referred to as the radiative transfer algorithm (RTA) (Strow et al., 2003), utilized by the retrieval. This is necessitated by the *measurement equation* (e.g., Taylor and Kuyatt, 1994), which, for a physical retrieval algorithm (e.g., OE), is given by (Rodgers, 1990)

$$\hat{\mathbf{x}} = I[F(\mathbf{x},\mathbf{b}),\mathbf{b},\mathbf{c}], \tag{1}$$

where F is the forward model with parameters $\mathbf{b}$ (e.g., spectroscopy), and I is the inverse model (i.e., retrieval), with parameters $\mathbf{c}$ not included in F (i.e., unrelated to the measurement). Eq. (1) thus includes forward and inverse operators to estimate the *measurand*, $\mathbf{x}$, on forward model layers, relating the profile retrieval (i.e., the *estimate*), $\hat{\mathbf{x}}$, to a correlative state $\mathbf{x}$. Rigorous validation therefore requires that high-resolution reference measurements (e.g., from a radiosonde) be reduced to correlative RTA/RTM layers (Nalli et al., 2013).

Analogous to the radiative transfer physics, we may obtain a correlative state **x** from a high-resolution measurement by taking the path integral of column densities from the top to the bottom of the observed atmospheric column. We may then interpolate those to the RTA layer boundaries (i.e., levels), then compute the RTA layer quantities from the interpolated level values.

Effective layer pressures, $\bar{p}_L$, are defined as $\bar{p}_L \equiv \int_{z_{l+1}}^{z_l} p(z)\, dz / \int_{z_{l+1}}^{z_l} dz$ where z is geopotential height, and subscripts l denote layer boundaries (i.e., levels). It can be shown that for the 101 standard RTA standard pressure levels (Strow et al., 2003), the $n = 100$ RTA layer pressures can then be derived as

$$\bar{P}_L = \frac{P_{l+1} - P_l}{\ln(P_{l+1}/P_l)}, \quad l = L = 1, 2, \ldots, n, \tag{2}$$

where uppercase P denote RTA pressures, and subscripts L and l denote layer and level, respectively (Nalli et al., 2013).

Generally speaking, in situ profile field data (e.g., radiosonde and ozonesonde) are at higher vertical resolution than the RTA layering, with the number of in situ measurements points much greater than the RTA pressure levels. Given a typical sounding measuring pressure, temperature, and humidity (p, T, Q), the number densities for constituent X, N_x (molecules cm^{-3}) may be determined. The integrated column abundances (from the top measurement z_t to the measurement level height z) for atmospheric constituent profiles $X(z)$ (layer abundances, molecules cm^{-2}) are then given by

$$\Sigma X(z) \equiv \int_{z_t}^{z} N_x(z')\, dz', \tag{3}$$

and in a similar manner, the temperature profile can be integrated, weighted by the airmass number density,

$$\Sigma T(z) \equiv \int_{z_t}^{z} T(z')\, N_a(z')\, dz'. \tag{4}$$

The calculations from Eqs. (3), (4) can then be linearly interpolated to the RTA pressure levels P_l (spanning only the measured column), noting that the summations include the observed surface and top levels. The top RTA level l_0 is defined simply as the level just below the top observed pressure level, p_{top} (e.g., at balloon burst or aircraft cruising altitude). Because of variable terrain and surface pressure, the bottom level, l_b, is defined to be consistent with the RTA calculation used in the retrieval algorithm. Given $\Sigma X(P_l)$ and $\Sigma T(P_l)$, the effective RTA layer quantities at the effective layer pressures $\bar{P}_L$ defined by Eq. (2) can be calculated as the molecular-weighted average using the levels defined by P_l.

3.2.4 Coarse-layer and fractional error statistics

Given correlative profiles on effective RTA layers, IR sounder profile error statistics may be computed on what are called "coarse layers" (Nalli et al., 2013). The reason for this pertains to the fact that passive IR sounders have fundamental limitations on vertical resolution. While the 100 RTA layers are usually coarser than the original high-resolution field measurements, they are nevertheless intentionally at a finer vertical resolution than the sounder's resolving capability. Therefore, even after high-resolution in situ profiles are reduced to 100 RTA layers, statistics computed on the 100 layers will contain *null-space errors* (i.e., errors caused by features theoretically unresolvable by the weighting functions of the remotely sensed measurement) that are not indicative of whether the sounder retrieval system is performing properly. Coarse-layer profile statistics are calculated based on the error analysis equations discussed in Chapter 1, usually ignoring mismatch error and uncertainties in the reference profiles (e.g., radiosonde error). Note that this conventional disregarding of the reference profile error within such analyses aliases those errors as retrieval errors, thereby artificially exaggerating them.

For temperature, the coarse layers are roughly 1–5 km thick, depending on the location in the troposphere. The coarse-layer temperatures $T_\mathcal{L}$ are computed from the RTA layer temperatures T_L as the pressure-weighted average within the coarse-layer boundaries according to

$$T_\mathcal{L} = \frac{\sum_L \ln\left(\frac{P_l}{P_{l-1}}\right) T_L}{\ln\left(\frac{P_\mathcal{L}}{P_{\mathcal{L}-1}}\right)}, \quad \mathcal{L} = 1, 2, \ldots, \mathcal{N}, \tag{5}$$

where L are the RTA layers that fall within the coarse layer $\mathcal{L}$, l is the RTA layer boundary (numerically equal to L), and $\mathcal{N}$ depends on the layering scheme (typical layering schemes have $\mathcal{N} = 30$ or 18). Given the coarse-layer values for both the correlative and retrieved profiles, $T_\mathcal{L}$ and $\hat{T}_\mathcal{L}$, respectively, we may use the statistical formulas in Chapter 1 to compute the temperature profile validation statistics.

For computing vertical profile statistics of gas concentration profile EDRs (namely H_2O and O_3) on coarse layers, both retrieval and correlative profiles (in RTA layer abundances) are summed over each coarse layer and converted to mass abundances in $g\,cm^{-2}$. For gas constituent X, the RTA layer abundances are denoted as X_L, and the coarse-layer mass abundances $X_\mathcal{L}$ are calculated as

$$X_\mathcal{L} = \frac{M_x}{N_A} \sum_L X_L, \quad \mathcal{L} = 1, 2, \ldots, \mathcal{N}, \tag{6}$$

where M_x is molecular mass, N_A is Avogadro's number, L are the RTA layers within the coarse-layer $\mathcal{L}$ as in Eq. (5), and $\mathcal{N}$ depends on layering scheme (currently for H_2O, $\mathcal{N} = 8$ or 12; for O_3, $\mathcal{N} = 9$). The bottom layer adjacent to the surface is typically a thinner layer (i.e., a partial RTA layer). To account for this, a bottom-layer multiplier factor must be included in the finite-difference summation calculation (e.g., Berndt et al., 2020; Nalli et al., 2013, 2020). Given coarse-layer abundances, $X_\mathcal{L}$, the fractional deviation is taken to be the absolute deviation divided by the correlative value

$$\Delta X_\mathcal{L} \equiv \frac{\hat{X}_\mathcal{L} - X_\mathcal{L}}{X_\mathcal{L}}, \quad \mathcal{L} = 1, 2, \ldots, \mathcal{N}. \tag{7}$$

Eq. (7) may then be applied to the statistical formulas in Chapter 1 to compute the water vapor or ozone profile validation statistics. However, the denominator in Eq. (7) can result in large $\Delta X_\mathcal{L}$ in dry atmospheres (e.g., middle to upper troposphere or polar regions) and thereby skew the statistics toward these cases. To avoid this, generalized weighted means (as opposed to simple arithmetic means) are used for water vapor statistics, the details of which can be found in a study by Nalli et al. (2013). For ozone, given a correlative ozone sounding (e.g., from an ozonesonde), statistics are calculated in a manner similar to the water vapor.

3.3 Validation highlights using field data

Beginning with the SNPP satellite, JPSS has directly and indirectly supported an ongoing dedicated radiosonde program leveraging a number of collaborating institutions. In addition, radiosondes from both operational synoptic sites as well as GCOS Reference Upper Air Network (GRUAN) sites have also been routinely leveraged by NPROVS (cf. Chapter 16). As mentioned earlier, dedicated radiosondes are optimally collocated and synchronous with satellite overpasses at various selected sites, including those from the US DOE Atmospheric Radiation Measurement (ARM) program (Mather and Voyles, 2013; Tobin et al., 2006) and other experimental sites (e.g., the Howard University Beltsville Campus [HUBC] site; cf. Chapter 12), along with NOAA Aerosols and Ocean Science Expedition (AEROSE) intensive campaigns of opportunity (Morris et al., 2006; Nalli et al., 2011). Likewise, global ozonesonde network sites (cf. Chapter 4) have been leveraged to include Southern Hemisphere Additional Ozonesonde (SHADOZ) (Thompson et al., 2004) and World Ozone and Ultraviolet Radiation Data Centre (WOUDC) network sites. Dedicated ozonesondes (Nalli et al., 2018b) have also been launched from ship-based campaigns (cf. Chapter 11) and more recently from the HUBC site. Using the previously mentioned methods and data, the NUCAPS $T/H_2O/O_3$ validation effort for the SNPP satellite (using CrIS nominal spectral resolution data) have been previously reported elsewhere (Nalli et al., 2018a, b; Sun et al., 2017). More recently, the validation of the NUCAPS carbon trace gases (CO, CH_4, and CO_2) for SNPP and NOAA-20 full-spectral resolution (FSR) CrIS has been performed and reported in a study by Nalli et al. (2020).

3.3.1 NOAA-20 temperature and moisture validation

NPROVS (Reale et al., 2012; Sun et al., 2017) (cf. Chapter 16) routinely compiles datasets of collocated radiosonde and satellite-derived geophysical profiles for conveniently comparing satellite profiles from over 20 products. These include IR-based temperature and moisture soundings from NOAA-20 NUCAPS, Aqua AIRS, and NOAA-19 ATOVS (the predecessor of the AIRS and CrIS/ATMS systems). This section highlights an example assessment of sounding products derived from NUCAPS, AIRS, and ATOVS versus global radiosonde field data as the reference. Radiosondes are from the conventional World Meteorological Organization network and the satellite products are the operational systems: IR + MW for NUCAPS, IR-only for AIRS (IR + MW not available after September 2016), and IR + MW for ATOVS. Since most of the radiosondes (and hence the collocations with the satellite data) are over the Northern Hemisphere mid-latitude land areas, the vertical statistics reflect sounding performance mainly for mid-latitude land regimes.

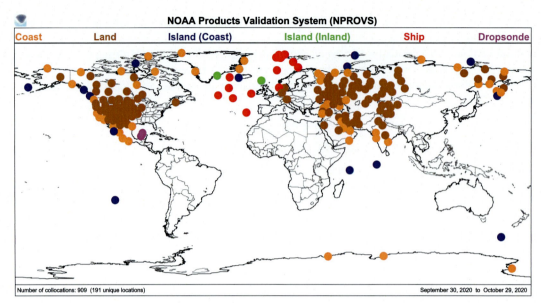

FIG. 3 NPROVS global distribution of conventional radiosonde collocations with NOAA-20 NUCAPS, Aqua AIRS, and NOAA-19 ATOVS for satellite IR-based profiles. Radiosonde locations are color coded as *brown* (land), *orange* (coast), *blue* (island), and *red* (ship); dropsonde locations are color-coded as *magenta*. Each radiosonde is collocated to all three products within ±3 h and 50 km over the period September 30 through October 29, 2020.

Fig. 3 shows the global distribution of collocated radiosondes with NUCAPS NOAA-20, Aqua-AIRS (v5), and the legacy ATOVS NOAA-19 from September 30 to October 29, 2020, which are used to estimate satellite product performance. At each radiosonde launch site, the radiosonde is collocated to all three products each within 3 h and 50 km. The locations are color coded, with brown, orange, blue, and red indicating radiosondes over land (inland), coast, islands, and ships, respectively (dropsondes denoted with magenta). Due to the time differences among the sun-synchronous polar satellite orbits and synoptic radiosonde observations, the 3-h collocations are located primarily over western North America, eastern Europe, and western Asia, with a small portion of observations from ships and islands. This highlights the utility of dedicated radiosondes as mentioned in Section 3.2.1. Furthermore, the NUCAPS and AIRS are typically 1–2 h before the radiosonde (AIRS about 30 min after NOAA-20) whereas ATOVS are 1–2 h after the radiosonde, given their respective local overpasses. Targeting all three systems to the same radiosonde is a unique approach available in NPROVS that allows an intercomparison of satellite products (methods 2 and 3 in the "hierarchy"; cf. Section 3.2.1) against the same sample of radiosondes.

The vertical profile statistics for temperature and water vapor fraction (%) for the sample in Fig. 3 are shown for each product suite in Fig. 4. The profile statistics are calculated consistent with the IR sounder validation approach summarized in Sections 3.2.3 and 3.2.4. Bias and root mean square error (RMSE) are computed in ≈1 km coarse layers for temperature and ≈2 km coarse layers for moisture. For atmospheric temperature (top panel), biases between NUCAPS (orange) and AIRS (green) are overall comparable to each other except near the surface where the NUCAPS appears warmer (≈0.5 K) and in the lower stratosphere (150 hPa) where the AIRS appears colder (≈1 K), otherwise an average cold bias <0.5 K is shown for both systems. ATOVS (blue) exhibits a cold bias of 0.5–1 K in the lower troposphere shifting to a warm bias (<0.5 K) in the middle-upper troposphere and a more prominent vertical sinusoidal bias pattern. RMSE in NUCAPS and AIRS are comparable and smaller by about 0.5–1.0 K than for ATOVS except above 100 hPa. For atmospheric moisture (bottom panel), all the products appear too moist, with AIRS (green) and NUCAPS (orange) about 5%–15% too moist, and ATOVS (blue) approaching 20% too moist (except near the surface). Again, NUCAPS appears overall to be comparable to AIRS except near the surface where the NUCAPS has an increased moist bias likely associated with the warm temperature bias. Part of the bias shown in the products could be attributable to dry bias in conventional radiosondes particularly in the upper troposphere (Sun et al., 2017). The RMSE for NUCAPS and AIRS are comparable to each other throughout the troposphere with ATOVS RMSE significantly larger. This assessment, based on 1 month of data, indicates that NUCAPS NOAA 20 IR-based retrievals are comparable to those of AIRS and show improvement over those of ATOVS for both atmospheric temperature and moisture, which is expected given the improved sensor capability.

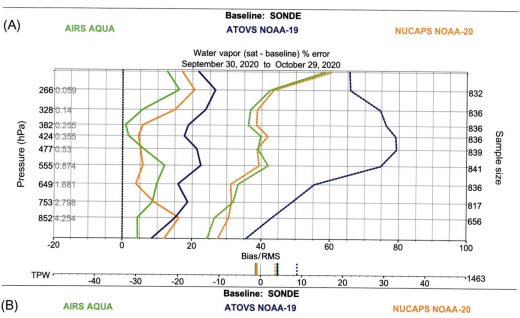

FIG. 4 IR sounder profile statistics for NUCAPS (*orange*), AIRS v5 (*green*), and ATOVS (*blue*) versus global conventional radiosonde collocations acquired via NPROVS: (*top*) atmospheric temperature mean bias (*solid*) and RMSE (*dashed*), and (*bottom*) water vapor mixing ratio percentage difference (%). Values in *gray* on the left y-axis are radiosonde temperature (K) and water vapor mixing ratio (g kg^{-1}) averaged from the collocations, the number of which is indicated on the right y-axis.

3.3.2 NOAA-20 IR ozone profile validation

Fig. 5 shows the ozonesonde sample utilized for the NOAA-20 CrIS-FSR IR ozone profile EDR validated maturity effort, with Fig. 6 summarizing the coarse-layer profile uncertainty statistics for the sample. For reference, the dashed lines show the JPSS requirements for RMSE and bias, from which it can be seen that the NUCAPS retrievals meet requirements. The red lines show the results of the climatological a priori (or "background") used for the OE physical retrievals (cf. Smith and Barnet, 2019) (blue lines). It can be seen that the OE retrievals improve upon the a priori, especially in terms of the random error (RMSE and σ) as would be expected. It is also seen that the CrIS sounder has good performance and sensitivity in the upper-troposphere/lower-stratosphere (UT/LS) ozone layer.

4. Applications

As discussed earlier, the NUCAPS algorithm is designed to meet NOAA/JPSS requirements and validation methods are employed to characterize its products and quantify the degree to which they meet NOAA requirements. These efforts, however, do not guarantee NUCAPS utility to operational forecasters who are usually focused on "difficult cases." A

330 PART | III Satellite applications

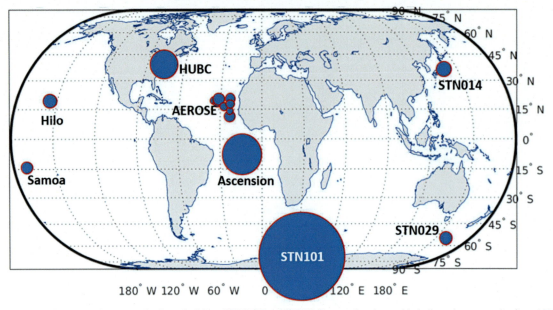

FIG. 5 Geographic histogram of ozonesondes (launched from SHADOZ and WOUDC networks, along with dedicated ozonesondes from AEROSE and HUBC) collocated with NOAA-20 overpasses for NUCAPS retrievals accepted by the quality flag within space-time criteria of $\delta x \leq 125$ km radius and $-240 \leq \delta t \leq +120$ min of launches. Circle sizes depict the relative ozonesonde collocation sample sizes for each ozonesonde launch location.

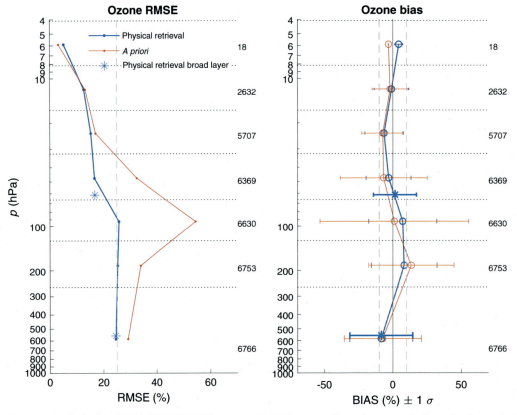

FIG. 6 Coarse-layer error analysis of the NOAA-20 NUCAPS IR ozone profile EDR versus collocated ozonesondes for retrievals accepted by the quality flag (under clear to partly cloudy conditions) within space-time collocation criteria indicated in Fig. 5. The *left and right plots* show the coarse-layer RMSE and bias ($\pm 1\sigma$ variability), respectively. NUCAPS IR physical retrieval and a priori performances are given in *blue* and *red*, respectively, with collocation sample size for each coarse-layer given in the right margins. The *black dashed lines* designate the JPSS global performance requirements for two broad atmospheric layers with *asterisks* denoting the calculated broad-layer averages for the physical retrievals.

modern-era forecaster is overwhelmed by the amount and diversity of available data and information. While careful algorithm design and rigorous validation studies are vital steps toward a successful research-to-operations (R2O) transition, NUCAPS soundings cannot reach and maintain NWS operational relevance if the quality and value of their information is not readily available and evident. The question forecasters ask is: What unique information do NUCAPS soundings provide that distinguishes it from other products? The answer to this question was not straightforward and was not to be found within the operational NetCDF output files, global statistics, NOAA technical reports, or peer-reviewed technical papers. One can only develop an approximate understanding of NUCAPS value in operations through the testimony of forecasters themselves. It is here where the JPSS Proving Ground and Risk Reduction (PGRR) program has been instrumental in establishing and maintaining open, interactive communication between forecasters and NUCAPS developers (remote sensing scientists) for the sake of an effective R2O pathway. The PGRR Sounding Initiative and the value in fostering collaboration between forecasters and product developers has since been documented in a number of published papers (Esmaili et al., 2020; Smith et al., 2018a, b; Berndt et al., 2020; Wheeler et al., 2018).

With feedback from forecasters, NUCAPS algorithm improvements have been implemented, its sounding product files have been tailored for visualization in the NWS Advanced Weather Interactive Product System (AWIPS-II) (Berndt et al., 2020), product latency has been improved so that forecasters receive NUCAPS soundings within 60 min of satellite overpass and ahead of severe convective events (Smith et al., 2019). New data applications exist, such as the forecasting of cold air aloft events (Weaver et al., 2019), tracking of long-range smoke plumes (Smith et al., 2020), and the characterization of extra tropical transition (Berndt and Folmer, 2018). Forecasters can also use NUCAPS to assess fire-weather potential, which is partly characterized by very dry atmospheric conditions (Gilbreath, L., "NUCAPS for Assessing the Fire Weather Environment," https://weather.msfc.nasa.gov/sport/training/rise/nucaps/intro_fire_weather/content/#/ (Accessed 15 December 2020)). Fig. 7A shows an annotated AWIPS-II terminal displaying available profiles from the NUCAPS swath (dots) overlaid with the storm prediction center (SPC) warning (magenta outline) for critical fire weather on December 7, 2020 from the 10:00 UTC NOAA-20 overpass. One can readily see the value of the vastly improved coverage offered by the satellite (passive IR) observations that forecasters can use to supplement operational in situ radiosonde data. Clicking on the colored dots (which indicate data quality) in AWIPS-II displays the skew-T profile (Fig. 7B). The profile helps the forecaster inspect the dry conditions in the low- and mid-levels of the column (below 500 hPa), which is indicated by the large gap between the temperature (red) and dewpoint temperature (green) line. While satellite water vapor imagery

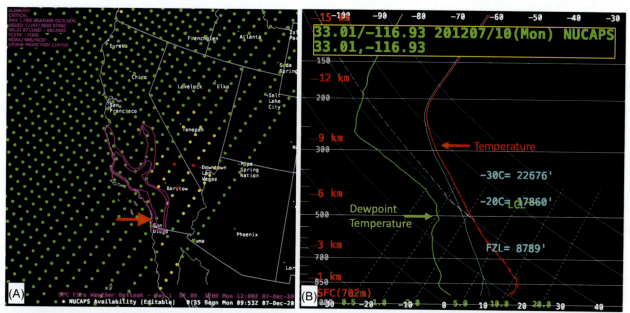

FIG. 7 Critical fire weather condition issued by SPC on December 7, 2020 at 10:00 UTC as a result of forecasted Santa Ana winds and dry conditions across Southern California as captured by the Advanced Weather Interactive Product System (AWIPS-II): (A) warning area (*magenta*) overlaid with available NUCAPS sounding profiles (*dots*) from NOAA-20 and (B) skew-T diagram from NUCAPS temperature and moisture retrievals near San Diego, California (indicated by the *red arrow* in (A)) indicating that the air is very dry below 500 hPa. The *dots* in (A) show a simplified NUCAPS quality flag scheme specifically developed for NWS forecasters, which indicate where the IR + MW retrieval passed (*green*), the MW-only retrieval passed (*yellow*), and where both failed (*red*).

can also show the spatial distribution of dry air, forecaster feedback has indicated that satellite sounders can complement these products with relatively high-resolution observations of the vertical distribution of dry air, during both day and night. Additional information can be found in studies by Berndt et al. (2020) and Esmaili et al. (2020), who present summaries of the lessons IR remote sensing scientists have since learned in making satellite sounding products more accessible to forecasters.

5. Summary and outlook

Satellite passive IR sounders have become an indispensable tool for weather forecasting, both in terms of numerical model assimilation and in providing real-time soundings for operational forecasters and nowcasting applications. While IR sounders were originally designed with the weather forecasting application in mind, the IR Earth emission spectra contain a plenitude of additional information about atmospheric trace gas concentrations. Furthermore, today's high spectral-resolution spectrometers (both grating and FTS types) include high-accuracy onboard blackbody calibration, thereby extending their utility for climate change applications. Because both NOAA and EUMETSAT have committed to maintaining LEO satellite constellations with high-resolution IR spectrometer sounders well into the 2020s and beyond, these instruments are now being exploited for such extended applications (e.g., climate) beyond their original mission scopes. Although there was an implicit emphasis in this chapter on hyperspectral thermal IR instruments onboard operational LEO satellites (polar orbiters), there is also a long history of narrowband IR sounders on GEO satellites (comparable to ATOVS), with more advanced ones on the way (e.g., the EUMETSAT Infrared Sounder, planned for the Meteosat Third Generation series, is a GEO hyperspectral FTS sounder).

In this chapter, we specifically discussed the development, validation, and application of NUCAPS, a NOAA operational retrieval system that generates satellite soundings for atmospheric monitoring and weather forecasting at the NOAA National Weather Service (NWS) and other environmental agencies. The success of IR sounder-retrieved products in operational applications depends strongly on knowledge of their accuracy and precision, which are determined using statistical intercomparison methods and accurate in situ field measurements, especially radiosondes, but also ozonesondes and other instrumentation discussed in earlier chapters of this book. In both assimilation and retrieval systems, these field data are needed for the forward problem (i.e., forward radiance calculations from the RTM or RTA), the inverse problem (i.e., the geophysical retrievals, which require training and/or tuning data), and last, but not least, the validation problem (i.e., the error characterization of the end-to-end sounding system). Knowledge of the three-dimensional nature and complexity of atmospheric processes across space and time is facilitated through the synergistic utilization of available instruments and observing systems available to us, both in situ and remotely sensed.

Acknowledgments

The authors acknowledge the support of Satya Kalluri and the STAR IR Soundings Team (K. Pryor, M. Divakarla, C. Tan, T. Zhou, et al.). We also express our appreciation to Prof. Peter Minnett (UM/RSMAS) for reviewing the chapter and providing expert constructive feedback. The scientific results and conclusions, as well as any views or opinions expressed herein, are those of the authors and do not necessarily reflect the views of NOAA or the Department of Commerce.

References

Aumann, H.H., Chen, X., Fishbein, E., Geer, A., Havemann, S., Huang, X., 2018. Evaluation of radiative transfer models with clouds. J. Geophys. Res. Atmos. 123 (11), 6142–6157. https://doi.org/10.1029/2017JD028063.

Barnet, C., 2009. NPOESS Community Collaborative Calibration/Validation Plan for the NPOESS Preparatory Project CrIS/ATMS EDRs (Tech. Rep. I30004, Ver. 1 Rev. B). Integrated Program Office (IPO), Silver Spring, MD.

Barnet, C.D., et al., 2021. NOAA Unique Combined Atmospheric Processing System (NUCAPS) algorithm theoretical basis document (ATBD). ATBD v3.1, NOAA/NESDIS/STAR Joint Polar Satellite System, College Park, MD.

Berndt, E., Folmer, M., 2018. Utility of CrIS/ATMS profiles to diagnose extratropical transition. Results Phys. 8, 184–185. https://doi.org/10.1016/j.rinp.2017.12.006.

Berndt, E., Smith, N., Burks, J., White, K., Esmaili, R., Kuciauskas, A., 2020. Gridded satellite sounding retrievals in operational weather forecasting: product description and emerging applications. Remote Sens. 12 (20), 3311. https://doi.org/10.3390/rs12203311.

Chahine, M.T., 1977. Remote sounding of cloudy atmospheres. II. Multiple cloud formations. J. Atmos. Sci. 34, 744–757.

Chahine, M.T., et al., 2006. AIRS: improving weather forecasting and providing new data on greenhouse gases. Bull. Am. Meteorol. Soc. 87 (7), 911–926.

Divakarla, M.G., Barnet, C.D., Goldberg, M.D., McMillin, L.M., Maddy, E., Wolf, W., Zhou, L., Liu, X., 2006. Validation of atmospheric infrared sounder temperature and water vapor retrievals with matched radiosonde measurements and forecasts. J. Geophys. Res. 111, D09S15. https://doi.org/10.1029/2005JD006116.

Divakarla, M., et al., 2008. Evaluation of atmospheric infrared sounder ozone profiles and total ozone retrievals with matched ozonesonde measurements, ECMWF ozone data, and ozone monitoring instrument retrievals. J. Geophys. Res. 113, D15308. https://doi.org/10.1029/2007JD009317.

Dole, R.M., et al., 2018. Advancing science and services during the 2015/16 El Niño: the NOAA El Niño rapid response field campaign. Bull. Am. Meteorol. Soc. 99 (5), 975–1001. https://doi.org/10.1175/BAMS-D-16-0219.1.

Esmaili, R.B., Smith, N., Berndt, E.B., Dostalek, J.F., Kahn, B.H., White, K., 2020. Adapting satellite soundings for operational forecasting within the hazardous weather testbed. Remote Sens. 12 (5), 886. https://doi.org/10.3390/rs12050886.

Fetzer, E., et al., 2003. AIRS/AMSU/HSB validation. IEEE Trans. Geosci. Remote Sens. 41 (2), 418–431.

Gambacorta, A., Barnet, C., 2013. Methodology and information content of the NOAA NESDIS operational channel selection for the cross-track infrared sounder (CrIS). IEEE Trans. Geosci. Remote Sens. 51 (6), 3207–3216. https://doi.org/10.1109/TGRS.2012.2220369.

Gambacorta, A., et al., 2014. An experiment using high spectral resolution CrIS measurements for atmospheric trace gases: carbon monoxide retrieval impact study. IEEE Geosci. Remote Sens. Lett. 11 (9), 1639–1643. https://doi.org/10.1109/LGRS.2014.2303641.

Goldberg, M.D., Qu, Y., McMillin, L.M., Wolf, W., Zhou, L., Divakarla, M., 2003. AIRS near-real-time products and algorithms in support of operational numerical weather prediction. IEEE Trans. Geosci. Remote Sens. 41, 379–389.

Inness, A., et al., 2019. The CAMS reanalysis of atmospheric composition. Atmos. Chem. Phys. 19, 3515–3556.

Jacobson, A.R., et al., 2020. CarbonTracker CT2019. NOAA Earth System Research Laboratory, Global Monitoring Division, https://doi.org/10.25925/39M3-6069.

Karion, A., Sweeney, C., Tans, P., Newberger, T., 2010. AirCore: an innovative atmospheric sampling system. J. Atmos. Ocean. Technol. 27 (11), 1839–1853. https://doi.org/10.1175/2010JTECHA1448.1.

Le Marshall, J., Jung, J., Goldberg, M., Barnet, C., Wolf, W., Derber, J., Treadon, R., Lord, S., 2008. Using cloudy AIRS fields of view in numerical weather prediction. Aust. Meteorol. Mag. 57, 249–254.

Li, J., Wolf, W.W., Menzel, W.P., Zhang, Z., Huang, H.L., Achtor, T.H., 2000. Global soundings of the atmosphere from ATOVS measurements: the algorithm and validation. J. Appl. Meteorol. 39 (8), 1248–1268.

Maddy, E.S., Barnet, C.D., Goldberg, M., Sweeney, C., Liu, X., 2008. CO_2 retrievals from the atmospheric infrared Sounder: methodology and validation. J. Geophys. Res. 113, D11301. https://doi.org/10.1029/2007JD009402.

Maddy, E.S., Barnet, C.D., Gambacorta, A., 2009. A computationally efficient retrieval algorithm for hyperspectral sounders incorporating a priori information. IEEE Geosci. Remote Sens. Lett. 6, 802–806. https://doi.org/10.1109/LGRS.2009.2025780.

Mather, J.H., Voyles, J.W., 2013. The ARM climate research facility: a review of structure and capabilities. Bull. Am. Meteorol. Soc. 94 (3), 377–392. https://doi.org/10.1175/BAMS-D-11-00218.1.

Membrive, O., Crevoisier, C., Sweeney, C., Danis, F., Hertzog, A., Engel, A., Bönisch, H., Picon, L., 2017. AirCore-HR: a high-resolution column sampling to enhance the vertical description of CH_4 and CO_2. Atmos. Meas. Tech. 10, 2163–2181.

Miloshevich, L.M., Vömel, H., Whiteman, D.N., Lesht, B.M., Schmidlin, F.J., Russo, F., 2006. Absolute accuracy of water vapor measurements from six operational radiosonde types launched during AWEX-G and implications for AIRS validation. J. Geophys. Res. 111, D09S10. https://doi.org/10.1029/2005JD006083.

Morris, V., et al., 2006. Measuring trans-Atlantic aerosol transport from Africa. Eos Trans. AGU 87 (50), 565–571.

Nalli, N.R., et al., 2006. Ship-based measurements for infrared sensor validation during aerosol and ocean science expedition 2004. J. Geophys. Res. 111, D09S04. https://doi.org/10.1029/2005JD006385.

Nalli, N.R., et al., 2011. Multi-year observations of the tropical Atlantic atmosphere: multidisciplinary applications of the NOAA aerosols and ocean science expeditions (AEROSE). Bull. Am. Meteorol. Soc. 92, 765–789. https://doi.org/10.1175/2011BAMS2997.1.

Nalli, N.R., et al., 2013. Validation of satellite sounder environmental data records: application to the cross-track infrared microwave sounder suite. J. Geophys. Res. Atmos. 118, 13628–13643. https://doi.org/10.1002/2013JD020436.

Nalli, N.R., et al., 2018a. Validation of atmospheric profile retrievals from the SNPP NOAA-unique combined atmospheric processing system. Part 1: temperature and moisture. IEEE Trans. Geosci. Remote Sens. 56 (1), 180–190. https://doi.org/10.1109/TGRS.2017.2744558.

Nalli, N.R., et al., 2018b. Validation of atmospheric profile retrievals from the SNPP NOAA-unique combined atmospheric processing system. Part 2: Ozone. IEEE Trans. Geosci. Remote Sens. 56 (1), 598–607. https://doi.org/10.1109/TGRS.2017.2762600.

Nalli, N.R., et al., 2020. Validation of carbon trace gas profile retrievals from the NOAA-unique combined atmospheric processing system for the cross-track infrared sounder. Remote Sens. 12 (19), 3245. https://doi.org/10.3390/rs12193245.

Neiman, P.J., et al., 2017. An analysis of coordinated observations from NOAA's Ronald Brown Ship and G-IV aircraft in a landfalling atmospheric river over the North Pacific during CalWater-2015. Mon. Weather Rev. 145, 3647–3669. https://doi.org/10.1175/MWR-D-17-0055.1.

Newman, S.M., et al., 2012. The joint airborne IASI validation experiment: an evaluation of instrument and algorithms. J. Quant. Spectrosc. Radiat. Transf. 113, 1372–1390.

Noh, Y.-C., Lim, A.H.N., Huang, H.L., Goldberg, M.D., 2020. Global forecast impact of low data latency infrared and microwave sounders observations from polar orbiting satellites. Remote Sens. 12 (14). https://doi.org/10.3390/rs12142193.

Oleson, J.P., 2008. Testing the waters: the role of sounding-weights in ancient Mediterranean navigation. In: Hohlfelder, R.L. (Ed.), The Maritime World of Ancient Rome. vol. 6. University of Michigan Press, Ann Arbor, MI, pp. 117–174.

Ralph, F.M., et al., 2016. CalWater field studies designed to quantify the roles of atmospheric rivers and aerosols in modulating U.S. west coast precipitation in a changing climate. Bull. Am. Meteorol. Soc. 97 (7), 1209–1228. https://doi.org/10.1175/BAMS-D-14-00043.1.

Reale, T., Sun, B., Tilley, F.H., Pettey, M., 2012. The NOAA products validation system. J. Atmos. Ocean. Technol. 29, 629–645. https://doi.org/10.1175/JTECH-D-11-00072.1.

Revercomb, H.E., Buijs, H., Howell, H.B., LaPorte, D.D., Smith, W.L., Stromovsky, L.A., 1988. Radiometric calibration of IR Fourier transform spectrometers: solution to a problem with the high-resolution interferometer sounder. Appl. Opt. 27 (15), 3210–3218.

Rodgers, C.D., 1990. Characterization and error analysis of profiles retrieved from remote sounding measurements. J. Geophys. Res. 95 (D5), 5587–5595.

Rodgers, C.D., 2000. Inverse Methods for Atmospheric Sounding: Theory and Practice, Atmospheric, Oceanic and Planetary Physics. vol. 2 World Scientific, Singapore.

Rodgers, C.D., Connor, B.J., 2003. Intercomparison of remote sounding instruments. J. Geophys. Res. 108 (D3), 4116. https://doi.org/10.1029/2002JD002299.

Smith, W.L., 1968. An improved method for calculating tropospheric temperature and moisture from satellite radiometer measurements. Mon. Weather Rev. 96 (6), 387–396.

Smith, W.L., 1991. Atmospheric soundings from satellites—false expectation or the key to improved weather prediction? Q. J. R. Meteorol. Soc. 117 (498), 267–297.

Smith, N., Barnet, C.D., 2019. Uncertainty characterization and propagation in the community long-term infrared microwave combined atmospheric product system (CLIMCAPS). Remote Sens. 11 (10), 1227. https://doi.org/10.3390/rs11101227.

Smith, N., Barnet, C.D., 2020. CLIMCAPS observing capability for temperature, moisture, and trace gases from AIRS/AMSU and CrIS/ATMS. Atmos. Meas. Tech. 13 (8), 4437–4459. https://doi.org/10.1109/TGRS.2002.808244.

Smith, W.L., et al., 2009. Technical note: Evolution, current capabilities, and future advance in satellite nadir viewing ultra-spectral IR sounding of the lower atmosphere. Atmos. Chem. Phys. 9, 5563–5574.

Smith, N., Shontz, K., Barnet, C.D., 2018a. What is a satellite measurement? Communicating abstract satellite science concepts to the world. In: 14th Annual Symposium on New Generation Operational Environmental Satellite Systems, AMS Annual Meeting, American Meteorological Society, Austin, TX.

Smith, N., White, K.D., Berndt, E.B., Zavodsky, B.T., Wheeler, A., Bowlan, M.A., Barnet, C.D., 2018b. NUCAPS in AWIPS—rethinking information compression and distribution for fast decision making. In: 22nd Conference on Satellite Meteorology and Oceanography, AMS Annual Meeting, American Meteorological Society, Austin, TX.

Smith, N., Berndt, E.B., Barnet, C.D., Goldberg, M.D., 2019. Why operational meteorologists need more satellite soundings. In: 15th Annual Symposium on New Generation Operational Environmental Satellite Systems, AMS Annual Meeting, American Meteorological Society, Phoenix, AZ.

Smith, N., Esmaili, R.B., Barnet, C.D., Frost, G.J., McKeen, S.A., Trainer, M.K., Francoeur, C., 2020. Monitoring atmospheric composition and long-range smoke transport with NUCAPS satellite soundings in field campaigns and operations. In: 22nd Conference on Atmospheric Chemistry, AMS Annual Meeting, American Meteorological Society, Boston, MA.

Strow, L.L., Hannon, S.E., Souza-Machado, S.D., Motteler, H.E., Tobin, D., 2003. An overview of the AIRS radiative transfer model. IEEE Trans. Geosci. Remote Sens. 41 (2), 303–313.

Sun, B., Reale, A., Seidel, D.J., Hunt, D.C., 2010. Comparing radiosonde and cosmic atmospheric profile data to quantify differences among radiosonde types and the effects of imperfect collocation on comparison statistics. J. Geophys. Res. 115, D23104. https://doi.org/10.1029/2010JD014457.

Sun, B., Reale, A., Tilley, F., Pettey, M., Nalli, N.R., Barnet, C.D., 2017. Assessment of NUCAPS S-NPP CrIS/ATMS sounding products using reference and conventional radiosonde observations. IEEE J. Sel. Top. Appl. Earth Observ. 10 (6), 2499–2509. https://doi.org/10.1109/JSTARS.2017.2670504.

Susskind, J., Barnet, C.D., Blaisdell, J.M., 2003. Retrieval of atmospheric and surface parameters from AIRS/AMSU/HSB data in the presence of clouds. IEEE Trans. Geosci. Remote Sens. 41 (2), 390–409.

Susskind, J., Blaisdell, J., Iredell, L., Keita, F., 2011. Improved temperature sounding and quality control methodology using AIRS/AMSU data: the AIRS science team version 5 retrieval algorithm. IEEE Trans. Geosci. Remote Sens. 49 (3), 883–907. https://doi.org/10.1109/TGRS.2010.2070508.

Taylor, B.N., Kuyatt, C.E., 1994. Guidelines for evaluating and expressing the uncertainty of NIST measurement results. US National Institute of Standards and Technology, Gaithersburg, MD. NIST Technical Note 1297.

Taylor, J.P., et al., 2008. EAQUATE: an international experiment for hyperspectral atmospheric sounding validation. Bull. Am. Meteorol. Soc. 89 (2), 203–218. https://doi.org/10.1175/BAMS-89-2-203.

Thompson, A.M., Witte, J.C., Oltmans, S.J., Schmidlin, F.J., 2004. SHADOZ—a tropical ozonesonde-radiosonde network for the atmospheric community. Bull. Am. Meteorol. Soc. 85 (10), 1549–1564. https://doi.org/10.1175/BAMS-85-10-1549.

Tobin, D.C., et al., 2006. Atmospheric radiation measurement site atmospheric state best estimates for atmospheric infrared sounder temperature and water vapor retrieval validation. J. Geophys. Res. 111, D09S14. https://doi.org/10.1029/2005JD006103.

Warner, J.X., Carminati, F., Wei, Z., Lahoz, W., Attié, J.L., 2013. Tropospheric carbon monoxide variability from airs under clear and cloudy conditions. Atmos. Chem. Phys. 13, 12469–12479. https://doi.org/10.5194/acp-13-12469-2013.

Weaver, G., Smith, N., Berndt, E.B., White, K.D., Dostalek, J.F., Zavodsky, B.T., 2019. Addressing the cold air aloft aviation challenge with satellite sounding observations. J. Oper. Meteorol. 7 (10), 138–152. https://doi.org/10.15191/nwajom.2019.0710.

Wheeler, A., Smith, N., Gambacorta, A., Barnet, C.D., 2018. Evaluation of NUCAPS products in AWIPS-II: results from the 2017 HWT. In: 14th Annual Symposium on New Generation Operational Environmental Satellite Systems, AMS Annual Meeting, American Meteorological Society, Austin, TX.

Wofsy, S.C., The HIPPO Science Team, Cooperating Modellers, and Satellite Teams, 2011. HIAPER pole-to-pole observations (HIPPO): fine-grained, global-scale measurements of climatically important atmospheric gases and aerosols. Philos. Trans. R. Soc. A 369, 2073–2086. https://doi.org/10.1098/rsta.2010.0313.

Wofsy, S., et al., 2018. Atom: merged atmospheric chemistry, trace gases, and aerosols. ORNL distributed active archive center., https://doi.org/10.3334/ornldaac/1581.

Wunch, D., et al., 2011. The total carbon column observing network. Philos. Trans. R. Soc. A 369, 2087–2112. https://doi.org/10.1098/rsta.2010.0240.

Xiong, X., Barnet, C., Maddy, E.S., Gambacorta, A., King, T.S., Wofsy, S.C., 2013. Mid-upper tropospheric methane retrieval from IASI and its validation. Atmos. Meas. Tech. 6 (9), 2255–2265.

Zhou, L., Divakarla, M., Liu, X., 2016. An overview of the joint polar satellite system (JPSS) science data product calibration and validation. Remote Sens. 8, 139. https://doi.org/10.3390/rs8020139.

Zhou, L., Divakarla, M., Liu, X., Layns, A., Goldberg, M., 2019. An overview of the science performances and calibration/validation of joint polar satellite system operational products. Remote Sens. 11 (6). https://doi.org/10.3390/rs11060698.

Chapter 19

Sea surface temperature validation and blended analysis

Christopher J. Merchant[a], Owen Embury[a], Chelle Gentemann[b], John J. Kennedy[c], Elizabeth C. Kent[d], Peter J. Minnett[e], and James While[c]
[a]Department of Meteorology and National Centre for Earth Observation, University of Reading, Reading, United Kingdom, [b]Farallon Institute, Petaluma, CA, United States, [c]Met Office, Exeter, United Kingdom, [d]National Oceanography Centre, Southampton, United Kingdom, [e]Department of Ocean Sciences, Rosenstiel School of Marine and Atmospheric Science, University of Miami, Miami, FL, United States

Chapter outline

1. Sea surface temperature products: An overview — 337
2. In situ measurements in SST validation — 342
3. Blended analysis of SST — 346
4. Outlook — 348
Acknowledgments — 349
References — 349

I can live with doubt and uncertainty and not knowing.

Richard Feynman

1. Sea surface temperature products: An overview

Blended analyses of sea surface temperature (SST) are data products made from multiple streams of measurements merged into a convenient form for users. The availability of such products as a sustained service is important for many reasons, related to the fundamental roles SST plays within Earth's biogeochemical and physical systems (O'Carroll et al., 2019). The atmosphere and ocean interact in complex ways that affect the surface marine environment directly, affect the development of weather systems, and ultimately influence the climate of Earth and its evolution. For shipping, exploitation of ocean resources, tourism, oceanographic and weather forecasting, environmental policy and management, energy systems, and many other users, operational SST products are crucial. Long-term, stable, and uncertainty-quantified climate data records of SST are also required within climate science and climate services.

This chapter focuses on two aspects of the use of in situ measurement of SST in connection to operational products: validation of satellite observations and use of in situ measurements in and with blended SST products (including climate data records).

Satellite observations are central to achieving worldwide, daily estimation of the SST at the spatial resolutions (1 km to 25 km) needed, which cannot feasibly be achieved by in situ measurements across the vast scale of the global oceans. Estimates of SST from satellites are, of course, indirect, based on inference from electromagnetic radiation measured hundreds or thousands of kilometers above the ocean surface (e.g., Anding and Kauth, 1970; Walton et al., 1998; Merchant et al., 2008). In situ measurements of SST have therefore been used both to formulate and validate the "retrievals" obtained by satellite remote sensing.

Measurements of SST from ships and buoys have variously been used directly within the creation of blended analyses too (e.g., Reynolds, 1988; Donlon et al., 2012; Fiedler et al., 2019). Over time, in situ measurements have become less directly influential on the quantitative SST values obtained in blended analyses, as satellite observations have increased in both accuracy and density of coverage. However, in situ observing systems remain crucial to anchor blended analyses via various means (to be discussed), in addition to their use for assessing marine climate change on centennial scales. Indeed, in situ observing systems have undergone a parallel evolution toward denser patterns of measurement, and smaller uncertainties as requirements by users have become more demanding.

Observations of SST have been made aboard ships with sampling in the current digital archives sufficient to generate estimates of the global mean since the 1850s (Kennedy et al., 2019). Measurements recorded in ship logbooks that are yet to be digitized have the potential to extend this record further back in time. The challenge of producing a homogeneous record spanning observations made from observing platforms ranging from sailing ships to moored and drifting buoys and with a range of different methods (Fig. 1) is significant (Kent et al., 2017) and ongoing. The earliest systematic measurements were made from sailing ships. Initially, the primary measurement was of air temperature, but over time the number of measurements of water temperature increased based on recording the temperature of seawater samples captured using buckets that were likely made of wood (Kent and Kennedy, 2021). Although detailed records of how the measurements were made (known as observational metadata) is not typically available for every ship, there is documentary evidence of a move away from wooden buckets to canvas buckets for ease of use, with an associated change in the characteristics of the measurements (Kennedy et al., 2011). The advent of steam ships meant that many ships could report the temperature of the pumped water used to cool the engines, the characteristics of these observations depends critically on where in the pumped system the measurement was taken. These measurements were made at greater depths than those from bucket samples, but despite this, these were often biased warm if the water sample was near the hot engine room. The tendency for these engine room intake (ERI) measurements to be too warm has decreased over time, probably due to improved technology for remote reading with an increased recognition of the potential biases. Despite this, significant variations between observations from different ships remain (Kennedy, 2014). Dedicated sensors measuring through the ship's hull became more common from the 1990s and showed smaller biases and less volatility than the ERI measurements. There has been substantial recent progress toward reconciling these differences largely due to a recognition of the importance of reconciling measurements from the different measurement types, and strongly motivated by the offsets becoming apparent between SSTs from ships and those from drifting buoys. The derivation of bias adjustments for ship-based SST requires the development of an error model for the observations (Kent et al., 2019) that ideally depends on the measurement method, protocols, and the ambient conditions (Kent et al., 2017). The availability of information on all these factors is extremely patchy in the International Comprehensive Ocean-Atmosphere Data Set (ICOADS) observational archive (Freeman et al., 2017). In the worst case, there is no information identifying the originating ship (Carella et al., 2017) or on the methods, protocols, or prevailing conditions. Observations either recorded or digitized from ship's logbooks in the past 20 years or so typically have more information available. Kennedy et al. (2019) provide a recent assessment of measurement methods for the ship observations

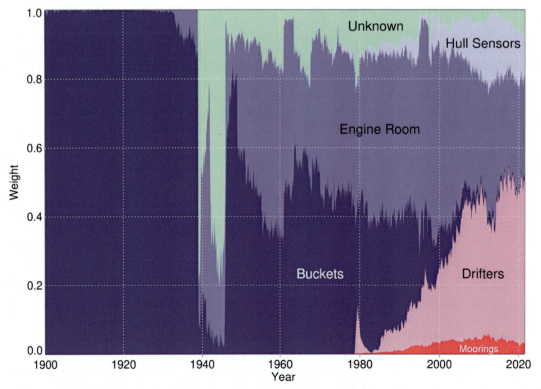

FIG. 1 Fraction of annual sea surface temperature in situ observations by type of observation.

in ICOADS since the 1950s. An initial, uncertain assignment of measurement method was made using available measurement method metadata and country information, which was then refined using observed offsets between groups and estimated biases. The results were broadly consistent with an independent method (Carella et al., 2018). This picture has since been extended by Chan et al. (2019) and Chan and Huybers (2021) who quantified offsets between different groups of observations linked to different types of buckets and identified a larger proportion of engine intakes around the World War 2 period, which helped to reduce a pervasive bias in this period when the characteristics of the observations changed rapidly.

In recent decades, the contribution of SST measurements from moored and drifting buoys to the global record has increased dramatically. Although ship-based in situ observations remain important in the calculation of global average SST up to the present day (Kennedy et al., 2019), arguably of most value to developers of satellite-derived data products would be an improved record of observations from ships in the 1980s and 1990s before coverage from drifting buoys became widespread. This will require the extension and application of methods such as those used by Carella (2017) who applied measurement method-dependent models for SST bias to observations made by individual ships. Complementing the long record of in situ SST observations from ships and buoys made at depths between a few cm and tens of meters below the surface is the smaller number of measurements of the radiometric skin temperature providing measurements more directly comparable to those from satellites (Section 3.1.2), albeit much more sparsely.

Blended SST analyses may also be referred to as "level 4 products," or "L4 SST." This jargon relates to conventional levels of satellite data processing in which higher levels (larger "L" numbers) mean data that are more highly processed. Higher levels are perhaps easier to use for users, but are farther removed from the original satellite and in situ measurements. Questions of validation of SST are somewhat dependent on which level is being discussed, and therefore it is worthwhile to give an overview of the sequence of processing that typically leads to a blended SST product.

Satellite operating agencies undertake the highly technical business of generating the level 1 (L1) satellite data from which SST is retrieved. The L1 data consist of values representing the electromagnetic radiation captured by the satellite sensor, typically at a number of different wavelengths for any given location. The wavelengths from which SST can be estimated are in the infrared (IR) or microwave (MW) portions of the electromagnetic spectrum. The ocean surface emits radiation with an intensity that depends on surface temperature across a wide range of IR and MW frequencies, which is the fundamental reason remote sensing of SST is possible. The wavelengths used for remote sensing must also have the property that a useful proportion of the at-satellite signal originates from the surface emission, which means, among other things, that the atmosphere has to be relatively transparent at those wavelengths. Finer spatial resolution (100 m to 1 km) is possible when using IR wavelengths, but clouds are not transparent to the IR, which limits the SST-relevant coverage achieved. MW observations are coarser ($\sim$50 km), but sensors can "see-through" non-raining clouds and detect the surface emission. The uncertainty of IR-based SSTs tends to be better than that of MW-based SSTs, but the value of near all-weather observation can be significant (Wentz et al., 2000).

The retrieval of SST therefore involves identifying the locations within the satellite imagery where SST can validly be obtained and then performing the inference of the SST from the satellite measurements for those locations. This process is shown for an IR example in Fig. 2. The lower panel of the figure displays the content of a level 2 (L2) product, which means that the SST locations are the same as the locations at which the satellite sensor measured radiance.

L2 data organized by satellite swaths are commonly then rendered more easily usable by averaging and/or sampling the SSTs to a fixed regular projection such as a latitude-longitude grid. This creates the "level 3" (L3) products that are often used as inputs to blended SSTs (although L2 products may also be used). L3 products may first be produced uncollated between orbits ("L3U"), while a common strategy is to collect a ay's worth of orbits into a collated product ("L3C"). The greater usability of the L3 products comes at the cost of degraded spatial fidelity and increased uncertainty in the SST of a specific location from resampling effects (Fig. 3).

Level 4 (L4) products are often referred to as analyses. Generally, at this stage, satellite and in situ data are both used, i.e., these are blended products, although satellite-only analyses are also made. SST analyses are gap-free estimates of temperature, which implies that data gaps within the analysis time window (see Fig. 3) must be filled by some useful form of interpolation. As discussed further below, useful interpolation methods are relatively sophisticated, making use of constraints from earlier analyses, climatology and accounting for the feasible scales of persistence of SST anomalies in time and horizontally. SST is not observed in regions of sea-ice, although for high ice concentration it can be taken to be close to $-1.8°C$. Often the observation density both from satellites and in situ measurements is low in the marginal ice zone and sea-ice periphery. An SST analysis method must therefore also handle the transition between observations of SST and the ice-covered regions where the SST is the freezing temperature of sea water. Typical contemporary SST analyses aim for resolution of order 10 km and daily estimates, although because of observation gaps, true feature resolution tends to be lower on average, even when the analyses are presented to users on grids as fine as $\sim$1 km

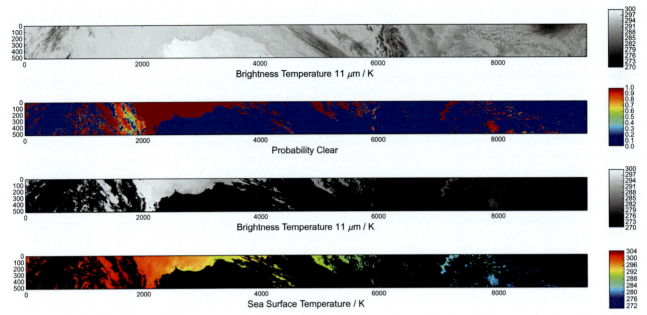

FIG. 2 The satellite measurements obtained from one quarter of an orbit round the Earth (from an Along Track Scanning Radiometer). The swath passes roughly northward *(from left to right)* along the North West African coast line and crosses the northern United Kingdom at around 7000 km. *Top panel*: clouds, being cold, are often obvious as dark features, and warm land can be seen *(bright)*. *Second panel*: an assessment of where in the swath the measurements are representative of clear-sky over ocean waters *(red areas)*. *Third panel*: as the top, except locations not good for SST are masked in black. *Lower panel*: SSTs obtained. *(Adapted and reproduced from Bulgin, C., Merchant, C.J., 2014. Steps in Obtaining Sea Surface Temperature from Infra-Red Imagery. figshare. Figure. https://doi.org/10.6084/m9.figshare.1060227.v1 under CCBY 4.0 license.)*

(Reynolds et al., 2013). This is because mean densities of observation are on average of order 10 to 100 km^{-2} yr^{-1} (Merchant et al., 2019), limited by cloud cover.

An issue that must be handled in blending data of different sources is that "SST" measurements may reflect different depths (Fig. 4). At the air-sea interface, the ocean skin effect changes the temperature within the upper ~1 mm compared to the subskin temperature below the surface sublayer. The magnitude and sign of this skin-to-sub-skin difference depends on the air-sea heat flux, and a value of order −0.2 K (e.g., Wong and Minnett, 2018) is considered to be typical (negative because the ocean generally is losing heat to the overlying atmosphere). MW measurements are more representative of subskin temperatures. Under windy conditions (>6 m/s), subskin temperature may be representative of SST for several meters or tens of meters in depth. With less wind-driven mixing and sustained solar warming, however, a stratified layer may form that is warmer than deeper SSTs by (typically) a few tenths of degree, and up to (rarely) 7 K (Gentemann et al., 2008). The prevalence of such sun-warmed layers has a diurnal cycle, with de-stratification occurring usually, although not always, overnight. Bucket, drifting buoy and mooring SSTs are usually obtained within the warm layer when such exists, although under more extremely calm conditions there can be significant stratification above these measurement depths. Ship SSTs would be considered typical of the upmost several meters of the ocean under most conditions, although the degree to which the passage of the ship causes near-surface mixing is likely quite variable.

Some blended SST product suites and analyses span up to four decades. Because SST is a recognized essential climate variable (Bojinski et al., 2014; Centurioni et al., 2019), this naturally raises the question as to whether such products can be considered as climate data records (CDRs). A CDR for SST must be constructed with due attention to the long-term stability of observation, because an objective is to be able to quantify slow climatic changes in temperature from data whose daily to decadal variability is relatively larger in magnitude (e.g., NRC, 2000). In the context of CDRs, therefore, greater concern is placed on questions of traceability to measurement standards (e.g., Minnett and Corlett, 2012) and quantification and reporting of uncertainty (e.g., Merchant et al., 2017). A CDR that blends in situ and satellite is likely effectively to rely on the in situ record to provide the long-term "calibration" of the record—i.e., the CDR's stability is that of the in-situ measurements used. Alternatively, by paying close attention to harmonization of SST across satellite missions (e.g., Merchant et al., 2020), it has proved possible to construct satellite-only CDRs for SST (e.g., Merchant et al., 2012, 2019) that closely corroborate the picture of marine climate change inferred independently from the in situ record (Hausfather et al., 2017).

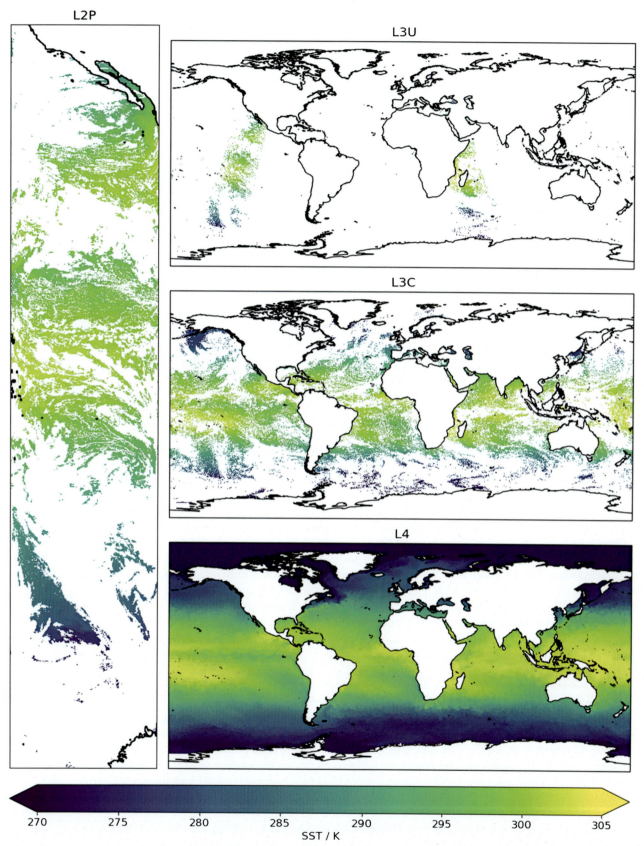

FIG. 3 Sea surface temperature content in data in different product levels. L2P data are on the original satellite viewing geometry, i.e., an orbital strip (here, from an AVHRR) with gaps in SST from cloud cover. The L2P data are gridded per orbit to L3U (uncollated gridded data). One day's worth of L3U data from a given sensor is collated to form the L3C product (collated gridded data). Data from multiple sensors are merged and interpolated to give the daily gap-free SST field of the L4 analysis. All these levels of data are generated within the European Space Agency's Sea Surface Temperature Climate Change Initiative (ESA SST CCI v2.1). *(Reproduced from Embury, O., Merchant, C.J., 2019. ESA SST CCI v2.1: Data in Different Product Types. figshare. Figure. https://doi.org/10.6084/m9.figshare.8305910.v1 under CCBY 4.0 license.)*

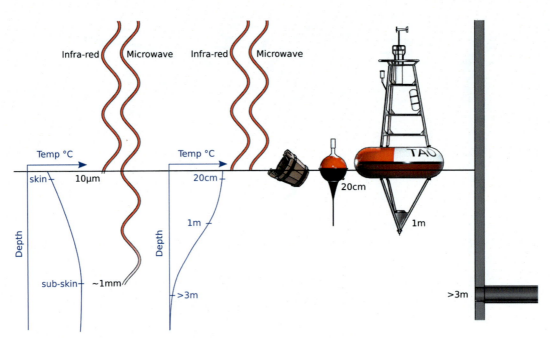

FIG. 4 Inter-related types of SST. From *left* to *right* are illustrated the depths of SST measured by IR and MW techniques (~10μm to ~1mm), buckets and drifting buoys (10–20cm), moored buoys (~1m), and automated ship measurements (~3m).

2. In situ measurements in SST validation

In-situ measurements of SST are obtained by a number of technologies, including some emerging technologies to be discussed in Section 4 below and shown in Fig. 5. Here, the focus is on the use of in situ measurement for satellite SST validation. Validation using temperatures measured by drifting buoys, Argo buoys and ship-borne radiometers will be considered.

It is useful to be explicit about the purposes of validation of SST. The most obvious purpose is to assess how closely the SSTs being obtained from a given satellite mission correspond to the true SST: are there any systematic differences ("bias") and how widely are the satellite SST dispersed ("noise")? There are a number of points to be clarified about such "validation," which are listed here and discussed in more depth below. First, the buoy or cruise data are not the true SST, but are themselves measurements of SST with their own error characteristics, which must be accounted for when interpreting comparisons. Second, there are sources of difference in such a comparison that are not errors, but the result of true geophysical differences arising because the various measurement technologies sample the ocean differently in the vertical, horizontal and time dimensions. Third, various aspects of the design of the validation may depend on the exact question we want to ask about the satellite SSTs. For example, do we wish to assess the SST retrieval method, or to assess the L2 SST product that results not only from an implementation of the retrieval but also from steps such as cloud screening and quality indication? Fourthly, the meaning and generality of the comparison of SSTs differ according to whether the sources being compared are or are not independent. If an SST retrieval relies on in situ data in some way, the conclusions that can be drawn from the comparison are rather less general than if the two SST data sets are highly independent.

A further purpose of SST validation is assessment of our degree of understanding of the uncertainty of the satellite SST. (We take it for granted here that every measurement should be provided with an evaluation of the uncertainty of the measurement, Merchant et al., 2017.) The uncertainty in satellite SST is highly variable (by a factor of several times). The uncertainty varies with respect to atmospheric and surface conditions, the satellite sensor, and observing conditions such as view angle and instrument state. For this reason, every SST should be accompanied by an uncertainty estimate specific to the context in which the retrieval was made. To do this, the variability in uncertainty needs to be understood and modeled, and the uncertainty needs to be validated.

The above point of view is in contrast to the practice of assuming that validation statistics provide or define uncertainty information. The latter, outdated approach is embedded in the long-standing community convention of single-sensor error statistics (SSES) discussed further below.

Similarly, if a quality indicator is supplied in an SST product, it is appropriate to validate that the quality indicator fulfills the purpose for which it is given. The purpose of quality indicators is not standardized, but a strong case can be

Sea surface temperature validation and blended analysis **Chapter | 19** 343

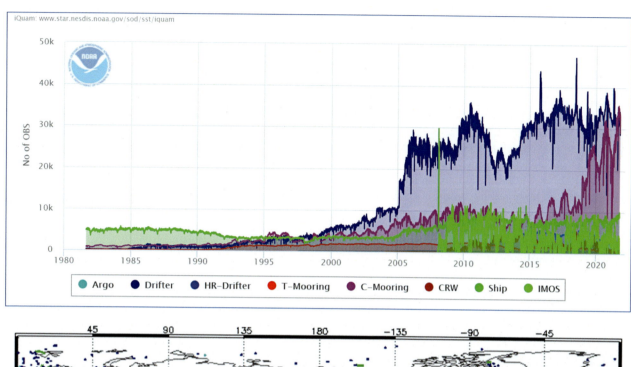

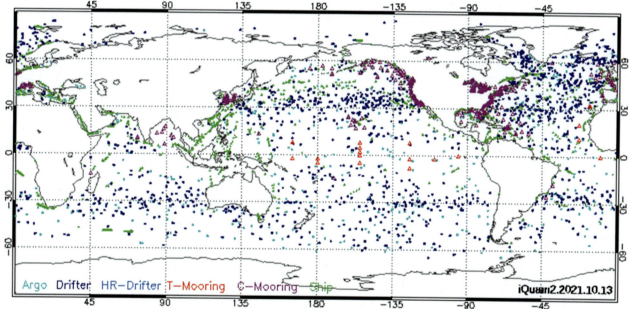

FIG. 5 In situ observations of SST have only been numerous since approximately 2005, and while there is fairly even global distribution of observations, the high latitudes remain under-sampled. *(Top)* Number of in situ SST observations over time and *(bottom)* spatial distribution on October 13, 2021. The Tropical Pacific TAO array was deployed in 1992, the Tropical Atlantic PIRATA in 1999. https://www.star.nesdis.noaa.gov/socd/sst/iquam/?tab=0&dateinput_year=2021&dateinput_month=10&dayofmoninput_day=13&dateinput_hour=00&dayofmon=daily&qcrefsst=_qcrey&qcrefsst=_qccmc&outlier=qced#qmap

made that it should tell users something different from uncertainty (Merchant et al., 2017). An SST with higher uncertainty is not lower quality if the higher uncertainty is correctly reflected in the uncertainty attributed to the SST measurement. A measurement should be flagged as lower quality; however, if there is evidence that the error in the SST may be greater than the uncertainty would suggest. This may arise if the retrieval context can lead to SST errors that are not accounted for in the uncertainty model. So, it is useful if a quality indicator flags conditions when the SST uncertainty estimate is more likely to be invalid (e.g., conditions where an assumption made in the retrieval or uncertainty evaluation is violated). Defined in this way, a quality indicator can be validated using in situ data by assessing whether data flagged with lower quality are more often disagreeing with the validation data by an amount that is not explicable by the attributed uncertainty.

Given these considerations regarding SST validation, it is important to have a clear validation strategy. The strategy should be tied to the specific, stated purpose(s) of the validation and the target properties of a particular SST product. The strategy should address the following aspects:

1. The specific purpose of the validation exercise. Does the validation assess an SST retrieval method, or an SST product?
2. The degree of independence of the satellite SST from the validation data, and how any dependence constrains the interpretation of the results. This is considered further below.
3. The SST-related quantities to be validated. Validation of the satellite SST, the uncertainty and the quality indicator attributed to the satellite SST can all be considered, and ideally will all be addressed.
4. The nature of the satellite SST measurand. Is the satellite SST a skin (IR) or subskin (MW) temperature retrieval, or is the measured value somehow tuned to represent on average an aspect of SST which the satellite does not actually observe (i.e., a satellite SST tuned to temperatures at some depth below the subskin)? Is the SST product intended to represent an instantaneous snapshot of surface temperature (e.g., L2 SST) or a spatio-temporal average (e.g., a daily L4 SST on a grid)?
5. The nature of the validation SST data. What area or track is observed, at what depth and with what temporal sampling? How best to use the available in situ data to correspond to the purpose of the validation and the nature of the satellite data? Can complementary use of multiple types of in situ data strengthen the validation (e.g., radiometer, drifting buoy and Argo measurements)?
6. Consideration of true geophysical differences. Given different location-depth-time representativity between the datasets, is a physical or statistical model to be used to estimate and reduce (some of) the true geophysical differences that arise? What uncorrected variability from representativity effects remains after any correction is applied? Where no correction is to be applied or the uncorrected variability remains significant, what steps will be taken in the interpretation of results to account for this?
7. The selection and preprocessing of matches between the validation and satellite data. Decisions such as spatio-temporal matching criteria are obvious considerations here and have been discussed in the literature; there is a trade-off between the looseness of matching and the statistical power (number of matches) to be considered whose balance likely varies over the satellite era because of the improving in situ observation density over time. Other decisions to be made in this category include: filtering or not of "outliers" in either data stream; filtering by observation condition (e.g., suppressing low-wind-speed matches if validating a foundation SST product); aggregation of data (e.g., averaging a full day of buoy observations if validating a daily L4 SST); and re-weighting the match-up distribution of opportunity to be more appropriate to the purpose of the validation (e.g., reducing the weighting or preponderance of less cloudy areas relative to cloudier regions so that results are more globally representative).
8. The quantitative metrics to be employed for assessment. Traditional vs. robust statistics (such as the trimmed mean or scaled median absolute deviation) emphasize different aspects of the distribution of difference. Beyond aggregate statistics, what potential geographical, temporal, and functional dependencies can be explored for systematic tendencies? Useful choices include looking for time-of-day, seasonal, geographical, and long-term temporal dependencies, along with functional factors such as wind speed, satellite zenith angle, instrument temperature, proximity to cloud, etc. For validating uncertainty and quality indicators, distributional metrics need to be chosen.

Once a sufficient number of comparisons between the satellite retrievals and surface measurements has been compiled, a statistical analysis can be made to assess the accuracy of the derived SST and the of the effectiveness of the retrieval algorithms, both for cloud screening and correcting the effects of the intervening atmosphere. A common approach has been to calculate the means and standard deviations of the differences and attribute these to the errors and uncertainties of the satellite retrieval, under the assumption that the surface measurement is more accurate than the satellite retrievals. A better approach is to compare the data sets in the light of their uncertainties and geophysical variability (Immler et al., 2010).

When comparing two data sets of matched SST measurements, 1 and 2, the following statistic may ideally be expected to be normally distributed with unit standard deviation:

$$(x_2 - x_1)/\sqrt{u_1^2 + u_2^2 + v^2}$$

where x_i represents an SST measurement, u_i is the corresponding uncertainty, and v is the expected real geophysical variability expected between the measurements. The geophysical variability arises from differences in depth, time of measurement and spatial representativity. While this approach is more effort than finding the mean and standard deviation of $x_2 - x_1$, since some knowledge of expected uncertainties and variability is required, it is far more instructive to consider measurement differences in this context. One could almost ask, what use is the standard deviation of difference in two data sets without this context? Not much. It is just a number. Deviations from normality of the above statistic tell us that our understanding is lacking somewhere: there are errors from unknown sources, or deficiencies in our understanding of uncertainty, or the SST variability is not what we thought it was.

Why is it important to test understanding of SST measurement and uncertainty so rigorously? One argument is that it is important to characterize the uncertainty and quality of satellite SSTs everywhere, not only where validation data are abundant. Equatorial and high latitude regions tend to have relatively less abundant drifting buoy SSTs available for validation purposes, and in earlier decades, there were large stretches of ocean with very few in situ measurements. Only by validating both the satellite SSTs and the uncertainty model of the satellite SSTs can there be confidence in areas without in situ measurements in the satellite measurements and their attributed uncertainty values.

The independence of the data to which satellite SSTs are compared in validation is important. For regression-based SST retrievals, these comparisons serve as the basis of the derivation of the coefficients used in the nonlinear atmospheric correction algorithm (Walton et al., 1998; Kilpatrick et al., 2015), a subset of in situ should be withheld to provide independent data for the accuracy assessment. The in situ subsets should be partitioned by in situ platform so that a given in situ sensor contributes to only one subset, otherwise true independence of errors and situational context between the subsets is not achieved.

In the framework of the Group for High Resolution Sea-Surface Temperature (GHRSST), the concept of Sensor Specific Error Statistics was developed (Donlon et al., 2007) with the intention that every SST retrieval should be accompanied by estimates of the mean and standard deviation of difference from a reference, the standard reference being drifting buoys. For example, one approach to satisfy this objective was developed for MODIS through the analysis of the comparisons with measurements from drifting and moored buoys in a database, the MarchUp DataBase (MUDB), which involved creating a seven-dimensional array of parameters, a hypercube, the variations of which are likely to influence the accuracy of the satellite SST retrievals: latitude, month, satellite zenith angle, SST, brightness temperature differences between 11 and 12 μm measurements as a proxy for water vapor, assigned quality flag, and whether it is a day or night measurement (Evans and Kilpatrick, 2007). The records in the MUDBs were assigned to the appropriate cell in the array and the mean and standard deviations were calculated. For each pixel, the relevant cell was identified and the mean, or median, and standard deviation, or robust standard deviation, in that cell were used as the relevant SSES. This approach has been used for MODIS on Terra and Aqua (Kilpatrick et al., 2015) and for the S-NPP VIIRS (Minnett et al., 2020), shown in Fig. 6.

There are issues, however, with the hypercube approach associated with the uneven population of the cells, with some being empty, and the risk of discontinuities at cell boundaries. A study by Kumar et al. (2021) has applied several Machine Learning techniques to the Aqua MODIS MUDB to characterize the SSES, with the result that the best, the Cubist model, identified eight Rule Sets into one of which any retrieved SST falls. It is the properties of these Rule Sets that determine the SSES.

The objective of the SSES is to provide guidance on how best to use each SST retrieval, such as how they should contribute to downstream processing to produce blended analyses; retrievals with poorer SSES should be given less weight than those with better SSES. The use to which the SSES for each retrieval is put depends on the application. Correcting the retrieved SST by the SSES bias estimate should improve the accuracy of the derived SST, but in so doing it eliminates information that can be used to refine the retrieval algorithms and correct systematic shortcomings in the algorithm formulations.

As discussed in Chapter 6, the SI-traceable calibration of shipborne radiometers provides the mechanism for merging SST fields from multiple satellite radiometers to generate multidecadal time series, and therefore the radiometer data should be reserved for validation of the satellite-derived SSTs and not lose the element of independence that would follow if they were to be used in algorithm generation. Similar arguments can be made for other sources of validation measurements, including Argo profilers and near-surface measurements from autonomous gliders.

The discussion above is relevant to validation L2 and L3 SST products from satellites. When it comes to the blended L4 analysis products discussed in the next section, additional considerations apply. Blended analyses generally use drifting buoy and moored buoy observations as part of the data stream. In analyses, such data may receive more or less weight as a constraint on the local blended SST outcome. This means it is impossible to make an objective comparison of L4 products against these in situ data sources. The reason is that it is not clear whether an L4 product that agrees closely with these data sets is objectively more truthful to the real-world SST, or is merely more heavily weighted toward in situ data than satellite products.

For this reason, the GHRSST recommendation has been to reserve the near-surface observations of Argo profiling floats as independent, low-uncertainty references for the purpose of validating blended L4 analysis products. Despite the immense and unique value of Argo temperature profiles for quantifying the subsurface ocean, the numbers of SSTs obtained from the network is too low to have a useful impact of L4 analysis products. Argo-based SSTs are, however,

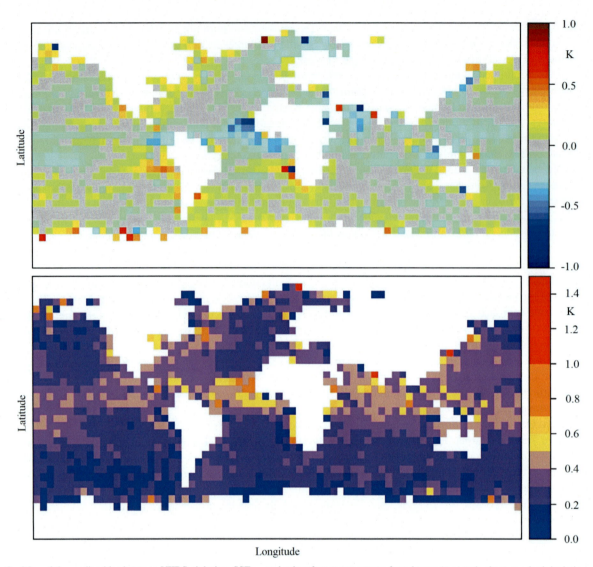

FIG. 6 Map of the median bias between VIIRS nighttime SST$_{skin}$ and subsurface temperatures from buoys *(top)* and robust standard deviation of the differences *(bottom)*. The buoy temperatures have been adjusted for the effects of the mean cool skin layer by subtracting 0.17 K from their measurements. *(From Minnett, P.J., Kilpatrick, K.A., Podestá, G.P., Evans, R.H., Szczodrak, M.D., Izaguirre, M.A., Williams, E.J., Walsh, S., Reynolds, R.M., Bailey, S. W., Armstrong, E.M., Vazquez-Cuervo, J., 2020. Skin sea-surface temperature from VIIRS on Suomi-NPP—NASA continuity retrievals. Remote Sens. 12, 3369. https://doi.org/10.3390/rs12203369, (Creative Commons CC BY 4.0).)*

numerous enough and well distributed enough to be immensely valuable as validation data, particularly for blended analyses for the reasons just stated. Not all L4 producers adhere to the GHRSST recommendation, but it remains good community practice.

3. Blended analysis of SST

For applications requiring continuous SST across the ocean, gap-free analyses (L4) are preferred products among users. The large majority of L4 analyses target daily-mean SST estimation (development of sub-daily analysis is discussed below). Daily analyses blending satellite and in situ data are typically made at 5 to 25 km grid resolution (a few are finer or coarser). Real feature resolution is, on average, typically coarser than this grid resolution in practice (>20 km, Reynolds et al., 2013) and varies with cloud coverage. Several near-real-time (NRT) global operational analyses are associated with GHRSST, facilitating common data standards and comparison through a multimodel ensemble. Yang et al., 2021, have compared six relatively long analyses from a climate perspective.

All L4 products, including those using near-all-weather microwave SST observations, must interpolate across data gaps from the combined observational networks. A variety of infilling techniques exist, but in general it can be said that (1) SST anomalies, not absolute values, constitute the interpolated field, (2) to preserve ocean spatial structure where possible in observational voids, an element of persistence of previous analyses is included, (3) L4 products are generally smoother (have smaller gradients) than the observations at L2 and L3 product levels that go into them, and (4) where multiple observations enable averaging-down of errors, L4 SSTs may be less uncertain than the individual observations contributing to the analysis.

Given the orders-of-magnitude numerical preponderance of satellite observations within the data flows into a blended analysis, the use of in situ data at all requires some comments. Other than the L4 product of Merchant et al. (2019), all analyses blend in situ data with satellite data (to our knowledge). Where there is a satellite data gap, such an in situ observation will naturally have significant influence on the interpolated field. Where the in situ observation coexists with satellite observations of the same locality, it is typical to privilege the influence of the in situ measurement of SST by some means. The in situ observation may be considered likely to have smaller systematic uncertainty (providing it has passed quality filters) or otherwise be directly upweighted in estimates around its vicinity. This practice anchors the analysis to in situ measurements (as discussed in the validation section above, regarding the preservation of Argo as an independent means of assessing analysis uncertainty), which may be valuable in maintaining stability of the analysis in the face of changes in the satellite observing constellation. On the other hand, some satellite data streams have been considered to be of sufficiently low bias also to be used as low-bias references for other components of the observing network (Good et al., 2020).

It is important for a given SST analysis to have a clearly stated measurand. The most common choices are analyses targeting foundation SST or SST representative of drifting-buoy depth. The foundation temperature is not observed by satellite observations or, most of the time, in situ observations (except in some circumstances by Argo or mooring profiles). For foundation SST analysis, data are used if physical reasoning suggests the difference of the measurement from the (hypothetical) foundation SST is adequately small. Thus, for example, satellite observations may be included at night and during the day wherever the wind speed is above a threshold that promotes vertical mixing (Good et al., 2020). Analyses for drifting-buoy depth SST target the daily mean at a nominal 20 cm depth, which therefore includes an element of near-surface diurnal cycle, and should therefore be typically a few tenths of kelvin warmer than a foundation analysis. In this case, all data tend to be used. Satellite observations that explicitly measure skin temperature should be adjusted to compensate for the skin effect and other surface stratification.

The skin temperature of the ocean can vary considerably on diurnal and subdiurnal time scales. Indeed, diurnal signals of several degrees have been observed on occasion (see, for example, Castro et al., 2014). The factors driving these diurnal variations are the loss of heat to the atmosphere, absorption of solar radiation in the top few meters of the water column and vertical mixing of water by the wind or stratification changes. More specifically, between the skin SST and sub-skin SST, heat losses dominate leading to a cool thermal skin layer (or cool skin) about 1 mm thick, while between the surface and the foundation depth absorption of solar radiation can cause the water to warm giving rise to a so called "warm layer."

Producing a blended analysis of skin SST from observations can be challenging. Almost all in situ observations are taken at a finite depth of a few 10's of cm (drifting buoys) or a meter or more (moorings and ships). Consequently, in situ observations only measure part of the diurnal warming signal. Only radiometers, principally mounted on satellites, can measure the skin and subskin temperatures. Infrared (IR) detectors directly sample the skin SST, while microwave instruments are more sensitive to the subskin. However, satellites have their own issues. Polar orbiting satellites will only observe a point on the Earth's surface twice a day, once during daylight, and once at night, which is far short of the sampling frequency needed to resolve any diurnal variation. Geostationary satellites can provide much better temporal coverage, but only observe part of the globe and cannot observe high latitudes. Satellite measurements can be blocked by cloud (IR) and rain (IR and microwave) and have their accuracy degraded by atmospheric aerosols. Furthermore, diurnal changes in SST can be underrepresented in satellite signals as retrieval algorithms can lack sensitivity to small changes in SST (Merchant et al., 2009).

Despite the challenges, there have been several attempts to create analyses of the diurnal component of SST. To fill in gaps in space and time such analyses generally contain a model component of the cool skin, warm layer, or both. Marullo et al. (2014) describe a diurnal analysis of the Mediterranean sea using an optimal interpolation technique to merge SST observations from the geostationary SEVIRI instrument and model outputs from the Mediterranean Forecasting System (Pinardi and Coppini, 2010). An implementation of the techniques described in Marullo et al. (2014) is used to generate hourly diurnal analyses of the Mediterranean and Black Seas; available from the Copernicus marine service (https://marine.copernicus.eu/). In Pimentel et al. (2008), a data assimilation methodology is proposed for combining gridded SST observations with a grid of 1D high-resolution mixed layer models. A more sophisticated 4DVar assimilation system is used in While et al. (2017) to assimilate various polar and geostationary satellite observations into a global implementation of the

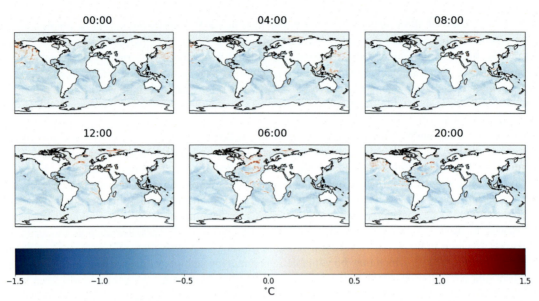

FIG. 7 Difference between skin SST and foundation SST for the June 26, 2021. The plot shows the signal for a particular time of day; time is given in UTC. Results are from the diurnal analysis system described in While et al. (2017). Units are degrees Celsius. Negative values are due to the cool skin component.

Takahashi warm layer (Takaya et al., 2002) and Artale cool skin (Artale et al., 2002) models. The system described in While et al. (2017) is used operationally to produce daily global analyses of the diurnal variation in SST with outputs available from the Copernicus marine service (https://marine.copernicus.eu/). An example of the diurnal variation in the While et al. (2017) system for one day is given in Fig. 7.

4. Outlook

In this section, relevant developments are discussed that are yet to be fully exploited (saildrones) or are in the future (satellite missions in development).

The existing in situ SST observations used for algorithm development and validation (Fig. 5) have provided valuable SST data for decades. Each observation type has its strengths and weaknesses. Moored buoys provide ancillary air-sea observations, but only measure at a single location and are expensive to deploy and maintain. Drifting buoy observations are able to fill in many remote regions and are relatively inexpensive to deploy but only measure SST and sea-level pressure. Ships are able to adaptively sample across fronts and other interesting features, collecting a full suite of air–sea measurements, but are extremely costly.

A new generation of air–sea autonomous uncrewed surface vehicles (USVs), saildrones, are able to provide a suite of air–sea measurements. Saildrones are solar and wind powered vehicles that can be deployed almost anywhere in the world and remain at sea for up to a year (Gentemann et al., 2020). An Iridium link to the vehicles allows for rapid course adjustments and uplinks of near-real-time data. These vehicles have been deployed on ~50 cruises, in locations around the world, the Arctic, across the Pacific, and around Antarctica. They are providing valuable data for SST algorithm development and validation in remote regions that are otherwise too expensive or difficult to sample routinely. These steerable vehicles can be tasked to explore SST features, including diurnal warming features and ocean fronts, helping scientists to better understand uncertainties in SST measurements.

As mentioned above, microwave radiometry from space obtains SST observations in nearly all weathers, irrespective of cloud. Regions of precipitation and within ~100 km of coasts and sea-ice have not been observable for SST from previous sensors. Other limitations have included relative coarse spatial resolution (>50 km feature resolution) and higher uncertainty than infrared SSTs. The ambitious Copernicus Microwave Imaging Radiometer in development by the European Space Agency will significantly address these limitations (Kilic et al., 2018). The mission's challenging technical specification will support feature resolution similar to the feature resolution of the best current SST analysis products (~30 km) and lower uncertainty than earlier microwave radiometers. Simulation studies suggest the impact of spatially resolved, all-weather coverage on SST analysis products will be significantly beneficial (Pearson et al., 2019), driving analysis uncertainty toward <0.25 K worldwide.

Acknowledgments

We would like to acknowledge Yuling Liu for reviewing our draft and providing constructive feedback.

References

Anding, D., Kauth, R., 1970. Estimation of sea surface temperature from space. Remote Sens. Environ. 1 (4), 217–220.

Artale, V., et al., 2002. Role of surface fluxes in ocean general circulation models using satellite sea surface temperature: validation of and sensitivity to the forcing frequency of the Mediterranean thermohaline circulation. J. Geophys. Res. 107, 1–24. https://doi.org/10.1029/2000JC000452.

Bojinski, S., Verstraete, M., Peterson, T.C., Richter, C., Simmons, A., Zemp, M., 2014. The concept of essential climate variables in support of climate research, applications, and policy. Bull. Am. Meteorol. Soc. 95, 1431–1443. https://doi.org/10.1175/bams-d-13-00047.1.

Carella, G., Kennedy, J.J., Berry, D.I., Hirahara, S., Merchant, C.J., Morak-Bozzo, S., Kent, E.C., 2018. Estimating sea surface temperature measurement methods using characteristic differences in the diurnal cycle. Geophys. Res. Lett. 45, 363–371. https://doi.org/10.1002/2017GL076475.

Carella, G., Morris, A.K.R., Pascal, R.W., Yelland, M.J., Berry, D.I., Morak-Bozzo, S., Merchant, C.J., Kent, E.C., 2017. Measurements and models of the temperature change of water samples in sea surface temperature buckets. Q. J. Roy. Meteorol. Soc. 143 (706), 2198–2209. https://doi.org/10.1002/qj.3078.

Carella, G., 2017. New Estimates of Uncertainty in the Marine Surface Temperature Record. University of Southampton. Doctoral Thesis, 173pp.

Castro, S.L., et al., 2014. Comparison of diurnal warming estimates from unpumped Argo data and SEVIRI satellite observations. Remote Sens. Environ. 140, 789–799. https://doi.org/10.1016/j.rse.2013.08.042.

Centurioni, L.R., Turton, J., et al., 2019. Global in situ observations of essential climate and ocean variables at the air-sea interface. Front. Mar. Sci. https://doi.org/10.3389/fmars.2019.00419.

Chan, D., Huybers, P., 2021. Correcting observational biases in sea surface temperature observations removes anomalous warmth during World War II. J. Climate 34 (11), 4585–4602. https://doi.org/10.1175/JCLI-D-20-0907.1.

Chan, D., Kent, E.C., Berry, D.I., Huybers, P., 2019. Correcting datasets leads to more homogeneous early 20th century sea surface warming. Nature 571, 393–397. https://doi.org/10.1038/s41586-019-1349-2.

Donlon, C.J., Martin, M., et al., 2012. The operational sea surface temperature and sea ice analysis (OSTIA) system. Remote Sens. Environ. 116, 140–158.

Donlon, C.J., Robinson, I., Casey, K.S., Vazquez-Cuervo, J., Armstrong, E., Arino, O., Gentemann, C., May, D., LeBorgne, P., Piollé, J., Barton, I., Beggs, H., Poulter, D.J.S., Merchant, C.J., Bingham, A., Heinz, S., Harris, A., Wick, G., Emery, B., Minnett, P., Evans, R., Llewellyn-Jones, D., Mutlow, C., Reynolds, R.W., Kawamura, H., Rayner, N., 2007. The Global Ocean data assimilation experiment high-resolution sea surface temperature pilot project. Bull. Am. Meteorol. Soc. 88, 1197–1213.

Evans, R.H., Kilpatrick, K., 2007. The MODIS hypercube. In: Proceedings of the 8th International GHRSST Science Team Meeting, Melbourne, Australia, pp. 99–100. Available from https://www.ghrsst.org/meetings/8th-international-ghrsst-science-team-meeting-ghrsst-viii/.

Fiedler, E.K., et al., 2019. Intercomparison of long-term sea surface temperature analyses using the GHRSST multi-product ensemble (GMPE) system. Remote Sens. Environ. 222, 18–33. https://doi.org/10.1016/j.rse.2018.12.015.

Freeman, E., Woodruff, S.D., Worley, S.J., Lubker, S.J., Kent, E.C., Angel, W.E., Berry, D.I., Brohan, P., Eastman, R., Gates, L., Gloeden, W., Ji, Z., Lawrimore, J., Rayner, N.A., Rosenhagen, G., 2017. ICOADS Release 3.0: A major update to the historical marine climate record. Int. J. Climatol. 37, 2211–2232. https://doi.org/10.1002/joc.4775.

Gentemann, C.L., Minnett, P.J., Le Borgne, P., Merchant, C.J., 2008. Multi-satellite measurements of large diurnal warming events. Geophys. Res. Lett. 35 (22), L22602. https://doi.org/10.1029/2008GL035730.

Gentemann, C.L., Scott, J.P., Mazzini, P.L.F., Pianca, C., Akella, S., Minnett, P.J., Cornillon, P., Fox-Kemper, B., Cetinić, I., Chin, T.M., Gomez-Valdes, J., Vazquez-Cuervo, J., Tsontos, V., Yu, L., Jenkins, R., De Halleux, S., Peacock, D., Cohen, N., 2020. Saildrone: adaptively sampling the marine environment. Bull. Am. Meteorol. Soc. 101, E744–E762. https://doi.org/10.1175/bams-d-19-0015.1.

Good, S., et al., 2020. The current configuration of the OSTIA system for operational production of foundation sea surface temperature and ice concentration analyses. Remote Sens. 12, 720. https://doi.org/10.3390/rs12040720.

Hausfather, Z., Cowtan, K., Clarke, D.C., Jacobs, P., 2017. Assessing recent warming using instrumentally homogeneous sea surface temperature records. Sci. Adv. 3 (1). https://doi.org/10.1126/sciadv.1601207.

Immler, F.J., Dykema, J., et al., 2010. Reference quality upper-air measurements: guidance for developing GRUAN data products. Atmos. Meas. Tech. 3 (5), 1217–1231.

Kennedy, J.J., 2014. A review of uncertainty in in situ measurements and data sets of sea surface temperature. Rev. Geophys. 52, 1–32. https://doi.org/10.1002/2013RG000434.

Kennedy, J.J., Rayner, N.A., Atkinson, C.P., Killick, R.E., 2019. An ensemble data set of sea surface temperature change from 1850: the Met Office Hadley Centre HadSST.4.0.0.0 data set. J. Geophys. Res. Atmos. 124, 7719–7763. https://doi.org/10.1029/2018JD029867.

Kennedy, J.J., Rayner, N.A., Smith, R.O., Parker, D.E., Saunby, M., 2011. Reassessing biases and other uncertainties in sea surface temperature observations measured in situ since 1850: 2. Biases and homogenization. J. Geophys. Res. 116, D14104. https://doi.org/10.1029/2010JD015220.

Kent, E.C., Kennedy, J.J., 2021. Historical estimates of surface marine temperatures. Ann. Rev. Mar. Sci. 13 (1), 283–311. https://doi.org/10.1146/annurev-marine-042120-111807.

Kent, E.C., Kennedy, J.J., Smith, T.M., Hirahara, S., Huang, B., Kaplan, A., Parker, D.E., Atkinson, C.P., Berry, D.I., Carella, G., Fukuda, Y., Ishii, M., Jones, P.D., Lindgren, F., Merchant, C.J., Morak-Bozzo, S., Rayner, N.A., Venema, V., Yasui, S., Zhang, H., 2017. A call for new approaches to quantifying biases in observations of sea surface temperature. Bull. Am. Meteorol. Soc. 98 (8), 1601–1616. https://doi.org/10.1175/BAMS-D-15-00251.1.

Kent, E.C., Rayner, N.A., Berry, D.I., Eastman, R., Grigorieva, V., Huang, B., Kennedy, J.J., Smith, S.R., Willett, K.M., 2019. Observing requirements for long-term climate records at the ocean surface. Front. Mar. Sci. 6, 441. https://doi.org/10.3389/fmars.2019.00441.

Kilic, L., et al., 2018. Expected performances of the Copernicus imaging microwave radiometer (CIMR) for an all-weather and high spatial resolution estimation of ocean and sea ice parameters. J. Geophys. Res. Oceans 123 (10), 7564–7580.

Kilpatrick, K.A., Podestá, G., Walsh, S., Williams, E., Halliwell, V., Szczodrak, M., Brown, O.B., Minnett, P.J., Evans, R., 2015. A decade of sea surface temperature from MODIS. Remote Sens. Environ. 165, 27–41. https://doi.org/10.1016/j.rse.2015.04.023.

Kumar, C., Podestá, G., Kilpatrick, K., Minnett, P., 2021. A machine learning approach to estimating the error in satellite sea surface temperature retrievals. Remote Sens. Environ. 255, 112227. https://doi.org/10.1016/j.rse.2020.112227.

Marullo, S., et al., 2014. Combining model and geostationary satellite data to reconstruct hourly SST field over the Mediterranean Sea. Remote Sens. Environ. 146, 11–23. https://doi.org/10.1016/j.rse.2013.11.001.

Merchant, C.J., Block, T., Corlett, G.K., Embury, O., Mittaz, J.P.D., Mollard, J.D.P., 2020. Harmonization of space-borne infra-red sensors measuring sea surface temperature. Remote Sens. 12 (6), 1048. https://doi.org/10.3390/rs12061048.

Merchant, C.J., Embury, O., Bulgin, C.E., Block, T., Corlett, G.K., Fiedler, E., Good, S.A., Mittaz, J., Rayner, N.A., Berry, D., Eastwood, S., Taylor, M., Tsushima, Y., Waterfall, A., Wilson, R., Donlon, C., 2019. Satellite-based time-series of sea-surface temperature since 1981 for climate applications. Sci. Data 6, 223. https://doi.org/10.1038/s41597-019-0236-x.

Merchant, C.J., Embury, O., Rayner, N.A., Berry, D.I., Corlett, G., Lean, K., Veal, K.L., Kent, E.C., Llewellyn-Jones, D., Remedios, J.J., Saunders, R., 2012. A twenty-year independent record of sea surface temperature for climate from along-track scanning radiometers. J. Geophys. Res. 117, C12013. https://doi.org/10.1029/2012JC008400.

Merchant, C.J., Harris, A.R., Roquet, H., Le Borgne, P., 2009. Retrieval characteristics of non-linear sea surface temperature from the Advanced Very High Resolution Radiometer. Geophys. Res. Lett. 36 (17), L17604. https://doi.org/10.1029/2009GL039843.

Merchant, C.J., Le Borgne, P., Marsouin, A., Roquet, H., 2008. Optimal estimation of sea surface temperature from split-window observations. Remote Sens. Environ. 112 (5), 2469–2484. https://doi.org/10.1016/j.rse.2007.11.011.

Merchant, C.J., Paul, F., Popp, T., Ablain, M., Bontemps, S., Defourny, P., Hollmann, R., Lavergne, T., Laeng, A., de Leeuw, G., Mittaz, J., Poulsen, C., Povey, A.C., Reuter, M., Sathyendranath, S., Sandven, S., Sofeiva, V.F., Wagner, W., 2017. Uncertainty information in climate data records from Earth observation. Earth Syst. Sci. Data 9 (2), 511–527. https://doi.org/10.5194/essd-9-511-2017.

Minnett, P.J., Corlett, G.K., 2012. A pathway to generating climate data records of sea-surface temperature from satellite measurements. Deep-Sea Res. II Top. Stud. Oceanogr. 77–80, 44–51. https://doi.org/10.1016/j.dsr2.2012.04.003.

Minnett, P.J., Kilpatrick, K.A., Podestá, G.P., Evans, R.H., Szczodrak, M.D., Izaguirre, M.A., Williams, E.J., Walsh, S., Reynolds, R.M., Bailey, S.W., Armstrong, E.M., Vazquez-Cuervo, J., 2020. Skin sea-surface temperature from VIIRS on Suomi-NPP—NASA continuity retrievals. Remote Sens. 12, 3369. https://doi.org/10.3390/rs12203369.

NRC, 2000. Issues in the Integration of Research and Operational Satellite Systems for Climate Research: II. Implementation 0–309–06994-7. National Academy of Sciences, Washington, DC, USA, p. 82.

O'Carroll, A.G., Armstrong, E.M., Beggs, H.M., Bouali, M., Casey, K.S., Corlett, G.K., Dash, P., Donlon, C.J., Gentemann, C.L., Hoyer, J.L., Ignatov, A., Kabobah, K., Kachi, M., Kurihara, Y., Karagali, I., Maturi, E., Merchant, C.J., Marullo, S., Minnett, P., Pennybacker, M., Ramakrishnan, B., Santoleri, R., Sunder, S., Saux Picart, S., Vazquez-Cuervo, J., Wimmer, W., 2019. Observational needs of sea surface temperature. Front. Mar. Sci. 6, 420. https://doi.org/10.3389/fmars.2019.00420.

Pearson, K., Good, S., Merchant, C.J., Prigent, C., Embury, O., Donlon, C., 2019. Sea surface temperature in global analyses: gains from the copernicus imaging microwave radiometer. Remote Sens. 11 (20), 2362. https://doi.org/10.3390/rs11202362.

Pimentel, S., et al., 2008. The assimilation of satellite-derived sea surface temperatures into a diurnal cycle model. J. Geophys. Res. 113. https://doi.org/10.1029/2007JC004608.

Pinardi, N., Coppini, G., 2010. Preface "Operational oceanography in the Mediterranean Sea: the second stage of development". Ocean Sci. 6, 263–267. https://doi.org/10.5194/os-6-263-2010.

Reynolds, R.W., 1988. A real-time global sea surface temperature analysis. J. Climate 1 (1), 75–86.

Reynolds, R.W., Chelton, D.B., Roberts-Jones, J., Martin, M.J., Menemenlis, D., Merchant, C.J., 2013. Objective determination of feature resolution in two sea surface temperature analyses. J. Climate 26 (8), 2514–2533. https://doi.org/10.1175/JCLI-D-12-00787.1.

Takaya, Y., et al., 2002. Refinements to a prognostic scheme of skin sea surface temperature. J. Geophys. Res. 115. https://doi.org/10.1029/2009JC005985.

Walton, C.C., Pichel, W.G., Sapper, J.F., May, D.A., 1998. The development and operational application of nonlinear algorithms for the measurement of sea surface temperatures with the NOAA polar-orbiting environmental satellites. J. Geophys. Res. 103, 27, 999–28,012.

Wentz, F.J., Gentemann, C., et al., 2000. Satellite measurements of sea surface temperature through clouds. Science 288 (5467), 847–850.

While, J., et al., 2017. An operational analysis system for the global diurnal cycle of sea surface temperature: implementation and validation. Quart. J. R. Meterol. Soc. 143, 1787–1803. https://doi.org/10.1002/qj.3036.

Wong, E.W., Minnett, P.J., 2018. The response of the ocean thermal skin layer to variations in incident infrared radiation. J. Geophys. Res. Oceans 123 (4), 2475–2493.

Yang, C., Leonelli, F.E., Marullo, S., Artale, V., Beggs, H., Bunogiorno Nardelli, B., Chin, T.M., De Toma, V., Good, S., Huang, B., Merchant, C.J., Sakurai, T., Santoleri, R., Vazquez-Cuervo, J., Zhang, H.-M., Pisano, A., 2021. Sea surface temperature intercomparison in the framework of the Copernicus Climate Change Service (C3S). J. Climate 34, 5257–5283. https://doi.org/10.1175/JCLI-D-20-0793.1.

Chapter 20

Satellite ocean color validation

Jianwei Wei[a,b], Menghua Wang[b], Michael Ondrusek[b], Alexander Gilerson[c], Joaquim Goes[d], Chuanmin Hu[e], Zhongping Lee[f], Kenneth J. Voss[g], Sherwin Ladner[h], Veronica P. Lance[b], and Nicholas Tufillaro[i]

[a]*Global Science and Technology, Inc., Greenbelt, MD, United States,* [b]*NOAA/NESDIS Center for Satellite Applications and Research (STAR), College Park, MD, United States,* [c]*The City College of New York, New York, NY, United States,* [d]*Lamont-Doherty Earth Observatory, Palisades, NY, United States,* [e]*University of South Florida, St. Petersburg, FL, United States,* [f]*University of Massachusetts Boston, Boston, MA, United States,* [g]*University of Miami, Coral Gables, FL, United States,* [h]*Naval Research Laboratory, Stennis Space Center, MS, United States,* [i]*Oregon State University, Corvallis, OR, United States*

Chapter outline

1. Introduction	351
2. Satellite ocean color products	352
2.1 Remote sensing reflectance	352
2.2 Water bio-optical properties	353
2.3 Quality measures	355
3. Field measurements of the radiance distribution	356
4. Field measurements of $R_{rs}(\lambda)$	357
4.1 Above-water approach	357
4.2 In-water approach	358
4.3 On-water approach	359
5. Field measurements of water inherent optical properties	360
5.1 Backscattering coefficient	360
5.2 Non-water absorption coefficient	361
5.3 Phytoplankton and CDOM absorption coefficients	362
6. Field measurements of phytoplankton pigments and functional types	363
6.1 Phytoplankton pigments	363
6.2 Phytoplankton cell counts and cell sizes (>10 μm)	363
6.3 Phytoplankton groups of small sizes (<10 μm)	365
6.4 Phytoplankton quantum efficiency of carbon fixation	365
7. Matchup analysis	366
8. Concluding remarks	368
Acknowledgments	369
References	369

There's nothing wrong with enjoying looking at the surface of the ocean itself, except that when you finally see what goes on underwater, you realize that you've been missing the whole point of the ocean.

Dave Barry

1. Introduction

The color of the ocean is a result of the sunlight interaction with water constituents. In the coastal ocean, microscopic algae or phytoplankton strongly absorb blue and red light while scattering back green light, rendering a greenish color overall. In nearshore or inland lakes, colored dissolved organic matter (CDOM) substantially absorbs light in the shorter wavelengths but not much in the longer wavelengths. Thus, as a rule of thumb, CDOM-rich waters often exhibit a brownish hue. Open ocean waters, on the other hand, usually have few particles or CDOM present. Pure seawater scatters the blue light, and the combination of this scattering with the strong absorption of red light by pure seawater makes the oceans appear blue. Such connections between color and water constituents are the foundation of the satellite ocean color remote sensing.

Polar-orbiting satellites image the oceans from a low earth orbit (LEO), allowing for repeat coverage of global waters. The LEO ocean color missions started with the proof-of-concept instrument Coastal Zone Color Scanner (CZCS) (1978–1986) (Gordon et al., 1980; Hovis, 1981). Almost a decade later, the US National Aeronautics and Space Administration (NASA) successfully launched the Sea-viewing Wide Field-of-view Sensor (SeaWiFS) (1997–2010) on the SeaStar satellite and then the Moderate Resolution Imaging Spectroradiometer (MODIS) onboard the Terra and Aqua satellites (since 1999 and 2002, respectively) (Esaias et al., 1998). The Visible Infrared Imaging Radiometer Suite (VIIRS) sensors aboard the Suomi National Polar-orbiting Partnership (SNPP, 2011–present) and NOAA-20 (2017–present) are recent members of

FIG. 1 Schematic of satellite remote sensing process and in situ ocean color observation.

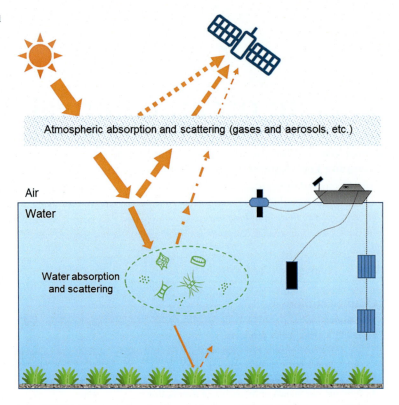

the US ocean color satellite constellation (Wang et al., 2013). Ocean color instruments have also been incorporated as a payload on geostationary (GEO) satellites, such as the Geostationary Ocean Color Imager (GOCI, 2010–present) operated by South Korea (Choi et al., 2012). GEO sensors permit high-frequency measurements but with regional coverage. Together with many other international missions, these ocean color satellites have provided essential measurements required for environmental monitoring and climate-related applications.

Satellite sensors measure a total signal scattered from the atmosphere, water surface, and water constituents (Fig. 1), of which the water-leaving radiance ($L_w(\lambda)$) only accounts for a small percentage (~10%) (IOCCG, 2010). To meet the required measurement accuracy, satellite sensors need post-launch on-orbit calibration. Over time, satellite sensors can also undergo drifting or degradation and require continuous monitoring and calibration. The $L_w(\lambda)$ retrieval from satellites is inevitably subject to uncertainties. The uncertainties in $L_w(\lambda)$ will eventually propagate to subsequently estimated water optical, biological, and biogeochemical products, such as chlorophyll-a (Chl-a) concentration. Thus, satellite ocean color products must be continuously validated with ground-truth data before wide application.

Ocean color validation centers on developing coincident matchup data of in situ measurements and satellite products. It requires practical knowledge of the derivation of radiometric properties from both satellites and in situ platforms, water inherent optical properties (IOPs), and water sampling and measurement of various biogeochemical quantities. It also demands appropriate data processing, quality control (QC), and quality assurance (QA) of various quantities. Since the very beginning of ocean color satellite missions, shipborne experimentation has been a critical element in offering in situ measurements of almost every optical, biological, and biogeochemical property needed for ocean color validation. In this work, we review the state-of-the-art instrumentation and measurements conventionally employed in field campaigns for satellite ocean color validation. In Section 2, we first recap the satellite ocean color products. We then introduce the observations of angular radiance distribution (Section 3). Following this, we describe and emphasize the observation of remote sensing reflectance ($R_{rs}(\lambda)$), an apparent optical property (AOP) of waters, and a critical satellite ocean color product (Section 4). In Section 5, we overview the measurements of water IOPs. Section 6 is dedicated to the problems of determining phytoplankton pigments and classification. In Section 7, a matchup analysis is demonstrated with concurrent satellite and in situ measurements. Finally, we provide concluding remarks for the chapter in Section 8.

2. Satellite ocean color products

2.1 Remote sensing reflectance

Satellites measure the top-of-atmosphere (TOA) radiance ($L_{toa}(\lambda)$), including $L_w(\lambda)$. $L_{toa}(\lambda)$ and $L_w(\lambda)$ are related by

$$L_{toa}(\lambda) = L_{path}(\lambda) + t(\lambda) \cdot L_w(\lambda), \quad (1)$$

where $L_{path}(\lambda)$ is the path radiance, and $t(\lambda)$ refers to the atmospheric diffuse transmittance from the water surface to the satellite. By this definition, $L_{path}(\lambda)$ includes radiance contributions from both atmospheric and surface reflection (Gordon and Wang, 1994; IOCCG, 2010). Also, it is implicitly assumed $L_w(\lambda) << L_{path}(\lambda)$, and thus the $L_w(\lambda)$ and $L_{path}(\lambda)$ terms are additive without coupling. To derive $L_w(\lambda)$ from $L_{toa}(\lambda)$, an atmospheric correction algorithm is required to model the path radiance and atmospheric transmittance accurately and subsequently to remove the path radiance from $L_{toa}(\lambda)$. The global algorithms, such as those for MODIS and VIIRS, follow similar procedures as Gordon and Wang (1994) to compute the radiance due to the Rayleigh scattering by air molecules, with the knowledge of air pressure. Estimating the radiance from the aerosol scattering is often based on the assumption that water-leaving radiance in the near-infrared (NIR) is negligible, thus facilitating the determination of aerosol properties in this domain. While this is mostly true for clear waters, for turbid coastal waters, this assumption is often violated due to the non-negligible backscattering of particles. Recent atmospheric correction algorithms have employed the shortwave infrared (SWIR)-based algorithm and the NIR-SWIR algorithm (Ibrahim et al., 2019; Wang and Shi, 2007) or a NIR iteration procedure (Bailey et al., 2010; Jiang and Wang, 2014) to determine the aerosol types and contributions over coastal and inland waters.

A common challenge with the operational algorithms is that they are designed for nonabsorbing or weakly absorbing aerosols. Over many coastal and inland water areas, the presence of strongly absorbing aerosols in the atmosphere may cause the standard algorithms to fail (Wang et al., 2020). Innovative approaches and algorithms are available for the $L_w(\lambda)$ retrievals when the absorption aerosols become a problem, such as the spectral optimization method (Chomko and Gordon, 2001), the multiscattering method (Antoine and Morel, 1999; Nobileau and Antoine, 2005), the neural network (Fan et al., 2017), the UV-blue scheme (He et al., 2012; Wang and Jiang, 2018), the two-angle atmospheric correction algorithm (TAACA) (Wang et al., 2020), and the blue-band estimation (BBE) model (Wei et al., 2020). Another challenge is the presence of extensive sun glitter patterns in the subtropical regions within a satellite's field of view. To recover the sun-contaminated ocean color data, a POLYnomial-based algorithm applied to MERIS Algorithm (POLYMER) has been found to be effective (Steinmetz et al., 2011). Recall that $L_w(\lambda)$ typically accounts for 10% or less of the $L_{toa}(\lambda)$ for most waters (IOCCG, 2010). The reliability of resultant $L_w(\lambda)$ is thus sensitive to the calibration of the satellite sensors. A small error in $L_{toa}(\lambda)$ can propagate to the estimated $L_w(\lambda)$ as a significant one.

The magnitude of $L_w(\lambda)$ varies with the illumination condition. Therefore, the $L_w(\lambda)$ data are often normalized by the downwelling plane irradiance at the water surface ($E_s(\lambda)$), leading to the commonly known remote sensing reflectance:

$$R_{rs}(\lambda) = \frac{L_w(\lambda)}{E_s(\lambda)}, \quad (2)$$

where $R_{rs}(\lambda)$ has a unit of sr^{-1}. In practice, the satellite-derived $L_w(\lambda)$ should also be corrected for the bidirectional reflectance distribution function (BRDF) effect (Gordon, 2005; Morel et al., 2002; Morel and Gentili, 1991, 1993, 1996; Wang, 2006). In Fig. 2A, an example of a satellite $R_{rs}(443)$ image is captured over the Mid-Atlantic Bight and adjacent coastal waters, demonstrating spatial variability. The $R_{rs}(\lambda)$ spectra are theoretically insensitive to the variation of cloudiness and solar zenith angles (<70 degrees) and can be directly used for ocean color inversion (see Section 2.2). The normalized water-leaving radiance ($nL_w(\lambda)$) is used interchangeably with $R_{rs}(\lambda)$ as $nL_w(\lambda) = R_{rs}(\lambda)F_o(\lambda)$, where $F_o(\lambda)$ is the mean extraterrestrial solar irradiance (a time-independent constant for each wavelength).

2.2 Water bio-optical properties

The principle of deriving water bio-optical properties from $R_{rs}(\lambda)$ is based on the relationship between $R_{rs}(\lambda)$ and the total backscattering coefficient ($b_b(\lambda)$) and total absorption coefficient ($a(\lambda)$) from every water constituent. According to radiative transfer simulation, $R_{rs}(\lambda)$ can be modeled as a simple function of $b_b(\lambda)$ and $a(\lambda)$ (Zaneveld, 1995), specifically:

$$R_{rs}(\lambda) \propto \frac{b_b(\lambda)}{a(\lambda) + b_b(\lambda)}. \quad (3)$$

The $a(\lambda)$ quantity can be decomposed into the absorption coefficients by phytoplankton ($a_{ph}(\lambda)$), CDOM ($a_{CDOM}(\lambda)$), detritus ($a_d(\lambda)$), and pure seawater ($a_w(\lambda)$). Similarly, the quantity $b_b(\lambda)$ can be divided into contributions from particles ($b_{bp}(\lambda)$) and pure seawater itself ($b_{bw}(\lambda)$). The goal of ocean color remote sensing is to retrieve various water components and their optical properties from satellite-measured $R_{rs}(\lambda)$ spectra.

Inversion algorithms are either semianalytical or empirical (IOCCG, 2000, 2006). The former lays emphasis on the semianalytical relationships between $R_{rs}(\lambda)$ and the optical properties of water components, including Eq. (3). They

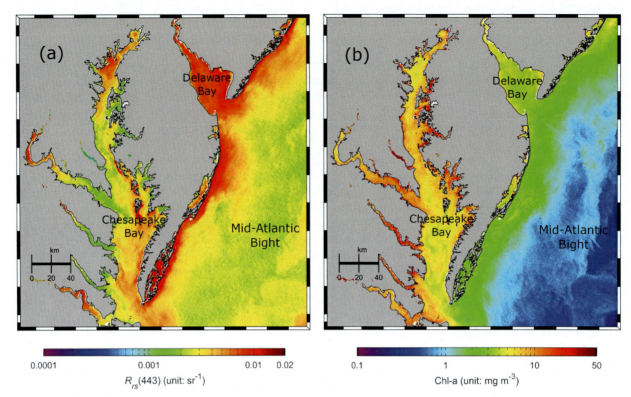

FIG. 2 Example satellite ocean color products for (A) $R_{rs}(443)$ and (B) Chl-a over the Mid-Atlantic Bight, Chesapeake Bay, and Delaware Bay (Satellite: VIIRS-SNPP; Date: November 20, 2020).

quantify the spectral variation of the IOPs of water components with analytical models. For example, $b_{bp}(\lambda)$ is usually described using a power-law model,

$$b_{bp}(\lambda) = b_{bp}(\lambda_0)[\lambda_0/\lambda]^\eta, \tag{4}$$

where λ_0 is a reference wavelength, often at ∼440 nm or 550 nm, and η is the power of this function. $a_d(\lambda)$ can be characterized with an exponential-decay function of wavelength similar to that of $a_{CDOM}(\lambda)$, which is explicitly

$$a_{CDOM}(\lambda) = a_{CDOM}(\lambda_0)\exp\left[-S_{CDOM}(\lambda - \lambda_0)\right], \tag{5}$$

where S_{CDOM} is the spectral slope of $a_{CDOM}(\lambda)$. The $a_{ph}(\lambda)$ spectra generally exhibit two peaks, one at the blue domain and the other at the red domain. Sometimes simplistic functions exist to describe the variation of $a_{ph}(\lambda)$ (Bricaud et al., 1995, 1998). The semianalytical algorithms, including the quasi-analytical algorithm (QAA) (Lee et al., 2002) and the Garver-Siegel-Maritorena (GSM) algorithm (Maritorena et al., 2002), usually estimate $a_{ph}(\lambda)$, $a_{dg}(\lambda)$ (which is the sum of $a_{CDOM}(\lambda)$ and $a_d(\lambda)$), and $b_{bp}(\lambda)$. The empirical algorithms are built upon the relationships between in situ $R_{rs}(\lambda)$ measurements and relevant bio-optical properties. Although they are data-dependent, empirical algorithms are also widely used in ocean color inversions for their simplicity and ease of regional tuning. Chl-a concentrations, for example, are derived from the spectral ratios of $R_{rs}(\lambda)$ at blue and green bands (O'Reilly and Werdell, 2019) or a color index algorithm based on the spectral difference between blue, green, and red bands (Hu et al., 2012).

Fig. 2B is an example of the daily Chl-a product derived from satellite $R_{rs}(\lambda)$ data using a combination of the blue/green algorithm and the ocean color index (OCI) algorithm (Wang and Son, 2016). The spatial variation of Chl-a concentrations can be ascribed to nutrient supply and light availability/limitation as well as variations in the phytoplankton community composition. In addition to the IOPs and Chl-a, many other bio-optical properties, including the diffuse attenuation coefficient for downwelling irradiance ($K_d(\lambda)$) (Wang et al., 2009a), particulate organic carbon (POC) (Stramski et al., 2008), suspended particulate matter (SPM) (Wei et al., 2021a), and phytoplankton functional type (PFT) (e.g., Xi et al., 2020), have also been generated as a part of the ocean color products.

2.3 Quality measures

Environmental interference from clouds and land, inaccurate modeling of the atmospheric and surface effects, or degradation of instruments can introduce errors in satellite radiometric products. It is imperative to devise QC and QA models to inspect the data quality of satellite radiance and reflectance products. A common practice is to use flags and masks to identify potentially problematic pixels. For instance, the QC flags indicate whether the radiance or reflectance value is possibly subject to atmospheric correction failure (ATMFAIL), land contamination (LANDADJ), cloud (CLOUD), high glint (HIGLINT), cloud shadow and stray light contamination (CLDSHDSTL), strongly absorbing aerosols (ABSAER), or moderate glint (MODGLINT) (Wang et al., 2017). They have been implemented as a part of the data stream from the ocean color instruments, including VIIRS and MODIS. The QA model of Wei et al. (2016) is a more recent addition to the quality measures, which examines the overall spectral shape of normalized $R_{rs}(\lambda)$ and assigns a score varying from zero to one to quantify the $R_{rs}(\lambda)$ quality. The QC and QA measures do not automatically identify "good" or "bad" pixels. The end-users have the freedom to decide which pixels to use for their own applications. Depending on specific applications, various combinations of flags can be used, as the final satellite data products are always a compromise between data quality and data quantity (Hu et al., 2020). Similarly, the QA scores provide a more quantitative perspective of the satellite data quality and different levels of the scores may be adopted in practice (Wei et al., 2020).

Fig. 3 gives an example of the QC flags and the QA scores for a satellite $R_{rs}(\lambda)$ image over the Mid-Atlantic Bight. In Fig. 3A, the pixels susceptible to LANDADJ, HIGLINT, ABSAER, and MODGLINT are highlighted, while the white pixels indicate none of these four flags being evoked. According to Fig. 3B, some pixels with low QA scores are coincident with either LANDADJ or ABSAER. A large patch of open water pixels in the east has low scores (<0.6), where the flag MODGLINT is evoked. On the other hand, the pixels along the nearshore region within the Chesapeake Bay have high QA scores, irrespective of potential contamination due to the land adjacency effect.

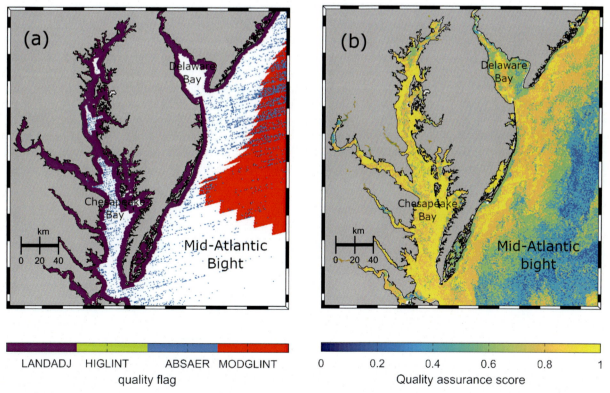

FIG. 3 Example satellite ocean color products for (A) quality control flags and (B) quality assurance scores over the Mid-Atlantic Bight, Chesapeake Bay, and Delaware Bay (Satellite: VIIRS-SNPP; Date: November 20, 2020).

3. Field measurements of the radiance distribution

The radiance distribution, $L(\theta,\varphi,\lambda)$, describes the variation of radiance with viewing direction, conventionally expressed by zenith angle (θ) and azimuth angle (φ). The shape of the radiance distribution varies with the incident solar angle and with the scattering and absorption properties of the water. Measurements of the radiance distribution are vital for remote sensing vicarious calibration for the following reasons:

First, many in situ measurements of the upwelling radiance are done with a nadir-viewing instrument, while satellites measure the surface at varying angles. To compare the in situ and satellite measurements, one needs to know the shape of the radiance distribution. Second, when comparing measurements from multiple satellites, the variation in the viewing geometries of the satellites will cause the radiance, or reflectance, measured by each instrument to be different and must be taken into account. Lastly, since all satellite algorithms are developed using nadir-viewing radiance measurements, the variation with sensor viewing angle must be considered when applying algorithms to satellite data to obtain accurate products. To account for these differences, the BRDF model has been developed for the upwelling radiance (Gordon, 2005; Morel et al., 2002; Morel and Gentili, 1991, 1993, 1996; Wang, 2006), in which the validation requires measurements of the radiance distribution.

Measuring $L(\theta,\varphi,\lambda)$ requires either pointing a single detector in many different directions (Aas and Højerslev, 1999; Tyler, 1960) or using a camera system with a fisheye lens system (Smith et al., 1970). The single detector has the advantage of enabling hyperspectral measurements but has the disadvantage of limited viewing angles, which increases the time needed to take measurements and risks changes to environmental conditions during the measurements. The fisheye method has the advantage of making measurements in all directions at once. It can do multispectral measurements by inserting interference filters in the optical train but cannot perform hyperspectral measurements. Multispectral instruments using the fisheye technique have been developed (Voss, 1989; Voss and Chapin, 1992).

One of the recently developed instruments is NuRads (Voss and Chapin, 2005). This instrument has a fisheye lens behind a glass dome window. In the optical train behind the fisheye lens is a spectral filter wheel with six bandpass interference filters to select the spectral region of interest, and then a charge-coupled device (CCD) camera. With the CCD camera and the fisheye lens (and domed window), a full hemisphere of radiance distribution data is obtained at once, although the spectral variation must be determined through sequential images. The system floats at the surface, pointing down (thus, for measuring upwelling radiance distribution), and is tethered to the ship by a neutrally buoyant signal/power cable. This allows the instrument to be deployed at some distance from the ship to avoid ship shadow and reflection. The instrument is small (0.3 m diameter, 0.3 m long) to reduce instrument self-shadow. When deployed, NuRads sequentially takes data at six wavelengths, completing a measurement cycle of all wavelengths in approximately 2 min. Typically, the instrument is deployed for 15 min or longer to acquire multiple sets of data. Data are averaged in 10-min segments to smooth out variations due to wave focusing, but in a short enough time to avoid significant changes in solar illumination conditions. Each image is manually inspected to remove images taken at times when the sun is occluded by clouds. Fig. 4 shows the normalized radiance for high and low Chl-a cases at 436 nm. The high Chl-a (1.4 mg m^{-3}) case has more variation in the normalized radiance due to the increased absorption. In both cases, along the principal plane (the plane containing the nadir and anti-solar direction), between −20 degrees and −40 degrees, one can see the dip in the normalized radiance caused by instrument self-shading.

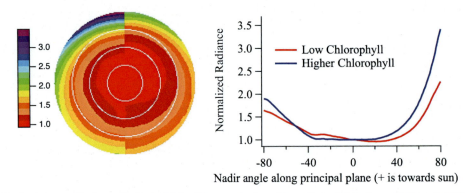

FIG. 4 Radiance distribution normalized to nadir. On the *left* is a fisheye projection of the radiance distribution, with the center being nadir and the white rings representing 20 degrees, 40 degrees, and 60 degrees nadir angles. The *top* of the image is toward the sun. High Chl-a ($= 1.4$ mg m^{-3}) on the *left side* of the image, low Chl-a ($= 0.06$ mg m^{-3}) on the *right side* of the image. The solar zenith angle is 48 degrees in both cases. On the right is a graph of the normalized radiance along the principal plane.

The fisheye design by Voss and colleagues inspired the development of new radiance cameras. This includes the complementary metal-oxide-semiconductor (CMOS) radiance camera system (Wei et al., 2012), which adopted nonlinear responsivity functions. As a result, the free-fall profiling CMOS camera system is able to measure the full radiance distribution at a faster rate (up to ~10 Hz) but at a single wavelength of 555 nm (Lewis et al., 2011). Shortly after the free-fall system was demonstrated, a miniaturized CMOS radiance camera was tested for observations at six bands (Antoine et al., 2013).

Recently, measurements of the spectral radiance distribution have been expanded to include the linear polarized spectral radiance distribution. The water-leaving radiance is partially linearly polarized. This can both provide additional information, in the case of satellite instruments that measure the polarization state (e.g., the future Plankton, Aerosol, Cloud, ocean Ecosystem (PACE) earth-observing satellite mission being planned by NASA), and interact with any polarization sensitivity in the satellite instrument and cause errors. The newest development in measuring the upwelling polarization is PixPol, which uses a camera with polarizers, in one of four orientations on each pixel in the camera, along with a fisheye lens, to capture the polarization state of the upwelling spectral light field. Earlier measurements have shown that the upwelling polarization depends on the water properties and the polarization state of the incident light fields (Bhandari et al., 2011).

4. Field measurements of $R_{rs}(\lambda)$

Three radiometric approaches are generally used for measuring $R_{rs}(\lambda)$ in field campaigns. Above- and in-water methods have been used for ocean optical measurements for decades. The on-water approach is a relatively recent development in the family of in situ optical radiometry. These three approaches are distinct in determining $L_w(\lambda)$, instrument setups, and data processing protocols.

4.1 Above-water approach

The above-water approach often has two radiometers deployed above water from a ship or fixed structure to observe the radiances. Precisely, one sensor points to the water surface and measures the total radiance $L_t(\lambda,\theta_v,\Delta\varphi)$ along a viewing direction (at zenith angle θ_v and relative azimuth angle $\Delta\varphi$). The other sensor measures the downward (sky) radiance $L_d(\lambda,\pi-\theta_v,\Delta\varphi)$ in the "mirror" direction, a portion of which reflects at the air-water interface into the water-viewing direction (Fig. 5). A third sensor looking upward can be added to record the downwelling irradiance right above the water surface, $E_s(\lambda)$. The water-leaving radiance in the water-viewing direction is estimated from the equation:

$$L_w(\lambda, \theta_v, \Delta\varphi) = L_t(\lambda, \theta_v, \Delta\varphi) - \rho_f(\lambda) \cdot L_d(\lambda, \pi - \theta_v, \Delta\varphi). \tag{6}$$

In Eq. (6), ρ_f is the reflectance coefficient representing the fraction of incident skylight reflected toward the water-viewing sensor at the air-water interface. The second part of the equation, $\rho_f \cdot L_d(\lambda,\pi-\theta_v,\Delta\varphi)$, estimates the reflected sky glint radiance, which cannot be measured independently.

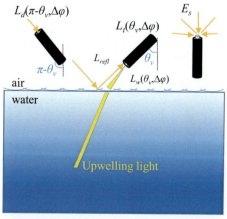

FIG. 5 Schematic of above-water radiometry. Two radiance sensors in *black* measure L_t and L_d, respectively; the third irradiance sensor also in *black* measures E_s. The *arrows* indicate the directions of light.

The above-water approach is adopted as a pragmatic way of estimating and removing the upwelling radiance that originates from reflection at the air-water interface. However, it is well understood that the reflected sky radiance may originate from portions of the sky dome other than the portion that is actually measured and may include reflection of direct sun glint. In reality, the water surface is not flat but a wavy interface, which complicates calculations of the reflectance coefficient and defines certain conditions where this coefficient is minimally affected by wind speed and solar angles, thus minimizing uncertainties associated with the sky glint removal. Such conditions were established as $\theta_v = 40$ degrees for the viewing angle and $\Delta\varphi = 90$–135 degrees for the azimuth angle (Mobley, 1999).

For measurements of $L_d(\lambda, \pi - \theta_v, \Delta\varphi)$, a small number of replicate measurements, usually three to five for the clear blue sky, should be sufficient. For measurements of $L_t(\lambda, \theta_v, \Delta\varphi)$, a much larger number of replicate measurements (about 10) is needed because of the rapid and significant temporal variations associated with surface waves. The temporal processing of $L_t(\lambda, \theta_v, \Delta\varphi)$ measurements should also depend on the integration time of each measurement and may be linked to the method for estimation of the reflectance coefficient ρ_f. Additional filtering of these measurements may be required to minimize sun glint contamination (Ruddick et al., 2019).

Such derived $L_w(\lambda, \theta_v, \Delta\varphi)$ need to be converted to $L_w(\lambda)$ from nadir with the BRDF function. BRDF is a function of water IOPs, viewing directions, sea surface roughness, and aerosol optical thickness (Gordon, 2005; Morel and Gentili, 1996; Wang, 2006).

Unique robotic instrument systems such as the SeaWiFS Photometer Revision for Incident Surface Measurements (SeaPRISM) and Hyperspectral Surface Acquisition System (HyperSAS) were developed to observe $L_t(\lambda, 0^+, \theta_v, \Delta\varphi)$ and $L_d(\lambda, 0^+, \pi - \theta_v, \Delta\varphi)$ based on the above-water radiometry. The SeaPRISM instruments are now installed at ~40 platforms in coastal areas in different parts of the world's oceans, estuaries, and lakes. In addition to estimating $L_w(\lambda)$ and $R_{rs}(\lambda)$, these instruments also measure several parameters of atmospheric aerosols and thus have become components of the Aerosol Robotic Network (AERONET) and AERONET-Ocean Color Network operated by NASA with data from the instruments transmitted to the NASA AERONET group through the Internet or communications satellites (Zibordi et al., 2021, 2009). The HyperSAS instrument is portable and can be installed as an automated underway system on a ship to collect high-frequency observations for validation. Automated, open-source data processing tools for these data also exist (e.g., NASA Community Processor for In Situ Radiometry; https://github.com/nasa/HyperInSPACE/). Hand-held radiometers are a class of portable above-water systems. The operator uses only one radiance sensor to measure $L_t(\lambda, \theta_v, \Delta\varphi)$ and $L_d(\lambda, \pi - \theta_v, \Delta\varphi)$, sequentially, for the derivation of $L_w(\lambda, \theta_v, \Delta\varphi)$. The same sensor can estimate the downwelling irradiance by measuring the light reflected from a Lambertian plaque.

The primary challenge of the above-water method is to remove the contaminations from residual sky glint. In addition, the variations in sky conditions and the sampling angles that may vary significantly among users can introduce non-negligible uncertainties in the radiometric products.

4.2 In-water approach

The in-water approach has a long history of use by the ocean optics community for observing various water optical properties. The in-water approach, together with the above-water approach, has been recommended for collecting $L_w(\lambda)$ and $R_{rs}(\lambda)$ for the validation of satellite ocean color products (Mueller et al., 2003c). As shown in Fig. 6, this method measures

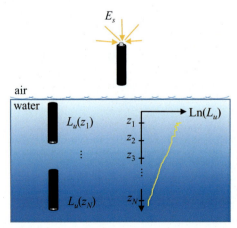

FIG. 6 Schematic of in-water radiometry. A free-fall radiance sensor in *black* descends from depths of z_1 to z_N, while continuously collecting $L_u(z)$ data. An irradiance sensor measures $E_s(\lambda)$ from the above. The *yellow curve* exemplifies the upwelling radiance profile between z_1 and z_N.

the upwelling radiance from the nadir direction at multiple depths, $L_u(z,\lambda)$. The $L_u(z,\lambda)$ values are then extrapolated to zero depth to obtain the upwelling radiance right below the water surface, $L_u(0^-,\lambda)$, with the following formula:

$$L_u(0^-, \lambda) = L_u(z, \lambda) \exp[z \cdot K_L(\lambda)], \tag{7}$$

where $K_L(\lambda)$ is the diffuse attenuation coefficient for $L_u(z,\lambda)$. Such derived $L_u(0^-,\lambda)$ can then propagate across the air-water interface to obtain $L_w(\lambda)$, with

$$L_w(\lambda) = L_u(0^-, \lambda)\frac{t_{wa}(\lambda)}{n_w^2(\lambda)}, \tag{8}$$

where the term $n_w(\lambda)$ refers to the refractive index of seawater and $t_{wa}(\lambda)$ is Fresnel transmittance of the air-water interface. The term $t_{wa}(\lambda)/n_w^2(\lambda)$ is the upwelling radiance transmittance (Austin, 1974; Wei et al., 2015). The $L_w(\lambda)$ data and the $E_s(\lambda)$ measurements enable the calculation of $R_{rs}(\lambda)$ (see Eq. (2)).

Hand-deployed profiling spectral radiometers, or simply profilers, is predominant in-water instruments deployable from research ships or small boats. Profiling radiometers have a downward-looking radiance sensor and an upward-looking irradiance sensor and are paired with a surface irradiance sensor mounted vertically aboard the vessel in a location with an unobstructed view of the sky. A telemetry cable is used to transmit the data back to an on-deck computer. Using these instruments requires avoiding shadow and reflection from the ship's hull (Shang et al., 2020) while collecting continuous data sets to account for wave focusing variations near the surface (Zibordi et al., 2004). With the buoyancy adjusted to be slightly negative, the profilers descend slowly (0.1 to 0.3 m s^{-1}) into the water column. The instrument continuously collects data while descending to the desired depth (a few meters to tens of meters depending on the water clarity) and then is quickly pulled back up to the surface. During deployment, enough slack is given to the cable to allow for a free fall through the water column with the minimal inclination of the radiometers. As the vessel is constantly drifting, it is necessary to use sufficiently long cables to ensure that the instrument can float far enough away from the ship to avoid shadowing and reflection.

For validation purposes, multicast deployments are often preferred, enabling the collection of hundreds of measurements per meter in depth. In the multicast mode, three to five casts in one logged file usually provide enough data points while leaving the total measurement duration short enough to minimize the impact of changing sky conditions. During processing, outliers of the $L_u(z,\lambda)$ measurements are first removed for high instrument tilts or for being outside a defined standard deviation. Then $L_u(0^-,\lambda)$ and $L_w(\lambda)$ are calculated as described above (Eq. (7) and Eq. (8), respectively). The $R_{rs}(\lambda)$ spectra can eventually be derived as the ratio of $L_w(\lambda)$ to measured $E_s(\lambda)$ (Eq. (2)).

The in-water, at depth, measurements can also utilize sensors at fixed depths. The legacy Marine Optical BuoY (MOBY) (Clark et al., 1997) off Lanai, Hawaii, is an excellent example of a permanently moored optical platform with fixed-depth sensors that measure upwelled radiance at three depths (see Chapter 5). The location of MOBY in clear oligotrophic waters makes the processing of ocean color data less complicated and is used as the primary source for vicarious calibration of ocean color satellites (Clark et al., 2003; Voss et al., 2010). In-water methods can also have sensors floating at the surface using a collar to measure the in-water upwelling radiance at a depth near the surface to approximate $L_u(0^-,\lambda)$. Both in-water measurements obtained using floaters and the at-depth profiling measurements must then be propagated to the sea surface $L_u(0^-,\lambda)$ and then through the air-sea interface to obtain $L_w(\lambda)$.

4.3 On-water approach

The on-water in situ radiometric approach was first proposed and tested by Ahn et al. (1999) and Tanaka et al. (2006). Lee et al. (2013) recently advanced data sampling and data processing. The paradigm in Fig. 7 shows the on-water setup that involves a skylight-blocking apparatus (SBA) (usually a cone) attached to the downward-looking radiance sensor. Ideally, the bottom rim of the cone touches the water surface and blocks off all possible stray light reflected from above. Since the radiance sensor is in the air, the reading from this instrument is equal to $L_w(\lambda)$ after correcting for the self-shading error from the instrument.

For each deployment, a collection of a large number of continuous measurements from the on-water instrument is recommended, usually with several hundred radiance spectra. Because the water surface is sometimes ruffled with big waves, the instrument package can be thrown off balance. As a result, the glass window of the radiance sensor could be instantly submerged in the water, or the cone could swing above the water surface so that the ambient light enters the fore optics of the radiance sensor. The radiance readings recorded under either of these situations are contaminated and should be excluded from the data processing. At present, the on-water data processing is performed with a probability distribution function (PDF)-based filtering procedure (Wei et al., 2021b). It uses the raw radiance data in red-NIR bands as a

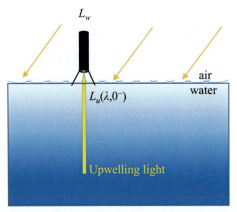

FIG. 7 Schematic of on-water radiometry. The downward-viewing radiance sensor is depicted in *black*, with the cone slightly immersed in the water.

reference, where the sky (and solar) radiance is much higher than $L_w(\lambda)$. The principle of PDF filtering is that the contaminated measurements only account for a nonsignificant portion of the raw data points. As readily observable from an experimentalist's view, most of the raw measurements represent the true $L_w(\lambda)$ values and can be considered "constant" for a short period of time under the assumption of a constant water property and incident irradiance. Thus, the true $L_w(\lambda)$ at a reference wavelength should be the first mode of the probability distribution of the radiance time-series data at that wavelength. If the objective is to measure $R_{rs}(\lambda)$ only, a practical routine is to measure $L_w(\lambda)$ simultaneously with $E_s(\lambda)$. This allows the filtering procedure to be carried out with $R_{rs}(\lambda)$ raw measurements, which are insensitive to the variation of the above-water light field.

The $L_w(\lambda)$ measurements include the radiance from the shadow of the instrument. The self-shading errors depend on the dimension of the cone, the water's IOPs, and the sun angles. Generally, the shading errors become less significant with a smaller cone, lower total absorption and backscattering coefficients, or larger solar zenith angles. Models have been developed for the correction of the self-shading effect (Shang et al., 2017; Yu et al., 2021).

The on-water method does not utilize $K_L(\lambda)$ as it does with the in-water approach. Neither does it need a surface correction procedure as employed in the above-water approach. Furthermore, it is not limited by sky conditions or water column stratification, and it is applicable to optically shallow water observations. In comparison, the on-water radiometry is more difficult to operate under rough sea conditions. The uncertainty budget for the on-water measurements has yet to be fully investigated. However, a recent study suggests that the measurement precision for $L_w(\lambda)$ and $R_{rs}(\lambda)$ from the on-water radiometry is only a few percent across various water types (Wei et al., 2021b).

5. Field measurements of water inherent optical properties

5.1 Backscattering coefficient

The backscattering or backward scattering coefficient $b_b(\lambda)$ is the fraction of incident light that is attenuated by scattering in the backward direction between 90 degrees and 180 degrees in units of m^{-1}. It can be expressed as the integral of the angular distribution of scattered light, or so-called volume scattering function ($\beta(\psi,\lambda)$),

$$b_b(\lambda) = 2\pi \int_{\pi/2}^{\pi} \beta(\psi, \lambda) \sin\psi \, d\psi, \qquad (9)$$

where ψ refers to the scattering angle. Measuring the $\beta(\psi,\lambda)$ function or the particulate contribution $\beta_p(\psi,\lambda)$ is a daunting task. Only a few prototype instruments have been developed for this type of observation (Lee and Lewis, 2003; Petzold, 1972). The commercial instruments for $b_b(\lambda)$ often measure $\beta(\psi,\lambda)$ at one to several fixed angles. Then $b_b(\lambda)$ is estimated by multiplying the measured $\beta(\psi,\lambda)$ by a scaling factor and by integrating the curve fit to $\beta(\psi,\lambda)$ from $\pi/2$ to π. Fig. 8A shows examples of $b_{bp}(\lambda)$ spectra measured by an ECO-BB9 instrument (WET Labs, Inc.) in the Mid-Atlantic Bight. For this data set, the $b_{bp}(\lambda)$ spectra between 443 nm and 700 nm largely follow the power-law distribution of Eq. (4), but with some variability. At 410 nm, however, some $b_{bp}(\lambda)$ spectra exhibit a substantial drop. The abnormal decrease in $b_{bp}(410)$ is not uncommon and was also observed in the Bristol Channel, United Kingdom (McKee et al., 2009) and Martha's Vineyard

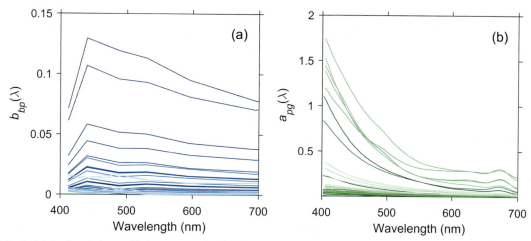

FIG. 8 (A) Particulate backscattering coefficients measured with ECO-BB9 in the Mid-Atlantic Bight in September 2019 and (B) nonwater absorption coefficients determined with ac-S in the South Atlantic Bight. The $a_{pg}(\lambda)$ data were corrected with the baseline^{+} model of Röttgers et al. (2013).

Coastal Observatory (MVCO) (Slade and Boss, 2015). The absorption effect of particles in the blue domain is partly responsible for this phenomenon.

5.2 Non-water absorption coefficient

The light absorption coefficient is composed of contributions from particulates ($a_p(\lambda)$), CDOM, and the water itself. The frequently used in situ approach is to measure the attenuation of the collimated light traveling through a reflective tube filled with water samples. Ideally, the inner wall of the absorption tube will reflect all light impinging upon it, and the collector accepts uncollimated light in a forward direction. Thus, the light extinction over the optical pathlength can be ascribed to light absorption. The reflective tube-based absorption meter is often accompanied by an attenuation meter, which has a dark chamber and a collector that only accepts collimated light. Commercial instruments, such as the ac-9 sensor with nine wavelengths and the ac-S sensor with hyperspectral resolution (WET Labs, Inc), are designed with two tubes for the absorption coefficient and the attenuation coefficient ($c(\lambda)$), respectively. The ac-9 and ac-S meters can be deployed from a winch to profile the water columns at discrete locations or work as a benchtop monitor to continuously record data on sample water pumped through a ship's underway flow-through system (IOCCG, 2019a). They require frequent calibration to monitor the optical alignment and lamp degradation. It is also necessary to account for instrument drift by removing the pure water background with Milli-Q water as a reference and to correct for the temperature and salinity differences in deployment (Sullivan et al., 2006). The resultant absorption values from the ac-9 and ac-S meters are thus representative of the non-water component of the total absorption coefficient, denoted as $a_{pg}(\lambda)$ ($= a(\lambda) - a_w(\lambda)$). These in situ instruments can also measure the absorptions due to particles and dissolved components by applying an in-line 0.2 μm capsule filter (see Section 5.3). Fig. 8B shows examples of hyperspectral $a_{pg}(\lambda)$ measurements determined for the surface waters in the South Atlantic Bight. The peaks in the red domain (675 nm) are caused by the phytoplankton absorption.

The absorption measurements from reflective-tube instruments are subject to significant scattering errors. Two methods for scattering corrections are described in the work of Zaneveld et al. (1994). The baseline method is to shift the raw (uncorrected) $a_{pg}(\lambda)$ values downward so that the spectrum at a NIR wavelength is equal to zero. The proportional method uses the $c(\lambda)$ measurements to make a wavelength-dependent correction. Despite questionable assumptions, these two methods have played an essential role in generating a large amount of $a_{pg}(\lambda)$ data for the ocean color community (Mueller et al., 2003a). In a recent effort, two approaches are upgraded to baseline^{+} and proportional^{+} methods, respectively, with improved performance (Röttgers et al., 2013).

The point-source integrating cavity absorption meter (PSICAM), initially proposed by Kirk (1997), is another type of instrument for light absorption measurement. It has a diffuse light source in an integrating cavity. The basic principle is that by measuring the diffuse light field in the cavity, scattering effects are reduced, and the multiple reflection and scattering of the light within the cavity extends the effective pathlength. Preliminary tests suggest that the PSICAM is a promising concept for measuring the absorption coefficients for discrete samples (Röttgers et al., 2007) and flow-through samples (Wollschläger et al., 2013). One advantage of discretized sampling with benchtop measurements over the ac-meters is that it enables the further partitioning of $a_{pg}(\lambda)$ into living, nonliving, and dissolved components (see Section 5.3).

5.3 Phytoplankton and CDOM absorption coefficients

Measurements of light absorption due to particulates and dissolved matter are useful for understanding and modeling the underwater light field, modeling primary production, and validating satellite-based estimates. One of the often used methods to determine the absorption coefficients of seawater particulate and dissolved matter is through water filtration (Kishino et al., 1985; Mueller et al., 2003a). A water sample of known volume, collected from the ocean, is filtered through a Whatman GF/F filter (25 mm diameter, 0.7 μm pore size). The particles on the filters are measured using a dual-beam spectrophotometer or in an integrating sphere to determine their absorbance, which is converted to their absorption coefficient in water after taking into account the filtered water volume, filter size, and pathlength elongation factor (Carder et al., 1999). After extracting the phytoplankton pigments from the particles using hot methanol or other reagents (Kishino et al., 1985; Roesler et al., 1989), the absorption coefficient of nonpigmented particles (detritus) is measured again with retaining particles on the filter. The difference is the phytoplankton pigment absorption coefficient, i.e., $a_{ph}(\lambda) = a_p(\lambda) - a_d(\lambda)$.

The filtrate from the filtration of seawater is further filtered using a Nuclepore 0.2 μm membrane filter, with the new filtrate being regarded as containing only dissolved matter as opposed to particulate matter. This filtrate is measured using a double-beam spectrophotometer to measure the absorption coefficient of CDOM.

In practice, duplicate water samples are often collected at each station. After filtration on the ship, if absorption measurements are not possible due to lack of instrumentation, the filters (with particles) are placed into containers and quickly frozen to −80°C using liquid nitrogen for later analysis in the lab. Likewise, the filtrate can be stored in dark glass bottles and kept refrigerated (∼4°C) until laboratory analysis.

In each of the NOAA dedicated calibration/validation (Cal/Val) cruises (Ondrusek et al., 2015, 2016, 2017, 2019) (Fig. 9A), water samples were collected to determine their $a_{ph}(\lambda)$, $a_{CDOM}(\lambda)$, and $a_d(\lambda)$ using the procedures described above, where data are presented in Fig. 9C–E, respectively. Clearly, these absorption values from different water

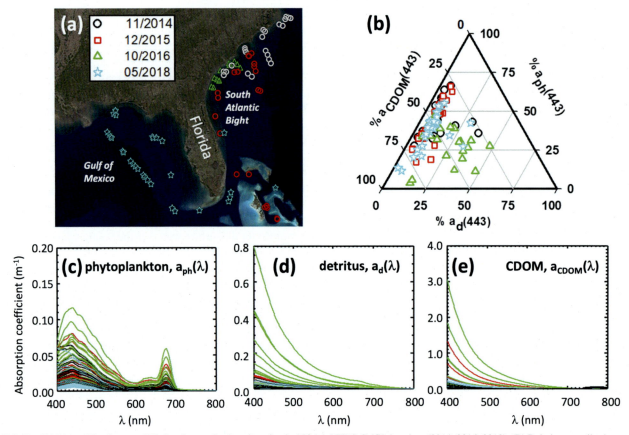

FIG. 9 (A) Map of Southeastern US showing station locations for the NOAA VIIRS Cal/Val cruises (2014–2016, 2018). (B) Relative contributions of $a_{ph}(443)$, $a_d(443)$, and $a_{CDOM}(443)$ to total nonwater absorption at 443 nm. (C–E) Phytoplankton, detrital, and CDOM absorption spectra. All samples were collected from near-surface waters only.

environments cover an extensive range (> one order of magnitude). This is important for validating satellite-based estimates of these absorption coefficients. For convenience, a reference wavelength (often at 440 nm) is used to represent these coefficients. Fig. 9B shows a ternary plot of the three absorption coefficients where their relative contribution to the total absorption coefficient can be visualized. Having information on the dominant factor(s) contributing to absorption can help distinguish unique optical water masses, some of which may impart more/less uncertainty on ocean color algorithms.

6. Field measurements of phytoplankton pigments and functional types

6.1 Phytoplankton pigments

Phytoplankton contain photosynthetic pigments, which are colored organic compounds strongly absorbing visible light. These photosynthetic pigments can be divided into three major classes: chlorophylls, carotenoids, and phycobilin pigments. The chlorophylls (usually a greenish color) can harvest light energy and pass electrons to molecules that manufacture sugars in the cell. The essential chlorophyll is Chl-a, which is present in almost all photosynthesizing cells. Chlorophyll-b and chlorophyll-c are not as universal. Other accessory pigments, such as carotenoids (usually yellow-red color), must transfer absorbed light energy to chlorophyll. Phycobilin pigments are soluble and colored pigments and are exceptionally efficient at absorbing green and yellow-red light that is not well absorbed by Chl-a. In biological oceanography, the Chl-a concentration has been a fundamental proxy for phytoplankton biomass. Thus, identifying Chl-a and other pigments is helpful in assessing the phytoplankton community structure and growth rate.

The pigment measurements start with water sampling and filtration. The seawater samples can be drawn with a Niskin bottle or from a flow-through system. As a general rule, for the fluorometric method, it is recommended that the following volumes be filtered for these water types: 0.5–1.0 L for oligotrophic, 0.2–0.5 L for mesotrophic, and 0.1 L and less for eutrophic water (Mueller et al., 2003b). Higher volumes are required for high-performance liquid chromatography (HPLC) analysis, depending on the method and the relative biomass present. The samples are filtered immediately after collection with glass-fiber filters and stored in liquid nitrogen or ultra-cold ($-80°C$) freezers.

The fluorometric technique is based on Chl-a's unique optical properties—fluorescing in red wavelengths when excited by blue light (Holm-Hansen and Riemann, 1978). It requires that the Chl-a pigments from the filters be extracted with a solvent, such as 90% acetone. Fluorescence signals of the nonacidified and acidified subsamples of the extracts are then measured with a fluorometer (Mueller et al., 2003b). The acidified measurement is to correct for the significant fluorescence by phaeopigments. This kind of measurement can be easily completed aboard a research vessel or in a land-based lab.

The HPLC technique separates various phytoplankton pigments in the order of polarity during passage through a column (Van Heukelem and Thomas, 2001). Commonly detected pigments from HPLC include but are not limited to Chl-a, chlorophyll-b, chlorophyll-c, carotene pigments, 19′-butanoyloxyfucoxanthin, 19′-hexanoyloxyfucoxanthin, diadinoxanthin, alloxanthin, diatoxanthin, zeaxanthin, and fucoxanthin. Compared with HPLC, the fluorometric method may be subjected to a relatively larger uncertainty because of overlapping excitation and emission wavelengths of Chl-a and other pigments. According to the archived measurements (Werdell and Bailey, 2005), the relative difference between fluorometric and HPLC Chl-a is ~11% in oligotrophic waters (Chl-a ≤ 0.1 mg m^{-3}), 13% in mesotrophic waters (0.1 < Chl-a ≤ 1 mg m^{-3}), and 13% in eutrophic waters (Chl-a > 1 mg m^{-3}). The differences associated with measurement techniques are usually far smaller than the satellite Chl-a product uncertainties.

The HPLC pigments can be used to identify phytoplankton classes. For example, divinyl-chlorophyll-a and divinyl-chlorophyll-b are unique to *Prochlorococcus* and alloxanthin to cryptophytes (IOCCG, 2014). However, many pigments are shared among taxonomic groups. Therefore, the plasticity of pigment composition and concentration between different ecological conditions and even between strains of the same phytoplankton species should be considered in interpretations of this data.

6.2 Phytoplankton cell counts and cell sizes (>10 μm)

One of the aspirational goals of satellite ocean remote sensing is to generate products of PFTs, which are defined as conceptual groupings of phytoplankton species with common ecological functionality (IOCCG, 2014). Some common examples of PFTs are silicifiers (diatoms), calcifiers (coccolithophores), DMS producers (*Phaeocystis* sp.), nitrogen fixers (*Trichodesmium sp.*), and toxin-producing harmful algae. PFTs are of considerable interest to ocean biogeochemists, fisheries and climate scientists, and coastal resource managers. Several approaches have been developed to derive PFTs from satellite ocean color observations with varying degrees of success (Alvain et al., 2006; Amin et al., 2009; Cannizzaro et al., 2019; Devred et al., 2011). These methods are based on the realization that changes in phytoplankton community composition and phytoplankton size class (Brewin et al., 2011; Hirata et al., 2009; Kostadinov et al., 2009; Mouw et al., 2017;

Mouw and Yoder, 2010) at any given location can have a significant influence on $a(\lambda)$ and $b_b(\lambda)$ derived from ocean color data. Satellite-derived PFTs, however, need to be validated using in situ data sets as with other satellite ocean color products (Neeley et al., 2021).

Until recently, one of the most common means of measuring PFTs and taking measurements in seawater was through microscopy. Microscopic measurements of PFTs are time consuming and are usually undertaken on samples preserved at sea either with buffered formaldehyde, Lugol's iodine, or glutaraldehyde, which are then brought back to the shore lab. Recent technological advances have included the development of sophisticated automated imaging systems, allowing the rapid analysis of the phytoplankton samples at sea. The Imaging Flow Cytobot (McLane Research Laboratories, Inc.) and the FlowCAM (Fluid Imaging Inc.) are among the most commonly used instruments.

Shipboard measurements of PFTs ($>10\,\mu m$) are often undertaken on board using FlowCam and microscopy. Together with the dimensions of each particle, the two methods allow estimations of phytoplankton community structure and particle size class (PSC) of both phytoplankton and detrital particles (Jenkins et al., 2016). In order to encompass the full-size range of PFTs, samples are run using a 300 μm field-of-view flow cell and a 4× objective (UPlan FLN, Olympus), which allows for taxonomic identification and sizing of phytoplankton over a broad $>10\,\mu m$ to 400 μm size range. The instrument can be operated in two modes: (1) manual mode for samples drawn from predetermined depths and (2) automatic mode, wherein the instrument is programmed to image phytoplankton from the ship's flow-through seawater system. The ability to sample seawater at fixed intervals along the cruise track allows the high-resolution (in space and time) measurements of PFTs and PSCs, in turn, enabling the observation of fine-scale oceanographic features (Fig. 10), which would be overlooked when sampling is restricted to fixed stations.

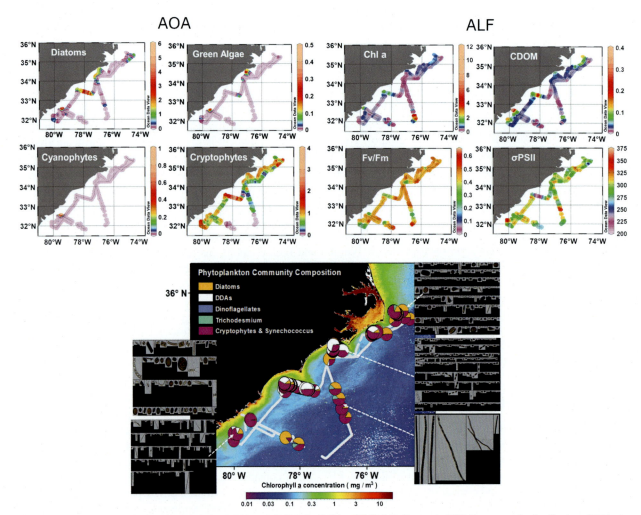

FIG. 10 *Upper panels*: distribution of AOA-derived PFTs and ALF-derived Chl-a, CDOM, Fv/Fm, and σPSII. *Lower panels*: distribution of PFTs along the 2014 NOAA VIIRS dedicated Cal/Val cruise track overlaid on a weekly $K_d(490)$ image over the cruise period from VIIRS-SNPP.

Continuous measurements of PFTs are also undertaken with a biological-biophysical-engineering (bbe) Moldaenke AlgaeOnlineAnalyser (AOA, Richardson et al., 2010) during the VIIRS Cal/Val cruises (Ondrusek et al., 2015, 2016). The AOA provides continuous seawater fluorescence measurements excited with light from light-emitting diodes at six wavelengths (370 nm, 470 nm, 525 nm, 570 nm, 590 nm, and 610 nm). The resulting excitation spectra are then utilized to estimate Chl-a and taxonomic discrimination of PFTs, i.e., cyanobacteria, green algae, brown algae (diatoms and dinoflagellates), and cryptophytes based on the shapes of the fluorescence spectra (see Fig. 10).

6.3 Phytoplankton groups of small sizes (<10 μm)

To account for PFTs such as cyanobacteria, chlorophytes, and cryptophytes that are <10 μm in diameter (Olson et al., 1990), one can use the automated laser fluorometer (ALF, WET Labs, Inc.). The ALF combines high-resolution spectral measurements of blue (405 nm) and green (532 nm) laser-stimulated fluorescence with spectral deconvolution techniques to quantify the fluorescence of Chl-a (peak at 679 nm) and three phycobilin pigment types: PE-1 (peak at 565 nm), PE-2 (peak 578 nm), and PE-3 (peak at 590 nm) (Chekalyuk et al., 2012; Goes et al., 2014a,b). Also measurable by the ALF are CDOM (peak at 508 nm) and variable fluorescence (Fv/Fm), a measure of photo-physiological performance of phytoplankton (see Fig. 10). All fluorescence values obtained are normalized to water Raman spectra and are generally expressed as relative fluorescence units (RFU), whereas Fv/Fm is unitless. PE-1 type pigments are associated with blue-water or oligotrophic cyanobacteria with high phycourobilin/phycoerythrobilin (PUB/PEB) ratios, PE-2 type phytoplankton with low-PUB/PEB ratios are generally associated with green water cyanobacteria that usually thrive in coastal mesohaline waters, and PE-3 tye are attributable to eukaryotic photoautotrophic cryptophytes. RFU values for Chl-a and phycobilin pigments can be converted into their concentrations using least-square regressions of HPLC or fluorometrically derived Chl-a in acetone extracts and phycobilin pigments in a phosphate buffer.

6.4 Phytoplankton quantum efficiency of carbon fixation

Measurements of phytoplankton quantum efficiencies are an additional goal for the ocean color Cal/Val activities as these data can be used to derive net phytoplankton productivity (NPP) using satellite ocean color data. Satellite-based NPP algorithms span a wide range of complexity from empirical (Eppley et al., 1985) to semi-analytical models (Platt et al., 1990). They can be generally categorized into three types based on the modeling strategy (Lee et al., 2015). Two of them are the biology-based models, of which one uses satellite-derived Chl-a (Arrigo et al., 1998, 2008; Behrenfeld and Falkowski, 1997; Carr et al., 2006; Platt and Sathyendranath, 1988), while the other, the Carbon-based Productivity Model (CbPM), uses phytoplankton carbon stock (C_{phyto}, mg m^{-3}) retrieved from satellite-derived $b_{bp}(443)$ (Behrenfeld and Boss, 2006; Behrenfeld et al., 2005). The third category is the absorption-based models (AbPM), which rely on the absorption coefficient of phytoplankton derived directly from $R_{rs}(\lambda)$. A recent model, Carbon Absorption Fluorescence Euphotic resolving (CAFE) (Silsbe et al., 2016), belongs to this category. Absorption-based models derive NPP as the product of $a_{ph}(\lambda)$, photosynthetically available radiation (PAR), and the quantum efficiency of carbon fixation by which absorbed energy is converted into carbon biomass (Hirawake et al., 2011; Kiefer and Mitchell, 1983; Lee et al., 2015; Zoffoli et al., 2018). Currently, the general use of AbPM models has been hampered by the lack of adequate in situ quantum efficiency measurements, forcing reliance on estimates that ignore large, temporal (diurnal, seasonal), and spatial (regional and vertically) variability (Hirawake et al., 2011; Marra et al., 2000, 2003; Ondrusek et al., 2001; Ostrowska et al., 2012; Sorensen and Siegel, 2001; Wu et al., 2022).

During recent NOAA dedicated Cal/Val cruises (Ondrusek et al., 2015, 2016), pilot studies were undertaken on the possibility of using the quantum efficiency estimated by the electron transport rate (ETR) with a Fast Repetition Rate Fluorometer (FRRf). Quantum efficiencies of phytoplankton at sea are typically derived from incubation-based methods involving the use of radioactive or stable isotope tracers. Although considered the gold standard for measurements of quantum efficiency, the method is time consuming and is generally incapable of providing the spatial and temporal coverage required for use in NPP algorithms. The FRRf allows the measurements of fluorescence-based photosynthetic ETR over a range of ambient irradiances that can be utilized to construct plots of ETR as a function of irradiance (similar to photosynthesis-versus-irradiance curves). The FRRf technique also eliminates the need for on-deck incubations, which are time consuming and often impeded by regulatory constraints for the handling of radioactive isotopes. Measurements are possible with both discrete seawater samples and continuous flow-through seawater samples. Currently, ETR data from several cruises are being used to improve understanding of the variability in quantum efficiencies using this method in relation to PFT, PSC, and environmental data collected during the cruise and for possible use in estimating NPP from space.

7. Matchup analysis

To validate satellite ocean color with in situ data, concurrent observations first need to be paired together ("matchups"). The satellite and field data matchups are often constructed following community-accepted protocols (e.g., Bailey and Werdell, 2006; Wang et al., 2009b). The interval between satellite overpass and in situ observations should be limited to a short period of time, ideally within several hours. Although the selection of field locations is not restricted, the waters should be spatially homogeneous. To extract the satellite measurement for a particular location, a box of 3×3 or 5×5 pixels, regardless of the satellite spatial resolution, is often preselected centered on the location of the field observation. The quality flags (e.g., HIGLINT, ATMFAIL, HISOLZEN, and CLDSHDSTL) are usually inspected to determine the satellite data quality (Barnes et al., 2019; Hu et al., 2020; Wang et al., 2017). The selection of these flags needs to be consistent with the way that flags are used to discard data in specific applications. At least half of these pixels within the pixel box must pass the quality check. Furthermore, the coefficient of variation of these pixel values should be reasonably small to assume that the pixels are representative of homogeneous waters. Failure in such a homogeneity test does not indicate low-quality data but that the data are simply not ideal for matchups. The final values, either of radiometric (e.g., $R_{rs}(\lambda)$) or bio-optical (e.g., Chl-a) properties, are then averaged to be evaluated against the near-coincident in situ data. The selection of specific criteria for constructing satellite and in situ matchups depends on specific applications. If a certain application requires one set of flags, the same flags should be used in the matchup selections. In general, the global study of long-term ocean changes requires higher data quality than applications tracing ocean color features and therefore requires more flags to be applied. Thus, individual researchers may adopt a different combination of quality flags, time constraints, and pixel boxes. The satellite ocean color processing procedures by NOAA, NASA, and NRL may use slightly different threshold values for certain flags (Hu et al., 2020; Lawson et al., 2019; Wang et al., 2017).

A range of metrics, including percentage difference (PD) and absolute difference (AD), can be used to quantify the matchup uncertainties. The percentage difference can be subdivided into absolute percentage difference (APD), unbiased absolute percentage difference (UAPD), and relative percentage difference (RPD). The absolute difference can also extend to the root mean squared difference (RMSD). These metrics can be calculated using the medians or means of the matchup data. For example, the median absolute percentage difference (MAPD) can be expressed as the following:

$$\text{MAPD} = \text{median}\left\{\left|\frac{Y_i - X_i}{X_i}\right|\right\} \times 100\%, i = 1, 2, \ldots N, \tag{10}$$

where Y and X refer to satellite and in situ measurements, respectively, and N is the total number of matchups. The corresponding median relative percentage difference (MRPD) or bias can be derived as

$$\text{bias} = \text{median}\left\{\frac{Y_i - X_i}{X_i}\right\} \times 100\%, i = 1, 2, \ldots N. \tag{11}$$

In addition, researchers have also started to use the QA scores as a direct measure of the $R_{rs}(\lambda)$ data quality (Chen et al., 2021; O'Shea et al., 2020).

Most existing satellite and in situ matchups are from the coastal ocean. In Fig. 11, the map shows the locations with coincident VIIRS $R_{rs}(\lambda)$ data and shipborne on-water radiometric measurements. These in situ $R_{rs}(\lambda)$ data were primarily observed where the water depths are less than 1000 m. The satellite $R_{rs}(\lambda)$ data represent the mean values from a 3×3 pixel box, where the pixels have passed the flags of ATMFAIL, HIGLINT, HILT, HISATZEN, and CLDSHDSTL. Fig. 12 presents the scatter plots of the $R_{rs}(\lambda)$ matchups for visual comparison. According to the quantitative measures, VIIRS-derived $R_{rs}(\lambda)$ at 410 nm and 443 nm are subjected to relatively high uncertainties where MAPD = 43% and 30%, respectively. VIIRS $R_{rs}(\lambda)$ uncertainties at 486 nm and 551 nm are minor, with MAPD = ~20%. The VIIRS $R_{rs}(\lambda)$ measurement uncertainty goes up at the red band, where MAPD = 30%. It reaches the maximum at the NIR band of 745 nm, partially due to the small reflectance values and noise at this band.

Table 1 compares the matchup statistics derived from boxes of 3×3 and 5×5 pixels. The results indicate no significant difference in the satellite $R_{rs}(\lambda)$ uncertainties, except that the larger box favors a higher number of matchups. Table 1 further divides the VIIRS and field matchup data into subgroups according to the QA scores of the VIIRS $R_{rs}(\lambda)$ spectra. The high-quality VIIRS data with QA ≥ 0.8 are found to be consistent with in situ measurements with MAPD $\approx 25\%$ for all visible bands. A small percentage of the VIIRS data has low QA scores (≤ 0.6); the corresponding MAPD values are much more significant, particularly at 410 nm and 443 nm.

Similar procedures may be followed to validate other satellite ocean color products. In the end, such field-based validations only examine pixel-wise data uncertainties from discrete data points, which inherently include uncertainties in both

Satellite ocean color validation **Chapter | 20 367**

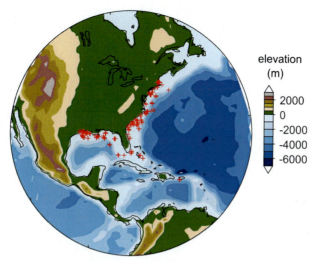

FIG. 11 Stations with concurrent VIIRS-SNPP and in situ $R_{rs}(\lambda)$ measurements. The team from the University of Massachusetts at Boston collected the in situ $R_{rs}(\lambda)$ data using the on-water approach. A detailed field measurement protocol is described in Wei et al. (2021b).

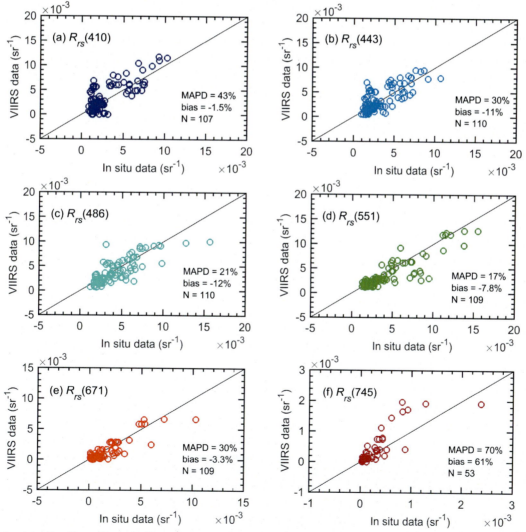

FIG. 12 Scatter plots of satellite and in situ $R_{rs}(\lambda)$ matchups. The VIIRS-SNPP data were determined from a box of 3×3 pixels, and the in situ data were measured with the on-water approach. The following quality flags were considered in selecting valid pixels: ATMFAIL, HIGLINT, HILT, HISATZEN, and CLDSHDSTL. The time constraint was ± 5 h. The maximum coefficient of variation of $R_{rs}(\lambda)$ values within the box was 20%.

TABLE 1 Statistics of the satellite $R_{rs}(\lambda)$ uncertainties from satellite and in situ matchups.

Range of satellite QA scores[a]	Wavelength (nm)	3×3 pixels MAPD	3×3 pixels Bias	3×3 pixels N	5×5 pixels MAPD	5×5 pixels Bias	5×5 pixels N
[0, 1]	411	43%	−1.5%	107	43%	−0.7%	120
	443	30%	−11%	110	30%	−9.7%	120
	486	21%	−12%	110	21%	−13%	122
	551	17%	−7.8%	109	18%	−8.8%	121
	671	30%	−3.3%	109	32%	1.9%	120
	745	70%	61%	53	69%	58%	56
[0.8, 1]	411	26%	−2%	79	26%	0%	85
	443	22%	−11%	80	20%	−8%	85
	486	19%	−12%	80	17%	−12%	85
	551	16%	−9%	80	15%	−9%	85
	671	25%	1%	80	24%	4%	84
	745	72%	72%	35	68%	65%	36
[0, 0.6]	411	95%	−14%	27	94%	−13%	34
	443	53%	−30%	27	52%	−31%	32
	486	29%	−23%	27	32%	−24%	34
	551	22%	−5%	27	28%	−6%	34
	671	50%	−20%	26	54%	−21%	33
	745	69%	−22%	16	70%	−21%	18

The quality flags used for the satellite data are similar to those in Fig. 12, except that two box sizes (3 × 3 and 5 × 5) are considered here and that the matchup data are further divided into two subgroups based on the quality assurance scores of the VIIRS-SNPP $R_{rs}(\lambda)$ spectra.
[a]The QA score is equal to 0, 0.2, 0.4, 0.6, 0.8, or 1; the QA score of 1 indicates the highest quality.

field data and satellite data (IOCCG, 2019b). When data are averaged spatially or temporally, uncertainties will be reduced dramatically (Qi et al., 2017) as most of the uncertainties come from data spread rather than bias, and therefore can be averaged out.

8. Concluding remarks

Satellite ocean color validation determines the spatial and temporal uncertainty of various ocean color products. Such uncertainty information is essential for the interpretation of observed optical, biological, and biogeochemical states of oceans. The feedback of ocean color validation, in turn, informs us of the future directions for algorithm improvement and sensor characterization. Validation of the ocean color products is also critical to ensure the continuity of long-term ocean color products from multiple satellite missions and to ensure the spatial consistency of ocean color products merged across different satellite measurements. The present work has focused on the state-of-the-art instruments and methods that offer in situ optical, biological, and biogeochemical measurements for satellite ocean color data validation. However, some quantities are missing from our discussion. For instance, the volume scattering function is only occasionally measured for lack of commercial instruments. In situ measurements of other properties such as POC and SPM are covered by NASA ocean optical protocols and IOCCG publications. Like satellite ocean color products, the field measurements are not devoid of errors. Future work should continue to explore this problem and account for them in matchup analysis. On the other hand, the selection of criteria for constructing matchup data is user and application dependent. Investigators may exercise varying constraints to construct matchups to suit specific validation purposes. Nonetheless, a community-wide effort remains necessary to address existing problems found with ocean color validation. Finally, note that commercial equipment, instruments, or materials are identified in this chapter to foster understanding. Such identification does not imply recommendation or endorsement by NOAA, nor does it imply that the materials or equipment identified are necessarily the best available for the purpose.

Acknowledgments

We thank Ryan Vandermeulen (NASA/GSFC) and Greg Foltz (NOAA/AOML/PHOD) for comments. This work was partially supported by the Joint Polar Satellite System (JPSS) funding for the NOAA ocean color calibration and validation (Cal/Val) project. Many people have contributed to the NOAA ocean color Cal/Val project since 2011, including Robert Arnone, Sam Ahmed, Curt Davis, Carol Johnson, Burt Jones, Charles Trees, Brian Barnes, Jennifer Cannizzaro, David English, Michael Feinholz, Stephanie Flora, Bill Gibson, Art Gleason, Helga do Rosario Gomes, Wesley Goode, Charles Kovach, Adam Lawson, Chunyan Li, Matthew Ragan, Zhehai Shang, Eric Stengel, Jinghui Wu, Mark Yarbrough, and Shuai Zhang. The in situ data used in this study were collected from the NOAA dedicated VIIRS ocean color Cal/Val cruises supported by the JPSS program and the NOAA Office of Marine and Aviation Operations. The scientific results and conclusions, as well as any views or opinions expressed herein, are those of the authors and do not necessarily reflect those of NOAA or the Department of Commerce.

References

Aas, E., Højerslev, N.K., 1999. Analysis of underwater radiance observations: apparent optical properties and analytic functions describing the angular radiance distribution. J. Geophys. Res. 104, 8015–8024.

Ahn, Y.-H., Ryu, J.-H., Moon, J.-E., 1999. Development of red tide and water turbidity algorithms using ocean color satellite. In: KORDI Report No. BSPE 98721-00-1224-01. KORDI, Seoul, Korea, p. 287.

Alvain, S., Moulin, C., Dandonneau, Y., Loisel, H., Bréon, F.M., 2006. A species-dependent bio-optical model of case I waters for global ocean color processing. Deep-Sea Res. I 53, 917–925.

Amin, R., Gilerson, A., Gross, B., Moshary, F., Ahmed, S., 2009. MODIS and MERIS detection of dinoflagellates blooms using the RBD technique. Proc. SPIE 7473. https://doi.org/10.1117/12.830631.

Antoine, D., Morel, A., 1999. A multiple scattering algorithm for atmospheric correction of remotely sensed ocean colour (MERIS instrument): principle and implementation for atmospheres carrying various aerosols including absorbing ones. Int. J. Remote Sens. 20, 1875–1916.

Antoine, D., Morel, A., Leymarie, E., Houyou, A., Gentili, B., Victori, S., Buis, J., Buis, N., Meunier, S., Canini, M., Crozel, D., Fougnie, B., Henry, P., 2013. Underwater radiance distributions measured with miniaturized multispectral radiance cameras. J. Atmos. Ocean. Technol. 30, 74–95.

Arrigo, K., Worthen, D., Schnell, A., Lizotte, M.P., 1998. Primary production in Southern Ocean waters. J. Geophys. Res. 103, 15,587–515,600.

Arrigo, K.R., Dijken, G.L.V., Bushinsky, S., 2008. Primary production in the Southern Ocean, 1997-2006. J. Geophys. Res. 113. https://doi.org/10.1029/2007JC004551. C008004.

Austin, R.W., 1974. Inherent spectral radiance signatures of the ocean surface. In: Duntley, S.Q., Austin, R.W., Wilson, W.H., Edgerton, C.F., Moran, S.E. (Eds.), Ocean Color Analysis. Scripps Institution of Oceanography, La Jolla, California, pp. 1–20.

Bailey, S.W., Franz, B.A., Werdell, P.J., 2010. Estimation of near-infrared water-leaving reflectance for satellite ocean color data processing. Opt. Express 18, 7521–7527. https://doi.org/10.1364/OE.18.007521.

Bailey, S.W., Werdell, P.J., 2006. A multi-sensor approach for the on-orbit validation of ocean color satellite data products. Remote Sens. Environ. 102, 12–23. https://doi.org/10.1016/j.rse.2006.01.015.

Barnes, B.B., Cannizzaro, J.P., English, D.C., Hu, C., 2019. Validation of VIIRS and MODIS reflectance data in coastal and oceanic waters: an assessment of methods. Remote Sens. Environ. 220, 110–123. https://doi.org/10.1016/j.rse.2018.10.034.

Behrenfeld, M.J., Boss, E., 2006. Beam attenuation and chlorophyll concentration as alternative optical indices of phytoplankton biomass. J. Mar. Res. 64, 431–451.

Behrenfeld, M.J., Boss, E., Siegel, D., Shea, D.M., 2005. Carbon-based ocean productivity and phytoplankton physiology from space. Glob. Biogeochem. Cycles 19, GB1006. https://doi.org/10.1029/2004GB002299.

Behrenfeld, M.J., Falkowski, P.G., 1997. A consumer's guide to phytoplankton primary productivity models. Limnol. Oceanogr. 42, 1479–1491.

Bhandari, P., Voss, K.J., Logan, L., 2011. An instrument to measure the downwelling polarized radiance distribution in the ocean. Opt. Express 19, 17609–17620.

Brewin, R.J.W., Devred, E., Sathyendranath, S., Lavender, S.J., Hardman-Mountford, N.J., 2011. Model of phytoplankton absorption based on three size classes. Appl. Opt. 50, 4535–4549. https://doi.org/10.1364/ao.50.004535.

Bricaud, A., Babin, M., Morel, A., Claustre, H., 1995. Variability in the chlorophyll-specific absorption coefficients of natural phytoplankton: analysis and parameterization. J. Geophys. Res. 100, 13321–13332.

Bricaud, A., Morel, A., Babin, M., Allali, K., Claustre, H., 1998. Variations of light absorption by suspended particles with chlorophyll a concentration in oceanic (case 1) waters: analysis and implications for bio-optical models. J. Geophys. Res. 103, 31033–31044. https://doi.org/10.1029/98JC02712.

Cannizzaro, J.P., Barnes, B.B., Hu, C., Corcoran, A.A., Hubbard, K.A., Muhlbach, E., Sharp, W.C., Brand, L.E., Kelble, C.R., 2019. Remote detection of cyanobacteria blooms in an optically shallow subtropical lagoonal estuary using MODIS data. Remote Sens. Environ. 231, 111227. https://doi.org/10.1016/j.rse.2019.111227.

Carder, K.L., Chen, F.R., Lee, Z.P., Hawes, S.K., Kamykowski, D., 1999. Semianalytic moderate-resolution imaging spectrometer algorithms for chlorophyll-a and absorption with bio-optical domains based on nitrate-depletion temperatures. J. Geophys. Res. 104, 5403–5421.

Carr, M.-E., Friedrichs, M.A.M., Schmeltz, M., Noguchi Aita, M., Antoine, D., Arrigo, K.R., Asanuma, I., Aumont, O., Barber, R., Behrenfeld, M.J., Bidigare, R., Buitenhuis, E.T., Campbell, J., Ciotti, A., Dierssen, H., Dowell, M., Dunne, J., Esaias, W., Gentili, B., Gregg, W., Groom, S., Hoepffner, N., Ishizaka, J., Kameda, T., Le Quéré, C., Lohrenz, S., Marra, J., Mélin, F., Moore, K., Morel, A., Reddy, T.E., Ryan, J., Scardi, M., Smyth, T., Turpie, K., Tilstone, G., Waters, K., Yamanaka, Y., 2006. A comparison of global estimates of marine primary production from ocean color. Deep-Sea Res. II Top. Stud. Oceanogr. 53, 741–770. https://doi.org/10.1016/j.dsr2.2006.01.028.

Chekalyuk, A.M., Landry, M.R., Goericke, R., Taylor, A.G., Hafez, M.A., 2012. Laser fluorescence analysis of phytoplankton across a frontal zone in the California current ecosystem. J. Plankton Res. 34, 761–777. https://doi.org/10.1093/plankt/fbs034.

Chen, S., Du, K., Lee, Z.P., Liu, J., Song, Q., Xue, C., Wang, D., Lin, M., Tang, J., Ma, C., 2021. Performance of COCTS in global ocean color remote sensing. IEEE Trans. Geosci. Remote Sens. 59, 1634–1644. https://doi.org/10.1109/TGRS.2020.3002460.

Choi, J.-K., Park, Y.J., Ahn, J.H., Lim, H.-S., Eom, J., Ryu, J.-H., 2012. GOCI, the world's first geostationary ocean color observation satellite, for the monitoring of temporal variability in coastal water turbidity. J. Geophys. Res. 117. https://doi.org/10.1029/2012JC008046.

Chomko, R.M., Gordon, H.R., 2001. Atmospheric correction of ocean color imagery: test of the spectral optimization algorithm with the sea-viewing wide field-of-view sensor. Appl. Opt. 40, 2973–2984. https://doi.org/10.1364/AO.40.002973.

Clark, D.K., Gordon, H.R., Voss, K.J., Ge, Y., Broenkow, W., Trees, C., 1997. Validation of atmospheric correction over the oceans. J. Geophys. Res. 102, 17209–17217. https://doi.org/10.1029/96JD03345.

Clark, D.K., Yarbrough, M.A., Feinholz, M., Flora, S., Broenkow, W., Kim, Y.S., Johnson, B.C., Brown, S.W., Yuen, M., Mueller, J.L., 2003. MOBY, a radiometric buoy for performance monitoring and vicarious calibration of satellite ocean color sensors: measurement and data analysis protocols. In: Mueller, J.L., Fargion, G.S., McClain, G.S. (Eds.), Ocean Optics Protocols for Satellite Ocean Color Sensor Validation, Revision 4, Volume VI: Special Topics in Ocean Optics Protocols and Appendices. NASA, Goddard Space Flight Center, Greenbelt, MD, pp. 138–170.

Devred, E., Sathyendranath, S., Stuart, V., Platt, T., 2011. A three component classification of phytoplankton absorption spectra: application to ocean-color data. Remote Sens. Environ. 115, 2255–2266. https://doi.org/10.1016/j.rse.2011.04.025.

Eppley, R.W., Stewart, E., Abbott, M.R., Heyman, U., 1985. Estimating ocean primary production from satellite chlorophyll. Introduction to regional differences and statistics for the Southern California bight. J. Plankton Res. 7, 57–70. https://doi.org/10.1093/plankt/7.1.57.

Esaias, W.E., Abbott, M.R., Barton, I., Brown, O.B., Campbell, J.W., Carder, K.L., Clark, D.K., Evans, R.H., Hoge, F.E., Gordon, H.R., Balch, W.M., Letelier, R., Minnett, P.J., 1998. An overview of MODIS capabilities for ocean science observations. IEEE Trans. Geosci. Remote Sens. 36, 1250–1265. https://doi.org/10.1109/36.701076.

Fan, Y., Li, W., Gatebe, C.K., Jamet, C., Zibordi, G., Schroeder, T., Stamnes, K., 2017. Atmospheric correction over coastal waters using multilayer neural networks. Remote Sens. Environ. 199, 218–240. https://doi.org/10.1016/j.rse.2017.07.016.

Goes, J.I., Gomes, H.D.R., Chekalyuk, A.M., Carpenter, E.J., Montoya, J.P., Coles, V.J., Yager, P.L., Berelson, W.M., Capone, D.G., Foster, R.A., Steinberg, D.K., Subramaniam, A., Hafez, M.A., 2014a. Influence of the Amazon River discharge on the biogeography of phytoplankton communities in the western tropical North Atlantic. Prog. Oceanogr. 120, 29–40. https://doi.org/10.1016/j.pocean.2013.07.010.

Goes, J.I., Gomes, H.D.R., Haugen, E.M., McKee, K.T., D'Sa, E.J., Chekalyuk, A.M., Stoecker, D.K., Stabeno, P.J., Saitoh, S.-I., Sambrotto, R.N., 2014b. Fluorescence, pigment and microscopic characterization of Bering Sea phytoplankton community structure and photosynthetic competency in the presence of a cold pool during summer. Deep-Sea Res. II Top. Stud. Oceanogr. 109, 84–99. https://doi.org/10.1016/j.dsr2.2013.12.004.

Gordon, H.R., 2005. Normalized water-leaving radiance: revisiting the influence of surface roughness. Appl. Opt. 44, 241–248.

Gordon, H.R., Clark, D.K., Mueller, J.L., Hovis, W.A., 1980. Phytoplankton pigments from the Nimbus-7 coastal zone color scanner: comparisons with surface measurements. Science 210, 63–66.

Gordon, H.R., Wang, M., 1994. Retrieval of water-leaving radiance and aerosol optical thickness over the oceans with SeaWiFS: a preliminary algorithm. Appl. Opt. 33, 443–452.

He, X., Bai, Y., Pan, D., Tang, J., Wang, D., 2012. Atmospheric correction of satellite ocean color imagery using the ultraviolet wavelength for highly turbid waters. Opt. Express 20, 20754–20770.

Hirata, T., Hardman-Mountford, N., Aiken, J., Fishwick, J., 2009. Relationship between the distribution function of ocean nadir radiance and inherent optical properties for oceanic waters. Appl. Opt. 48, 3129–3138.

Hirawake, T., Takao, S., Horimoto, N., Ishimaru, T., Yamaguchi, Y., Fukuchi, M., 2011. A phytoplankton absorption-based primary productivity model for remote sensing in the Southern Ocean. Polar Biol. 34, 291–302. https://doi.org/10.1007/s00300-010-0949-y.

Holm-Hansen, O., Riemann, E., 1978. Chlorophyll-a determination: improvements in methodology. Oikos 30, 43–47.

Hovis, W.A., 1981. The Nimbus-7 coastal zone color scanner (CZCS) program. In: Gower, J.F.R. (Ed.), Oceanography from Space. Springer US, Boston, MA, pp. 213–225.

Hu, C., Barnes, B.B., Feng, L., Wang, M., Jiang, L., 2020. On the interplay between ocean color data quality and data quantity: impacts of quality control flags. IEEE Geosci. Remote Sens. Lett. 17, 745–749. https://doi.org/10.1109/LGRS.2019.2936220.

Hu, C., Lee, Z.P., Franz, B., 2012. Chlorophyll a algorithms for oligotrophic oceans: a novel approach based on three-band reflectance difference. J. Geophys. Res. 117, 2156–2202. https://doi.org/10.1029/2011JC007395.

Ibrahim, A., Franz, B.A., Ahmad, Z., Bailey, S.W., 2019. Multiband atmospheric correction algorithm for ocean color retrievals. Front. Earth Sci. 7. https://doi.org/10.3389/feart.2019.00116.

IOCCG, 2000. In: Sathyendranath, S. (Ed.), Remote Sensing of Ocean Colour in Coastal, and Other Optically-Complex, Waters. International Ocean Colour Coordinating Group, Dartmouth, Nova Scotia, Canada, p. 140. doi: 10.25607/OBP-95.

IOCCG, 2006. In: Lee, Z.P. (Ed.), Remote Sensing of Inherent Optical Properties: Fundamentals, Tests of Algorithms, and Applications. International Ocean Color Coordinating Group, Dartmouth, Nova Scotia, Canada, p. 126. doi: 10.25607/OBP-106.

IOCCG, 2010. In: Wang, M. (Ed.), Atmospheric Correction for Remotely-Sensed Ocean Color Products. International Ocean Color Coordinating Group, Dartmouth, Nova Scotia, Canada, p. 78. doi: 10.25607/OBP-101.

IOCCG, 2014. Phytoplankton Functional Types from Space. International Ocean-Color Coordinating Group, Dartmouth, Canada. doi: 10.25607/OBP-106.

IOCCG, 2019a. Inherent optical property measurements and protocols: best practices for the collection and processing of ship-based underway flow-through optical data. In: Neeley, A.R., Mannino, A. (Eds.), IOCCG Ocean Optics and Biogeochemistry Protocols for Satellite Ocean Colour Sensor Validation. International Ocean Colour Coordinating Group in conjunction with the National Aeronautics and Space Administration (NASA), Dartmouth, NS, Canada, p. 22.

IOCCG, 2019b. In: Mélin, F. (Ed.), Uncertainty in Ocean Color Remote Sensing. International Ocean Color Coordinating Group, Dartmouth, NS, Canada, p. 164. https://dx.doi.org/10.25607/OBP-696.

Jenkins, C., Goes, J., McKee, K., Gomes, H.D.R., Arnone, R., Wang, M., Ondrusek, M., Nagamani, P., Preethi Latha, T., Rao, K., Dadhwal, V., 2016. High-resolution shipboard measurements of phytoplankton: a way forward for enhancing the utility of satellite SST and chlorophyll for mapping microscale features and frontal zones in coastal waters. Proc. SPIE 9878. https://doi.org/10.1117/12.2225875.

Jiang, L., Wang, M., 2014. Improved near-infrared ocean reflectance correction algorithm for satellite ocean color data processing. Opt. Express 22, 21657–21678. https://doi.org/10.1364/OE.22.021657.

Kiefer, D.A., Mitchell, B.G., 1983. A simple, steady state description of phytoplankton growth based on absorption cross section and quantum efficiency. Limnol. Oceanogr. 28, 770–776.

Kirk, J.T.O., 1997. Point-source integrating-cavity absorption meter: theoretical principles and numerical modeling. Appl. Opt. 36, 6123–6128. https://doi.org/10.1364/AO.36.006123.

Kishino, M., Takahashi, M., Okami, N., Ichimura, S., 1985. Estimation of the spectral absorption coefficients of phytoplankton in a thermally stratified sea. Bull. Mar. Sci. 37, 634–642.

Kostadinov, T.S., Siegel, D.A., Maritorena, S., 2009. Retrieval of the particle size distribution from satellite ocean color observations. J. Geophys. Res. 114, C09015. https://doi.org/10.1029/2009JC005303.

Lawson, A., Ladner, S., Crout, R., Wood, C., Arnone, R., Bowers, J., Martinolich, P., Lewis, D., 2019. Establishing optimal matchup protocols between ocean color satellites and ground truth AeroNET-OC radiance. Proc. SPIE 11014. https://doi.org/10.1117/12.2521062.

Lee, M.E., Lewis, M.R., 2003. A new method for the measurement of the optical volume scattering function in the upper ocean. J. Atmos. Ocean. Technol. 20, 563–571.

Lee, Z.P., Carder, K.L., Arnone, R., 2002. Deriving inherent optical properties from water color: a multi-band quasi-analytical algorithm for optically deep waters. Appl. Opt. 41, 5755–5772.

Lee, Z.P., Marra, J., Perry, M.J., Kahru, M., 2015. Estimating oceanic primary productivity from ocean color remote sensing: a strategic assessment. J. Mar. Syst. 149, 50–59. https://doi.org/10.1016/j.jmarsys.2014.11.015.

Lee, Z.P., Pahlevan, N., Ahn, Y.-H., Greb, S., O'Donnell, D., 2013. Robust approach to directly measuring water-leaving radiance in the field. Appl. Opt. 52, 1693–1701.

Lewis, M.R., Wei, J., van Dommelen, R., Voss, K.J., 2011. Quantitative estimation of the underwater radiance distribution. J. Geophys. Res. 116, C00H06. https://doi.org/10.1029/2011JC007275.

Maritorena, S., Siegel, D.A., Peterson, A.R., 2002. Optimization of a semianalytical ocean color model for global-scale applications. Appl. Opt. 41, 2705–2714.

Marra, J., Ho, C., Trees, C.T., 2003. An Alternative Algorithm for the Calculation of Primary Productivity from Remote Sensing Data. Lamont-Doherty Earth Observatory of Columbia University, p. 27.

Marra, J., Trees, C., Bidigare, R.R., Barber, R.T., 2000. Pigment absorption and quantum yield in the Arabian Sea. Deep-Sea Res. II 47, 1279–1299.

McKee, D., Chami, M., Brown, I., Calzado, V.S., Doxaran, D., Cunningham, A., 2009. Role of measurement uncertainties in observed variability in the spectral backscattering ratio: a case study in mineral-rich coastal waters. Appl. Opt. 48, 4663–4675.

Mobley, C.D., 1999. Estimation of the remote-sensing reflectance from above-surface measurements. Appl. Opt. 38, 7442–7455.

Morel, A., Antoine, D., Gentili, B., 2002. Bidirectional reflectance of oceanic waters: accounting for Raman emission and varying particle scattering phase function. Appl. Opt. 41, 6289–6306.

Morel, A., Gentili, B., 1991. Diffuse reflectance of oceanic waters: its dependence on sun angle as influenced by the molecular scattering contribution. Appl. Opt. 30, 4427–4438.

Morel, A., Gentili, B., 1993. Diffuse reflectance of oceanic waters. II. Bidirectional aspects. Appl. Opt. 32, 6864–6879.

Morel, A., Gentili, B., 1996. Diffuse reflectance of oceanic waters, III, implications of bi-directionality for the remote sensing problem. Appl. Opt. 35, 4850–4862.

Mouw, C.B., Hardman-Mountford, N.J., Alvain, S., Bracher, A., Brewin, R.J.W., Bricaud, A., Ciotti, A.M., Devred, E., Fujiwara, A., Hirata, T., Hirawake, T., Kostadinov, T.S., Roy, S., Uitz, J., 2017. A consumer's guide to satellite remote sensing of multiple phytoplankton groups in the Global Ocean. Front. Mar. Sci. 4. https://doi.org/10.3389/fmars.2017.00041.

Mouw, C.B., Yoder, J.A., 2010. Optical determination of phytoplankton size composition from global SeaWiFS imagery. J. Geophys. Res. 115. https://doi.org/10.1029/2010jc006337.

Mueller, J., Fargion, G., McClain, C.R., 2003a. Ocean Optics Protocols for Satellite Ocean Color Validation, Revision 4, Volume IV: Inherent Optical Properties: Instruments, Characterizations, Field Measurements and Data Analysis Protocols. NASA, Greenbelt, Maryland, p. 76.

Mueller, J.L., Fargion, G.S., McClain, C.R., 2003b. Ocean Optics Protocols for Satellite Ocean Color Sensor Validation, Revision 5, Volume V: Biogeochemical and Bio-Optical Measurements and Data Analysis Protocols. NASA Goddard Space Flight Space Center, Greenbelt, Maryland, p. 36.

Mueller, J.L., Fargion, G.S., McClain, C.R., 2003c. Radiometric measurements and data analysis protocols. In: Ocean Optics Protocols for Satellite Ocean Color Sensor Validation, Revision 4. NASA Goddard Space Flight Center, Greenbelt, MD, p. 84.

Neeley, A., Beaulieu, S.E., Proctor, C., Cetinić, I., Futrelle, J., Soto Ramos, I., Sosik, H.M., Devred, E., Karp-Boss, L., Picheral, M., Poulton, N., Roesler, C.S., Shepherd, A., 2021. Standards and practices for reporting plankton and other particle observations from images. In: WHOAS: Woods Hole Open Access Server. WHOI, Woods Hole, MA, p. 38.

Nobileau, D., Antoine, D., 2005. Detection of blue-absorbing aerosols using near infrared and visible (ocean color) remote sensing observations. Remote Sens. Environ. 95, 368–387. https://doi.org/10.1016/j.rse.2004.12.020.

Olson, R.J., Chisholm, S.W., Zettler, E.R., Armbrust, E.V., 1990. Pigments, size, and distributions of Synechococcus in the North Atlantic and Pacific oceans. Limnol. Oceanogr. 35, 45–58. https://doi.org/10.4319/lo.1990.35.1.0045.

Ondrusek, M., Lance, V.P., Wang, M., Arnone, R.A., Ladner, S., Goode, W., Vandermeulen, R., Freeman, S., Chaves, J.E., Mannino, A., Gilerson, A., Ahmed, S., Carrizo, C., El-Habashi, A., Foster, R., Ottaviani, M., Goes, J.I., Gomes, H.D.R., McKee, K., Hu, C., Kovach, C., English, D., Cannizzaro, J., Johnson, B.C., Lee, Z., Wei, J., Wang, Q., Lin, J., Tufillaro, N., Nahorniak, J., Davis, C.O., Voss, K.J., 2015. Report for Dedicated JPSS VIIRS Ocean Color Calibration/Validation Cruise. National Environmental Satellite, Data, and Information Service, Washington, DC, United States. doi: 10.7289/V52B8W0Z.

Ondrusek, M., Lance, V.P., Wang, M., Arnone, R.A., Ladner, S., Goode, W., Vandermeulen, R., Freeman, S., Chaves, J.E., Mannino, A., Gilerson, A., Ahmed, S., Carrizo, C., El-Habashi, A., Foster, R., Ottaviani, M., Goes, J.I., Gomes, H.D.R., McKee, K., Hu, C., Kovach, C., English, D., Cannizzaro, J., Johnson, B.C., Lee, Z., Wei, J., Wang, Q., Lin, J., Tufillaro, N., Nahorniak, J., Davis, C.O., Voss, K.J., 2016. Report for Dedicated JPSS VIIRS Ocean Color December 2015 Calibration/Validation Cruise. National Environmental Satellite, Data, and Information Service, Silver Spring, Maryland, United States. doi: 10.7289/V5/TR-NESDIS-148.

Ondrusek, M.E., Bidigare, R.R., Waters, K., Kar, D.M., 2001. A predictive model for estimating rates of primary production in the subtropical North Pacific Ocean. Deep-Sea Res. II 48, 1837–1863.

Ondrusek, M., Lance, V.P., Wang, M., Stengel, E., Kovach, C., Arnone, R.A., Ladner, S., Goode, W., Gilerson, A., Ahmed, S.A., El-Habashi, A., Foster, R., Ottaviani, M., Goes, J.I., Gomes, H.D.R., McKee, K., Kang, J.W., Hu, C., Cannizzaro, J., Sun, S., English, D., Johnson, B.C., Lee, Z., Zoffoli, L., Lin, J., Tufillaro, N., Lalovic, I., Nahorniak, J., Davis, C.O., Twardowski, M., Stockley, N., Voss, K.J., 2017. Report for Dedicated JPSS VIIRS Ocean Color Calibration/Validation Cruise, October 2016. National Environmental Satellite, Data, and Information Service, Washington, DC, United States. https://doi.org/10.7289/V5/TR-NESDIS-151.

Ondrusek, M., Lance, V.P., Wang, M., Stengel, E., Kovach, C., Arnone, R.A., Ladner, S., Goode, W., Gilerson, A., El-Habashi, A., Carrizo, C., Herrera, E., Ahmed, S.A., Goes, J.I., Gomes, H.D.R., McKee, K., Hu, C., Cannizzaro, J., Zhang, Y., Huang, C.-W., English, D., Johnson, B.C., Lee, Z., Yu, X., Shang, Z., Tufillaro, N., Lalovic, I., Voss, K.J., 2019. Report for Dedicated JPSS VIIRS Ocean Color Calibration/Validation Cruise May 2018. National Environmental Satellite, Data, and Information Service, Washington, DC, United States. https://doi.org/10.25923/scyb-qf42.

O'Reilly, J.E., Werdell, P.J., 2019. Chlorophyll algorithms for ocean color sensors – OC4, OC5 & OC6. Remote Sens. Environ. 229, 32–47. https://doi.org/10.1016/j.rse.2019.04.021.

O'Shea, R.E., Laney, S.R., Lee, Z., 2020. Evaluation of glint correction approaches for fine-scale ocean color measurements by lightweight hyperspectral imaging spectrometers. Appl. Opt. 59, B18–b34. https://doi.org/10.1364/ao.377059.

Ostrowska, M., Woźniak, B., Dera, J., 2012. Modelled quantum yields and energy efficiency of fluorescence, photosynthesis and heat production by phytoplankton in the world ocean. Oceanologia 54, 565–610. https://doi.org/10.5697/oc.54-4.565.

Petzold, T.J., 1972. Volume Scattering Functions for Selected Natural Waters. Scripps Institute of Oceanography, San Diego, California, pp. 72–78.

Platt, T., Sathyendranath, S., 1988. Oceanic primary production: estimation by remote sensing at local and regional scales. Science 241, 1613–1620.

Platt, T., Sathyendranath, S., Ravindran, P., 1990. Primary production by phytoplankton: analytic solutions for daily rates per unit area of water surface. Proc. Biol. Sci. 241, 101–111.

Qi, L., Lee, Z., Hu, C., Wang, M., 2017. Requirement of minimal signal-to-noise ratios of ocean color sensors and uncertainties of ocean color products. J. Geophys. Res. 122, 2595–2611. https://doi.org/10.1002/2016JC012558.

Richardson, T.L., Lawrenz, E., Pinckney, J.L., Guajardo, R.C., Walker, E.A., Paerl, H.W., MacIntyre, H.L., 2010. Spectral fluorometric characterization of phytoplankton community composition using the algae online analyser®. Water Res. 44, 2461–2472. https://doi.org/10.1016/j.watres.2010.01.012.

Roesler, C.S., Perry, M.J., Carder, K.L., 1989. Modeling in situ phytoplankton absorption from total absorption spectra in productive inland marine waters. Limnol. Oceanogr. 34, 1510–1523. https://doi.org/10.2307/2837036.

Röttgers, R., Häse, C., Doerffer, R., 2007. Determination of the particulate absorption of microalgae using a point-source integrating-cavity absorption meter: verification with a photometric technique, improvements for pigment bleaching, and correction for chlorophyll fluorescence. Limnol. Oceanogr. Methods 5, 1–12. https://doi.org/10.4319/lom.2007.5.1.

Röttgers, R., McKee, D., Woźniak, S.B., 2013. Evaluation of scatter corrections for ac-9 absorption measurements in coastal waters. Methods Oceanogr. 7, 21–39. https://doi.org/10.1016/j.mio.2013.11.001.

Ruddick, K.G., Voss, K., Boss, E., Castagna, A., Frouin, R., Gilerson, A., Hieronymi, M., Johnson, B.C., Kuusk, J., Lee, Z., Ondrusek, M., Vabson, V., Vendt, R., 2019. A review of protocols for fiducial reference measurements of water-leaving radiance for validation of satellite remote-sensing data over water. Remote Sens. 11, 2198.

Shang, Z., Lee, Z.P., Dong, Q., Wei, J., 2017. Self-shading associated with a skylight-blocked approach system for the measurement of water-leaving radiance and its correction. Appl. Opt. 56, 7033–7040. https://doi.org/10.1364/AO.56.007033.

Shang, Z., Lee, Z.P., Wei, J., Lin, G., 2020. Impact of ship on radiometric measurements in the field: a reappraisal via Monte Carlo simulations. Opt. Express 28, 1439–1455. https://doi.org/10.1364/OE.28.001439.

Silsbe, G.M., Behrenfeld, M.J., Halsey, K.H., Milligan, A.J., Westberry, T.K., 2016. The CAFE model: a net production model for global ocean phytoplankton. Glob. Biogeochem. Cycles 30, 1756–1777. https://doi.org/10.1002/2016GB005521.

Slade, W., Boss, E., 2015. Spectral attenuation and backscattering as indicators of average particle size. Appl. Opt. 54, 7264–7277.
Smith, R.C., Austin, R.W., Tyler, J.E., 1970. An oceanographic radiance distribution camera system. Appl. Opt. 9, 2015–2016.
Sorensen, J.C., Siegel, D.A., 2001. Variability of the effective quantum yield for carbon assimilation in the Sargasso Sea. Deep-Sea Res. II 48, 2005–2035.
Steinmetz, F., Deschamps, P.-Y., Ramon, D., 2011. Atmospheric correction in presence of sun glint: application to MERIS. Opt. Express 19, 9783–9800.
Stramski, D., Reynolds, R.A., Babin, M., Kaczmarek, S., Lewis, M.R., Röttgers, R., Sciandra, A., Stramska, M., Twardowski, M.S., Franz, B.A., Claustre, H., 2008. Relationships between the surface concentration of particulate organic carbon and optical properties in the eastern South Pacific and eastern Atlantic oceans. Biogeosciences 5, 171–201.
Sullivan, J.M., Twardowski, M.S., Zaneveld, J.R.V., Moore, C.M., Barnard, A.H., Donaghay, P.L., Rhoades, B., 2006. Hyperspectral temperature and salt dependencies of absorption by water and heavy water in the 400-750 nm spectral range. Appl. Opt. 45, 5294–5309.
Tanaka, A., Sasaki, H., Ishizaka, J., 2006. Alternative measuring method for water-leaving radiance using a radiance sensor with a domed cover. Opt. Express 14, 3099–3105.
Tyler, J.E., 1960. Radiance distribution as a function of depth in an underwater environment. In: Bulletin of the Scripps Institution of Oceanography of the University of California, La Jolla, California, 7, pp. 363–411.
Van Heukelem, L., Thomas, C.S., 2001. Computer-assisted high-performance liquid chromatography method development with applications to the isolation and analysis of phytoplankton pigments. J. Chromatogr. A 910, 31–49. https://doi.org/10.1016/S0378-4347(00)00603-4.
Voss, K.J., 1989. Electro-optic camera system for measurement of the underwater radiance distribution. Opt. Eng. 28, 384–387.
Voss, K.J., Chapin, A.L., 1992. Next generation in-water radiance distribution camera system. In: Gilbert, G.D. (Ed.), Ocean Optics XI. SPIE, San Diego, CA, United States, pp. 384–387.
Voss, K.J., Chapin, A.L., 2005. Upwelling radiance distribution camera system, NURADS. Opt. Express 13, 4250–4262.
Voss, K.J., McLean, S., Lewis, M.R., Johnson, C., Flora, S., Feinholz, M., Yarbrough, M., Trees, C.C., Twardowski, M.S., Clark, D.K., 2010. An example crossover experiment for testing new vicarious calibration techniques for satellite ocean color radiometry. J. Atmos. Ocean. Technol. 27, 1747–1759.
Wang, J., Lee, Z.P., Wei, J., Du, K., 2020. Atmospheric correction in coastal region using same-day observations of different sun-sensor geometries with a revised POLYMER model. Opt. Express 28, 26953–26976. https://doi.org/10.1364/OE.393968.
Wang, M., 2006. Effects of ocean surface reflectance variation with solar elevation on normalized water-leaving radiance. Appl. Opt. 45, 4122–4128.
Wang, M., Jiang, L., 2018. Atmospheric correction using the information from the short blue band. IEEE Trans. Geosci. Remote Sens. 56, 6224–6237. https://doi.org/10.1109/TGRS.2018.2833839.
Wang, M., Liu, X., Jiang, L., Son, S., 2017. VIIRS Ocean Color Algorithm Theoretical Basis Document (ATBD)—Visible Infrared Imaging Radiometer Suite Ocean Color Products. NOAA/NESDIS/STAR, College Park, MD, p. 68.
Wang, M., Liu, X., Tan, L., Jiang, L., Son, S., Shi, W., Rausch, K., Voss, K.J., 2013. Impacts of VIIRS SDR performance on ocean color products. J. Geophys. Res. 118, 10347–10360. https://doi.org/10.1002/jgrd.50793.
Wang, M., Shi, W., 2007. The NIR-SWIR combined atmospheric correction approach for MODIS ocean color data processing. Opt. Express 15, 15722–15733.
Wang, M., Son, S., 2016. VIIRS-derived chlorophyll-a using the ocean color index method. Remote Sens. Environ. 182, 141–149. https://doi.org/10.1016/j.rse.2016.05.001.
Wang, M., Son, S., Harding, L.W., 2009a. Retrieval of diffuse attenuation coefficient in the Chesapeake Bay and turbid ocean regions for satellite ocean color applications. J. Geophys. Res. 114. https://doi.org/10.1029/2009JC005286.
Wang, M., Son, S., Shi, W., 2009b. Evaluation of MODIS SWIR and NIR-SWIR atmospheric correction algorithm using SeaBASS data. Remote Sens. Environ. 113, 635–644.
Wei, J., Lee, Z.P., Lewis, M., Pahlevan, N., Ondrusek, M., Armstrong, R., 2015. Radiance transmittance measured at the ocean surface. Opt. Express 23, 11826–11837.
Wei, J., Lee, Z.P., Shang, S., 2016. A system to measure the data quality of spectral remote sensing reflectance of aquatic environments. J. Geophys. Res. 121, 8189–8207. https://doi.org/10.1002/2016JC012126.
Wei, J., Van Dommelen, R., Lewis, M.R., McLean, S., Voss, K.J., 2012. A new instrument for measuring the high dynamic range radiance distribution in near-surface sea water. Opt. Express 20, 27024–27038.
Wei, J., Wang, M., Jiang, L., Yu, X., Mikelsons, K., Shen, F., 2021a. Global estimation of suspended particulate matter from satellite ocean color imagery. J. Geophys. Res. 126, e2021JC017303. https://doi.org/10.1029/2021JC017303.
Wei, J., Wang, M., Lee, Z.P., Ondrusek, M., Zhang, S., Ladner, S., 2021b. Experimental analysis of the measurement precision in spectral water-leaving radiance in different water types. Opt. Express 29, 2780–2797. https://doi.org/10.1364/OE.413784.
Wei, J., Yu, X., Lee, Z.P., Wang, M., Jiang, L., 2020. Improving low-quality satellite remote sensing reflectance at blue bands over coastal and inland waters. Remote Sens. Environ. 250, 112029. https://doi.org/10.1016/j.rse.2020.112029.
Werdell, P.J., Bailey, S.W., 2005. An improved bio-optical data set for ocean color algorithm development and satellite data product validation. Remote Sens. Environ. 98, 122–140.
Wollschläger, J., Grunwald, M., Röttgers, R., Petersen, W., 2013. Flow-through PSICAM: a new approach for determining water constituents absorption continuously. Ocean Dyn. 63, 761–775. https://doi.org/10.1007/s10236-013-0629-x.
Wu, J., Goes, J.I., do Rosario Gomes, H., Lee, Z., Noh, J.-H., Wei, J., Shang, Z., Salisbury, J., Mannino, A., Kim, W., Park, Y.-J., Ondrusek, M., Lance, V.P., Wang, M., Frouin, R., 2022. Estimates of diurnal and daily net primary productivity using the Geostationary Ocean Color Imager (GOCI) data. Remote Sens. Environ. 280, 113183. https://doi.org/10.1016/j.rse.2022.113183.

Xi, H., Losa, S.N., Mangin, A., Soppa, M.A., Garnesson, P., Demaria, J., Liu, Y., d'Andon, O.H.F., Bracher, A., 2020. Global retrieval of phytoplankton functional types based on empirical orthogonal functions using CMEMS GlobColour merged products and further extension to OLCI data. Remote Sens. Environ. 240, 111704. https://doi.org/10.1016/j.rse.2020.111704.

Yu, X., Lee, Z., Shang, Z., Lin, H., Lin, G., 2021. A simple and robust shade correction scheme for remote sensing reflectance obtained by the skylight-blocked approach. Opt. Express 29, 470–486. https://doi.org/10.1364/OE.412887.

Zaneveld, J.R.V., 1995. A theoretical derivation of the dependence of the remotely sensed reflectance of the ocean on the inherent optical properties. J. Geophys. Res. 100, 13135–13142.

Zaneveld, J.R.V., Kitchen, J.C., Moore, C.C., 1994. Scattering error correction of reflecting-tube absorption meters. In: Ocean Optics XII. SPIE, Bergen, Norway.

Zibordi, G., D'Alimonte, D., Berthon, J.F., 2004. An evaluation of depth resolution requirements for optical profiling in coastal waters. J. Atmos. Ocean. Technol. 21, 1059–1073.

Zibordi, G., Holben, B.N., Talone, M., D'Alimonte, D., Slutsker, I., Giles, D.M., Sorokin, M.G., 2021. Advances in the ocean color component of the aerosol robotic network (AERONET-OC). J. Atmos. Ocean. Technol. 38, 725–746. https://doi.org/10.1175/jtech-d-20-0085.1.

Zibordi, G., Mélin, F., Berthon, J.-F., Holben, B., Slutsker, I., Giles, D., D'Alimonte, D., Vandemark, D., Feng, H., Schuster, G., Fabbri, B.E., Kaitala, S., Seppälä, J., 2009. AERONET-OC: a network for the validation of ocean color primary products. J. Atmos. Ocean. Technol. 26, 1634–1651. https://doi.org/10.1175/2009JTECHO654.1.

Zoffoli, M.L., Lee, Z., Marra, J.F., 2018. Regionalization and dynamic parameterization of quantum yield of photosynthesis to improve the ocean primary production estimates from remote sensing. Front. Mar. Sci. 5. https://doi.org/10.3389/fmars.2018.00446.

Chapter 21

Land surface temperature validation

Yuling Liu[a], Yunyue Yu[b], Heshun Wang[a], and Peng Yu[a]
[a]Earth System Science Interdisciplinary Center of University of Maryland, College Park, MD, United States, [b]NOAA/NESDIS Center for Satellite Applications and Research (STAR), Silver Spring, MD, United States

Chapter outline

1. Land surface temperature product introduction	375	3.2 Strategies and methodologies	377
2. Methodology of LST retrieval	376	3.3 Ground LST estimates	378
2.1 LST algorithm derivation	376	3.4 Validation results and highlights	378
2.2 Simulation study and theoretical analysis	376	**4. Summary**	**387**
3. LST product validation	377	**Acknowledgments**	**387**
3.1 VIIRS LST product	377	**References**	**388**

Look deep into nature, and then you will understand everything better

Albert Einstein

1. Land surface temperature product introduction

Land surface temperature (LST) is defined as a measure of how hot or cold the surface of the Earth would feel to the touch (Guillevic et al., 2018). Surfaces over inland water and coastline are included in the LST domain. For ground-based, airborne, and space satellite remote sensing instruments, it is the aggregated radiometric surface temperatures of all components within the sensor's field of view in the direction of observation (Yu et al., 2017a). The knowledge of LST provides critical information on the temporal and spatial variations of the surface equilibrium state and is of fundamental importance in the weather and climate system controlling surface heat and water exchange between land and the atmosphere (Li et al., 2013; Yu et al., 2012). LST has been widely used in a variety of fields such as numerical weather prediction models and data assimilation systems (Meng et al., 2009; Zheng et al., 2012; Trigo et al., 2015), evapotranspiration evaluation (Fang et al., 2019; Sun et al., 2012; Galleguillos et al., 2011), irrigation and hydrological cycle particularly agricultural drought monitoring (Anderson et al., 2012), and urban heat island monitoring (Rajasekar and Weng, 2009; Weng et al., 2004). In 2016, LST was listed as one of the essential climate variables (ECVs) by the Global Climate Observation System (GCOS) of the World Meteorological Organization (WMO) (GCOS 200).

The thermal signal of the satellite remote sensing provides the unique advantage in measuring the LST at regional and global scale with advanced capability in spatial and temporal resolutions. The LST product as one of the important components in the environmental data records has been developed for both low earth orbit (LEO) satellites such as the Advanced Very High Resolution Radiometer (AVHRR), MODIS and VIIRS, and geostationary orbit (GEO) satellites, e.g., the Spinning Enhanced Visible and InfraRed Imager (SEVIRI) and Geostationary Operational Environmental Satellite (GOES) missions for over four decades. Therefore, it is very important to quantify the LST product performance through various validation practices. In this study, we will focus on the validation of the VIIRS LST product from JPSS satellite series including NOAA20 and the Suomi National Polar-orbiting Partnership (SNPP). The method used in this study can be adapted to validate similar LEO LST products, e.g., future JPSS-2 LST, Meteorological imager (METimage) LST, etc.

The chapter is organized as follows: the LST retrieval method is briefly described in the second section, followed with the LST product validation including strategy and methodology, ground LST estimate methods, and the validation results and highlights. The summary and concluding remarks are given in Section 4.

2. Methodology of LST retrieval

2.1 LST algorithm derivation

The LST is derived based on the physics of radiative transfer process from the Earth's surface to the remote sensor. It can be sensed by infrared signals from the thermal infrared (TIR) channels, particularly those at around 10 to 12 μm spectrum where atmospheric transmittance reaches the maximum. However, the infrared channels cannot penetrate the cloud so it can only be used for LST retrieval under cloud-free condition.

Under clear sky condition, the radiance at-sensor received within the TIR spectral range (8–14 μm) can be described by

$$I(\lambda) = I_s(\lambda) + I_{atm}(\lambda)^\uparrow + I_{atm}(\lambda)^\downarrow \tag{1}$$

where $I_s(\lambda)$, $I_{atm}(\lambda)^\uparrow$, and $I_{atm}(\lambda)^\downarrow$ represent the radiance contributions from the surface emission, the atmospheric upwelling, and the reflected downwelling sky irradiance, respectively; λ is the wavelength of the sensor channel.

Contribution of the surface emitted radiance $I_s(\lambda)$ is a function of the land surface temperature and emissivity and gets attenuated along the atmospheric path as shown in Eq. (2), in which $B(\lambda, T_s)$ can be calculated using Planck function (Eq. 3):

$$I_s(\lambda) = \varepsilon(\lambda)\tau_0(\lambda)B(\lambda, T_s) \tag{2}$$

$$B(\lambda, T_s) = \frac{1}{\lambda^5 \left(\exp\left(\frac{c_2}{\lambda T}\right) - 1\right)} \tag{3}$$

where c_1 and c_2 are constants ($c_1 = 1.191 \cdot 10^8 \text{W} \cdot \mu\text{m}^4 \cdot \text{sr}^{-1} \cdot \text{m}^{-2}$ and $c_2 = 1.439 \cdot 10^4 \text{W} \cdot \mu\text{m} \cdot \text{K}$).

The LST algorithm development is to derive the LST T_s from the sensor measured radiance $I(\lambda)$, in which the coupling of T_s with the land surface emissivity (LSE) and the atmospheric impact of absorption and attenuation due to the water vapor absorption makes it hard to derive an analytic solution to the above problem. Besides, the radiance is measured at N channels so there will be N+1 unknowns, i.e., N emissivities from the N channels and one unknown LST for N equations. Many algorithms have been proposed to deal with the characteristics of various sensors with different assumptions and approximations for the radiative transfer equation and LSEs (e.g., McMillin, 1975; Gillespie et al., 1996; Hook et al., 1992). These approaches can be roughly grouped into three categories: single-channel methods, multichannel methods, and multiangle methods provided that the LSEs are known a priori. If the LSEs are not known, then the algorithms can be categorized into three types: stepwise retrieval method, simultaneous retrieval of LSEs and LST with known atmospheric information, and simultaneous retrieval (Li et al., 2013). The most widely used approach is the linear regression split window (SW) technique, i.e., the atmospheric effects are compensated using data from two or more adjacent TIR channels (typically at 10–12.5 μm) (Becker and Li, 1990; Atitar and Sobrino, 2009; Prata, 1994; Price, 1984; Sun and Pinker, 2003). This method is computationally efficient and does not require accurate atmospheric profiles so that it is widely used for producing the operational satellite LST estimates for VIIRS, MODIS, SEVIRI, and Advanced Baseline Imager (ABI) sensors (Becker and Li, 1990; Prata, 1994; Wan and Dozier, 1996; Sun and Pinker, 2003; Yu et al., 2009; Trigo et al., 2009). For the VIIRS sensor, the surface-type dependent SW algorithm is firstly used for LST retrieval in which the look up table (LUT) is stratified by the 17 International Geosphere-Biosphere Programme (IGBP) surface types (Liu et al., 2015), and then it was replaced by the emissivity explicit algorithm so called enterprise algorithm which utilized the spectral emissivity explicitly in the retrieval formula. Liu et al. (2019) introduce the enterprise LST algorithm determination through analysis of the theoretical accuracy and sensitivity of the candidate SW algorithms using a comprehensive simulation data set. The enterprise LST algorithm is defined as follows:

$$T_s = C + A_1 T_{11} + A_2(T_{11} - T_{12}) + A_3 \varepsilon + A_4 \varepsilon (T_{11} - T_{12}) + A_5 \Delta\varepsilon \tag{4}$$

where T_{11} and T_{12} represent the top-of-atmosphere brightness temperatures at ~11 μm and 12 μm, respectively, $\varepsilon = (\varepsilon_{11} + \varepsilon_{12})/2$ and $\Delta\varepsilon = (\varepsilon_{11} - \varepsilon_{12})$, where ε_{11} and ε_{12} are the spectral emissivity values of the land surface at ~11 μm and 12 μm channels, respectively. C, A_1, A_2, A_3, A_4, and A_5 are algorithm coefficients stratified by the day/night condition, precipitable water vapor, and satellite viewing zenith angle.

2.2 Simulation study and theoretical analysis

A simulation data set is used for algorithm training and evaluation. The MODerate resolution atmospheric TRANsmission (MODTRAN) (Berk et al., 2003) was used to compute the at-sensor radiation with a comprehensive collection of atmospheric profiles, which represents a wide range of surface and atmospheric conditions for global land coverage. Several collections of profiles were used to build up the simulation database. The first collection consists of 60 daytime and 66 nighttime cloud-free radiosonde from the CrIS F98-Weather Products Test Bed Data Package (NOAA88, Rev. 1.0, M. Goldberg, personal communication, 1998). The second collection is composed of 354 cloud-free profiles selected from the version 3 Thermodynamic Initial Guess Retrieval (TIGR) database. The third collection is from Seebor V5.0 that consists of 15,704 global profiles of temperature, moisture, and ozone at 101 pressure levels for clear sky conditions (Borbas et al., 2005). The Seebor profiles over land represent a variety of atmospheric conditions, spanning a column water vapor range from 0.2 to 7.5 g/cm^2, a surface air temperature range from 200 K to 319 K, and the LST range from 204 K to 337 K. Quality check and additional cloud screening were conducted to reduce the impact from cloud/foggy residue. The simulation data are separated into two groups: one group is training database and the remaining profiles are used for the independent evaluation. The overall evaluation result indicates a bias of 0.19 K and 0.34 K and a standard deviation (STD) of 0.48 K and 0.69 K for nighttime and daytime, respectively (Liu et al., 2019).

3. LST product validation

3.1 VIIRS LST product

JPSS is the Nation's new generation polar-orbiting operational environmental satellite system and consists of five satellites: S-NPP, NOAA20, JPSS-2, 3, and 4. The onboard VIIRS sensor provides observations for measuring earth surface parameters including LST. The SNPP VIIRS LST product has been operationally generated since 2012. The VIIRS moderate resolution channels M15 and M16 centered at 10.76 μm and 12.01 μm, respectively, are utilized in the LST algorithm.

Both NOAA20 and SNPP VIIRS LST products have achieved the validated maturity level and have been archived and distributed by NOAA Comprehensive Large Array-data Stewardship System (CLASS) (CLASS data portal). In this study, the purpose is to validate the VIIRS LST generated using the enterprise algorithm therefore the SNPP VIIRS LST for the time period between 2012 and May 2019 is reprocessed with the enterprise algorithm, while the NOAA20 VIIRS LST was originally produced with the enterprise algorithm since 2018.

3.2 Strategies and methodologies

LST validation is a challenge because of the spatial, temporal, and directional variabilities of land surface temperature. It is nearly impossible to find the exact measurements to match with the satellite pixel LST. Three approaches are widely used for the LST product validation: temperature-based (T-based) validation, radiance-based validation (R-based), and cross-satellite comparison method. Besides, the time series intercomparison method, widely used in sea surface temperature validation (e.g., Goela et al., 2016), is also adopted for LST validation such as detection of the instrument problems (Hook et al., 2007) or unrealistic outliers due to undetected clouds (Merchant et al., 2013). However, this approach requires relatively long-time series of observations over temporally highly stable targets, e.g., inland water bodies (Hook et al., 2007), therefore it greatly limits its application in the LST validation. These methods are complementary and provide different levels of information about the accuracy of the satellite-derived LST products (Guillevic et al., 2018).

T-based validation uses the ground in situ LST measurements as a reference to match the satellite pixel LST. There are several requirements for the ground sites to be suitable for the LST validation. First of all, the ground site should be able to provide long-term high quality in situ LST observations. The instrument calibration, data quality control, and the site maintenance should be performed on a regular basis. Second, the ground site is able to provide stable and sufficiently high sampling rate in- situ measurements, e.g., every 1 min to have a good temporal match up with the satellite LST. Third, the ground site should be relatively homogeneous within its surrounding areas with the size comparable to the satellite pixel. Most of the ground LST is measured within a small area with limited field of view comparing the relatively large satellite pixel size. In addition, the polar satellite pixel size varies over the satellite zenith angle, which results in the edge pixel two or three times of the nadir pixel size. Therefore, the site needs to have a homogenous surface cover over a relatively large area to be able to spatially represent the pixel LST. Unfortunately, such in-situ measurements are rare, limiting the statistical significance and the seasonal and global representativeness of the results (Yu et al., 2017a).

The R-based method applies a numerical radiative transfer model to reversely estimate the reference LST using the satellite sensed BTs. It does not rely on ground LST measurements, but it requires accurate atmospheric temperature and water vapor profile concurrent with the satellite overpass time and the surface emissivity information. The advantage of the R-based method is that it can be applied to the LST validation during both nighttime and daytime period over large number of sites at global scale where the atmospheric condition and emissivity are known (e.g., from field measurements or emissivity products). The LST error, or uncertainty in the LST retrieval, is simply found by taking the difference between the retrieved LST product and the R-based LST. (Guillevic et al., 2017). This method has been used for MODIS LST (Wan and Li, 2008; Coll et al., 2009a,b; Duan et al., 2018) and SEVIRI LST validation (Niclòs et al., 2011).

The cross-satellite comparison method uses a well-calibrated satellite LST as a reference to evaluate the target LST, therefore it is a relative comparison not the absolute quantification of the error. This method has been widely used in the satellite LST evaluation (Hulley and Hook, 2009; Liu et al., 2015, 2019). Considering that the MODIS LST has been successfully in operation for over 20 years and VIIRS was designed to extend and improve upon its predecessor MODIS, the cross comparison between VIIRS LST and MODIS LST shall provide good evaluation of the VIIRS LST retrieval performance with respect to characterization of the differences, spatial pattern, systematic error budget, etc. It is recommended to select data with view angles lower than 45 degrees, a satellite angular separation limit of ±10 degrees, and a satellite time separation limit of ±10 min (Guillevic et al., 2017). In this chapter, the VIIRS LST is assessed using both the T-based method and the cross-satellite comparison method.

The quality control (QC) of both the ground data and satellite data is critical in order to obtain reliable comparison results. Liu et al. (2015, 2019) described a procedure of the QC control for the matchup between the satellite LST and ground in situ LST measurements, and it is also used in this study as follows:

(1) The temporal difference is less than 86 s, which is the typical duration of a single granule
(2) Spatially closest pixel is used for the comparison
(3) Confidently clear indicated by the cloud mask product
(4) To reduce the nearby cloud impact, the 3*3 neighboring pixels are all marked as confidently clear
(5) The standard deviation of the band 15 brightness temperature in the neighboring 3 by 3 box is less than the threshold, which is set as 1.5 K for all sites.
(6) The standard deviation of the 30 min (centered at the matchup time) downwelling radiation from in situ observations is less than a predetermined threshold, which is set as 1.2 for all sites.

Principally, the quality control of the ground measurements is to exclude bad or suspicious observations by using the quality indicator provided in the data set. In addition, the temporal variation within 30 min interval is considered to exclude the noisy and cloud contaminated data. The main purpose for satellite data quality control is to reduce the impact from the cloud contamination and suboptimal atmospheric conditions.

3.3 Ground LST estimates

The in situ LST is usually measured from the infrared thermometer that provides the surface upwelling and downwelling radiations of the scene. In this study, the ground measurements from SURFRAD network, BSRN and ARM were used to validate the VIIRS LST. The ground LST is derived from its upwelling and downwelling radiance flux by Stefan-Boltzmann law:

$$F^\uparrow = \varepsilon \sigma T_s^4 + (1-\varepsilon) F^\downarrow \tag{5}$$

where $F^\uparrow$ and $F^\downarrow$ are upwelling and downwelling longwave radiation flux, respectively; ε is the broadband surface emissivity; σ is the Stefan-Boltzman constant ($\sigma = 5.67051 \times 10^{-8}$ Wm^{-2}k^{-4}); and T_s is the surface skin temperature. T_s is then obtained by inverting Eq. (5):

$$T_s = \left[\frac{F^\uparrow - (1-\varepsilon)F^\downarrow}{\varepsilon \sigma} \right]^{\frac{1}{4}} \tag{6}$$

Here, the broadband emissivity ε is from NOAA VIIRS LSE product (Wang et al., 2020).

From Eq. (6), the accuracy of the ground LST depends on the accurate measurements of the upwelling and downwelling radiation and the broadband emissivity. Different types of instruments are used in in situ networks. The infrared radiometer, e.g., pyrgeometers measuring both downwelling and upwelling radiation has an uncertainty about ±5 W·m^{-2} (Augustine and Dutton, 2013). The resulting overall LST uncertainty lies in a range between 0.6 and 2 K for all stations, which translates to a relative uncertainty smaller than 1% of the LST value. The NOAA broadband emissivity for the spectral range of 8–13.5 μm has an uncertainty of 0.012 based on the comparison with in situ emissivity measurements (Yu et al., 2017b), which might give rise to an uncertainty in the ground LST of less than 1 K.

3.4 Validation results and highlights

3.4.1 Validation with SURFRAD data

NOAA SURFRAD network site locations were chosen with the intent of best representing the diverse climates of the continental United States. It provides long-term high-quality downwelling and upwelling infrared radiation, along with other

TABLE 1 Geolocation and surface type of the seven SURFRAD stations.

No.	Site location	Station acronyms	Lat(N)/Lon(W)	Surface type
1	Bondville, IL	BON	40.05/88.37	Crop Land
2	Fort Peck, MT	FPK	48.31/105.10	Grass Land
3	Table Mountain, CO	TBL	40.13/105.24	Grass/Crop Land
4	Desert Rock, NV	DRA	36.63/116.02	Shrub Land
5	Pennsylvania State University, PA	PSU	40.72/77.93	Mixed Forest
6	Sioux Falls, SD	SFX	43.73/97.49	Cropland

meteorological parameters (Augustine et al., 2000). SURFRAD consists of seven stations in which The Goodwin Creek (GWN) site was removed in the LST validation due to the onsite thermal heterogeneity, which caused the ground LST colder than the satellite LST at daytime while warmer at nighttime (Liu et al., 2019). The other six sites with geolocation and surface-type information are given in Table 1. The data are with 3 min interval before 2009 and 1 min thereafter so it meets the temporal requirement for ground observations. SURFRAD has been widely used for satellite LST validation such as GOES, MODIS, and VIIRS (Liu et al., 2015, 2019; Li et al., 2014; Yu et al., 2012).

Besides the QC procedure described in Section 3.2, additional temporal filter is applied over the site in Bondville, IL. During late spring and early summer, an obvious daytime discrepancy between the satellite LST and in situ measurement has been reported from previous studies (Liu et al., 2015; Guillevic et al., 2014; Li et al., 2014). The satellite LST is 6 K to 10 K warmer than the ground in situ observations. Guillevic et al. (2014) mentioned that validation results obtained for stations surrounded by croplands present strong seasonal dependency: station observations may be closer/deviate more from the temperature of surrounding fields, according to the crop maturity phases. Considering its great impact on the validation, the matchups in May and June were removed from the results over BON site.

Eight years (between December 2012 and December 2020) of the in situ LST data from SURFRAD were used to validate the SNPP VIIRS LST, and 3 years (between Jan. 2018 to Jan. 2021) of data were used to validate the NOAA20 VIIRS LST product. Fig. 1 illustrates the comparison results between the VIIRS LST and ground LST (left subfigure for and middle subfigure for NOAA20). Similar comparison between the AQUA/MODIS LSTs and the ground LST (right subfigure) is presented as a reference in which 2 years between January 2019 and December 2020 of MYD11A1 LST was used. Daytime matchups are in red color and nighttime in blue color. The daytime and nighttime validation statistics, i.e., the number of matchups, bias, and STD are also provided in lower portion of Fig. 1 with the same color. The overall results show a bias of −0.3 K, −0.43 K, and −0.5 K and a root mean square error (RMSE) of 1.81 K, 1.84 K, and 2.26 K for SNPP VIIRS LST, NOAA20 VIIRS LST, and MODIS LST, respectively. The nighttime results indicate a bias of −0.25 K, −0.26 K, and −0.32 K and STD of 1.61 K, 1.59 K, and 1.82 K, while daytime LST yields a bias of −0.36 K, −0.78 K, and −0.82 K for SNPP VIIRS LST, NOAA20 VIIRS LST, and MODIS LST, respectively. The better nighttime performance is expected considering stronger thermal heterogeneity usually during daytime and a dryer atmosphere at night (Li et al.,

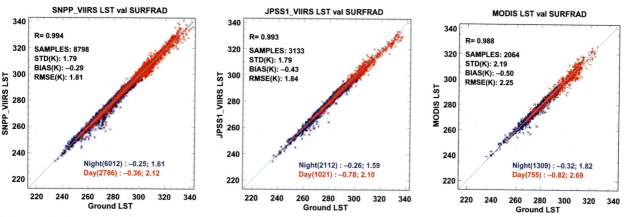

FIG. 1 Validation results against SURFRAD observations.

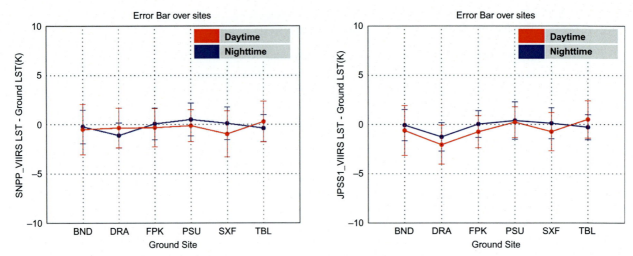

FIG. 2 Site wide validation results against SURFRAD observations for SNPP *(left)* and NOAA20 LST product *(right)*.

2014). Outliers were observed in both VIIRS and MODIS LST validation, which might be attributed to the cloud effect or suboptimal atmosphere conditions. In addition, the mismatch of the field of view by the in situ instrument and satellite sensor might cause large deviation under specific conditions. SURFRAD sites are mostly vegetated sites as shown in Table 1. The site itself is mostly grassland, but the surrounding area might be cropland with seasonal growth and harvest cycles. The results vary among the sites as shown in Fig. 2 for site-wide results where LST underestimation is observed over DRA site and SXF site particularly for the NOAA20 VIIRS LST validation.

To characterize the spatial representativeness of the ground site LST, ASTER global emissivity data set (GED) climatology at 100 m resolution is used to form a 1 km area centered at each station to illustrate surroundings as shown in Fig. 3. STD of the emissivities within the surrounding area is calculated as an indicator of the surface heterogeneity. It is found that Desert Rock site has the stable landscape in an annual cycle with a STD of emissivity around 0.007, which indicates it is the most homogeneous site in the SURFRAD network. The systematic underestimation of LST at Desert Rock, NV (Arid shrub land) has been attributed to uncertainties associated with land surface emissivities and other input data sets (Zhou et al., 2019).

The degree of agreement between VIIRS satellite LSTs and ground measurements denoted by RMSE shows fairly strong seasonal variation as shown in Table 2. The seasonality appears more pronounced at daytime (d) than at nighttime (n). The best agreement at nighttime occurs in fall with a bias of −0.19 K and an RMSE of 1.40 K, while the worst agreement occurs in winter with a bias of −0.81 K and a RMSE of 2.15 K. The best agreement at daytime occurs in winter with a bias of −0.42 K and an RMSE of 1.53 K, while the worst agreement occurs in summer with a bias of −0.96 K and a RMSE of 3.32 K. The worse performance in winter nighttime might be related to the in pixel low cloud or warm cloud, which is not excluded from the cloud screening procedure.

The difference between VIIRS LST and ground LST is associated with dry/wet atmosphere conditions and satellite viewing angle as shown in Fig. 4. The cold LST bias increases significantly when the view angle is greater than 50 degree for both daytime and nighttime, but the magnitude at nighttime is smaller than that during the daytime. With respect to the total column water vapor, it shows no clear trend at nighttime while the cold bias at daytime first rises slowly and then decreases gradually thereafter.

3.4.2 Validation with BSRN data

The radiometric network BSRN provides typically 1-min averaged short-wave and long-wave surface radiation fluxes of the best possible quality currently available. Currently, a total of over 7000 station-month data sets from 58 stations available in the World Radiation Monitoring Center (WRMC) (GCOS-174). However, only two stations are selected for LST validation for several reasons: the longwave upwelling and downwelling data availability, temporal overlap with VIIRS measurements, and the site homogeneity requirements for LST validation. Particularly, some sites were excluded because they were too close to water bodies which results in the mixture of land and water in the match-up satellite pixel. The selected two sites are located at Gobabeb, Namibia (GOB) and Cabauw, The Netherlands (CAB). Fig. 5 shows the site

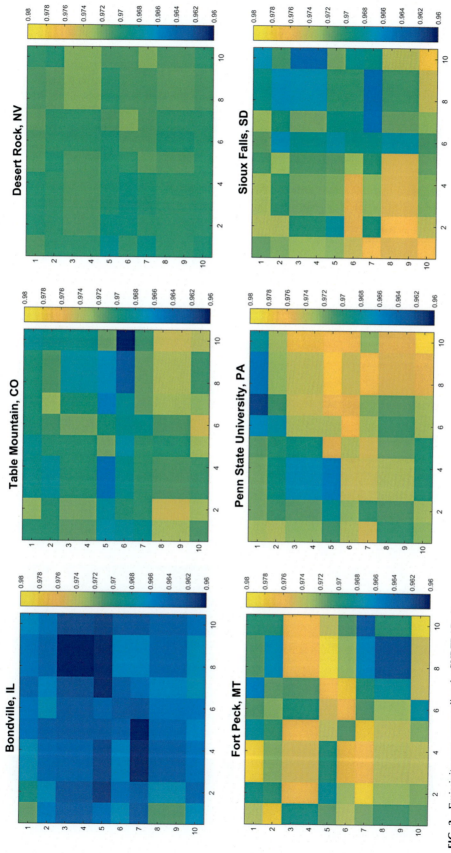

FIG. 3 Emissivity map surrounding the SURFRAD sites.

382 PART | III Satellite applications

TABLE 2 Seasonal analysis of the validation performance.

Season	Count	Bias	RMSE	Count (d)	Bias (d)	RMSE (d)	Count (n)	Bias (n)	RMSE (n)
Spring	1919	−0.28	1.81	541	0.17	2.20	1378	−0.45	1.63
Summer	2219	−0.04	1.98	457	−0.96	3.32	1762	0.21	1.45
Fall	2669	−0.24	1.58	945	−0.33	1.86	1724	−0.19	1.40
Winter	1991	−0.64	1.91	843	−0.42	1.53	1148	−0.81	2.15

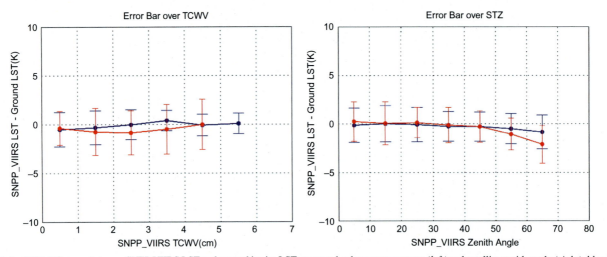

FIG. 4 LST difference between SNPP VIIRS LST and ground in situ LST over total column water vapor *(left)* and satellite zenith angle *(right)*, *blue* for nighttime and *red* for daytime.

FIG. 5 BSRN site map: GOB *(left)* and CAB *(right)*.

map. It is noticed that a couple of new sites, e.g., PAY site in Mexico have been put into running since early 2020. The suitability of these sites for the LST validation will be investigated with regard to the data quality.

GOB site is located in the hyper-arid climate of the Namibia desert. Note that for GOB site, the downwelling radiation sensor and upwelling radiation sensor are mounted at different locations with about 10 km distance between them. The downwelling sensor is located near Gobabeb Research and Training Center (latitude 23.56°S, longitude 15.04°E), where the nearby Kuiseb river forms a natural boundary between large gravel plains (>900 km^2) and the sand dunes of the Namib Desert (Göttsche et al., 2013). The upwelling sensor is located at latitude 23.519°S and longitude 15.083°E, covered by homogeneous gravel plains with very sparsely grass coverage. The matchups between in situ and satellite data are set at the homogeneous upwelling sensor location for revealing the ground conditions. Five years of the BSRN data from September 2015 to August 2020 were acquired for the SNPP VIIRS LST validation, and about 3 years of the BSRN data from January 2018 to November 2020 were acquired for the NOAA20 VIIRS LST validation. Two years of data in 2019 and 2020 were used for the MODIS LST validation as a reference. The same quality control procedure described in the Section 3.2 was implemented, and the validation results are shown in Fig. 6. The VIIRS LST from both SNPP and NOAA 20 are in good agreement with the in situ LST. For the GOB site, the bias is −0.37 K, −0.24 K, and 0.89 K, and the RMSE is 1.72 K, 1.74 K, and 2.37 K for SNPP, NOAA20, and MODIS LST, respectively. For the CAB site, the bias is 0.02 K, −0.02 K, and −0.78 K, and the RMSE is 1.92 K, 2.04 K and 3.22 K for SNPP, NOAA20, and MODIS LST, respectively. Some outliers due to subpixel clouds can be observed in which the station LST is higher than VIIRS LST. VIIRS LST is about 0.5 K colder than ground LST measurements at nighttime but has nearly no bias at daytime, while MODIS LST is about 0.9 K colder and 1.6 K warmer than ground LST for nighttime and daytime, respectively. Over CAB site, a (Fang et al., 2019; Galleguillos et al., 2011; Gillespie et al., 1996; Goela et al., 2016; Guillevic et al., 2018; Hook et al., 1992, 2007; Hulley and Hook, 2009; Li et al., 2013, 2014, Liu et al., 2015, 2019; McMillin, 1975; Meng et al., 2009; Merchant et al., 2013; Niclòs et al., 2011; Prata, 1994; Price, 1984; Rajasekar and Weng, 2009; Sun et al., 2012; Trigo et al., 2009, 2015; Wan and Dozier, 1996; Wan and Li, 2008; Weng et al., 2004; Zheng et al., 2012; Martin and Göttsche,

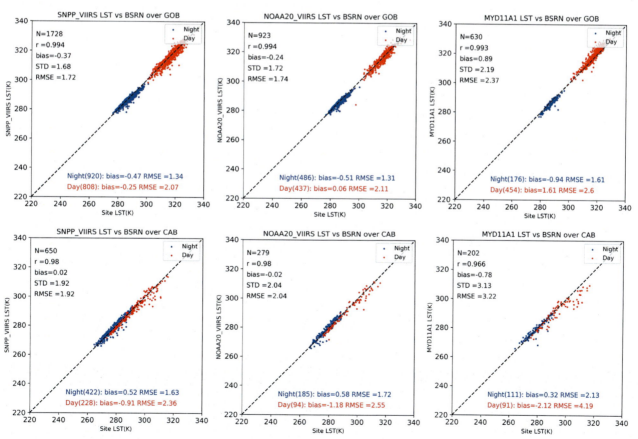

FIG. 6 BSRN validation results: GOB site *(top)* and CAB site *(bottom)*.

2016) warm bias about 0.6 K at nighttime and a cold bias about 1 K at daytime and a warm bias of 0.3 K at nighttime and a cold bias of −2.1 K at daytime are found for VIIRS and MODIS LST, respectively.

The seasonal variation of VIIRS LST is also observed in the validation with BSRN data as shown in Table 3. The seasonality appears more pronounced at daytime than at nighttime. The best agreement at nighttime occurs in winter with the bias of −0.04 K and RMSE of 1.24 K, while the worst agreement occurs in spring with a bias of 0.24 K and RMSE of 1.53 K. The best agreement at daytime occurs in winter with a bias of −0.27 K and RMSE of 1.75 K, while the worst agreement occurs in summer with a bias of 0.42 K and RMSE of 2.60 K.

3.4.3 Validation with ARM data

The Atmospheric Radiation Measurement (ARM) operates three fixed-location atmospheric observatories including Southern Great Plains (SGP), North Slope of Alaska (NSA), and Eastern North Atlantic (ENA) that represent a broad range of conditions to gather massive amounts of atmospheric data. The Southern Great Plains (SGP) atmospheric observatory, the first field measurement site established by ARM, is the world's largest and most extensive climate research facility (ARM website). ARM provides longwave downwelling and upwelling radiance as 60 s averages with pyrgeometer instrument similar to the instrument used in BSRN stations. The ARM observations from the SGP station were used to validate GOES LST (Sun et al., 2012) and Glob temperature LST products (Martin and Göttsche, 2016). In addition to the SGP central station, i.e., sgpsirsC1, fifteen extended SGP sites have been included in this study with geolocation and surface type information are given in Table 4.

One year of ARM data from October 2019 to October 2020 were obtained to validate the SNPP VIIRS LST and NOAA20 VIIRS LST products and the same time period of data was used to validate the MODIS LST as a reference. The validation results are shown in Fig. 7. It indicates a bias of 0.04 K, 0.17 K, and −0.48 K and RMSE of 1.71 K,

TABLE 3 Seasonal analysis of the BSRN validation performance.

Season	Count	Bias	RMSE	Count (d)	Bias (d)	RMSE (d)	Count (n)	Bias (n)	RMSE (n)
Spring	1057	−0.06	1.82	421	−0.53	2.20	636	0.24	1.53
Summer	409	−0.19	1.98	169	0.42	2.60	240	−0.63	1.39
Fall	495	−0.83	1.71	234	−0.85	1.97	261	−0.81	1.42
Winter	417	−0.16	1.52	212	−0.27	1.75	205	−0.04	1.24

TABLE 4 Geolocation and surface type of the ARM stations.

ID	Name	Longitude	Latitude	Altitude (M)	Surface type
1	sgpsirsC1	−97.485	36.605	318	Rangeland (Sandy)
2	sgpsirsE9	−97.266	37.133	386	Pasture
3	sgpsirsE11	−98.285	36.881	360	Pasture
4	sgpsirsE12	−96.427	36.841	331	Native Prairie
5	sgpsirsE13	−97.485	36.605	318	Pasture and Wheat
6	sgpsirsE15	−98.284	36.431	418	Pasture
7	sgpsirsE31	−98.362	37.1509	412.1	Pasture
8	sgpsirsE32	−97.8199	36.819	328	Pasture
9	sgpsirsE33	−97.0817	36.9255	357	grassy field
10	sgpsirsE34	−96.7606	37.0694	417	Pasture
11	sgpsirsE35	−97.0695	35.8615	294.1	Pasture

TABLE 4 Geolocation and surface type of the ARM stations—cont'd

ID	Name	Longitude	Latitude	Altitude (M)	Surface type
12	sgpsirsE36	−97.5112	36.1166	336.8	Pasture
13	sgpsirsE37	−97.928	36.3106	378.9	grass
14	sgpsirsE38	−98.1728	35.8797	371.2	pasture
15	sgpsirsE40	−96.76203	36.31937	247	Pasture
16	sgpsirsE41	−97.08645	36.87956	340	Grassy field

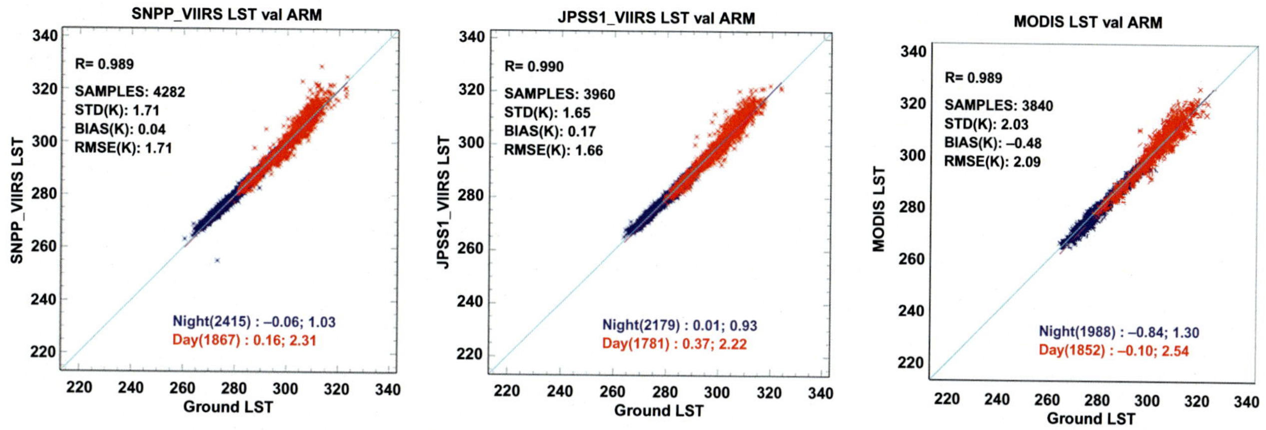

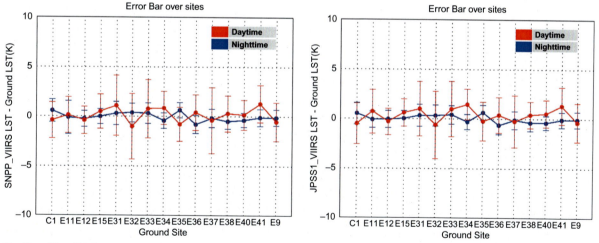

FIG. 7 Overall ground validation results against ARM observations for SNPP VIIRS LST *(left)*, NOAA20 LST *(middle)*, and MODIS LST *(right)*.

FIG. 8 Site-wide validation results against ARM observations for SNPP VIIRS LST *(left)* and NOAA20 LST *(right)*.

1.66 K, and 2.09 K for SNPP VIIRS LST (left), NOAA20 VIIRS LST(middle), and AQUA/MODIS LSTs (right), respectively. A fairly significant overestimation of up to 2 K is observed over several sites such as E31, E34, and E41 at daytime from both NOAA20 and SNPP VIIRS LST, as shown in Fig. 8.

3.4.4 Cross-satellite comparison with MODIS LST

Previous cross-satellite validation effort of VIIRS LST was performed at granule level using Simultaneous Nadir Overpasses (SNOs) tool (SNO predictions) to search for the "just-miss" scenes from the granule data set of both satellite

(Liu et al., 2015, 2019; Yu et al., 2017b). In this chapter, the cross-satellite comparison between the VIIRS LSTs and the MODIS LSTs is conducted at global scale using the gridded LST data. The gridded VIIRS LST is a daily global product at 1 km spatial resolution with daytime and nighttime LST in separate data set. It was developed based on the best selection method, in which the granule pixel LST that meet the criteria was selected to represent the grid cell LST. The gridded MODIS LST in its latest version 6 consists of two LST data sets, i.e., MYD11A1 and MYD21A1 providing daily per-pixel LST&LSE with 1 km spatial resolution in sinusoidal projection. The MYD11A1 product is derived from the MYD11_L2 swath product that uses the split-window technique in the algorithm, whereas the MYD21A1 as a new suite of the MODIS LST products is based on the ASTER Temperature and Emissivity Separation (TES) algorithm.

The global data from March to October 2020 were used for the cross comparison between L3 NOAA20 VIIRS LST and AQUA MODIS LST. For the comparison between SNPP VIIRS LST and MODIS LST, the global data in May, July, and August 2020 were used in the comparison with MYD11A1 LST and the data in March and April 2020 were used in the comparison with MYD21A1 LST. The matchup procedure limits the temporal difference to be within 12 min and chooses the cloud clear LST only from both satellite LSTs products. Considering the coverage representativeness, only the dates with good global coverage were selected for the comparison. As shown in Fig. 9, the cross comparisons between NOAA20 VIIRS LST and MYD11A1 LST indicate a consistent warm bias at nighttime about 0.5 K and STD below 1.5 K, and the absolute daytime bias is mostly below 0.5 K. The comparison results with MYD21A1 LST indicate an average cold bias of 1.5 K at daytime and nearly no bias at nighttime. Similar results are observed for the cross comparison between SNPP VIIRS LST and MODIS LST for both MYD11A1 and MYD21A1 LSTs. The comparison result is affected by the cloud, contamination, temporal difference, field of view difference, and the data composition method employed in each LST product. The multiple daily mean LST difference between NOAA20 VIIRS LST and MODIS LST results as shown in Table 5 indicate that the NOAA20 VIIRS LST is in between the MYD11A1 and MYD21A1 for both daytime and nighttime, i.e., NOAA20 LST is 0.4 K and 0.7 K warmer than MYD11A1 but 1.6 K and 0.2 K colder than MYD21A1 LST for daytime and nighttime, respectively.

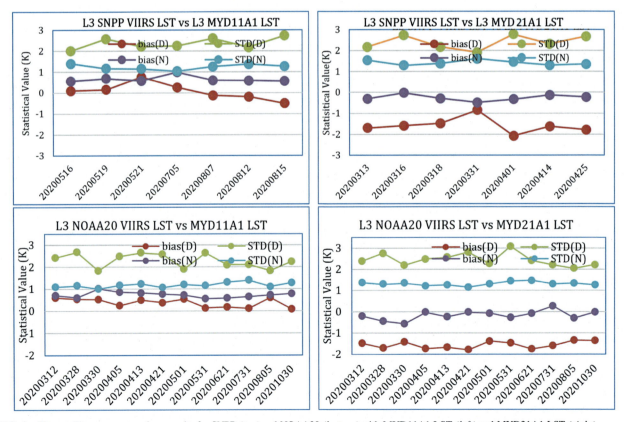

FIG. 9 The satellite cross comparison results for SNPP *(top)* and NOAA20 *(bottom)* with MYD11A1 LST *(left)* and MYD21A1 LST *(right)*.

TABLE 5 Multiple daily mean LST difference statistics.

	Bias(D)	STD(D)	Bias(N)	STD(N)
VIIRS LST-MYD11A1	0.37	2.29	0.73	1.18
VIIRS LST-MYD21A1	−1.56	2.45	−0.17	1.32

4. Summary

The VIIRS LST product has been developed for both SNPP and NOAA20 satellites based on the enterprise LST algorithm, which utilizes the split window technique as the theoretical basis. A reliable and high-quality emissivity product is developed to support both LST production by providing the spectral emissivity for LST retrieval as well as the broadband emissivity for LST validation.

Two validation approaches are used in the VIIRS LST validation, namely the T-based method to compare the satellite LST with ground LST observations and the cross-satellite comparison with AQUA MODIS LST product. These two methods are complementary to each other and both are useful to quantify and characterize the accuracy of the satellite LST product and help refine the LST retrieval algorithms. The comparisons with ground measurements indicate a similarly close agreement between SNPP and NOAA20 VIIRS LST products. Over SURFRAD network (Fig. 1), the VIIRS LST EDR yields a reasonable accuracy with a bias of −0.29 K and −0.43 K and RMSE of 1.81 K and 1.84 K for SNPP and NOAA20, respectively. The accuracy at nighttime is much better than that in daytime. In addition, the VIIRS LST data show significant seasonality with the best performance in fall and the worst in summer over SURFRAD sites. The comparisons with ground measurements from BSRN (Fig. 6) demonstrate a good agreement with a bias of −0.37 K and −0.24 K and RMSE of 1.72 K and 1.74 K in Gobabeb, Namibia, for SNPP and NOAA20, respectively, a bias close to zero and RMSE of 1.92 K and 2.04 K in Cabauw, Netherlands, for SNPP and NOAA20, respectively. The seasonal analysis over BSRN sites shows the best performance in winter and the worst also in summer. The validation over 17 ARM sites shows a bias of 0.04 K and 0.17 K and RMSE of 1.71 K and 1.66 K for SNPP and NOAA20, respectively. During the validation practice, the mismatch exists between the satellite pixel size observation and the ground spot size observations, which restrict the LST validation only at the relatively homogeneous ground sites. It is still a challenge how to characterize the ground site with strong heterogeneity and make it more comparable to the satellite retrieval (Yu et al., 2017b). The mismatch is often related to the variation of ground surface cover with seasonal or annual feature such as cropland growth and harvest, crop rotation, seasonal green/dry grass, etc. The performance of the VIIRS LST strongly depends on the atmospheric conditions, and it degrades under hot and wet weather conditions and large view zenith angles.

The cloud contamination is found to have a great impact on the validation results though the satellite cloud mask and additional cloud screening procedures were applied. Outliers can still be found from the comparison between the satellite retrievals and their in situ counterpart particularly at nighttime. This is often due to failure in identifying cloudy pixels. In addition, the cloud mask quality varies among different sensors attributed to the algorithm difference and threshold used in the cloud classifications.

The cross-satellite comparison with AQUA MODIS LST product at global scale generally exhibits a difference between daytime and nighttime. In daytime, VIIRS LST is closer to the MYD11A1 LST with an average bias of 0.4 K while closer to MYD21A1 LST with a bias of −0.2 K at nighttime. A cold bias about 1.5 K on average is observed between the VIIRS and MYD21A1 LST at daytime. Because VIIRS LST is on average warmer than the MYD11A1 LST with a bias of 0.5 K, which suggests that the MYD21A1 is about 2 K warmer than MYD11A1 LST. The result is affected by the temporal difference, viewing angle difference, cloud residue and the composition method. The angle difference is not considered in the cross comparison because the viewing angle information is not available in the L3 VIIRS LST yet.

Acknowledgments

This study was supported by NOAA grant NA19NES4320002 (Cooperative Institute for Satellite Earth System Studies-CISESS) at the University of Maryland/ESSIC. We would also like to acknowledge Jingjing Peng for reviewing the draft and providing constructive feedback. The manuscript contents are solely the opinions of the authors and do not constitute a statement of policy, decision, or position on behalf of NOAA or the U. S. Government.

References

Anderson, M.C., Allen, R.G., Morse, A., Kustas, W.P., 2012. Use of Landsat thermal imagery in monitoring evapotranspiration and managing water resources. Remote Sens. Environ. 122, 50–65.

Atitar, M., Sobrino, J.A., 2009. A split-window algorithm for estimating LST from Meteosat 9 data: test and comparison with in situ data and MODIS LSTs. IEEE Geosci. Remote Sens. Lett. 6, 122–126.

Augustine, J.A., DeLuisi, J.J., Long, C.N., 2000. SURFRAD—a national surface radiation budget network for atmospheric research. Bull. Am. Meteorol. Soc. 81, 2341–2357.

Augustine, J.A., Dutton, E.G., 2013. Variability of the surface radiation budget over the United States from 1996 through 2011 from high-quality measurements. J. Geophys. Res. Atmos. 118 (1), 43–53.

Becker, F., Li, Z.-L., 1990. Towards a local split window method over land surfaces. Int. J. Remote Sens. 11, 369–393.

Berk, A., Anderson, G.P., Acharya, P.K., Hoke, M.L., Chetwynd, J.H., Bernstein, L.S., et al., 2003. MODTRAN4 Version 3 Revision 1 user's Manual. Hanscom Air Force Base. Mass: Air Force Res. Lab.

Borbas, E.E., Seemann, S.W., Huang, H.-L., Li, J., Paul Menzel, W., 2005. Global profile training database for satellite regression retrievals with estimates of skin temperature and emissivity. In: Proceedings of the XIV. International ATOVS Study Conference, Beijing, China. University of Wisconsin-Madison, Space Science and Engineering Center, Cooperative Institute for Meteorological Satellite Studies (CIMSS), Madison, WI, pp. 763–770.

Coll, C., Hook, S.J., Galve, J.M., 2009a. Land surface temperature from the advanced along-track scanning radiometer: validation over inland waters and vegetated surfaces. IEEE Trans. Geosci. Remote Sens. 47, 350–360.

Coll, C., Wan, Z., Galve, J.M., 2009b. Temperature-based and radiance-based validations of the V5 MODIS land surface temperature product. J. Geophys. Res. 114, D20102.

Duan, S., Li, Z., Wu, H., Leng, P., Gao, M., Wang, C., 2018. Radiance-based validation of land surface temperature products derived from collection 6 MODIS thermal infrared data. Int. J. Appl. Earth Obs. Geoinf. 70, 84–92. https://doi.org/10.1016/j.jag.2018.04.006.

Fang, L., Zhan, X., Schull, M., Kalluri, S., Laszlo, I., Yu, P., Carter, C., Hain, C., Anderson, M., 2019. Evapotranspiration data product from NESDIS GET-D system upgraded for GOES-16 ABI observations. Remote Sens. (Basel) 11 (22), 2639. https://doi.org/10.3390/rs11222639.

Galleguillos, M., Jacob, F., Prévot, L., Frenchc, A., Lagacheriea, P., 2011. Comparison of two temperature differencing methods to estimate daily evapotranspiration over a Mediterranean vineyard watershed from ASTER data. Remote Sens. Environ. 115, 1326.

Gillespie, A.R., Rokugawa, S., Hook, S.J., Matsunaga, T., Kahle, A.B., 1996. Temperature/Emissivity Separation Algorithm Theoretical Basis Document, Version 2.4. NASA/GSFC, Maryland, USA, pp. 1–64.

Goela, P.C., Cordeiro, C., Danchenko, S., Icely, J., Cristina, S., Newton, A., 2016. Time series analysis of data for sea surface temperature and upwelling components from the southwest coast of Portugal. J. Mar. Syst. 163, 12–22.

Göttsche, F.-M., Olesen, F.-S., Bork-Unkelbach, A., 2013. Validation of land surface temperature derived from MSG/SEVIRI with in situ measurements at Gobabeb. Namibia. Int. J. Remote Sens. 34, 3069–3083.

Guillevic, P.C., Biard, C.J., Hulley, G.C., Privette, J.L., Hook, S.J., Olioso, A., Göttsche, F.M., Radocinski, R., Román, M.O., Yu, Y., Csiszar, I., 2014. Validation of land surface temperature products derived from the Visible Infrared Imaging Radiometer Suite (VIIRS) using ground-based and heritage satellite measurements. Remote Sens. Environ. 154, 19–37.

Guillevic, P., Göttsche, F., Nickeson, J., Hulley, G., Ghent, D., Yu, Y., Trigo, I., Hook, S., Sobrino, J.A., Remedios, J., Román, M., Camacho, F., 2018. Land surface temperature product validation best practice protocol. Version 1.0. In: Guillevic, P., Göttsche, F., Nickeson, J., Román, M. (Eds.), Best Practice for Satellite-Derived Land Product Validation. Land Product Validation Subgroup (WGCV/CEOS), p. 58, https://doi.org/10.5067/doc/ceoswgcv/lpv/lst.001.

Hook, S.J., Gabell, A.R., Green, A.A., Kealy, P.S., 1992. A comparison of techniques for extracting emissivity information from thermal infrared data for geologic studies. Remote Sens. Environ. 42, 123–135.

Hook, S.J., Vaughan, R.G., Tonooka, H., Schladow, S.G., 2007. Absolute radiometric in-flight validation of mid infrared and thermal infrared data from ASTER and MODIS on the Terra spacecraft using the Lake Tahoe, CA/NV, USA, automated validation site. IEEE Trans. Geosci. Remote Sens. 45, 1798–1807.

Hulley, G.C., Hook, S.J., 2009. Intercomparison of versions 4, 4.1 and 5 of the MODIS land surface temperature and emissivity products and validation with laboratory measurements of sand samples from the Namib desert, Namibia. Remote Sens. Environ. 113, 1313–1318.

Li, S., Yu, Y., Sun, D., Tarpley, D., Zhan, X., Chiu, L., 2014. Evaluation of 10 year AQUA/MODIS land surface temperature with SURFRAD observations. Int. J. Remote Sens. 35, 830–856.

Li, Z., Tang, B., Wu, H., Ren, H., Yan, G., Wan, Z., Trigo, I.F., Sobrino, J.A., 2013. Satellite-derived land surface temperature: current status and perspectives. Remote Sens. Environ. 131, 14–37.

Liu, Y., Yu, Y., Yu, P., Göttsche, F.M., Trigo, I.F., 2015. Quality assessment of S-NPP VIIRS land surface temperature product. Remote Sens. (Basel) 7 (9), 12215–12241. https://doi.org/10.3390/rs70912215.

Liu, Y., Yu, Y., Yu, P., Wang, H., Rao, Y., 2019. Enterprise LST algorithm development and its evaluation with NOAA 20 data. Remote Sens. (Basel) 11, 2003.

Martin, M.A., Göttsche, F.-M., 2016. Satellite LST Validation Report (DEL-13). ESA DUE Glob Temperature Project, Reference GlobT-WP4-DEL-12. www.globtemperature.info.

McMillin, L.M., 1975. Estimation of sea surface temperatures from two infrared window measurements with different absorption. J. Geophys. Res. 80 (C36), 5113–5117.

Meng, C.L., Li, Z.-L., Zhan, X., Shi, J.C., Liu, C.Y., 2009. Land surface temperature data assimilation and its impact on evapotranspiration estimates from the common land model. Water Resour. Res. 45, W02421.

Merchant, C.J., Matthiesen, S., Rayner, N.A., Remedios, J.J., Jones, P.D., Olesen, F., et al., 2013. The surface temperatures of Earth: steps towards integrated understanding of variability and change. Geosci. Instrum. Methods Data Syst. 2, 305–321. https://doi.org/10.5194/gi-2-305-2013.

Niclòs, R., Galve, J.M., Valiente, J.A., Estrela, M.J., Coll, C., 2011. Accuracy assessment of land surface temperature retrievals from MSG2-SEVIRI data. Remote Sens. Environ., 2126–2140.

Prata, A.J., 1994. Land surface temperatures derived from the advanced very high resolution radiometer and the along-track scanning radiometer 2. Experimental results and validation of AVHRR algorithms. J. Geophys. Res. 99, 13025–13058.

Price, J.C., 1984. Land surface temperature measurements from the split window channels of the NOAA 7 AVHRR. J. Geophys. Res. 89, 7231–7237.

Rajasekar, U., Weng, Q., 2009. Urban heat island monitoring and analysis by data mining of MODIS imageries. ISPRS J. Photogramm. Remote Sens. 64, 86–96.

Sun, D., Pinker, R.T., 2003. Estimation of land surface temperature from a geostationary operational environmental satellite (GOES-8). J. Geophys. Res. 108, 4326.

Sun, J., Salvucci, G.D., Entekhabi, D., 2012. Estimates of evapotranspiration from MODIS and AMSR-E land surface temperature and moisture over the Southern Great Plains. Remote Sens. Environ. 127, 44–59.

Trigo, I., Freitas, S., Bioucas-Dias, J., Barroso, C., Monteiro, I., Viterbo, P., 2009. Algorithm Theoretical Basis Document for Land Surface Temperature (LST) Products: LSA-4(MLST).

Trigo, I.F., Boussetta, S., Viterbo, P., Balsamo3 G., Beljaars A., Sandu I., 2015. Comparison of model land skin temperature with remotely sensed estimates and assessment of surface-atmosphere coupling. J. Geophys. Res. Atmos. 120 (23). https://doi.org/10.1002/2015JD023812.

Wan, Z., Dozier, J., 1996. A generalized split-window algorithm for retrieving land-surface temperature from space. IEEE Trans. Geosci. Remote Sens. 34, 892–905.

Wan, Z., Li, Z.-L., 2008. Radiance-based validation of the V5 MODIS land-surface temperature product. Int. J. Remote Sens. 29, 5373–5395.

Wang, H., Yu, Y., Yu, P., Liu, Y., 2020. Land surface emissivity product for NOAA JPSS and GOES-R missions: methodology and evaluation. IEEE Trans. Geosci. Remote Sens. 58 (1), 307–318. https://doi.org/10.1109/TGRS.2019.2936297.

Weng, Q., Lu, D., Schubring, J., 2004. Estimation of land surface temperature vegetation abundance relationship for urban heat island studies. Remote Sens. Environ. 89, 467–483.

Yu, Y., Liu, Y., Yu, P., 2017a. Land surface temperature product development for JPSS and GOES-R missions. In: Reference Module in Earth Systems and Environmental Sciences., https://doi.org/10.1016/B978-0-12-409548-9.10522-6. December.

Yu, Y., Tarpley, D., Privette, J., Goldberg, M., Raja, M., Vinnikov, K., Xu, H., 2009. Developing algorithm for operational GOES-R land surface temperature product. IEEE Trans. Geosci. Remote Sens. 47, 936–951.

Yu, Y., Tarpley, D., Privette, J.L., Flynn, L.E., Xu, H., Chen, M., Vinnikov, K.Y., Sun, D., Tian, Y., 2012. Validation of GOES-R satellite land surface temperature algorithm using SURFRAD ground measurements and statistical estimates of error properties. IEEE Trans. Geosci. Remote Sens. 50, 704–713.

Yu, Y., Wang, H., Liu, Y., Yu, P., 2017b. Enterprise Land Surface Emissivity Algorithm Theoretical Basis Document, Version 2.0. September,.

Zheng, W., Wei, H., Wang, Z., Zeng, X., Meng, J., Ek, M., Mitchell, K., Derber, J., 2012. Improvement of daytime land surface skin temperature over arid regions in the NCEP GFS model and its impact on satellite data assimilation. J. Geophys. Res. Atmos. 117, D06117.

Zhou, J., Liang, S., Cheng, J., Wang, Y., Ma, J., 2019. The GLASS land surface temperature product. IEEE J. Sel. Top. Appl. Earth Obs. Remote Sens. 12, 493–507.

Chapter 22

Heterogeneity of smoke from fires: Evaluation of VIIRS smoke detection using FIREX-AQ field campaign data

Pubu Ciren[a,b] and Shobha Kondragunta[b]

[a]I.M. Systems Group, Inc., College Park, MD, United States, [b]NOAA/NESDIS Center for Satellite Applications and Research (STAR), College Park, MD, United States

Chapter outline

1. Introduction — 391
2. NOAA JPSS enterprise processing system aerosol detection product — 393
   2.1 Deep-blue aerosol detection algorithm — 394
   2.2 IR-visible aerosol detection algorithm — 396
   2.3 eMAS measurements during FIREX-AQ 2019 — 398
   2.4 Application of EPS ADP algorithm to eMAS measurements — 401
3. Comparison of ADPe with ADPv — 404
4. Conclusions — 407
   Acknowledgment — 408
   Disclaimer — 408
   References — 409

Scientific truth is universal, because it is only discovered by the human brain and not made by it, as art is.

Konrad Lorenz

1. Introduction

Atmospheric aerosols impact human health and the economy by influencing weather, climate, air quality, and ecosystems (Nicholson, 2000; Prospero and Lamb, 2003; Yoshioka and Mahowald, 2007; Kim et al., 2010; Al-Saadi et al., 2005; Marlier et al., 2015; Price et al., 2018). Among the many sources of aerosols, outbreaks of smoke and dust aerosol events are evolving as major sources for poor air quality (Schoennagel et al., 2017); they are becoming more frequent because of increased warming, drought, and insect outbreaks, which are all caused by or linked to climate change (Krawchuk et al., 2009; Westerling et al., 2011). To understand and mitigate the impacts of dust and smoke, routine monitoring of their global distribution, including sources and transport, is crucial (IPCC, 2007). To meet this need, satellite remote sensing provides the ability to monitor the spatial and temporal variability of both smoke and dust events.

Various smoke/dust classification approaches have been developed in recent years. The first category is human-based smoke/dust plume identification, such as the NOAA Hazard Mapping System (HMS) smoke product, in which human analysts hand-draw smoke plumes by observing satellite images (McNamara et al., 2004; Rolph et al., 2009). However, besides inherent subjectivity, this approach is labor-intensive and time-consuming, rendering it unsuitable for global-scale applications, especially in the era of rapid refresh geostationary satellite imagery available at 5-min intervals (Schroeder et al., 2008; Schroeder and Giglio, 2017; Tosca et al., 2011; Nelson et al., 2013; Williamson et al., 2013; Mota and Wooster, 2018; Lu et al., 2019). The second category is multi-spectral threshold-based smoke/dust detection (Li et al., 2001; Wang et al., 2007; Xie et al., 2007; Tosca et al., 2011; Ismanto et al., 2019; Zhao et al., 2010). Bearing the advantage of being fast and easily applied from regional and global scales, the performance of this approach, however, is hindered by the optimality of the fixed spectral thresholds based on the region of interest (Chrysoulakis and Cartalis, 2003; Wang et al., 2007; Xie et al., 2007; Lu et al., 2021; Ciren and Kondragunta, 2014). The third category is smoke/dust detection based on machine learning (ML) theory, a quickly evolving technology (Li et al., 2001, 2015; Ba et al., 2019; Chacon et al., 2011; Boroughani et al., 2020; Lee et al., 2021). Unlike the other techniques, the ML approach classifies a scene into smoke, dust, cloud, or clear background land/ocean but with a continuous value representing the probability of these conditions. However, a training database, which must well represent various scenarios of the truth data, is the core for the success of ML techniques. For smoke and dust detection, limited by the lack of real truth and the unavailability of a large volume of such datasets, the global scale application of the ML classification for smoke and dust aerosols is currently unfeasible.

Ciren and Kondragunta (2014) proposed a new smoke/dust classification approach based on satellite remote sensing observations, which falls into the category of the multi-spectral threshold-based techniques. This new technique first identifies both smoke and dust as absorbing aerosol using the spectral dependence of absorption by both smoke and dust in the deep-blue part of the visible spectrum. Then, dust and smoke are further separated by using the spectral contrast between two wavelengths: one in the deep-blue and the other in a much longer wavelength range, such as shortwave infrared (SWIR). This separation is possible because dust, owing to its larger size, scatters light in the SWIR, whereas smoke and small particles present in smog do not. Compared to measurements from both the Aerosol Robotic Network (AERONET) (Holben et al., 1998) and the Vertical Feature Mask (VFM) product from the Cloud-Aerosol Lidar with Orthogonal Polarization (CALIOP), the probability of correct detection (POCD) of dust using the new smoke/dust classification approach applied to Aqua Moderate Resolution Imaging Spectroradiometer (MODIS) measurements was found to be ~70%–80%. Following this initial work on Aqua MODIS, this new approach was refined to not only detect dust but also identify smoke (including smog) as well. This updated approach became the foundation of the NOAA JPSS Enterprise Processing System (EPS) Aerosol Detection Product (ADP) algorithm (Joint Polar Satellite System (JPSS), 2016a). An operational version of the ADP has been generated from the Visible Infrared Imaging Radiometer Suite (VIIRS) on the Suomi NPP (SNPP) satellite since 2012 and from the NOAA-20 satellite since 2018. Validation against both AERONET measurements and CALIOP VFM products indicates the POCD is 80% and 78%, respectively for dust and smoke.

Validation of ADP is very challenging, however, since there is no "real" truth data in terms of smoke and dust masks. In addition, aerosol type is mostly present as a mixture. As an indirect strategy, ADP is validated against smoke/dust type classified from AERONET observations, typically by using thresholds of the observed Aerosol Optical Depth (AOD) and its spectral dependence, which can be characterized by the Ångström exponent (Holben et al., 2001; Giles et al., 2012). However, to match with satellite-derived ADP, the AERONET data are spatially and temporally averaged, which creates uncertainty for binary (yes/no) products such as ADP. Another indirect strategy is to validate ADP against the aerosol-type product from active space-borne sensors, such as the CALIOP VFM; however, the narrow track of space-borne lidars not only substantially reduces the number of match-ups with ADP from other satellite platforms but also creates uncertainty due to the large difference in spatial coverage between the two products.

Measurements from field campaigns that are targeted on specific aerosol events are valuable data sources for validation of satellite-based ADP and also for deep-dive analyses to improve satellite-based retrieval algorithms. This is because a suite of instruments, both ground-based and airborne, are deployed to characterize better both the surface and the atmosphere during the aerosol event. In addition, airborne sensors have much higher spatial resolution than satellite sensors (e.g., 50m versus 750m for VIIRS) and thus can provide unique information on the small-scale distribution of various geophysical parameters.

The Fire Influence on Regional to Global Environments and Air Quality (FIREX-AQ) field campaign, a joint venture led by NOAA and NASA, made comprehensive observations to investigate the impact on air quality and climate from wildfires and agricultural fires across the continental United States (Warneke et al., 2018). During the 2019 fire season, FIREX-AQ made measurements of several wildfires and prescribed burns over the Western and Eastern US from July 22 to August 31. A suite of instruments onboards the NASA DC-8, NOAA Twin Otter, and NASA ER-2 aircraft, and ground-based instruments at multiple sites, made intensive measurements during wildfire smoke and haze episodes. These measurements were coordinated with the overpass times of several satellites, including SNPP, NOAA-20, GOES-16 and -17, and Sentinel 5-Precursor (S5P). A key instrument onboard the NASA-ER2, the enhanced MODIS Airborne Simulator (eMAS) is a multispectral scanning spectrometer configured to approximate the Moderate-Resolution Imaging Spectrometer (MODIS). Therefore, eMAS has a spectral band coverage similar to VIIRS onboard both SNPP and NOAA-20. Data acquired by the eMAS have been helping to define, develop, test, and refine MODIS algorithms (King et al., 1996), and it is expected to be valuable for VIIRS as well.

In this chapter, we demonstrate how field campaign measurements are used to validate and improve the VIIRS ADP by using the eMAS observations during FIREX-AQ 2019 as an example. In Section 2, the JPSS EPS ADP algorithm is described and the process of applying the algorithm to the eMAS measurements is explained. Comparisons between ADP generated from three sets of data, i.e., the VIIRS observations, the matchup of the eMAS observations at their original resolution (which is about 50m) with the VIIRS pixels, and the matchup of the eMAS observations with the eMAS pixels aggerated to the resolution of the VIIRS pixels (750m), are presented in Section 3. In addition, the subpixel variability, which is represented by the standard deviation of the VIIRS M5/eMAS band 3 in a 3×3 box, was examined to characterize smoke heterogeneity. The results are discussed in Section 4.

2. NOAA JPSS enterprise processing system aerosol detection product

The EPS ADP algorithm attempts to separate cloudy and clear pixels from those with smoke or dust. The detection of smoke or dust relies on the distinctive signature of smoke or dust which is often expressed in terms of spectral variations of the observed brightness temperature (BT) or solar reflected energy. The spectral variation of the refractive index (i.e. the aerosol composition) plays an important role in the success of these methods. In addition, the scattering and absorption properties of aerosols also depend on the particle size distribution and the particle shape. Several aerosol remote sensing techniques have been developed using observations from the Advanced Very High-Resolution Radiometer (AVHRR) (e.g., Barton et al., 1992). Similar to dust plumes, volcanic ash plumes often generate negative brightness temperature differences (BTD) between 11 μm and 12 μm. Prata (1989) has demonstrated the detection of volcanic aerosols using two infrared channels, while Ackerman and Strabala (1994) applied observations at 8.6, 11, and 12 μm from the HyperSpectral Infrared Sound (HIRS) instrument to study the Mt. Pinatubo stratospheric aerosol. Recently, Ciren and Kondragunta (2014) developed a simple and fast technique to detect dust based on the MODIS deep-blue (412 nm), blue (440 nm), and shortwave-IR (2130 nm) bands. This technique utilizes the spectral dependence of dust absorption, surface reflectance, and differences in absorbing/scattering properties between small and large particles.

The essence of the NOAA EPS is to have one set of algorithms working on observations from multiple sensors; the algorithm path is determined by the spectral coverage of the observations, with uniform input and output. As for the EPS ADP algorithm, it consists of three algorithm paths, i.e., the Deep-Blue path, the IR-Visible path, or both paths. The detailed flow chart of the EPS ADP algorithm is given in Fig. 1. As seen from Fig. 1, for retrievals over land and over the ocean, the ADP algorithm has path 1 (Deep-Blue path), path 2 (IR-Visible path), and path 3 (combined Deep-Blue with IR-Visible path). The choice of algorithm path depends on the spectral coverage of the sensor, for example, path 3 for SNPP/NOAA-20 VIIRS, path 2 for GOES-R ABI and Himawari AHI, and path 1 for S5P TROPOMI.

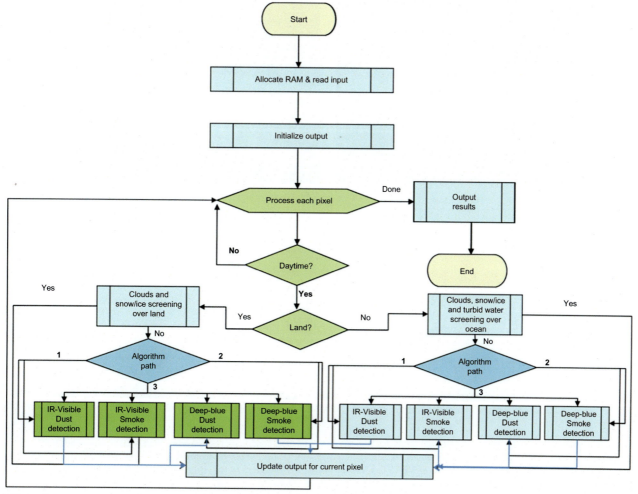

FIG. 1 High-level flowchart of the EPS ADP algorithm, illustrating the main processing sections.

2.1 Deep-blue aerosol detection algorithm

The advantage of using measurements in the blue wavelength region (410 to 490 nm) to retrieve aerosol optical properties has been demonstrated by Hsu et al. (2004, 2006). Because the sensitivity of reflectance to dust in the atmosphere decreases with increasing wavelength, the spectral contrast between two neighboring wavelengths can be used as an indicator for the presence of dust. The spectral shape of the reflectance at the Top of the Atmosphere (TOA) for a cloud-free atmosphere is determined by three main processes: Rayleigh scattering, absorption and scattering by aerosols, and reflection by the underlying surface. The strong wavelength-dependent Rayleigh scattering creates a strong contrast between two neighboring wavelengths. The presence of dust, however, reduces this contrast as a result of the increased absorption with decreasing wavelength. As for the underlying surfaces, their effects on spectral contrast depend on the spectral variability of surface reflectance. For surface types such as water, the surface reflectance is nearly independent of wavelength within deep-blue to blue regions, except for areas with sunglint; therefore, its effect on the spectral contrast is minimal. However, arid or desert surfaces show an increased reflection with the increasing wavelength (Hsu et al., 2004), similar to absorbing aerosol; reflection from these surfaces will also reduce the spectral contrast.

The reduction in the spectral contrast of Rayleigh scattering when dust is present in the atmosphere is also seen in satellite observations. Fig. 2a and b shows the regions highlighted by boxes where pixels are identified as clear (red box) and as dusty (yellow box), respectively, over land and ocean. Fig. 2c and d shows the ratio of TOA reflectance at 412 nm and

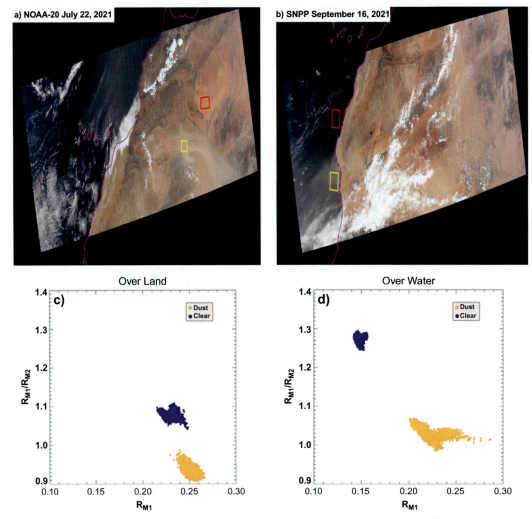

FIG. 2 VIIRS true color (RGB) images (*red box* and *yellow box* indicate dusty and clear conditions, respectively): NOAA-20 granules on July 22, 2021, ~13:43-13:44 UTC (a) and SNPP granules on September 16, 2021, ~13:42-13:43 UTC (b); for pixels identified as clear (*blue*) and dust-laden (*orange*), scatter plots of the observed TOA reflectances (R) of VIIRS bands: ratio of R_{M1} (M1, 412 nm) to R_{M2} (M2, 440 nm) vs. R_{M1} (M1, 412 nm) over land (c) (corresponding to colored boxes in (A)) and over water (d) (corresponding to colored boxes in (B)).

440 nm as a function of TOA reflectance at 412 nm for pixels from the selected boxes over land and ocean. The ratio of 412 nm to 440 nm for pixels with dust is distinct from clear pixels over the ocean, showing the reduced contrast between 412 nm and 440 nm (Fig. 2d). Such separation is also seen for dust over the desert (Fig. 2c).

This analysis based on theory and observations suggested that the effect of dust in reducing the spectral contrast between 412 nm and 440 nm could be used as a way to detect the presence of dust in the atmosphere. However, using the spectral contrast under clear conditions as a reference to detect dust requires knowledge of the spectral reflectance of the underlying surface. To bypass this problem, the spectral contrast from pure Rayleigh scattering is used as a reference, and different thresholds for the reduction in spectral contrast are chosen over land and water. Consequently, an index similar to the Aerosol Index and absorbing aerosol index used in SeaWiFS (Hsu et al., 2000), called Absorbing Aerosol Index (AAI), has been developed:

$$\text{AAI} = -100 \left[\log_{10}\left(\frac{R_{412\,nm}}{R_{440\,nm}}\right) - \log_{10}\left(\frac{R'_{412\,nm}}{R'_{440\,nm}}\right) \right] \quad (1)$$

In Eq. (1), R is the TOA reflectance, R' is the reflectance from Rayleigh scattering as computed by 6S (Radiative transfer code) (Kotchenova et al., 2006; Kotchenova and Vermote, 2007) for a given location and satellite viewing geometry.

However, as shown in Fig. 3a, other absorbing aerosols, such as smoke, also have an effect on the spectral contrast between 412 nm and 440 nm that is similar to dust. Due to the fact that the particle size of dust is considerably larger than that of smoke, dust extends its scattering signature to shortwave IR wavelengths, whereas smoke is mostly transparent in this range (Kaufman et al., 2005). By using observations at shortwave IR wavelengths, it is possible to separate dust from other absorbing aerosols. For example, Fig. 3b shows a scatter plot between TOA reflectance at 412 nm and 2250 nm from VIIRS for pixels identified as smoke, dust, and clear sky over the ocean. Dust pixels have higher reflectance at 2250 nm and are well separated from clear sky pixels and smoke pixels. To this end, a second index, named the Dust Smoke Discrimination Index (DSDI), is computed as follows:

$$\text{DSDI} = -10 \left[\log_{10}\left(\frac{R_{412\,nm}}{R_{2250\,nm}}\right) \right] \quad (2)$$

In Eq. (2), R is the observed TOA reflectance. The DSDI is applied to pixels that pass through the AAI threshold test.

As shown in the above section, by using the spectral contrast between the deep-blue and blue wavelengths, absorbing aerosols can be detected, including smoke and dust. Furthermore, utilizing the spectral contrast between the deep-blue and shortwave IR wavelengths, dust can be separated from other absorbing aerosols, including smoke, which indicates that this technique can be used for both smoke and dust detection.

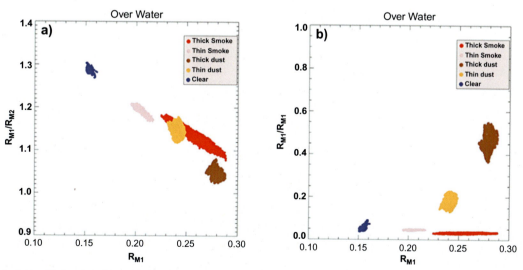

FIG. 3 For pixels identified as clear *(blue)*, thick smoke-laden *(red)*, thick dust-laden *(brown)*, thin smoke-laden *(pink)*, and thin dust-laden *(orange)* over water, scatter plots of the observed TOA reflectances (R) of VIIRS bands: ratio of R_{M1} (M1, 412 nm) to R_{M2} (M2, 440 nm) vs. R_{M1} (M1, 412 nm) (a) and ratio of R_{M11} (M11, 2250 nm) to R_{M1} (M1, 412 nm) vs. R_{M1} (M1, 412 nm) (b).

2.2 IR-visible aerosol detection algorithm

In addition to the deep-blue wavelengths, which are used to identify the signatures of the presence of smoke/dust aerosols, other wavelengths are also used for smoke/dust detection. The bulk transmittances of dust and volcanic ash display a strong spectral variation in the 8–10 μm and 10–12 μm regions. This is also a spectral region over which the atmosphere is fairly transparent. For these reasons, techniques have been developed which successfully employ satellite radiance measurements at 11 and 12 μm to detect dust and volcanic ash. These split-window IR techniques have primarily been applied to volcanic aerosols, particularly those from sulfur-rich eruptions (e.g. Prata, 1989; Barton et al., 1992) as well as dust outbreaks (Legrand et al., 1992, 2001; Evan et al., 2006).

Positive $BT_{11\mu m} - BT_{12\mu m}$ values are usually associated with clear sky atmospheres since water vapor both absorbs and emits in the 11 and 12 μm channels, and the weighting function for the 11 μm channel peaks lower in the atmosphere than the 12 μm channel. However, the presence of dry air, often associated with dust events, will tend to reduce the positive $BT_{11\mu m} - BT_{12\mu m}$ values. In addition, dust has a larger absorption at 12 μm than at 11 μm, so dust plumes generally have higher emissivity and lower transmissivity in the 12 μm channel (Ackerman, 1997; Dunion and Velden, 2004). For more elevated dust layers, the increased temperature separation between the dust layer and the surface, and coincident reduction of dry air closer to the peak of the 11 μm weighting function, make the split window brightness temperature difference even less positive. However, this difference has also been observed to be affected by the optical thickness of a given dust plume, so that for thick optical depths, the $BT_{11\mu m} - BT_{12\mu m}$ difference becomes more negative. Darmenov and Sokolik (2005) further explored the brightness temperature difference technique using MODIS data applied to dust outbreaks from different regions of the globe. In general, $BT_{8\mu m} - BT_{11\mu m}$ becomes less negative and $BT_{11\mu m} - BT_{12\mu m}$ becomes more negative with increasing dust loading (Fig. 4). However, in the EPS ADP algorithm, the 3.7 μm band is chosen instead of 8 μm because 3.7 μm has less water vapor absorption and also in order to eliminate false alarms from low-level clouds (often towering cumulus).

The NOAA-20 VIIRS RGB image in Fig. 5A shows a dust plume with different regions of heavy dust (yellow box), thin dust (green box), and clear sky (red box) identified. For these different regions, the relationship between different visible reflectance and IR BTD is plotted in the four panels of Fig. 5B(a)–(d). Clear sky pixels have low reflectance at both 0.49 μm (M3 band) and 0.67 μm (M5 band), thin dust has elevated reflectances at these channels, and thick dust pixels have 20% or greater reflectance at these channels. The BTD between 3.7 μm (M12 band) and 10.76 μm (M15 band) plotted against the BTD between 10.76 μm (M15 band) and 12.0 μm (M16 band) shows a clear separation of thick dust pixels from both thin dust and clear pixels. In addition, the advantage of using 3.7 μm (M12 band) instead of 8.55 μm (M14 band) to separate thin dust from thick dust is evident.

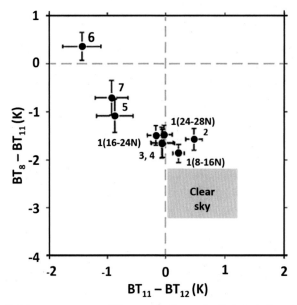

FIG. 4 Combined trispectral diagram of brightness temperature differences for "heavy dust" pixels, indicated by the numbers 1 to 7 (for different geographic locations), and for clear sky. *(From Darmenov, A., Sokolik, I.N., 2005. Identifying the regional thermal-IR radiative signature of mineral dust with MODIS. Geophys. Res. Lett. 32, L16803. https://doi.org/10.1029/2005GL023092.)*

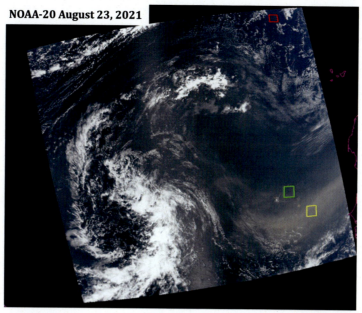

FIG. 5A NOAA-20 VIIRS true color (RGB) image on August 23, 2021, at 15:20–15:26 UTC for a dust event over the west coast of Africa. The *yellow*, *green*, and *red boxes* represent heavy dust, thin dust, and clear conditions.

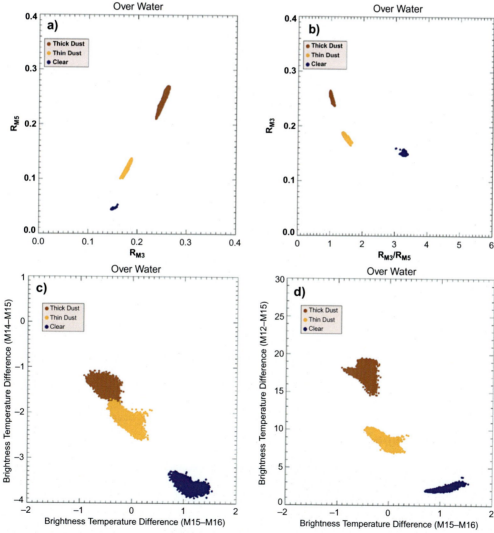

FIG. 5B For pixels identified as clear (*blue*; corresponding to *red box* in Fig. 5A), thick dust-laden (*brown*; corresponding to *yellow box* in Fig. 5A), and thin dust-laden (*orange*; corresponding to *green box* in Fig. 5A), scatter plots of the relationship between various combinations of VIIRS bands: observed TOA reflectances R_3 (M3, 488 nm) vs. R_5 (M5, 672 nm) (a), observed TOA reflectances R_3 (M3, 488 nm) vs. ratio of R_3 (M3, 488 nm) to R_5 (M5, 672 nm) (b), brightness temperature difference M14 (8.55 μm)-M15 (10.76 μm) vs. brightness temperature difference M15 (10.76 μm)-M16 (12.01 μm) (c), and brightness temperature difference M12 (3.70 μm)-M15 (10.76 μm) vs. brightness temperature difference M15 (10.76 μm)-M16 (12.01 μm) (d).

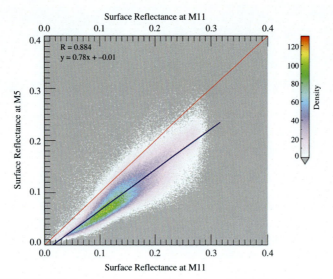

FIG. 6 Surface reflectance of S-NPP VIIRS M5 (0.67 μm) vs. that of M11 (2.25 μm) over AERONET stations. The source of the data set is the JPSS EPS AOD algorithm Joint Polar Satellite System (JPSS), 2016b.

For smoke detection over land, fire spots are detected first by looking at pixels with BTs at 4.05 μm (M13 band) greater than 350 K and a BTD between 4.05 μm and 10.76 μm greater than or equal to 10 K. Pixels that pass these fire tests are assumed to have thick smoke. The smoke tests over land also take advantage of a linear relationship between the reflectance at the visible band (0.67 μm for VIIRS M5) and shortwave IR band (2.25 μm for VIIRS M11). Fig. 6 shows this relationship with the corresponding bands of VIIRS. The surface reflectance at M5 is generally around 80% of the surface reflectance at 2.25 μm. However, because the size of smoke particles is relatively small, the signal from smoke will be extremely small in the shortwave IR wavelength range; therefore, there is a larger increase in R_{M5} than R_{M11} for an atmosphere that contains smoke. Spatial variability tests for the M5 (0.67 μm) band are further used over land to separate smoke from clouds, since clouds show large variability in this band compared to smoke, and also the surface is relatively darker. The spatial variability test is defined as the standard deviation of the reflectance at M5 in a box of 3×3 pixels centered in the pixel for which the detection is performed. In the current ADP algorithm, the threshold for this spatial variability test is set at 0.04.

Spatial variability tests also help in avoiding the misclassification of clouds as smoke for smoke detection over water. Since clear pixels, pixels loaded with thick smoke, and cloudy pixels are more uniform than pixels with partial cloud or thin dust, by using the standard deviation of reflectance at 0.86 μm (VIIRS M7), where both aerosol and clouds effects are moderate, pixels which contain thick smoke can be separated from those containing clouds/thin smoke. A threshold value of 0.008 is used in the current version of ADP algorithm. Smoke in visible channels looks brighter than the water surface but darker than a cloud. However, it is very difficult to completely separate them by only using the reflectance test. Therefore, because reflection from clouds is spectrally independent while reflection from smoke has a strong wavelength dependence, spectral contrast tests are combined to separate clouds, smoke, and the water surface. First, the ratio between R_{M3} and R_{M10} is used; these two channels are used because the aerosol effect is larger at 0.49 μm (VIIRS M3) but water is darker at 1.61 μm (VIIRS M10). Second, the ratio between R_{M11} and R_{M10} is combined to enhance the separation of smoke from clouds. Third, by constraining R_{M3} and R_{M10}, thick smoke can be identified. Scatter plots of the ratio of R_{M3} to R_{M10} and the ratio of R_{M11} to R_{M10} against R_{M3} and R_{M10} are shown in Fig. 7, respectively, for clear pixels, pixels loaded with thick smoke, thin smoke, and cloudy pixels.

2.3 eMAS measurements during FIREX-AQ 2019

FIREX-AQ field campaign was carried out to study fires and smoke as related to their influence on air quality. The experiment flew many instruments on aircrafts to study smoke aerosol properties including composition, physical and optical properties, and changes to these properties in fresh smoke versus aged smoke, etc. The eMAS instrument was one of the sensors flown to make smoke reflectance observations at multiple visible and infrared bands. eMAS made measurements on a total of 10 flights, encompassing 168 flight tracks during the period August 2–21. These flights included 16 wildfire events. Fig. 8 shows examples of eMAS images from uncalibrated band 3 (0.65 μm), band 10 (1.61 μm), band 20

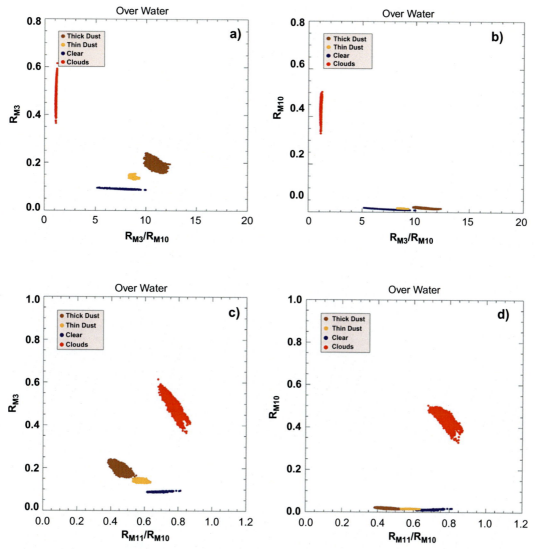

FIG. 7 For pixels identified as clear *(blue)*, thick smoke-laden *(dark brown)*, thin smoke-laden *(light brown)*, and cloudy *(red)* over water, scatter plots of the observed TOA reflectances (R) of VIIRS bands: R_3 (M3, 488 nm) vs. ratio R_3 (M3, 488 nm) to R_{10} (M10, 1061 nm) (a); R_{10} (M10, 1061 nm) vs. ratio of R_3 (M3, 488 nm) to R_{10} (M10, 1061 nm) (b); R_3 (M3, 488 nm) vs. ratio of R_{11} (M11, 2250 nm) to R_{10} (M10, 1061 nm) (c); and R_{10} (M10, 1061 nm) vs. ratio of R_{11} (M11, 2250 nm) to R_{10} (M10, 1061 nm) (d).

(2.13 μm), and band 22 (10.99 μm), and a false-color RGB image (combination of bands 20, 10 and 2) from flight #19–911, track #12 over the Williams Flats fire, WA on August 7. A huge smoke plume is observed in the upper right corner of the track, indicated by a high signal at 0.65 μm (white shading in Fig. 8a) and a blue color in the false-color RGB image (Fig. 8e). This smoke plume is nearly transparent in the shortwave-IR band (Fig. 8c).

In contrast, the calibrated reflectances for flight #19–191, track #12, made from eMAS L1b data for 0.65, 0.55, and 0.47 μm and the RGB image from the combined bands are shown in Fig. 9; these radiances cover only the region of the smoke plume shown in Fig. 8. In Fig. 9, the smoke plume is evident; the signal from the smoke plume diminishes with increasing wavelength, with the strongest signal at 0.47 μm, weak signal at 1.61 μm, and essentially no signal (nearly transparent) at 2.23 μm. As discussed in Section 2, all three of these channels are used in the ADP algorithm for smoke detection over land. The very high spatial resolution eMAS data set, observed directly over smoke plumes, provides great potential for evaluating the EPS ADP algorithm. Two focus areas of our validation are the agreement between VIIRS smoke detection and eMAS smoke observations and to characterize the heterogeneity of smoke as observed by the two instruments, VIIRS and eMAS. One of the critical elements of VIIRS EPS ADP algorithm is the spatial variability test that screens smoke from cloud. Understanding smoke heterogeneity from high-resolution eMAS data will help validate the spatial standard deviation test thresholds we employ in the VIIRS ADP algorithm.

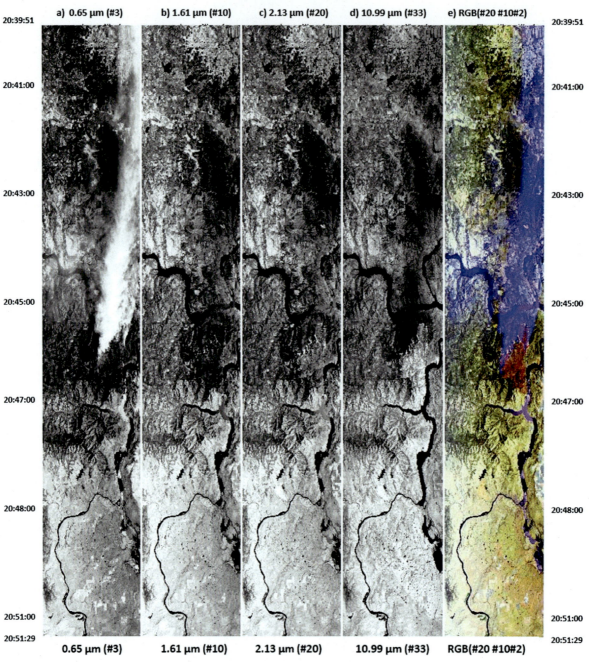

FIG. 8 eMAS images for bands 3 (a), 10 (b), 20 (c), and 33 (d), and the RGB image (e) from flight #19–191, track #12 over the Williams Flat fire, WA on August 7, 2019. *(Credit: https://mas.arc.nasa.gov/data/deploy_html/firex-aq_home.html#satpass.)*

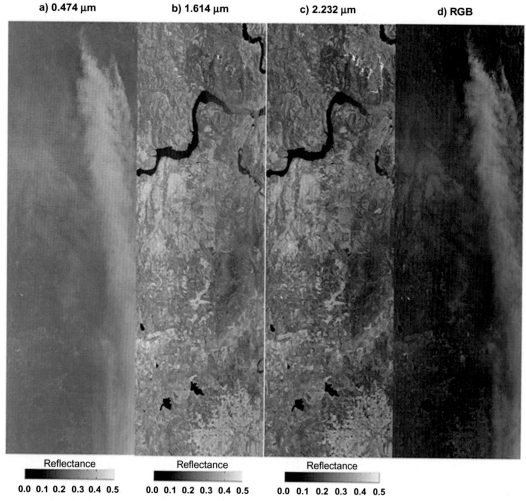

FIG. 9 Observed eMAS L1b reflectances at 0.47 μm (a), 1.61 μm (b) and 2.25 μm (c) over the Williams Flat fire, WA on August 7, 2019. The corresponding RGB image (*Red*—0.65 μm, *Green*—0.55 μm, *Blue*—0.47 μm) (d).

2.4 Application of EPS ADP algorithm to eMAS measurements

The spectral band configuration of the eMAS that flew during FIREX-AQ 2019 is given in Table 1, and the corresponding bands from VIIRS are shown for comparison. Most of the bands used for smoke detection in the VIIRS ADP IR-Visible algorithm overlaps with bands from the eMAS; the much finer spatial resolution of the eMAS, which is about 50 m versus 750 m for VIIRS, provides more precise characterization of the spatial variability inside VIIRS pixels, which is one of the tests used to screen out residual clouds in the VIIRS ADP algorithm.

During FIREX-AQ campaign in 2019, several eMAS flight tracks were coordinated with the overpass time of satellites, including GOES-16 and -17, SNPP, NOAA-20, and S5P. This eMAS-satellite coordination allowed us to apply the EPS ADP algorithm to eMAS observations and to make comparisons of airborne-derived ADP with satellite-derived ADP. To demonstrate this approach, the eMAS flight tracks that coincided with SNPP and NOAA-20 were selected for this study. For the case shown in Fig. 8, eMAS flight #19-911 and track #12, the range of the eMAS observations over the smoke plume is around 20:39–20:45 UTC, which is close to the NOAA-20 VIIRS granule overpass time of 20:40 UTC. The first step in making a temporally matched pair of measurements is to spatially match eMAS pixels with VIIRS pixels. Due to the finer resolution of eMAS, about 225 to 800 eMAS pixels fall into each VIIRS pixel. The exact number of eMAS pixels in each VIIRS pixel varies with its location along the scanline.

Fig. 10 shows the RGB images created from the collocated eMAS bands (Red-band 3, Green-band 2, Blue-band 1; Fig. 10a) and the collocated VIIRS bands (Red-M5, Green-M4, Blue-M3; Fig. 10b). Owing to the higher resolution of the eMAS sensor, the eMAS RGB image (Fig. 10a) has more detail for both the smoke plumes and the surface, whereas

TABLE 1 List of eMAS bands configured during FIRX-AQ 2019 and the corresponding VIIRS bands.

	eMAS			VIIRS	
Band No.	Central wavelength (μm)	Spectral range (μm)	Band No.	Central wavelength (μm)	Spectral range (μm)
1	0.4716	0.4518–0.4914	M3	0.488	0.478–0.498
2	0.5535	0.5324–0.5746	M4	0.555	0.545–0.565
3[a]	0.6595	0.6338-0.6852	M5[a]	0.672	0.662-0.682
4	0.7056	0.6849–0.7263			
5	0.7466	0.7257–0.7674	M6	0.746	0.739–0.754
6	0.8279	0.8070–0.8489			
7	0.8691	0.8487–0.8896	M7	0.865	0.846–0.885
8	0.9096	0.8894–0.9299			
9	0.9497	0.9295–0.9699			
10[a]	1.6163	1.5881-1.6445	M10[a]	1.61	1.580-1.640
11	1.6715	1.6445–1.6985			
12	1.7272	1.7012–1.7531			
13	1.7798	1.7545–1.8051			
14	1.8328	1.8082–1.8573			
15	1.885	1.8605–1.9096			
16	1.9361	1.9116–1.9606			
17	1.9862	1.9619–2.0105			
18	2.0378	2.0129–2.0627			
19	2.0873	2.0637–2.1109			
20	2.1372	2.1132–2.1612			
21	2.1862	2.1626–2.2097			
22[a]	2.2357	2.2113-2.2601	M11[a]	2.25	2.225-2.275
23	2.2862	2.2626–2.3097			
24	2.337	2.3132–2.3608			
25	2.386	2.3622–2.4098			
26[a]	3.7315	3.6424-3.8206	M12[a]	3.70	3.660-3.840
27	6.6372	6.5315–6.7429			
28	7.2576	7.1232–7.3921			
29	8.2011	8.0561–8.3462			
30	8.5018	8.3645–8.639	M14	8.55	8.400–8.700
31	9.6871	9.5413–9.8329			
32	10.1622	10.036–10.288			
33	11.0018	10.870–11.132	M15	10.76	10.263–11.263
34	11.993	11.872–12.113	M16	12.01	11.538–12.488
35	12.5673	12.4447–12.69			
36	13.3142	13.1894–13.43			
37	13.6049	13.479–13.730			
38	13.9055	13.786–14.025			

[a]The band used in smoke detection over land in the IR-Visible path of EPS ADP.

the VIIRS RGB image (Fig. 10b) shows a similar pattern, but with fewer details. Since both data sets are collocated, the variability of each VIIRS pixel at the subpixel level, which may affect the detectability of smoke plumes, can be examined with the help of the eMAS measurements.

As shown in the flowchart of the EPS ADP algorithm (Fig. 1), smoke and dust detection are not retrieved for any pixel that is identified as cloudy or covered by snow/ice. In the operational VIIRS EPS ADP algorithm, cloud information from the VIIRS Enterprise Cloud Mask (ECM) product is applied; however, the VIIRS cloud mask is at the VIIRS pixel resolution, so it is not suitable for application to eMAS measurements. Since the purpose of our study is to examine subpixel variability in ADP, it is intuitive to not use cloud masks at the VIIRS pixel level. Fortunately, the extremely fine spatial resolution of eMAS measurements enables us to select cases and regions which are cloud-free, such as the one shown in Fig. 10, so we can avoid the need for a cloud mask. The same rationale is applied for snow/ice masks.

Given the spectral coverage of eMAS, the IR-visible path of the EPS ADP algorithm is selected for smoke detection. From the collocated data between eMAS and VIIRS, three data sets are generated: (1) ADP_e: eMAS L1b data at its original spatial resolution (50 m), (2) ADP_{ev}: eMAS L1b data aggregated through averaging of pixels falling into each VIIRS pixel (750 m), and (3) ADP_v: VIIRS reflectances/radiances at their original resolution (750 m). The EPS ADP algorithm is then applied to these three data sets. Subsequently, the smoke mask from these three data sets is compared. In addition, an inhomogeneity (spatial variability) test, i.e., the standard deviation of $0.65\,\mu m$ ($0.67\,\mu m$ for VIIRS) in a box of 3×3 pixels is calculated from all original eMAS pixels, aggregated eMAS pixels and VIIRS pixels; the standard deviation is used to screen out any pixels potentially contaminated by residual clouds or large variability of surface features/reflectance.

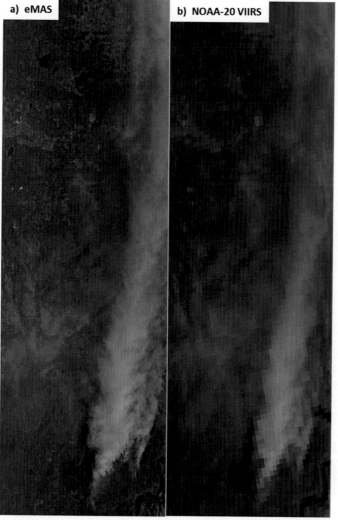

FIG. 10 True color (RGB) images of the Williams Flat fire, WA on August 7, 2019, from eMAS (20:39–20:45 UTC) (a) and NOAA-20 VIIRS (20:40 UTC) (b).

3. Comparison of ADP$_e$ with ADP$_v$

During FIREX-AQ 2019, a total of 76 wildfires were observed by eMAS flights (Dibb et al., 2020). The JPSS EPS ADP algorithm as discussed in Section 2 was first applied to eMAS observations for all flight tracks over wildfire events that produced visible smoke plumes. The results showed that for all flight tracks, ADP$_e$ was able to identify smoke-filled pixels; however, the coverage of ADP smoke-flagged pixels varied with the thickness of the plumes. For example, Fig. 11 shows ADP$_e$ for two extreme cases: one with a very small and thin smoke plume on August 6 observed by eMAS flight #19-910, track #19 (Fig. 11a and b), and one with a large thick smoke plume, observed by eMAS flight #19-912, track #19 (Fig. 11c and d). Fig. 11 shows that the ADP$_e$ smoke flags (red shading in Fig. 11b and d) generally coincided with the smoke plumes seen in the RGB images (Fig. 11a and c), indicating ADP$_e$ can capture the most densely smoke-filled pixels. However, there are differences in coverage for the two cases. For the small-scale smoke plumes in Fig. 11a and b, the ADP$_e$ algorithm only flags the thickest part of the smoke plume, whereas the thinner parts of the smoke plume, where surface features are visible in the RGB image, are missed by the ADP$_e$, which accounts for about 80% of the whole smoke plume. In contrast, for the large thick smoke case in Fig. 11c and d, nearly all of the smoke plume pixels evident in the RGB image are identified as smoke by the ADP$_e$, except for the portion that seems to have underlying clouds (bright white in features in Fig. 11c). These two cases demonstrate the challenge for smoke detection over land related to the large variability of the underlying surface when smoke plumes are thin. One of the assumptions for smoke detection over land is that the smoke plume is much more uniform than the underlying surface. When smoke is present, the spatial variability, i.e., the standard deviation of 0.65 µm, is smaller. Any locations with large variability values are considered as "no smoke" pixels or pixels contaminated by residual clouds.

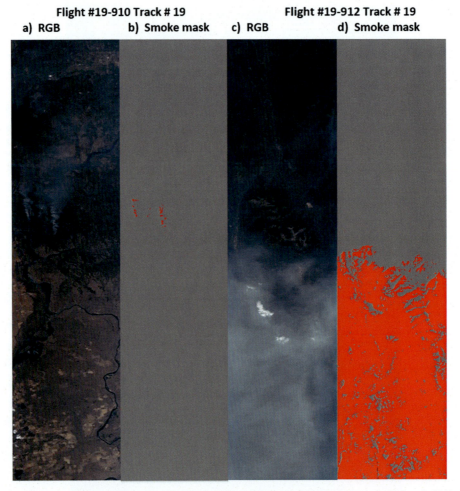

FIG. 11 True color (RGB) image (a) and ADP$_e$ smoke mask (b) for eMAS Flight #19-910, Track # 19 on August 6, 2019; RGB image (c) and ADP$_e$ smoke mask (d) for eMAS Flight #19-912 Track # 19 on August 15, 2019. In the ADP$_e$ images, smoke is represented by the red color and the gray color represents no smoke.

Besides spatial variability, the dynamics of smoke plumes also varies in temporal space. This phenomenon is demonstrated in Fig. 12, which shows three eMAS observations of the Williams Flats fire from flight tracks spanning about an hour and half on August 7. Fig. 12a and b show the flight track for 18:27 UTC, Fig. 12c and d are the observations for 19:36 UTC, and the final set of observations from 19:55 UTC are shown in Fig. 12e and f. The thick smoke plumes around fire sources are captured well by the ADP_e smoke mask (red shading in Fig. 12b, d and e) for all three eMAS flight tracks, and all exhibit a similar spatial pattern. However, there is a difference in ADP_e retrievals for downwind smoke plumes among the three flight tracks, where the smoke plumes are dispersed and thinner. For track #5 in Fig. 12a and b, most of the smoke plume is retrieved by ADP_e, with only a small portion missing for the thin smoke plume at the top part of the track. For track #8 in Fig. 12c and d, which was observed about 1 h after track #5, the smoke plume has become more diffuse and spread out, partially due to the difference in eMAS viewing angle. Fig. 12c and d indicates that the ADP_e smoke mask failed to retrieve more of the top part of the plumes in track #8 compared to track #5. As for track #9, Fig. 12e and f shows that the eMAS viewing angle at 19:55 UTC is significantly different from the other two earlier tracks, as indicated by the sun-glint over the river (white/silver colored features in Fig. 12e), which is not present in Fig. 12a or c. In Fig. 12f, the ADP_e accurately captured most of the smoke plumes in great detail, which was likely promoted by the favorable viewing geometry. To give an estimate on the percentage of the missed smoke detection, we used the reflectance at eMAS Band1 (0.47 μm) > 0.2 as a proxy for smoky pixels in the images. It is found that about 54%, 25%, and 70% of pixels with reflectance at 0.47 μm > 0.2 are identified as smoke for ADP_e at track #5, track #8, and track #9, respectively.

One of the benefits from a field campaign such as FIREX-AQ is the coordination of aircraft observations with satellite overpasses, which enables the evaluation of satellite-retrieved ADP. On August 7, one of the eMAS flight tracks over the Williams Flats fire occurred at 20:39–20:45 UTC, which was very close to the NOAA-20 overpass time. Fig. 13 shows the results from this analysis, including the smoke flag indicated by red shading from ADP_e, ADP_v, and ADP_{ev} (Fig. 13b, e and h), the corresponding RGB images (Fig. 13a, d, and g), and the standard deviation of the 0.65 μm band reflectance in a box containing 3 × 3 pixels (Fig. 13c, f and i). Comparing the ADP and RGB tiles in Fig. 13 indicates that there are no

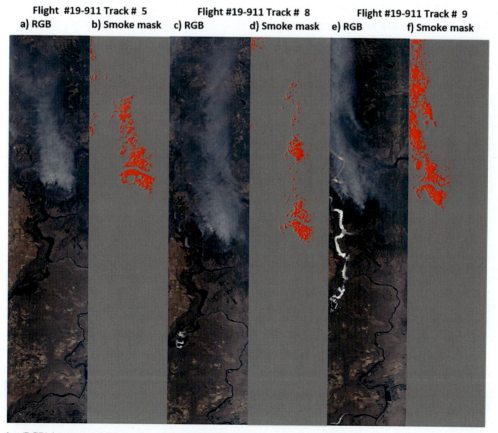

FIG. 12 True color (RGB) image and ADP_e smoke mask pairs for three tracks of eMAS Flight #19-911 over the Williams Flat fire, WA on August 7, 2019: Track # 5 (18:27 UTC, 20,201 MSL) (a and b), Track #8 (19:36 UTC, 20,319 MSL) (c and d), and Track #9 (19:55 UTC, 20,186 MSL) (e and f). In the ADP_e images, smoke is represented by the *red color*, and the *gray color* represents no smoke.

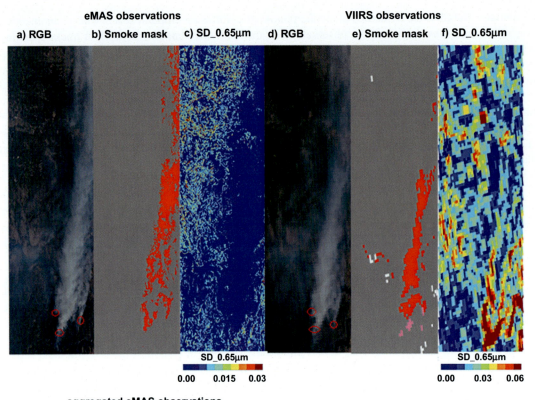
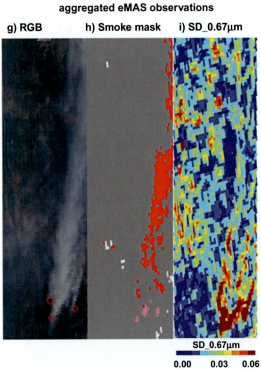

FIG. 13 Comparison of the true color (RGB) image, ADP smoke mask, and the standard deviation (SD) of 0.65 μm (0.67 μm for VIIRS) from eMAS observations in their original resolution (a, b, and c), the collocated NOAA VIIRS pixel-level observations from the NOAA-20 granule (20:39–20:45 UTC) (d, e, and f), and eMAS observations aggregated to VIIRS pixels (g, h, and i) for part of eMAS Flight # 19911, Track #12 over the Williams Flat fire, WA on August 7, 2019. The *red circles* on the RGB images indicate smoke sources.

false smoke detections in areas where smoke plumes are not visible in the RGB images. For areas covered by smoke, Fig. 13b shows that the extremely high native spatial resolution of eMAS produces ADP_e that resolves the smoke plumes in detail and with greater coverage compared to the ADP_v and ADP_{ev} at the coarser resolution of 750 m (Fig. 13e and h). The number of the detected smoke pixels (in eMAS original pixel resolution) is 150,112, 115,976, and 139,270, respectively, for ADP_e, ADP_v, and ADP_{ev}, which is about 7% (20%) difference between ADP_e and ADP_{ev} (ADP_v) and about 15% difference between ADP_{ev} and ADP_v. By comparing all three retrievals, the smoke plumes over the fire source points, which are marked with red circles in Fig. 13a, d and g, are clearly shown as smoke-filled pixels in all versions of ADP. Besides its finer spatial resolution, Fig. 13c shows that the spatial variability in a box of 3×3 pixels in the original resolution of eMAS is very small, with a value of <0.03 for the smoke-covered areas. In contrast, for ADP from the coarse resolution observations (Fig. 13d–i), the coverage of smoke plumes is much smaller and more compact. Larger variability is also indicated from the standard deviation of 0.65 μm in a box 3×3 pixels (Fig. 13f and i), especially over the edge of smoke plumes, which usually are associated with regions where smoke detection is missed. However, the very similar pattern of ADP and standard deviation of 0.65 μm indicate that the EPS ADP algorithm can produce similar results from two independent observations that have a similar spatial resolution.

Fig. 14a shows a scatterplot of the aggregated reflectance of eMAS band 3 (0.65 μm) versus the reflectance of VIIRS band M5 (0.67 μm) for eMAS flight track #5, shown in Fig. 12. Even though the central wavelength of the two sensors is slightly different, the reflectances from the bands agree well, with a correlation coefficient of 0.82; this result gives us confidence in comparing ADP derived from these two data sets. Fig. 14b shows the variability of the three data sets, represented by the standard deviation of 0.65 μm (0.67 μm for VIIRS) from these three data sets, plotted against the reflectance of VIIRS band M5 (0.67 μm). It indicates that for high reflectance values, which correspond mostly to pixels laden with smoke, the variability is much smaller for eMAS in its original resolution (blue empty triangles), leading to more smoke being detected by the ADP. The standard deviation values for the other two data sets at 750 m are very similar and display no significant dependence on the 0.67 μm reflectances. Among all the pixels shown in Fig. 14b, there are about 0.1%, 13%, 15% pixels above the threshold value (0.04) for ADP_e, ADP_{ev}, and ADP_v, indicating the amount of potential missing smoke pixels due to the spatial variability test. However, it should be noted that the spatial variability test is not the sole reason for missing true smoke detection. As shown in Section 2.2, the surface reflectance relationship (in Fig. 6) is based on VIIRS observations, it may contribute to some miss-detection of smoke, as seen in the difference between ADP_{ev} and ADP_v. But the difference between ADP_e and ADP_{ev} can be considered to be dominated by the difference in spatial variability associated with pixel resolution (50 m vs. 750 m), due to the fact that both are from eMAS observations.

4. Conclusions

In this chapter, we demonstrated the utility of field campaign measurements for validation of satellite ADP by using observations from eMAS onboard the NASA ER-2 aircraft during FIREX-AQ 2019. The JPSS EPS ADP algorithm, which is used to generate operational ADP from VIIRS on SNPP and NOAA-20, was first applied to eMAS measurements for flight tracks over wildfire smoke plumes. Results showed that ADP_e can detect moderate to thick smoke plumes, but detection of thin and diffuse smoke plumes is more difficult. Second, for eMAS flight tracks that coincided with satellite overpasses, eMAS pixels (50 m resolution) were spatially matched with VIIRS satellite pixels (750 m resolution). In general, detected smoke plumes are shown in much more detail at the eMAS original resolution and benefit from less pixel variability, indicated by the lower standard deviation of 0.65 μm in a box of 3×3 pixels. For both the eMAS pixels aggregated to and VIIRS observations at 750 m resolution, the detected smoke plume coverage is smaller by about 20% compared to those for the eMAS original resolution data, but the two data sets show good agreement, indicating that EPS ADP has a similar performance for two sets of independent observations with similar resolution. However, all three detection results suggest that the ability of smoke detection over land in the EPS ADP algorithm may be limited for thin or diluted smoke, since the surface variability plays a major role in misclassifying smoke-filled pixels as smoke free. Treating ADP_{ev} as truth, we calculated the POCD for ADP_v and find that the accuracy of VIIRS detection is at 85% for very thick smoke. This is similar to what we found to be the accuracy of VIIRS ADP when it was compared to CALIOP VFM and AERONET smoke classification. When ADP algorithm was applied to eMAS 50 m resolution data, the accuracy of the ADP_e product is found to be 93%. Only when either smoke plumes are thin and diffuse or viewing angles are not favorable, the ADP_v is found to have a lower accuracy (25% to 70%).

Field campaigns, such as FIREX-AQ, make measurements from multiple instruments on various airborne, ground-based, and space-borne platforms targeted at specific locations and events. These measurements complement each other spatially and spectrally, providing more complete observations to characterize the atmosphere and the underlying surface. Combining all measurements from these various platforms will benefit not only satellite-based retrievals but also retrievals from other platforms, such as retrievals from ground-based instruments and airborne instruments.

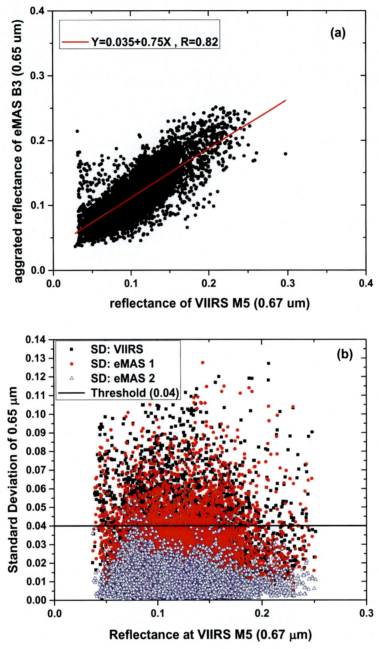

FIG. 14 Scatter plots of: the standard deviation of the aggregated reflectance of eMAS band 3 (0.65 μm) vs. the reflectance of VIIRS band M5 (0.67 μm) (a) and the standard deviation of the reflectance of eMAS band 3 (0.65 μm) and of VIIRS band M5 (0.67 μm) in a box of 3 × 3 pixels vs. the reflectance of VIIRS band M5 (0.67 μm) in a box of 3 × 3 VIIRS pixels (b). In plot (b), "eMAS1" *(red diamonds)* indicates eMAS observations aggregated to VIIRS pixels (750 m), "eMAS2" *(blue empty triangles)* indicates eMAS observations in their original resolution (50 m), and "VIIRS" *(black squares)* indicates VIIRS observations in their original resolution (750 m).

Acknowledgment

We gratefully acknowledge the crew, logistical personnel, science team, and science leadership who facilitated the FIREX-AQ mission for making valuable measurements available. We would like to thank Dr. Amy Huff (I.M. Systems Group, Inc.) for her editorial work on the manuscript, and Dr. Carlos Toledano (University of Valladolid) for his valuable comments.

Disclaimer

The scientific results and conclusions, as well as any views or opinions expressed herein, are those of the author(s) and do not necessarily reflect those of NOAA or the Department of Commerce.

References

Ackerman, S.A., 1997. Remote sensing aerosols using satellite infrared observations. J. Geophys. Res. 102, 17069–17079.

Ackerman, S.A., Strabala, K.I., 1994. Satellite remote sensing of H_2SO_4 aerosol using the 8–12 µm window region: application to Mount Pinatubo. J. Geophys. Res. 99, 18,639–18,649.

Al-Saadi, J., Szykman, J., Pierce, R.B., Kittaka, C., Neil, D., Chu, D.A., Remer, L., Gumley, L., Prins, E., Weinstock, L., Macdonald, C., Wayland, R., Dimmick, F., Fishman, J., 2005. Improving national air quality forecasts with weather satellite aerosol observations. Bull. Am. Meteorol. Soc. 86, 1249–1261.

Ba, R., Chen, C., Yuan, J., Song, W., Lo, S., 2019. SmokeNet: satellite smoke scene detection using convolutional neural network with spatial and channel-wise attention. Remote Sens. (Basel) 11, 1702.

Barton, I.J., Prata, A.J., Watterson, I.G., Young, S.A., 1992. Identification of the Mount Hudson volcanic cloud over SE Australia. Geophys. Res. Lett. 19 (12), 1211–1214.

Boroughani, M., Pourhashemi, S., Hashemi, H., Salehi, M., Amirahmadi, A., Asadi, M., Berndtsson, R., 2020. Application of remote sensing techniques and machine learning algorithms in dust source detection and dust source susceptibility mapping. Eco. Inform. 56, 101059. https://doi.org/10.1016/j.ecoinf.2020.101059.

Chacon, M., Quezada-Holguín, Y., Pablo, R.P., Cabrera, S., 2011. Dust storm detection using a neural network with uncertainty and ambiguity output analysis. In: Pattern Recognition. vol. 6718., ISBN: 978-3-642-21586-5, pp. 305–313.

Chrysoulakis, N., Cartalis, C., 2003. A new algorithm for the detection of plumes caused by industrial accidents, based on NOAA/AVHRR imagery. Int. J. Remote Sens. 24, 3353–3368.

Ciren, P., Kondragunta, S., 2014. Dust aerosol index (DAI) algorithm for MODIS. J. Geophys. Res.-Atmos. 119 (8), 4770–4792. https://doi.org/10.1002/2013JD020855.

Darmenov, A., Sokolik, I.N., 2005. Identifying the regional thermal-IR radiative signature of mineral dust with MODIS. Geophys. Res. Lett. 32, L16803. https://doi.org/10.1029/2005GL023092.

Dibb, J.E., Crawford, J.H., Schwarz, J.P., Warneke, C., Fahey, D.W., Lefer, B.L., Ryerson, T.B., 2020. Overview of the Fire Influence on Regional to Global Environments and Air Quality (FIREX-AQ) 2019 Field Campaign. American Geophysical Union. Fall Meeting.

Dunion, J.P., Velden, C.S., 2004. The impact of the Saharan air layer on Atlantic tropical cyclone activity. Bull. Am. Meteorol. Soc. 90, 353–365.

Evan, A.T., Heidinger, A.K., Pavolonis, M.J., 2006. Development of a new over-water advanced very high-resolution radiometer dust detection algorithm. Int. J. Remote Sens. 27, 3903–3924. https://doi.org/10.1080/01431160600646359.

Giles, D.M., Holben, B.N., Eck, T.F., Sinyuk, A., Smirnov, A., Slutsker, I., Dickerson, R.R., Thompson, A.M., Schafer, J.S., 2012. An analysis of AERONET aerosol absorption properties and classifications representative of aerosol source regions. J. Geophys. Res. 117, D17203. https://doi.org/10.1029/2012JD018127.

Holben, B.N., Eck, T.F., Slutsker, I., Tanré, D., Buis, J.P., Setzer, A., Vermote, E., Reagan, J.A., Kaufman, Y., Nakajima, T., Lavenu, F., Jankowiak, I., Smirnov, A., 1998. AERONET—a federated instrument network and data archive for aerosol characterization. Remote Sens. Environ. 66, 1–16.

Holben, B.N., et al., 2001. An emerging ground-based aerosol climatology: aerosol optical depth from AERONET. J. Geophys. Res. 106 (D11), 12067–12097. https://doi.org/10.1029/2001JD900014.

Hsu, N.C., Robinson, W.D., Bailey, S.W., Werdell, P.J., 2000. The description of the SeaWiFS absorbing aerosol index. In: SeaWiFS NASA Technical Memorandum 2000-206892. vol. 10, pp. 3–5.

Hsu, N.C., Tsay, S.-C., King, M.D., Herman, J.R., 2004. Aerosol properties over bright reflecting source regions. IEEE Trans. Geosci. Remote Sens. 42 (3), 557–569. https://doi.org/10.1109/TGRS.2004.824067.

Hsu, N.C., Tsay, S.-C., King, M.D., Herman, J.R., 2006. Deep blue retrievals of Asian aerosol properties during ACE-Asia. IEEE Trans. Geosci. Remote Sens. 44 (11), 3180–3195. https://doi.org/10.1109/TGRS.2006.879540.

IPCC, 2007. In: Solomon, S., Qin, D., Manning, M., Chen, Z., Marquis, M., Averyt, K.B., Tignor, M., Miller, H.L. (Eds.), Climate Change 2007: The Physical Science Basis. Contribution of Working Group I to the Fourth Assessment Report of the Intergovernmental Panel on Climate Change. Cambridge University Press, Cambridge, United Kingdom and New York, NY, USA.

Ismanto, H., Hartono, H., Marfai, M.A., 2019. Smoke detections and visibility estimation using Himawari_8 satellite data over Sumatera and Borneo Island Indonesia. Spat. Inf. Res. 27, 205–216.

Joint Polar Satellite System (JPSS), 2016a. Aerosol Detection Product Algorithm Theoretical Basis Document (ATBD), April 15. https://www.star.nesdis.noaa.gov/jpss/documents/ATBD/ATBD_EPS_Aerosol_ADP_v1.1.pdf.

Joint Polar Satellite System (JPSS), 2016b. EPS Aerosol Optical Depth (AOD) Algorithm Theoretical Basis Document (ATBD), June 28. https://www.star.nesdis.noaa.gov/jpss/documents/ATBD/ATBD_EPS_Aerosol_AOD_v3.0.1.pdf.

Kaufman, Y., Koren, I., Remer, L., Rosenfeld, D., Rudich, Y., 2005. The effect of smoke, dust, and pollution aerosol on shallow cloud development over the Atlantic Ocean. Proc. Natl. Acad. Sci. U. S. A. 102, 11207–11212. https://doi.org/10.1073/pnas.0505191102.

Kim, K.M., Lau, W.K.-M., Sud, Y.C., Walker, G.K., 2010. Influence of aerosol-radiative forcings on the diurnal and seasonal cycles of rainfall over West Africa and Eastern Atlantic Ocean using GCM simulations. Climate Dynam. 26. https://doi.org/10.1007/s00382-010-0750-1.

King, M.D., Menzel, W.P., Grant, P.S., Myers, J.S., Arnold, G.T., Platnick, S.E., Gumley, L.E., Tsay, S.-C., Moeller, C.C., Fitzgerald, M., Brown, K.S., Osterwisch, F.G., 1996. Airborne scanning spectrometer for remote sensing of cloud, aerosol, water vapor and surface properties. J. Atmos. Oceanic Tech. 13, 777–794.

Kotchenova, S.Y., Vermote, E.F., 2007. Validation of a vector version of the 6S radiative transfer code for atmospheric correction of satellite data. Part II. Homogeneous Lambertian and anisotropic surfaces. Appl. Optics 46 (20), 4455–4464.

Kotchenova, S.Y., Vermote, E.F., Matarrese, R., Klemm Jr., F.J., 2006. Validation of a vector version of the 6S radiative transfer code for atmospheric correction of satellite data. Part I: path radiance. Appl. Optics 45 (26), 6762–6774.

Krawchuk, M.A., Moritz, M.A., Parisien, M.A., Van Dorn, J., Hayhoe, K., 2009. Global pyrogeography: the current and future distribution of wildfire. PLoS One 4, e5102. https://doi.org/10.1371/journal.pone.0005102.

Lee, J., Shi, Y.R., Cai, C., Ciren, P., Wang, J., Gangopadhyay, A., Zhang, Z., 2021. Machine learning based algorithms for global dust aerosol detection from satellite images: inter-comparisons and evaluation. Remote Sens. (Basel) 13, 456. https://doi.org/10.3390/rs13030456.

Legrand, M., Cautenet, G., Buriez, J.-C., 1992. Thermal impact of Saharan dust over land, part II: application to satellite IR remote sensing. J. Appl. Meteorol. 31, 181–193.

Legrand, M., Plana-Fattori, A., N'Doume, C., 2001. Satellite detection of dust using IR imagery of Meteosat 1. Infrared difference dust index. J. Geophys. Res. 106, 18,251–18,274.

Li, X., Song, W., Lian, L., Wei, X., 2015. Forest fire smoke detection using back-propagation neural network based on MODIS data. Remote Sens. (Basel) 7, 4473–4498.

Li, Z., Khananian, A., Fraser, R.H., Cihlar, J., 2001. Automatic detection of fire smoke using artificial neural networks and threshold approaches applied to AVHRR imagery. IEEE Trans. Geosci. Remote Sens. 39, 1859–1870.

Lu, X., Zhang, X., Li, F., Cochrane, M.A., 2019. Investigating smoke aerosol emission coefficients using MODIS active fire and aerosol products—a case study in the CONUS and Indonesia. Eur. J. Vasc. Endovasc. Surg. 124, 1413–1429.

Lu, X.M., Zhang, X.Y., Li, F.J., Cochrane, M.A., Ciren, P., 2021. Detection of fire smoke plumes based on aerosol scattering using VIIRS data over global fire-prone regions. Remote Sens. (Basel) 13, 2. https://doi.org/10.3390/rs13020196.

Marlier, M.E., DeFries, R.S., Kim, P.S., Koplitz, S.N., Jacob, D.J., Mickley, L.J., Myers, S.S., 2015. Fire emissions and regional air quality impacts from fires in oil palm, timber, and logging concessions in Indonesia. Environ. Res. Lett. 10, 085005.

McNamara, D., Stephens, G., Ruminski, M., Kasheta, T., 2004. The Hazard mapping System (HMS)—NOAA'S multi-sensor fire and smoke detection program using environmental satellites. In: Conference on Satellite Meteorology and Oceanography.

Mota, B., Wooster, M.J., 2018. A new top-down approach for directly estimating biomass burning emissions and fuel consumption rates and totals from geostationary satellite fire radiative power (FRP). Remote Sens. Environ. 206, 45–62.

Nelson, D.L., Garay, M.J., Kahn, R.A., Dunst, B.A., 2013. Stereoscopic height and wind retrievals for aerosol plumes with the MISR INteractive eXplorer (MINX). Remote Sens. (Basel) 5, 4593–4628.

Nicholson, S.E., 2000. Land surface processes and Sahel climate. Rev. Geophys. 38, 117–139.

Prata, A.J., 1989. Observations of volcanic ash clouds in the 10-12-MU-M window using AVHRR/2 data. Int. J. Remote Sens. 10 (4–5), 751–761.

Price, O.F., Purdam, P.J., Williamson, G.J., Bowman, D.M., 2018. Comparing the height and area of wild and prescribed fire particle plumes in south-East Australia using weather radar. Int. J. Wildland Fire 27, 525–537.

Prospero, J.M., Lamb, P.J., 2003. African droughts and dust transport to the Caribbean: climate change implications. Science 302, 1024–1027.

Rolph, G.D., Draxler, R.R., Stein, A.F., Taylor, A., Ruminski, M.G., Kondragunta, S., Zeng, J., Huang, H., Manikin, G., McQueen, J.T., Davidson, 2009. Description and verification of the NOAA smoke forecasting system: the 2007 fire season. Weather Forecast. 24, 361–378. https://doi.org/10.1175/2008WAF2222165.1.

Schoennagel, T., Balch, J.K., Brenkert-Smith, H., Dennison, P.E., Harvey, B.J., Krawchuk, M.A., Mietkiewicz, N., Morgan, P., Moritz, M.A., Rasker, R., Turner, M.G., Whitlock, C., 2017. Adapt to more wildfire in western North American forests as climate changes. Proc. Natl. Acad. Sci. U. S. A. 114, 4582–4590.

Schroeder, W., Giglio, L., 2017. Visible Infrared Imaging Radiometer Suite (VIIRS) 750 m Active Fire Detection and Characterization Algorithm Theoretical Basis Document 1.0.

Schroeder, W., Ruminski, M., Csiszar, I., Giglio, L., Prins, E., Schmidt, C., Morisette, J., 2008. Validation analyses of an operational fire monitoring product: The Hazard Mapping System. Int. J. Remote Sens. 29, 6059–6066.

Tosca, M.G., Randerson, J.T., Zender, C.S., Nelson, D.L., Diner, D.J., Logan, J.A., 2011. Dynamics of fire plumes and smoke clouds associated with peat and deforestation fires in Indonesia. J. Geophys. Res. 116, 1–14.

Wang, W., Qu, J.J., Hao, X., Liu, Y., Sommers, W.T., 2007. An improved algorithm for small and cool fire detection using MODIS data: a preliminary study in the southeastern United States. Remote Sens. Environ. 108, 163–170.

Warneke, C., Schwarz, J.P., Ryerson, T., Crawford, J., Dibb, J., Lefer, B., Roberts, J., Trainer, M., Murphy, D., Brown, S., Brewer, A., Gao, R.-S., Fahey, D., 2018. Fire Influence on Regional to Global Environments and Air Quality (FIREX-AQ): A NOAA/NASA Interagency Intensive Study of North American Fires: 2018. National Oceanic and Atmospheric Administration, Boulder, CO, USA. Available online https://www.esrl.noaa.gov/csd/projects/firex/whitepaper.pdf. (Accessed 14 June 2020).

Westerling, A.L., Bryant, B.P., Preisler, H.K., Holmes, T.P., Hidalgo, H.G., Das, T., Shrestha, S.R., 2011. Climate change and growth scenarios for California wildfire. Clim. Change 109, 445–463. https://doi.org/10.1007/s10584-011-0329-9.

Williamson, G.J., Price, O.F., Henderson, S.B., Bowman, D.M., 2013. Satellite-based comparison of fire intensity and smoke plumes from prescribed fires and wildfires in South-Eastern Australia. Int. J. Wildland Fire 22, 121–129.

Xie, Y., Qu, J., Xiong, X., Hao, X., Che, N., Sommers, W., 2007. Smoke plume detection in the eastern United States using MODIS. Int. J. Remote Sens. 28, 2367–2374.

Yoshioka, M., Mahowald, N.M., 2007. Impact of desert dust radiative forcing on Sahel precipitation: relative importance of dust compared to sea surface temperature variations, vegetation changes, and greenhouse gas warming. J. Climate 20, 1445–1467.

Zhao, T.X.-P., Ackerman, S., Guo, W., 2010. Dust and smoke detection for multi-channel imagers. Remote Sens. (Basel) 2, 2347–2368. https://doi.org/10.3390/rs2102347.

Chapter 23

Downburst monitoring and prediction studies

Kenneth L. Pryor
NOAA/NESDIS Center for Satellite Applications and Research (STAR), College Park, MD, United States

Chapter outline

1. Introduction	411	4.1 27 April 2020 South Texas severe thunderstorm downbursts	416
2. Theoretical background of severe convective windstorms: Genesis and evolution	412	4.2 29 June 2012 North American derecho	421
3. Field measurement application methodology	414	5. Summary	427
4. Case studies of field measurement applications	416	Acknowledgments	428
		References	428

And suddenly a sound came from heaven like the rush of a mighty wind, and it filled all the house where they were sitting.

Acts 2:2 Revised Standard Version Bible

1. Introduction

Downbursts are strong downdrafts that induce an outburst of damaging winds at or near the ground, and a microburst as a very small downburst with an outflow diameter of less than 4 km and a lifetime of less than 5 min (Fujita, 1985; Wakimoto, 1985). The dangers posed by convective storm-generated downbursts have been extensively documented. Since 2000, the National Transportation Safety Board (NTSB) has recorded 48 downburst-related accidents over CONUS with 42 fatalities (National Transportation Safety Board 2021) that involved personal or instructional aircraft. Severe windstorms (i.e., widespread convective wind gusts $>25.7 \text{ m s}^{-1}$ (50 kt)) resulting from mesoscale convective systems (MCS) cause significant disruption to society, including widespread power outages, tree and structural damage, and transportation accidents that affect multistate regions and metropolitan areas along their track. Among them, a derecho, defined as a long-lived, widespread severe convective windstorm, is composed of numerous downbursts that are organized into clusters or families of clusters. Derechos can produce winds above hurricane force along a track that may exceed several hundred ($\sim$400) kilometers. Between 1987 and 2002, severe convective windstorms resulted in a total property loss of over $3 billion in the United States, with an average loss per event of $96 million. Also, between 1986 and 2003, severe convective windstorms were responsible for a total of 153 deaths and 2605 injuries, proving to be more deadly and hazardous than the low-end (F-0/F-1 intensity) tornado outbreaks that occurred during the same period and resulted in only 71 deaths (Ashley and Mote, 2005). Because these events are severe, it is important to understand the factors that lead to the downbursts and utilize all available observations to monitor and forecast their development.

Proctor (1989) and Pryor (2015) noted that convective windstorm potential has been traditionally expressed as a grouping of stability parameters relevant for downburst generation. These include the lower-to-mid-tropospheric temperature and equivalent potential temperature (theta-e) lapse rates, vertical relative humidity differences, and the amount of convective available potential energy (CAPE) in the troposphere. Some factors increase the likelihood of severe convective winds, which are (1) an elevated mixed layer that promotes instability by generating powerful storm updrafts and downdrafts (Banacos and Ekster, 2010) and (2) a rear-inflow jet into an MCS (Smull and Houze Jr., 1987; Weisman, 1992) which channels unsaturated mid-tropospheric air into the leading convective storm line. The establishment of an elevated, ascending front-to-rear flow originating from deep, moist convection, overlying a strong and deep outflow-induced cold pool has been found to generate and sustain a robust rear inflow jet (Weisman, 1992). Other factors documented by Proctor

(1989) and Srivastava (1985) can reduce the likelihood of severe convective winds, such as the presence of a lower-tropospheric temperature inversion and a surface-based layer of unsaturated air that reduces virtual temperature.

Meteorological satellite measurement data, especially brightness temperature measured at the surface and convective cloud tops, have been previously exploited to study the troposphere's thermodynamic structure and convective storms' physical structure. Fujita and Wakimoto ("FW," 1981) demonstrated one of the earliest studies that identified infrared imagery from geostationary satellites showing definite cloud-top signatures associated with large and robust downbursts on the ground. Ellrod (1989) first applied geostationary sounder instrument data to the study of downburst potential assessment for the August 2, 1985, Dallas-Ft. Worth (DFW), Texas microburst storm. Since then, new technology and instrumentation have improved monitoring severe windstorms.

This chapter presents a discussion of the science of operational forecasting of severe windstorms through examples of employing new satellite and ground-based microwave and vertical wind profile data. Accordingly, this chapter is organized as follows: Section 2 is a detailed background of severe convective windstorm theory that includes a discussion of windstorm genesis and evolution. Section 3 is a summary of instrumentation and measurement application methodology. Finally, as an example of the coordinated use of surface- and satellite-based observational instrumentation, Section 4 presents case studies of downburst events from local to regional scale. The objective of this instruction is to build a multistep procedure for operational convective storm downburst monitoring and prediction.

2. Theoretical background of severe convective windstorms: Genesis and evolution

The outline of convective windstorm theory begins with the requirement that severe storms are highly organized. Construction of a model of severe convective winds from a simplified vertical momentum equation is derived from Newton's 2nd Law and expressed in Eqs. (1a), (1b):

$$a = F/m = (-1/\rho)(dp/dz) + g\left(T_{ve} - T_{vp}\right)/T_{ve} \tag{1a}$$

$$a = F/m = (-1/\rho)(dp/dz) + g\left(\rho_o - \rho_f\right)/\rho_o \tag{1b}$$

where a is the acceleration, ρ is the density of the medium, p is the pressure, z is the altitude in the atmosphere, and T_{ve} and T_{vp} are virtual temperatures of the environment (atmosphere) and the parcel, respectively.

Downdraft initiation proceeds as a departure from hydrostatic equilibrium. For a volume of air with a high concentration of ice phase precipitation that develops within a convective storm, the resultant force F on the precipitation volume is downward and imparts negative buoyancy. This physical basis can be extended to more complex convective systems in which updrafts, downdrafts, and outflow foster a rear inflow jet's development.

The standard NOAA/National Weather Service (NWS) definition of a severe thunderstorm includes damaging winds with gusts of $26\,\mathrm{m\,s^{-1}}$ (50 kt) or greater and hail with a diameter of 2.5 cm (1 in.) or greater. Severe thunderstorms are most identifiable in weather radar imagery, in which a large concentration of ice-phase precipitation within a volume results in high reflectivity resulting from increased backscattering. Downdraft severity is governed by phase change and the loading of ice-phase precipitation. Loading, in effect, refers to the mass of a collection of hydrometeors. When gravity is imposed upon the volume of hydrometeors, downward acceleration occurs due to precipitation's weight.

Numerous studies through the 1980s and 1990s provided observational analyses and conceptual models for severe convective wind generation that addressed environmental and storm microphysical and dynamic attributes. A prototypical conceptual model of a deep moist convective (DMC) storm is shown in Fig. 1. Fujita and Wakimoto ("FW," 1981) noted that damaging winds induced by thunderstorms were classified into a tornado and straight-line winds and further identified a subclassification of straight-line winds into two categories: the downburst and the gust front. FW formulated the definition of the downburst as stated at the beginning of Section 1. This study identified that infrared imagery from geostationary satellites showed definite cloud-top signatures associated with large and strong downbursts on the ground.

Srivastava (1987) found that precipitation in the form of ice increases the convective downdraft intensity. This effect increases with precipitation content and the stability of the environmental lapse rate of temperature. The power of the downdraft also increases in proportion to the relative concentration of smaller particles. Condensate loading (Srivastava, 1987), sometimes combined with subsaturated air entrainment in the storm middle level (Knupp, 1989), initiates the convective downdraft. The subsequent melting of frozen hydrometeors and subcloud evaporation of liquid precipitation, in conjunction with precipitation loading, result in the cooling and negative buoyancy that accelerate the downdraft in the unsaturated layer (Srivastava, 1987). The melting of ice-phase precipitation, subsequent evaporative cooling, and the resulting downdraft strength are enhanced by sizeable liquid water content and the related water surface available for evaporation. A large

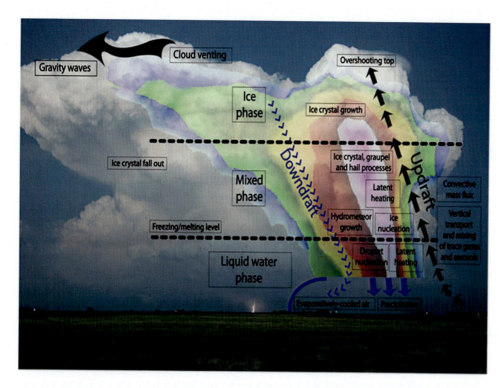

FIG. 1 Conceptual model of a deep convective storm with the potential to generate intense downdrafts and damaging downburst winds. *(Courtesy of Rob Seigel and Susan C. van den Heever, Global Precipitation Measurement (GPM, available online at https://gpm.nasa.gov/GPM, accessed on 7 July 2020)).*

lapse rate maintains negative buoyancy as the downdraft descends in the subcloud layer (Srivastava, 1987). As the lapse rate's stability is increased, higher precipitation contents, precipitation in the form of ice, and relatively higher concentrations of small precipitation particles are required to force an intense downdraft. At this point in the downdraft initiation process, potential energy resulting from temperature deficit between the precipitation-infused parcel and the ambient environment is converted to downward air parcel motion, which, collectively, comprises the convective downdraft. As the lapse rate becomes even more stable, only wet downbursts having substantial precipitation in the form of ice are possible. A downburst can be driven solely below the cloud base where melting and evaporation of precipitation and precipitation loading below the cloud base are sufficient to produce wet downbursts.

Knupp (1989, 1996) refined the understanding of the downburst generation's physical and dynamic processes. The author noted that low-level downdrafts are closely controlled by the arrival of precipitation at low levels. In the storm middle levels, air flows quasi-horizontally around the updraft flanks and converges into the downshear flank, referred to as the wake. Within the wake region, where entrainment reduces positive buoyancy and associated updraft strength, precipitation at middle levels (where it is grown most effectively) is then allowed to descend to lower levels. The intrusion of drier air into the wake's precipitation region also enhances the evaporation/sublimation process. Diabatic cooling from melting and evaporation is most effective at levels below the melting level. Convergence within the downshear wake is thus instrumental in transporting precipitation into the downshear flank. Therefore, a comprehensive understanding of the downdraft initiation process is closely related to the precipitation initiation and transport process within clouds and is observable in passive MW imagery, as shown in the following case studies. Such processes depend not only on vertical profiles of temperature and moisture but also on vertical environmental wind profiles. Knupp (1996) identified the protrusion echo produced by settling hydrometeors from a line of weak updraft that formed in association with low-level confluence located east of the storm core. The protrusion appears to be partly forced by existing mesoscale convergence, while the updraft within its lower levels represents the weak low-level ascent along the up-down downdraft branch. In this regard, the protrusion is indirectly connected to the strong core downdraft. Initial bowing of the echo (Przybylinski, 1995) is associated with the early microburst activity, a characteristic observed in other case studies. The inference of downburst occurrence can be successfully applied by the synergistic use of satellite-based passive MW and ground-based Doppler radar data and imagery.

From Weisman et al. (1988) as a departure point, Weisman (1992) explored the role of vertical wind shear and buoyancy in the generation of a rear inflow jet and visualized the associated conceptual model of this process. Weisman (1992) noted that rear inflow is generated in response to the development of an upshear-tilted updraft, as the horizontal buoyancy gradients along the back edge of the expanding system create a circulation that draws midlevel air in from the rear. The rear inflow jet system can take two forms, descending or elevated. For a descending-jet system, the convective circulation is characterized by an updraft current that ascends gradually above a spreading surface cold pool, with light-to-moderate convective and stratiform rainfall extending well behind the leading edge of the cold pool. This structure is often associated with a decaying system. The gust-front lifting is not strong or deep enough to regenerate new convective cells, and the mesoscale circulation slowly weakens. However, for an elevated-jet system, the circulation is dominated by strong, erect updrafts along the leading edge of the surface cold pool, with the updraft current spreading rapidly rearward above 7–8 km above ground level. Moderate-to-heavy convective rainfall exists at the system's leading edge, with lighter rainfall extending to the rear. This structure tends to be longer lived than the descending-jet case, as the deeper gust-front lifting regularly regenerates strong convective cells. The rear-inflow jet represents a new, potentially significant horizontal vorticity source that must be included when diagnosing various circulation sources' relative importance. Specifically, a rear-inflow jet that descends and spreads along the surface is characterized by the same sign of horizontal vorticity generated by the cold pool, thereby accentuating the cold pool circulation.

In contrast to a descending jet, an elevated rear-inflow jet is characterized by the opposite sign of horizontal vorticity generated by the cold pool (up to jet level), thereby accentuating the ambient vertical shear effects. Since significant rear-inflow characteristically develops after the cold-pool circulation overwhelms the ambient shear, a surface jet's development reinforces the upshear-tilting process that tends to weaken the system. However, an elevated rear-inflow jet's development reverses this process, promoting powerful, upright convective cells along the cold pool's leading edge. Johns (1993) built on the basis established by previous observational and modeling studies of environmental conditions associated with the development and maintenance of bow echo-induced damaging winds, focused on parameters related to storm outflow and updraft strengths. Specifically, wind speeds and relative humidity values in the mid-levels (related to outflow strength) and instability (related to updraft strength) were examined. The results indicated that these parameters exhibit a wide range of values when considering all bow echo situations in which damaging winds are reported. Further, combinations of wind speeds in the mid-levels and instability tend to vary with the season and the synoptic situation. For example, as detailed by Johns (1993) and Moller (2001), when powerful winds are present in the mid-levels, bow echo development has been observed in only marginally unstable environments. Bow echo events associated with the powerful wind-marginal instability combination typically occur with strong, rapidly moving low-pressure systems ("dynamic" synoptic pattern) in the colder months of the year. On the other hand, events associated with the relatively weak wind-extreme instability combination typically occur along a quasi-stationary thermal boundary in relatively stagnant weather regimes ("warm season" synoptic pattern) in the late spring or summer. Many bow echo wind events are associated with wind-instability combinations between the extremes. Some of these events are related to synoptic patterns that do not sufficiently match either prototypical pattern.

As will be demonstrated in the following case studies, passive microwave (MW) observations from polar-orbiting satellites are instrumental in identifying convective storms with dense ice-phase precipitation cores that can generate intense downdrafts by the processes of loading, melting, and evaporation. At a 12.5 km horizontal resolution, the 91 GHz channel on the Special Sensor Microwave Imager Sounder (SSMIS) effectively views the rainband structure in which both 91 GHz polarization (horizontal, "H" and vertical, "V") generate images where cirrus cloud decks are transparent. The SSMIS is a conical scanner with a 53.1 degrees zenith angle, swath width 1700 km, and scan rate of 31.9 scans/min, which equals 12.5 km/scan. Ferraro et al. (1998) introduced an expression to quantify brightness temperature derived from Schwarzchild's equation:

$$TB = T_u + \tau \cdot \left[\varepsilon \cdot T_s + (1 - \varepsilon) \cdot T_d \right] \tag{2}$$

where T_u is the upwelling atmospheric emission, τ is the transmittance, ε is the emissivity, T_s is the surface temperature, and T_d is the downwelling atmospheric emission. Scattering by large precipitation particles, especially by graupel, hail, and ice crystal aggregates (i.e., snow) above the freezing level, causes 91 GHz brightness temperatures (TB) to be low, referred to as a TB depression (Ferraro et al. (2015), Laviola et al. (2020)). Thus, convective rainbands tend to have very low TB, often below 200 K. Time trends in cloud top TB, as measured in MW spectrum window channels, and spatial patterns of cloud top TB, especially the geometry of TB gradients, can be exploited to infer downburst generation.

3. Field measurement application methodology

Surface-based measurements in the radio and microwave regions of the electromagnetic spectrum provide important environmental parameters for monitoring atmospheric stability and mesoscale and microphysical processes associated with convective storm development. Over CONUS, traditional data sets applied to both operational downburst monitoring and prediction, as well as product validation, include surface-based observations of atmospheric parameters (i.e., temperature, humidity, wind speed/direction, sky condition, precipitation accumulation, etc.) from NWS/FAA aviation routine meteorological reports (METAR) stations, mesonetwork (mesonet) stations, radiosonde observations (RAOBs; see Chapter 2), and meteorological Doppler radar reflectivity and velocity measurements. The highest quality networks in CONUS include the Oklahoma Mesonet and West Texas Mesonet (Brock et al., 1995; Schroeder et al., 2005). These networks have a spatial density of weather stations suitable for observing patterns such as outflow boundaries and temperature perturbations resulting from convective storms. In effect, surface weather observation and analysis represent a primary important step in convective storm diagnosis process.

An ultra-high frequency (UHF) boundary layer profiler (BLP) is capable of identifying mesoscale features such as low-level jets, rear-inflow jets, and convective storm outflow and can supplement the Next Generation Weather Radar (NEXRAD) velocity azimuth display wind profile (VWP) product at longer distances (>30 km) from the adjacent NEXRAD site. The BLP is a Doppler radar system operating at a frequency of 915 MHz that provides high-resolution observations with enhanced sensitivity to hydrometeors (Ecklund et al., 1988) in which backscattered signals from turbulence-induced refractive index variations are detected by the radar (Martner et al., 1993). The BLP retrieves horizontal wind speed and direction up to an elevation of four kilometers above ground level at a vertical resolution of 60 m. In selected regions of CONUS, an expanded data set incorporates lower tropospheric vertical wind profile data. For example, in this chapter, we will utilize a BLP in Beltsville, Maryland (United States), which is near the capital, Washington, DC. The BLP at Howard University Beltsville Campus (HUBC) provides the most representative wind conditions in the Washington, DC metropolitan area. These BLPs are component of the Cooperative Agency Profilers (CAP) network, which compiles data in real-time, applies quality control, and distributes the data online. Current and archived BLP wind data are available on the CAP website: https://madis-data.ncep.noaa.gov/cap/

In addition, vertical temperature and moisture sounding data sets generated by the surface-based microwave radiometer profiler (MWRP) provide routine monitoring of thermodynamic patterns in both the preconvective and storm environments (Westwater et al., 2005). In the Washington, DC–Baltimore, Maryland corridor, the HUBC MWRP, manufactured by Radiometrics Corporation, observes atmospheric brightness temperatures in 12 frequency bands from 22 to 59 GHz and retrieves temperature and humidity soundings up to 10 km height with a vertical resolution of 50 m below 500 m AGL and a resolution of 100 m between 500 and 2000 m AGL. The MWRP exploits the 30 to 50 GHz transmission window to retrieve water vapor profiles, while exploiting the absorption band near 60 GHz for temperature sensing. In addition, the MWRP can obtain retrievals at zenith and a 15° elevation angle (Cimini et al., 2015). The HUBC MWRP employs the neural network (NN) inversion method of retrieval, as described in Cimini et al. (2015), trained with a large data set of profiles generated from historical data sets of operational radiosondes. Vertical temperature and humidity profiles are often applied to calculate CAPE, temperature lapse rates, and other atmospheric stability indices to determine the presence of conditional instability (CI) and potential instability. The parcel choice for CAPE computation is an important consideration. The most unstable parcel CAPE (MUCAPE) was selected due to its universality and versatility as an estimator of positive buoyancy for both elevated and surface-based convection (Bunkers and Klimowski, 2002), and thus applicable to the range of thermodynamic environments observed during the severe wind events. This chapter will demonstrate the thunderstorm downburst potential applications of the microburst windspeed potential index (MWPI, Pryor, 2015) as calculated from MWRP and satellite sounding data sets.

In the next section, we present two case studies, one for April 2020 in South Texas and another the June 2012 North American Derecho. These studies were selected because they (1) demonstrate the physical process of downburst generation described in Section 2 and they were (2) observed simultaneously by the microwave sensors onboard polar-orbiting meteorological satellites, vertical sounding profiles generated from the Infrared Atmospheric Sounding Interferometer (IASI) and the NOAA Unique Combined Atmospheric Processing System (NUCAPS), NEXRAD, and BLPs. NUCAPS (Nalli et al., 2020) is a NOAA enterprise algorithm that retrieves atmospheric profile environmental data records and is described in more detail in Chapters 14 and 18. The IASI instrument and its applications for vertical atmospheric sounding are also highlighted in Chapter 18. Data collection, processing, and visualization follow the methodology of Pryor (2015, 2017). We use microwave sensors on Meteorological Operational (METOP) and Defense Meteorological Satellite Program (DMSP) satellites. DMSP Special Sensor Microwave Imager Sounder (SSMIS) and METOP Microwave Humidity Sounder (MHS) 89–91 GHz window channel data sets were obtained from the NOAA Comprehensive Large Array-data Stewardship System (CLASS) and the EUMETSAT Data Centre, respectively. Dual-polarized 91 GHz brightness temperature data sets allow for the calculation of polarization-corrected temperature (Liu et al., 1995) as presented in the study of the April 2020 South Texas downburst case. Vertical temperature and wind profile data, up to 5 km above ground level, from the CAP network are applied to further study the favorable environment for severe convective storm winds.

The Next Generation Weather Radar (NEXRAD) level-II reflectivity and differential reflectivity factor ZDR are obtained from the National Center for Environmental Information (NCEI) and used to verify that observed wind gusts are associated with downbursts originating from high reflectivity factor storms and are not associated with other types of convective wind phenomena (i.e., gust fronts). Plan-view images of radar reflectivity and ZDR are constructed from the lowest elevation angle scan (0.46 degrees). An additional application of radar reflectivity factor imagery is to infer microscale physical properties of downburst-producing convective storms. Differential reflectivity factor is employed in case studies to analyze the vertical precipitation composition in convective storms and thereby indicate the presence of graupel and hail and discuss their role in the enhancement of convective downdrafts. Particular reflectivity signatures, such as bow echoes (Przybylinski, 1995) and protrusion echoes (Knupp, 1996), are effective indicators of downburst occurrence. Downburst occurrence can be further confirmed by calculating a surface ΔT value, where $\Delta T \equiv T(\text{downburst}) - T(\text{ambient})$ and represents the peak temperature departure from ambient at ground level (Proctor, 1989). ΔT can therefore serve as a proxy variable for the surface density perturbation through the ideal gas law. In summary, a comprehensive approach of observational data analysis involves both surface- and satellite-based instrumentation. Because this approach utilizes operational products available to weather service forecasters, it can feasibly be used for monitoring and forecasting downburst occurrence. Compared to other ground-based microwave imagery sources, such as Doppler radar, spatial patterns in TB can also infer airflow characteristics and circulation patterns surrounding the convective storm of interest.

4. Case studies of field measurement applications

4.1 27 April 2020 South Texas severe thunderstorm downbursts

A cluster of thunderstorms developed west of the dryline over the Davis Mountains of southwestern Texas during the afternoon of April 27, 2020, while a dryline extended from east of Midland, Texas to the Big Bend area as shown in the 2100 UTC surface analysis in Fig. 2. The thunderstorm cluster then merged to form an intense MCS with a leading bow echo (not shown) near the westward bulge in the dryline. As noted by Schaefer (1986) and Ziegler et al. (1997), the dryline is often a focus of differential heating, boundary layer convergence, and solenoidal circulation with an attendant release of instability and convective storm intensification. After dryline interaction, the MCS tracked rapidly southeastward toward the Gulf Coastal Plain. Fig. 3A shows a mid-afternoon (1946 UTC) NUCAPS physical retrieval sounding profile over Del Rio, Texas, which can be compared to a RAOB released from the same location 4h later, at 0000 UTC on 28 April (Fig. 3B). The earlier NUCAPS Skew-T diagram (Fig. 3A) shows considerable CAPE (> 4000 J/kg), a large lower tropospheric temperature lapse rate (>9°C/km), and a prominent mid-tropospheric unsaturated layer. These features indicate an elevated probability of severe thunderstorm downburst occurrence, roughly 6h before the onset of the severe windstorm. A

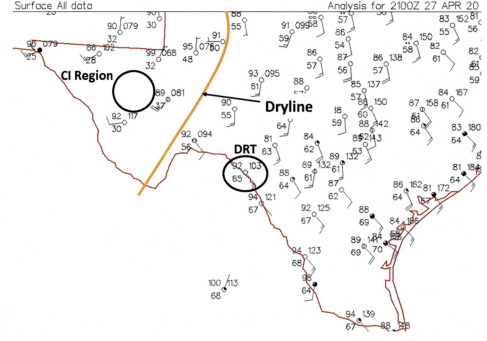

FIG. 2 Surface analysis over the southern Great Plains region at 2100 UTC 27 April 2020. The *black circled* region labeled "DRT" represents the location of the Del Rio, Texas and the location of the NOAA-20 sounding retrieval. "CI" represents the convective initiation region over the Davis Mountains of Texas. *(Courtesy of Plymouth State Weather Center.)*

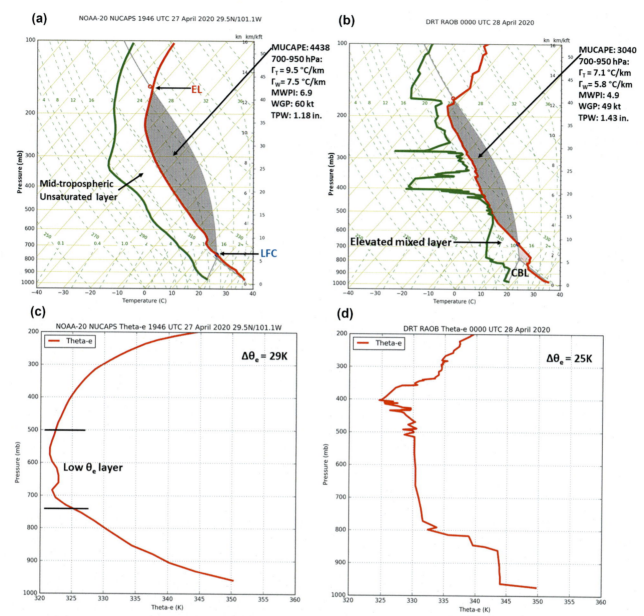

FIG. 3 (A) A NOAA-20 NUCAPS sounding profile retrieved during the afternoon of April 27, 2020, as compared to (B) a radiosonde observation at Del Rio, Texas (DRT) at 0000 UTC April 28, 2020; (C) A NOAA-20 NUCAPS theta-e profile retrieved during the afternoon of April 27, 2020, as compared to (D) a radiosonde observation theta-e profile at Del Rio, Texas (DRT) at 0000 UTC April 28, 2020. In (A) and (B), *red curves* and *green curves* represent the temperature and dewpoint soundings in degrees Celsius (°C), respectively. "MUCAPE" is most unstable parcel CAPE in J kg^{-1}, "MWPI" represents the Microburst Windspeed Potential Index (Pryor, 2015), "WGP" represents wind gust potential derived from the MWPI in knots (kt), and "TPW" represents total precipitable water in inches (in.). Γ_T and Γ_w represent dry-bulb temperature and wet-bulb temperature lapse rates, respectively.

MWPI value of 6.9 indicated thunderstorm wind gust potential of 60 knots. MWPI values greater than 5 indicate a high probability of severe winds greater than 50 knots in magnitude. In addition, the corresponding equivalent potential temperature (theta-e) profile signified potential instability with a decreasing value from the surface upward to a minimum in the middle troposphere. The calculated surface to 500 mb theta-e difference of 29 K (Fig. 3C) significantly exceeds the 20 K threshold for downburst occurrence as documented by Atkins and Wakimoto (1991). Comparing the 1945 UTC (2:45 pm local time) NUCAPS sounding profile (Fig. 3A), to the 0000 UTC (7:00 pm local time) 28 April RAOB from Del Rio (Fig. 3B), with a distance between retrieval locations of 22 km (12 n mi), shows that the risk for severe weather persisted into the evening.

Previous evaluations of NUCAPS for stability index calculation and convective weather forecasting applications (Bloch et al., 2019; Esmaili et al., 2020) described the necessity and resultant implementation of a modification technique to mitigate temperature and moisture biases in the boundary layer. This boundary layer correction replaces NUCAPS surface temperature and dewpoint temperature measurements with adjacent data from the Real Time Mesoscale Analysis (RTMA). The modification technique for the NUCAPS sounding profile in this case study, as shown in Fig. 4, was applied by adding the surface temperature and dew point measurements from Del Rio International Airport (DRT) to the retrieval. This procedure resulted in a stronger signal for severe outflow wind generation as evidenced by increased wind gust potential to 65 knots. Accordingly, the National Weather Service/Storm Prediction Center, in mesoscale discussion (MCD) #0481 issued at 2333 UTC 27 April, placed the Del Rio area in the southeastern periphery of a threat region for isolated strong/severe wind gusts (Fig. 5A). A Laughlin AFB NEXRAD reflectivity image highlighted the linear structure of the MCS with an array of protrusion echoes pointing downshear (Fig. 5B). This MCS likely generated a downburst cluster, resulting in wind gusts of 67 and 65 knots, recorded at Del Rio International Airport at 0135 UTC and at Laughlin Air Force Base at 0150 UTC 28 April, respectively. Calculated gust factors near 1.6 and ΔT values of −12 to −13°C for both wind events are consistent with downburst occurrence (Choi and Hidayat, 2002; Proctor, 1989). In addition, there was an unofficial report of a 90-knot wind gust from a personal weather station in the Del Rio area (D. Schreiber, Personal communication, April 30, 2020) as well as reports of major structural (i.e., roof) damage, uprooted trees, and power line and grid damage.

Defense Meteorological Satellite Program (DMSP) F-17 and MetOP-A overpasses were optimal for retrieving cloud microphysical properties before and shortly after the downburst in the Del Rio area, as inferred from TB measurements shown in Fig. 6. Imagery from Special Sensor Microwave Imager Sounder (SSMIS) and Microwave Humidity Sounder (MHS) displayed in Fig. 6 show remarkably cold cloud tops (white shading) and large graupel water path values (>1 mm) that indicated significant storm severity. As shown in Fig. 6A and B, the 91 GHz channel is an atmospheric window in the microwave spectrum. Scattering by ice-phase precipitation particles, especially graupel, hail, and snow above the freezing level, causes the TB depression (Ferraro et al., 2015; Laviola et al., 2020) around the time of peak storm intensity. For this case, TB near the storm centroid was remarkably low (~120 K) and corresponded to a maximum in graupel water path (GWP) values (>10 mm), indicating the presence of a dense core of graupel/hail. A large ice content and prominent dry-air notches on the lateral and downshear (forward) flanks of the storm indicated favorability for strong thunderstorm downdraft generation (Srivastava, 1987; Knupp, 1989). MetOP-A MHS imagery shown in Fig. 6C and D

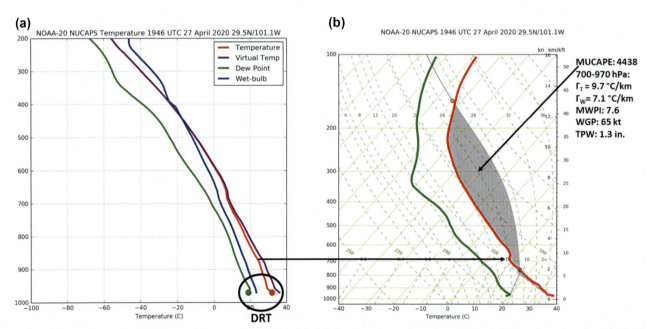

FIG. 4 (A) The NOAA-20 NUCAPS temperature sounding profile marked with surface observations of temperature and dew point retrieved at 1946 UTC near Del Rio, Texas (DRT); (B) corresponding sounding profile over Del Rio modified with the surface temperature and dew point observations at DRT. In (B), *red curves* and *green curves* represent the temperature and dewpoint soundings in degrees Celsius (°C), respectively. "MUCAPE" is most unstable parcel CAPE in J kg^{-1}, "MWPI" represents the Microburst Windspeed Potential Index (Pryor, 2015), "WGP" represents wind gust potential derived from the MWPI in knots (kt), and "TPW" represents total precipitable water in inches (in). Γ_T and Γ_w represent dry-bulb temperature and wet-bulb temperature lapse rates, respectively.

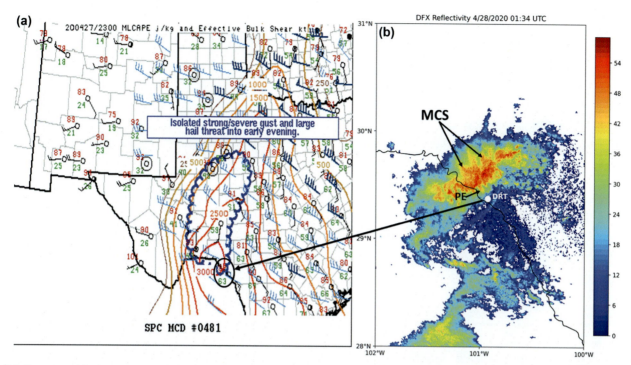

FIG. 5 (A) NWS/SPC mesoscale convective discussion (MCD) at 2300 UTC April 27, 2020. (B) Laughlin AFB, Texas NEXRAD reflectivity at 0134 UTC April 28, 2020. *Black-circled* region in (A) marks the location of Del Rio ("DRT"). "PE" represents the location of a protrusion echo.

displayed remarkably cold cloud tops (white shading) and large graupel water path values (>1 mm) that indicated significant storm severity.

This intensive study of a severe convective wind event demonstrates the value of a synergistic analysis of satellite and ground-based sensor data. This is illustrated by the afternoon NUCAPS vertical profile pointing to the potential for convection and verification by the appearance of severe convective storms in SSMIS and MHS imagery. Successive overpasses of DMSP F-17 and EPS MetOP-A satellites shown in Fig. 7 provided microwave imagery that effectively visualized the severe MCS evolution. Fig. 8 demonstrates the application of brightness temperature differencing to more effectively extract signatures associated with storm severity and intense convective downdraft generation. A new algorithm is the differential brightness temperature (BTDR) between the horizontal and vertical polarization channels that is analogous to Doppler radar-derived differential reflectivity:

$$\text{BTDR} = 100^{*}(\log_{10}(TB91_H/TB91_V)) \qquad (3)$$

where $TB91_H$ and $TB91_V$ represent horizontally and vertically polarized 91 GHz brightness temperatures, respectively. Small negative values of SSMIS-derived BTDR correspond to large graupel water path values (>10 mm), and thus to a mixture of rain, hail, and graupel. The calculation of BTDR is compared to the established polarization corrected temperature (PCT, Spencer et al. (1989)), defined as.

$$\text{PCT} = (1.818^{*}TB_V) - (0.818^{*}TB_H) \qquad (4)$$

where TB_H and TB_V represent horizontally and vertically polarized brightness temperatures, respectively. PCT values below 200 K typically correspond to high precipitation rates and large ice-phase precipitation content.

As shown in Fig. 9, the velocity azimuth display (VAD) wind profile (VWP) from Laughlin Air Force Base (AFB) NEXRAD, this multicellular storm tracked east-southeastward with weak low-level shear that was nearly parallel to the storm line, which allowed for a short episode of severe downburst winds over the Del Rio area, followed by gradual weakening of the system and cessation of severe winds. Also apparent were inward-directed V-shaped TB gradients on the downwind (eastern) flank of the storm that suggests the occurrence of wake entrainment of sub-saturated air and subsequent downdraft acceleration by the process detailed in Knupp (1989). The absence of an apparent rear-inflow jet was likely a

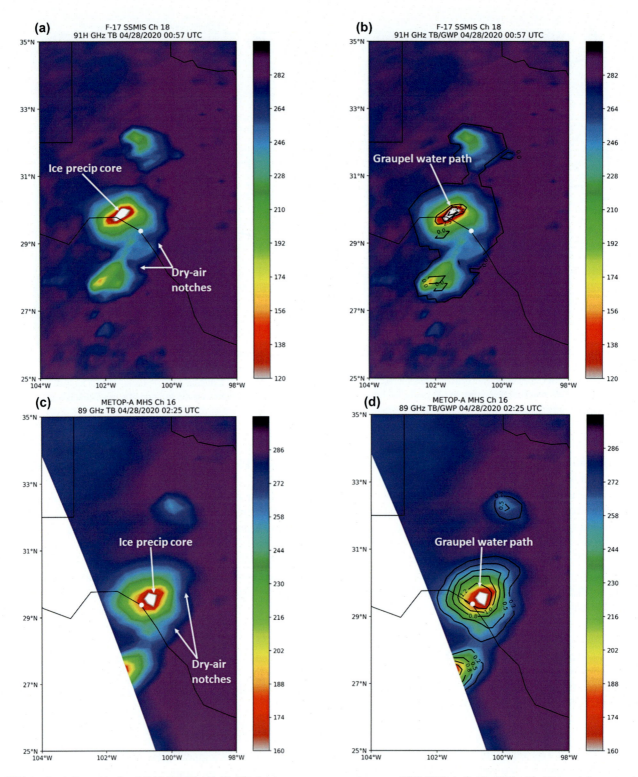

FIG. 6 South Texas Regional F-17 SSMIS (A) 91 GHz brightness temperature image at 0057 UTC April 28, 2020 with (B) overlying graupel water path ("GWP") measurements (in mm); (C) MetOP-A Microwave Humidity Sounder (MHS) 89 GHz brightness temperature image at 0225 UTC April 28, 2020 with (D) overlying graupel water path ("GWP") measurements (in mm). GWP contours are every 2.5 mm. *White circle* marks the location of Del Rio.

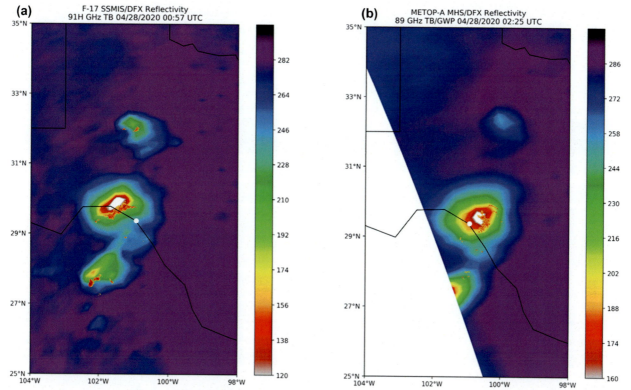

FIG. 7 South Texas Regional F-17 SSMIS (A) 91 GHz brightness temperature image at 0057 UTC April 28, 2020 and (B) MetOP-A MHS 89 GHz brightness temperature image at 0225 UTC April 28, 2020 with overlying NEXRAD reflectivity measurements (in dBZ). *White circle* marks the location of Del Rio.

factor in the episodic and short-lived occurrence of downburst winds at Del Rio with a duration only 15 min, from 0135 to 0150 UTC 28 April. However, these storms still produced significant structural damage and power outages on a local scale.

4.2 29 June 2012 North American derecho

During the morning of June 29, 2012, an area of convective storms over Iowa organized into a quasi-linear convective system (QLCS) as it tracked into northern Illinois. The system then evolved into a bow echo (Przybylinski, 1995) during the afternoon and tracked southeastward over the Ohio Valley to the Mid-Atlantic coast by late evening. What would eventually become the June 29, 2012, North American Derecho, this QLCS produced its first significant severe downburst, with winds measured over 65 knots, at Michigan City, Indiana during the early afternoon. This extraordinary derecho-producing convective system (DCS) event resulted in 22 deaths and nearly a thousand severe wind reports from northern Illinois to the Atlantic Coast. This system was more typical of a warm-season progressive derecho, as shown in Fig. 10, associated with a major heat wave and an elevated mixed layer (Banacos and Ekster, 2010). This DCS was generated and then propagated within a mid-tropospheric ridge synoptic pattern as identified by Johns (1993) and Coniglio et al. (2004). In addition, the derecho system maintained a type 2 echo pattern, as described by Przybylinski (1995), through most of its track through the Ohio Valley and Mid-Atlantic regions. The type 2 echo is characterized by "a short, solid bowing convective line segment of between 80 and 100 km in length. A band of scattered to broken convective elements is associated with a surface frontal boundary or warm advection zone and typically extends downwind (eastward) from the northern end of the bulging line echo" Przybylinski (1995). During the evening of 29 June, the derecho tracked rapidly eastward across the mountains of West Virginia (WV), western Virginia (VA), southwest Pennsylvania (PA), and western Maryland (MD) during mid-evening. The derecho's effects were particularly formidable in the Washington, DC–Baltimore, MD corridor, where measured wind gusts of 60–70 knots severed numerous overhead electrical feeders.

Comparison of the late evening MetOP-A Infrared Atmospheric Sounding Interferometer (IASI) sounding near Salisbury, Maryland to the HUBC microwave radiometer (MWR) sounding at 0200 UTC 30 June in Figs. 11 and 12 exhibits a transition to a moist and highly unstable profile favorable for severe wet microbursts. Fig. 11 illustrates the sounding

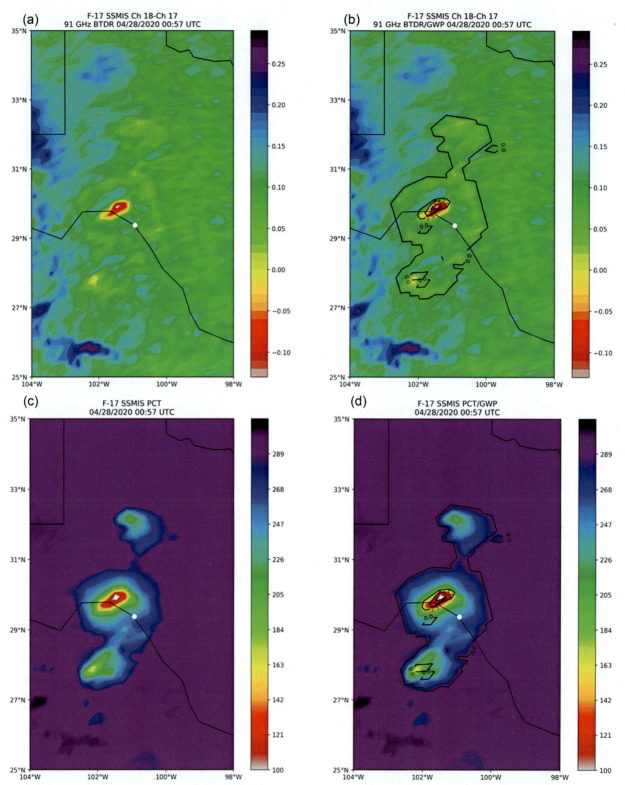

FIG. 8 F-17 SSMIS (A) differential brightness temperature (BTDR) imagery and (C) polarization corrected temperature (PCT) near 0100 UTC 28 April 2020. (B) and (D) display the product imagery with overlying graupel water path ("GWP", in mm). GWP contours are every 2.5 mm. *White circle* marks the location of Del Rio.

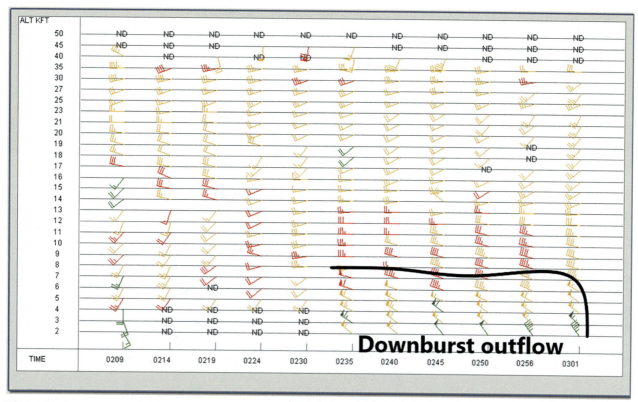

FIG. 9 Velocity azimuth display (VAD) wind profile (VWP) from Laughlin Air Force Base (AFB) NEXRAD, near Del Rio, Texas between 0200 and 0300 UTC April 28, 2020.

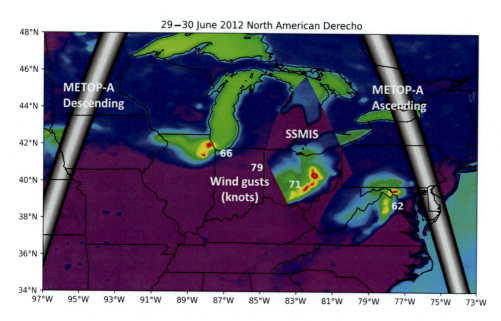

FIG. 10 Summary composite image of the June 2012 North American Derecho displaying the 29 June descending node and 30 June ascending node METOP-A orbit (nadir) tracks, MHS and SSMIS 89–91 GHz channel brightness temperature (TB, degrees Kelvin (K)) and significant wind reports (kt) along the storm track.

modification process to further enhance the signal for severe deep convective storm development. In Fig. 11A, the IASI thermodynamic profile indicated modest convective storm potential. Incorporating the 0154 UTC surface temperature and dew point observation from Salisbury Regional Airport (27 km (14 n mi) north of the IASI retrieval site), as shown in Fig. 11B, results in significantly larger CAPE. Finally, substituting the dry bulb temperature data set with calculated virtual temperature yields the strongest signal for severe downburst generation with wind gust potential of 57 knots, comparable to

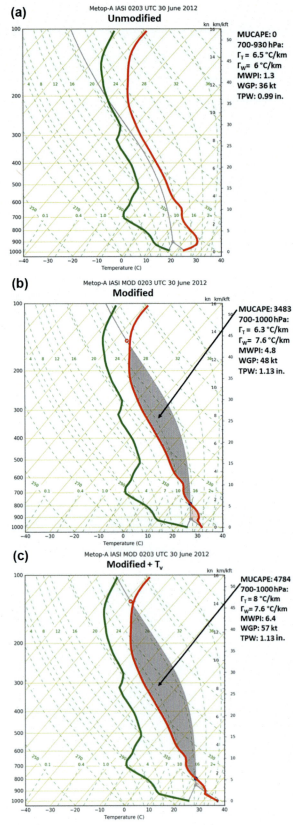

FIG. 11 MetOP-A IASI retrievals near Salisbury, Maryland during the evening of June 29, 2012 (0203 UTC 30 June): (A) IR+MW sounding profile; (B) IR+MW sounding profile modified by observed surface temperature and dew point at Salisbury Regional Airport; (C) modified IR+MW sounding profile plotted with virtual temperature. *Red curves* and *green curves* represent the temperature and dewpoint soundings in degrees Celsius (°C), respectively. "MUCAPE" is most unstable parcel CAPE in J kg^{-1}, "MWPI" represents the Microburst Windspeed Potential Index (Pryor, 2015), "WGP" represents wind gust potential derived from the MWPI in knots (kt), and "TPW" represents total precipitable water in inches (in). Γ_T and Γ_w represent dry-bulb temperature and wet-bulb temperature lapse rates, respectively.

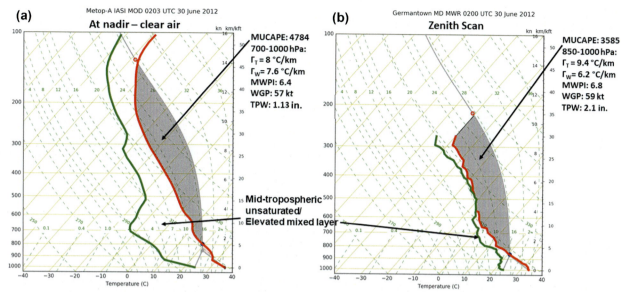

FIG. 12 (A) Modified MetOP-A IASI IR+MW sounding profile retrieved near Salisbury, Maryland at 0203 UTC June 30, 2012, as compared to (B) a ground-based sounding profile retrieval from the Germantown, Maryland microwave radiometer (MWR). *Red curves* and *green curves* represent the temperature and dewpoint soundings in degrees Celsius (°C), respectively. "MUCAPE" is most unstable parcel CAPE in J kg^{-1}, "MWPI" represents the Microburst Windspeed Potential Index (Pryor 2015), "WGP" represents wind gust potential derived from the MWPI in knots (kt), and "TPW" represents total precipitable water in inches (in). Γ_T and Γ_w represent dry-bulb temperature and wet-bulb temperature lapse rates, respectively.

the potential derived from the Germantown, Maryland MWR sounding as demonstrated in Figs. 11C and 12. Accordingly, the MWPI increased in magnitude prior to the onset of the derecho to eventually indicate convective wind gust potential of 57–59 knots with an hour of lead time. Since this author is not aware of routine use of this sounding modification technique in weather forecasting operations, the dissemination of this procedure to the operational meteorology community is paramount. The value added to the application of hyperspectral sounding analysis is demonstrated by the results of this technique. Between 0000 and 0200 UTC, as shown in Fig. 13, the DCS evolved into a double-bow echo pattern with a "warm advection wing" (Smith, 1990) over Frederick County (near latitude 39.5°N/longitude 77.4°W) that developed in an east-west oriented region of weak surface convergence over central Maryland.

Fig. 13 exhibits a type 2 derecho echo pattern with a warm advection wing (Przybylinski, 1995) that extended downwind (eastward) from the northern end of the bulging line echo. Microbursts occurred in Frederick County within the warm advection wing of the derecho. MetOP-A MHS, with overlying Sterling, VA (LWX) NEXRAD reflectivity, revealed the presence of the warm advection wing. A dry air notch, displayed as an inward (eastward) pointing TB gradient, likely indicated the presence of a rear-inflow jet ("RIJ") that sustained the MCS and the generation of downburst clusters in the DC-Baltimore corridor during the following hour. A shown in Figs. 14 and 15, the RIJ was apparent and distinguishable from surface-based outflow in the Sterling, Virginia NEXRAD VWP and Beltsville, Maryland 915 MHz Boundary Layer Profiler (BLP) wind observation time series, respectively. Near 0250 UTC, a downburst cluster tracking over downtown Washington, DC produced measured wind gusts of 61 and 47 knots at Reagan National Airport and the Washington Physical Oceanographic Real-Time System (PORTS) station, respectively. A gust factor of 1.42 and ΔT values of −10 to −11°C were also consistent with downburst occurrence embedded in the larger scale DCS.

In retrospect, NWS/Storm Prediction Center (SPC) adequately indicated the likelihood of scattered severe winds over the Washington, DC–Baltimore, MD corridor as the derecho tracked east of the Appalachian Mountains during the late evening. However, the density and magnitude of severe wind events, and associated impacts, over the Washington, DC metropolitan area, including the adjacent Maryland and Virginia suburbs, was not anticipated by neither SPC nor the NWS Office Baltimore-Washington. Furthermore, the MCD did not document any use of information from satellite-based sounding profilers or imagers, or ground-based profilers such as network MWRPs. Thus, science value added with this study of this derecho event entails the coordinated application of evening IASI and MWRP sounding profiles and derived parameters that will provide more insight into the evolution of the nocturnal convective lower troposphere. MW window channel data will more effectively interrogate evolving DCSs and reveal greater detail of storm structure, especially pertaining to convective wind generation. An important outcome of this study will be to formulate a correlation between MW parameters and signatures, and severe convective wind occurrence. The results of the evaluation of this derecho event

426 PART | III Satellite applications

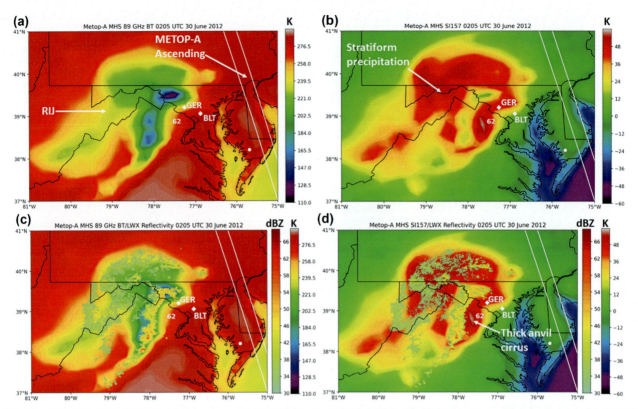

FIG. 13 METOP-A MHS 89 GHz brightness temperature (TB, K) image at 0200 UTC June 30, 2012 with (A) overlying Sterling, Virginia (LWX) NEXRAD radial velocity (kt) and (B) reflectivity (dBZ) measurements. (C, D) as in (A, B) with overlying Sterling, Virginia (LWX) NEXRAD reflectivity (dBZ) measurements. "GER" and "BLT" mark the location of the Germantown and Beltsville, Maryland MWRPs, respectively, The *white circle* marks the location of the IASI retrieval over Salisbury, Maryland, and "62" is the location of the first severe wind report in the Washington, DC metropolitan area (31.7 m s^{-1} (62 kt)) recorded at Dulles International Airport, Virginia. *White lines* mark the 30 June ascending node METOP-A orbit (nadir) tracks.

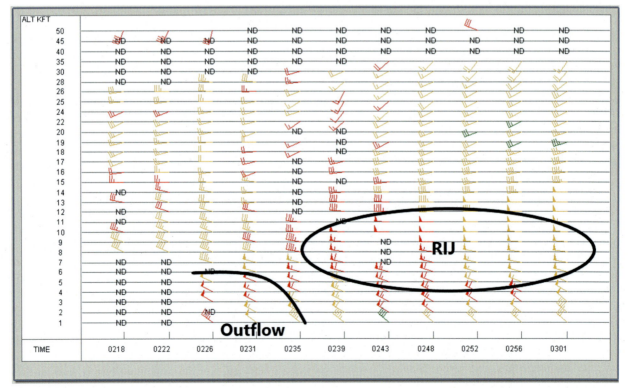

FIG. 14 Velocity azimuth display (VAD) wind profile (VWP) from Sterling, Virginia NEXRAD between 0200 and 0300 UTC June 30, 2012.

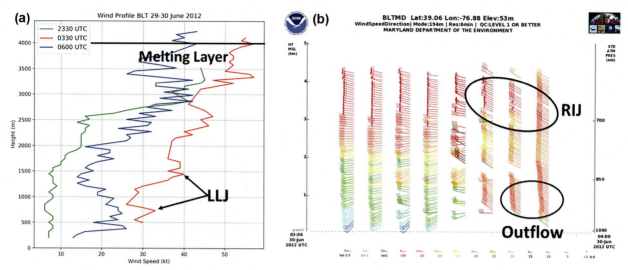

FIG. 15 Time series of HUBC 915 MHz Boundary Layer Profiler (BLP) (A) wind speed vs height (meters) between 2330 UTC 29 June and 0600 UTC June 30, 2012, and (B) wind speed and direction vs height between 0300 and 0400 UTC June 30, 2012.

necessitate the development and implementation of a nationwide NOAA ground-based microwave profiler network to provide the operational meteorology community near real-time access to high temporal resolution vertical temperature and moisture soundings.

5. Summary

Convective storm-generated downbursts are an operational forecasting challenge due to the spectrum of time, space, and intensity scales in which they occur. This chapter assembled the governing physical theory essential for development of downburst prediction algorithms that proceeds from vertical momentum equations and the aggregate of thermodynamical and microphysical processes of precipitation. Accordingly, downburst monitoring and subsequent prediction is a three-step process with an objective to build a three-dimensional model of the thermodynamic structure of the ambient environment and conceptual model of downburst-producing convective storms:

1. Collection and exploitation of surface-based observations including measurements from tower platforms and Doppler radar-measured reflectivity and wind velocity. This step promotes the enhanced use of the network of private and university-partnered ground-based MWRPs, as well as the archival of profiler datasets.
2. Ground-based microwave and radio profiler instruments, including MWRPs and BLPs, to obtain vertical profiles of temperature, humidity, and wind velocity. This step continues encouragement of algorithm development and multi-instrument synergistic interpretation of existing data assets.
3. Satellite-based 2-D plan view images of brightness temperature and vertical profiles of temperature and humidity. Modification of sounding profiles with surface observations of temperature and humidity is an additional step that results in improved representation of the ambient environment. Modifying sounding profiles with surface observations of temperature and humidity is an additional step that improves the representation of the ambient environment especially when performed with co-located MWRP sounding retrievals.

The case studies demonstrate how both ground-based and satellite-based observational data for convective storms can be combined for monitoring and forecasting applications. The strategic application of polar-orbiting meteorological satellite datasets and ground-based MWRP datasets allow for the comprehensive tracking of severe convective windstorms and DCSs through most of their lifecycles. With the advent of geostationary-satellite based hyperspectral infrared sounders, such as the InfraRed Sounder (IRS) to be deployed on Meteosat Third Generation (MTG) upon launch in 2024 (Iturbide-Sanchez et al., 2022), these observational techniques can be readily applied in near real-time to future convective windstorm events. With the expected improvement in humidity observations, the MTG IRS should satisfy a growing interest in severe convective windstorms that occur in Europe, in consideration of recent derecho events that impacted Germany in 2002 (Gatzen, 2004) and 2014 (Mathias et al., 2017). The IRS coverage area over the Atlantic Ocean, Europe, and Africa will suit preliminary assessment of this instrument in severe convective storm monitoring and prediction.

Passive microwave sounder imagery, as generated by low-Earth-orbit (LEO) satellite platforms, is vital for the inference of precipitation physical process and the resultant generation of convective outflow winds and confirms the hypothesis that convective windstorms are driven by unique aspects of the thermodynamic structure of the ambient environment and precipitation characteristics of the parent MCS. A challenge of the operational application of the derived microwave product imagery is the low temporal resolution (i.e., retrieval frequency) of LEO satellite passes over a region of interest, especially during the development phase of an MCS. Latency minimization procedures would mitigate a lower temporal resolution by providing derived products in near real-time to operational forecasting agencies. For MCSs with longer lifetimes (>6 h), successive overpasses with microwave image retrievals can effectively detect changes in storm intensity and the potential for widespread severe convective wind occurrence. The NASA Short-term Prediction Research and Transition (SPoRT) project disseminates microwave sounder imagery in near real-time through the web-based SPoRT Viewer (available online at https://weather.msfc.nasa.gov/sport/viewer/), which should serve as a benchmark to provide these products to operational weather forecasters. Field measurements are the cornerstone for remote sensing techniques and are essential for understanding deep convective storms and associated downburst occurrence, phenomena that encompass the vertical dimension of the troposphere.

Acknowledgments

The Author would like to acknowledge Rebekah Esmaili (STC) for reviewing the draft and providing constructive feedback. The author thanks Randolph Ware of Radiometrics Corporation for providing a comprehensive dataset of microwave radiometer profiles for analysis in this research effort and also thanks Dr. Belay Demoz of University of Maryland, Baltimore County, and Dr. Laurie Rokke of NOAA/NESDIS/STAR for their guidance and thorough internal review of the manuscript.

References

Ashley, W.S., Mote, T.L., 2005. Derecho hazards in the United States. Bull. Am. Meteorol. Soc. 86, 1577–1592.

Atkins, N.T., Wakimoto, R.M., 1991. Wet microburst activity over the southeastern United States: implications for forecasting. Weather Forecast. 6, 470–482.

Banacos, P.C., Ekster, M.L., 2010. The association of the elevated mixed layer with significant severe weather events in the Northeastern United States. Weather Forecast. 25, 1082–1102.

Bloch, C., Knuteson, R.O., Gambacorta, A., Nalli, N.R., Gartzke, J., Zhou, L., 2019. Near-real-time surface-based CAPE from merged hyperspectral IR satellite sounder and surface meteorological station data. J. Appl. Meteorol. Climatol. 58, 1613–1632.

Brock, F.V., Crawford, K.C., Elliott, R.L., Cuperus, G.W., Stadler, S.J., Johnson, H.L., Eilts, M.D., 1995. The Oklahoma Mesonet: a technical overview. J. Atmos. Oceanic Tech. 12, 5–19.

Bunkers, M.J., Klimowski, B.A., 2002. The importance of parcel choice and the measure of vertical wind shear in evaluating the convective environment. Preprints, 21st Conf. Severe Local Storms, San Antonio. Am. Meteorol. Soc. J117–J120, 11–16.

Choi, E.C., Hidayat, F.A., 2002. Gust factors for thunderstorm and non-thunderstorm winds. J. Wind Eng. Ind. Aerodyn. 90 (12–15), 1683–1696.

Cimini, D., Nelson, M., Güldner, J., Ware, R., 2015. Forecast indices from a ground-based microwave radiometer for operational meteorology. Atmos. Meas. Tech. 8 (1), 315–333.

Coniglio, M.C., Stensrud, D.J., Richman, M.B., 2004. An observational study of derecho-producing convective systems. Weather Forecast. 19, 320–337.

Ecklund, W.L., Carter, D.A., Balsley, B.B., 1988. A UHF wind profiler for the boundary layer: brief description and initial results. J. Atmos. Oceanic Tech. 5 (3), 432–441.

Ellrod, G.P., 1989. Environmental conditions associated with the Dallas microburst storm determined from satellite soundings. Weather Forecast. 4, 469–484.

Esmaili, R.B., Smith, N., Berndt, E.B., Dostalek, J.F., Kahn, B.H., White, K., Barnet, C.D., Sjoberg, W., Goldberg, M., 2020. Adapting satellite soundings for operational forecasting within the hazardous weather testbed. Remote Sens. (Basel) 12, 886. https://doi.org/10.3390/rs12050886.

Ferraro, R., Beauchamp, J., Cecil, D., Heymsfield, G., 2015. A prototype hail detection algorithm and hail climatology developed with the advanced microwave sounding unit (AMSU). Atmos. Res. 163, 24–35.

Ferraro, R.R., Kusselson, S.J., Colton, M., 1998. An introduction to passive microwave remote sensing and its applications to meteorological analysis and forecasting. Natl. Weather Dig. 22, 11–23.

Fujita, T.T., 1985. The Downburst, Microburst and Macroburst. Satellite and Mesometeorology Research Paper 210. University of Chicago. 122 pp.

Fujita, T.T., Wakimoto, R.M., 1981. Five scales of airflow associated with a series of downbursts on 16 July 1980. Mon. Weather Rev. 109, 1438–1456.

Gatzen, C., 2004. A derecho in Europe: Berlin, 10 July 2002. Weather Forecast. 19, 639–645. https://doi.org/10.1175/1520-0434(2004)019<0639:ADIEBJ>2.0.CO;2.

Iturbide-Sanchez, F., Wang, Z., Kalluri, S., Chen, Y., Lynch, E., Divakarla, M., Tan, C., Zhu, T., Cao, C., 2022. Exploration of a future NOAA infrared sounder in geostationary earth orbit. IEEE J. Sel. Top. Appl. Earth Observ. Remote Sens. 15, 1543–1561. https://doi.org/10.1109/JSTARS.2022.3142069.

Johns, R.H., 1993. Meteorological conditions associated with bow echo development in convective storms. Weather Forecast. 8, 294–299.

Knupp, K.R., 1989. Numerical simulation of low-level downdraft initiation within precipitating cumulonimbi: some preliminary results. Mon. Weather Rev. 117, 1517–1529.

Knupp, K.R., 1996. Structure and evolution of a long-lived, microburst producing storm. Mon. Weather Rev. 124, 2785–2806.

Laviola, S., Levizzani, V., Ferraro, R.R., Beauchamp, J., 2020. Hailstorm detection by satellite microwave radiometers. Remote Sens. (Basel) 12 (4), 621.

Liu, G., Curry, J.A., Sheu, R.W., 1995. Classification of clouds over the western equatorial Pacific Ocean using combined infrared and microwave satellite data. J. Geophys. Res. Atmos. 100 (D7), 13811–13826.

Martner, B.E., Wuertz, D.B., Stankov, B.B., Strauch, R.G., Westwater, E.R., Gage, K.S., Dabberdt, W.F., 1993. An evaluation of wind profiler, RASS, and microwave radiometer performance. Bull. Am. Meteorol. Soc. 74 (4), 599–614.

Mathias, L., Ermert, V., Kelemen, F.D., Ludwig, P., Pinto, J.G., 2017. Synoptic analysis and hindcast of an intense bow echo in Western Europe: the 9 June 2014 storm. Weather Forecast. 32, 1121–1141. https://doi.org/10.1175/WAF-D-16-0192.1.

Moller, A.R., 2001. Severe local storms forecasting. In: Doswell, C.A. (Ed.), Severe Convective Storms. 28. American Meteorological Society, Boston, MA, pp. 433–480.

Nalli, N.R., Tan, C., Warner, J., Divakarla, M., Gambacorta, A., Wilson, M., Zhu, T., Wang, T., Wei, Z., Pryor, K., Kalluri, S., Zhou, L., Sweeney, C., Baier, B.C., McKain, K., Wunch, D., Deutscher, N.M., Hase, F., Iraci, L.T., Kivi, R., Morino, I., Notholt, J., Ohyama, H., Pollard, D.F., Té, Y., Velazco, V.A., Warneke, T., Sussmann, R., Rettinger, M., 2020. Validation of carbon trace gas profile retrievals from the NOAA-unique combined atmospheric processing system for the cross-track infrared sounder. Remote Sens. (Basel) 12, 3245. https://doi.org/10.3390/rs12193245.

Proctor, F.H., 1989. Numerical simulations of an isolated microburst. Part II: sensitivity experiments. J. Atmos. Sci. 46, 2143–2165.

Pryor, K.L., 2015. Progress and developments of downburst prediction applications of GOES. Weather Forecast. 30, 1182–1200. https://doi.org/10.1175/WAF-D-14-00106.1.

Pryor, K.L., 2017. Advances in downburst monitoring and prediction with GOES-16. In: 17th Conference on Mesoscale Processes, San Diego, CA, Amer. Meteor. Soc. Paper No. 10.6.

Przybylinski, R.W., 1995. The bow echo. Observations, numerical simulations, and severe weather detection methods. Weather Forecast. 10, 203–218.

Schaefer, J.T., 1986. The Dryline. In: Ray, P.S. (Ed.), Mesoscale Meteorology and Forecasting. American Meteorological Society, pp. 549–572.

Schroeder, J.L., Burgett, W.S., Haynie, K.B., Sonmez, I., Skwira, G.D., Doggett, A.L., Lipe, J.W., 2005. The West Texas Mesonet: a technical overview. J. Atmos. Oceanic Tech. 22, 211–222.

Smith, B.E., 1990. Mesoscale structure of a derecho-producing convective system: the southern Great Plains storms of May 4 1989. In: Preprints, 16th Conf on Severe Local Storms, Kananaskis Park, AB, Canada, Amer. Meteor. Soc, pp. 428–433.

Smull, B.F., Houze Jr., R.A., 1987. Rear inflow in squall lines with trailing stratiform precipitation. Mon. Weather Rev. 115, 2869–2889.

Spencer, R.W., Goodman, H.M., Hood, R.E., 1989. Precipitation retrieval over land and ocean with the SSM/I: identification and characteristics of the scattering signal. J. Atmos. Oceanic Tech. 6 (2), 254–273.

Srivastava, R.C., 1985. A simple model of evaporatively driven downdraft: application to microburst downdraft. J. Atmos. Sci. 42 (10), 1004–1023. https://doi.org/10.1175/1520-0469(1985)042<1004:ASMOED>2.0.CO;2.

Srivastava, R.C., 1987. A model of intense downdrafts driven by the melting and evaporation of precipitation. J. Atmos. Sci. 44, 1752–1773.

Wakimoto, R.M., 1985. Forecasting dry microburst activity over the high plains. Mon. Weather Rev. 113, 1131–1143.

Weisman, M.L., 1992. The role of convectively generated rear inflow jets in the evolution of long-lived mesoconvective systems. J. Atmos. Sci. 49, 1826–1847.

Weisman, M.L., Klemp, J.B., Rotunno, R., 1988. Structure and evolution of numerically simulated squall lines. J. Atmos. Sci. 45, 1990–2013.

Westwater, E.R., Crewell, S., Matzler, C., 2005. Surface-based microwave and millimeter wave radiometric remote sensing of the troposphere: a tutorial. IEEE Geosci. Remote Sens. Soc. Newslett. 134, 16–33.

Ziegler, C.L., Lee, T.J., Pielke, R.A., 1997. Convective initiation at the Dryline: a modeling study. Mon. Weather Rev. 125, 1001–1026.

Index

Note: Page numbers followed by *f* indicate figures, *t* indicate tables and *b* indicate boxes.

A

Absolute difference, 366
Absolute percentage difference, 366
Absorbing Aerosol Index (AAI), 395
Absorption-based models, 365
Advanced Along-Track Scanning Radiometer (AATSR), 112
Advanced Baseline Imager (ABI), 13, 112, 166
Advanced Earth Observing Satellite (ADEOS), 210
Advanced Infrared Radiometry and Imaging Facility (AIRI), 164
Advanced Meteorological Imager (AMI), 190
Advanced Microwave Sounding Unit (AMSU), 29–30, 221–222, 247, 286, 297–299
Advanced Scatterometer (ASCAT), 190–191
Advanced Technology Microwave Sounder (ATMS), 14, 246, 283, 297–299, 320
Advanced TIROS Observational/Operational Vertical Sounder (ATOVS), 29–30, 286, 319–320
Advanced Very High-Resolution Radiometer (AVHRR), 11, 87, 112, 375, 393
Advanced Weather Interactive Processing System (AWIPS), 246, 320
Aerosol detection product, 392
Aerosol optical depth (AOD), 8, 122–123, 254, 392
Aerosol Robotic Network (AERONET), 4, 121, 358, 392
Aerosols, 132–135
  quality assurance, control, 132–135, 134*f*
  types, 132, 133*f*
Aerosols and Ocean Science Expeditions (AEROSE), 205–206, 283, 324
Aggregating aircraft observations, regional intensives, 243
Airborne Sounder Testbed Interferometer (NAST-I), 189
Airborne Vertical Atmospheric Profiling System (AVAPS), 185
AirCore atmospheric sampling system, 141–149
  flight string components accompany, 142–144, 143*f*
  operating principle, 141–142, 142*f*
  profile retrieval, vertical resolution, 143*f*, 144–145
  sampler configurations, associated platforms, 145–146
    Active AirCore sampler, 146
    balloon-borne dual-AirCore, 146, 147*f*
    balloon-borne, high-resolution passive sampler, 145–146, 145–146*f*
AirCore balloon-borne measurement sites, 149, 149*f*, 150*t*
AirCore, high-altitude uncrewed aircraft systems, 152–153
AirCore remote sensing retrievals evaluation, 149–152
  column trace gas abundance calculation, 150
  vs. satellite XCO$_2$, 150, 151*f*
  WMO standard scales, 151–152, 152*t*
AirCore retrieved profile uncertainty, reproducibility, 146–149
  profile measurement accuracy, 147–148
  profile reproducibility, 148–149, 148*f*
Air-sea interface, 86–87
AIRS processing system, 286–287
AIRS Water Vapor Experiment-Ground (AWEX-G), 324
AlgaeOnlineAnalyser, 365
Analytical Spectral Device, 213–214
Angle of incidence, 267
Angstrom exponent (AE), 122–123
Angular distribution models, 271–272
Apparent optical property, 352
Aqua satellite, 2
ARM Cloud Aerosol and Precipitation Experiment (ACAPEX), 105, 207–208, 324
Atlantic Regional Campaign (ATReC), 189
Atlantic Tradewind Ocean-Atmosphere Mesoscale Interaction Campaign (ATOMIC), 83, 209
Atmospheric correction failure, 355
Atmospheric Emitted Radiance Interferometer (AERI), 103–104, 160–162, 204
Atmospheric Infrared Sounder (AIRS), 30, 189–190, 221–222, 237–238, 247, 319–320
Atmospheric Laser Doppler Instrument (ALADIN), 31, 192
Atmospheric motion vectors (AMV), 31, 192
Atmospheric Radiation Measurement (ARM), 103–104, 160–162, 204, 220, 283, 327, 384
Atmospheric rivers (ARs), 207–208
Atmospheric Sounding Processing Environment (ASPEN), 188
Atmospheric Tomography Mission (ATom), 239, 325
Atmospheric trace gases, 139–141
  vertical profiling, 140–141, 141*f*
At-sea instrument, sensor deployments, 198–203
  atmospheric, 198–201
    in situ atmospheric surface measurements, 201
    radiosonde, ozonesonde launches, 197*f*, 198–199, 199*f*
    shipboard radiometers, 199–201, 200*b*
  oceanographic, 201–203
    CTD casts, 203
    drifting buoys, 203
    expendable ocean profilers, 203
    moored buoys, 197*f*, 201–202, 202*f*
Averaging kernel matrix (AKM), 251

B

Balloon Experiment on Standards for Ozone (BESOS), 63
2-Balloon launch strategy, 224
Beam splitter, 103
Best estimate, 223
Bidirectional reflectance distribution function (BRDF), 353
Binary Universal Form for the Representation of meteorological data (BUFR), 27, 189
Biological-biophysical-engineering, 365
Blackbody, 264
Blackbody curves, 7
Blended analyses, 337
Blue-band estimation, 353
Boundary Layer Profiler, 415
Brightness temperature, 7, 414
Brightness temperature differences, 393
British Aerospace (BAe), 189–190

C

Calibration, 235–236
Calibration/validation (cal/val), 16–17, 79, 320, 362–363
  error analysis, 16–17
  intensive cal/val, long-term monitoring, 17
  measurement correlation, 16
  campaigns, 189–192
    clouds, precipitation, 192
    surface winds, 190–191, 191*f*
    temperature, humidity, 189–190
    wind profiles, 192
  satellite, 83

Capping inversions, 249–250
Carbon Absorption Fluorescence Euphotic resolving (CAFE), 365
Carbon-based Productivity Model (CbPM), 365
Carbon cycle, 140–141
Carbon dioxide (CO_2), 139, 235
Carbon monoxide (CO), 14, 144, 239
Carbon Observatory version 2 (OCO-2), 141
Cavity ring-down spectrometer, 144
Center of Satellite Applications and Research (STAR), 281
Charge-coupled device, 356
Chlorophyll-a (Chl-a), 352
Climate data records (CDRs), 113, 263–264, 340
Climate Forecast System Reanalysis (CFSR), 287
Closure experiments, 2–3
Cloud-Aerosol Lidar with Orthogonal Polarization (CALIOP), 392
Cloud-cleared radiances (CCRs), 206
Cloud-clearing, 322
Cloud liquid water (CLW), 299–300
Cloud Physics Lidar (CPL), 176–177, 189
Cloud shadow and stray light contamination, 355
Coastal Marine Automated Network (C-MAN), 87
Coastal Zone Color Scanner (CZCS), 210, 351–352
Collaborative remote sensing campaigns, 205–209
  CalWater/ACAPEX, 207–208, 207f
  EUREC4A/ATOMIC, 209
  PNE/AEROSE, 205–207, 206f
    SAL, cross-sectional observations, 206–207
  SHEBA/MOSAiC, 208, 209f
Colored dissolved organic matter (CDOM), 351
Column CO_2 (XCO_2), 150
Combined ASTER MODIS Emissivity over Land (CAMEL), 173
Combined Sensor Program (CSP), 203–204
Community Long-Term Infrared Microwave Combined Atmospheric Product System (CLIMCAPS), 253–254, 320
Community Radiative Transfer Model (CRTM), 15, 286, 299–300
Community satellite processing package (CSPP), 246
Compact Optical Profiling System (COPS), 213–214
Complementary metal-oxide semiconductor (CMOS), 357
Complete Hemispherical Infrared Laser-based Reflectometer (CHILR), 164
Comprehensive Large Array-data Stewardship System (CLASS), 248, 266, 377, 415
Conductivity-temperature-depth (CTD), 80–81, 197
Confidence intervals, 275
Convection, 246
Convective available potential energy (CAPE), 411–412
Convective storms, 416

Cooperative Agency Profilers, 415
Coordinated universal time (UTC), 198
Copernicus Atmosphere Monitoring Service (CAMS), 325
Correlative data, 212–213
Coupled Ocean Atmosphere Response Experiment (COARE), 90–91
Cross-track Infrared Sounder (CrIS), 14, 30, 190, 222–223, 239–240, 246, 283, 319–320
Cryogenic FPH (CFH), 42
Cyclone Global Navigation Satellite System (CYGNSS), 191

D
Dark signals, 43
Data Buoy Cooperation Panel (DBCP), 80
Day Night Band (DNB), 264
Days since launch (DSL), 274
Deep convective cloud (DCC), 264
Deep moist convective (DMC) storm, 412
Defense Meteorological Satellite Program (DMSP), 415
Degrees of freedom (DOFs), 70
Department of Energy (DOE), 103–104, 204, 220, 283
Derecho-producing convective system (DCS), 421
Derived motion winds (DMWs), 14
Dew point, 37–38
Differential brightness temperature (BTDR), 419
Differential optical absorption spectroscopy (DOAS), 121
Direct solar irradiance, 122
Distributed Active Archive Center (DAAC), 286–287
Diurnal heating, 102
Diurnal thermocline, 205
DOE ARM campaigns, 220–226
  AIRS Cal/Val, 221–222, 222f
  JPSS CrIS Cal/Val, 222–226, 223–226f
  water vapor intensive operating periods, 220–221, 221f
Doppler Wind Lidar (DWL), 192
Downburst monitoring, field measurement, 414–427
  North American derecho, 421–427, 423–427f
  South Texas severe thunderstorm downbursts, 416–421, 416–423f
Dropsondes, measurement, 186–189, 186t, 186f
  data processing, quality control, 187–189, 188f, 188t
  launchers, 187, 187f
Dual Regression Retrieval with De-Aliasing (DRDA), 162
Dust Smoke Discrimination Index (DSDI), 395

E
Early infrared remote sensing campaigns, 203–205
  Combined Sensor Program (CSP), 205
  Ocean Temperature Interferometric Survey (OTIS), 204
Earth observation (EO), 32

Earth Observing System (EOS), 221–222, 286
Earth's surface area, 195, 196f
Earth System Research Laboratory (ESRL), 248
Earth view radiance, 268
Electrochemical concentration cell (ECC), 60
Electromagnetic radiation, 7–8
Electron transport rate (ETR), 365
El Nino Rapid Response (ENRR), 245, 324
El Niño Southern Oscillation (ENSO), 47–48
Emission, 376
Empirical orthogonal functions (EOFs), 299–300
Engine room intake, 338–339
Enterprise algorithm, 376
Enterprise Cloud Mask, 403
Enterprise Processing System (EPS), 392
Environmental data record (EDR), 15, 248, 285–286, 320
Environmental satellites, 319–320
Equator crossing times (EXT), 11
Error analysis, 16–17
Essential climate variables (ECVs), 375
Essential ocean variables, 84–85
European Aqua Thermodynamic Experiment (EAQUATE), 189–190, 324
European Brewer Network (EUBREWNET), 129
European Centre for Medium-Range Weather Forecasts (ECMWF), 247–248, 301
European Meteorological Operational (Metop) series, 297–299
European Organization for the Exploitation of Meteorological Satellites (EUMETSAT), 11
European Polar System-Second Generation (EPS-SG), 297–299
Exclusive economic zone (EEZ), 80, 196
Expendable bathythermographs (XBTs), 198
Expendable CTDs (XCTDs), 198

F
Fast Repetition Rate Fluorometer (FRRf), 365
Field campaigns, 2–3, 392
Field-of-regard (FOR), 285–286
Field of view (FOV), 4, 166–169, 285–286, 322
Fire Influence on Regional to Global Environments and Air Quality (FIREX-AQ), 254–256, 255–256f, 392
Fluorometric technique, 363
Fourier-transform infrared spectrometer (FTIR), 102–103, 121
  calibration, 103–104
  principles, 103, 103f
Fourier transform spectrometer (FTS), 13, 160, 204, 241–242, 319–320
French Polarization and Directionality of the Earth's Reflectances (POLDER), 210
Frequency, 7
Frost point hygrometers (FPHs), 37–46
  balloon-borne, 39–41, 42f
  development, 41–44
    cryogen implementation, 44
    frost control stability, 42

Index

photodetection system, 42–43, 43f
radiosonde improvements, 44
thermistor calibration technique, 43–44
history, 37–39, 38f, 40f
measurement uncertainties, 44–46
calibration uncertainties, 45, 45f
environmental uncertainties, 46
frost control uncertainties, 44–45, 45f
radiosonde measurement errors, 45–46, 45f
measurement records, 46–48
record profile, 47–48, 48f
sounding site, 46–47, 47t, 47f
satellite-based remote sensors comparisons, 47t, 48–53, 50–52f
Full-spectral resolution (FSR), 327

G

GCOS Reference Upper Air Network (GRUAN), 27, 63, 220, 226–229, 227f, 327
data products, 227
radiosondes, 283
RS92-RS41 radiosonde intercomparisons, 228, 229f
sites, 227–228, 228f
WMO radiosonde intercomparisons, 228–229, 229t, 230f
Geopotential altitudes, 46
Geostationary Coastal and Air Pollution Event Airborne Simulator (GCAS), 176–177
Geostationary Earth orbit (GEO), 10, 166, 319–320, 375
Geostationary Lightning Mapper (GLM), 176–177
Geostationary Ocean Color Imager (GOCI), 351–352
Geostationary Operational Environmental Satellite (GOES), 252, 375
Global AirCore network, 149f, 152
Global climate models (GCMs), 203–204
Global Climate Observing System (GCOS), 27, 46–47, 80, 375
Global Drifter Program (GDP), 81, 203
Global intensive profiling campaigns, 236–240
Atmospheric Tomography Mission (ATom), 239–240, 240–241f
HIPPO, HIaper pole-to-pole observations, 236–239, 237–239f, 238t
Global Monitoring Laboratory (GML), 140–141
Global Navigation Satellite System (GNSS), 26, 185, 281
Global Navigation Satellite System Radio Occultation (GNSS-RO), 304
Global Ocean Data Assimilation Experiment (GODAE), 87
Global Ocean Observing System (GOOS), 80
Global Positioning Satellites/System (GPS), 26, 185, 198, 304
Global precipitation measurement, 192
Global Space-based Inter-Calibration System (GSICS), 265
Global telecommunications system (GTS), 27, 95–96, 189
Global Tropical Moored Buoy Array, 84–85

GNSS-derived precipitable water vapor column (GNSS-PW), 227
GNSS RO profiles, 287
Goddard Space Flight Center (GSFC), 126, 286–287
Graupel water path (GWP), 418–419
Greenhouse gases (GHGs), 140–141, 240–241
Greenhouse gases Observing Satellite (GOSAT), 237–238
Ground truth, 248
Group for High Resolution Sea-Surface Temperature (GHRSST), 87, 345
Group of Atmospheric Optics (GOA), 126

H

Hazard Mapping System (HMS), 391–392
Hazardous Weather Testbed (HWT), 249–251, 250f
HIAPER pole-to-pole observations (HIPPO), 325
High-performance liquid chromatography (HPLC), 212–213, 363
High-Resolution Infrared Radiation Sounder (HIRS), 29–30, 286, 319–320
High-Resolution Interferometer Sounder (HIS), 160–162
Howard University Beltsville Campus (HUBC), 327, 415
Hurricane and Severe Storm Sentinel (HS3) mission, 162
Hyperspectral Infrared Sounder (HIRS), 393
Hyperspectral Profilers (HyperPro), 213–214
Hyperspectral Surface Acquisition System (HyperSAS), 358
Hyperspectral Tethered Spectral Radiometer Buoy (HTSRB), 213–214

I

Indium-gallium-arsenide (InGaAs), 124
In-flight Calibration Target (ICT), 169
Infrared (IR), 7, 30, 195–196, 245, 319
Infrared Atmospheric Sounding Interferometer (IASI), 30, 247, 319–320, 415
Inherent optical properties, 212–213, 352
In situ measurement, 4
Integrate Carbon Observation System (ICOS), 148
Integrated cal/val and long-term monitoring (ICVS-LTM), 17
Integrated water vapor (IWV), 192
Intensity Forecast Experiment (IFEX), 251–254, 253f
Intensive Ozonesonde Network Studies (IONS), 68–69
Intergovernmental Oceanographic Commission (IOC), 80
International Comprehensive Ocean-Atmosphere Data Set (ICOADS), 338–339
International Geosphere-Biosphere Programme (IGBP), 376
International Ozone Commission (IOC), 63
International Polar Year (IPY), 208

International Space Station (ISS), 50–52, 70
Inversion algorithms, 353–354
Isotropic, 7–8

J

Japan Aerospace Exploration Agency (JAXA), 270
Joint airborne IASI Validation Experiment (JAIVEx), 190
Joint Center for Satellite Data Assimilation (JCSDA), 15, 299–300
Joint Polar Satellite System (JPSS), 11, 30, 125, 212, 222–223, 281, 319–320
Jülich Ozone Sonde Intercomparison Experiment (JOSIE), 63

K

Korean Multi-Purpose Satellite-5 (KOMPSAT-5), 287
Korea's Geostationary Multi-Purpose Satellite-2A (GK2A), 190

L

Laboratory of Atmospheric Optics (LOA), 126
Land-based field campaigns, 220
Land surface emissivity (LSE), 376
Land surface temperature (LST), 375
product validation, 377–386
ground LST estimates, 378
LST validation results, highlights, 378–386
strategies, methodologies, 377–378
VIIRS LST product, 377
retrieval methodology, 376–377
simulation study, theoretical analysis, 376–377
Lightweight stratospheric air sampler, 147–148
Limb Infrared Monitor of the Stratosphere (LIMS), 30
Line-by-Line Radiative Transfer Model (LBLRTM), 173
Liquid water path (LWP), 192
Look up table (LUT), 376
Low-earth orbit (LEO), 10, 59, 166, 204, 222–223, 263–264, 319–320, 351–352, 375
Lunar calibration, 264

M

Machine learning (ML), 391–392
MarchUp DataBase (MUDB), 345
Marine-Atmospheric Emitted Radiance Interferometer (MAERI), 87, 104–105, 104–106f, 203–204
applications, 112–115
atmospheric profiles, 113–114, 115f
greenhouse gas heating, 102f, 114–115, 115f
satellite SSTskin retrievals validation, 112–113, 113t, 114f
ARM, 105
geophysical variables, 106–112
atmospheric profiles, 110–112

Marine-Atmospheric Emitted Radiance
    Interferometer (MAERI) *(Continued)*
  near-surface, air-sea temperature
    differences, 110, 111*f*
  sea surface emissivity, 106–107,
    108–109*f*
  SSTskin, 107–108, 109*f*
  thermal skin layer profiles, 109
  RSMAS, 105
Marine boundary layer (MBL), 206–207
Marine measurements, 195–196
Marine Optical Buoy (MOBY), 84–85, 210, 359
Marine Optical Characterization Experiments
    (MOCE), 210
Matchups, 366
Maxwell equations, 4
Mean sea level (MSL), 140
Measurement correlation, 16
Measurement principles, 3–4, 3*f*
  in situ vs. remote sensing, 4
Measurements of Aerosols, Radiation and
    CloUds over the Southern Ocean
    (MARCUS), 105
Measurements of Pollution in the Troposphere
    (MOPITT), 241
Median absolute percentage difference (MAPD),
    366
Medium Resolution Infrared Radiometer
    (MRIR), 29
Memorandums of understanding (MOUs),
    80
Mesoscale convective systems (MCS), 411
Meteorological Research Flight, 38
Meteosat Second Generation (MSG), 11
Methane (CH_4), 139, 235, 320
Michelson interferometer, 103
Microburst windspeed potential index (MWPI),
    415
Microwave (MW), 7, 195–196, 245, 319, 414
Microwave Humidity Sounder (MHS), 286,
    297–299, 415
Microwave Integrated Retrieval System (MiRS),
    299–300, 300*f*, 301*t*
  product validation, 301–315
    temperature profile validation, 304–307,
      306–309*f*
    total precipitable water validation, 303*f*,
      311–315, 313–315*f*
    validation background, 301–304, 302–305*f*,
      309*f*
    water vapor profile validation, 308–310,
      310–312*f*
Microwave Limb Sounder (MLS),
    50–52, 325
Microwave radiometer (MWR), 421–425
Microwave radiometer profiler (MWRP), 415
Microwave sounder data, 297–299, 298*t*, 299*f*
Microwave Sounding Unit (MSU), 29–30
MODerate resolution atmospheric
    TRANsmission (MODTRAN),
    376–377
Moderate Resolution Imaging
    Spectroradiometer (MODIS), 94, 112,
    166, 351–352, 392

MODIS/ASTER airborne simulator (MASTER),
    169
Monitoring Atmospheric Composition and
    Climate (MACC), 237–238

N

NASA African Monsoon Multidisciplinary
    Analyses (NAMMA), 190
NASA bio-Optical Marine Algorithm Dataset
    (NOMAD), 212–213
NASA scatterometer (NSCAT), 190–191
National Aeronautics and Space Administration
    (NASA), 208, 249, 286–287, 351–352
National Center for Atmospheric Research
    (NCAR), 185
National Center for Environmental Information
    (NCEI), 416
National Data Buoy Center (NDBC), 84–85
National Ecological Observatory Network
    (NEON), 126
National Hurricane Center (NHC), 189
National Institute of Science and Technology
    (NIST), 94, 105, 267
National Meteorological Services (NMHs),
    227–228
National Oceanic and Atmospheric
    Administration (NOAA), 11, 124,
    140–141, 185, 240–241, 297–299
National Physical Laboratory (NPL), 105
National Polar-orbiting Operational
    Environmental Satellite System
    (NPOESS), 189
National Science Foundation (NSF), 208,
    236
National Transportation Safety Board (NTSB),
    411
National Weather Service (NWS), 246, 412
Naval Research Laboratory (NRL), 38–39
Near-infrared (NIR), 12, 175–176, 352–353
Near-real-time (NRT), 346
Net phytoplankton productivity, 365
Network for the Detection of Atmospheric
    Composition Change (NDACC),
    46–47, 129
Neural network (NN), 415
Next Generation Weather Radar (NEXRAD),
    415
Nitrous oxide (N_2O), 139
NOAA-20 satellites, 166–169, 246, 283,
    297–299
NOAA Center for Atmospheric Sciences
    (NCAS), 205–206
NOAA JPSS Enterprise Processing System,
    aerosol detection, 393–403, 393*f*
  deep-blue aerosol detection algorithm,
    394–395, 394–395*f*
  EPS ADP algorithm application, 401–403,
    402*t*, 403*f*
  FIREX-AQ 2019, eMAS measurements,
    398–400, 400–401*f*
  IR-visible aerosol detection algorithm,
    396–398, 396–399*f*
NOAA National Weather Service
    (NOAA/NWS), 320
NOAA Products Validation System (NPROVS),
    18, 190, 282, 282*f*, 324

  collocation compiling, 287–288
  data sources, 282–287
    conventional radiosondes, 282–283,
      284*f*
    handling file, formatting product
      differences, 287
    numerical weather prediction, 287
    satellites, 285–287
    special sondes, 283–285, 285*f*
  vertical statistics computation, 288–289
NOAA-Unique Combined Atmospheric
    Processing System (NUCAPS), 14,
    206–207, 285, 320, 322–323,
    329–332, 331*f*, 415
North Atlantic Waveguide and Downstream
    Experiment (NAWDEX), 192
Northwest Tropical Atlantic Station for air-sea
    flux measurement (NTAS), 209
NUCAPS algorithm validation, 323–329
  field data, validation highlights, 327–329
    NOAA-20 IR ozone profile validation, 329,
      330*f*
    NOAA-20 temperature, moisture
      validation, 327–328, 328–329*f*
  objectives, 323
  strategies, methodologies, 323–327
    coarse-layer, fractional error statistics,
      326–327
    high-resolution field data reduction,
      correlative layers, 325–326
    temperature, moisture, ozone hierarchy,
      324
    trace gas hierarchy, 324–325
Null-space errors, 326
Numerical weather prediction (NWP), 15, 27,
    189, 195, 263–264, 281, 319

O

Observing Air-Sea Interactions Strategy
    (OASIS), 95
Ocean Color and Temperature Scanner (OCTS),
    210
Ocean color campaigns, 210–214, 210–211*t*,
    212–213*f*, 214*t*
Ocean color index (OCI), 354
Ocean color matchup analysis, 366–368, 367*f*,
    368*t*
Ocean expeditions, 196
Ocean Observations Panel for Climate (OOPC),
    80
Oceanographic buoy platforms, 80–85
  moored buoys, 83–85, 83–85*f*
  surface drifting buoys, 81–83, 81–82*f*
Oceanographic research vessel (RV)/ship
    platforms, 196–198, 197*f*
Olympic Mountains Experiment (OLYMPEX),
    192
On-board calibrators (OBCs), 264
On-orbit radiometric calibration, reflective solar
    band, 267–272
  DCC technique, 271–272
  lunar F-factor calculation algorithm, 270–271
  scheduled lunar calibration collection, 265*f*,
    268–269, 268–269*f*, 270*t*

simultaneous nadir overpass (SNO) trend derivation algorithm, 272
VIIRS RSB calibration coefficient, 266f, 267–268, 267t
Optimal estimation (OE), 15, 322
Optimal estimation method (OEM), 286
Optimum interpolation sea surface temperature (OISST), 87
Orbiting Carbon Observatory version 2 (OCO-2), 150
Organization of Tropical East Pacific Convection (OTREC), 187
Ozone (O_3), 320
Ozone partial pressure, 63
Ozonesonde, 59, 199
  global ozone measurement, 59–60, 59–60f
  networks, 64–69
    global, long-term sites, 64–67, 65–66f
  operation, data quality control, 60–64
    Assessment of Standard Operating Procedures (SOPs) for OzoneSondes (ASOPOS), 63–64
    ECC, operating principle, 61–63, 61–62f
    electrochemical ozonesondes, 60
    homogenization, 64, 65–66f
    quality assurance, 63–64
    strategic networks, global campaign operations, 68–69, 68f, 69t

P

Particle size class, 364
Particulate organic carbon, 354
Passive remote sensing, 297–299
Photosynthetically active radiation (PAR), 365
Phycourobilin/phycoerythrobilin (PUB/PEB), 365
Physical Oceanographic Real-Time System (PORTS), 425
Physical retrieval, 15, 247–248, 322
Phytoplankton functional type (PFT), 354
Phytoplankton pigments, field measurements, 363–365
  cell counts, cell sizes (10 μm), 363–365, 364f
  cell sizes (10 μm), 365
  pigments, 363
  quantum efficiency, 365
Planck function, 7
Plankton, Aerosol, Cloud, Ocean Ecosystem (PACE), 357
Point-source integrating cavity absorption meter, 361
Polarization corrected temperature (PCT), 419
Polar Operational Environmental Satellite (POES), 12, 297–299, 319–320
Post launch test, 176–177, 266
Potassium chloride (KCL), 88
Potassium iodide (KI), 61
Precision Filter Radiometer (PFR), 129
Pre-Depression Investigation of Cloud-Systems in the Tropics (PREDICT), 192
Principal component (PC), 247–248
Probability distribution function (PDF), 359–360
Probability of correct detection, 392

Proving Ground and Risk Reduction (PGRR), 246–247, 329–331
Pseudo Invariant Calibration Site, 264

Q

Quality assurance (QA), 60, 352
Quality control (QC), 283, 352
Quasi-analytical algorithm, 353–354
Quasi-biennial oscillation (QBO), 47–48
Quick Scatterometer (QuikSCAT), 190–191

R

Radiance/radiant intensity, 7, 356–357, 356f
Radiative transfer algorithm (RTA), 325
Radiative transfer equation (RTE), 7, 319
Radiometer, 11–12
Radiometric uncertainty (RU), 164
Radio occultation (RO), 281
Radio Occultation Meteorology (ROM), 287
Radiosonde Intercomparison and Validation (RIVAL) experiment, 283
Radiosonde observations (RAOBs), 414–415
Radiosondes, 23–27, 198, 302–303, 415
  cal/val considerations, 28–32
    clouds, 31–32
    NWP data assimilation fields, 32
    temperature, humidity profiling, 29–31
    winds, 31
    launching, 23, 24f
  measurement technology, 23–28, 24–25f
    humidity, 26
    pressure, 26–27
    temperature, 25–26
    winds, 26
  operations, 27–28, 28f
  receiving system, data format, 27
Rain rate, 299–300
Rain water path, 192
Raw data records (RDRs), 15
Real Time Mesoscale Analysis (RTMA), 418
Rear-inflow jet, 425
Reflectance ($R_{rs}(\lambda)$), field measurements, 210, 352–353, 357–360
  above-water approach, 357–358, 357f
  in-water approach, on-water approach, 358–360, 358f, 360f
Reflected solar band (RSB), 176–177
Relative fluorescence units (RFU), 365
Relative humidity (RH), 31, 37, 190, 319
Relative percentage difference, 366
Remotely operated vehicle (ROV), 197
Remotely sensed measurement, 4
Remote sensing primer, 4–9, 4–6b, 5f
  radiative transfer equations, 7–9
    passive MW, 9
    solar spectrum (VIS), 7–8
    thermal IR, 8
  terrestrial radiation, 6f, 7
Research-to-operations (R2O), 329–331
Research vessels (RVs), 196
Retrieval(s), 344
  accuracy, 113
  algorithms, 322
Robotic Lunar Observatory (ROLO), 265

Root mean square (RMS), 190
Root mean squared difference (RMSD), 366
Root mean square error (RMSE), 328
Rosenstiel School of Marine and Atmospheric Science (RSMAS), 104
Rotating telescope assembly, 267
R-Series Geostationary Operational Environmental Satellites (GOES-R), 11

S

Saharan air layer (SAL), 114, 190, 205–206, 252
Salinity, 88
Salinity Process in the Upper-Ocean Regional Study (SPURS), 83
Satellite Application Facility (SAF), 287
Satellite broadcast network (SBN), 246
Satellite data assimilation, 208
Satellite evaluation, 243
Satellite infrared remote sounding, 319–321, 320b, 321f
Satellite Infrared Spectrometer (SIRS), 29
Satellite ocean color products, 352–355, 354f
  quality measures, 355, 355f
  remote sensing reflectance, 352–353
  water bio-optical properties, 353–354, 354f
Satellite orbits, 288
Satellite product validation, 135
Satellite remote sensing, 9–15, 9t
  applications, 11–15, 12–14f
  environmental satellites, 10–11, 10f
    GEO, 11
    LEO, 10f, 11
  methods, 13–14f, 15
Satellite remote sensors, 352, 352f
Satellite sounders, 281
Scanning High-Resolution Interferometer Sounder (S-HIS), 160–163, 160b, 161f, 162b, 249
  high-altitude airborne calibration validation, 166–180
    Advanced Baseline Imager (ABI), cal/val, 175–180, 177–179f
    Cross-Track Infrared Sounder (CrIS), cal/val, 166–175, 170–172f, 174–176f
    radiometric calibration, verification, traceability, 163–166, 163b, 165f, 166t, 167–168f
Scattering, 7–8
Scientific Community on Oceanographic Research (SCOR), 95
Sea and Land Surface Temperature Radiometer, 112
Seabird Hyperspectral Surface Acquisition System (HyperSAS), 213–214
Sea ice concentration, 299–300
Sea-level pressure, 83, 91–92
Sea surface salinity (SSS), 87–89
Sea-surface temperature (SST), 15, 83, 85–87, 86f, 101, 102f, 337
  blended analysis, 346–348, 348f
  method-dependent model, 339
  validation, in situ measurements, 342–346, 343f, 346f

Sea-surface temperature (SST) *(Continued)*
  products, 337–341, 338f, 340–342f
Sea-viewing Wide Field-of-view Sensor
    (SeaWiFS), 94, 210, 351–352
  Bio-Optical Archive and Storage System
      (SeaBASS), 212–213
  Photometer Revision for Incident Surface
      Measurements (SeaPRISM), 358
Selective chopper radiometers, 29
Sensing solution types, 63
Sensor data record (SDR), 15, 174, 320
Severe convective windstorms, 412–414, 413f
Ship-based intensive field campaigns, 196
Short-wave infrared (SWIR), 124, 352–353, 392
Signal-to-noise ratio (SNR), 322
Simultaneous nadir overpasses (SNOs), 264,
    385–386
Single-scattering approximation, 7–8
Single-sensor error statistics, 342
Skin temperature, 347
Skylight-blocking apparatus, 213–214, 359
Snowfall rate, 299–300
Snow water equivalent, 299–300
Soil moisture active passive, 87–88
Soil moisture and ocean salinity, 87–88
Solar Diffuser, 263–264
Solar Diffuser Stability Monitor (SDSM),
    263–264, 267
Solar spectrum (VIS), 7–8
Soundings, 322
Soundings of Ozone and Water in the Equatorial
    Region (SOWER), 46–47
Southern Hemisphere Additional Ozonesonde
    (SHADOZ), 327
Space Science and Engineering Center (SSEC),
    104, 160–162, 205, 283
Space View, 264
Special Sensor Microwave Imager (SSM/I),
    190–191
Special Sensor Microwave Imager Sounder
    (SSMIS), 414
Spectral band adjustment factors, 272
Spectral response functions (SRFs), 173
Spectrometer, 9–10, 358
Spherical integrating sphere, 267
Spinning Enhanced Visible and Infrared Imager
    (SEVIRI), 375
Split-window technique, 376
Stepped Frequency Microwave Radiometer
    (SFMR), 191
Storm Prediction Center (SPC), 331–332,
    425–427
Stratosphere, 37, 59
Stratosphere-troposphere exchange (STE),
    68–69
Stratospheric Aerosol and Gas Experiment
    (SAGE III), 50–52
Stratospheric sampling, 146
Stratospheric Sounding Unit (SSU), 29–30
Sun glint, 212
Sun photometers, measurement networks,
    124–130
  aerosol robotic network (AERONET),
      125–126, 125t, 126f
  brewer networks, 129–130, 130f
  global atmosphere watch, precision filter
      radiometer network (GAW-PFR),
      127–129, 128–129f
  Maritime Aerosol Network (MAN), 126–127,
      127f
  other networks, 130
Sun photometry, 122–125
  calibration, uncertainty, 131–132, 131f
  instrument specifications, 124–125
  theoretical background, 122–124, 122–123f
Suomi National Polar-orbiting Partnership
    (SNPP), 30, 94, 112–113, 166–169,
    222–223, 239–240, 264, 319–320,
    375, 392
Surface Heat Budget of the Arctic Ocean
    (SHEBA), 208
Surface heat fluxes, 90–91
Suspended particulate matter, 354
Systematic error/bias, 16

T

Temperature profile, 326
The Observing Research and Predictability
    EXperiment (THORPEX), 189
Thermal and Near Infrared Sensor for Carbon
    Observation-Fourier Transform
    Spectrometer (TANSO-FTS),
    166
Thermal emissive band (TEB), 176–177, 264
Thermal infrared (TIR), 376
Thermal infrared radiance spectra, 166–169
Thermal Infrared Transfer Radiometer (TXR),
    164
Thermal skin layer, 102
Thermodynamic Initial Guess Retrieval (TIGR),
    376–377
THORPEX-Pacific Asian Regional Campaign
    (T-PARC), 190
Three-mirror collimator (TMC), 267
Tipping calibration (tip-cal), 201
TIROS Operational Vertical Sounder (TOVS),
    29–30, 281
Top-of-atmosphere (TOA), 7, 92, 352–353,
    394
Total Carbon Column Observation Network
    (TCCON), 235–236, 325
Total column ozone (TCO), 63
Total precipitable water (TPW), 299–300
Traceability, 160–162
Trace gas, 235
  sustained observations, 240–243, 242f
Transmittance, 414
Tropical cyclones, 251–252
Tropical Rainfall Measuring Mission (TRMM),
    192
Tropical tropopause layer, 68
Tropospheric column ozone, 69
Tropospheric Emission Spectrometer (TES),
    166, 237–238
Tropospheric Monitoring Instrument
    (TROPOMI), 240, 325
Turbulent heat fluxes, 90
Two-angle atmospheric correction algorithm,
    353

U

Ultra-high frequency (UHF), 415
Ultraviolet (UV), 9–10, 122–123
Unbiased absolute percentage difference, 366
United Nations Educational, Scientific and
    Cultural Organization (UNESCO), 80
University Corporation for Atmospheric
    Research (UCAR), 287
Unmanned aircraft vehicle (UAV), 160–162
Upper tropopause/lower stratosphere (UTLS),
    239, 329

V

Validation, 16, 27, 69, 101, 281
Variable fluorescence, 365
Velocity azimuth display (VAD), 419–421
Velocity azimuth display wind profile (VWP),
    415
Vertical Feature Mask (VFM), 392
Vertical profiles, 48, 236, 328
VIIRS prelaunch calibration, 266–267
Viscous sublayer, 102
Visible (VIS), 12, 124
Visible Infrared Imaging Radiometer Suite
    (VIIRS), 12, 94, 112, 169, 254,
    351–352, 392

W

Warm layer, 347
Water inherent optical properties, field
    measurements, 360–363
  backscattering coefficient, 360–361, 361f
  non-water absorption coefficient, 361
  phytoplankton, CDOM absorption
      coefficients, 362–363, 362f
Water-leaving radiance, 92–94, 95f, 352
  characterization, calibration, maintenance, 94
  data processing, 93–94
  instruments, measurements, 92–93, 93f
  location selection, 92
Water vapor (H_2O), 37–38, 144, 320
Water Vapor Intensive Operating Periods,
    220–221
Water vapor transmittance (T_w), 124
Wavelength, 122
Wavenumber, 102–103
Weather balloons, 198
Weather forecast offices (WFOs), 320
Wet downbursts, 412–413
Wind speed, direction, 89–90
Windstorms, 411
World Calibration Centre for Ozonesondes
    (WCCOS), 63
World Climate Research Program (WCRP), 80
World Meteorological Organization (WMO), 27,
    63, 80, 139–140, 189, 196–197,
    227–228, 236, 282, 375
World Ocean Circulation Experiment (WOCE),
    80–81
World Optical Depth Research Calibration
    Center (WORCC), 129
World Ozone and Ultraviolet Radiation Data
    Centre (WOUDC), 129, 327
World Radiation Monitoring Center (WRMC),
    380–383

Printed in the United States
by Baker & Taylor Publisher Services